Physical Geology

Physical Geology

Brian J. Skinner
Yale University

Stephen C. Porter
University of Washington

John Wiley & Sons

New York
Chichester
Brisbane
Toronto
Singapore

Production supervised by Linda R. Indig
Cover and text design by Sheila Granda
Cover photo Galen Rowell/Mountain Light
Photo researched by Safra Nimrod
Illustrations by Dennis Tasa
Manuscript edited by J. Nestor under the supervision of Deborah Herbert

Library of Congress in Publication Data:

Skinner, Brian J., 1928–
  Physical geology.

  Includes index.
  1. Physical geology. I. Porter, Stephen C.
II. Title.
QE28.S65 1987   550   86-32533
ISBN 0-471-05668-5

Printed and bound by Von Hoffmann Press, Inc.

10 9 8

Preface

Our understanding of the Earth is built on centuries of patient observations and careful insights. Geology is a major branch of science that deals with those observations and insights. It is a special kind of science, because the geological laboratory is the world in which we live. Sometimes we can test geological conclusions by controlled experiments, but under most circumstances we cannot carry out experiments in the geological laboratory. Either the scales of space and time needed for such experiments are too large to deal with, or the experiments would change our living environment in some unfortunate way. We must therefore make careful and systematic observations of the Earth, and then draw conclusions about the phenomena from the assembled evidence. A recent conclusion which has been drawn from such observations is that each and every one of us plays a part in the changes that ceaselessly alter the face of the Earth. Although our individual contributions are small, the sum is large. We influence the atmosphere, streams, lakes, and oceans; we affect rates of erosion, and the way deserts expand or contract; we cover the surface of the land with roads and cities, and we redistribute the Earth's materials by digging them up and then using them for the multitude of things we need for living in a complex society. We humans have become a vital force in the shaping of our own environment.

There are many questions that can be asked concerning human interactions with the environment for which answers are not yet available. For example, we are still uncertain of the degree to which the continual burning of fossil fuels, and the consequent increase of carbon dioxide in the atmosphere, will affect global climates in the coming centuries. How a climatic change might influence the world's agricultural productivity, the distribution of polar ice, or the position of global sea level, is not known. Study of the Earth and its multitude of dynamic interacting systems is an expanding, evolving, and tremendously exciting science. It is a science in which each of us inevitably plays a part. It is one to which we can all contribute and, in our turn, bequeath increased understanding to those who will follow us.

Revolutionary advances in the breadth and depth of our knowledge of the Earth have occurred over the past thirty-five years. At no previous time during human history have so many dramatic discoveries been made within such a short time. The revolutionary discoveries continue unabated. Earth science is a field in ferment, a subject laced with challenging excitement; new discoveries, new insights, and new theories heighten the excitement every day. A few short years ago the suggestion that the outer layer of the Earth moves laterally at rates of up to 10 cm/yr was embodied in a theory called *plate tectonics.* Evidence to support the theory was discovered in rocks on the sea floor, and while the evidence was compeling that movement had happened in the past, it did not prove that such movements are occurring today. Now, in 1986, measurements using satellites and lasers have demonstrated that continents really are moving. Plate tectonics is no longer just a theory; it's a fact, and it's happening now.

The revolutionary advances in the Earth sciences can be grouped into three categories. The first advance concerns our understanding of the way the Earth works; plate tectonics is one of the products of that understanding. The advance has come about through such seemingly unrelated studies as exploration of the ocean floor, seismic studies of the Earth's core, and long-term measurements of the strength of the magnetic field. The studies are not unrelated. The Earth's magnetism arises in the core, and rocks of the sea floor are influenced by the magnetic field in distinctive ways. The realization that all of the Earth's processes, whether large or small, interact in many unsuspected ways is forcing geologists to reexamine every scrap of evidence, and to rethink every conclusion. One important result has been a refined understanding of plate tectonics. As continents are moved around on plates, small fragments of continents can be sliced off and rafted away; volcanic mountains arise in the sea, and limestone plateaus like those in the modern Bahamas can form on the volcanic ramparts in warm tropical seas. The fate of all such small fragments of crust is to be swept up and eventually accreted on to the margins of larger continental masses. Four hundred million years ago small segments of exotic and far-traveled crust were accreted to the eastern border of North America. A hundred million years ago the same process added exotic blocks of crust to the western margin of the continent. Extraordinary though the notion seems, there are pieces of California, Oregon, and Washington that were once separated from North America by 6000 kilometers of ocean, and other pieces that were once attached to Asia.

The second advance springs from space exploration, and in particular from systematic investigations of the Moon, Mars, Mercury, Venus, and the rocky moons of the giant planets Jupiter and Saturn. All planets and moons in the Solar System, plus the asteroids, comets, and meteorites, have a common birthright, and although each planet has evolved differently, common threads run through their histories. By unraveling those common threads a new discipline called comparative planetology has arisen. Comparative planetology is helping to provide answers to why the Earth exists at all, why it is like it is, why other rocky bodies in the Solar System are not suitable for human habitation, and whether there are likely to be hospitable, earthlike planets anywhere else in the Universe.

The third advance involves an increasing awareness, particularly in industrial nations, of the effect of human activities on environments at the surface of the Earth. With that awareness has come a realization that attempts to analyze and modify changes are fraught with difficulty because the multitude of natural processes operating at the Earth's surface interact in innumerable and often complex ways. We have finally come to understand that people are not just one of the minor forces of nature—they are a major force. What the Earth will be like in the future depends very much on how we act today.

This book is a successor to a long line of textbooks of physical geology. The book has been written in sharp realization of our heritage of knowledge and of the need to integrate with the corpus of classical knowledge the fruits of current research and thinking. We have tried to do so in a cohesive form that is intelligible to students who have not previously studied geology. The present book has been almost completely rewritten, and more than 80 percent of the illustrations are new. One chapter, devoted to the evolution of landscapes, is entirely new and is unique. The chapter uses a number of well-known places, such as the Himalaya, the Andes, and the Dead Sea, to illustrate how the Earth's internal and external forces interact to produce distinctive topographies. Another chapter, devoted to climatic changes through geologic time, and how those changes have influenced surface processes and environments, is also unique. Inevitably, the flavor of the book differs from its predecessors. More attention has been paid to the role of people, and to their involvement with and reliance on resources from the Earth. Plate tectonics is integrated throughout the book rather than being treated as an isolated topic, and the total coverage has been greatly increased. The early history of continents is discussed more fully. The geographic coverage in the present volume is more global than in previous editions. Geology involves the whole Earth, and we have therefore consciously tried to introduce examples from many corners of our planet.

We have continued features of earlier editions that have proved to be well received. Among these are the review summaries at the ends of chapters and the unique scheme of defining technical terms within the body of the text. A term defined in the text appears in boldface italic type, and the defining phrase or clause appears in lightface italics. The page on which the term is defined is indicated by boldface type in the index. The definition thus occurs in context, in the presence of additional information, and perhaps also in one or more clarifying illustrations. Thus, in some instances, the reader can see how a definition is built and even why some definitions are difficult to frame. In addition, a glossary is included for those who may wish to use it. We have employed Standard International and metric units throughout the book. For those wishing to convert from metric to English units, we have included an appendix of conversion factors.

Many people have helped with the preparation of the volume. Some have helped with discussions, others have proferred ideas or illustrations, some have read and offered comments on specific chapters, and several brave souls critically reviewed the entire manuscript. To all of these people we are extremely grateful. The number is too large and our memories too short to list everyone who has helped, but special thanks must be given to the following: Glen Allcott, Richard Allmendinger, John Allock, Robert Anderson, Robert Berner, David Best, Francis Birch, Shelly Boardman, J. Allan Cain, Eric Cheney, Stanley Chernicoff, Gregory Davis, Richard Enright, Rolfe Erickson, Irving Friedman, Richard Fiske, Roy Gill, Bruce Goodwin, Richard Harvey, George Haselton, Robin Holcomb, Glen Izett, Peggy Keating, Michael Kimberley, Richard Krueger, Anthony Lasaga, Steve Laubach, John Longhi, Bruce Marsh, Charles Naeser, Jack Oliver, Anne Porter, Raymundo Punongbayan, Emmanuel Ramos, Nicholas Rast, Jill Schneiderman, John Shelton, Ronald Shreve, Brenda Sirois, Peter Stifel, Lynn Topinka, Robert Tracy, George Turner, and James Wilson.

We particularly wish to thank Donald Deneck, the editor under whose direction the planning and preparation of this volume was commenced. His encouragement and vision are greatly appreciated.

Brian J. Skinner
Stephen C. Porter

Contents

Physical Geology

The interaction of the Earth's mobile lithosphere with its fluid atmosphere and hydrosphere leads to the formation of varied landscapes, including that of the spectacular Monument Valley country of the American Southwest.

Planet Earth
and Its
Materials

C H A P T E R 1

The Human Planet

Distinctively shaped sand-dunes, called barchans, advance across irrigated fields in the Danakil Depression, Egypt.

OUR PLACE IN THE WORLD

Step by slow step we are beginning to understand how the Earth works. Much remains to be learned, but we are starting to understand how and why new mountain chains are thrust up while old chains are eroding away. The map of the globe is ever changing because the continents are continually moving and being rearranged, sometimes splitting into smaller pieces, sometimes colliding and forming super continents. Surprisingly, it has been discovered that seemingly unrelated things, such as the saltiness of seawater and the composition of the atmosphere, are influenced by the slow movement of continents. Indeed, it seems more and more likely that everything that happens on the Earth influences everything else, either directly or indirectly. The story that is being unraveled is one of great fascination and it has helped us to understand many of the major changes that have occurred during the Earth's long history. This understanding has led us, finally, to a realization that even the actions of the human race play an important role in shaping the face of the Earth. We can now perceive that Planet Earth is a place of great complexity, of innumerable interactions between all the component parts—such as the ocean, the atmosphere, the soil, the solid rocks, and the living plants and animals. The Earth is a planet where nothing is static and change is ever present. The story of these changes, past and present, is one full of surprises and unexpected discoveries; it is the story discussed in this book, the story of the planet we call home.

Life and the Earth

Perhaps the most startling part of the emerging story is the discovery that the Earth is the way it is because there is life on Earth. The converse is also true—all living things on the Earth (microorganisms, plants, and animals) are the way they are because the Earth is the way it is. The compositions of the atmosphere, the ocean, and even the rocks beneath our feet are strongly influenced by the activities of living matter; and living things, in turn, are influenced and controlled by the atmosphere, ocean, rocks, and soils.

Within the solar system our planet is unique. No other planet is known to support life. None has such a friendly atmosphere nor such a comfortable range of climates. None has an ocean of water. So far as we can tell, no other planet can supply the wide range of minerals and fuels we humans have come to need in order to support our complex society.

The human race is completely dependent on the resources of the Earth. Those resources include air, water, and soil, as well as the materials we grow and dig from the ground. We have come to depend on such a great many things in order to live that we seem to teeter on the brink of a crisis. Because there are so many human beings there is a possibility that the friendly balance between man and nature could be upset. At this point it helps to gain perspective on the human race as a geological force by looking back into history in order to see how the crisis in the Earth's development has come about. As we look back, we can see that the crisis had to happen at some time, simply because of the presence of our species. The only uncertainty was when it would happen.

The crisis came about in this way. Through billions of years of evolution, the Earth's surface has been inhabited by various kinds of living things that gradually became more and more complex. The fossils found in the Earth's layers of rock provide evidence that as long ago as 3.5 billion years microscopic organisms lived in the oceans. These tiny organisms were responsible for slowly changing the composition of the atmosphere by producing oxygen. As the atmosphere changed, the water in the ocean and the rocks in contact with the atmosphere were also changed and slowly the Earth became hospitable for more complex and larger organisms. One billion years ago small but distinct animals as well as very primitive plants were living in the ocean. Four hundred million years ago both plants and animals had invaded the land, and animals with backbones had developed. One hundred fifty million years ago reptiles (including dinosaurs) dominated the land and primitive kinds of mammals had already appeared. Sixty million years ago mammals had displaced reptiles as the dominant land animals, and three million years ago (and perhaps even earlier)· humans had evolved from the mammal stock and had begun to make tools from stone. The fossil human bones are there to prove it.

The Dominant Species

A hundred thousand years ago *Homo sapiens*, our own species, had evolved and was making tools with increasing skill. *Homo sapiens* has an exceptionally large and complex brain and this enabled our ancestors to make tools and to use them in ever more sophisticated ways. Throughout their long existence, Stone Age people were skillful hunters of wild game. So far as we can tell, Stone Age people hunted both large and small animals, and they supplemented their diet with birds, fish, and wild seeds and berries. As the number of Stone Age hunters increased, people hunted game to the far corners of the Earth.

The Origin of Agriculture and Cities

The *Homo sapiens* population kept increasing, so that between 9000 and 12,000 years ago a worldwide agricultural revolution was underway. Agriculture and the domestication of animals began to supplant hunting as the principal sources of food. Agriculture made other things happen. Instead of roving as small bands of hunters and sleeping in caves or temporary shelters, people began to build more substantial dwellings, and gathered into villages. In this way they turned to a more settled life. Gradually, cities containing thousands of inhabitants were formed. Because agriculture could provide much more food than hunting ever did, forests were cut down and replaced by fields. Gradually, streams were diverted, rivers were dammed, and hillslopes were terraced to provide more favorable growing conditions (Fig. 1-1). Such changes began early. Evidence in Iraq shows that in order to irrigate crops, people had begun to interfere with the Tigris and Euphrates Rivers as early as 7000 years ago. In many parts of the world the most ancient evidence of agricultural communities is found beside rivers; the Indus in Pakistan, the Yellow in China, and the Nile in Egypt are examples.

The effect of this tremendous change was to replace natural groupings of wild animals and wild plants with artificial groupings of domesticated plants and animals. But the changes had many side effects. One side effect was that people were removed a long step from participation in a wholly natural economy. Agriculture and city living pushed people toward a life in a world of their own creation. A second side effect was that unexpected and often unappreciated changes were wrought in the environment. It has recently been suggested, for example, that the decline of the Mayan civilization about a thousand years ago occurred because the Mayan farmers did not appreciate and counteract the excessive soil erosion that denuded their lands. It is probable, too, that the Mesopotamian societies of the Tigris and Euphrates valleys, and also the complex society of the Indus valley, were weakened by environmen-

FIGURE 1.1 These mountain slopes in the northeastern part of the island of Luzon, Philippines, were cleared and terraced more than 2000 years ago for growing rice. They are sill being used today.

tal stresses and eventually could not repel outside invaders.

Energy Consumption and the Expanding Population

Agriculture continued to be the principal basis of the economy as well as the main occupation of the human population for many thousands of years. Energy came largely from human muscles or from animals. Starting about a thousand years ago, an important change occurred when increasing use was made of wind and water power. Wind energy was used to pump water, and the energy of running water was used to grind grains and to power machines. Then, beginning in the eighteenth century on both sides of the Atlantic Ocean, a marked change in the use of energy began to make itself felt. This was the start of the age of intensive use of energy, also called the Industrial Age because it is the age of the high-energy industry.

Stone Age people had an industry too—they made weapons and tools of stone. But the energy needed to make stone tools was only the small amount supplied by human muscle. Other sources of energy remained to be tapped. Six thousand years ago people discovered metals and learned to smelt them from their ores, and to work the metals into implements. The smelting required energy, most of which came from the burning of wood. Still, until the eighteenth century, human industry drew most of its energy from the wind and running water, or from the muscles of men, horses, and cattle, and so production remained small.

The eighteenth century brought with it the invention of steam engines and the ability to convert the energy locked up in wood, coal, and other fuels into other forms of energy. Machines could do work formerly done by people and by horses, so that, in theory, all individuals could have machine "servants" and machine "animals" working for them. Coal came into wide use as a fuel because it yields, on an equal weight basis, far more energy than wood. Then, with the invention of internal-combustion engines, petroleum became a widely used fuel. From our vantage point in the last quarter of the twentieth century, it appears that solar or nuclear energy could eventually become the successor to wood, coal, and petroleum. The use of high-energy fuels enabled a given number of people to exploit more plants, animals, and inorganic materials and to produce much more food and many more products than had been possible in the days of predominant agriculture. However, high-energy fuels also led to huge industrial plants, giant cities, commuters, and huge apart-

ment buildings. Machines and cities are built with cement, metals, and other materials won from the Earth. Every year we seem to need more and more, until today we use the Earth's materials in such vast amounts that it is hard to imagine where they all go (Fig. 1.2).

This huge change affected everyone who participated in a high-energy economy. It surrounded people with a whole range of artificial things. In doing so it removed them a further, long step from direct participation in a natural economy; and it resulted in major changes in the environments in which energy use was highest. Stone Age people had made little more change in their environments than did other species of animals then living; they fitted well into their natural surroundings. But with the turn to agriculture, things began to be different. The most obvious changes were the destruction of forests and the altered shape of the land due to irrigation and terracing. Less obvious changes occurred in and near cities where streams became polluted. The changes were local and small enough so that the entire ocean and atmosphere remained essentially unaffected. With the age of intensive energy use, the changes became far greater. There were many more people, many more cities, and new, powerful machines. The re-

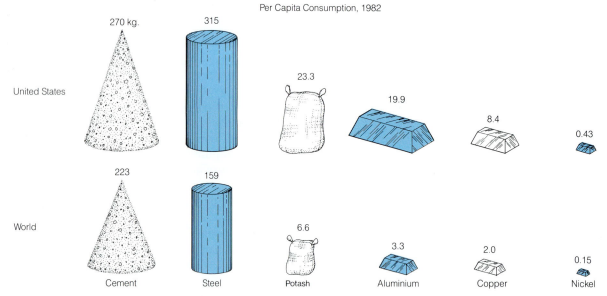

FIGURE 1.2 The amount of material dug from the Earth grows larger as the population grows. Use rates vary greatly around the world. The average amount of a material consumed, per person, is called the per capita consumption. For almost all of the materials won from the Earth, per capita consumptions are larger in industrially advanced countries such as the United States, than they are for the whole world. (*Source:* After data from U.S. Bureau of Mines.)

FIGURE 1.3 Overgrazing during years of drought killed most of the vegetation around wells in the Azaouak Valley, Mali. Without vegetation, soil blows away and the desert advances.

sult was gradual destruction of the plant cover, devastation of whole landscapes, pollution of the air, of streams and lakes, and even of the ocean. Gradually, parts of the Earth's lands have approached a state in which they will no longer be habitable (Fig. 1.3). Today, even the composition of the atmosphere is being changed by human activities. Yet the growth of the Earth's human population continues.

The history of the human population from about 8000 B.C. was one of steady growth until about the sixteenth century (Fig. 1.4). A few breaks did interrupt the steady growth, the most severe of which was the Black Death, an epidemic of bubonic plague that started in Italy in 1248 A.D. when the Crusaders inadvertently brought back rats bearing plague-carrying fleas from the Middle East. By the time the Black Death had run its course, it had killed between a quarter and a third of Europe's population! About the end of the sixteenth century the world's population started to rise more rapidly; the reasons were probably an increase in medical care and hygiene in cities, and a marked improvement in the diets of Europeans through the in-

troduction of corn and potatoes from the Americas. The rate of population growth became even faster during the eighteenth century when industrialization led to ever higher standards of living.

The world's population reached 1 billion about 1800 A.D. By 1930, the population had doubled to 2 billion. Despite a disastrous depression and the most destructive war ever fought, the next doubling of the population took only 45 years, so that by 1975 the population was 4 billion. The world's population is still growing, but important changes can now be discerned. Through the 1960s and 1970s, the population grew at a rate of 2.2 percent a year. By the 1980s, however, it was observed that, worldwide, the population was (and still is) growing only at a rate of 1.7 percent a year. Even so, and despite projections of continued slowing, the world's population is likely to reach about 7 billion people by the year 2000.

What the population picture might be in the more distant future remains uncertain. Both the densities of populations and the rates at which those populations grow vary greatly around the

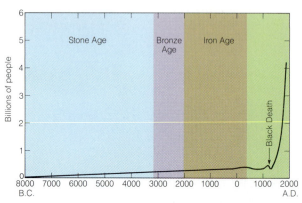

FIGURE 1.4 Population growth from ancient times to the present. The slow growth up to the Industrial Age was steady except for the setback caused by the Black Death and other less drastic disasters. (After Population Reference Bureau.)

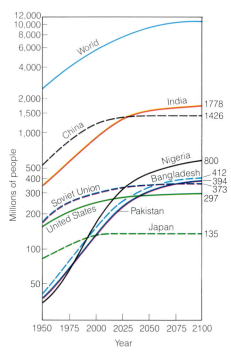

FIGURE 1.5 Projected population growth curves for the world and selected countries to 2100 A.D. (*Source:* data by Demeny in Report of the Population Council.)

world. The two most populous countries in the world today are China (with more than a billion people) and India (with about 700 million); these countries, plus the United States, the Soviet Union, and Indonesia account for more than half the world's population. Demographers working for the United Nations perceive that the picture is changing rapidly. Birthrates in technologically developed countries such as the United States, France, and Canada are slowing down so much that population sizes of these countries may soon be static. Birthrates in less-developed countries are still high, however. Even though projections suggest that changing birth and death rates will eventually bring all populations to static sizes, the demographers suggest it may be as long into the future as 2100 A.D. before the world's population stops growing. By that time, they project, people will probably number between 11 and 12 billion (Fig. 1.5), and the seven most populous countries will be, in order of decreasing size, India, China, Nigeria, Bangladesh, Pakistan, Indonesia, and the Soviet Union. These seven countries will contain 50 percent of the world's population at that time. Or, to say the same thing in a different way, the population of the seven most populous countries 110 years in the future will be larger than the entire world's population today!

Population and the Changing Environment

Can the Earth supply all the food and materials needed for 11 billion people? It is not yet widely appreciated that despite the great farming advances wrought by the so-called "green revolution" of the 1950s and 1960s, gains in agricultural output from 1970 to 1984 failed to keep pace with food needs of the expanding population. Nor is it widely realized that today only a small handful of countries consistently produce sufficient food to feed their populations and have a large enough excess remaining to export to the hungry of the world. The fortunate few countries are Australia, Canada, New Zealand, South Africa, and the United States. Not all of the food problems are environmental. In some countries political and social problems interfere with agriculture. Even so, as populations grow larger, environmental limitations on food production become more and more important.

The demand for food is already pushing forward ever more strongly the exploitation of plants and animals, and will tend to increase the existing intensities of pollution. That is to say, this demand will subject the delicately balanced economy to additional pressure. To put the case differently, the human species is becoming less and less well-adapted to its proper place in nature. That this is true is shown by three basic indicators: (1) serious overcrowding; (2) decrease in per capita consumption of food along with increase in per capita consumption of energy; and (3) widespread destruction of terrain and pollution of air and water.

With a limited view confined to the natural community around us, we sometimes fail to realize the magnitude of the changes that human activities create in the Earth's natural features. A major example is the group of changes being wrought by strip mining of coal in the Appalachian region of the United States. Destruction of terrain and pollution of water are clearly evident in the huge continuous gashes along hillsides. Broad horizontal shelves floored with broken rock waste and backed by sheer rock walls are the most conspicuous feature of many Appalachian landscapes (Fig. 1.6). Such shelves now aggregate more than 30,000 km in length, a distance that is more than one half the circumference of the Earth. As stripping proceeds, it destroys forests and soils, reduces the amount and the quality of the water in the ground, and pollutes streams by pouring into them quantities of debris and acid chemicals released from the coal. Landslides are created and this causes erosion that leads to the piling up of debris on valley

floors. Slopes once forested become dangerous eyesores; entire stream systems can become clogged and lifeless. In short, the human population has become a major geological force.

The extent of human interference with natural systems is evident also in the recent history of the Caspian Sea in the southwestern Soviet Union, the world's largest salt lake (about the size of the state of California). Since about 1930, its water surface has lowered by nearly 2.5 m, leaving many port facilities high and dry. The water has increased in salinity and the commercial catch of fish has decreased by half and is still falling. Also, because there are fewer fish to eat mosquito larvae, there have been epidemics of malaria.

The principal source of water for the Caspian is the huge Volga River. Because the lake has no outlet, the inflow of water is balanced by evaporation under a warm, dry climate. Since about 1930, contribution of water by the Volga has decreased markedly. The decrease is influenced in part by a decrease in rainfall and a warming trend in the climate, but it has been influenced considerably, too, by withdrawal of water from the Volga for irrigation. Engineers are now drawing up plans for the diversion of water from rivers that flow north

FIGURE 1.6 Coal mining in the Appalachians, West Virginia. Benches cut in the steep hillsides expose coal seams and change the shape of the land.

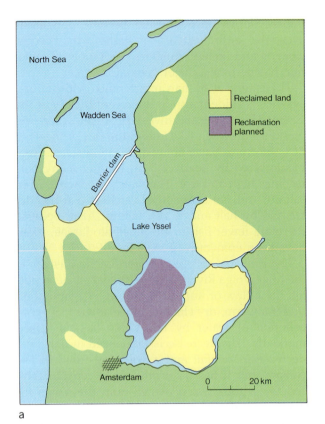

a

b

FIGURE 1.7 Land reclaimed from the sea in the Netherlands. (*a*) Map of the northern part of the country. In 1932 a barrier dam sealed off the Zuider Sea. Sea water was pumped out and fresh water from the River Yssel flowed in to create Lake Yssel. Bit by bit portions of Lake Yssel are being dammed, pumped dry and converted to farmland. (*b*) False color image of the same area of northern Holland. Green plants appear red. Fallow fields are blue.

into the Arctic Ocean in order to replenish the Volga. What the effects of such a massive restructuring of a natural system might be is still being debated, and may not be known until the diversion scheme is actually installed.

We should not give the impression that human interference with nature is always bad. Terrain sometimes can be rearranged on a very large scale without obvious unfavorable consequences. Large scale farming, with careful attention paid to soil conservation, changes the natural landscape but in the process the production of edible food is greatly increased. Another favorable example is the conversion of shallow seafloor into usable land (Fig. 1.7). Since about 1200 A.D., the Dutch people have been reclaiming land from the sea by building dikes and draining lakes and marshes. Today the reclamation process has become part of the high-energy economy, and the resulting dikes and other engineering works are huge. The reclaimed land adds up to more than 10 percent of the present area of The Netherlands. However, the further dikes are built into the sea, the harder it is to maintain them. Dutch engineers have not been able to maintain all of their winnings. Since the year 1200, nearly 5000 km² of formerly reclaimed land have been lost along the Dutch coast.

Two vital lessons can be learned from the examples described, and from the many other examples given in this book. The first is that any artificial change in a dynamic natural system will probably lead to unforeseen side effects, and these side effects are sometimes undesirable. The second lesson is nicely illustrated by the Dutch reclamation example: The more a natural system is changed—the further it is pushed out of balance with nature—the greater becomes the effort that must be expended to maintain it.

From this brief review of the history of human interactions with the surrounding environment, we can return with more understanding to the crisis we face. We have become complacent with the feeling that simply because the Earth has always provided the human population with more land whenever land was needed, and more food and minerals when known supplies ran out, that somehow new land and new resources will always be available through a future of indefinite length. In short, there has developed a belief that a limitless cornucopia of riches will continue. That comfortable point of view made it easy in the past to abandon land damaged by poor agriculture practices, by the cutting down of forests, or by mining, and to shift to another site. There was not obvious need to feel responsibility for the Earth because the habitable lands had not been exhausted. As long as there were unoccupied regions, people could and did move into them. However, nearly all of the easily habitable lands are now occupied, and only a handful of countries have space left to accommodate their anticipated population growth comfortably. In order to meet future needs, we must appreciate how all activities are energized. Some activities occur rapidly, others very slowly. We must understand how human activities directly change the speeds of other activities, and indirectly influence many more.

To the minds of some, the possibility of inhabiting other nearby planets offers an escape from future problems. However, the space-age examination of the Moon, Mars, Venus, Mercury, and the moons of Saturn and Jupiter has made it abundantly clear that free habitation is impossible. If humans are to live in such hostile places, it will be as inhabitants of small, artificially controlled environments. Like it or not, we humans must not only live on Planet Earth, we must also learn to repair the damage of the past and to keep future damage to an absolute minimum. In order to do so, we must understand as carefully and as exactly as possible how all the parts of the Earth interact.

The Earth, Small But Intricate

Although small, Planet Earth is made up of different materials, all of them involved in a wonderful array of interlocking movements, some slow, some fast. There is movement in the various layers that form the Earth's interior. The Earth's outermost layer is continually being destroyed and renewed as rocks form soil, then streams and winds transport the soil particles to new resting places. At the surface of the Earth all living things interact with each other and with the Earth's nonliving parts such as the atmosphere, seawater, and soils. It is especially necessary to understand and appreciate the repeating, cyclic pattern of activities at the Earth's surface.

To you, the reader, the facts of physical geology and related facts about the world of plants and animals offer a graphic explanation of these Earth systems. Your generation has now become responsible for their maintenance. The Earth sciences, of which physical geology is the foundation, are the study of our environment. To the average educated man or woman, regardless of occupation, this knowledge is necessary as never before. A broad view of the Earth's natural activities is an essential beginning to the person who aims at a professional career in any of the many aspects of environmental studies. The realization of the Earth's finite limits is creating a social revolution. However, an understanding of the limits can only come through scientific study of the Earth itself and how it works. This is the domain of *Geology, the science of the Earth.*

SUMMARY

1. All parts of the Earth interact. Rocks, soils, seawater, the atmosphere, and all living things influence each other in many complex ways.

2. The human population has become so large that it is now an important geological force.

3. The Earth has finite limits. The study of those limits and of the way all of the Earth's components interact is the domain of geology.

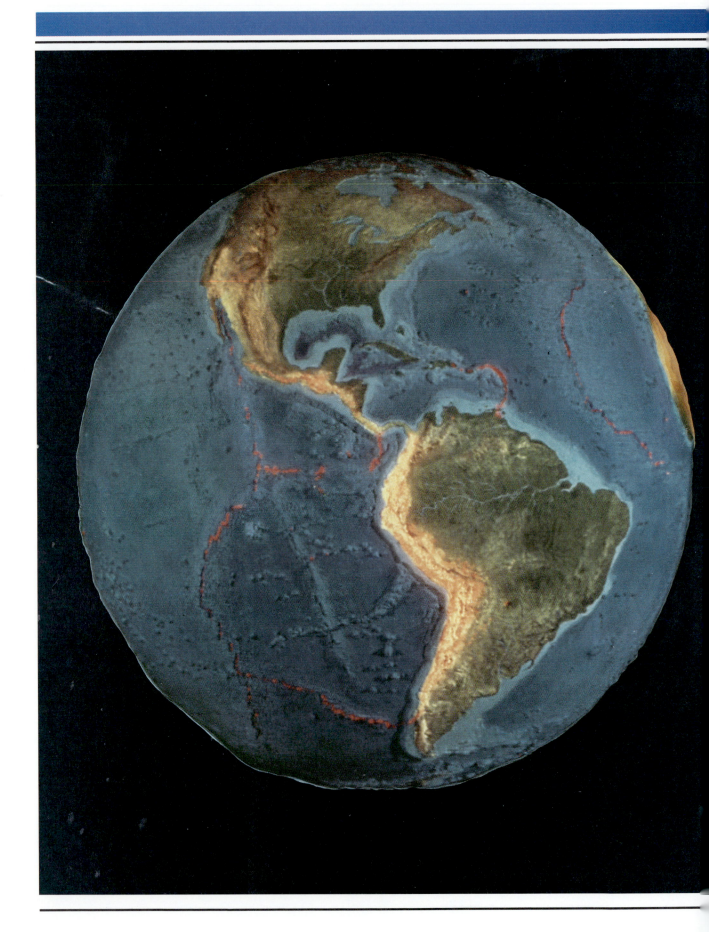

CHAPTER 2

The Earth: Inside and Out

Planet Earth and its distinctive topography. Much of the volcanism, and most earthquakes, are localized along the margins between tectonic plates.

MATERIALS AND PROCESSES

Living at the Earth's solid surface, we are aware of *materials* and *processes*. The *materials* are common substances such as rock, sand, clay, and organic matter. The *processes* change materials and move them around. Examples of processes are: the breakdown of materials to form soil; water flowing in a river and carrying mud or sand; surf pounding against a shore and making a beach; and air flowing (wind), carrying sand and dust. These and other processes expend energy derived from the Sun's heat as they work on, and wear down, the materials they continually move from place to place. The solid particles they are moving are mostly bits of rock. *Rock is any naturally formed, firm, and coherent aggregate mass of solid matter that constitutes part of a planet.* Note that the definition specifies a coherent aggregate of solid matter. A pile of loose sand grains is not a rock because the grains are not coherent. Nor is a tree a rock, even though it is solid. But coal, which is a compressed and coherent aggregate of twigs, leaves, and other bits of plant matter, is a rock.

Moving water and the solid particles in it are a very small sample of a great chain of processes that operate all over the Earth's surface. The water that flows off through gutters and rivers comes from rain, which comes from clouds, which in turn form from moisture evaporated from the ocean. As rivers empty into the ocean, the traveling water, ocean to land and land to ocean, completes a cycle and is ready to begin anew. This cycle, called the *hydrologic cycle*, is discussed in greater detail in Chapter 10. The bits of rock carried in the flowing water are part of another cycle, one that interlocks with the cycle traveled by water. The particles of rock came originally from the firm rock that forms all continents and that is continually being broken into particles. Like the water itself, the rock particles are on their way to the ocean, where they will be spread out and deposited, and will eventually form new rock. *The complex group of related processes by which the products of rocks that have been broken down physically and chemically, and then moved, is called* **erosion** ("wearing away").

All the activities involved in erosion, and also in the transport and deposition of the eroded materials, are together called **external processes** because they operate at or near the Earth's solid surface. The energy to drive external processes comes from the Sun.

It is not possible to see, directly, the interior regions of the Earth. Nevertheless, it is possible to reason from evidence such as earthquakes, volcanic eruptions, and the thrusting up of mountain ranges, that activities must also be occurring inside the Earth. *All activities involved in movement or chemical and physical change of rocks in the Earth's interior are called* **internal processes.** The energy that drives internal processes comes from the Earth's internal heat.

THE INTERNAL LAYERS

By measuring things indirectly, it is possible to infer a good deal about the interior portions of the Earth. One of the indirect measurements is the way the density of rock changes with depth; this can be calculated by measuring the speeds with which earthquake waves pass through the Earth. From such indirect measurements it can be deduced that the solid Earth does not consist of one single material, but must instead consist of distinct layers, like the layers of an onion. Unlike the onion, however, each layer has a different composition.

There are three such compositional layers (Fig. 2.1). At the Earth's center is the most dense of the three layers, the *core*. The core is a *spherical mass, largely of metallic iron, with admixtures of nickel, sulfur, silicon, and other elements.* It can be deduced that the admixed materials must be present because pure metallic iron would make the density of the core too high for the overall density of the Earth. *The thick shell of dense, rocky matter that surrounds the core* is called the **mantle.** The mantle is less dense than the core, but it is more dense than the outermost layer. Above the mantle lies the thinnest and outermost layer, the *crust,* which consists of *rocky matter that is less dense than the rocks of the mantle below.*

The core and the mantle have roughly constant thicknesses. The crust, on the other hand, is far from uniform and differs in thickness from place to place by a factor of nine. The crust beneath the oceans, called the oceanic crust, has an average thickness of about 8 km, whereas the continental crust, which is the crust that comprises the continents, ranges from 20 to 70 km in thickness (Fig. 2.2). The mantle and the core have very different compositions and the boundary between them is sharp. There are almost certainly compositional variations within the mantle, but we still know little about them. The crust, however, can be observed to be quite varied in its composition and in places seems to have a composition that differs little from the mantle below. Even so the boundary between the crust and mantle is distinct.

There is another kind of layering inside the Earth besides the compositional layering. This is a layering of physical properties, like the layering of ice floating on the water of a partly frozen lake. Just as H_2O can exist, according to its temperature, in different states (ice, water, gaseous vapor), so do the materials in the Earth change their physical

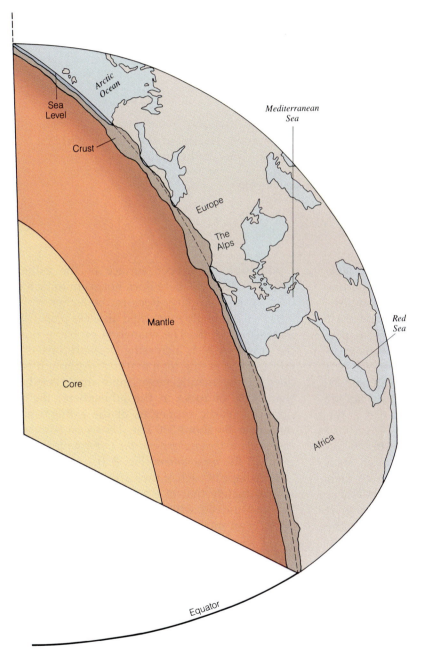

FIGURE 2.1 A slice of the Earth, revealing layers with distinctly different composition. The core is composed largely of metallic iron with an admixture of nickel, sulfur, silicon, and other elements. The mantle is a zone of dense rocky matter, and the crust (here shown with exaggerated thickness), is much less dense rocky matter with a composition distinctly different from that of the mantle. The slice cuts through the North Pole. Note that crust is thicker under the continents and thinner under the ocean floor.

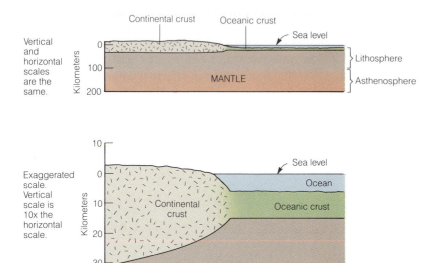

FIGURE 2.2 Cross section of the crust and upper part of the mantle. The crust is of two different kinds. Continental crust underlies the continents and is 20 to 70 km thick with an average of about 40 km; oceanic crust underlies the oceans and is only about 10 km thick.

properties as the temperature and pressure change. The regions where physical properties change do not coincide exactly with the compositional boundaries between core, mantle, and crust (Fig. 2.3). Within the core there is an inner region where pressures are so great that the metals present form a solid. Surrounding the inner core is a zone where temperature and pressure are so balanced that the metals melt and exist as a liquid. Analogous changes occur in the upper part of the mantle.

Starting at a depth of about 100 km below the Earth's surface, rocks in the mantle reach such high temperatures that they lose much of their strength. *The region of the mantle where rocks become plastic, like toffee or tar, and are easily deformed is* called the **asthenosphere** (a word meaning the "weak sphere"). The asthenosphere continues to a depth of about 350 km. Above the asthenosphere, and corresponding approximately to *the outer 100 km of the solid Earth*, is a region *where rocks are harder and more rigid than those in the plastic asthenosphere.* This hard outer region is called the **lithosphere** (meaning the "rock sphere").

The boundary between the lithosphere and the asthenosphere is distinct, but it does not correspond to a sudden change of composition. The composition at the base of the lithosphere and the top of the asthenosphere is, so far as we can tell, identical. The boundary between them is merely one where rising temperature causes the physical properties of rock to change rapidly.

Although the upper surface of the asthenosphere is distinct, there does not seem to be a distinct lower boundary. Rock in the asthenosphere becomes gradually more rigid and less plastic with increasing depth. The change is probably due to the way in which increasing pressure offsets the effect of increasing temperatures. Although rocks deep in the mantle never quite reach the hard, rigid state of rocks in the lithosphere, the pronounced plastic properties of the asthenosphere disappear by about 350 km. *The region between the base of the asthenosphere and the core-mantle boundary is called the* **mesosphere** (Fig. 2.3).

The shape of each of the Earth's internal layers is approximately that of a sphere. Why this should be, and indeed why the Earth has the shape it does, is an interesting question.

THE SHAPE OF THE EARTH

The Earth is an approximately spherical body, 12,756 km in diameter, that rotates on its axis once a day; the axis is inclined at an angle of 23.5° to *the plane of the Earth's orbit around the Sun*, known as the **plane of the ecliptic.** All of the other planets rotate on their axes too, each is nearly spherical and each moves in an orbit around the Sun that is approximately coplanar with the Earth's orbit (Fig. 2.4).

Everyone knows that an object on a spinning disc will fly off unless it is held securely in place.

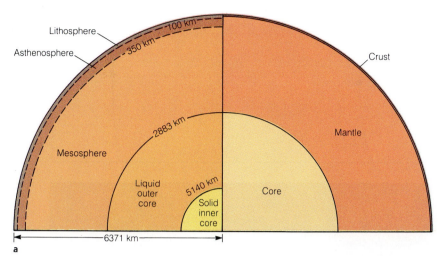

a

FIGURE 2.3 Layering of physical properties in the Earth. (*a*) Right side shows the compositional layering of crust, mantle, and core. Left side depicts changing physical properties with depth. Note that compositional changes and physical property changes do not coincide. (*b*) Expanded view of the upper portion of the left side of (*a*). Lithosphere is hard and rigid by comparison with the asthenosphere. The asthenosphere is a region in the upper mantle where increased temperature makes the rock weak, plastic, and easily deformed. The mesosphere, lying below the asthenosphere, is stronger and more rigid than the asthenosphere because even though it is hot, increased pressure offsets the effects of temperature.

The force that pulls outward on a spinning body is called the centrifugal force. An easy way to sense the centrifugal force is to tie a small weight on the end of a string and then spin the string around your head. The faster you spin and the longer the string, the greater is the centrifugal pull. The great English scientist, Sir Isaac Newton, pointed out in 1666 that objects on a spinning Earth must be subjected to a centrifugal force. Objects on the Earth would necessarily fly off into space, he reasoned, unless a force stronger than the centrifugal force held them in place. This was the starting point of the reasoning that led him to enunciate the universal law of gravitational attraction. Newton expressed his law in the following manner:

$$F \text{ is proportional to } \frac{M_1 \times M_2}{d^2}$$

where F is the force of gravitational attraction, M_1 and M_2 are the masses of two attracting bodies, and d is the distance between the center of M_1 and the center of M_2.

Clearly, the larger M_1 and M_2 are, and the smaller d is, the greater F (the force of attraction) will be. Nevertheless, unless the attracting bodies are very large, we cannot easily sense the attract-

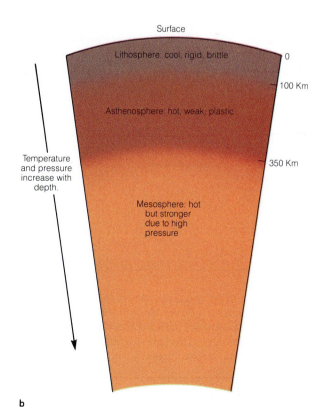

b

a

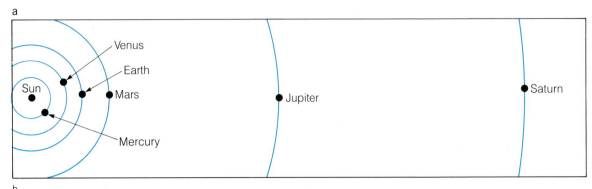

b

c

	Mercury	Venus	Earth	Mars	Jupiter	Saturn	Uranus	Neptune	Pluto
Diameter (km)	4880	12,104	12,756	6787	142,800	120,000	51,800	49,500	6000?
Mass (Earth = 1)	0.06	0.81	1	0.11	317.9	95.2	14.6	17.2	?
Density (water = 1)	5.4	5.2	5.5	3.9	1.3	0.7	1.2	1.7	?
Number of moons	0	0	1	2	13	10	5	2	0
Length of day (in Earth hours)	1416	5832	24	24.6	9.8	10.2	11	16	153
Period of one revolution around Sun (in Earth years)	0.24	0.62	1.00	1.88	11.99	29.5	84.0	165	248
Average distance from sun (millions of kilometers)	58	108	150	228	778	1427	2870	4497	5900

FIGURE 2.4 Orbits and properties of the planets. (*a*) The orbits of the planets around the Sun are all very close to the same plane. (*b*) Arranged in order of their relative positions, outward from the Sun, the planets are shown in their correct relative sizes. The Sun, 1.6 million km in diameter, is 13 times larger than Jupiter, the largest planet.

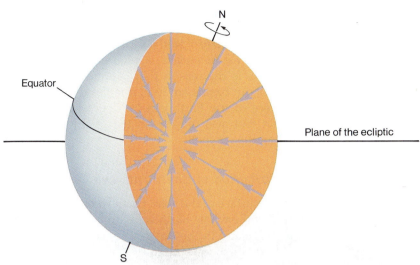

FIGURE 2.5 Cut-away view of the Earth, demonstrating how the force of gravity constantly pulls every object toward the center of mass. The direction of pull is radial and the result tends to give Earth a spherical form. Note that the Earth's axis of rotation is not perpendicular to the plane of the ecliptic—the plane of the Earth's orbit around the Sun—but is tilted 23.5° off the vertical. It is the tilt of the axis that causes the seasonal climatic cycles.

ing force between them because gravity is not a very strong force. We cannot readily sense the attraction between small objects like knives and forks, or even somewhat larger objects like automobiles and houses. However, if a body is very large, as the Earth is, its gravitational attraction becomes very large too. Even a small object will fall to the Earth's surface because the gravitational pull is so great. Indeed, were it not for the Earth's strong gravitational attraction holding the air in place, the gases of the atmosphere would simply leak away into space.

The Earth's gravity is an inward-acting force (Fig. 2.5) which tends to pull all objects toward the center of mass. The Earth's gravitational pull is radial, which means that regardless of where an object lies on the Earth, the gravitational pull is along the straight line between the object and the center of the Earth. Where all particles in a deformable mass of material are pulled equally toward the center, the shape of the mass becomes a sphere. Gravity, and the fact that the seemingly solid Earth can be deformed, is the reason the Earth is round. The same two factors are the reasons that the internal layers are spherical—or at least approximately spherical.

Newton realized that the balance between the centrifugal force, due to the Earth's rotation, and the gravitational force must distort the shape of the Earth from a sphere to an ellipsoid. This is so because while gravity is a radial force that pulls all objects toward the Earth's center, the opposing centrifugal force acts in a direction perpendicular to the axis of rotation (Fig. 2.6). The centrifugal force is greater the farther away the object is from the axis of spin; that is, the centrifugal force is greatest at the equator, and least at the poles. The interaction of the two forces produces an Earth that is slightly flattened at the poles and bulged out at the equator. As a result, the Earth's radius is 6378 km at the equator, but only 6357 km at the poles. This departure from a perfect sphere has an interesting effect on weight; a man who weighs 90 kg at the equator, weighs 90.5 kg at the North Pole. But we should not get a wrong picture of the Earth's departure from a perfect sphere. If we could shrink our planet to the size of a basketball and keep its exact shape, it would seem to be a perfect and highly polished sphere.

The ellipsoidal shape of the Earth corresponds approximately to the surface of sea level. Indeed, it is possible to calculate that if the entire Earth were composed of water, it would still be an ellipsoid of almost identical shape. This in turn is proof that the seemingly solid Earth, and all its compositional layers, must, over time, be capable of flowing like

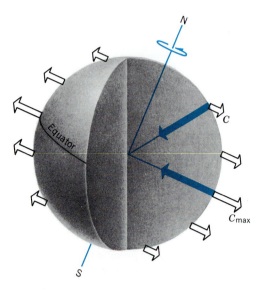

FIGURE 2.6 The force of gravity, shown in blue arrows, acts along a radius. The centrifugal force caused by Earth's rotation (white arrows) acts in a direction perpendicular to the axis of rotation. The centrifugal force is a maximum at the equator (C max) and zero at the poles. As a result of the opposing forces, Earth bulges out at the equator but is flattened at the poles.

liquid. Planet Earth is not all water, however, and its actual shape departs from a mean sea-level ellipsoid because of continents, mountains, volcanoes in the ocean, and other features that cause irregularities. These irregularities are due to an important property—rocks may be able to flow, but they also have considerable strength. Indeed, without changing either the temperature or pressure, a rock can be both a hard, brittle solid and a weak plastic one. This seeming enigma has a straightforward explanation. Rock has properties akin to a cold stick of butter. If the stick is bent quickly, it will break. However, if it is bent very slowly it will not break, but instead will remain bent. Similarly, when rocks are subjected to a rapid buildup of force, they behave as brittle solids, and they break. When the force is slowly applied, rocks will flow plastically. The subject of rock strength and rates of deformation is a complicated one and is further discussed in Chapter 15.

The fluid-like properties of the Earth have many important consequences. One consequence is that when a mass of low-density rocks is piled up to form a mountain, the mountain range has a low-density root beneath it, just as a floating iceberg has a large mass of ice beneath the sea to balance the small tip above water. Indeed, all portions of the crust and the lithosphere are, in a sense, floating. *The property of ideal flotational balance among segments of the lithosphere* is referred to as **isostasy**. Isostasy is the reason that continents stand high and ocean basins are low; it is the reason that the surface of the Earth can be depressed by an ice sheet, but also the reason that it slowly rises up when the ice melts.

Distribution of Continents and Oceans

If the surface of the solid Earth were smoothed out so that the whole Earth were uniformly covered by all the water in the ocean, the depth of the ocean would be 2.6 km. In reality the oceans cover only 71 percent of the world's surface, so the average depth must be greater than 2.6 km. In fact, the average depth is 3.8 km. The depth is very irregular, however; the greatest depth of 11 km is reached near the island of Guam, in the western Pacific. The remaining 29 percent of the world's surface is occupied by land, the average height of which is 0.8 km above mean sea level.

If, by some mysterious means, it were possible to remove briefly all the water from the oceans and then to view the dry Earth from a spaceship, it would be observed that the continents rise abruptly and stand about 4.6 km above the average

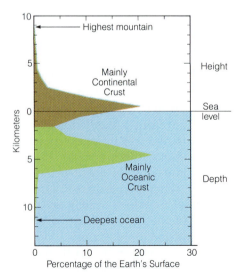

FIGURE 2.7 Distribution of the areas of the Earth's solid surface expressed as a percentage. Note that the surface of the continental crust is considerably higher than the surface of the oceanic crust. (*Source:* After Wyllie, 1976.)

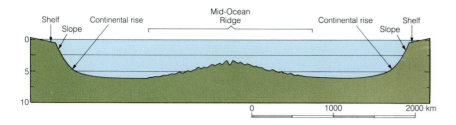

FIGURE 2.8 Idealized diagram of a section across portion of the Atlantic Ocean showing the major topographic features.

depth of the ocean floor. The reason this is so is that continental crust has a density of about 2.7 g/cm³, while oceanic crust has a density closer to 3.2 g/cm³. In a sense, both kinds of crust can be considered to be "floating" on the mantle. Continental crust is light, so it stands high. Oceanic crust is dense, so it sits lower (Fig. 2.7). If, while the water were still removed from the ocean, the margin of a continent were examined at some place, such as the southern part of the Atlantic (Fig. 2.8), it would immediately be seen that some of the ocean water must spill out of the ocean basin onto the continent because shorelines do not coincide with the steep continental margin. As a result there is *a submerged*

platform of variable width that forms a fringe around a continent known as the **continental shelf.** The actual edge of the ocean basin is the **continental slope,** *a pronounced slope beyond the seaward margin of the continental shelf.* Considering the edge of the continents to be marked by the continental slope rather than by the shoreline, only 60 percent of the Earth's surface is occupied by ocean basins, while 40 percent is occupied by continents. About a quarter of the continental crust, or 11 percent of the Earth's surface, is covered by water (Fig. 2.9). At the base of the continental slope there is *a region of gently changing slope where the floor of the ocean basin meets the margin of the continent; this is referred*

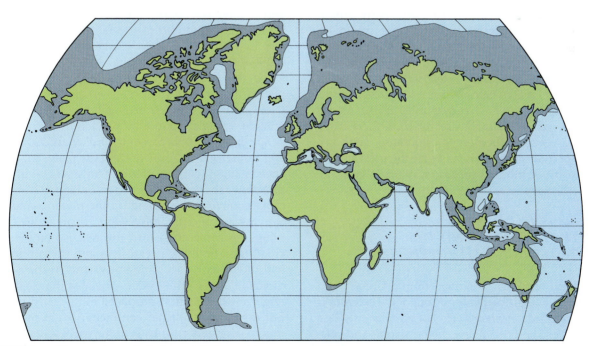

FIGURE 2.9 The continental shelves and slopes (shown in gray), together, add about 15 percent to the areas of the continents. The map projection exaggerates the shelves in the arctic region, which are nevertheless very extensive.

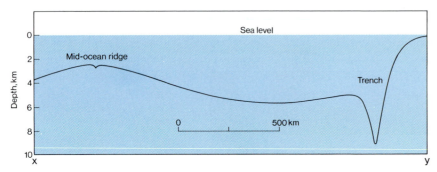

FIGURE 2.10 Profile of the sea floor, along the line *X-Y* in Figure 2.11, comparing the shapes of a mid-ocean ridge and a trench. Where a trench is flanked by a continent, the continental side is steeper (to the right), than the oceanic side.

to as the ***continental rise.*** The rise is actually part of the floor of the ocean basin, but it is a distinctive part because it is covered by a thick pile of sedimentary debris shed from the adjacent continent.

Beyond the continental slope and rise, lies the strange, rarely seen world of the deep ocean floor. With newly perfected devices for sounding the sea-bottom, for making submarine dives of limited time, and for sampling its sediment, teams of oceanographers and seagoing geologists have explored and mapped the ocean floor so that almost as much is now known about the seafloor as is known about the land surface. Particularly prominent features include great oceanic mountain chains, called ***mid-ocean ridges,*** which are *continuous rocky ridges on the ocean floor, many hundred to a few thousand kilometers wide, with a relief of more than 0.6 km* and *long, narrow, very deep and arcuate basins in the seafloor,* called ***trenches*** (Fig. 2.10). As we shall see, all of the major features on the Earth's surface, whether submerged or on land, arise as a result of internal processes. By far the most important result of internal processes is the lateral motion of lithosphere.

PLATE TECTONICS AND THE EVER-CHANGING LITHOSPHERE

The word tectonics is derived from a Greek word *tekton* that means carpenter or builder. ***Tectonics*** is *the study of movements and deformation of the crust on a large scale.* The lithosphere, being strong and relatively rigid, can move and slide as a coherent layer over the much weaker and more easily deformed asthenosphere. ***Plate tectonics*** is *the special branch of tectonics that deals with the processes by which the lithosphere is moved laterally over the asthenosphere.*

The very idea that a process such as plate tectonics might occur is only about 20 years old. Because it is so new many of the details concerning the process are still being investigated. The discoveries and new understandings that have come from plate tectonics are so profound, however, that the concept has given rise to a modern revolution in the earth sciences. The semirigid rocky lithosphere not only moves, but it does so in a series of plate-like pieces; it is movement of the plates that causes continents and ocean basins to be where they are and to have the shapes they do. The plates range from several hundred to several thousand kilometers in width. Trying to visualize them, we can think of the skin of an orange. Imagine the orange to be the Earth, its skin the lithosphere. If we peeled the orange, then replaced all the pieces of peel, each piece, large or small, would be analogous to a single plate of lithosphere.

The lithosphere is presently broken into six large and numerous small plates and the plates move at speeds ranging from 1 to 12 cm a year—in the directions shown by the arrows in Figure 2.11. As a plate moves, everything on the plate, including the capping of crust, moves too. If part of the capping is oceanic crust and the rest is continental crust, then both the ocean floor and the continent will move with the same speed and in the same direction as the entire plate. The idea that seafloor might move was first proposed in the early 1960s. But the realization that continents might move is much older and goes back to the early years of the present century. The idea of continental movement was most forcefully and carefully proposed by a German scientist, Alfred Wegener. The concept came to be called *continental drift.* When first proposed the idea did not receive widespread support because there was no adequate explanation as to how it could happen. Plate tectonics has provided the answer.

The original suggestion for continental drift was that continents must somehow slide across the seafloor. It was soon realized that friction and the strength of rocks would not allow this to happen. Eventually, following the discovery that the seafloor also moves, and that the asthenosphere is weak and easily deformed, geologists realized that the entire lithosphere was in motion, not just the continents.

Yet there are many unanswered questions concerning plate tectonics. One of the questions concerns the shapes and sizes of all plates from past ages. There is convincing evidence to demonstrate that at times in the past there have sometimes been fewer and at other times, apparently, more plates than at present. Plates change both in size and shape because new, smaller plates can form through the breaking-up of larger plates. Also, larger plates can form by the collision and welding

together of smaller plates. As discussed in later chapters, there are ways by which past breakups and weldings can be inferred from the geological record, but the evidence is sometimes difficult to find, and rarely easy to decipher. The complete deciphering of the Earth's tectonic history will be a long and complex task and possibly it may never be completed.

When we examine how plates move, we think of rafts, but a better analogy is conveyor belts. In a conveyor, the belt continually appears from below, moves along the length, then turns down, and passes temporarily from sight as it completes its circuit. Although broad and irregular rather than long and narrow, a plate of lithosphere acts like the top of a slowly moving conveyor belt. Along one edge (not the side) of most plates is a long, clearly defined fracture in the oceanic crust that coincides with the mid-ocean ridge. The plate moves away from the ridge just as if it were a continuous belt rising up the fracture from the mantle below. The analogy is only partly correct, because the plate is not rising as a solid ribbon. It is being built, or rather added to, as it rises.

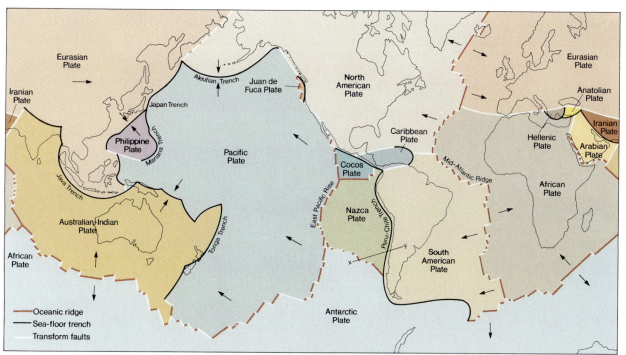

FIGURE 2.11 Six large plates of lithosphere and several smaller ones cover the Earth's surface and move continuously, in the direction shown by arrows. Plates have three kinds of margins: (1) growing or spreading margins, delineated by oceanic ridges, (2) subduction margins, delineated by sea-floor trenches, and (3) transform faults. The profile shown in Figure 2.10 lies along the line X-Y.

It is not possible to see into the mantle beneath the mid-ocean ridges, but it is possible to infer what must be happening. Hot, plastic rock in the asthenosphere must rise toward the surface beneath the ridges. As a result, the boundary between the lithosphere and the asthenosphere is very shallow near the ridges. Also, some small portion of the asthenosphere must melt, giving rise to magma. *Magma* is defined as *molten rock material, together with any suspended crystals and dissolved gases, that forms when temperatures rise and melting occurs in the mantle or crust.* The magma that forms beneath the mid-ocean ridges rises upward to the top of the lithosphere where it cools and crystallizes to form new oceanic crust (Fig. 2.12).

Another reason that the analogy between a plate and a conveyor is only partly correct is that two plates form and move away in opposite directions from the mid-ocean ridge. The narrow depression that marks the center of a mid-ocean ridge is actually the surface expression of the join between the two plates. It is also the top of the new oceanic crust. Because plates move, or spread outward, away from the mid-ocean ridge, *the new, growing edge of a plate* is called a **spreading edge.**

Formation of new oceanic crust is a continuous process. Movement of lithosphere away from the oceanic ridge, like the movement of a conveyor belt, is a continuous process also. Near the spreading edge the lithosphere is thin and it has a low density because it is heated and expanded by the rising magma. As the lithosphere moves away from the spreading edge it cools, contracts, and becomes denser. Also, the depth of the boundary between the lithosphere and the asthenosphere moves down. Finally, at a distance of a thousand or more kilometers from the spreading edge, the lithosphere and its capping of oceanic crust is so cool it is more dense than the hot, weak asthenosphere below and it starts to sink downward. Like a conveyor belt, old lithosphere disappears back into the mantle. *The edges along which plates of lithosphere turn down into the mantle,* called **edges of consumption** or **subduction zones,** are marked by deep trenches in the seafloor.

As the moving strip of lithosphere passes from view and sinks slowly through the asthenosphere into the depths of the mantle it can no longer be seen. Consequently, what happens next is still largely conjecture. On one point we can be quite certain: The lithospheric plate does not turn under, as a conveyor belt does, and reappear at the

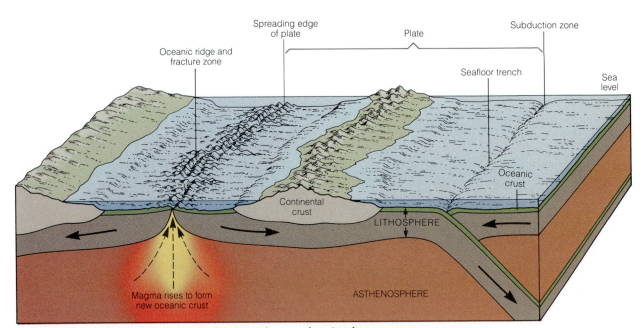

FIGURE 2.12 Cross section of the Earth's outer layers, showing how magma (dotted arrows) moves from the mantle upward into spreading centers in the ocean floor, and cools there to form new oceanic crust. To acommodate the new materials, the lithosphere (solid arrows) moves away from the fracture zone and eventually sinks slowly down into the mantle again, where it is reheated, and eventually mixed again with the mantle.

spreading edge; rather, it is reheated and, presumably, slowly remixed with the material of the mantle.

The trip we have described is one for plates of lithosphere that are capped by oceanic crust. As a result, all oceanic crust is geologically young because old crust has all returned to the mantle. Unlike oceanic crust, continental crust is not recycled into the mantle; it takes a shorter trip that ends more suddenly. Continental crust is lighter and less dense than even that part of the mantle in the hot asthenosphere. As a result, continental crust is too buoyant to be dragged downward on top of the sinking lithosphere. So, in continent-sized pieces, such crust moves from place to place on the Earth's surface, much as a log of driftwood, partly embedded in an ice floe, floats on a lake or river. Movement stops or a plate changes direction when something happens, such as one continental mass colliding with another or a new spreading edge splitting the continental crust apart. Because continental crust does not sink down into the mantle, most of the remaining evidence concerning ancient plates and their motions is recorded in the scars carried by ancient continental rocks. Some of the evidence is remarkable.

In parts of the Sahara Desert convincing evidence indicates that some 500 million years ago that region, hot and dry today, was covered by an enormous ice sheet, similar to the one that now covers the Antarctic continent. The Sahara region then lay near the South Pole, and the plate of which that portion of Africa is a part has since traveled slowly northward.

Substantiation of an entirely different kind can be found in the long, more-or-less linear belts of highly deformed rocks called mountain ranges. The belts are found along, or close to, the edges of continental masses that have been involved in continental collisions. There is an old term, *orogeny,* that refers to *the tectonic processes by which large regions of the crust are deformed and uplifted to form mountains.* The causes of orogenies were obscure before plate tectonics provided an answer.

Further evidence can be seen in the rocks on both sides of the Atlantic Ocean. When a new spreading edge forms and splits a large mass of continental crust, a new strip of growing oceanic crust separates the two pieces. Africa and Europe on one side with the Americas on the other, provide an example. Two hundred fifty million years ago there was no Atlantic Ocean. Instead, the continents that now border it were joined together into a single huge continent. The place where New

York now stands was then a dry, arid place as far from the sea as central Mongolia is today. About 200 million years ago, for reasons that we do not yet fully understand but that presumably involved changes in the mantle below, new spreading edges formed. They split the ancient continent into the pieces we see today. These fragments then drifted slowly into their present positions. At first the Atlantic Ocean was a narrow body of water that separated North America from Europe and North Africa, but as movement continued, the ocean widened and lengthened, splitting South America from Africa and then growing to its present form. The Atlantic is still growing wider, by about 5 cm each year. There is abundant evidence to mark where the torn edges formerly fitted together. In one region, pieces of mountain ranges that once formed a long, narrow mountain belt, like the Rocky Mountains of today, have been pulled apart so that now they lie on the two sides of the Atlantic. If these pieces are fitted back together, the now deeply eroded mountains fit like matched pieces of a jigsaw puzzle (Fig. 2.13).

A consequence of moving plates of lithosphere arises from the geometry of a sphere. It is impossible to cover a spherical Earth with more than two rigid plates bounded only by edges of expansion and consumption. There has to be a third kind of edge, analogous to the sides of a conveyor belt, along which plates simply slip past each other. These edges of slipping are great vertical fractures—or to use a term defined in Chapter 15, *transform faults*—that cut right down through the lithosphere. One transform fault, much in the public eye because it continually threatens earthquakes, is the San Andreas Fault in California. This fault separates the American Plate, on which San Francisco sits, from the Pacific Plate, on which Los Angeles sits. As the two plates slide past each other, Los Angeles is slowly moving northwest toward San Francisco; some millions of years in the future, should the cities still exist, Los Angeles and San Francisco will become joined as a single large city.

THE EXTERNAL LAYERS

Above the crust sits the ocean, the lakes, the streams, and other bodies of water, and above them the atmosphere. The interface between the crust and the water + air is a region of intense activity, for it is here that erosion occurs. Erosion

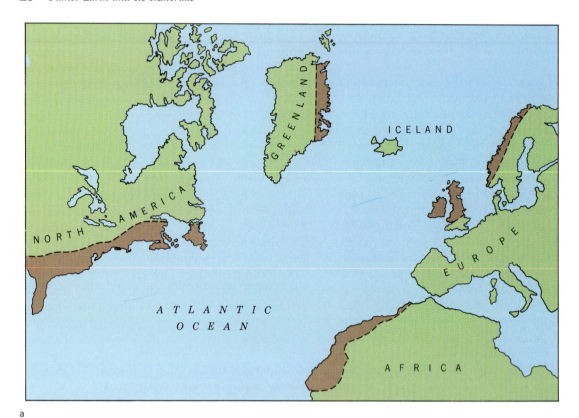

a

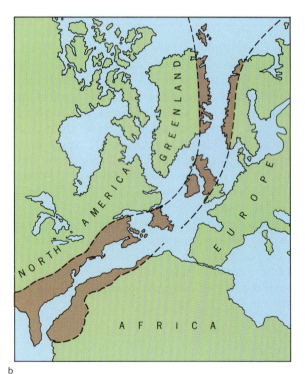

b

FIGURE 2.13 Opening of the Atlantic Ocean. (*a*) Rock in eroded fragments of similar mountain belts (brown)—each 350 to 470 million years old—is found on both sides of the Atlantic Ocean. (*b*) When continents are moved and fitted together as they were 200 million years ago, the fragments are seen to form a continuous belt. The reconstruction provides evidence that the present continents were once part of a larger land mass broken up by the moving lithosphere. Note that Iceland is not present in the reconstruction. It is a young land mass, and is a piece of the ocean ridge that marks the line along which the continental separation occurred. (*Source:* Adapted from P. M. Hurley, 1968.)

either on the blanket of loose rock debris or close to the water–air interface. It is helpful to think of the water, the air, living matter, and the blanket of eroded rock particles as layers or envelopes that surround the solid Earth, analogous to the internal layers. These four external layers are termed the *hydrosphere*, the *atmosphere*, the *biosphere*, and the *regolith*, respectively.

Hydrosphere

The **hydrosphere** is the "water sphere," embracing *the world's oceans, lakes, streams, water underground, and all the snow and ice, including glaciers.* Most of the water in the hydrosphere resides in the oceans,

continually breaks down rock and moves the broken particles around. As a result, the crust is mantled by an irregular blanket of loose rock debris. Most of the world's plants and animals live

but important parts are also to be found in lakes, streams, groundwater, and ice sheets. The hydrosphere is discussed in more detail in Chapter 10. The **cryosphere** is a small part of the hydrosphere and is defined as *that portion of the hydrosphere that is ice, snow, and frozen ground.* The cryosphere is discussed in detail in Chapter 13.

Atmosphere

The **atmosphere** is *the air sphere, consisting of the mixture of gases that together we call air.* It penetrates into the ground, filling the openings, small and large, that are not already filled with water. The atmosphere is the busiest sphere and is never quiet. Some part of it is always moving, as we are well aware when we feel a wind blowing.

Regolith

Where the atmosphere and the hydrosphere are in contact with rocks, reactions take place between them. This is the beginning of erosion. For example, the atmosphere with its content of water vapor acts upon **bedrock,** *the continuous mass of solid rock that makes up the crust,* breaking it up mechanically and causing it to decay chemically. We can observe the process at any **exposure** (or **outcrop**), *a place where solid rock is exposed at the earth's surface.* As a result, the surface rock, although still a solid, is no longer a continuous solid. It has been subdivided into many pieces, most of them very small particles. This broken-up part of the crust has a separate name, the **regolith** ("blanket rock"), because it lies like a blanket draped over (and in many places grading into) the continuous solid rock beneath. It is defined as *the blanket of loose, noncemented rock particles that commonly overlie bedrock.* We would generally have to use a hammer or even a high-speed drill to collect a sample of bedrock, but a shovel or pick would ordinarily be enough for sampling regolith.

Not all regolith has been broken down and left in place. Some of it has been moved and set down

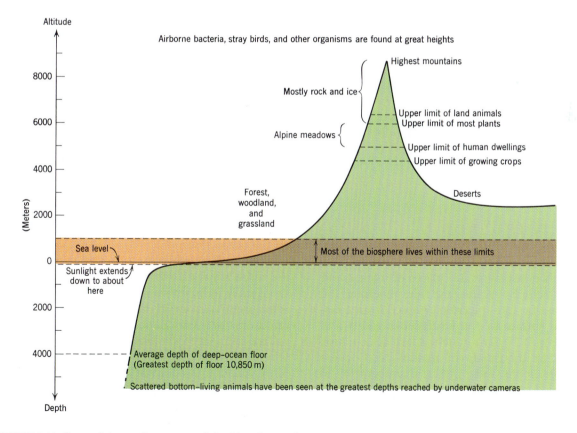

FIGURE 2.14 Some data on the extent of the biosphere, shown in a schematic way.

in a new site. It is on its way (discontinuously) to the ocean. *Regolith that has been transported by any of the external processes* is called **sediment** (a word that means "settling").

In some places, where bare bedrock is exposed at the surface, there is no regolith at all. In other places regolith is 100 m or more thick. In most land areas the upper part of the regolith forms soil (Chapter 9) in which plants grow, including the crops that are the principal food of human populations.

Biosphere

The **biosphere** (the "life sphere"), as its name implies, is *the totality of the Earth's organisms and, in addition, organic matter that has not yet been completely decomposed.* It embraces innumerable living things, large and small, grouped into millions of different species. We humans are one of the species. The composition of the biosphere is distinctive; its chief constituents are compounds of carbon, hydrogen, and oxygen, although it includes other chemical elements as well. All of these materials are drawn from other spheres, but they are fixed in patterns peculiar to the biosphere. Whether they form parts of living organisms or dead ones, the organic compounds remain a part of the biosphere until they have been destroyed by chemical alteration.

The biosphere extends through, or at least into, each of the outer envelopes (Fig. 2.14). Although living things manage to exist through an enormous vertical range—nearly 20,000 m—they are very sparse at great heights and great depths. Most of them are crowded into a narrow zone that extends from a little below sea level to 1000 m or so above it.

CYCLES: INTERACTIONS BETWEEN THE LAYERS

The Earth's outer layers are places of intense and continual activity. Water and air penetrate the regolith and far into the crust. Viewed from a distance, the surface of the Earth looks sharp and distinct (Fig. 2.15), but in actual fact, it is a zone hundreds to thousands of meters deep where chemical reactions and physical disintegration of rock proceed continually.

Rock in the crust crumbles slowly and becomes regolith. Driven by energy from the Sun, water and air move as flowing currents. In very cold places, water in the air freezes and falls as snow.

FIGURE 2.15 Satellite image of the delta at the mouth of the Mississippi River. The apparent sharpness of the line that separates the lithosphere from the hydrosphere is misleading because the two spheres blend gradually. Off the river's mouth are plumes of turbid water, loaded with suspended particles of rock derived from erosion of the land. When the suspended particles settle, they build up a delta. Compare Figure 11.31.

The snow in turn forms glaciers. Air currents, water currents, and glaciers pick up particles and may move them long distances. The biosphere penetrates air, water, and regolith so thoroughly that living things can be found in almost every available nook and cranny within the other spheres, filling them all with additional activity. Partly through physical processes, but more importantly through biochemical reactions, organisms help to crumble rock and also to deposit solid substances on the ocean floor.

When rock weathers to form sediment, some of the more soluble constituents from the rock dissolve and eventually concentrate in the ocean. This is the origin of many of the salts in seawater. When raindrops form they dissolve gases from the atmosphere and carry them down to the Earth's surface where they react to form new minerals in the soil. Since material is continually transferred between the Earth's spheres, why should the composition of the atmosphere be constant? Why doesn't the sea become saltier, or fresher? Why does rock two billion years old have the same composition as rock only two million years old? The answers to these questions are the same. The chemical elements and rock particles follow cyclic paths. "Wheels within wheels" is an apt way to describe the movement.

Cycle is a term that describes *a sequence of recurring events.* The rotation of the Earth on its axis, the circling of the Earth around the Sun, and the circling of the Moon around the Earth are all cycles. One result of the first cycle is the daily heating and cooling of the Earth's surface. An important effect of the first and third cycles is the tide, the recurring rise and fall of the ocean surface. Cycles of one sort or another control almost everything that happens on the Earth. Even the growth, the development, and the death of human beings is part of a cycle of events—the life cycle. Our human economy involves many kinds of cycles. For instance, we cycle glass (but we call it *recycling*), a process in which bottles are ground up and melted down to form a liquid from which new bottles are made. We "recycle" paper, which is pulped and made into new paper. Thus, any single particle of glass or paper substance may pass through the same series of states or forms again and again.

The operation of all the Earth's natural processes, both internal and external, and the interactions between the various compositional layers, all involve distinct cycles. Chemical cycles may have many local circuits, and the circuits may interact in many ways; however, if we carefully measure the circuits, we find that in any sphere (such as the atmosphere) the chemical elements added are balanced by those removed. Thus, as the cycles roll onward, the spheres maintain their compositions, or, if changes occur, they may happen exceedingly slowly, often over millions of years. The spheres are large reservoirs and between the reservoirs there are flows, or fluxes, of materials that balance out and keep the reservoir compositions nearly constant. Some of the most important questions to be answered about human interactions with the environment concern the way we are changing the natural fluxes through activities such as the burning of fossil fuels, the clearing of forests, intensive agriculture with massive additions of fertilizers to the soil, and the mining of ever-larger amounts of mineral resources. How will the reservoirs respond to the changing fluxes brought about by human activities?

The Rock Cycle

Although some products of weathering are soluble and are carried away in solution by streams and rivers, most are subdivided, loose particles that are carried away in suspension. Both the dissolved and suspended materials can be later deposited as sediment and eventually become *sedimentary rock.* The definition of sedimentary rock is *any rock formed by chemical precipitation or by sedimentation and cementation of mineral grains transported to a site of deposition by water, wind, or ice.* Sedimentary rocks constitute one of the three major rock families. The second major rock family is *igneous rock* (named from the Latin word *igneus,* meaning fire). It is *rock formed by the cooling and consolidation of magma.* The final major rock family is *metamorphic rock* (named from the Greek words *meta,* meaning change, and *morphe,* meaning form; hence, change of form). Metamorphic rock is *rock whose original compounds, or textures, or both, have been transformed to new compounds and new textures by reactions in the solid state as a result of high temperature, high pressure, or both.* Metamorphism, the process that forms metamorphic rocks, is analogous to cooking. When meat is put in the oven, it undergoes a series of chemical reactions as a result of the increased temperature. As a result, cooked meat looks and tastes very differently from raw meat. When sedimentary or igneous rocks are buried deep in the crust, they undergo chemical reactions too and are said to have been metamorphosed.

Most rock in the crust has formed, initially, from magma. It is estimated, for example, that 95 percent of all rock in the crust is either igneous rock or metamorphic rock derived from rock that was originally igneous. However, we can see in Figure 2.16 that most of the rock we actually see at the Earth's surface is sedimentary. The difference arises because sediments are products of external weathering processes, and as a result they are draped as a thin veneer over the largely igneous crust below.

The internal processes that form magma, and that in turn lead to the formation of igneous rock, interact with external processes through erosion. When a body of rock is subjected to erosion, the eroded particles form sediment. The sediment may eventually become cemented, usually by substances carried in ground water, and thereby converted into new sedimentary rock. In places where such rock subsides, it can reach depths at which pressure and heat cause new compounds to grow, thus forming metamorphic rock. Sometimes great enough depths can be reached so that the internal heat is capable of melting the rock, converting it to magma. The new magma can then move upward through the crust, where it can cool and form another body of igneous rock. If uplift occurs, due to internal processes, erosion can gradually wear down the surface of the land to a depth so great that the top of the new body of igneous rock can be uncovered and begin to be eroded. Once again the igneous rock is attacked by weathering and broken up, its particled waste starting once more on its way to the sea.

So far as we can tell, the sequence of events described above has occurred again and again throughout the Earth's long history. The events are just one of many sequences by which materials in the Earth's internal layers interact with materials in the external layers. Indeed, it seems possible that the compositions and balances of each layer, whether internal or external, depend to some extent on all the other layers. Many details of the interactions between internal and external layers are still subjects of research. An important part of the interactions involves tectonics. The internal processes that move the Earth's surface up and down and cause plates of lithosphere to move laterally are tectonic activities. Without tectonism, rock would not continually be uplifted and exposed to erosion.

The cycle that most directly connects internal and external layers is the rock cycle depicted in Figure 2.17. There are two parts to the cycle, one involving continental crust, the other involving oceanic crust. The example involving igneous rock and sedimentary rock, described above, is simply one possible flux among a great many that occur in the rock cycle of the continental crust. As shown in Figure 2.17, the cycle can be interrupted by short circuits at several places. For example, some bodies of sedimentary rock are never metamorphosed, never melted, or never even deeply buried before they are uplifted and eroded. Whether the circuits are long or short, the continental crust is continually being recycled. Because the mass of the continental crust is large, the average time a rock takes to complete the cycle, from erosion, through rock, to erosion again, is long. Time estimates vary, and they are difficult to make, but the average age of all rock in the continental crust seems to be about 650 million years.

By contrast with the cycle of the continental crust, the oceanic crustal cycle is rapid. The magma that rises to form new oceanic crust forms hot igneous rocks and these in turn react with seawater. In the reaction process, some of the constituents in the hot rock, such as calcium, are dissolved in the seawater. Other constituents already in the seawater, such as magnesium, are deposited in the igneous rock. This is one way through which the mantle interacts directly with the hydrosphere and indirectly with the continental crust.

When sinking lithosphere carries old oceanic crust back down into the mantle, the crust is eventually remixed into the mantle. The most ancient oceanic crust of the ocean basins is only about 180 million years old. The average age, by contrast, is about 60 million years. This means that the time taken for a circuit through the oceanic rock cycle, from magma, through igneous rock, to remixing in the mantle, is very much less than the time for a circuit through the continental crust.

The Carbon Cycle

The rock cycle acts continually, but slowly. By contrast, movement of materials between the hydrosphere, the atmosphere, and the biosphere is very rapid. The chemical element carbon, which is essential to all forms of life, provides an example of how rapid and complicated a chemical cycle can be. The example is also interesting because it is one that human activities are distorting at a measurable rate.

Carbon occurs in four reservoirs: (1) in the atmosphere it occurs in carbon dioxide (CO_2); (2) in the biosphere it occurs in organic compounds; (3) in

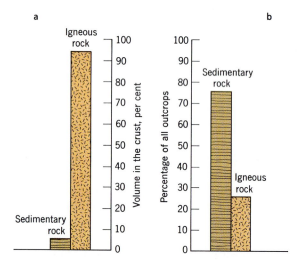

FIGURE 2.16 Relative amounts of sedimentary and igneous rock in the Earth's crust. Pyroclastic rocks are considered to be igneous; metamorphic rocks are considered to be either sedimentary or igneous, depending on their origin. (*a*) The great bulk of the crust consists of igneous rock (95 percent), while sedimentary rock (5 percent) forms a thin covering at and near the surface. (*b*) The extent of sedimentary rock at the surface is much larger than that of igneous rock, so 75 percent of all rock seen at the surface is sedimentary and only 25 percent is igneous. (*Source:* Data from Clarke and Washington, 1924.)

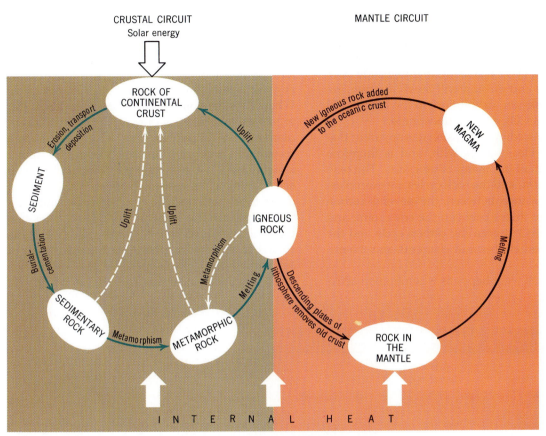

FIGURE 2.17 The rock cycle, an interplay of internal and external processes. Rock material in the continental crust can follow any of the arrows from one phase to another. At one time or another it has followed all of them. Within the mantle circuit, magma rises from a depth and forms new igneous rock in the lithosphere. The old lithosphere descends again to the mantle where it is eventually remixed. Reservoirs are labeled in capital letters; paths representing processes are labeled in lowercase letters.

the hydrosphere it occurs as carbon dioxide dissolved in lakes, rivers, and seawater; and (4) in the crust it occurs both in the calcium carbonate ($CaCO_3$) that forms limestone, and in buried organic matter such as coal and petroleum. Each of the reservoirs is involved in the carbon cycle (Fig. 2.18).

The key to the carbon cycle is the biosphere, where plants continually extract CO_2 from the atmosphere, then break the CO_2 down by the process of photosynthesis to form organic compounds. When plants die, or are consumed by animals, the organic compounds decay again by combining with oxygen from the atmosphere to reform CO_2. The passage of material through the biosphere reservoir is so rapid that the entire content of CO_2 in the atmosphere "turns over" every 4.5 yr. However, the biosphere and atmosphere

circuits interact with circuits in the hydrosphere and crust too.

Not all dead plant matter in the biosphere decays immediately back to CO_2. A small fraction is transported and deposited as sediment; then some is buried in sedimentary rock and thus protected from the oxygen in the atmosphere. The buried organic matter joins the slower moving rock cycle, and will only reenter the atmosphere when uplift and erosion again have exposed the rock in which it is trapped.

CO_2 from the atmosphere also dissolves in the waters of the hydrosphere. There it is used by aquatic plants in the same way that land plants use CO_2 from the atmosphere. Additionally, aquatic animals extract calcium and carbon dioxide from the water to make shells of $CaCO_3$. When the animals die the shells accumulate on the seafloor, mixing with any $CaCO_3$ that may have been precipitated as a chemical sediment. When compacted and cemented, the $CaCO_3$ forms limestone. In this way also, some carbon joins the rock cycle. Eventually the rock cycle will bring the limestone back to the surface and erosion will break it down, returning the calcium, in solution, to the ocean, and the carbon, as CO_2, to the atmosphere.

From the preceding discussion it is apparent that the carbon cycle interacts with, and is therefore partly dependent on, many other cycles. The oxygen, the calcium, and the rock cycles are examples. There is a balance between the several carbon reservoirs, the fluxes between the reservoirs, and the interactions between the carbon cycle and the cycles of the other chemical elements and materials.

We might ask what happens when human activ-

ities become so great that they start to change the sizes of the reservoirs and influence the fluxes between the reservoirs? When we dig coal or pump oil from the crust, and burn them, we convert organic matter to CO_2 and thereby speed up the rate at which carbon moves through the rock cycle. When we clear vast areas of forests, or allow deserts to expand because of overgrazing that kills off plant life, we are reducing the size of the biosphere reservoir. While an individual event may appear insignificant, the cumulative effects of all human activities have a measurable effect. Both the burning of fossil fuels and the clearing of the land cause CO_2 to pass to the atmosphere at faster rates than dictated by natural causes. Unless the CO_2 is dissolved in the hydrosphere, or buried rapidly in sediments, the CO_2 content of the atmosphere must inevitably increase. The rate at which CO_2

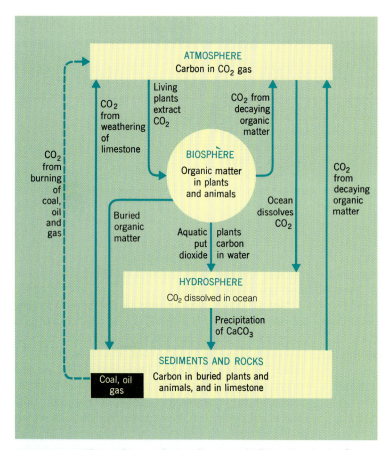

FIGURE 2.18 The carbon cycle involves interlocking circuits in the biosphere, hydrosphere, atmosphere, and crust. Boxes show main reservoirs of carbon, and arrows denote the paths along which carbon moves. Circuits and reservoirs are in balance except for human interference. As we burn coal and petroleum we speed up the circuit of carbon from sedimentary rock to the atmosphere (dashed arrow). As a consequence, the CO_2 content of the atmosphere is rising.

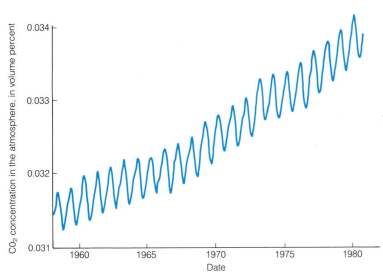

FIGURE 2.19 The steady upward trend of CO_2 in the atmosphere is recorded in daily measurements at an observatory on Mauna Loa, a mountain in Hawaii. The saw-toothed shape reflects the annual change in the rate of photosynthesis in the northern hemisphere where Mauna Loa is located. During summer months, plant growth is rapid and CO_2 respiration is high; during winter months growth is slow and respiration is low. The seasonal cycles are not changing, even though the average CO_2 content of the atmosphere is clearly rising. (*Source:* From National Oceanic and Atmospheric Administration, 1985.)

dissolves in the hydrosphere is known to be slower than the rate at which human activities are adding it to the atmosphere. As a consequence, the CO_2 content of the atmosphere is increasing (Fig. 2.19). This has led to much research. One of the concerns, as discussed in Chapter 20, is that increased concentrations of CO_2 in the atmosphere will cause the global climate to warm up. This in turn could have unpredictable effects on food production, and might lead to rapid decrease in volume of polar ice and a consequent rise in sea level.

The carbon cycle reinforces the point that "everything is connected to everything else." Every cycle, no matter how seemingly far removed from interactions with other cycles, can be shown to interact with and be dependent on numerous other cycles.

The Phosphorus Cycle

One of the nutrients needed for the healthy growth of plants and animals in the biosphere is the chemical element phosphorus. Plants cannot grow without phosphorus, and animals cannot grow their calcium-phosphate skeletons without phosphorus. The principal reservoir of phosphorus is the crust. Only a tiny fraction of the crust contains phosphorus-rich, mineable ores, but all

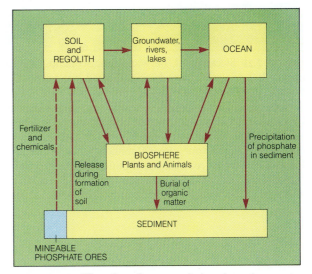

FIGURE 2.20 The phosphorus cycle involves the crust, the biosphere, and the hydrosphere, but the atmosphere does not play an important role. Yellow boxes show the main reservoirs of phosphorus; arrows denote paths along which phosphorus moves. Circuits and reservoirs are in balance except for mining activities and the use of phosphorus for fertilizers and other chemicals. (*Source:* Adapted from Richey, 1983.)

FIGURE 2.21 Siccar Point, Berwickshire, Scotland. Layers of sedimentary rock, originally horizontal, were bent and tilted into vertical layers during uplift. Erosion developed a new land surface which, on submergence, became the surface on which younger sediments were laid. The interface between the two rock units is called a surface of unconformity. The rocks standing vertical are about 450 million years old. Those lying above the unconformity, named the Old Red Sandstone, are 370 million years old. At this locality, in 1788, James Hutton first realized the significance of unconformity and the fact that it records a cycle of sedimentation, uplift, and erosion that is repeated again and again. The realization led him, in turn, to the Principle of Uniformity.

rocks contain at least a trace amount of phosphorus. As weathering breaks down rock to form the regolith, phosphorus becomes available for extraction by plants. When the plants die, or when animals that eat the plants die, the phosphorus returns to the soil, to groundwater, or to the ocean. The system is in balance because the reservoirs—biosphere, rocks, soil, groundwater, and ocean—and the fluxes between the reservoirs are in balance (Fig. 2.20).

Human activities are now disturbing this bal-ance through the mining of phosphate rock and the conversion of phosphorus compounds to soluble substances that make the phosphorus rapidly available to plants. The flux from rock and regolith to the hydrosphere is greatly increased. Much of the phosphorus added as fertilizer finds its way via groundwater into lakes, rivers, and coastal marine waters. There the soluble phosphorus compounds cause greatly increased biological activity through massive growth of certain algae. This massive algal growth suppresses other life-forms that play major roles in the food chains of many higher animals. The higher animals die and, thus, a portion of the biosphere is upset. Research continues on the effects of human interaction with the phosphorus cycle, but it is already clear that our activities are leading to unexpected consequences in unexpected places.

Uniformity

The first person to realize and enunciate the deep significance of the cycles, and in particular the rock cycle, was James Hutton (1726–1797), a Scotsman. Hutton examined the evidence in the rocks (Fig. 2.21) and concluded that there was neither evidence of a beginning to the cycles nor signs of an end. Rather, he pointed out, there was abundant evidence that the same processes we observe today have all operated in the past. During the early years of the nineteenth century, Hutton's findings were developed into what we call the *Principle of Uniformity.* It says that *the same external and internal processes we recognize in action today have been operating unchanged throughout most of the Earth's history.* This principle provides us with an important capability. We can examine any rock, however old, and compare its characteristics with those of similar rock forming today in a particular environment. We can then infer that the old rock very likely formed in the same sort of environment. Therefore, the Principle of Uniformity provides a first and very significant step in understanding the Earth's history.

Uniformity has played a tremendously important role in the development of geology as a science. However, there have been occasions when it led to misinterpretations. During the nineteenth century, geologists tried to estimate the duration of the rock cycle by estimating the thickness of all sediments laid down through geological time. By assuming that rates of deposition remained constant and equal to today's rate of deposition, it was a simple calculation to estimate the time needed to produce all the sediments. The results (which are discussed further in Chapter 8) we now know were greatly in error. One of the principal reasons

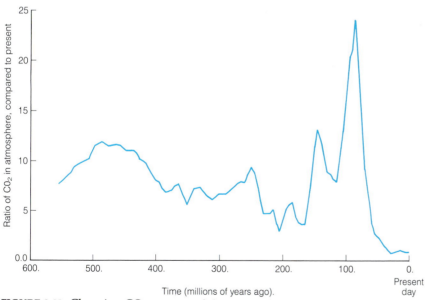

FIGURE 2.22 Changing CO_2 content of the atmosphere for the past 570 million years taking the present CO_2 content as 1. A CO_2 content of 10 means 10 times greater than today's value. It is apparent that about 90 million years ago, the CO_2 content was 24 times higher, and that the present and recent past has been a period of very low CO_2 levels. This curve is calculated from the estimated changes in certain flux rates and reservoir sizes based on the geological record. The high CO_2 contents calculated for the periods about 100 and from 400 to 500 million years ago, coincide with the times when the geological record suggests the Earth's climate was very warm. (*Source:* Personal communication from Berner, Lasaga, and Garrels.)

for the error was the assumption of the constancy of geological rates. So widely was the idea of constancy accepted that it even became incorporated into the statement of the Principle of Uniformity. The more that is learned of Earth's history, and the more extensively the timing of past events are determined through radiometric dating (Chapter 8), the clearer it becomes that the rates of the cycles have not always been the same. The evidence is strongly against constancy; some rates were once more rapid, others much slower. One reason that the rate of the rock cycle has probably changed through time is that the Earth is very slowly cooling down as its internal heat is lost. The Earth's internal temperature is maintained, in part, by the heat given off whenever an atom of uranium, thorium, or any other radioactive atom decays. Early in the Earth's history there must have been many more radioactive atoms, so more heat must have been produced than is produced at present. Internal processes, which are all driven by the Earth's internal heat, must have been more rapid than they are today. It is possible that three billion years ago oceanic crust was created at a faster rate than it is now, and that continental crust was uplifted and eroded at a faster rate. Either or both actions would cause the rock cycle to speed up. It is probable that even though the cycles have been continuous, none has maintained a constant rate through time.

When human influence on the carbon cycle was discussed earlier in this chapter, it was pointed out that an increase in the flux of CO_2 from the ground to the atmosphere caused the reservoir of atmospheric CO_2 to grow larger also. Whenever the flow of material between two reservoirs is changed, all parts of a cycle must adjust to accommodate that change. In particular, the sizes of the reservoirs must change. A striking example of this can be seen in the way the CO_2 content of the atmosphere has varied during the past 500 million years (Fig. 2.22). The effects that brought about the big changes in CO_2 content included such things as varying rates at which magma created new oceanic crust along mid-ocean ridges, and what percentage of the continental crust was covered by seawater.

There would be no changes of any kind, past or present, if it were not for the existence of energy and of materials for the energy to work on. Motions, processes, activities, and changes of any kind require the expenditure of energy. We turn next, therefore, to a closer look at the sources of energy that drive the Earth's internal and external processes, and to an investigation of the materials involved in those processes.

SUMMARY

1. The solid Earth is layered. It consists of a core, mantle, and crust, each differing in composition.

2. The crust consists of two parts: oceanic crust with an average thickness of 6 km, and continental crust with an average thickness of about 40 km.

3. The lithosphere, approximately the outer 100 km of the solid Earth, consists of rock that is hard and relatively rigid. Beneath the lithosphere, down to a depth of 350 km, is the asthenosphere, a region of the mantle where high temperatures make rock weak and easily deformed.

4. The lithosphere slides slowly over the asthenosphere.

5. The lithosphere is divided into six large plates and several small ones.

6. Moving plates of lithosphere can cause continents to move and ocean basins to open and close. Moving plates are the most important factors shaping the face of the Earth.

7. Outside the solid Earth are the external layers called the hydrosphere, atmosphere, biosphere, and regolith.

8. The Earth's active processes can be divided into internal processes, driven by internal heat, and external processes, driven by the Sun's heat.

9. The internal and external layers continually interact leading to repeated cycles in which materials flow from one layer to another.

10. The rock cycle in the continental crust begins with magma, which solidifies and forms igneous rock. The rock is eroded, creating sediment, which is deposited in layers that become sedimentary rock. Deep burial leads to metamorphism and eventually temperatures and pressures so high that rock melts and forms new magma.

11. Chemical elements follow cyclic paths between the crust, mantle, hydrosphere, atmosphere, and biosphere.

12. On a time scale of human observation, cycles do not change through natural causes, so the compositions of layers remain approximately in balance. Over periods of thousands or millions of years, or due to massive human activity, cycles can change.

13. One cycle in which human activity can be detected is the CO_2 cycle. Burning fossil fuel increases the flux of CO_2 from the crust to the atmosphere. This flux is raising the CO_2 content of the atmosphere.

SELECTED REFERENCES

Broecker, W. S., 1983, The ocean: Scientific American, v. 249, no. 3, p. 146–161.

Cloud, P., 1983, The biosphere: Scientific American, v. 249, no. 3, p. 176–189.

Garrels, R. M., Mackenzie, F. T., and Hunt, C., 1975, Chemical cycles and the global environment: Los Altos, Calif., Wm. Kaufman Inc.

Ingersall, A. P., 1983, The atmosphere: Scientific American, v. 249, no. 3, p. 162–174.

Siever, R., 1983, The dynamic Earth: Scientific American, v. 249, no. 3, p. 46–55.

Energy, Matter, and Minerals

Opal, a mineraloid with the approximate formula $SiO_2 nH_2O$, owes its striking color patterns to diffraction caused by geometric arrays of submicroscopic spheres of colloidal silica. This specimen was found in Australia.

ENERGY

Turn a page of this book. You are using energy. Walking outside, driving a car, or merely turning on a light—whatever the activity, you are using energy. Activities and energy are so intimately related that we may define **energy** as *the capacity to produce activity*. Energy is vital for our existence, and vital, also, for the Earth's existence. Without energy, the Earth would be a lifeless planet.

Energy takes many forms, each of which produces characteristic activities. We speak of kinetic energy, meaning the energy of a moving body, and heat energy, meaning the energy of a hot body. Other forms of energy are electrical, chemical, radiant, and atomic. Each of these kinds of energy is important for some of the Earth's activities (Table 3.1), but the four principal kinds of energy are kinetic, atomic, heat, and radiant.

Kinetic Energy

Every moving body has *kinetic energy,* named from the Greek word *kinetikos* (to move). The movements of the Earth around the Sun and that of the Earth spinning on its axis mean that our planet possesses kinetic energy. A ball moving in a tennis game, a boulder rolling downhill, a stream flowing down a valley, and tiny rock particles in a dust storm each have kinetic energy. Besides running water and rolling boulders, winds, waves, icebergs drifting in the ocean, and moving glaciers are other common examples. A boulder on a cliff—or any other object that could move if pushed or freed for movement—is said to possess *potential energy* because it has the potential for acquiring kinetic energy.

Atomic Energy

Atom bombs and hydrogen bombs release vast amounts of the energy locked within atoms. Such energy can be very destructive. But natural processes also release energy from certain kinds of atoms, and the processes go on continuously—fortunately, in a nonthreatening manner. To discuss these processes it is necessary, first, to talk about atoms. Appendix A contains a more detailed discussion; here we mention only the essential points.

Asked to analyze a rock, a chemist would report the kinds and amounts of the chemical elements present, because **chemical elements** are *the most fundamental substances into which matter can be sepa-*

TABLE 3.1   *Activities Produced by Common Forms of Energy*

Energy	Common Activity
Kinetic	Flowing water, wind, waves, landslides
Heat	Volcanoes, hot springs, rainstorms
Chemical	Decaying vegetation, forest fires, rusting, burning coal
Electrical	Lightning, aurora
Radiant	Daylight, sunburn
Atomic	Heating Earth's interior

rated by chemical means. At present 104 elements are known, and 88 of them occur naturally. Each is separately named and identified by a symbol, such as H for hydrogen, Ag for silver, Si for silicon, and U for uranium. The known elements and their symbols are listed in Appendix A.

If the chemist were asked what an element is made of, the answer would be that each elemental substance consists of a large number of identical particles called atoms. An **atom** is *the smallest individual particle that retains all the properties of a given chemical element*. We cannot actually see an atom of hydrogen, or an atom of lead, or an atom of any other element, even with the most powerful microscopes, because individual atoms are so small. When we handle a pure chemical element we are seeing instead an aggregation of a vast number of identical atoms. A cube of pure silver 1 cm on an edge contains about 58×10^{21} atoms, a number so large that its significance is almost impossible to grasp. A faint idea of its magnitude may be conceived by imagining that we had somehow spread the cube of silver thinly and evenly over the entire face of the Earth. Each square centimeter of the Earth's surface would then contain approximately 10,000 atoms.

Everything on the Earth is composed of atoms, but atoms in turn are built up from still smaller subatomic particles. The principal subatomic particles are *protons* (which have positive electrical charges), *neutrons* (which, as their name suggests, are electrically neutral), and *electrons* (which have negative electrical charges that balance exactly the positive charges of protons). Protons and neutrons are dense but very tiny particles and they aggregate together to form the core or *nucleus* of an atom. Protons give a nucleus a positive charge, and we call *the number of protons in the nucleus of an atom* the **atomic number.** Electrons are even tinier particles; they move, like a distant and diffuse cloud, in orbits around the nucleus (Fig. 3.1).

Elements are catalogued systematically by

atomic number, beginning with hydrogen which has an atomic number of one because it has one proton. Hydrogen is followed by helium which has two protons—and so the list continues up to the heavier elements such as uranium, which has 92 protons. In order for an atom to be electrically neutral, the number of orbiting electrons must balance the number of protons in the nucleus. If the number of orbiting electrons is smaller or larger than the number of protons, a net positive or negative charge results.

All atoms having the same atomic number are atoms of the same element. The number of neutrons that accompany the protons in a nucleus can vary, within small limits, without markedly affecting the chemical properties. Therefore, any chemical element may have several *isotopes, atoms having the same atomic number but differing numbers of neutrons.* For one element ten isotopes have been discovered, and all elements have at least two isotopes. However, not all combinations of protons and neutrons are completely stable, so that some isotopes break down spontaneously, forming in the process new isotopes and different elements. *The decay process by which an unstable atomic nucleus spontaneously transforms to another nucleus is called* **radioactivity.** As we shall see in Chapter 8, the rate at which a given radioactive isotope decays is constant; the rate is a characteristic property of that isotope. Because the decay rate is constant, it can be used as a kind of geological clock.

The most common radioactive isotopes in the Earth are potassium-40 (written ^{40}K, meaning an isotope with a total of 40 neutrons and protons in the nucleus), uranium-235 (^{235}U), uranium-238 (^{238}U), and thorium-232 (^{232}Th). When an atom decays by radioactive disintegration to form a new atom, a tiny amount of energy is released as heat. It is this heat energy that we refer to as the Earth's atomic energy.

Radioactive isotopes of potassium, thorium, and uranium (together with the isotopes of a few less common elements such as rubidium) are widely distributed in tiny amounts through the crust and mantle, and their rates of disintegration are very slow. Nevertheless, they generate sufficient heat to keep the Earth's interior exceedingly hot.

Atomic energy can also be released by another and somewhat different process. When very light atoms, such as those of hydrogen or helium, are raised to temperatures of millions of degrees, the atomic nuclei combine, or *fuse,* to form heavier atoms, and as they do so a great deal of atomic energy is released as heat. The combination of hydrogen atoms to form heavier helium atoms is the process that produces heat in the Sun. This process is called *fusion.*

Heat Energy

We can think of heat as a special form of kinetic energy. It is the energy possessed by the motions of atoms. All atoms move constantly. The faster they move, the more heat energy they have, and the hotter a body feels. In a solid object atoms move within confined spaces—in a sense they rattle around their assigned spaces in the solid structure—but if they move fast enough (become hot enough), they break out of their fixed positions and the solid is said to have melted. With faster movement still, which means more heat, atoms move with complete freedom, and the liquid is then said to have vaporized.

Hotness, or degree of heat, is a cumbersome term; so we have the term *temperature* instead. Temperature is an indication of the average speed of the moving atoms and is measured by arbitrary scales. A common one is the *Celsius scale,* in which we select as 100° C the temperature (or speed of the atoms) just sufficient to boil water at sea level, and 0° C as the freezing of water.

Heat energy is transmitted in three ways. The first is by *conduction,* which occurs when atoms pass on some of their motions to adjacent atoms. Conduction is the process by which heat is transmitted through the side of a cup of hot coffee; it is also the way by which most of the Earth's internal heat reaches the surface. **Conduction,** therefore, is *the means by which heat is transmitted through solids;* it is a slow process because the transfer occurs atom by atom. The second and much faster way by

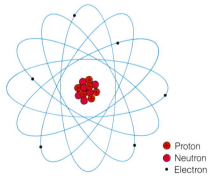

● Proton
● Neutron
• Electron

FIGURE 3.1 Schematic illustration of an atom of carbon. The nucleus contains 6 protons and 6 neutrons. Moving in orbits around the nucleus are 6 electrons.

which heat can be transmitted is *convection.* Convection occurs in liquids and gases in which the distribution of heat is uneven. When a liquid or gas is heated it expands and its density (mass per unit volume) decreases. Hot and less dense material floats upward, while colder and more dense material sinks to replace it, thereby setting up a convection cell or *convection current.* The third way by which heat is transmitted is by radiation as discussed in the following section.

Radiant Energy

The forms of energy discussed so far involve matter. How then is it possible for energy to pass through empty space? It is possible because energy can be transmitted by energy waves, commonly called *electromagnetic waves.* Radiant energy, therefore, is electromagnetic waves: cosmic rays, visible light, X rays, and radio waves are well-known forms of radiant energy. Radiant heat is also electromagnetic waves.

The distance between the crests of two adjacent waves is a **wavelength.** The wavelengths of electromagnetic waves range from less than 10^{-9} cm (cosmic rays) to more than 10^6 cm (radio waves). The names do not indicate any fundamental difference among electromagnetic waves. In a vacuum, all such waves move at the speed of light; the names are merely convenient ways of designating various ranges of wavelengths. Cosmic rays and X rays are the shortest wavelengths. Light has wavelengths

ranging from 10^{-5} to 10^{-4} cm, and heat has still longer wavelengths.

Radiant energy and its transmission through space by electromagnetic waves are vitally important for the Earth. It is by this means that the Earth receives the atomic energy the Sun generates by nuclear fusion.

Conversions Between Forms of Energy

When we burn gasoline in a car, we convert chemical energy to heat energy. When the car moves, the heat energy in turn is converted to kinetic energy. Conversion of heat energy to kinetic energy in a controlled manner was first accomplished in the eighteenth century, when the steam engine was invented, thereby inaugurating the era of intensive energy use in which we now live.

In all natural processes, energy is continually converted from one form to another. As radiation arrives from the Sun it strikes the Earth's surface, where the rays are absorbed, and the ground is heated. Air that is in contact with the ground is warmed by conduction and this causes it to expand and to rise. The rising air is replaced by cooler and denser air from above and wind results (Fig. 3.2). The process of lifting warm air from the Earth's surface to greater altitudes is an example of convection. Thus, atomic energy that was generated as heat in the Sun and was radiated to the Earth by electromagnetic waves can eventually be transformed to kinetic energy in a blowing wind.

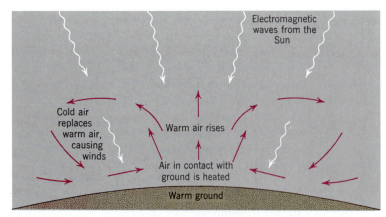

FIGURE 3.2 Energy in the form of electromagnetic radiation from the Sun becomes the kinetic energy of wind. Radiation heats the Earth's surface. Air in contact with the surface is warmed; it expands as a consequence of this heating and rises. Cold air flows in to replace the rising hot air, and winds are the result. The cycle of rising warm air and inflowing cold air is called a convection cell.

TABLE 3.2 *Amounts of Energy Available from Different Sources. The Amount of Energy per Unit Mass from Nuclear Burning Is Vastly Greater Than the Chemical Energy Resulting from Ordinary Burning*

Reaction	Amount of Heat Energy
1 g of hydrogen fuses to form helium	6.3×10^{11} J
1 g of ^{238}U decays by radioactivity	0.8×10^{11} J
1 g of oil burns	4.2×10^{4} J
1 g of coal burns	3.3×10^{4} J
1 g of sugar is eaten	17.6×10^{3} J
1 g mass moving at a velocity of 1000 cm/s[a]	5.0×10^{2} J

[a] 1000 cm/s = ~ 23 mph.

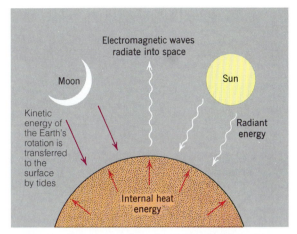

FIGURE 3.3 Energy reaches the Earth's surface from three sources—two are internal, one external. The two internal sources are the Earth's internal heat, and the kinetic energy of the Earth's rotation. The external energy source is the Sun. To maintain a thermal balance, long-wavelength electromagnetic waves radiate energy into space.

The multitude of internal and external processes operating in and on the Earth result from the fact that vast numbers of paths exist through which energy conversions can happen.

Before people realized that motion and heat were different expressions of energy, and that the different forms could be converted back and forth, they had already begun to use different measurement units for each form. To avoid the confusion of multiple units, we will use International System Units, for which the unit of energy is the joule. A *joule* (J) is defined as *the work done when a force of one newton is displaced a distance of one meter.* A *newton* (N), in turn, is defined as *that force which gives a mass of one kilogram an acceleration of one meter per second per second.* The joule is possibly an unfamiliar unit. A more familiar unit of energy to most people is the *calorie* (C), which is *the amount of heat energy needed to raise the temperature of one gram of water by one degree Celsius.* One calorie is equal to 4.183 J.

In Table 3.2 the amounts of energy derived from different sources are compared. It is apparent that an incomparably greater amount of energy comes from fusion and fission reactions than from burning.

SOURCES OF ENERGY

Energy reaches the Earth's surface from three principal sources: (1) radiant energy arrives from the Sun; (2) kinetic energy arrives from the rotations of the Moon, Earth, and Sun and appears as tides; and (3) energy reaches the Earth's surface by continuous outflow of the Earth's internal heat. Because the surface receives energy from three sources, we might reason that it is heating up.

However, this is not the case. The average temperature of the Earth's surface does not vary from year to year, so some sort of energy balance must be maintained. The reason for the balance is not difficult to find: The Earth's surface not only receives energy, it also radiates energy back into space as long-wavelength electromagnetic waves. The Earth's surface is said to be in a **steady state,** or state of **dynamic equilibrium,** by which is meant *a condition in which the rate of arrival of energy equals the rate of escape of energy* (Fig. 3.3).

Energy from the Sun

Radiant energy reaches the Earth from all the stars in the universe, but the total amount is tiny by comparison with the Sun's radiant energy. When the Sun's rays reach the Earth's atmosphere, approximately 40 percent is simply reflected back into space without any change. It is this reflected radiation that astronauts see when, standing on the Moon, they look back at the Earth (Fig. 3.4). The remaining 60 percent is absorbed, partly by the atmosphere (which becomes heated in the process) and partly by the land and the sea. The energy absorbed by the sea warms the water and causes evaporation. The resulting water vapor forms clouds and eventually rain, snow, and all other forms of precipitation. The energy absorbed by the land eventually warms the air, causes convection, and creates winds which, blowing over

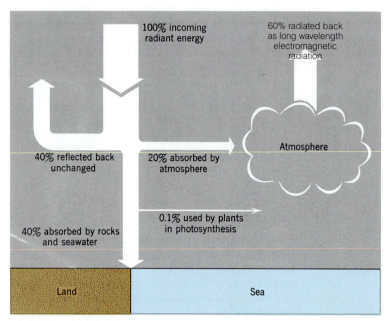

FIGURE 3.4 Paths followed by incoming radiant energy from the Sun. The energy used in heating the land and sea causes winds plus the evaporation of water to form clouds which in turn causes rain, snow, and hail.

the sea, create waves. Thus, all the major agents of transportation of sediment that operate at the Earth's surface—rain, ice, streams, winds, waves, and glaciers—are driven by the Sun's energy.

Energy from the Tides

One does not always think of tides in the context of energy. Nevertheless, tides are the mechanism by which some of the kinetic energy from the motions of the Moon, Earth, and Sun reach the Earth's surface. The principal effect arises from interactions between the Moon and Earth; tides are principally lunar effects.

Gravitational attraction by the Moon on the Earth creates tidal bulges in the ocean (Fig. 3.5). No significant amount of energy would be involved were it not for the Earth's rotation about its axis and the Moon's movement in orbit around the Earth. As a consequence, the positions of the bulges move continually, and at most places on the sea two high tides and two low tides can be observed each day. However, as Figure 3.5 shows, the Sun also affects the tides, sometimes opposing the Moon by pulling at right angles, and sometimes aiding it by pulling in the same direction. The Sun's effect is smaller than the Moon's, so the two effects never exactly cancel each other. The actual heights of high and low tides, therefore,

vary on a cycle of approximately 14 days, matching the enhancement and opposition of tides by the Sun and Moon.

Tidal bulges cannot move around the Earth unhindered because continents get in the way. Therefore, water piles up against the continental margins whenever a tidal bulge arrives. Water from a high tide flows back to the ocean basin as the Earth rotates and the tidal bulge passes. The piling-up effect is the reason why tides are much higher along coasts than in the open ocean. Piling-up of water masses in coastal tides, and prevention of the movement of water in a tidal bulge, means in effect that the coastline runs into a mass of water at every high tide. This in turn means that kinetic energy must be taken from the Earth's store of kinetic energy of rotation.

If the kinetic energy of rotation is transferred to the Earth's surface to overcome tides, what is happening to the Earth's rate of rotation? The Earth's spinning motion, and therefore its kinetic energy of rotation, remains from the days of its formation more than 4.5 billion years ago (Chapter 23); we know of no new sources of kinetic energy that are being added. Therefore, removing kinetic energy can have only one effect. The tides are acting as weak but steady brakes, and the Earth is gradually slowing down. The rate of slowing is, fortunately, not great because the rate of energy transfer is

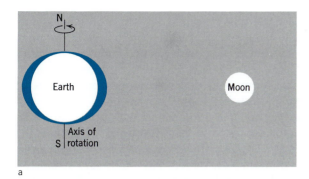

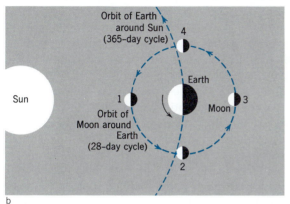

FIGURE 3.5 The gravitational attractions of the Moon and Sun on the Earth raise tidal bulges in the ocean. (*a*) Idealized diagram of the tidal bulges, relative to the Earth's axis of rotation, and the position of the Moon. (*b*) When the Moon and Sun attract in the same direction (Moon positions 1 and 3), highest tides are observed. When the Moon and Sun have opposing positions (Moon positions 2 and 4), lowest tides are experienced.

small. Astronomers have measured the exact length of the day over the past three centuries and find that it is increasing by 0.002 seconds each century. Over hundreds of millions of years the effect of this small slowing can become very large. About 570 million years ago, it is estimated that the Earth spun so fast, a day was about 20.6 h long. At a time some billions of years in the future, the Earth will probably stop rotating completely.

Energy from the Earth's Interior

Anyone who has been down in a mine realizes that rock temperatures increase with depth. Measurements made in deep drillholes and mines show that the rate of temperature increase (the *geothermal gradient*) varies in different parts of the world from 15–75°C/km. Direct temperature measurements cannot be made beyond the deepest drillholes, which are little more than 10 km deep.

Therefore, indirect means must be used to estimate temperatures in the Earth's interior. From physical properties that vary with temperature, such as the speeds of earthquake waves, it is possible to estimate that temperatures continue to increase toward the earth's center and eventually reach values of 5000° C or more in the core. When the Earth's size is considered, it is obvious that a vast amount of energy must be stored as heat within it. As we have already seen, internal temperatures are maintained by the slow but continually operating processes of radioactive transformation. Some of the original heat energy from the Earth's earliest processes of formation must remain, in part, because internal temperatures are maintained by these radioactive transformations.

The average value of the heat flow that reaches the Earth's surface is about 6.3×10^{-6} J/cm²/s. To sense how small this amount of heat energy really is, imagine that we have an instrument that can catch and use all the heat reaching 1 m² of the Earth's surface. We would have to catch heat for approximately 14 days and nights before enough heat energy was gathered to raise the temperature of a cup of water to 100° C so that boiling would commence. Because the area of the Earth's surface is enormous, the total amount of internal heat energy that reaches the surface and is radiated to outer space is enormous—it is estimated to be 2700 billion J/s! The heat loss is not constant everywhere. Just as the geothermal gradient varies from place to place, so does the heat flow, which is greatest near young volcanoes and active hot springs, and least where the crust is oldest and least active. Variations in surface heat flow tell us a great deal about activities down below and (as we shall see in later chapters) they allow us to infer that convection may be occurring in the seemingly solid rock of the mesosphere and asthenosphere (Chapter 18).

Comparisons of the Earth's Energy Sources

Table 3.3 shows the amounts of energy reaching the Earth's surface every 24 hours from the three

TABLE 3.3 *Comparison of the Total Amount of Energy Reaching the Earth's Surface Every 24 h from the Three Principal Energy Sources*

Energy Source	Energy Each 24 h
Solar radiation	15.5×10^{21} J
Flow of internal heat	27.6×10^{17} J
Tides	2.5×10^{17} J

different energy sources. Because solar energy is vastly larger than the energy of either tides or internal heat flow, it is clear that the temperature at the Earth's surface must be controlled by the Sun. Changes in the atmosphere can influence the incoming radiation and produce a change in the surface temperature of the Earth. If the atmosphere is filled with dust and other pollutant particles, more of the Sun's incoming radiation will be reflected back into space. Similarly, if the composition of the atmosphere is changed through the addition of more carbon dioxide, it will be a better thermal blanket because escape of long wavelength radiation is slowed down. Other energy sources are too small to offset the effects on the environment that changes in the atmosphere can cause.

Solar energy does not contribute to the Earth's internal activities. Just as water cannot flow uphill, heat cannot flow from a cold body to a hot one. The geothermal gradient prevents heat energy from the Sun flowing more than a meter or so from the surface down into the interior. In a real sense, therefore, the Earth's surface is a surface of conflict between different energy sources. Internal activities, driven by internal heat energy, raise mountains and cause irregularities on the Earth's surface. External activities, driven by solar and tidal energy, continually erode and abrade the surface irregularities. Therefore, next we must examine the consequence of the great energy conflict. To do so, it is necessary to know something about the materials from which the Earth is constructed—minerals and the aggregates of minerals that we call rocks.

MINERALS

Pick up a stone, or if you cannot find a stone, some sand, soil, or gravel; you will be holding a handful of minerals. The world is made of minerals, so wherever you look you will either see minerals or materials made from minerals.

Most minerals are common and have neither commercial value nor any particular use. A few, like diamonds and rubies, are rare and prized for their beauty. Others, though still few in number, are the raw materials for industry and the basis for national wealth. Empires have been won and lost for minerals, and powerful countries have collapsed when deposits of valuable minerals became exhausted. The Romans conquered most of Europe and the Near East in their search for miner-

als containing copper, gold, tin, iron, silver, and lead. They built a great empire by using the mineral wealth they found. When the mineral deposits became exhausted, or were captured by local tribes, Rome was deprived of its great sources of wealth and its empire slowly died. During the eighteenth and nineteenth centuries, England prospered because rich resources of coal, iron, copper, tin, and other minerals aided its Industrial Revolution. However, as mineral resources waned, so did England's power. The United States, also, has prospered as a result of its abundant mineral wealth. Yet, as some of its sources are now facing exhaustion, there are those who wonder if history is about to repeat itself. Others believe that such fears are not justified. They argue that, as some supplies run out, we should be able to develop alternative and presently unused sources and that in some cases it may be possible to substitute other materials.

Because society has become so dependent on substances made from minerals, it is hardly surprising that many people have a natural curiosity about how minerals form and how rich deposits occur. But there are many other reasons to be curious about minerals. As stated in Chapter 2, rocks are aggregates of mineral matter which carry the record of the Earth's history. To decipher that history and to understand how the Earth works it is necessary to investigate two important subfields of geology; *mineralogy,* which is *the branch of geology that deals with the classification and properties of minerals;* and *petrology, the branch of geology that deals with the occurrence, origin, and history of rocks.*

So far, the term *mineral* has been used as if everyone knows precisely what it means. Unfortunately, this is not true because the word is often used loosely and in many different ways. For example, it is possible to read advertisements for plant foods that provide "minerals" for plant growth, or advertisements for vitamin pills to provide "minerals" for healthy human bodies. To avoid confusion we will use mineral in a specific, scientific way. In order to do so, we will give it an exact definition. However, before attempting a definition it is helpful to examine the two most important characteristics of minerals. The first is composition, by which is meant the proportions of the various chemical elements present. The second is the structure, meaning the way that the atoms of the chemical elements are packed together. Because most minerals contain several kinds of atoms, it is helpful to start by discussing the way in which atoms combine.

Compounds and Ions

A few minerals, such as metallic gold and platinum, are single chemical elements. Most minerals are *combinations of atoms of different chemical elements that are bonded together* to form **compounds.** (The reasons atoms combine, or bond, are discussed more fully in Appendix A.) This combination depends on how atoms transfer, or share, orbiting electrons. *An atom that has excess positive or negative charges caused by electron transfer* is called an **ion.** When the charge is positive (meaning that the atom gives up electrons), the ion is called a *cation;*

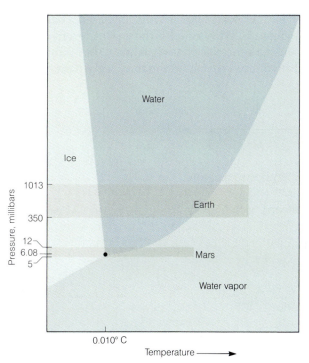

FIGURE 3.6 The control of temperature and pressure on the state of H_2O. In the region marked water, H_2O is in the liquid state; within the ice region it is solid and in the vapor region it is in the gaseous state. At temperatures and pressures defined by the lines separating regions, two states can coexist. For example, along the line separating ice and water, the solid and liquid states coexist—that is, along the line water freezes to ice. At one specific temperature and pressure the three lines of coexistence meet, and at that triple point, ice, water, and water vapor coexist. The triple point for H_2O is at 0.010°C and 6.08 millibars pressure. The International Standard unit of pressure, the pascal, is equal to 10^7 (ten million) millibars. The pressures at sea level and at the top of Mt. Everest bound the range of pressures at the Earth's surface. By comparison, the range of pressures at the surface of Mars is much smaller.

when negative, an *anion.* The convenient way to indicate ionic charges is to record them as superscripts. For example, Ca^{+2} is a cation that has given up two electrons, while S^{-2} is an anion that has accepted two electrons. Compounds contain one or more elements that are cations and one or more that are anions. For a compound to be stable the sum of the positive charges on the cations and the negative charges on the anions must equal zero.

Sometimes two different atoms form such strong bonds that they seem to act as a single atom. An especially strongly bonded pair is said to form a *complex ion.* Complex ions act in the same way as single ions, forming compounds by bonding with other elements. For example, carbon and oxygen combine to form the very stable carbonate anion $(CO_3)^{-2}$. Other important examples are the sulfate $(SO_4)^{-2}$, nitrate $(NO_3)^{-1}$, and silicate $(SiO_4)^{-4}$ anions.

Two broad classes of compounds are recognized. *Organic compounds* are made from carbon and hydrogen, with or without other elements such as nitrogen and oxygen. Organic compounds can form by direct combination of carbon and hydrogen, but most come directly or indirectly from the activities of living organisms. Mixtures of organic compounds are called *organic matter.* All other matter is said to be *inorganic* and its compounds are *inorganic compounds.* Most minerals are inorganic compounds.

One property of minerals, therefore, is that they are chemical compounds or (in a few cases such as metallic gold and copper, or sulfur) single chemical elements. However, a single property is insufficient to define a mineral uniquely; other properties must be considered as well. The most important of the additional properties is crystal structure.

Structure

Compounds and elements can exist in any of the three *states of matter*—solid, liquid, or gas. H_2O is a familiar example of a compound that can form a *liquid* (water), a *solid* (ice), and a *gas* (water vapor) under suitable conditions at the surface of the Earth. The two factors that control the state of H_2O are temperature and pressure. It is clear from Figure 3.6 that at high temperatures and low pressures water vapor is the stable state for H_2O, while ice forms at low temperatures and high pressures. The liquid state is found at intermediate temperatures and pressures. The temperatures and pressures over which different compounds change from one state to another differ greatly, but the same general statement is true for all substances— low temperatures and high pressures favor the solid state, high temperatures and low pressures favor the gaseous state, while the liquid state occurs in intermediate ranges of temperature and pressure.

With minerals we are dealing entirely with the solid state because all minerals are solids. Whereas atoms in gases and liquids are randomly jumbled, the atoms in almost all solids are organized in regular, geometric patterns, like eggs in a carton. *The geometric pattern that atoms assume in a solid* is called the **crystal structure.** The crystal structure of a mineral is a unique property, and all specimens of a given mineral have identical structures.

The packing of atoms in the mineral galena, PbS, the most common lead mineral, is shown in Figure 3.7. Notice that the sulfur atoms are larger than the lead atoms. The reason for this is that the lead in galena is the cation, Pb^{+2}, because it transfers two electrons to the sulfur, S^{-2}. Electrons are held in their orbits by the electrostatic attraction of the protons in the nucleus. The attracting power of the nucleus is fixed because the number of protons is fixed. Thus, when two additional electrons are added to the orbiting group around each sulfur nucleus, the electrostatic force exerted by the nucleus on each orbiting electron is slightly reduced. As a result, each of the electrons is held a little less strongly and it moves a bit further away from the nucleus, or, as the effect is more commonly stated, the **ionic radius,** which is *the distance from the center of the nucleus to the outermost orbiting electrons,* is increased. With cations the opposite effect happens and the electrons remaining after transfer are drawn more closely to the nucleus. Anions, there-

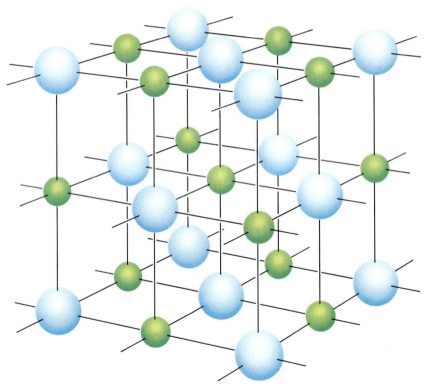

FIGURE 3.7 Arrangement of atoms in galena (PbS), the most common mineral containing lead. The packing arrangement is repeated continuously throughout a crystal, and atoms are so small that a cube of galena one centimeter on edge contains 10^{22} atoms each of lead and sulfur. The atoms are shown pulled apart along the marked lines to demonstrate how they fit together. Pb (green) is a cation with a charge of +2, S (blue) is an anion with a charge of −2. To maintain a charge balance between the atoms, there must be an equal number of Pb and S atoms in the structure.

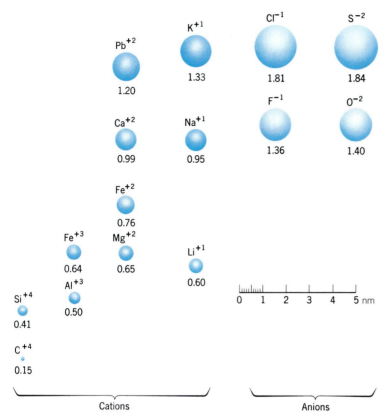

FIGURE 3.8 Radii of some common ions range from C^{+4} at lower left to S^{-2} at upper right. Ions are arranged in vertical groups based on charge, from $^+4$ at left to $^-2$ at right. Ions in each of the pairs Si^{+4} and Al^{+3}, Mg^{+2} and Fe^{+2}, and Na^{+1} and Ca^{+2} are about the same size and commonly substitute for each other in crystal lattices. Radii expressed in nanometers (nm), where 1 nm equals 10^{-8} cm. (*Source:* Data from Fyfe, 1964.)

fore, tend to have large radii; cations tend to be small. Most of the volume of a crystal structure is taken up by the large anions, and as a result the crystal structures of minerals are determined largely by the packing arrangements of anions. The radii of some common ions are shown in Figure 3.8 and are given in nanometers (abbreviated nm), a unit of length commonly used for atomic measurement (see Appendix E).

It is apparent from Figure 3.8 that certain ions have the same electrical charge and are nearly alike in size. For example, Fe^{+2} and Mg^{+2} have radii of 0.83 nm and 0.78 nm, respectively. Because of their similarity in size and charge, ions of Fe^{+2} are often found substituting for ions of Mg^{+2} in magnesium-bearing minerals. The structures of the magnesium minerals are not changed as a result of the substitution, but of course the compositions of the minerals are affected. *The substitution of one atom for another in a random fashion throughout a crys-*

*tal structure is **ionic substitution** or, to use an older term, **solid solution**.* There is a special way of writing ionic substitutions in chemical formulas. When Fe substitutes for Mg in the mineral olivine, Mg_2SiO_4, for example, the formula is written $(Mg,Fe)_2SiO_4$ which indicates that the Fe substitutes for the Mg, but not for any other atoms in the structure. Variations in mineral composition caused by solid solution are often large and, as will become apparent, they are important in the formation of common minerals and rocks.

Each mineral has a unique crystal structure. Some compounds are known to form two or more different minerals because the atoms can be packed to form more than one kind of crystal structure. The compound $CaCO_3$, for example, forms two different minerals. One is *calcite*, which is the mineral of which marble is composed; the other is *aragonite*, which is most commonly found in the shells of clams, oysters, and snails. Calcite and

TABLE 3.4 *Minerals with Identical Compositions But Different Crystal Structures Are Called Polymorphs. Some Well-Known Polymorphs Are Listed Below, with the Most Common Variety Listed First*

Composition	Mineral Name
C	Graphite
	Diamond
$CaCO_3$	Calcite
	Aragonite
FeS_2	Pyrite
	Marcasite
SiO_2	Quartz
	Cristobalite
	Tridymite
	Coesite
	Stishovite

aragonite have identical compositions, but entirely different crystal structures. *A compound that occurs in more than one crystal structure* is called a **polymorph** (a word meaning many forms). Some common polymorphs are listed in Table 3.4.

Definition of a Mineral

Minerals are naturally occurring substances. If the definition were not so limited, a vast number of man-made substances would have to be included. We are now ready to consider an exact definition. A **mineral** is *any naturally formed, solid, chemical substance having a definite chemical composition and a characteristic crystal structure.* Because most mineral substances are inorganic substances, some people include the word inorganic in the definition. There are, however, several natural organic substances not formed by living cells that meet the restriction of a definite composition and characteristic crystal structure. One example is the mineral whewellite, $CaC_2O_4 \cdot 2H_2O$, which is found in many different geologic environments.

The definition of a mineral does not include all naturally occurring solid compounds because some mineral-like substances do not fulfill either, or both, of the requirements that a mineral have a definite composition and a characteristic crystal structure. Examples are natural glasses and resins, both of which have wide and variable composition ranges and are *amorphous* (meaning they lack crystal structure). Another example is opal, which has a more-or-less constant composition, but is amorphous. The term *mineraloid* is used to describe such mineral-like substances.

Composition

Approximately 3000 minerals are known. Most have been found in the crust, because that is the only part of the Earth accessible to us. A few minerals have been identified only in meteorites, and two new ones were discovered in the Moon rocks brought back by the astronauts. The total number of minerals may seem large, but it is tiny by comparison with the astronomically large number of ways that all the naturally occurring elements can combine to form compounds. The reason for the disparity between observation and theory becomes apparent when the abundances of the chemical elements are considered—only 12 are sufficiently abundant so that they comprise 0.1 percent by weight, or more, of the Earth's crust. The 12 abundant elements (Table 3.5), collectively, make up 99.23 percent of the mass of the crust. The crust is constructed, therefore, of a limited number of minerals containing 2 or more of the 12 abundant elements.

Rather than forming minerals, the scarcer chemical elements tend to occur by ionic substitution; they substitute for more abundant elements in common minerals. For example, if a grain of the mineral olivine is analyzed, it would be observed that not only does the grain contain Mg and Fe in addition to Si and O, but that trace amounts of Cu, Ni, Co, Mn, and many other elements are also present. Each of the trace elements would be present as ionic substitutes for Mg in the structure. Minerals containing less common elements than the abundant 12 certainly do occur, but only in

TABLE 3.5 *The Most Abundant Chemical Elements in the Continental Crust*

Element	Weight (%)
Oxygen (O)	45.20
Silicon (Si)	27.20
Aluminum (Al)	8.00
Iron (Fe)	5.80
Calcium (Ca)	5.06
Magnesium (Mg)	2.77
Sodium (Na)	2.32
Potassium (K)	1.68
Titanium (Ti)	0.86
Hydrogen (H)	0.14
Manganese (Mn)	0.10
Phosphorus (P)	0.10
All other elements	0.77
Total	100.00

Source: After K. K. Turekian, 1969.

small amounts. Most of the very scarce elements only form minerals under special and restricted circumstances. In fact, hafnium, rhenium, and a few other chemical elements are so rare that they are not known to form minerals under any circumstances—they only occur by ionic substitution.

Referring to Table 3.5, it is apparent that two elements, oxygen and silicon, make up more than 70 percent of the crust. Oxygen forms a simple anion, O^{-2}, and silicon forms a simple cation, Si^{+4}, but oxygen and silicon together form an exceedingly stable, complex ion, the *silicate anion*, $(SiO_4)^{-4}$. Minerals that contain the silicate anion are called silicate minerals. In view of the abundances of silicon and oxygen, it is not surprising that silicate minerals are the most common, naturally occurring inorganic compounds. Compounds that contain simple O^{-2} anions, called oxides, are the second most abundant group of minerals. Other natural compounds, although less common than silicates and oxides, are sulfides, chlorides, carbonates, sulfates, and phosphates, for which the anions are, respectively, S^{-2}, Cl^{-1}, $(CO_3)^{-2}$, $(SO_4)^{-2}$, and $(PO_4)^{-3}$.

Properties

The properties of minerals are controlled by their compositions and crystal structures. Compositions can be determined by any of several methods of chemical analysis. Once the compositions are determined, chemical formulas can be calculated by balancing the number of anions and cations.

Determination of crystal structure is a more demanding process than measurement of composition because indirect methods have to be used. The way most structures are determined is by measurement of the way atoms scatter X rays. How this is done can be illustrated by a simple experiment. When a streetlight is viewed through the wire mesh of a screen door, the light appears to have a series of rays surrounding a central spot. The rays appear at regular angles around the bright spot, and close examination shows that each ray is made up of a series of closely spaced light spots. The pattern of rays is the result of reflections from the regular geometric array of the wires in the mesh. By careful and suitably controlled experiments, the geometry of a wire mesh can be determined by measuring the pattern of light rays and the distance of the streetlight from the mesh. So, too, can the geometric placement of atoms in a crystal structure be determined by observing the passage, and scattering, of electromagnetic radia-

tion of a suitable wavelength through crystals. The property employed is not that of reflection, as in the case of the wire mesh, but rather that of *diffraction*, by which rays are scattered from a regularly spaced series of points. In certain directions the wave crests of the rays are coincident and come through as bright spots; in all other directions, the wave crests do not coincide and destroy each other (Fig. 3.9).

The scattering points in a crystal are the tiny clouds of electrons that surround each atom. The

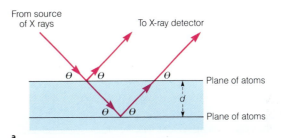

FIGURE 3.9 The way atoms pack to form a crystal structure can be determined with X rays. (a) Atoms in a crystal lattice lie on planes. The distance between adjacent, parallel planes can be determined through the Bragg equation, $2d = \lambda \sin \theta$, where d is the distance between adjacent planes, λ is the wavelength of the X ray, and θ is the angle of incidence of the incoming radiation. λ can be controlled and θ can be measured experimentally so that d can be calculated; when all the d's between all the planes have been measured for a given crystal, the structure can be calculated. (b) Example of one kind X-ray diffraction pattern prepared by directing a beam of X rays at a garnet crystal. Each of the spots recorded on the film is produced by one set of planes in the crystal structure.

intensity of scattering is proportional to the number of electrons, so atoms such as silver and uranium with large numbers of electrons give a very strong signal, while atoms such as hydrogen and lithium with very few electrons give a very

weak signal. The wavelengths of the electromagnetic rays—X rays—used to determine crystal structures are between 1 and 2 nm, about the diameter of the electron clouds that cause the scattering. The scattering process is called *X-ray diffrac-*

a

b

FIGURE 3.10 Distinctive and characteristic crystal forms of two minerals. (*a*) Pyrite, FeS_2, is commonly found in cube-shaped crystals with characteristic striations on the cube face. The largest crystals in the photograph are 3 cm on an edge. The specimen is from Bingham Canyon, Utah. (*b*) Elongate crystals of rutile, TiO_2, embedded in a crystal of quartz. The specimen comes from Alexander County, North Carolina. The largest crystal is 4 cm long.

a

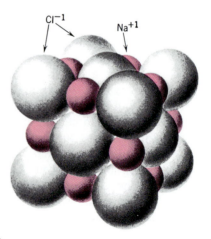

b

FIGURE 3.11 Relation between crystal structure and cleavage. (*a*) Halite, NaCl, has well-defined cleavage planes; it breaks into fragments bounded by perpendicular faces. (*b*) The crystal structure, in the same orientation as the cleavage fragments, shows that the plane of breakage is a plane in the crystal in which sodium and chlorine atoms occur in equal numbers.

tion, and it is one of the most powerful properties used in the identification and investigation of minerals.

Once all the properties of a mineral are known, and it has been determined which are uniquely characteristic of that particular mineral, it is possible to use these properties to identify the mineral. Where common minerals are concerned, it is rare indeed that it is necessary for identification to analyze a mineral chemically or to determine its crystal structure. Because more easily determined properties, such as hardness and color, are controlled by the composition and the crystal structure, it is possible in most cases to use a combination of simple properties to identify the minerals. The characteristics most often used in identifying minerals are the obvious properties, such as color, shape of crystal (Fig. 3.10), and hardness, plus some less obvious ones, such as the way a mineral breaks (Fig. 3.11) and its density. Each property is discussed in detail in Appendix B, which also contains a table of common minerals together with the most important physical properties used in their identification.

THE ROCK-FORMING MINERALS

A few minerals—no more than 20 kinds—are so common that they account for more than 95 percent of all the minerals in the continental and oceanic crust. These are the *rock-forming minerals,* so called because essentially all common rocks contain one or more of them. We have already seen that silicates and oxides are the most common minerals of the crust, but that silicates are more abundant than oxides. Most rock-forming minerals, therefore, are silicates.

The Silicates

Silicon and oxygen form the complex anion $(SiO_4)^{-4}$. The four oxygen atoms in the complex anion are tightly bound to the single silicon atom. Oxygen is a large ion (Fig. 3.8), while silicon is a small ion. The oxygens pack into the smallest space possible for four large spheres. As can be seen in Figure 3.12, the four oxygens sit at the corners of a tetrahedron and the small silicon sits in the hole between the oxygens at the center of the tetrahedron. Therefore, the shape of the complex silicate anion is a tetrahedron and the structures and properties of silicate minerals are deter-

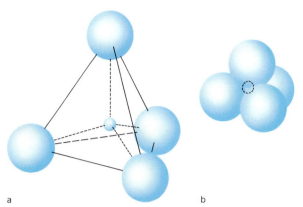

a   b

FIGURE 3.12 Silicate tetrahedron. (*a*) Expanded view showing large oxygen ions at the four corners, equidistant from a small silicon ion. Dotted lines show bonds between silicon and oxygen ions; solid lines outline the tetrahedron. (*b*) Tetrahedron with oxygen ions touching each other in natural positions. Silicon (dashed circle) occupies central space.

mined by the manner in which *silicate tetrahedra* pack together.

Each silicate tetrahedron has four unsatisfied negative charges. This is so because the silicon ion (Si^{+4}) has a charge of $+4$ while oxygen (O^{-2}) has a charge of -2. Each oxygen satisfies one of its negative charges by a bond to the silicon ion at the center of the tetrahedron, leaving each oxygen with one unsatisfied negative charge. To make a stable compound, therefore, each oxygen must satisfy its remaining negative charge. This can happen in two ways.

The first way for the charges to be satisfied is for the oxygens of the tetrahedra to form bonds with cations, just as if the tetrahedra were simple anions. In such a structure, the complex ions are separated from each other because they are completely surrounded by cations. An example of this is found in olivine, in which Mg^{+2} cations balance the charges. Each oxygen is bonded to one silicon and three magnesium atoms (Fig. 3.13).

The second, and entirely different, way by which oxygen charges can be satisfied is for two adjacent tetrahedra to share an oxygen. A shared oxygen is bonded thereby to two silicons, and its charges are balanced as a consequence; the two tetrahedra, now joined at a common apex, form an even larger anion unit. As shown in Figure 3.14, when two silicate tetrahedra share a single oxygen, the result is a large, complex anion with the formula $(Si_2O_7)^{-6}$. There is only one common mineral, epidote, plus a few rare minerals, in which

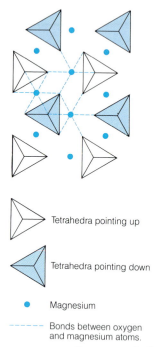

Tetrahedra pointing up

Tetrahedra pointing down

● Magnesium

----- Bonds between oxygen and magnesium atoms.

FIGURE 3.13 Structure of olivine. Isolated silicate tetrahedra, here shown in a geometric form such that an oxygen atom would sit at each apex and a silicon at the center of each tetrahedron. By viewing the structure from above it is apparent that one-half of the tetrahedra point up and one-half point down. The magnesium ions lie between the tetrahedra, and each is bonded to 6 oxygens. Each oxygen is bonded to 3 magnesiums. Oxygen has a charge of -2, one-half of which is balanced by a silicon atom; magnesium has a charge of $+2$ so each of the 6 oxygens to which it is bonded receives a charge of $+1/3$. Because each oxygen is bonded to three magnesiums, the structure is electrostatically balanced and the formula is Mg_2SiO_4.

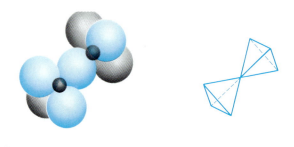

a   b

FIGURE 3.14 Two silicate tetrahedra share an oxygen, thereby satisfying some of the unbalanced electrical charges and, in the process, forming a larger and more complex anion group. (*a*) The arrangement of oxygen and silicon atoms in a double tetrahedra, giving the anion $(Si_2O_7)^{-6}$. (*b*) A geometric representation of the double silicate tetrahedra.

this simplest form of oxygen sharing occurs. However, just as more and more beads can be strung on a necklace, so can **polymerization,** *the process of linking silicate tetrahedra into large anion groups,* be extended to form huge units.

Most common silicate minerals have anions with complicated polymerizations. If a tetrahedron shares more than one oxygen with adjacent tetrahedra, large circular groups, endless chains, sheets, and three dimensional networks of tetrahedra can all be formed. There is, however, an important restriction to the polymerization process that must always be met—adjacent tetrahedra can never share more than one oxygen. Stated another way, tetrahedra only join at their apexes, never along the edges or faces. The common polymerizations, together with rock-forming minerals forming them, are shown in Figure 3.15 and are discussed below in order of increasing complexity of polymerization.

Olivine

Two important minerals contain isolated silicate tetrahedra. The first is *olivine,* a glassy-looking mineral that is usually pale green in color. As previously discussed, olivine has a range of compositions because Fe^{+2} can substitute readily for Mg^{+2}, giving rise to the general formula $(Mg,Fe)_2SiO_4$. Olivine sometimes occurs in such flawless and beautiful crystals that it is used as the gem, *peridot,* but it is also one of the most important minerals in the Earth, being a very common constituent of rocks of the oceanic crust and upper part of the mantle.

Garnet

The second important mineral with isolated silicate tetrahedra is *garnet.* As with olivine, garnets have a range of compositions due to ionic substitution. Garnet has the complex formula $A_3B_2(SiO_4)_3$, where A can be any of the cations Mg^{+2}, Fe^{+2}, Ca^{+2}, and Mn^{+2}, or any mixture of them, while B can be either of the triply charged cations Al^{+3} or Fe^{+3}. Garnet is characteristically found in rocks of the continental crust. One of the most striking features of garnet is its tendency to form beautiful crystals. The iron-rich variety of garnet, called *almandine,* is deep red and is well known as a gemstone. Another important property of garnet is its hardness, which makes it useful as an abrasive for grinding and polishing.

Energy, Matter, and Minerals **55**

Pyroxene and Amphibole

Pyroxenes and *amphiboles* are two silicate minerals groups that contain continuous chains of silicate tetrahedra. They differ in that pyroxenes are built from a polymerized chain of single tetrahedra, each of which shares two oxygens, while the amphiboles are built from double chains of tetrahedra equivalent to two pyroxene chains, in which half the tetrahedra share two oxygens and the other

	Arrangement of silica tetrahedra	Formula of the complex anions	Typical mineral	
			Name	Composition
Isolated tetrahedra		$(SiO_4)^{-4}$	Olivine	$(Mg, Fe)_2SiO_4$
Isolated polymerized groups		$(Si_2O_7)^{-6}$	Lawsonite	$CaAl_2Si_2O_7(OH)_2H_2O$
		$(Si_3O_9)^{-6}$	Benitoite	$BaTiSi_3O_9$
	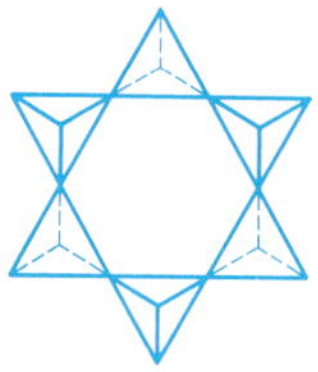	$(Si_6O_{18})^{-12}$	Beryl	$Be_3Al_2Si_6O_{18}$
Continuous chains	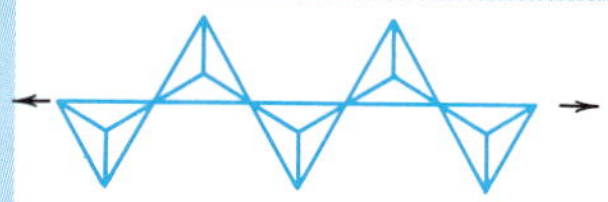	$(SiO_3)_n^{-2}$	Pyroxene	$CaMg(SiO_3)_2$ (Variety; diopside)
	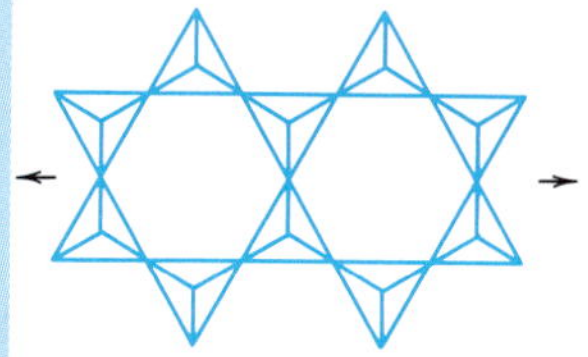	$(Si_4O_{11})_n^{-6}$	Amphibole	$(Ca_2Mg_5(Si_4O_{11})_2(OH)_2$ (Variety; tremolite)
Continuous sheets	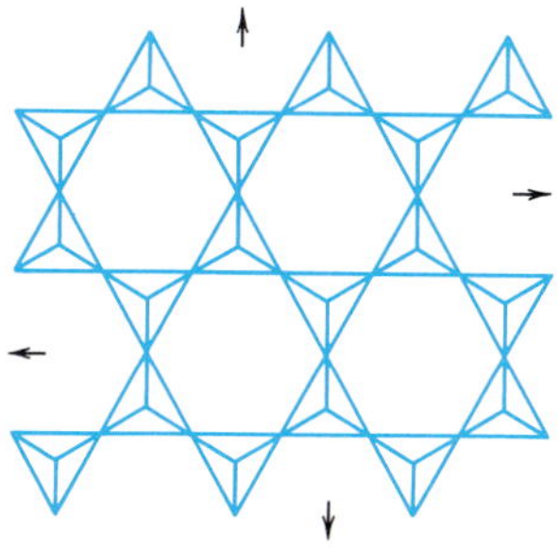	$(Si_4O_{10})_n^{-4}$	Mica	$(KAl_2(Si_3Al)O_{10}(OH)_2$ (Variety; muscovite)
Three-dimensional networks	Too complex to be shown by a simple two-dimensional drawing	(SiO_2)	Quartz	SiO_2

FIGURE 3.15 The way in which silicate tetrahedra polymerize by sharing oxygens determines the structures and compositions of the rock-forming silicate minerals. Polymerizations other than those shown are known, but do not occur in common minerals. The most important polymerizations are those that produce chains, sheets, and three-dimensional networks.

half share three oxygens. These relations can be seen clearly in Figure 3.15.

Both the pyroxene and amphibole chains are bonded together by cations such as Ca, Mg, and Fe. The cations bond to two or more oxygens in adjacent chains that have unsatisfied charges. Bonds between silicon and oxygen are stronger than the bonds between the other cations and oxygen. When a pyroxene or amphibole fractures, therefore, it is the weaker bonds that tend to be broken, and as a result there are prominent breakage surfaces parallel to the length of the polymerized chains (Fig. 3.16).

The general formula for pyroxene is $AB(SiO_3)_2$, where A and B can be any of a number of cations, the most important of which are Mg^{+2}, Fe^{+2}, Ca^{+2}, and Mn^{+2}. The pyroxenes are most abundantly found in rocks of the oceanic crust and mantle, but they also occur in many rocks of the continental crust. The most common pyroxene is a shiny black variety called *augite* which has the complex formula $Ca(Mg,Fe,Al)[(Si,Al)O_3]_2$.

Amphiboles have, perhaps, the most complicated formula of all the rock-forming minerals. The general formula is $A_2B_5(Si_4O_{11})_2(OH)_2$, in which A is most commonly either Ca^{+2} or Mg^{+2}, and B is usually Mg^{+2} or Fe^{+2}. Even this complicated formula does not completely describe the composition of the most abundant variety of amphibole, *hornblende*, a dark green to black mineral that looks very much like augite and, because of ionic substitution, has the approximate formula $Ca_2Na(Mg,Fe)_4[Si_6(Al,Fe,Ti)_3O_{22}](OH)_2$.

Clay, Mica, and Chlorite

Mica, *chlorite*, and *clay* are related minerals that have a polymerized sheet of silicate tetrahedra as their basic building unit. The sheet is formed by each tetrahedron sharing three of its oxygens with adjacent tetrahedra. This leaves a single, unbalanced oxygen in each tetrahedron, leading to the general anion formula $(Si_4O_{10})_n^{-4}$. In the simplest case, the electrical charges that remain are balanced by Al^{+3} cations, leading to the formula $Al_4Si_4O_{10}(OH)_8$ for the clay mineral *kaolinite*.

The Al^{+3} ions in kaolinite hold the polymerized sheets together by bonding to the oxygens with unsatisfied charges. In the cases of the micas and chlorites, other ions besides aluminum are present to hold the sheets together. As with the chain-structure minerals, the bonds between silicon and oxygen in the tetrahedra are stronger than the other cation–oxygen bonds that hold the sheets together. As a result, the clays, micas, and chlorites all display a very pronounced direction of breakage parallel to the sheets (Fig. 3.17).

A new principle must be mentioned to explain the composition of mica. Al^{+3} ions can replace the Si^{+4} ions in silicate tetrahedra by ionic substitution

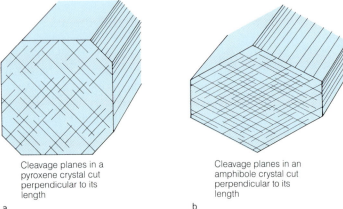

Cleavage planes in a pyroxene crystal cut perpendicular to its length

Cleavage planes in an amphibole crystal cut perpendicular to its length

a   b

FIGURE 3.16 Cleavage directions in pyroxenes and amphiboles. Bonds within polymerized chains of silicate tetrahedra are stronger than the bonds holding adjacent chains together. Cleavage breaks adjacent chains apart. (*a*) Pyroxenes have two directions of cleavage at approximately 90°. (*b*) Amphiboles have two directions of cleavage at approximately 120°.

FIGURE 3.18 Characteristic six-sided crystals of quartz. Crystals such as these form by slow growth when SiO_2 precipitates from a hot aqueous solution.

FIGURE 3.17 Perfect cleavage of the mica mineral, muscovite, shown by thin, plane flakes into which this six-sided crystal has been split. The cleavage flakes suggest leaves of a book, a resemblance embodied in the name "books of mica" for crystals elongated in a direction perpendicular to the cleavage flakes.

without affecting the property of polymerization. Because Al^{+3} has a smaller charge than Si^{+4}, a substituted tetrahedron has an extra negative charge to be satisfied. The charge cannot be fully satisfied by polymerization, so extra cations must be added to the crystal structure. Approximately one quarter of the tetrahedra in micas contain Al^{+3} instead of Si^{+4} ions. To make up the charge imbalance, cations such as K^{+1}, Mg^{+2}, and even some extra Al^{+3} must be added outside of the tetrahedra. The variety of mica called *muscovite*, for example, has the formula $KAl_2(Si_3Al)O_{10}(OH)_2$.

Chlorite is a complex sheet structure mineral that is usually green in color. It derives its name from a Greek word meaning green. The unbalanced charges of the polymerized sheet are satisfied by bonding to Mg^{+2}, Fe^{+2}, and Al^{+3} to give the general formula $(Mg,Fe,Al)_6(Si,Al)_4O_{10}(OH)_8$. Chlorite is a common alteration product from other minerals that contain iron and magnesium—such as biotite, hornblende, and augite.

Quartz

The only common mineral composed exclusively of silicon and oxygen is quartz. It is the classic example of a crystal structure that has all its charges satisfied by polymerization of the tetrahedra into a three-dimensional network—meaning that all the oxygens are shared.

Quartz characteristically forms six-sided crystals (Fig. 3.18), and is found in many beautiful colors. It is one of the most widely used gem and ornamental minerals. Common names for some gemstone varieties of quartz are *rock crystal* (colorless), *citrine* (yellow), and *amethyst* (violet). Quartz is a particularly abundant mineral in rocks of the continental crust. Indeed, it is so abundant that certain sedimentary rocks are composed entirely of quartz.

Certain varieties of quartz, formed by precipitation from water solution, are so finely grained they almost appear amorphous and can only be shown to have the internal crystal structure characteristic of minerals through the use of high-powered microscopes, X-ray machines, and other research tools. The common name given to these microcrystalline forms of quartz is *chalcedony* (Fig. 3.19).

FIGURE 3.19 Chalcedony, a microcrystalline variety of quartz formed by precipitation of SiO_2 from cool, aqueous solutions. Color banding is due to minute amounts of impurities. The photograph is 10-cm across.

Varietal names are *agate,* if the chalcedony has color banding; and *flint* (gray) and *jasper* (red), if the color is uniform.

Feldspars

The name feldspar is derived from two Swedish words, *feld* (field) and *spar* (mineral). Early Swed-ish miners were familiar with feldspar in their mines, and found the same mineral in the abundant rocks they had to pick from the fields around their homes. They were so struck by the abundance of feldspar that they chose a name to indicate that their fields seemed to be growing an endless crop. Of course they were mistaken about the fields growing feldspar, but they were not mistaken about the abundance of feldspar: it is the most common mineral in the crust. Feldspar accounts for about 60 percent of all minerals in the continental crust, and together with quartz comprises about 75 percent of the volume of the continental crust. Unlike quartz, which is rare in rocks of the oceanic crust, feldspar is also abundant in rocks of the seafloor.

Like quartz, feldspar has a structure formed by polymerization of all the oxygen atoms in the silicate tetrahedra. Unlike quartz, however, some of the tetrahedra contain Al^{+3} substituting for Si^{+4}—so, as in mica, other cations must be added to the structures to balance the charge.

Feldspar is a complex mineral that has a wide range of compositions. The varietal names and limiting (or ideal) compositions of common feldspar are: *potassium feldspar*, $K(Si_3,Al)O_8$, *albite* $Na(Si_3,Al)O_8$, and *anorthite* $(Ca(Si_2,Al_2)O_8)$. Potassium feldspar has several polymorphs—*orthoclase, microcline,* and *sanidine*—but the structural differences between them are subtle.

The most important ionic substitution in

FIGURE 3.20 Perthite, an intergrowth of irregular laths of plagioclase enclosed in potassium feldspar. The intergrowth forms when a high-temperature feldspar solid solution cools so that the limit of atomic substitution of plagioclase in potassium feldspar is exceeded. The plagioclase precipitates inside the crystal of potassium feldspar in rough, wormlike masses. The specimen is 6 × 8 cm.

feldspar involves the substitution of Ca^{+2} for Na^{+1}. That is possible because, as can be seen in Figure 3.8, the two ions are much closer in size than either is to the size of K^{+1}. However, Ca^{+2} and Na^{+1} have different charges, so the actual substitution scheme involves the coupled substitution of two ions: $(Na^{+1} + Si^{+4})$ for $(Ca^{+2} + Al^{+3})$. The substitution is so effective that *plagioclase* is *a variety of feldspar with an unbroken range of composition from albite to anorthite.*

To a limited extent, Na^{+1} and Ca^{+2} do substitute for K^{+1}, leading to compositions between potassium feldspar and plagioclase. The extent of this ionic substitution is strongly temperature-dependent, being much greater at high than at low temperatures. Just as a solution of sugar in hot water will precipitate sugar grains as the solution cools, so will a potassium feldspar formed at high temperature precipitate grains of plagioclase as it cools, and the limits of ionic substitution are exceeded. Potassium feldspar crystals often contain small, precipitated grains of plagioclase, giving an intergrowth called *perthite* (Fig. 3.20).

Other Minerals

Although silicates are the most abundant minerals on the Earth, a number of others—principally oxides, sulfides, carbonates, phosphates, and sulfates—are common enough to be called rock-forming minerals.

Some common oxide minerals are the compounds of iron [*magnetite* (Fe_3O_4) and *hematite* (Fe_2O_3)], the oxide of titanium [*rutile* (TiO_2)], and of course *ice*, the oxide of hydrogen [(H_2O)]. Oxides are important as ore minerals. The principal sources of iron, chromium, manganese, uranium, tin, niobium, and tantalum are oxide minerals.

The most common sulfide minerals are *pyrite* (FeS_2), *pyrrhotite* (FeS), *galena* (PbS), *sphalerite* (ZnS), and *chalcopyrite* $(CuFeS_2)$. The sulfide minerals are exceedingly important as ore minerals, being the principal sources of copper, lead, zinc, nickel, cobalt, mercury, molybdenum, silver, and many other metals.

The complex carbonate anion $(CO_3)^{-2}$ forms three important and common minerals: *calcite, aragonite,* and *dolomite*. We have already seen that calcite and aragonite have the same composition, $CaCO_3$, and are polymorphs. Calcite is much more common than aragonite. Dolomite has the formula $CaMg(CO_3)_2$.

The single important phosphate mineral contains the complex anion $(PO_4)^{-3}$. It is the mineral *apatite*, $Ca_5(PO_4)_3(F,OH)$ which is the substance from which our bones and teeth are made. It is also a common mineral in many varieties of rocks and the main source of phosphorus used for making phosphate fertilizers.

Sulfate minerals contain the complex anion $(SO_4)^{-2}$. Although many sulfates are known, only two are common, and both are calcium sulfate minerals: *anhydrite*, $CaSO_4$; and *gypsum*, $CaSO_4 \cdot 2H_2O$. Both form when seawater evaporates; they are the raw materials used for making plaster of all kinds. Plaster of paris got its name from a quarry near Paris where a very desirable, pure-white form of gypsum was mined centuries ago.

TABLE 3.6 *The Common Rock-Forming Minerals*

Silicates	Oxides	Sulfides	Carbonates	Sulfates	Phosphates
Olivines	Hematite	Pyrite	Calcite	Anhydrite	Apatite
Pyroxenes	Magnetite	Sphalerite	Aragonite	Gypsum	
Augite	Rutile	Galena	Dolomite		
Amphiboles	Ice	Chalcopyrite			
Hornblende					
Garnet					
Quartz	The common ferromagnesian minerals are:				
Feldspars					
Potassium feldspar	Augite				
Plagioclase	Biotite				
Micas	Chlorite				
Muscovite	Hornblende				
Biotite	Olivine				
Chlorites					
Clays					
Kaolinite					

SUMMARY OF MINERALS

Minerals are naturally occurring chemical elements or inorganic chemical compounds, and each mineral has distinctive properties that arise from its crystal structure and composition. All rocks are made from minerals, but only a few kinds of minerals form the mass of the crust. The common rock-forming minerals are listed in Table 3.6, which can be referred to as we proceed to discuss rocks and, in later chapters, how rocks are put together to form the Earth. One subgrouping of silicate minerals is commonly referred to as the *ferromagnesian minerals,* indicating they contain iron and/or magnesium as essential constituents. The ferromagnesian minerals include olivine, pyroxene, amphibole, chlorite, and the variety of mica called biotite. As we shall see in the next chapter, some of the ferromagnesian minerals are important in the characterization of igneous rocks.

MINERALS AS INDICATORS OF ENVIRONMENT

Minerals should not be regarded merely as objects of beauty or sources of economic materials. Contained within their makeup are the keys to the conditions under which they—and the rocks they are in—have formed. The study of minerals, therefore, can provide invaluable insight into the chemical and physical conditions in regions of the Earth that are inaccessible to direct observation and measurement.

An understanding of the growth environments of minerals has come very largely through studying minerals in the laboratory. By suitable experiments, for example, scientists have been able to define the temperatures and pressures at which a diamond forms rather than its polymorph graphite (Fig. 3.21). Because it is possible to infer how temperature and pressure increase with depth in the Earth, we can state with certainty that rocks in which diamonds are found are samples of the mantle from at least 145 km below the Earth's surface.

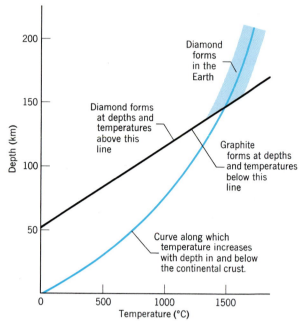

FIGURE 3.21 Line separating conditions of temperature and pressure of overlying rock (here plotted as depth) at which the two polymorphs of carbon, diamond and graphite, grow. At a pressure equal to that at a depth of 145 km, the diamond-graphite line intersects the curve depicting the way the Earth's temperature changes with depth. Diamond can only form 145 km or more below the surface—at depths well into the mantle.

The use of minerals to get information about environments is widely possible. Past climates, for example, can be deciphered from the kinds of minerals formed during erosion. The composition of seawater in past ages can be determined from the minerals formed when seawater evaporated and deposited its salts. Rather than elaborating many examples, we will turn to an examination of rocks, for rocks are after all simply assemblages of minerals. The kinds of minerals that group to form rocks, and the ways in which the groupings occur, are even more informative than individual minerals.

SUMMARY

Energy and Matter

1. Energy is the capacity to produce activity. Energy appears in many forms: kinetic, heat, chemical, radiant, and atomic.

2. Energy can be converted from one form to another. All of the Earth's activities and processes involve the conversion of energy.

3. Energy reaches the Earth's surface from three sources. Energy arrives from the Sun as electromagnetic radiation. Tidal energy comes from the kinetic energy of the Earth's rotation about its axis. Natural radioactive decay produces internal heat energy which flows to the surface.

4. Atoms form 104 chemical elements, and 88 of these occur naturally on the Earth.

5. Each chemical element has two or more isotopes. Isotopes are varieties of atoms that have the same chemical properties but differ in their masses because they contain different numbers of neutrons in their nuclei.

6. Some isotopes are not stable and change spontaneously by radioactive transformation. During transformation, energy is emitted and new isotopes and elements are produced.

Minerals

7. Minerals are naturally formed, solid, chemical substances having a definite chemical composition and a characteristic crystal structure. A crystal structure is the geometric array of atoms.

8. Approximately 3000 minerals are known, but of these about 20 make up more than 95 percent of the Earth's crust and are called the *rock-forming minerals*.

9. Silicates are the most common minerals, followed by oxides, carbonates, sulfides, sulfates, and phosphates.

10. The basic building block of silicate minerals is the silicate tetrahedron, a complex anion in which an Si^{+4} ion is bonded to four O^{-2} ions. The four O^{-2} ions sit at the apexes of a tetrahedron, with the Si^{+4} at its center. Adjacent silicate tetrahedra can bond together to form larger complex anions by sharing an oxygen. The process is called polymerization.

SELECTED REFERENCES

Berry, L. G., Mason, B., and Dietrich, R. V., 1983, Mineralogy, 2nd ed.: San Francisco, W. H. Freeman and Company.

Dietrich, R. V., and Skinner, B. J., 1979, Rocks and rock minerals: New York, John Wiley & Sons.

Klein, C., and Hurlburt, C. S., Jr., 1985, Manual of mineralogy: New York, John Wiley & Sons.

Walton, A. J., 1983, Three phases of matter, 2nd ed.: Oxford, Oxford University Press.

CHAPTER 4

Magma, Volcanoes, and Igneous Rock

Mount Fuji, Japan, a famous stratovolcano.

ROCK

At first glance rocks seem confusingly varied. Some appear platy or distinctly layered and display pronounced, flat crystals of mica. Others are coarse, evenly grained, and lack layering; yet, they may still contain the same kinds of minerals present in the platy, micaceous rock. By studying a large number of rock specimens it soon becomes clear that most of the differences between samples can be described in terms of two kinds of small-scale features. The first feature is *texture,* by which is meant *the overall appearance that a rock has because of the size, shape, and arrangement of its constituent particles.* For example, the mineral grains may be flat and parallel to each other, giving the rock a pronounced platy, or flaky, texture—like a pack of playing cards. In addition, the various minerals may be unevenly distributed and concentrated into specific layers. The rock texture is then both layered and platy.

The second small-scale feature in a rock is the assemblage of minerals present. A few kinds of rock contain only one mineral, but most rocks contain two or more of the rock-forming minerals. The specific minerals, and the amounts present, reflect the composition of the rock. *The varieties and abundances of minerals present in rocks,* commonly called **mineral assemblages,** are important pieces of information for interpreting a rock record. Mineral assemblages indicate, for example, whether a rock is formed in the crust or in the mantle. The composition of the crust and the mantle are distinctly different and their temperatures and pressures also differ. As a result, mineral assemblages of rocks formed in the mantle are quite different from those formed in the crust. *The systematic description of rocks in terms of mineral assemblage and texture is* termed *lithology.*

Two useful terms, *megascopic* and *microscopic,* are used to describe the textures and mineral assemblages of rocks. *Megascopic* refers to *those features of rocks that can be perceived by the unaided eye, or by the eye assisted by a simple lens that magnifies up to 10 times. Microscopic* refers to *those features of rocks that require high magnification in order to be viewed.* Commonly, examination of a microscopic texture requires the preparation of a special *thin section* that must be viewed through a microscope. A thin section is prepared by first grinding a smooth, flat surface on a piece of rock. The flat surface is then glued to a glass slide and the rock fragment is ground to a slice so thin that light passes through it

easily. The appearance of the same rock on a polished surface and in a thin section is shown in Figure 4.1.

The rock families (igneous, sedimentary, and metamorphic) were defined in Chapter 2. Igneous rocks are the most abundant of the three, and we therefore discuss them first. At the same time it is convenient to discuss an important subgroup of rocks that is transitional between igneous and sedimentary rocks. This group consists of **pyroclastic rocks** (named from the Greek words *pyro,* meaning heat, or fire, and *klastos,* meaning broken; hence, hot, broken fragments). They are *rocks comprised of fragments of igneous material ejected from a volcano, then sedimented, and either cemented or welded to a coherent aggregate.*

MAGMA

The formation of both igneous and pyroclastic rocks involves the cooling and solidification of magma. From direct observations of magma being erupted from volcanoes, and by studying the rock formed through the solidification of magma, three important conclusions can be drawn. The first conclusion concerns composition; magma is characterized by a range of compositions in which silica (SiO_2) is almost always predominant. The only exceptions are a few rare examples of carbonate and sulfide magmas. The second conclusion is that magmas are characterized by high temperatures, and that the properties of magmas are controlled both by composition and by temperature. The third conclusion is that magma has the properties of a liquid, including the ability to flow. This is true, even though some magma is almost as solid as window glass. Most magma is a mixture of crystals and liquid (often referred to as melt). As the temperature falls, the percentage of crystals increases and the percentage of melt decreases. The amount of liquid remaining when flow ceases varies from one magma to another, but is probably never less than 10 to 15 percent. So long as flow is possible, the mixture is referred to as magma.

Composition

Six minerals comprise the great bulk of all igneous and pyroclastic rocks—the minerals are quartz, feldspar, mica, amphibole, pyroxene, and olivine. Therefore, the chemical elements contained in these minerals are the principal chemical elements in magmas. They are Si, Al, Ca, Na, K, Fe, Mg, H, and O. By tradition, and because O is the most abundant anion, it is usual to express compositional variations in terms of oxides, such as SiO_2, Al_2O_3, CaO, and H_2O. The most abundant oxide component, and the most important one for controlling the properties of magma, is SiO_2.

Chemical analyses of all kinds of igneous and pyroclastic rock indicate that three specific compositions predominate. This suggests that three compositionally distinct types of magma must be more common than all others. The first type of magma contains about 50 percent SiO_2, the second about 60 percent, while the third contains about 70 percent SiO_2. The names of the common igneous rocks derived from these magma types are, respectively, basalt, andesite, and rhyolite (Fig. 4.2). However, the three most common varieties of magma are not found in equal abundance. Of all the igneous rock in the crust (oceanic and continental crust combined), approximately 80 percent forms from basaltic magma, 10 percent from andesitic magma, and 10 percent from rhyolitic magma.

The gases dissolved in magma are usually not major constituents by weight, but they are very important in determining the properties of the

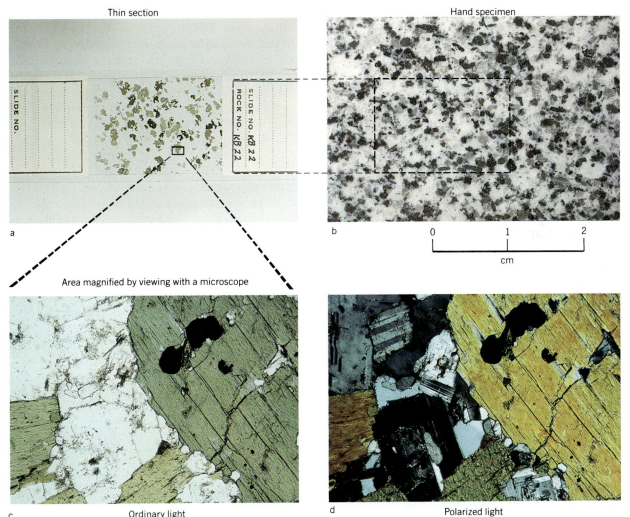

Thin section
Hand specimen

a
b

0   1   2
cm

Area magnified by viewing with a microscope

c
Ordinary light
d
Polarized light

FIGURE 4.1 In the study of rocks, polished surfaces and thin slices reveal textures and mineral assemblages to great advantage. The specimen here is an igneous rock containing quartz, plagioclase, hornblende, and biotite.

magma. Despite their importance, however, the amounts and compositions of dissolved gases are difficult to determine. As magma rises toward the Earth's surface, the confining pressure due to the weight of overlying rock must decrease. As pressure drops, the dissolved gases bubble out of solution—much as carbon dioxide bubbles out of an open bottle of soda. The principal gas is apparently water vapor, which, together with carbon dioxide, accounts for more than 90 percent of all gases emitted from volcanoes. Others are nitrogen, chlorine, sulfur, and argon which are rarely present in amounts exceeding 1 percent. Although vast quantities of gases are emitted by volcanoes, it is rarely possible to determine exactly how much has actually been released by the magma, and how much is merely gas that leaked into the magma from adjacent rock and has been boiled off. For example, at the height of its activity in May 1945, the volcano Parícutin, in Mexico, erupted an estimated 116,000 metric tons of material a day, of which 16,000 tons (14 percent) were water vapor, and 100,000 were lava and pyroclastic fragments. However, the maximum amount of water it is possible to dissolve in laboratory melts having the same composition as Parícutin magma, is only 10 percent. Therefore, we must conclude that the gases emitted by Parícutin magma contained not only the gases released from the magma, but also an unknown quantity of groundwater heated by the rising magma.

Despite uncertainties, estimates of both the compositions and the quantities of gases dissolved in magma can sometimes be obtained from analyses of glassy volcanic rock. Such rock is erupted and chilled so quickly that not all of the dissolved gases can escape completely. Typical volcanic glasses have water contents ranging from 0.2 to about 3 percent. Many scientists conclude that the dissolved gas contents of magmas rarely reach values as high as 10 percent and most are probably not much greater than the quantities observed in the most water-rich volcanic glasses.

Temperature

The temperature of lava as it comes out of a volcano is also difficult to measure. Volcanoes are dangerous places and scientists who study them are not anxious to be roasted alive. Measurements must be made some distance away from an active vent. This can be done with a pyrometer, which is an optical measuring device. Using pyrometers, scientists have determined magma temperatures ranging from 1040° to 1200° C. Once magma is extruded, it starts to cool. Temperatures as low as 800° C have been measured in some lavas that had nearly ceased to move. Experiments on synthetic magmas in the laboratory suggest that under some conditions, flow might even be possible at still lower temperatures—possibly down to 600° C.

Mobility

Although magma may seem to be very fluid, many of its properties are more akin to solids than to liquids; lava is more like asphalt than it is like water. With rare exceptions, lava flows slowly. *The internal property of a substance that offers resistance to flow is called* **viscosity.** The more viscous a lava,

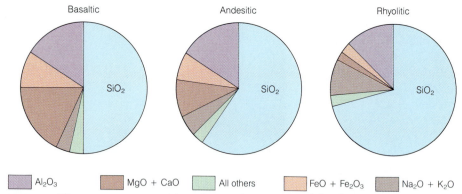

FIGURE 4.2 The average compositions (in weight percent) of the three primary magmas types. The compositions are derived from analyses of the most common types of igneous rocks. (*Source:* After Ronov and Yaroshevsky, 1969, and McBirney, 1969.)

FIGURE 4.3 The way a lava flows is controlled by viscosity. Slowly moving, very viscous basaltic lava from an eruption of Kilauea Volcano, Hawaii in 1972, broke into hot, rubbly fragments. The resulting rock is a rough mass of fragments (locally called *aa*). Underneath the aa flow is an older, smooth, ropy surfaced lava, locally called *pahoehoe*. Both lavas have the same composition. The pahoehoe flow formed from a rapidly moving, high fluid lava.

the less fluid it is. Viscosity of a lava depends on composition (especially the SiO_2 and dissolved-gas contents) and temperature.

The effect of temperature is simple to understand. As with asphalt or thick oil, high temperature leads to greater fluidity. The higher the temperature, therefore, the lower the viscosity and the more readily a magma flows. A very hot magma erupted from a volcano may flow readily, but it soon begins to cool, becoming more viscous and eventually slowing to a complete halt (Fig. 4.3).

The effect of silica content on viscosity is not so obvious as the effect of temperature. The same $(SiO_4)^{-4}$ tetrahedra that occur in silicate minerals (Chapter 3) occur also in magmas. Just as they do in minerals, the tetrahedra link together by sharing oxygens. However, unlike the tetrahedra in minerals, those in magma form irregular groupings of chains, sheets, and networks. As the average number of tetrahedra in the polymerized groups becomes larger, the magma is more and more resistant to flow and behaves increasingly like a solid. The number of tetrahedra in the groups depends on the silica content of the magma. The higher the silica content, the larger the polymerized groups. Therefore, magma composition exerts a direct and strong control on viscosity (Fig. 4.4). Magma with the composition of basalt (50 percent SiO_2) sometimes flows rapidly. Basaltic

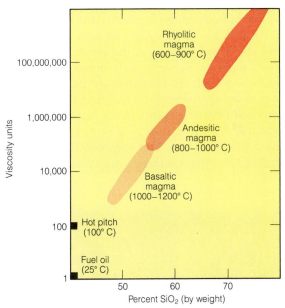

FIGURE 4.4 Viscosity of magma is strongly controlled by the silica (SiO_2) content. The more viscous a magma, the less tendency it has to flow. Viscosities of fuel oil at 25°C and pitch at 100°C are plotted for comparison.

magma moving down a steep slope on Mauna Loa in Hawaii, during an eruption in 1850, was clocked at an average speed of 16 km/h. However, such fluidity is very rare. Flow rates are more commonly measured in meters per hour or even meters per day.

Magmas that have rhyolitic compositions, containing 70 percent or more SiO_2, are so viscous and flow so slowly, that their movement can hardly be detected. Their high viscosity even makes it difficult for gas bubbles to escape. If such a viscous magma cools at a shallow depth, where the pressure of overlying rock is low, the pressure of the trapped gases may become so great that the confining pressure is exceeded and the whole mass of sticky magma simply explodes into billions of tiny, hot, glassy pyroclasts. The mass of tiny pyroclasts, or as they are often called, volcanic shards (Fig. 4.5), together with an expanding mass of hot gases, then erupts as volcanic ash, widely covering the surrounding country. When the volcanic ash becomes cemented, an important variety of pyroclastic rock is the result.

Origin

The making of a magma and the cooking of a meal in the kitchen are similar—both require ingredients and a source of heat. First consider the source of heat. In the kitchen it is the stove. In the Earth it is the great supply of heat in the mantle and the lower regions of the crust. But temperatures have to be very high indeed before rock melts—so high, in fact, that we usually think of rock as being fireproof. So high too, that one of the important turning points in the history of geology was the demonstration by a Scot, James Hall, almost 200 years ago, that common rocks could be melted. No longer is there a question as to whether

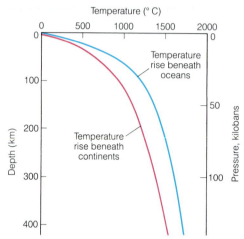

FIGURE 4.6 The geothermal gradient, the way temperature changes with depth, differs beneath oceanic crust and continental crust. The differences arise because radioactive elements, the principal sources of the Earth's heat, are distributed differently in the two regimes. The graph is drawn so that depth and pressure on the vertical axis increase downward, as they do in the Earth. The unit of pressure, a kilobar, is equal to 1000 bars.

or not a rock will melt; rather, the questions are, how does temperature increase with depth, and at what depths does melting occur? The answers are not straightforward and in order to approach them, it is necessary first to consider the *geothermal gradient,* which is *the rate of increase of temperature downward in the Earth.*

Geothermal gradients in the continental crust, and beneath it in the mantle, differ from those in and beneath the oceanic crust. This is so because the distribution of radioactive elements, the principal sources of heat, differ in the two regions, and because the rocks in the oceanic and continental crusts differ significantly in their capacities to serve as thermal blankets to the mass of hot mantle rocks below. Nevertheless, as we see in Figure 4.6, temperatures in both cases rise to about 1000° C at rather shallow depths. We already know that some magmas are fluid at 1000° C, so an immediate question is, "Why isn't the Earth's mantle entirely molten?" The answer is that the pressures are too great. Pressures increase with depth due to the increasing load of rock above. The vertical axis in Figure 4.6 could just as reasonably have units of pressure as units of depth. As the pressure rises, the temperature at which a compound melts also rises. For example, albite ($NaAlSi_3O_8$) melts at 1104° C at the Earth's surface, where the pressure

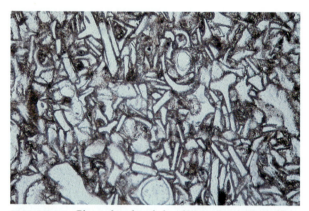

FIGURE 4.5 Glass shards of rhyolitic composition. The glass fragments, seen here in a microscopic thin section, are the main constituent of the Obispo Tuff in California. The photograph is 6 mm across.

is 1 bar,[1] but at a depth of 100 km, where the pressure is 35,000 times greater, the melting temperature is 1440° C. Therefore, whether a given rock melts and forms a magma at a specified place and depth in the Earth depends both on the geothermal gradient and on the effect that pressure has on the melting properties of the rock.

The effect of pressure on melting is straightforward provided the mineral is dry. When water or water vapor is present, however, a complication enters and another effect occurs that is similar to the effect salt has on an icy road. Salt causes ice to melt by forming a salty solution that can freeze only at temperatures below the freezing point of pure water. We say that salt depresses the freezing point of pure water. Similarly, the presence of water dissolved in magma depresses the freezing point of the magma. Or, to say it another way, wet minerals will melt at lower temperatures than dry minerals of the same composition because water dissolves in the melt. Furthermore, as the pressure rises, the effect of water also rises. This is so, because the higher the pressure, the greater the amount of water that will dissolve in the melt. Therefore, increasing pressure decreases still fur-

ther the temperature at which a wet mineral starts to melt. This is exactly opposite to the effect of pressure on the melting of a dry mineral. The effect of water on the melting of albite can be judged by comparing Figures 4.7*a* and *b*. The effects of pressure and water on the melting properties of rocks are very similar to the effects of pressure and water on the melting of albite.

If a rock melts completely, the resulting magma must have the same chemical composition as its parent. However, rock is a mixture of minerals, and it does not melt at one specific temperature as albite does. Rather, a rock melts over a temperature interval that may be as much as 500° C (Fig. 4.8). Once a rock reaches the temperature at which melting starts, a small quantity of liquid forms. The liquid has a different composition from the unmelted residue of minerals. As the temperature rises, first one mineral melts, and then another. At any instant, the melt has a composition that differs from the aggregate of unmelted crystals. Suppose now that the liquid from the partially melted aggregate is squeezed out of the remaining pile of unmelted crystals. The result will be a rock consisting of a residue of the unmelted crystals and a magma. Both the magma and the residual rock will have compositions that differ from the composition of the parent rock. *The process of forming magmas with differing compositions through the incomplete melting of rocks is known as* **magmatic differentia-**

---

[1] A bar is a unit of pressure equal to a force of one million (10⁶) newtons per square centimeter. The unit pressure in the International System of Units is the pascal, which is equal to a newton per square meter. Thus, a bar is equal to 10^{10} Pa. A bar is numerically equal to 1.02 kg/cm², and to 0.987 atm.

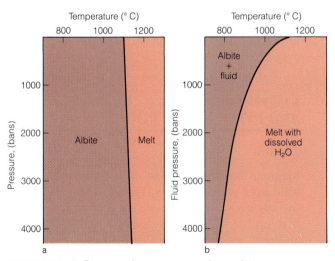

FIGURE 4.7 Influence of pressure on the melting temperature of albite ($NaAlSi_3O_8$). (*a*) Dry melting curve. Increasing pressure raises the melting temperature. (*b*) Wet melting curve. When an aqueous fluid is present, H_2O dissolves in the melt and decreases the melting temperature. The amount of H_2O that can dissolve, and hence the magnitude of the decrease in the melting temperature, increases with increased pressure.

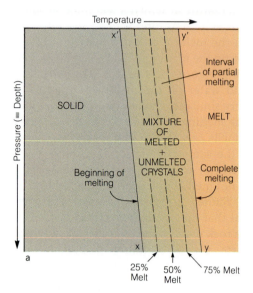

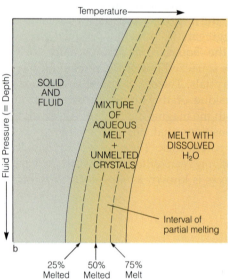

FIGURE 4.8 Diagrams illustrating the melting of rock. Even though rocks are composed of several kinds of minerals, the pressure effects on melting are very similar to those shown in Figure 4.7 for a single mineral. (*a*) Dry melting. Curve *X–X'* marks the onset of melting, curve *Y–Y'* the completion of melting. Between the two curves is a region in which melt and a mixture of unmelted crystals coexist. The melting interval can be 500°C or more wide. (*b*) Wet melting. As in the case of a single mineral, H_2O depresses the temperature of the onset of melting; the influence of H_2O increases at high pressure.

tion by partial melting (Fig. 4.9). It is not difficult to see how the composition of a magma that develops by partial melting depends both on the composition of the parent magma and the percentage of the rock that melts. Basaltic magma forms, appar-

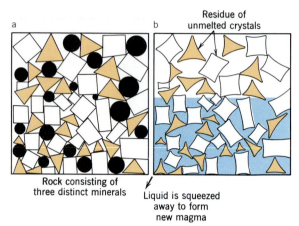

FIGURE 4.9 Creation of magma by partial melting of rock. (*a*) Rock consisting of three distinct minerals. (*b*) The first mineral that starts to melt will dissolve a small portion of the other minerals. The composition of the newly formed liquid, therefore, differs from the bulk composition of the remaining minerals. When the liquid is squeezed out, it forms magma of one composition and leaves a residue of unmelted crystals having a different composition.

ently, by partial melting of rock in the mantle—some people estimate that as much as 40 percent of the parent rock in the mantle must melt in order to produce magma of basaltic composition. Rhyolitic magma, too, can form by partial melting, but by partial melting of rock in the continental crust rather than in the mantle. Before discussing the process of partial melting any further, it is helpful to discuss first the kinds of rocks that form from magmas.

KINDS OF IGNEOUS AND PYROCLASTIC ROCK

Texture

Magma, like most liquids, is less dense than the solid from which it forms. Therefore, once formed, the low density magma will exert an upward pressure on the enclosing rocks, and will slowly flow and push its way upward through the mantle and crust. As a magma rises it will cool and start to solidify; eventually, when completely solidified an igneous rock is the result. Most magmas solidify below the surface, and we refer to *any igneous rock formed by solidification of magma below the Earth's surface* as an **intrusive igneous rock.** Some magma does, of course, reach the surface where it flows out to form **extrusive igneous rock** which we define

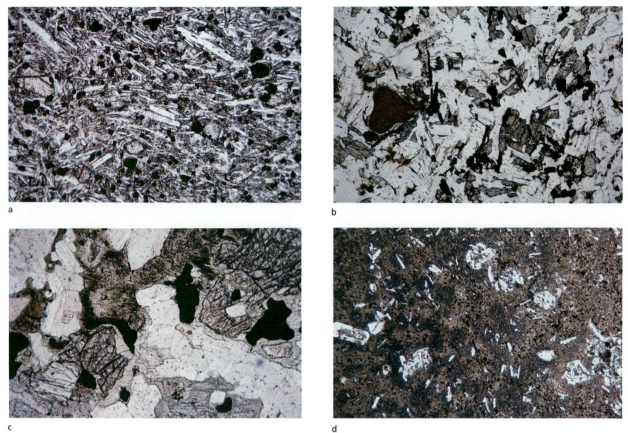

FIGURE 4.10 The textures visible in these thin sections of (*a*) basalt, (*b*) diabase, (*c*) gabbro, and (*d*) basalt porphyry, reflect the cooling history of each magma. The rocks all have the same composition (plagioclase, pyroxene, and olivine). Basalt, an aphanite, is a fine-grained lava that cooled very rapidly. Gabbro, a phanerite, is a coarse-grained, deep-seated, intrusive igneous rock that cooled very slowly. Diabase is a term used for a fine- to medium-grained gabbro, intruded at shallow depths and cooled at intermediate rates. Basalt porphyry contains phenocrysts of plagioclase set in a matrix that is so fine the grains can barely be resolved with a microscope. The coarse phenocrysts formed during slow cooling at depth; the matrix formed when the partly crystallized magma was suddenly extruded as a lava. The field of view is the same in each case: it is 7 mm across.

as *rock formed by the solidification of magma poured out onto the Earth's surface.*

In general, intrusive igneous rocks tend to be coarse-grained and referred to as **phanerites,** by which we mean that they are *igneous rocks in which the component mineral grains are distinguishable megascopically.* This is so because magma that solidifies below the surface tends to cool slowly and, thus, to have sufficient time to form large, clearly visible mineral grains, a millimeter or larger in diameter. By contrast, magma that cools rapidly forms very fine-grained and even glassy rocks. When lava cools and solidifies too rapidly for its atoms to or-

ganize themselves into minerals, the result is a *natural glass,* a substance called *obsidian.* Most extrusive igneous rocks are crystalline but fine-grained, or they are a mixture of glass and fine-grained minerals. Such rocks are called **aphanites** which we define as *igneous rocks in which the component grains cannot be readily distinguished with the naked eye or even with the aid of a simple hand lens.*

Most igneous rocks are either phanerites or aphanites, but one special textural class has a distinctive mix of coarse and fine grain sizes. A rock containing such a texture is called a **porphyry,** meaning *any igneous rock consisting of coarse mineral*

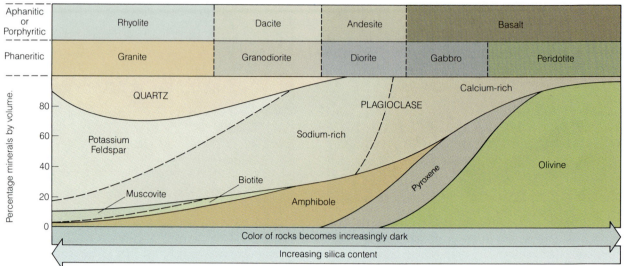

FIGURE 4.11 The textures and proportions of minerals in the common igneous rocks. Boundaries between rock types are not abrupt but gradational, as suggested by the broken lines. In granites, for example, there is a wide range in the proportions of minerals present: those with nearly 75 percent potassium feldspar belong at the left side of the diagram; others with only 20 percent are near the boundary with granodiorite. To determine the composition range for any rock type, project the broken lines vertically downward, then estimate the percentages of the minerals by means of the numbers at the edge of the diagram. (*Source:* Modified after Dietrich.)

grains scattered through a mixture of fine mineral grains (Fig. 4.10). *The isolated large crystals in a porphyry are* called **phenocrysts,** and they form in the same way that coarse crystals in phanerites do—by slow cooling of magma deep in the crust. The fine-grained, commonly aphanitic, groundmasses that enclose phenocrysts provide evidence that the partly cooled and crystallized magmas were moved rapidly upward. In their new settings, the magmas cooled more rapidly, and the later mineral grains, which now form the groundmass, are all tiny as a result. Most phenocrysts have well-developed crystal forms because they grow within a fluid, and do not encounter interference from other crystals growing near them, as we see in Figure 4.10c. The shapes of mineral grains in phanerites, in aphanites, and in the groundmass of porphyries are irregular and intensely interlocked, as can be seen in Figures 4.10a, b, and c. This is so, because during the final stages of mineral growth and crystallization, all the mineral particles are crammed against each other, thus preventing the formation of smooth crystal faces, and creating instead an interlocking network of grain boundaries

that act like a giant, three-dimensional jigsaw that makes an igneous rock coherent and solid.

Composition

Having determined whether an igneous rock has a phaneritic, aphanitic, or porphyritic texture, a name can be given to any specimen on the basis of its mineral assemblage. To see how this is done it is convenient to employ the diagram shown in Figure 4.11.

As mentioned earlier in this chapter, all common igneous rocks are comprised of one or more of these six minerals: quartz, feldspars (both potassium feldspar and plagioclase), mica, amphibole, pyroxene, and olivine. When the percentage of each mineral in a rock specimen has been estimated, the correct place in Figure 4.11 is located and the corresponding rock name determined. The aphanitic or phaneritic name is selected, whichever is appropriate. If a rock has a porphyritic texture, we use the name determined by the mineral assemblage as an adjective, using the grain size of the groundmass as an indication. For example, if the groundmass is aphanitic, we would refer to a rhyolite porphyry; but if it is phaneritic, we refer to a granite porphyry.

Varieties

The kinds of common igneous rocks, together with suggestions as to how to identify them, are discussed more fully in Appendix C. Here we mention only those points about the common rocks that are needed to discuss their origins. First, consider the *granites* and *granodiorites,* both of which are intrusive igneous rocks found only in the continental crust. Both contain abundant quartz and feldspar—granites are rich in potassium feldspar, granodiorites are rich in plagioclase. While huge intrusive masses of granite and granodiorite are common, masses of extrusive igneous rock with the same compositions—*rhyolite* and *dacite,* respectively—are less common and form much smaller masses than their intrusive equivalents. As we shall see, these facts provide important clues concerning the origin of the magmas that form the rocks.

Second, consider basalt, the fine-grained, generally aphanitic, extrusive igneous rock that everywhere underlies the thin skin of sediments that floor the oceans. Basalt contains, as major minerals, olivine, pyroxene, and plagioclase feldspar. As mentioned previously, rhyolite is less common than the compositionally equivalent intrusive igneous rock, granite. The situation is reversed with basalt which very commonly makes its way to the surface as a lava. Basalt is by far the most common extrusive igneous rock on earth. Intrusive igneous rock formed from basaltic magma, on the other hand, while common, seems to be less common than basalt. The intrusive products of basaltic magma also have a rather wide range of mineral assemblages and grain sizes. The

a

b

c

FIGURE 4.12 Tephra. (*a*) Large tephra fragments are called bombs. These spindle-shaped bombs, up to 50 cm in length, cover the surface of a cinder cone on Haleakala volcano, Maui. (*b*) Intermediate-sized tephra particles are called lapilli. Two layers of tephra are separated by a soil layer. The upper layer was erupted from Mount St. Helens in 1800 A.D., the lower one in 1480 A.D. (*c*) The finest tephra is called ash. Because the volcanic fragments are so small, they weather rapidly. Three ash layers are visible in a peat bog at Snoqualmie Pass, Washington. The thin, upper layer was erupted from Mount St. Helens 500 years ago; the middle layer also came from Mount St. Helens, 3,400 years ago. The bottom yellowish-colored layer was deposited as a result of the great eruption that destroyed Mount Mazama and produced Crater Lake 6,700 years ago.

TABLE 4.1 *Names for Tephra and Pyroclastic Rock*

Average Particle Diameter (mm)	Tephra (unconsolidated material)	Pyroclastic Rock (consolidated material)
> 64 mm	Bombs	Agglomerate
2–64 mm	Lapilli	Lapilli tuff
< 2 mm	Ash	Ash tuff

phaneritic equivalent of basalt is called *gabbro* if olivine, pyroxene, and feldspar are present in roughly equivalent amounts and the grain size is coarse. It is called fine-grained gabbro, or *diabase*, if the grain size is small but the rock is not aphanitic. Some phaneritic variants of gabbro contain 90 percent or more olivine, in which case we refer to them as *peridotites*; others contain 90 percent or more plagioclase, and are called *anorthosites*.

The third and final group of common igneous rocks is intermediate in composition between those formed from rhyolitic and basaltic magmas. The most distinctive rock of the class is *andesite*, an extrusive igneous rock that is commonly aphanitic and that consists largely of plagioclase and amphibole. The phaneritic equivalent of andesite is *diorite*. Not surprisingly, since diorite is intermediate in composition and properties between granites and gabbro, it is observed that andesites are just about as common as diorites.

Pyroclasts

When magma reaches the Earth's surface, it does so through a ***volcano,*** *a vent from which igneous matter, solid rock debris, and gases are erupted.* The term volcano comes from *Vulcan,* the Roman god of fire, and it immediately conjures up visions of sheets of lava pouring out over the landscape. In fact, volcanoes, and the eruption process, are much more varied than most of us imagine. ***Lava,*** to be sure, is *magma that reaches the Earth's surface through a volcanic vent, and flows as hot streams or sheets.* Most lava tends to cool quickly and to be fine-grained; some cools so rapidly that obsidian

FIGURE 4.13 Tuff, a pyroclastic rock formed by cementation of lapilli and ash. Note the fragments of volcanic rock. The scale marks are 1 mm. From Clark County, Nevada.

FIGURE 4.14 Welded tuff (ignimbrite) from the Jemez Mountains, New Mexico. The dark patches are fragments of obsidian flattened during welding. Note the fragments of other rocks in the specimen. The scale marks are 1 mm.

results. However, not all magma is erupted as a smooth-flowing liquid.

When gases escape from a volcanic vent, they sometimes do so in such a violent fashion that they rip pieces of solid rock off the walls of the vent and splatter and shatter the sticky magma into small, hot fragments. *Fragments extruded violently from a volcano are called **pyroclasts;** loose assemblages of pyroclasts are called **tephra,*** and as was discussed at the beginning of this chapter, rock formed through the cementation or welding of tephra is termed pyroclastic rock. Tephra and pyroclastic rocks are named on the basis of the fragment sizes they contain (Table 4.1; Fig. 4.12).

The word ash, applied to tephra, is somewhat misleading but is nevertheless a word in common use. It is misleading because ash means, strictly, the solid that is left after something inflammable, such as wood, has burned. But the fine particles thrown out by volcanoes look so like true ash, that it has become a convenient custom to use the word for these particles also. As seen in Table 4.1, the terms *agglomerate, lapilli tuff,* and *ash tuff* are used to describe the textures of pyroclastic rocks. Mineral assemblages are used to determine the composition of pyroclastic rocks just as they are with other igneous products. For example, we refer to a rock of appropriate composition as an andestic lapilli tuff if it is aphanitic, and dioritic lapilli tuff if it is phaneritic.

Conversion of tephra to pyroclastic rock can come about in two ways. The first, and most common way, is through the addition of a cementing agent introduced by groundwater (Fig. 4.13). The most common cementing agents are calcite and quartz. The second way tephra is transformed to pyroclastic rock is through the welding of hot, glassy ash particles. *Welded tuff* (also called ***ignimbrite***) is the name applied to *pyroclastic rocks, the glassy fragments of which were plastic and so hot when deposited that they fused to form a glassy rock.* Very commonly the glassy fragments became flattened and compacted during the fusion process, giving welded tuffs a rather distinctive texture that resembles a flow pattern (Fig. 4.14).

ORIGINS OF THE THREE PRIMARY MAGMAS

Basaltic Magma

Basalt and gabbro, both formed from basaltic magma, are the only important igneous rocks of the oceanic crust. The oceanic crust is thin and immediately below lies the mantle; therefore, we must conclude that the mantle is the source of basaltic magma. Basalts and gabbros are also found in the continental crust, which leads to the further conclusions that the mantle beneath continents must be the same, or at least very similar to that beneath the oceans, and that basaltic magma must be able to rise upward from the mantle and penetrate the thick continental crust.

The dominant minerals found in basalts and gabbros are olivine, pyroxene, and feldspar. Each of them is water-free. This fact, plus observations that basaltic magma in Hawaii, Iceland, and other active volcano sites contains little water, suggests that basalt is essentially a dry, or water-poor magma. Indeed, all evidence suggests that the water content of basaltic magma rarely exceeds 0.1 percent. It must be concluded, therefore, that basaltic magma originates by some sort of dry, partial-melting process in the mantle.

Much scientific debate has centered on the question of the composition of the mantle. Rarely can we actually observe mantle rocks. In a few places it is possible to find fragments carried up and ejected from volcanoes. From such evidence it appears that the upper portion of the mantle resembles a peridotite in composition, but a peridotite that contains garnet in addition to the major mineral, olivine. Laboratory experiments on the melting properties of *garnet peridotite* show that at pressures equivalent to those reached at depths of 100 to 350 km below the surface—that is, at the pressures of the asthenosphere—a 30 to 40 percent partial melt will yield a magma of basaltic composition (Fig. 4.15). While this leaves unanswered, for the moment, the question of a heat source, and why basaltic magma should develop in some places but not others, we can nevertheless confidently accept the conclusion that basalt forms by dry partial melting of rocks in the mantle.

Once a body of basaltic magma has formed, it

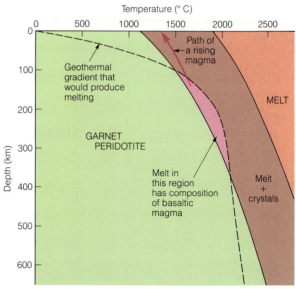

FIGURE 4.15 Dry melting properties of garnet peridotite and the generation of basaltic magma. The composition of garnet peridotite is believed to be similar to that of the upper mantle. If the geothermal gradient in some region of the mantle is like that shown by the dashed line, approximately 30 percent of the peridotite will melt at depths between 250 and 400 km. The melt that forms has the composition of basaltic magma. Once a highly fluid basaltic melt has formed it will start to rise toward the surface, following a depth–temperature curve like that shown by the dashed line. The magma will reach the surface without solidifying.

will start to rise because it is less dense than the rocks around it. Because basaltic magma is very fluid, the rate of rising can be quite rapid. One line of evidence comes from earthquakes. As basaltic magma rises, it widens fractures and sometimes causes new ones in the rock it is passing through. Movement of fractures causes small earthquakes that can be detected by sensitive devices on the surface. Evidence from Hawaii indicates that magma can rise at rates as high as many kilometers a day. Under such circumstances, the rate of rising is much faster than the rate of cooling. The depth–temperature path followed by a rising basaltic magma will, therefore, be something like that shown in Figure 4.15. The higher a magma rises, the further away it will be from its solidification temperature. Therefore, not surprisingly, a lot of basaltic magma manages to rise through the crust without solidifying. When it reaches the surface, it is erupted as lava.

Rhyolitic Magma

Two important observations suggest a unique origin for rhyolitic magma and its close relative, dacitic magma. Rhyolites and dacites are similar in composition, differing only in the relative proportions of potassium feldspar and plagioclase. So too are their respective phaneritic equivalents, granite and granodiorite. For this reason it is common practice to refer, loosely, to the magmas from which rhyolites and dacites form as being rhyolitic, and the magmas from which granites and granodiorites form, as being granitic. The first observation concerning origin is that modern volcanoes that extrude rhyolitic magmas are confined to regions of continental crust. Similarly, the distribution of ancient rhyolites and their equivalent intrusive rocks, is also confined to the continental crust. The second observation bearing on origin, concerns the water content of rhyolitic magma. Rhyolitic volcanoes give off a great deal of water vapor, and granitic rocks contain significant quantities of water-bearing minerals such as mica and amphibole. These two points of evidence suggest (1) that the sources of granitic magmas lie within the continental crust, and (2) that the origin involves some sort of wet partial melting. Laboratory experiments bear this suggestion out. When, in the laboratory, water-bearing rocks of the continental crust start to melt, the composition of the liquid that forms is rhyolitic. The average composi-

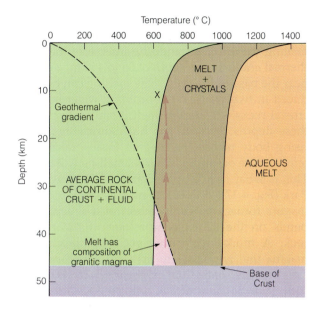

FIGURE 4.16 Wet melting properties of crustal rock, and the generation of granitic magma. If the geothermal gradient is as shown, it will intersect the region where melting commences at a pressure equivalent to a depth of about 35 km. The melt that forms has the composition of granitic magma. As the highly viscous magma slowly rises, it will follow a depth–temperature path like that shown by the red arrows, and it will continually approach the freezing curve; at point *X* the magma reaches the freezing curve and will therefore be completely solidified and will not reach the surface.

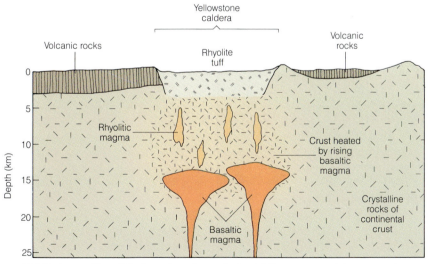

FIGURE 4.17 Simplified section through the Yellowstone caldera. Basaltic magma rising from the mantle fills magma chambers and heats and locally melts crustal rocks to form magma of rhyolitic composition. Eruption of the magma to form rhyolitic lava flows and welded tuffs leaves the magma chamber partly empty so the roof collapses to form a caldera. (Modified after Smith and Braile.)

tion of the continental crust is approximately that of an andesite. As seen in Figure 4.16, the wet-melting curve for an andesite intersects the geothermal gradient at a depth of 35 to 45 km, a depth near the base of the crust. The melting can be produced in two ways. First, ordinary water-bearing crustal rocks will start to melt if simply buried to these depths. Second, if a heat source such as a rising body of basaltic magma locally raises the temperature of the crust, partial melting will occur. This is what has happened beneath Yellowstone Park (Fig. 4.17). It does not matter which way the magma forms, the composition of a rhyolite depends on the actual composition of the parent rock. A muscovite-bearing rock will produce a magma that is potassium-rich and which crystallizes to form potassium feldspar. A calcium-rich rock, such as an andesite, will produce a magma that is calcium-rich and which crystallizes to form plagioclase as the predominant feldspar.

Although granite and granodiorite are very common (and granodiorite is more abundant than granite), rhyolite and dacite are not very common rocks. This is exactly the opposite of the relation between gabbro and basalt, where the extrusive rock, basalt, is more abundant than its intrusive equivalent. Scientists were long puzzled by this observation; Figure 4.16 illustrates a possible explanation. Once a rhyolitic magma has formed, it starts to rise. However, it rises slowly because it is very viscous, and as it rises the pressure on it decreases. As discussed earlier, the effectiveness of water in reducing the melting temperature is diminished by reduced pressure. A rising magma formed by wet partial melting must therefore increase in temperature or it will solidify because of reduced pressure, thereby forming an intrusive igneous rock. But a rising magma traverses bodies of cool rock, and as a consequence, there is no source of heat to cause an increase in temperature. As a result, the depth–temperature path of a rising body of rhyolitic magma comes closer and closer to its solidification temperature. Therefore, most rhyolitic magmas solidify and form granites rather than being extruded at the surface as a rhyolitic lava.

Andesitic Magma

The chemical composition of andesite is close to the average composition of the continental crust. Andesite and its equivalent intrusive rock, diorite, are commonly found in the continental crust. From these two facts we might suppose that andesitic magma forms simply by the complete melting of a portion of continental crust. Some andesitic magma may indeed be generated in this way, but this cannot be the origin of all such magma. Because andesitic magma is extruded from some volcanoes that are far from continental masses, it must, in those cases, be developed either from the mantle or the oceanic crust. Laboratory experiments provide a possible answer.

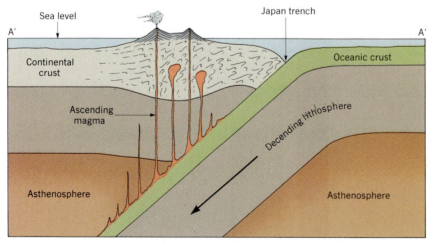

FIGURE 4.18 Cross section through the lithosphere and asthenosphere demonstrating the probable way by which andesitic magma is formed and erupted from volcanoes in Japan. At depths between 100 and 150 km, oceanic crust that has been carried downward by descending lithosphere undergoes wet partial melting. Andesitic magma rises, creating volcanoes on the Japanese islands.

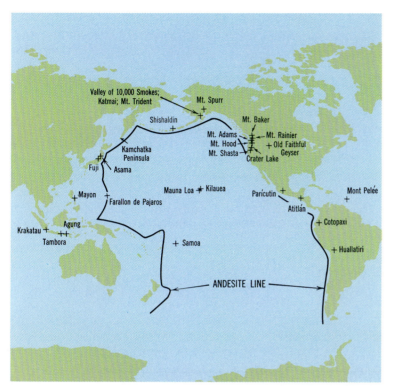

FIGURE 4.19 The Andesite Line surrounds the Pacific Ocean basin and separates areas within the basin where andesitic rocks are not found from areas where andesites are common. Volcanoes such as Mauna Loa, that are inside the line, erupt basaltic magma but not andesitic magma. Those outside the line, such as Mt. Shasta, may erupt basaltic magma too, but they also erupt andesitic magma.

In laboratory experiments, partial melting of wet basalt yields, under suitably high pressure, a magma of andesitic composition. An interesting hypothesis suggests how this might happen. When a moving plate of lithosphere plunges back into the asthenosphere, it carries a capping of wet oceanic crust with it. Oceanic crust is largely basalt and gabbro. The plate heats up and eventually the wet basalt starts to melt. A small degree of wet partial melting produces a liquid having the composition of andesitic magma (Fig. 4.18).

There are a number of details concerning the melting process of wet basalt that remain to be deciphered, but two pieces of evidence are very supportive of the idea that much of the andesitic magma forms in this manner. The first concerns the distribution of andesite. When the locations of active volcanoes in and around the Pacific Ocean are plotted, it is apparent that a well-defined line separates regions where andesite occurs from regions where it does not occur. The line is called the Andesite Line (Fig. 4.19). Inside this line, and inside the main ocean basin, andesite is unknown.

All active volcanoes inside the line erupt basaltic magma, and all the volcanic rock associated with dormant volcanoes formed from basaltic magma. Outside the line, andesite is common. The Andesite Line coincides closely with subduction zones, which, as we learned in Chapter 2, mark the very places where plates of lithosphere sink back into the asthenosphere. Therefore, if andesites form by the partial melting of wet basalt, the process occurs at the very places on the Earth where wet basalt is found at suitable depths and temperatures for partial melting to occur.

The second line of evidence comes from the distribution of andesitic volcanoes with respect to the subduction zone. On the upper surface of a plate of lithosphere, a subduction zone is marked by the presence of a deep-sea trench (Fig. 2.7). Beyond the subduction zone, the lithosphere sinks into the asthenosphere, carrying with it its capping of wet basalt. At a depth of about 80 km or more, melting of the wet basalt commences; the magma rises and forms clusters of volcanoes that form an arcuate belt parallel to the trench (Fig. 4.20).

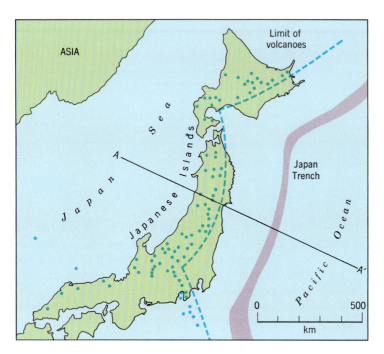

FIGURE 4.20 Relations between ocean trenches and arcs of andesitic volcanoes. Arc-shaped Japanese islands are parallel to the Japan Trench. Volcanoes, most of them andesitic and active during the last million years, are also confined behind arcuate boundaries. The cross section depicted in Figure 4.18 is approximately along the line *A–A'*. (*Source:* Adapted from Matsuda and Eyeda, 1971.)

SOLIDIFICATION OF MAGMA

Although only three common families of magma exist, there are literally hundreds of different kinds of igneous rocks. Most are rare, but the fact that they exist emphasizes an important point: A single magma can crystallize into many different kinds of igneous rock. This is true because magma is a complicated liquid. It does not solidify into a single compound, as water freezes to form ice. Solidifying magma forms several different minerals, and again unlike water, the minerals crystallize at different temperatures. The process is just the opposite of partial melting. As the temperature slowly falls and the magma freezes, first one mineral crystallizes, then another. Therefore, a freezing magma soon consists of a mixture of already-crystallized minerals and still unfrozen liquid. The combination is like a partly frozen bottle of cider. When cider cools, crystals of ice form from the water it contains. All the other ingredients—sugar, alcohol, and flavorings—become concentrated in the remaining liquid. Similarly, in cooling magma, the first minerals that crystallize have

compositions different from that of the remaining liquid. Because different minerals begin to crystallize at different temperatures, the composition of the remaining liquid changes continually as the temperature changes. If at any time during crystallization the remaining liquid becomes separated from the crystals, the liquid can continue to cool as a magma with a brand-new composition, while the crystals left behind form an igneous rock with an entirely different composition. One way by which liquid becomes separated from crystals occurs when early formed crystals are much more dense than the liquid. The crystals will then sink to the bottom of the magma chamber where they will form a separate layer with a distinctive composition. Further cooling, crystallization, and settling can produce layers of igneous rocks with widely differing compositions (Fig. 4.21).

There are a number of ways by which crystal–liquid separations can occur. For example, compression can squeeze liquid out of a crystal–liquid mixture. However a separation occurs, it leads to the formation of igneous rocks with compositions that differ from the compositions of the parent magma. We call the *compositional changes that occur in magmas by the separation of early-formed minerals from residual liquids* ***magmatic differentiation by fractional crystallization.***

The person who first recognized the importance of magmatic differentiation by fractional crystallization was the Canadian-born scientist, N. L. Bowen. Beginning in the early 1900s, Bowen and his colleagues at the Geophysical Laboratory in Washington, D.C., investigated the melting and crystallization properties of minerals, and in particular the order in which minerals crystallize in cooling magmas of different compositions.

Bowen recognized that several important features present in the textures and mineral assem-

blages of igneous rocks could be explained by his experimental data. For example, he knew that plagioclases in basalts and gabbros are usually calcium-rich (anorthitic), while those in granodiorites are usually sodium-rich (albitic). Andesites, he observed, tend to have plagioclases of intermediate composition. Bowen also knew that plagioclases in many igneous rocks have concentric zones of differing compositions such that the innermost, and therefore earliest formed core, is anorthitic in composition, and that successive layers are more and more albite-rich (Fig. 4.22). Bowen's experiments provided a common explanation for these two observations. He discovered that the composition of the first plagioclase that crystallizes from a magma of basaltic composition is calcium-rich, but that as crystallization proceeds, and the ratio of crystals to melt increases, so does the composition of the plagioclase in contact with the melt change toward a more sodic composition. This means that all the plagioclase crystals, even the earliest ones formed, should continually change their compositions as the magma cools. The plagioclase solid-solution series involves a coupled substitution in which $Ca^{+2} + Al^{+3}$ are replaced in the structure by $Na^{+1} + Si^{+4}$. Bowen referred to such a continuous change of mineral composition

FIGURE 4.21 Compositional layering in a phaneritic igneous rock. Light-colored layers of anorthosite contain calcium-rich plagioclase, dark-colored layers are nearly pure chromite ($FeCr_2O_4$). The layering was formed by magmatic differentiation in a parent magma of basaltic composition; if differentiation had not occurred during crystallization, the magma would have solidified to form a gabbro. From the Bushveld Igneous Complex, South Africa. The specimen is 8 cm wide.

FIGURE 4.22 Zoned crystal of plagioclase in a granite. The photograph was taken in polarized light in order to enhance the zoning. Each band in the crystal is a slightly different composition, progressing from anorthite-rich in the center, to albite-rich along the rim. The crystal is about 1 cm across.

as a *continuous reaction series,* by which he meant that the composition changed continually, but that the crystal structure remained unchanged. The process by which this occurs is controlled by the rates at which the four ions, Ca^{+2}, Al^{+3}, Si^{+4}, and Na^{+1}, can diffuse through the plagioclase structure—they are exceedingly slow processes. A complete chemical balance (commonly referred to as chemical equilibrium) is rarely attained because cooling rates are too fast. As a result, zoned plagioclase crystals are formed. The inner zones are out of chemical equilibrium with the outer zones and the residual magma. Bowen pointed out that the existence of zoned crystals has important implications. If anorthite-rich cores are present, the remaining liquid is necessarily richer in albite than it would have been if equilibrium had been maintained. The anorthite-rich cores are, he pointed out, another example of magmatic differentiation by fractional crystallization. If, in a partially crystallized magma containing zoned crys-

tals, the liquid were somehow squeezed out of the crystal mush, the result would be an albite-rich magma, and the residue would be an anorthite-rich rock.

Bowen's studies identified several sequences of reactions besides the continuous reaction series of the feldspars. For example, one of the earliest minerals to form in a cooling basaltic magma is olivine. Continued cooling will change the olivine composition slightly by solid solution, but eventually a point is reached where the olivine reacts with silica in the melt to form a more silica-rich mineral, pyroxene. An idealized example of such a reaction is:

$$Mg_2SiO_4 + SiO_2 \rightarrow 2MgSiO_3$$

Olivine + silica in $\rightarrow$ pyroxene
magma

At still lower temperatures, pyroxene reacts to form amphibole, which contains more silica than pyroxene, and then the amphibole reacts in turn to form an even more siliceous mineral, biotite. Such

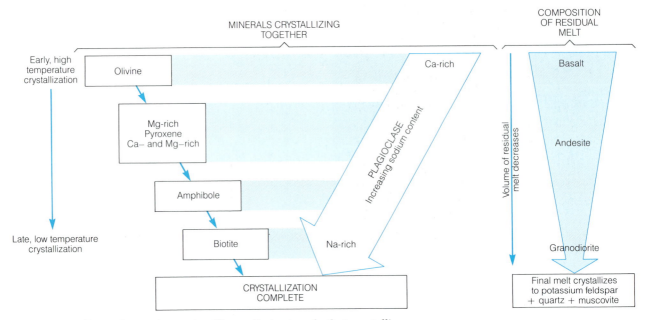

FIGURE 4.23 Bowen's reaction series. The earliest minerals that crystallize from a magma of basaltic composition are olivine and calcium-rich plagioclase (anorthite). As cooling and crystallization proceed, olivine (upper left) reacts with the remaining liquid to form a new mineral, pyroxene. Pyroxene in turn reacts to form amphibole, and amphibole forms biotite. The early plagioclase that co-crystallizes with olivine is calcium-rich, but as cooling proceeds, the early plagioclase reacts with the residual melt and continually changes its composition, becoming more and more sodium-rich. The composition of the residual melt in contact with the crystallized minerals becomes increasingly silica-rich, and eventually the final small fraction of melt has the composition of a granitic magma.

FIGURE 4.24 Microscopic thin section, 1 cm wide, showing effects of chilling at the contact of a basalt porphyry (right) and a sandstone (left). The basaltic magma was chilled against the cold sandstone at the time of intrusion. The phenocrysts of plagioclase in the basalt porphyry were present before the magma was chilled.

a series of reactions, where early formed minerals form entirely new compounds through reactions with the remaining liquid, is called a *discontinuous reaction series*. Bowen observed that if the conversion of olivine to pyroxene did not proceed to completion, and a core of olivine were shielded from further reactions by a rim of pyroxene, the remaining liquid would be more silica-rich than it would be if equilibrium were maintained and all the olivine were converted to pyroxene. If partial reactions occurred in both continuous and discontinuous reaction series, Bowen reasoned that differentiation by fractional crystallization in a basaltic magma could, under some circumstances, even produce a residual magma with a rhyolitic composition (Fig. 4.23). It is now known that Bowen's conclusions are correct so far as the reaction series are concerned, and that the reactions are vitally important in producing rock types of many intermediate compositions. However, neither large volumes of rhyolitic magma, nor large bodies of granite, apparently form by fractional crystallization. The main evidence against Bowen's idea is simply that, even under ideal conditions, no more than 10 percent of the volume of basaltic magma can be differentiated to rhyolitic magma. Yet granite bodies formed by crystallization of granitic magma are often of immense size—hundreds of thousands of cubic kilometers of rock. Basaltic

magma chambers where the differentiation could occur are simply not large enough to have produced such huge masses of granite. Another line of evidence comes from the distribution of granite. It always occurs in the continental crust. But if granite formed by direct differentiation of basaltic magma we would surely expect to find some in the oceanic crust, for it is there that basalts are most common. The principal manner by which granites form, therefore, must be through partial melting of continental crust.

Intrusion of Magma

Beneath every volcano there lies a complex of chambers and channel ways through which magma reaches the surface. The magmatic channels of an active volcano can not be seen, but ancient channel ways can be examined when they have been unroofed and laid bare by erosion. The ancient channel ways are filled with intrusive igneous rocks because they are the underground sites where magma solidified. *All bodies of intrusive igneous rock, regardless of shape or size*, are called **plutons** after the Greek god of the underworld, *Pluto*.

The magma that forms a pluton did not originate where we now find an intrusive igneous rock. Rather, the magma was intruded from the place where it was generated by partial melting. Proof that this is so lies in the composition of intrusive igneous rocks—they are usually completely different from the rocks that enclose them (Fig. 4.24). Commonly, too, it is possible to find evidence of chemical reactions between the magma and the intruded rocks.

Plutons are given special names depending on their shapes and sizes (Fig. 4.25). Dikes and sills are tabular bodies. *Dikes* are discordant with the layering in the enclosing rocks, by which we mean they are *tabular sheets of intrusive igneous rock cutting across the layering of the intruded rock.* *Sills* on the other hand are concordant, and are defined as *tabular sheets of intrusive igneous rock that are parallel to the layering of the intruded rock.* Dikes may occur at any depth; they are commonly steeply inclined or vertical, and they mark the ancient channel ways of upward-rising magma. Sills, by contrast, only tend to form at shallow depths and occur within piles of layered rocks. Intrusion of a sill requires that all the overlying rocks be lifted upward by an amount equal to the thickness of the sill. Thus, sills must be intruded at shallow depths, where the weight of overlying rock is low, and they must

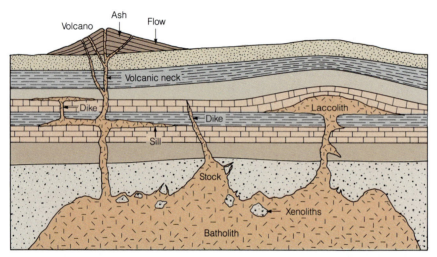

FIGURE 4.25 Diagrammatic section through part of the crust to show the various forms assumed by plutons. Many plutons were once connected with volcanoes, and there is a close relationship between intrusive and extrusive igneous rocks. A granitic pluton rises upward through the overlying rocks by a process called stoping; fragments of country rock are wedged off and sink through the magma to the floor of the chamber.

FIGURE 4.26 Shiprock, New Mexico, the eroded remains of a volcanic neck 400 m high. The prominent ridges radiating outward are dikes. The view was taken from the air, looking southwest.

follow planes of weakness in the intruded rock. Commonly, dikes and sills occur together as part of a network of plutons, as shown in Figure 4.25.

Both dikes and sills can be very large. For example, the Great Dike in Zimbabwe is a tabular body of gabbro nearly 500 km in length and about 8 km wide, with essentially vertical walls. An example of a large and well-known sill can be seen in the cliffs of the Palisades that line the Hudson River opposite New York City. The Palisades sill reaches a thickness of about 300 m, and like the Greak Dike, is a gabbro. Like the Great Dike also, the magma that formed the Palisades sill differentiated by fractional crystallization. In both masses it is possible to observe distinct compositional layers formed by the settling of early formed crystals. A variation of a sill is a *laccolith,* which is *a concordant, lenticular intrusive body along which the layers of the invaded country rock have been bent upward to form a dome.* A variant of a dike, because it is discordant, is a *volcanic neck, the approximately cylindrical conduit of igneous rock forming the feeder pipe immediately below a volcanic vent.* A famous example of a volcanic neck, together with associated dikes, can be seen at Shiprock, New Mexico (Fig. 4.26).

Large plutons are enormous bodies in comparison with dikes, sills, and other small plutons. The largest is a *batholith,* by which we mean *a very large, discordant, intrusive igneous body of irregular shape.* Most batholiths are composite masses that comprise a number of separate intrusive bodies of slightly differing composition. The differences possibly reflect variations in the crustal rocks from which the magma formed. Some batholiths exceed 1000 km in length and 250 km in width—the largest in North America is the Coast Range Batholith of British Columbia and northern Washington, which has a length of about 1500 km (Fig. 4.27). Geologists commonly reserve use of the term batholith for bodies of intrusive igneous rock that have outcrop exposures in excess of 100 km². When the outcrop exposure is less than 100 km², a discordant pluton is called a *stock,* which we define as *a small, discordant body of intrusive igneous rock.* It is apparent from Figure 4.25 that a stock may merely be a satellitic body to a batholith or even the top of a partly eroded batholith. What is not apparent in Figure 4.25, however, is what the bottom of a batholith looks like. Where it is possible to see them, the walls of batholiths tend to be very steep-sided or even vertical. This had led to a commonly held perception that batholiths extend downward to great depths—possibly even to the base of the crust. Geophysical measurements and

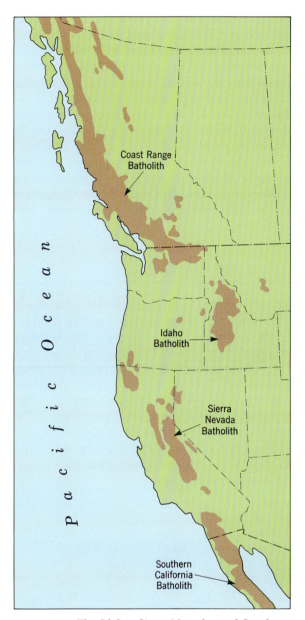

FIGURE 4.27 The Idaho, Sierra Nevada, and Southern California Batholiths, largest in the United States, are dwarfed by the Coast Range Batholith, in Southern Alaska, British Columbia, and Washington. Each of these giant batholiths is believed to have formed from magma generated by the partial melting of continental crust, and each intrudes metamorphosed rocks.

studies of very deeply eroded bodies of igneous rock suggest that this perception is incorrect. Batholiths seem to have floors, and to be 20 to 30 km thick, which is rather small compared to their great widths and lateral extents. Despite their huge sizes, batholiths do move upward. Even though intruded rocks can be pushed upward by the slowly rising magma, some other process must also operate. The rising magma can also dislodge fragments of the overlying country rock by a process known as *stoping*. Dislodged blocks are more dense than the rising magma and will therefore sink. As sinking proceeds, the fragments may react with and be partly dissolved by the magma. Most fragments apparently do not dissolve, but instead they sink all the way and reach the floor of the magma chamber. Those *fragments of country rock still enclosed in a magmatic body when it solidifies* are known as **xenoliths** (Fig. 4.28) (from the Greek words *xenos*, meaning stranger, and *lithos*, meaning stone).

MAGMA AT THE SURFACE

We noted earlier in the chapter that a volcano is a vent from which molten igneous matter, solid rock debris, and gases erupt. The vents are commonly surrounded by distinctive cone-shaped, or sheet-like, piles of tephra and volcanic rock. Most people refer to both the vent and the volcanic pile as a volcano, and we will do so also, although, strictly speaking, this is not correct in terms of the definition. The shapes of many volcanoes are so similar, and the processes that build them so nearly alike, that special names have been given to the common ones.

Shield Volcanoes

The volcano easiest to visualize is one built up of successive flows of lava. Such volcanoes are characteristically built by very fluid lavas, capable of flowing great distances down gentle slopes, and of forming thin sheets of nearly uniform thickness. Eventually the pile built up in this fashion develops a shape resembling a shield, convex-side up. A **shield volcano,** then, is *a volcano that emits fluid lava and builds up a broad, dome-shaped edifice (convex upward) with a surface slope of only a few degrees* (Fig. 4.29). The slope of a shield volcano is less near the summit than on the flanks. This happens because most of the magma is erupted near the summit, and because it is hot and very fluid it will readily run down a very slight slope. The further the lava flows, the cooler and less fluid it becomes, and the steeper a slope must be in order for it to flow. Slopes typically range from less than 5° near the summit, to 10° on the flanks.

Shield volcanoes are characteristically formed by the eruption of basaltic lava—the proportions of ash and other fragmental debris are small. Because basalt is the igneous rock of the ocean basins, shield volcanoes are characteristically oceanic but they are found on the continental crust as well. Hawaii, Tahiti, Samoa, the Galapagos, and many other oceanic islands are the upper portions of large shield volcanoes.

a

b

FIGURE 4.28 Xenoliths dislodged from overlying rock by an intruding body of magma. (*a*) Xenoliths of metamorphic rocks and gabbro in a body of phaneritic igneous rock at Crystal Cascade, Ascutney, Vermont. The marks on the hammer handle are in inches. (*b*) Xenolith of biotite gneiss in the Petersburg Granite is exposed in the bed of the James River, Richmond, Virginia. The size of the xenolith can be judged from the hammer.

Pyroclastic Cones

The lavas of some volcanoes, particularly those of rhyolitic, dacitic, and andesitic compositions, are so highly viscous that gas bubbles can only escape from them with great violence, ejecting quantities of pumice, cinders, volcanic ash, and other pyroclasts. As the debris showers down, a cone is built around the vent. The slope of the cone is determined by the angle of repose of the debris (Fig. 4.30). Fine ash will stand at a slope angle of 30° to 35°, while cinders generally stand at an angle of about 25°. Near the base of the cone, slumping and gradual decrease in the volume of fallout material away from the vent lead to more gentle slopes. Therefore, *pyroclastic cones* are *cones consisting entirely of pyroclastic debris surrounding a volcanic vent.* They are common in many areas of active volcanism, and have distinctive steep-sided profiles.

Stratovolcanoes

Large, long-lived volcanoes of andesitic, dacitic, and rhyolitic composition tend to emit a combina-

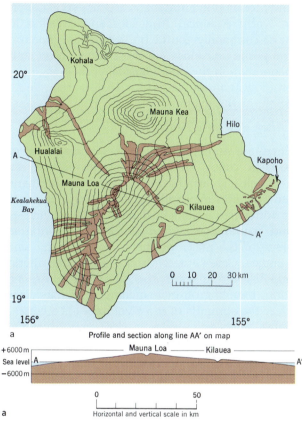

a

FIGURE 4.29 Shield Volcanoes. (*a*) The island of Hawaii is composed of five overlapping shield volcanoes, of which the largest are Mauna Loa and Mauna Kea. The topographic map shows the major volcanic centers and the longest lava flows erupted since 1750. The contour interval is 300 m. A profile along the section *A–A'* (lower diagram) illustrates the gentle slope characteristic of shield volcanoes. (After Sterns and MacDonald, 1946). (*b*) Mauna Kea as seen from Mauna Loa. The view is almost directly north. A pahoehoe flow on the northeast flank of Mauna Loa can be seen in the foreground.

b

FIGURE 4.30 Puu Hau Kea, Hawaii. The steep slopes of pyroclastic cones result from the high angle of repose of the pile of pyroclastic fragments. Mauna Loa is visible in the background.

FIGURE 4.31 Two steep-sided, towering stratovolcanoes in the Aleutians. The larger cone (rear) is Pavlov; the near one is Pavlov's Sister.

tion of lava flows and pyroclastic debris. The volume of pyroclastic material generally exceeds the volume of the lava, and so the slopes of the large volcanic cones, which may be thousands of meters high, are steep like those of pyroclastic cones. We define *stratovolcanoes* (also called *composite volcanoes*) as *volcanoes that emit both fragmental material and viscous lava, and that build up steep conical mounds.* Near the summit of a stratovolcano, the

slope is about 30°, like that near the summit of a pyroclastic cone. Toward the base, the slopes of stratovolcanoes flatten to about 6° to 10°. The beautiful steep-sided cones of composite volcanoes are among Earth's most picturesque sights (Fig. 4.31). The snow-capped peak of Mt. Fuji in Japan has inspired poets and writers for centuries. Mount Rainier and Mount Baker in Washington and Mount Hood in Oregon are majestic examples in North America. Andesites and rhyolites are most commonly found on continents; therefore, composite volcanoes are more common on continents than in the ocean basins.

Craters, Calderas, and Other Volcanic Features

There are numerous features associated with pyroclastic cones, shield, and stratovolcanoes that give volcanic terrains a special and unique character. Fractures may split the cone so that lava, ash, or both emerge along its flanks. Small, satellitic ash and spatter cones then develop, peppering the slope of the main mountain like so many small pimples. Gases emerge from small vents, altering and discoloring nearby rocks, and hot springs may form, bubbling off evil-smelling, sulfurous gases. Near the summits of most volcanoes is a *crater, a funnel-shaped depression from which gases, fragments of rock, and lava are ejected.*

Many volcanoes are marked near their summits by a striking and much larger depression than a crater. This is a **caldera,** *a roughly circular, steep-walled basin several kilometers or more in diameter.* Calderas originate through collapse following eruption and the partial emptying of a magma chamber. Most commonly, rapid ejection of magma involves large pyroclastic eruptions. Following eruption, the chamber from which the volcanic ash was emitted is empty or partly empty. The now-unsupported roof of the chamber slowly sinks under its own weight, like a snow-laden roof on a shaky barn, dropping downward on a ring of steep vertical fractures. Subsequent volcanic eruptions commonly occur along these fractures, thus creating roughly circular rings of small cones. Crater Lake, Oregon occupies a circular caldera 8 km in diameter (Fig. 4.32), formed after a great eruption about 6600 years ago. The volcano that erupted has been posthumously called Mount Mazama. What remained of the roof after a pyroclastic outpouring of about 75 km^3 of magma then collapsed into the partly empty magma chamber.

Large as the Crater Lake caldera may seem, it is

FIGURE 4.32 Crater Lake, Oregon, occupies a caldera 8 km in diameter that crowns the summit of a once lofty composite volcano, posthumously called Mount Mazama. Wizard Island is a small pyroclastic cone and lava flow that formed after the collapse which created the caldera.

FIGURE 4.33 A lava dome forming in the crater of Mount St. Helens, Washington. The photo was taken in May 1982. The plume rising above the dome is steam.

tiny by comparison with calderas observed in some regions of widespread volcanism. There, complexes of volcanoes, lava flows, and sheets of pyroclastic debris may cover 200,000 km² or more. Yellowstone National Park is a striking example.

Following several earlier periods of volcanic activity, a catastrophic eruption occurred in Yellowstone about 600,000 years ago. Approximately 1000 km³ of rhyolitic lava and hot ash were rapidly erupted from a shallow magma chamber, whose roof, no longer supported from below and suddenly loaded with the newly erupted lava above, collapsed to form a caldera 70 km long and 45 km wide. Most of the hot springs and other features for which Yellowstone is now so famous lie within the caldera. Scientists have recently discovered that a few kilometers beneath its floor a huge mass of rhyolitic magma still occupies the remains of the magma chamber. The scientists are divided on whether future eruptions are in store.

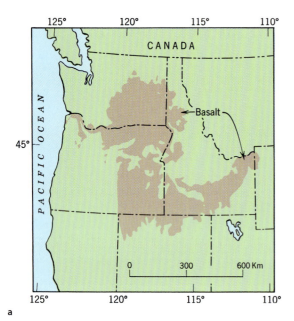

FIGURE 4.34 Plateau basalt. (*a*) Map of one of the world's largest basalt plateaus, the Columbia Plateau, northwestern United States. Areas of basalt (brown) are only remnants of the original flows, reduced by erosion and covered by younger sediments. (*b*) A stacked sequence of flows, part of the Columbia Plateau basalts exposed in the walls of the Snake River, Idaho. The light brown layers between are layers of volcanic ash, now largely altered to clay.

A volcano does not necessarily cease activity following the formation of a caldera. After collapse, it is common for magma to start reentering a magma chamber, and in the process cause *the uplifting of the collapsed floor of a caldera to form a structural dome.* Such a feature is called a **resurgent cauldron.** Subsequently, small pyroclastic and lava cones build up in the interior of the caldera and above the fractures that bound the caldera. Wizard Island, in Crater Lake, is a cone having such an origin. When lava, rather than pyroclastic debris, is extruded following a major eruption and collapse of the roof, the lava tends to be extremely viscous because it has very little dissolved volatile material in it. The sticky lava then squeezes out to form a **plug dome,** sometimes also called a *lava dome* (Fig. 4.33), which is *a volcanic dome characterized by an upheaved, consolidated conduit filling of lava.*

Fissure Eruptions on Land

Some lava does not reach the surface through pipelike conduits, but instead does so via elongate fractures which, when the walls are spread apart, become fissures. *Extrusion of volcanic materials along an extended fracture* is a **fissure eruption.** Such eruptions are characteristically associated with fluid basaltic magma, and the lavas that emerge from the fissure tend to spread widely and to build up flat plains. The fissure eruption at Laki in Iceland in 1783 occurred along a fracture 32 km long. Lava flowed 64 km outward from one side of the fracture and nearly 48 km outward from the other side. Altogether it covered an area of 588 km². The volume of the lava extruded has been estimated at 12 km³, making this the largest single, observed lava flow in recorded history. It was also one of the most deadly lava flows in history because an estimated 10,000 people were killed. There is good evidence to prove that larger eruptions have occurred in the past. The Roza flow, a great sheet of basalt in eastern Washington, can be traced over 22,000 km² and shown to have a volume of 650 km³.

When fissure eruptions occur on the continental crust, they produce distinctive, flat lava plateaus. The term *plateau basalt* is used for these features. A spectacular example of a plateau basalt is shown in Figure 4.34. Plateau basalts seem to develop when a plate of lithosphere carries continental crust over a source of basaltic lava in the mantle. The crust becomes thermally distended and as a result develops vertical fractures which serve as fissures for the escape of the basalt.

Fissure Eruptions Beneath the Sea

The most extensive and volcanically active system of fissures on the Earth lies beneath the sea. These are the midocean ridges. The fractures that split the center of the ridges serve as channel ways for the rising basaltic magma that forms new oceanic crust at the edges of plates of lithosphere. Not surprisingly, the water of the ocean cools the basaltic magma so rapidly after extrusion that plateau basalts do not form. Rather, two very distinctive, but much smaller, lava forms are observed. Close to a vent or fissure, where the lava temperature is highest, *thin sheets of lava with rapidly quenched, glassy surfaces* develop. These are called **sheeted flows** (Fig. 4.35). They build up piles of lava in which each sheet may only be 20 cm or so thick. Farther away from a vent, where the temperature has decreased, a very distinctive and common form of lava develops. The form is referred to as **pillow lava,** which describes a structure characterized by *discontinuous, pillow-shaped masses of lava, ranging in size from a few centimeters to a meter or more in greatest dimension* (Fig. 4.36).

Pillow structure forms when the surface of the viscous lava is quickly chilled. The brittle, chilled surface cracks, making an opening for the still-molten magma inside to ooze out like a strip of toothpaste. In turn, the newly oozed strip chills, its surface cracks, and the process continues. The end result is a pile of lava pillows, each with a quickly chilled, glassy skin that resembles a jumbled pile of sand bags.

Because oceanic crust is 10 km or more thick, we

FIGURE 4.35 Sheeted flows of basalt in the central part of the Galapagos Rift. The photo was taken during a cruise in 1979.

a

b

FIGURE 4.36 Pillow lava is one of the characteristic forms of lava extruded beneath the sea. (*a*) Tubular-shaped pillows of basalt photographed in the central rift of the East Pacific Rise, at a water depth of about 500 m. (*b*) Spectacular pile of basaltic pillow lavas exposed in the Sultanate of Oman, where an ophiolite complex has been thrust up from the sea floor.

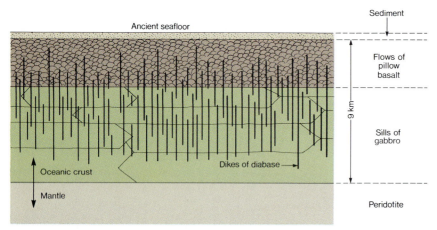

FIGURE 4.37 Idealized section through an ophiolite complex. Once part of the seafloor, ophiolite complexes are believed to have formed at a mid-ocean ridge, moved away from the ridge, and eventually been preserved for examination when part of the seafloor was caught in a collision between two continents moving on their rafts of lithosphere. Ophiolite complexes consist of flows of pillow basalt overlying a thick pile of gabbro sills. Both are intruded by large numbers of nearly vertical, diabase dikes (black). The boundary between the oceanic crust and the mantle below is the place where gabbro comes into contact with peridotite. (*Source:* Adapted from data by Moores and Vine, 1971.)

cannot see whether it is all made of pillow basalt. Nor has it yet been possible to sample the entire crust by drilling and dredging. But nature has provided a few samples of old oceanic crust that can be examined on land. They exist because two masses of continental crust, each riding on its plate of lithosphere, have collided with each other. During the collision, fragments of oceanic crust were broken off and caught up in the crumpled edges of the continental crust. One of the best preserved fragments of ancient oceanic crust, called an *ophiolite complex*, is exposed on the island of Cyprus. A generalized diagram of an ophiolite complex is shown in Figure 4.37. At the top is a thin veneer of sediment deposited after the igneous activity ceased. Beneath the sediment are layers of basaltic pillow lavas and beneath the lavas are many sills consisting of gabbro. The sills commonly contain *cumulate layering* formed by magmatic differentiation. The word cumulate is generally used to describe a layer formed by the accumulation of a single mineral. Cutting through the sills and lavas are thousands of vertical dikes of gabbro. The dikes form extensive, parallel sheets. Basalt and gabbro have the same composition and presumably formed from identical magma. Beneath the gabbro sills are rocks of different composition (rocks such as peridotite) that are characteristic of the upper mantle. Therefore, the whole array of rocks includes not only a sample of oceanic crust but also a small sample of the upper mantle as well.

VOLCANOES AND PEOPLE

We may think that all aspects of volcanoes are dangerous and life-threatening, but some aspects of volcanism are actually helpful to mankind. For example, weathering converts volcanic ash, with great rapidity, to exceptionally fertile soils. In some parts of the world, crops can be grown as soon as one year after an eruption of volcanic ash. In other places, such as Italy, Iceland, Mexico, New Zealand, and California, volcanic steam is tapped by deep drill holes and is used to drive electrical generators. Volcanic power of this sort is *geothermal power*. Despite these helpful aspects, volcanoes are most commonly thought of as danger spots. Underground movement of magma can trigger destructive earthquakes. Flowing lava, such as that shown in Figure 4.3, destroys everything in its path, riding over fields and villages alike. Yet, flowing lava usually moves so slowly that one can easily get out of its way. For example, in spite of the frequency of lava flows on Hawaii, loss of life is exceedingly rare.

The violent pyroclastic eruptions associated with andesitic, dacitic, and rhyolitic volcanism present a very different situation from the easily avoided basaltic lava flows—loss of life is more common. The most deadly and inescapable disasters occur when great incandescent ash flows are erupted ex-

a

b

FIGURE 4.38 A nueé ardente roars down the steep slopes of Mt. Mayon in the Philippines. (*a*) A mass of hot pyroclastic debris flows down a valley searing plants and animals and sending up a dense cloud of smoke and steam. (*b*) Viewed at night, the red-hot glow of a nueé ardente can be seen through the smoke.

FIGURE 4.39 Indicator of future disasters? A thick sequence of young, pyroclastic flow deposits on which part of Guatemala City is built.

plosively, and, buoyed by superheated steam and other volcanic gases, form deadly avalanches that roll down the steep slopes of stratovolcanoes at high speed (Fig. 4.38). This is the kind of eruption that occurred at Katmai in 1912, but fortunately no people were nearby when the eruption took place. Parts of Guatemala City that are built on such deposits may some day be destroyed by new ash flows (Fig. 4.39). A devastating eruption of this kind occurred on Mont Pelée, Martinique, in 1902. After 50 years of quiescence, Mont Pelée burst into life, emitting steam and erupting small amounts of ash for several months. On May 8, a swiftly flowing gaseous cloud, in places incandescent and containing ash and other pyroclasts, roared down the mountainside at 60 m/s, searing everything in its path. Such a cloud is called a *nuée ardente*, which is the French expression for "glowing cloud." Eight kilometers away from Mont Pelée lay St. Pierre, capital of Martinique, a community of 30,000 people and one of the biggest cities in the Caribbean. The nuée ardente roared through St. Pierre, engulfing everything and killing all but two persons, one a prisoner in a dungeon. All buildings were destroyed (Fig. 4.40), ships in the harbor burned and capsized, and the sea was converted to a boiling mass in which fish and other creatures were scalded to death.

Some of the most devastating eruptions do not give rise to nuées ardentes, but instead blast the already solidified top and sides of the volcanic cone into a mass of dust and rubble, which, mixed with volcanic ash, then expands at great speed both outward and upwards engulfing everything in its path. The cataclysmic eruption of Mount St. Helens on May 18, 1980, was of this kind. A similar destructive explosion, the largest of modern times, occurred in 1815 when Tamboro, an Indonesian volcano, decapitated itself. After a long period of apparent dormancy, Tamboro sprang to life, and with a massive eruption blew 150 km^3 of volcanic ash and pulverized rock debris into the air. The debris was thickly scattered over a roughly circular area 550 km in diameter, the mountain lost 1300 m in height, and after the top of the mountain collapsed into the empty magma chamber, a caldera nearly 11 km in diameter remained as mute evidence of the event. The vast number of people killed in heavily populated Indonesia has been estimated at 50,000 (Table 4.2), but is probably much higher. People around the world felt the effect of Tamboro. Fine dust from the eruption was blasted so high into the outer atmosphere that some of it remained for years. Suspended volcanic dust of this sort prevents so much sunlight from reaching the surface that it lowers temperatures worldwide through periods several years long. In the case of the Tamboro eruption, the following year was so cold that many places had frost during the summer, crops failed to ripen, and it is likely that a large portion of the world's population indirectly suffered some discomfort.

The geologic record is filled with evidence of eruptions of the Mount St. Helens, Pelée, Katmai, and Tamboro types. During the last few million years, for example, eruptions have occurred repeatedly in California, Oregon, Washington, Nevada, New Mexico, Alaska, and Montana. The eruption of Mount St. Helens in 1980 released between 1 and 2 km^3 of pyroclastic debris. What would be the magnitude of the human disaster if Yellowstone were to erupt again, as it did 2.2 mil-

TABLE 4.2 *The Most Disastrous Volcanic Eruptions in Recorded History*

Place	Date of Eruption	Estimated Deaths
Kelut, Indonesia	1586	10,000
Laki, Iceland	1783	10,000
Unzen, Japan	1792	10,000
Nevado del Ruiz, Colombia	1985	25,000
Mont Pelée, Martinique	1902	30,000
Krakatau, Indonesia	1883	36,000
Tamboro, Indonesia	1815	50,000

lion years ago, when 2500 km³ of fragmental material was ejected (much of it now preserved in the Huckleberry Ridge Tuff), or even if it repeated the eruption of the Lava Creek Tuff, 600,000 years ago, when 100 km³ of pyroclastic debris was ejected?

During the last thousand years about 520 volcanoes have been observed in eruption. They are widely distributed around the Earth's surface; a volcano is always erupting somewhere on the Earth. Many others that have not erupted for the past thousand years will almost certainly erupt during the next thousand. The timing and frequency of eruptions is irregular, making predictions uncertain. Yet, promising developments may help. Scientists have noted that prior to many eruptions, swarms of small earthquakes occur, and slight warping of the ground occurs in the immediate vicinity of the volcano. Through arrays of seismometers, the earthquakes can be monitored from a safe distance. With films sensitive to radiation in the infrared region, cameras mounted on orbiting satellites can post a continuous watch for developing hot spots, warning of areas of likely activity so that closer appraisals can be made and the endangered populations warned.

THERMAL SPRINGS AND GEYSERS

When volcanism finally ceases, igneous rock in the old magma chamber remains very hot for possibly a million years or more. Descending groundwater that comes into contact with the hot igneous rock becomes heated and tends to rise again toward the surface, along a fault or other avenue, where it forms a thermal spring.

Although most thermal springs are associated with volcanism, some are not. Groundwater sometimes descends so deeply into the earth that it is warmed by the general internal heat. Altogether there are more than 1000 thermal springs in the United States, most of them in volcanic regions in the western states. Even larger numbers exist in other parts of the world.

Water temperatures in thermal springs range all the way up to the boiling point. Because dissolution is more rapid in warm water than in cold, thermal springs are likely to be unusually rich in mineral matter dissolved from rocks with which they have been in contact. In some springs the mineral content is said to have medicinal properties.

FIGURE 4.40 Devastated remains of St. Pierre, Martinique. On May 8, 1902, Mont Pelée, 8 km away, erupted a fast-moving, sheetlike, cloud of hot volcanic ash and gases. The cloud removed all vegetation, reduced the town to utter ruin, and killed 30,000 people.

FIGURE 4.41 The great Geysir, Iceland, from which all other geysers take their name. The word means "to gush" in Icelandic.

A hot spring equipped with a system of plumbing and heating that causes intermittent eruptions of water and steam is a *geyser.* The name comes from the Icelandic word meaning to *gush,* for Iceland is the home of many geysers (Fig. 4.41). Most of the world's geysers that are not in Iceland are in New Zealand or in Yellowstone National Park. In both, there is evidence of volcanic activity late in geologic time, and the heat source for the geysers probably consists of masses of hot rock down below the surface.

The feature that marks a geyser is that it erupts not continuously, but intermittently. No two geysers behave in exactly the same way, and we cannot observe and study the system of underground passages that supplies any one of them. However, they are probably all alike in that they are fed by ordinary groundwater derived from rainfall. The water occupies a natural tube, probably crooked, that extends downward from the surface. It is heated by contact with hot rock until its temperature is nearly at the boiling point. In a straight tube, convection would occur as it does in a teakettle, and would equalize temperature throughout the tube. However, in a crooked tube convec-tion cannot occur effectively, and so from top to bottom of the tube the water is at its boiling point. Here we come to a basic principle, which probably explains the on-again off-again character of a geyser: Pressure increases with depth and the boiling point rises with increasing pressure. At the bottom of the tube, pressure is greatest and temperature is highest. When the condition is reached where all the water in the tube is at its boiling point, a very slight decrease of pressure or an increase of temperature will make the bottom water boil. The resulting steam pushes the overlying water up through the tube and a little is forced out at the top. The loss of water reduces the pressure below, so that through most or all of the tube the water is suddenly converted into steam and a violent eruption occurs at the surface.

Old Faithful in Yellowstone, the most famous American geyser, erupts for a few minutes about once an hour, throwing a jet of steam and water high in the air. During intervals between eruptions the emptied tube is refilled with water, which is then heated to the critical point at which the next eruption is triggered off.

SUMMARY

1. Igneous rock forms by the solidification and crystallization of magma.

2. Magma forms by the complete or partial melting of rock. There are three predominant kinds of magma—basaltic, andesitic, and rhyolitic.

3. Basaltic magma forms by dry partial melting of rock in the mantle. Andesitic magma forms by wet partial melting of oceanic crust. Rhyolitic magma forms by wet partial melting of rock in the continental crust.

4. The principal controls on the physical properties of magma are their contents of SiO_2 and H_2O. High temperature and low SiO_2 content result in fluid magma, such as basaltic magma. Lower temperatures and higher SiO_2 contents result in viscous magma, such as andesitic and rhyolitic magma.

5. The sizes and shapes of volcanic edifices depend on the kind of material extruded, viscosity of the lava, and explosiveness of the eruptions.

6. Volcanoes that dispense sluggish lava, generally rich in SiO_2, tend to be explosive. Those that extrude fluid lava, generally low in SiO_2, erupt less violently.

7. Widespread sheets of basaltic rock have resulted from fissure eruptions of fluid lava.

8. Processes that separate remaining liquid from already-formed crystals in a cooling magma lead to the formation of a wide diversity of igneous rocks.

9. Igneous rock may be intrusive (meaning it formed within the crust) or extrusive (meaning it formed on the surface). The texture and grain size of igneous rock indicate how and where the rock cooled.

10. Igneous rock rich in quartz and feldspar, such as granite, granodiorite, and rhyolite, is characteristically found in continental crust. Basalt, rich in pyroxene and olivine derived from the mantle, is common beneath the ocean basins.

SELECTED REFERENCES

Blong, R. J., 1985, Volcanic hazards: A sourcebook on the effects of eruptions: Sydney, Academic Press.

Decker, R. and Decker, B., 1981, Volcanoes: San Francisco, W. H. Freeman and Co.

Dietrich, R. V. and Skinner, B. J., 1979, Rocks and rock minerals: New York, John Wiley & Sons.

Eaton, G. P., Christiansen, R. L., Iver, H. M., Pitt, A. M., Mabey, D. R., Blank, H. R. Jr., Zietz, I., and Gettings, M. E., 1975, Magma beneath Yellowstone National Park: Science, v. 188, p. 787–796.

Ehlers, E. G. and Blatt, H., 1982, Petrology: Igneous, sedimentary and metamorphic: San Francisco, W. H. Freeman and Co.

Hamilton, W. and Myers, W. B., 1967, The nature of batholiths: U.S. Geol. Survey, Prof. Paper 554-C, p. C1–30.

Simkin, T. and Fiske, R. S., 1983, Krakatau 1883. The volcanic eruption and its effects: Washington, D.C., Smithsonian Institution Press.

Smith, R. L., 1960, Ash flows: Geol. Soc. America Bull., v. 71, p. 795–841.

Tabor, R. W. and Crowder, D. F., 1969, On batholiths and volcanoes—Intrusion and eruption of late Cenozoic magmas in the Glacier Peak area, North Cascades, Washington: U.S. Geol. Survey, Prof. Paper 604.

Williams, H. and McBirney, A. R., 1979, Volcanology: San Francisco, Freeman, Cooper, and Co.

CHAPTER 5

Sediments and Sedimentary Rocks

Colorful stratified sedimentary rocks in the Arizona desert.

FROM SEDIMENT TO SEDIMENTARY ROCK

Like a perpetually restless housekeeper, nature is ceaselessly sweeping regolith off the non-weathered rock below, carrying the sweepings away, and depositing them as sediment in river valleys, lakes, seas, and innumerable other places. We can see sediment being transported by trickles of water after a rainfall and by every wind that carries dust. The mud on a lake bottom, the sand on the beach, even the dust on a windowsill is sediment. Because erosion and deposition of rock particles take place almost continuously, we find sediment nearly everywhere.

When a thick pile of sediment accumulates, the particles near the base of the pile become compacted due to the weight of the overlying deposit. Over time, they become cemented together to form a solid aggregate rock. Most commonly the cementation occurs as films of new mineral sub-stance are deposited from water percolating slowly through the spaces between particles of sediment. By compaction and cementation, therefore, sediment is gradually transformed into sedimentary rock.

Sedimentary rock is the rock most commonly found at the Earth's surface, where it forms a thin but extensive blanket over igneous and metamorphic rocks beneath. Its most obvious feature is sedimentary layering, often strikingly exposed on mountainsides, in the walls of canyons, or in artificial cuts along highways or railroads (Fig. 5.1). Such layers drew the attention of observant people even in ancient times. The most thoughtful observers realized that many sedimentary layers are composed of fragments of other rocks, spread out as loose sediment and eventually cemented to form new rock. Writings by Greek philosophers long before the Christian era show that the meaning of sedimentary layers was understood even at that early time. Later, in the fifteenth century,

FIGURE 5.1 Multicolored layered sedimentary rocks in Capitol Reef National Park, Utah.

FIGURE 5.2 Examples of clastic sedimentary rocks. (*a*) Round pebbles of igneous and metamorphic rocks cemented in a matrix of sand form a coarse pebble conglomerate. (*b*) Grains of quartz sand cemented by silica produce a coarse, hard sandstone. (*c*) Platy fragments of shale are composed of lithified muds. (*d*) Broken fragments of sea shells cemented together with calcite makes a clastic sedimentary rock called *coquina*.

Leonardo da Vinci wrote an explicit statement about the connection between erosion, sediment, and sedimentary rock. In his notebooks he recorded the close similarity between sedimentary rock high in the mountains of northern Italy and the sand and mud he observed along the seashore.

In sedimentary rock can be seen some of the same kinds of features that are visible in igneous rock: its constituent minerals, its texture, and its color, for example. However, because sedimentary rock is clearly different than igneous rock, the questions we ask about it differ from those we might ask about igneous rock. Some of these questions are: Where did the sediment come from? How was it transported? What led to its deposition? What can it tell us about ancient environ-

ments at the Earth's surface? The terms used to describe and classify the kinds of sedimentary rock reflect such questions.

SEDIMENTS AND SEDIMENTATION

Clastic Sediment

A close look at a sediment shows that pebbles or sand grains are simply bits of rocks and minerals. A magnifying glass discloses that the finer sedimentary particles, too, are derived from broken-up rock, but that generally the particles have undergone chemical changes. Feldspars, for instance, have been partly altered to clay. All sediment of this kind is known as *detritus* or *clastic sediment* from the Greek word *klastos* (broken), meaning *the accumulated particles of broken rock and of skeletal remains of dead organisms* (Fig. 5.2). There is, of course, a continuous gradation of particle size, from the largest boulder down to submicroscopic clay particles. This range of particle size is embodied in the classification of Table 5.1.

If a sedimentary rock is made up of mineral particles derived from the erosion of igneous rock, how can we tell it is sedimentary and not igneous? Besides obvious clues such as sedimentary layering, there are clues in texture as well. A typical clastic sedimentary rock contrasts strongly with igneous rock in the shapes and arrangements of the grains (Fig. 5.2). In igneous rock the grains are irregular and are interlocked. In sedimentary rock the particles are commonly rounded and show signs of the abrasion they received during transport. Clastic sedimentary rock also reveals cement that holds the particles together, whereas igneous rock consists of interlocking crystals. Another important feature of many sedimentary rocks is the presence of fossils. Life cannot tolerate the high temperatures under which igneous rocks form, so the presence of ancient shells or similar evidence of past life is an excellent clue to sedimentary origin. Similarly, the presence of features such as ripple marks, marks of erosion by fast-flowing water, and mud cracks tell us that a rock was once a sediment.

Clastic sedimentary rock is classified mainly on the basis of size and shape of clastic particles, as outlined in Table 5.1. When distinctive minerals are abundant in the rock, these too will aid in classification (Appendix C).

Chemical Sediment

Certain kinds of rock contain fossils and other sedimentary characteristics, yet seem to be free of clastic sediment. The origin of such rock might be a puzzle were it not possible to find places such as the Bahama Banks and the Persian Gulf, where similar rock is forming today. The rock is indeed sedimentary, and the material composing it has been transported. However, the sediment is not clastic because its components were dissolved, transported in solution, and precipitated chemically instead of mechanically. *Sediment formed by precipitation of minerals from solution in water is chemical sediment.* It forms in two principal ways.

One way consists of biochemical reactions resulting from the activities of plants and animals within the water. For example, tiny plants living in seawater can decrease the acidity of the surrounding water and so cause calcium carbonate to precipitate.

Chemical sediment also forms as a result of inorganic reactions within the water. When the water of a hot spring cools, it may precipitate opal or calcite. Another common example is simple evaporation of seawater or lake water. As the water evaporates, thereby concentrating dissolved matter that is in solution, salts begin to precipitate out and remain as a residue of chemical sediment (Fig. 5.3). Precipitation commonly takes place in a definite sequence. The least soluble minerals, like gypsum and anhydrite, are first to settle out, followed by more soluble salts such as halite. The table salt we eat comes mainly from sedimentary rock formed in this way.

Most chemical sedimentary rocks contain only

TABLE 5.1 *Definition of Clastic Particles, Together with the Sediments and Sedimentary Rocks Formed from Them*

Name of Particle	Range Limits of Diameter (mm)[a]	Name of Loose Sediment	Name of Consolidated Rock
Boulder	More than 256	Gravel	Conglomerate and
Cobble	64 to 256	Gravel	sedimentary
Pebble	2 to 64	Gravel	breccia
Sand	1/16 to 2	Sand	Sandstone
Silt	1/256 to 1/16	Silt	Siltstone
Clay[b]	Less than 1/256	Clay	Claystone, mudstone, and shale

Source: After Wentworth, 1922.

[a]Note that size limits of sediment classes are powers of 2, just as are memory limits in microcomputers (for example, 2K, 64K, 128K, 256K, 512K).

[b]Clay, used in the context of this table, refers to particle size. The term should not be confused with clay minerals, which are definite mineral species.

FIGURE 5.3 Precipitated salts encrust the surface of Searles Lake playa in southeastern California. The chemical plant on the lake shore extracts and refines saline minerals.

one important mineral and this is used as a basis for classification. The most common rocks formed by chemical precipitation are *limestone,* formed chiefly of the mineral calcite through both organic and inorganic processes, and *dolostone,* formed mainly by replacement of calcite by the mineral dolomite.

Transport and Deposition of Sediment

Sediment is transported in many ways. It may simply slide down a hillside or may be carried by the wind, by a glacier, or by flowing water. In each case, when transport ceases, the sediment is deposited in a fashion characteristic of the transporting agent. When sediment is transported by sliding or rolling downhill, the result is generally a mixture of particles of all sizes. Much of the sediment carried by a glacier is deposited either beneath the ice or at the glacier's edge. Such sediment also is a mixture of sedimentary particles of all sizes.

In the transport of sedimentary particles by wind or water, deposition occurs when the flowing water or moving air slows down to a speed at which particles can no longer be moved. In a general way, therefore, the size of the grains in sediment carried by wind or water tells us something about the speed of the transporting medium. Coarse-grained sediment indicates deposition from fast-flowing wind or water; fine-grained sediment indicates either that the wind or water was slow-moving, or that only fine sediment was available for transport.

The size of particles in sedimentary rock and the way they are packed together, as well as other distinctive features, provide evidence about the environment in which the original sediment was deposited. The existence of ancient oceans, coasts, lakes, streams, swamps, and all the other places where sediment accumulates, can be demonstrated from clues in sedimentary rock. Fossils within a sedimentary layer may also provide information about former climates. Some animals and plants are restricted to warm, moist climates whereas others are associated only with cold, dry climates. By using the climatic ranges of modern plants and animals as guides and invoking the Principle of Uniformity, one can infer the general nature of the climate in which similar ancestral forms lived.

Diagenesis

Diagenesis is a term for the *changes that affect sediment after its initial deposition, and during and after its slow transformation into sedimentary rock;* however, it does not encompass surface weathering and soil formation, nor metamorphism. The first and simplest change is *compaction,* which occurs as the weight of the accumulating sediment forces the rock and mineral grains together, thereby reducing the pore space and eliminating some of the contained water. Precipitation of dissolved substances in circulating groundwater then bonds the grains together through *cementation.* Calcium carbonate is one of the most common cements (Fig. 5.4), but silica may also bond grains together forming a particularly hard cement. After burial, less stable minerals may change to more stable forms through *recrystallization,* a process that is especially common, for example, in porous reef limestone.

Important chemical alterations also affect sediments. In the presence of oxygen (an *oxidizing environment*), organic remains are quickly converted into carbon dioxide and water. If oxygen is lacking (a *reducing environment*), the organic matter does

FIGURE 5.4   A thin section of a sandstone formation from central Washington showing sand grains that are bonded together by calcite cement. Light-colored grains are plagioclase, brownish grains are pyroxene, and the large dark grain is a volcanic rock fragment. Largest plagioclase grain is 1 mm long.

FIGURE 5.5   Alternating layers of sandstone and shale along the coast of the Gaspé Peninsula in eastern Canada. Differences in grain size, color, and erosional characteristics emphasize the differences among the strata.

not decay but instead may be slowly transformed into solid carbon in the form of peat or coal. Similarly, organic oils and fats may be converted into carbon-rich residues (*hydrocarbons*).

FEATURES OF SEDIMENTS AND SEDIMENTARY ROCKS

Stratification and Bedding

Sedimentary stratification results from *a layered arrangement of the particles that constitute sediment or sedimentary rock*. Each **stratum** (plural = **strata**) is *a distinct layer of rock that accumulated at the Earth's surface*. While layering is an obvious feature of most sedimentary rocks, it is seen also in some volcanic rocks (lava flows, pyroclastic deposits) and in many metamorphic rocks. Looking closely at sedimentary rocks that are distinctly stratified,

we can see that the strata differ from one another because of differences in some characteristic of the constituent particles or in the way in which the particles are arranged (Fig. 5.5). Very commonly one stratum consists of particles of different diameter from those in another. In a clastic rock, such changes of diameter result from fluctuations of energy in a stream, in surf, in wind, in a lake current, or in whatever agent is responsible for the deposit. The energy changes, usually small, are not the exception but the rule.

Sorting

A conspicuous result of the transport of particles by flowing water or flowing air is *sorting* of sediments as they are deposited. Sorting according to *specific gravity* (ratio of the weight of a given volume of material to the weight of an equal volume of water) is evident in mineral *placers* (Chapter 22). Particles of unusually heavy minerals such as gold, platinum, and magnetite are deposited quickly and concentrated on stream beds or on beaches, whereas lighter particles are carried onward. Most of the particles transported by water or wind, however, consist of common rock-forming minerals such as quartz and feldspar having similar specific gravity. Therefore, such particles typically are not sorted according to specific gravity but rather according to size (Fig. 5.6). In a stream, gravel is deposited first, whereas sand and silt are carried farther before deposition. Thin, flat particles are carried farther than spherical particles of similar weight because their shape causes them to settle out more slowly. Long-continued handling of particles by turbulent water and air results in gradual destruction of the weaker particles. In this way rock particles and mineral grains that have pronounced cleavage or are otherwise easily broken down may be eliminated, leaving behind the particles that can better survive in the turbulent environment. Very commonly the survivor is quartz, because it is hard and lacks cleavage. In this case sorting is based on durability.

Sorting is the chief cause of stratification, but it is not the sole cause. Successive layers that do not differ from each other in grain size or composition may still be separated by surfaces representing minor intervals when no deposition occurred; commonly rocks tend to split most easily along such surfaces. Two adjacent layers, not otherwise distinct, can also differ from each other because of the kind or abundance of cement they contain.

Each stratum nearly always possesses definite characteristics by which it differs from the stratum beneath or above it. With this in mind, we can describe two chief kinds of stratification and then examine the particles within a stratum.

Parallel Strata

Layers of sediment fall into two classes according to the geometric relation between successive units. One class consists of **parallel strata,** *strata whose individual layers are parallel* (Fig. 5.1). Parallelism indicates that deposition probably occurred in water, and that the activity of waves and (except for the special case of graded layers, described below) currents was minimal. Indeed, sediment deposited on lake floors and in the deep sea occurs rather commonly in parallel layers.

A distinctive variety of parallel strata consists of repeated alternations of layers of unlike grain size or differing mineral composition. Such alternation suggests the influence of some naturally occurring rhythm, such as the rise and fall of the tide or the

FIGURE 5.7 Rhythmically laminated glacial lake sediments exposed in a seacliff along the Strait of Magellan in southernmost Chile. Each pair of layers in the upper part of the section constitutes a varve; in each couplet, light-colored silt deposited during summer months grades upward into darker clay deposited during winter.

FIGURE 5.6 Well-rounded grains of quartz sand from the St. Peter Sandstone of Wisconsin have been sorted by size and polished by constant shifting and abrasion in surf along an ancient shoreline.

FIGURE 5.8 Outcrop of Green River Formation at Green River, Wyoming, showing parallel-layered varved sediments deposited in an ancient lake that covered a large area in the northern Colorado Plateau region.

annual change of seasons. *A pair of sedimentary layers deposited during the seasonal cycle of a single year* is a *varve* (Swedish for "cycle"). Varves are forming today in many mountain and high-latitude lakes. They were also deposited in many lakes near the margins of ice-age glaciers (Fig. 5.7). Such paired layers are generally very distinct, because close to an ice sheet the contrast between summer and winter weather markedly affects the rate of melt-ing of ice. This, in turn, influences seasonal fluctuations in the amount of water and sediment entering a nearby lake. Similar pairs of laminae occur in ancient rocks. Certain claystones in South Africa have been interpreted as varves deposited in ice-dammed lakes during a glaciation more than 200 million years ago.

Varves of a different origin are seen in a formation of sedimentary rocks that underlies the vast Green River Basin in Wyoming, Colorado, and Utah (Fig. 5.8). In each varve one lamina consists of calcium carbonate, whereas the other includes dark-colored organic matter. The rhythmic stratification is explained as follows. The sediments were deposited in a lake which, as it warmed in summer, precipitated calcium carbonate from solution. During the same warm season floating microscopic organisms reached a peak of abundance. The relatively heavy carbonate sank promptly to form a light-colored summer layer. As the organisms died at the end of their life cycle, the less-dense organic matter sank more slowly to form an overlying dark-colored winter layer. Each pair of layers thus forms a varve. Because an estimated five to eight million varves are present in the formation, it is reasoned that an immense lake occupied the Green River region for many millions of years under remarkably uniform conditions.

FIGURE 5.9 Cross-stratified sandstone near Kanab, Utah consists of ancient sand dunes that have been converted to sedimentary rock.

Cross-Strata

By contrast, ***cross-strata*** are *strata that are inclined with respect to a thicker stratum within which they occur* (Fig. 5.9). All such strata consist of particles coarser than silt and are the work of a turbulent flow of water or air, as in streams, in wind, or in waves along a shore. As they move along, the particles tend to collect in ridges, mounds, or heaps in the form of ripples and waves which may migrate slowly forward in the direction of the current. As particles continually accumulate on the downcurrent slope of the pile, they result in strata having inclinations as great as 30° to 35°. The direction in which cross-strata are inclined, then, is the direction in which the related current of water or air was flowing at the time of deposition.

Cross-strata are commonly seen in river deltas (Fig. 11.29), sand dunes (Fig. 12.25), beaches (Fig. 14.31), and stream sediments. It must be kept in mind that although the dipping layers in deltas and in dunes are usually parallel with one another, they nevertheless are cross-strata because they are not parallel with the top and base of the larger stratum within which they lie.

Under some conditions, however, no larger enclosing strata are present. Instead, all the layers deposited at a locality are inclined. For example, pyroclastic layers characteristically accumulate in conical piles surrounding their source vent and have steep inclinations (Fig. 5.10). Because all such layers are inclined, they are not cross-strata.

Arrangement of Particles Within a Stratum

In addition to the relationships of layers to each other, several kinds of particle arrangements within a single layer are possible. Each kind gives information about the conditions under which the sediment was deposited.

Uniform Layers. A layer that consists of particles of about the same diameter is called a *uniform layer.* A uniform layer of clastic rock implies deposition of particles of a single size, with little change in the velocity of the transporting agent. A uniform layer of nonclastic rock implies uniform precipitation from solution, which produces crystalline particles of a single size. A layer that is subdivided into still thinner layers marked off from each other by differences in grain size suggests a transporting agent having a fluctuating velocity.

Graded Layers. If a quantity of small solid particles having different diameters and about the same specific gravity is placed in a glass of water, shaken vigorously, and then allowed to stand, the particles will settle out and form a deposit on the bottom of the glass. The largest particles settle first, followed by successively smaller ones. If small enough, the finest may stay in suspension and keep the water turbid for hours or days before finally settling out. Thus, particle size in the deposit decreases from the bottom upward. This arrangement characterizes a ***graded layer,*** defined as *a layer in which the particles grade upward from coarse to finer* (Fig. 5.11).

Although a graded layer produced by simple shaking of sediment and water in a jar is graded only in the vertical dimension, those produced in nature are graded laterally also because they are made by moving currents. Because the heaviest particles settle first, grading occurs laterally in the downcurrent direction.

Processes that produce graded sediments include stream flow, as velocity subsides and trans-

FIGURE 5.10 Layers of volcanic ejecta that were deposited at a steep angle on the slope of a pyroclastic cone on the flanks of Mauna Kea volcano, Hawaii. The oldest layers are at lower right, the youngest at upper left. Each layer represents a separate eruptive pulse.

FIGURE 5.11 Graded bedding in ancient sedimentary rock from Adelaide, Australia. Each layer of rock grades upward from coarse sand at the base to fine sand at the top.

when the agitation stopped, such processes involve rapid and continuous loss of energy, an essential condition for the creation of graded layers.

Nonsorted Layers. The particles in some sedimentary rocks are not sorted at all. Instead the rocks are a mixture of particles of different sizes arranged chaotically, without any obvious order. Such sediments are created, for example, by rockfalls, slow movement of debris down hillslopes, slumping of loose deposits on the seafloor, mudflows, and deposition of debris by glaciers and by floating ice. Such *nonsorted sediment, regardless of origin*, is referred to as a **diamicton;** the equivalent rock is a *diamictite*. If the origin of a nonsorted rock is known, then it may be given a genetic name (for example, **tillite**, *a diamictite of glacial origin*) (Fig. 5.12).

porting power diminishes, currents of turbid water in the ocean or in lakes that lose speed and drop their sediment load (see below), falls of volcanic ash during explosive eruptions, and dust storms as they die down. As in the case of the jar experiment, in which settling of sediment commenced

Rounding

Particles broken from bedrock by mechanical weathering and other processes tend to be angular, because breakage commonly occurs along grain boundaries, joints (Chapter 6), and surfaces

FIGURE 5.12 Stony tillite of an ancient glaciation overlies a glacially striated surface of older rock at Nooitgedacht, near Kimberly, South Africa.

of stratification. The same particles tend to become smooth and rounded as they undergo transport by water or air, and are abraded by impact with other rock fragments. At the same time, they become progressively better sorted. Figure 5.13 shows what can happen to pebbles on a beach, while Figure 5.14 indicates how sand grains become shaped during transport. The degree of rounding gives some idea of the distance or length of time involved in transport of particles by flowing water or air.

FIGURE 5.13 Rounded cobble gravel of a shingle beach on the Olympic coast of Washington. Pieces of locally derived sandstone bedrock, up to 20 cm in diameter, have been tossed, sorted, and abraded in the surf zone until they assume a flattened shape, like a discus.

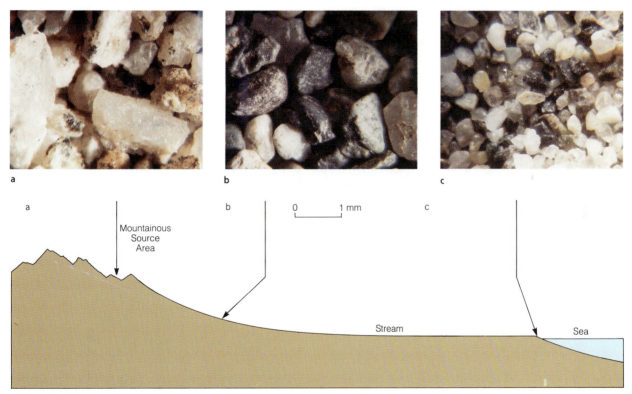

FIGURE 5.14 Rounding and sorting of mineral grains during transport. (*a*) Mineral grains loosened and separated from igneous and metamorphic rocks by mechanical and chemical weathering before transport. The angular shapes of the individual grains, slightly altered by weathering, are the forms they assumed when the minerals crystallized in the parent rock. The aggregate of grains is a nonsorted sand. (*b*) Sand carried downslope by streams undergoes abrasion and rounding. Some of the less durable mineral grains have been broken up and lost, leaving a larger proportion of the durable mineral quartz. (*c*) By the time the sediment reaches the mouth of the stream it has traveled a great distance and the grains have become well rounded. They now consist almost entirely of durable quartz.

a   b

FIGURE 5.15 Modern and ancient ripple marks. (*a*) Ripple marks forming in shallow water near the shore of Ocracoke Island, North Carolina. (*b*) Ripple marks on the bedding surface of an ancient quartzite bed in the Baraboo Syncline of south-central Wisconsin.

Glaciers, by contrast, tend to produce irregularly shaped particles by crushing, grinding, and abrasion. The faceted shapes of many coarse glaciated rock fragments, and the scratches on them, are distinctive and characteristic of this mode of transport (Fig. 13.25*b*). Sand and dust carried by wind can likewise abrade facets on coarse rock particles (Figs. 12.21 and 12.22), but the facets are more distinct and generally meet each other more sharply than do those made by glacial action. Large wind-shaped rocks are not rounded because wind can

not move them; they lie motionless while fine particles driven by the flowing air abrade them.

Surface Features on Sedimentary Layers

Various features found preserved on the surfaces of strata provide clues about the origin of sedimentary rocks and the environments in which they formed. Bodies of sand that are being moved by wind, by streams, or by coastal waves are often rippled, and such ripples are frequently preserved

a   b

FIGURE 5.16 Modern and ancient mudcracks. (*a*) Mudcracks formed at the surface of dry lake floor in Death Valley, California. (*b*) Ancient mudcracks preserved on bedding plane of sandstone bed exposed at Ausable Chasm, New York.

FIGURE 5.17 Tracks of a three-toed dinosaur exposed on the bedding plane of a sandstone bed in the Painted Desert near Cameron, Arizona. All of the tracks in the picture belong to a single species.

in sandstones and siltstones as *ripple marks* (Fig. 5.15). Some claystones and siltstones contain layers that are cut by polygonal markings. By comparison with sediments forming today, such as those in roadside puddles following a rain, we infer that these are **mud cracks,** *cracks caused by shrinkage of wet mud as its surface becomes dry* (Fig. 5.16). The presence of mud cracks in a rock generally implies at least temporary exposure to air and, therefore, suggests tidal flats, exposed stream beds, playa lakes, and similar environments. Occurring with some ripple marks and mud cracks, and preserved in a similar manner, are the footprints and trails of animals (Fig. 5.17). Even the impressions of large raindrops made during brief, hard showers can be preserved in strata.

Concretions

Enclosed in some sedimentary strata are unusual bodies called *concretions* (Fig. 5.18). A **concretion** may be defined as *a hard, localized body having distinct boundaries, enclosed in sedimentary rock, and consisting of a substance precipitated from solution, commonly around a nucleus.* They range in diameter from less than a centimeter to two meters or more. Their shape may be spherical or may range through a variety of bizarre forms, many with remarkable symmetry, to elongate bodies that parallel the stratification of the rock. Concretions are composed of many different substances, including

FIGURE 5.18 Round concretions the size of cannon balls litter the surface near Lake Powell, Arizona, where they have weathered out of the enclosing Navajo Sandstone.

calcite, silica, hematite, limonite, siderite (iron carbonate), and pyrite. Small concretions are dredged up from the seafloor, which shows that they are forming there today as sediments are deposited. This origin, contemporaneous with the enclosing sediments, is indicated also by the shapes of some concretions and by their relation to the stratification of the surrounding sedimentary rock. Others can be shown to have formed during diagenesis, as they retain the primary stratification of the surrounding rock.

The substances of which concretions are made indicate that these objects are the result of localized chemical precipitation of dissolved matter from seawater, lake water, or groundwater. Once precipitation starts around an organism or other object that differs from the enclosing rock, the concretion thus formed continues to grow. In some rocks, perfectly preserved fossils are found at the centers of concretions.

Fossils

The remains of animals and plants that were buried with sediments, protected against oxidation and erosion, and preserved through the slow process of conversion to rock constitute the fossils (Fig. 5.19) that form a very important element of the geologic record. Not only do fossils provide significant clues about former environments, but they are the chief basis for the correlation of strata and the construction of the geologic column (Chapter 7).

Color

The colors of sedimentary rocks vary considerably. Some rocks exposed in cliffs display a surface coloring imparted by products of chemical weathering. For example, a sandstone that is pale gray on freshly fractured surfaces may have a surface coating of yellowish-brown limonite, developed by the oxidation of sparse iron-rich minerals included among the grains of quartz sand.

The color of fresh rock is the combination of the colors of the minerals that compose it. Iron sulfides and organic matter, buried with a sediment, are responsible for most of the dark colors in sedimentary rocks. Microscopic examination of reddish and brownish rocks shows that their colors result mainly from the presence of iron oxides which occur as powdery coatings on grains of quartz and other minerals, or as very fine particles mixed with clay.

FIGURE 5.19 Fossil of a trilobite (*Olenellus getzi*) exposed on the bedding plane of an ancient marine sandstone.

Iron oxides get into strata in at least two different ways. In most reddish and brownish rocks the colored oxides probably are secondary; that is, they are created in the strata by alteration of ferromagnesian minerals, such as hornblende and biotite, through diagenesis and weathering. Such chemical alteration can take place in any warm climate, dry or moist.

In some strata the colored oxides are believed to be primary, having been deposited as components of the sediments themselves. Such oxides may have been derived from the erosion of reddish clayey soils like those forming today in the warm climates of the tropics and subtropics. Washed down from uplands into rivers, some of the reddish (ferric) iron oxide is deposited on swampy floodplains where decaying plant matter creates strong reducing conditions that transform it into dark-gray (ferrous) iron oxide. Rivers carry the rest of the reddish-colored sediment to the sea, where organic matter on the seafloor likewise reduces it.

Probably the reddish color can be preserved only if the sediments escape reduction after deposition. This is possible when deposition occurs in basins where, for one reason or another, organic matter does not accumulate in amounts sufficient to develop a reducing environment. An ancient reddish-colored stratum, therefore, tells us that at the time of accumulation the climate was warm, either at the site of deposition or in the region from which the sediment was derived.

DEPOSITIONAL ENVIRONMENTS

Nonmarine Environments

Sediment derived from the mechanical and chemical breakdown of rocks is moved inexorably toward the sea, but enroute, as it is transported by one or several processes, it may come to rest temporarily on land. We will explore these agencies of transport and the resulting deposits and landforms more fully in the following chapters, but the principal environments of sediment deposition are introduced here.

Alluvial Sediments

Streams constitute the principal agency for transporting sediment across the land. Their deposits, which can be seen nearly everywhere, are called **alluvium,** the general name given to *sediment deposited by streams in nonmarine environments.* The sediment differs from place to place depending on the type of stream, the energy available for work, and the nature of the sedimentary load. A smoothly flowing, meandering stream will deposit well-sorted layers of coarse and fine particles as it swings back and forth across its valley. Fine silt and clay will be deposited on the floodplain during spring floods, and organic sediments will accumulate in abandoned reaches of the channel. By contrast, a stream issuing from the front of a mountain glacier may divide into an intricate system of interconnected channels that change radically in size and direction as the volume of water fluctuates and the stream copes with an overabundance of debris. The resulting sediments will consist of cross-cutting channel deposits that are distinctly different from the broad and lenticular sedimentary layers left by a meandering stream.

Such differences in the texture and structure of sediments, and of rocks formed from them, can tell us a great deal about past environments of deposition. So too can associated plant and animal fossils

sometimes tell us whether the enclosing sediments represent, for example, the floodplain of a subtropical river or a desert alluvial fan.

Lake Sediments

Lakes occur in basins that can originate in many ways. Some may owe their origin to tectonic movements of the land, others to landslides, to glacial activity, or to solution of soluble rocks. Lake sediments accumulate chiefly in two environments: the lake shore and lake floor. Lakeshore deposits consist of generally well-sorted sand and gravel that form beaches as well as bars and spits across the mouths of bays. Where a stream enters a lake its sediment load will be dropped as both speed and transporting ability are suddenly checked. The resulting deposit, called a *delta* (chapters 11 and 14), consists of nearly horizontal stream-channel sediments that cap inclined and generally well-sorted layers of the delta front which in turn pass outward into thinner, finer, and even-laminated beds on the lake floor.

While deltaic strata provide a means of identifying a lake environment in the sedimentary record, often more can be found out by studying the mineralogy of the sediments, as well as the contained fossils. A lake in a desert basin, for example, may evaporate to such an extent that soluble salts begin to precipitate out and to form a distinctive mineral component of the lake-bottom sediments. Fossil pollen grains, which are generally abundant in lake sediments, can help identify the plants that lived in and around a former lake and also can tell something about the climatic setting in which the lake existed.

Glacial Sediments

Although glaciers now cover only about 10 percent of the world's land areas, in former times they were far more extensive and left a record of their presence both in landforms and sediments. Sediment eroded and transported by a glacier is either deposited at the bed of the glacier or is released at its front as melting occurs and is then subjected to further reworking by meltwater. Sediment deposited directly from ice commonly forms a heterogenous mix of particles ranging in size from clay up to large boulders and consisting of all the rock types over which the ice has passed (Fig. 13.25a). Such sediment characteristically is nonsorted and nonstratified, in contrast to the other nonmarine sediments discussed thus far. The rock fragments typically are angular (Fig. 5.20*a*) unless they were rounded prior to being picked up and reworked by the glacier.

Nonsorted sediment is characteristic of glacial deposition, but it also can result from other depositional agencies such as landslides and mudflows. This frequently makes positive identification of such sediments difficult. There are many examples of nonglacial sediments which have mistakenly been attributed to glacial activity. One feature that helps determine the glacial origin of a nonsorted deposit is the occurrence beneath it of a smoothed and abraded rock surface. While such features are known to be produced by erosion at the glacier bed, they are not likely to be found associated with sediment of nonglacial origin.

Sediments deposited adjacent to or in contact with glacier ice by meltwater tend to be sorted, just as waterlaid sediments from nonglacial environments are. Thus, we might find both stream sediments and lake sediments associated with ice-laid deposits in glaciated terrains. However, such sediments often are discontinuous and may change character abruptly, both vertically and horizon-

tally, reflecting the melting out of blocks of ice which caused sudden changes in local topography and in directions of meltwater flow.

Wind-Transported Sediments

Sediment carried by the wind tends to be finer-grained than that moved by other erosional agencies, because air is much less dense than water or ice. Sand grains are easily moved where strong winds occur, as along seacoasts and in deserts, and where there is little vegetation to stabilize the surface. The sand piles up to form dunes consisting of well-sorted sand grains and, characteristically, having inclined bedding similar to that in a delta. In addition, individual grains typically have a frosted appearance (Fig. 5.20*b*) due to the repeated impacts they receive as they bounce along. Such dune sands can be easily identified in the rock record and are especially common in the American West where they are exposed in bold cliffs in the scenic canyon lands (Fig. 5.9).

Silt and clay which are picked up and moved by the wind tend to settle out as a thin uniform blan-

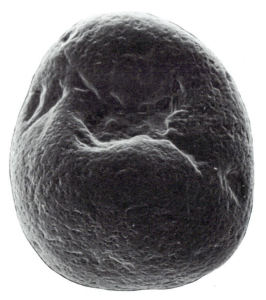

a   b

FIGURE 5.20 Surface features of sand grains, seen on enlarged pictures taken with a scanning electron microscope, aid in differentiating among transporting agencies. (*a*) Surface of a quartz grain (0.1 mm diameter) that has been crushed and abraded during transport at the bed of a Swiss glacier displays distinctive concoidal fractures. (*b*) Surface of a wind-transported quartz grain (0.5 mm diameter) from south-central Libya has a distinctive pitted appearance caused by mechanical chipping as grains impact one another during strong sand storms.

ket of sediment called *loess* (Chapter 12). It is thickest and coarsest near its source and becomes progressively thinner and finer with increasing distance downwind. It generally is massive, lacks bedding (Fig. 12.32), and may contain fossils of land plants and animals that help to identify its origin. Although common as a sediment in many parts of the world, it is virtually unknown as a sedimentary rock, probably because it is easily eroded from the landscape and therefore is unlikely to be widely preserved.

Marine Environments

Sediments of the Continental Shelves

The world's rivers continuously transport detritus to the edges of the land where much of it then accumulates on the continental shelves, sometimes to great thicknesses. Trapped within the sediments are large reservoirs of hydrocarbons that are increasingly being exploited to supply energy for the industrialized world. In part spurred on by a search for these undersea riches, geologists have learned a great deal about the sediments accumulating on the shelves.

Near-Shore Sediments. Stream sediments that reach the edge of the sea accumulate near the mouth of streams or are carried either seaward or laterally along the coast by currents. Much of the load transported by a large river may be trapped in an **estuary,** *a semi-enclosed body of coastal water within which seawater is diluted with fresh water.* Coarse sediment settles close to land and may quickly fill an estuary unless active subsidence is taking place, in which case a thick body of estuarine sediment may accumulate.

The outward flow of fresh water through an estuary is often substantial and may extend out across the submerged continental shelf as a distinct layer overlying denser, salty marine water. Some fine-grained sediment thereby reaches the outer shelf in suspension. The fine clay-sized particles tend to settle out in one of two ways. The first way is through *flocculation*, a process whereby many minute suspended particles form clotlike accumulations that become massive enough to settle. Fine particles, secondly, are ingested by organisms which aggregate the particles in fecal pellets that fall to the bottom.

Tidal processes control estuarine sedimentation. Where tidal fluctuations are large, turbulent mixing occurs, sediment is redistributed, and tidal flats and salt marshes are common depositional environments. However, where tidal range is small or sediment input is so high that estuaries are filled, sediment accumulates at a river mouth to form a delta (Chapter 14). Major deltas are built by large streams where rates of sedimentation are higher than rates of coastal erosion. Waves and near-shore currents move sediment laterally along beaches and also supply sediment for the construction of offshore bars and barrier islands (Chapter 14).

Offshore Sediments. Through the action of currents, sediment not trapped in deltas and estuaries is redistributed along the coast and onto the continental shelf where it slowly accumulates to form a thick prism of sediment. The nature of this sediment is now well known as a result of extensive sampling and geophysical investigations, largely carried out in the search for offshore oil and gas. A section through the continental shelf off eastern North America shows clearly how sediment has piled up behind banks and reefs that formed the ancient edge of the continental shelf (Fig. 5.21). In some places sediment as much as 14 km thick has accumulated over a period of some 70 to 100 million years. To build the whole pile, an average of less than a millimeter of sediment need have been deposited each year.

If sediment carried onto the shelf comes to rest at depths of less than half the distance between the crests of incoming waves, it will be stirred up and redeposited in deeper water until it lies too deep to be moved by even the greatest storm waves. Thereafter it can be transported across the shelves only by bottom currents.

When world sea level stands high against the continents, the submerged shelves trap most of the detritus that reaches them from the adjacent land and prevent it from reaching the deep ocean. Of course some detritus never reaches the shelf. It is trapped by inland basins, where it is buried and preserved. However, analysis of ancient strata shows that the amount of sediment thus retained on land is much less than the amount carried to the shelf. Only an estimated 10 percent of the sediment reaching the shelf remains in suspension long enough to eventually get to the deep sea, so it is clear that the great bulk of the Earth's sedimentary strata is shelf strata whose sediment originated within the continents themselves. The shelves, in effect, conserve continental crust which is continually recycled within the continental realm.

Because most sediment on the shelves was derived from the land, we would expect it to become gradually finer in the seaward direction. Samples of surface sediment collected from some areas on the shelves support this expectation. However, on most shelves the distribution of particle sizes is irregular. In general, modern coarse sediment seldom is carried more than a few tens of kilometers from the shore. Most of it is deposited within 5 or 6 km of the land and is dispersed by longshore currents. Coarse sediment is also found at greater distances offshore as far as the seaward limits of the shelves. Its observed patchy distribution may be partly the work of localized currents, but to a greater extent it is the result of changing sea level. At times when sea level had fallen below its present level, the shoreline migrated seaward across the shelves, exposing new land. Bodies of coarse sediment, deposited near shore or on the land itself at such times, subsequently were submerged as sea level rose and drowned the shelves. Such sediment is referred to as *relict* sediment, for it is not in equilibrium with present environments, but rather is a relict of past conditions. It has been estimated that as much as 70 percent of the sediment cover on the continental shelves can be classified as relict sediment.

Carbonate Shelves and Reefs. Carbonate sediments of biogenic origin accumulate where the influx of land-derived sediment is minimal and where the climate and sea-surface temperatures are warm enough to promote the abundant growth of carbonate-secreting organisms. Most carbonate sediment on carbonate shelves consists of sand-sized skeletal debris, together with inorganic precipitates. Coarser debris is found mainly near reefs or in areas of turbulence and strong currents. Carbonate sediments near the landward margins of such shelves often are mixed with clastic debris from the land.

Carbonate shelves are mainly of two kinds. One type consists of a wide protected lagoon in which fine carbonate muds accumulate. The lagoon is bordered on its seaward margin by a protective reef inhabited by reef-building corals and coralline algae (Fig. 5.22a). The shelf off eastern Australia with its Great Barrier Reef is a good example (Fig. 5.22b). Such reef-bordered carbonate shelves are well-developed only on the western side of ocean basins, for cold upwelling waters and surface currents make the eastern margins of oceans largely unfavorable for reef growth. A second type, of which the shelves of Yucatan and western Florida are examples, is an open carbonate platform that lacks a bordering reef. On such a platform, surface currents tend to winnow fine particles and move them to deeper water, leaving a coarse lag of carbonate sediment on the shelf. The surface deposits of carbonate shelves, like those of most other continental shelves, include both modern sediment and relict sediment that formed when sea level was lower than today.

Marine Evaporite Basins. Restricted basins of marine water located in areas of high evaporation can generate deposits of **evaporite,** *sedimentary rock composed chiefly of minerals precipitated from a saline solution through evaporation.* The Mediterranean Sea is such a basin. Were it not for continuous inflow

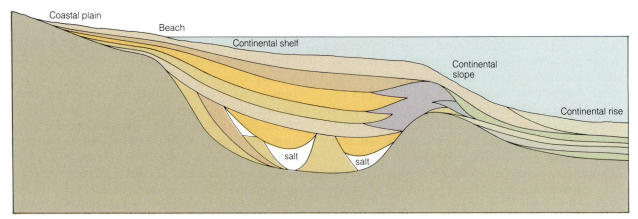

FIGURE 5.21 Section through the continental margin of the eastern United States showing how a thick prism of sediments has piled up behind banks and reefs at the ancient shelf edge, which now is buried by younger sediments of the continental slope. (*Source:* After McGregor, 1984.)

FIGURE 5.22 Tropical coral reefs. (*a*) Outer reef zone of Rongelap Atoll located in the Marshall Islands of the tropical Pacific Ocean. Corals and coralline algae form a hard interlocking reef framework that can withstand the vigorous assault of ocean waves. (*b*) Reef flat and rocky edge of Heron Island on the Great Barrier Reef off the coast of Australia.

of Atlantic water at its western end, the Mediterranean would gradually decrease in volume due to evaporation, which is especially high in its eastern half. It has been estimated that if deprived of new water, evaporation would cause the landlocked sea to dry up completely in about 1000 years, in the process precipitating a layer of salt about 70 m thick. Extensive evaporites underlying the Mediterranean Basin are evidence of former periods when high evaporation, together with a continuous inflow of Atlantic water to supply the necessary salt, produced evaporite deposits as much as 2 to 3 km thick.

Sediments of the Continental Slope and Rise

Along most of their lengths, the shallowly submerged continental shelves pass abruptly into continental slopes which plunge steeply to depths of several kilometers. Sediments that reach the shelf edge are poised for further transport down the slope and onto the adjacent continental rise.

Turbidity Currents. At the foot of the continental slope, and in places far beyond it on the deep-sea floor, thick bodies of coarse sediment of continental origin lie at depths as great as 4 to 5 km. Such occurrences were an unsolved mystery until it appeared that they could be explained by ***turbidity currents,*** *gravity-driven currents consisting of dilute mixtures of sediment and water having a density greater than the surrounding water.*

During the 1930s, researchers at the California Institute of Technology were pouring streams of

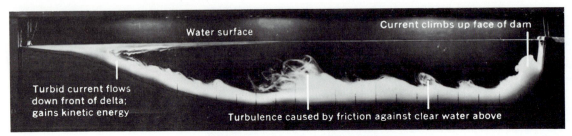

FIGURE 5.23 Turbidity current flowing from left to right, seen through the glass wall of a water-filled laboratory tank.

water, densified by dissolved salts or suspended silt and clay, into a water-filled tank in their laboratory. They found that the dense water, after flowing down the front of a small delta, possessed enough momentum to travel along the bottom on an extremely gentle slope beneath the clear water above (Fig. 5.23). Energy not expended in friction enroute enabled the current to climb part way up the face of a dam at the far end of the tank. The current had distinct boundaries and appeared to one researcher like "a dust storm under water." Indeed, such a current is analogous, mechanically, to a dust storm.

In 1935, soon after completion of Boulder (Hoover) Dam on the Colorado River, engineers were surprised to find that from time to time clear water discharging through the lowest outlet pipes became muddy. It soon was realized that muddy water entering Lake Mead far upstream from the dam formed dense sediment-laden currents that swept along the floor of the lake until they rose up the side of the dam and flowed out through the discharge pipes. Observations in this and other reservoirs confirmed that the suspended loads of rivers were the source of the currents of dense turbid water, which became known as turbidity currents.

Further laboratory experiments showed that turbidity currents could perform a surprising amount of erosion. The densities of natural turbidity currents on gently inclined lake floors are rarely as great as 1.02 g/cm^3, and their velocities generally are less than 30 cm/s. However, submarine turbidity currents flowing down a continental slope can develop velocities greater than that of the swiftest streams on land. Some reach a velocity of more than 90 km/h and transport up to 3 kg/m^3 of sediment, spreading it as far as 1000 km from the source.

Breaks in Transtlantic Cables. On November 18, 1929, a severe earthquake on the continental slope off Nova Scotia caused the breakage of 13 transatlantic cables in 28 places. At the time it was supposed that the breaks were the direct result of movements of the seafloor, the usual cause of submarine earthquakes. Turbidity currents were then unknown, and the event was all but forgotten. Many years later the evidence, including a thick file of measurements made by a repair ship, was reviewed and studied. There were two odd things about the cable breaks. First, although all the cables on the continental slope and deep-sea floor were broken, not one of the many cables that crossed the continental shelf was damaged. Second, the breaks occurred in sequence over a period of 13 hours, in order of increasing depth, and through a distance of 480 km from the earthquake center. Repair ships found that each damaged cable had been broken at two or three points more than 160 km apart. The detached segments of cable between the breaks had been carried partway down the continental slope or buried beneath sediment beyond its base.

The whole event was like a vast laboratory experiment with times and distances controlled by measurement. Each break was timed by the equipment that automatically records the messages transmitted through the cables, and was accurately located after the quake by the electric-resistance measurements always made to enable repair ships to locate breaks and repair damage.

The only hypothesis that can reasonably explain all these facts is that the quake triggered a large submarine landslide on the continental slope. As it developed, a series of slumps may have quickly formed turbidity currents that flowed downslope, breaking each cable as it was reached. Eight cables on the slope were broken instantaneously at the moment the quake occurred, and five others were broken at times ranging from 59 minutes to 13 hours 17 minutes after the quake (Fig. 5.24). The area affected was about 320 km wide and the currents traveled well over 720 km from their source.

From the times of the breaks and the distances between them the velocities of the inferred turbidity currents could be calculated. Velocities on the continental slope are believed to have been at least 30 km/h and may have reached 40–55 km/h, or about 6 to 8 times the average velocity of the lower Mississippi River.

Within the past 75 years, similar events have occurred at more than 40 localities around the world. Some were related to earthquakes. Others occurred off the mouths of large rivers, suggesting that the postulated turbidity currents were set off by large floods or by a slump of unconsolidated deposits. Still others may have resulted from major storms off the coast. This evidence is impressive, and it leads us to believe that turbidity currents are very effective geologic agents on continental slopes.

Turbidites. As the inferred turbidity currents off Nova Scotia lost energy, the sediment carried down the continental slope was dropped onto the ocean floor. The repair ships reported that some of the long, broken pieces of cable were buried beneath deposits of "sand and small pebbles." Two sediment cores obtained downslope from the cable breaks showed that the surface is blanketed with a layer of silt and muddy sand in the form of a graded layer (Fig. 5.11). In one core the layer is 70 cm thick, whereas in the other it is 128 cm. In all, the layer covers at least 100,000 km^2, an area larger than that of Austria or the state of Maine.

As noted previously, a graded layer is the result of rapid, continuous loss of energy in the transporting agent. This would occur in a spent turbidity current, but it is not expected in most other sorts of marine currents. A graded layer of this kind is termed a **turbidite**, defined as *sediment deposited by a turbidity current.*

Samples of turbidites of various ages from the seafloor obtained by deep-ocean coring suggest that at any site on the continental rise or adjacent deep-sea plain a turbidite is deposited once every 1000 to 20,000 years. In these places, far-distant from the source, the deposits are mainly thin-bedded layers a few millimeters to 30 cm thick. Although deposition is infrequent, over millions of years the turbidites can slowly accumulate to form vast deposits beyond the continental realm.

Deep-Sea Fans. Some large canyons that are cut into continental slopes are aligned with the mouths of big rivers such as the Hudson, Mississippi, Amazon, Congo, Ganges, and Indus. The mouths of most such canyons lead into *huge fan-shaped features at the base of the continental slope that spread downward and outward to the deep-sea floor* (Fig. 5.25). These features are **deep-sea fans.** The surfaces of some are marked by distributary channels as much as 200 m deep that have a radial pattern resembling those of an alluvial fan on land.

The Amazon deep-sea fan (Fig. 5.25a), one of the largest known, is about 350,000 km^2 in area,

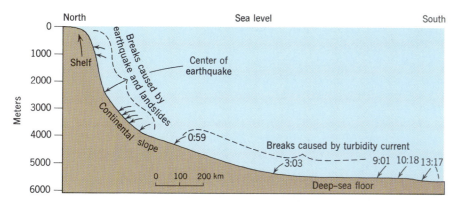

FIGURE 5.24 Profile of sea floor off Nova Scotia showing events of the earthquake of November 18, 1929. Short arrows point to the locations of breaks in transatlantic cables. The numbers show the times of breaks in hours and minutes after the earthquake. The vertical scale is exaggerated greatly. (*Source:* After Heezen and Ewing, 1952.)

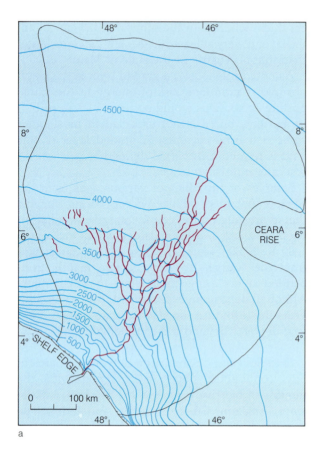

a

extends to a depth of about 4700 m, and is at least 1 to 5 km thick. It is thought to have been built largely during the last 8 million years, and mainly at times when sea level stood so low that the shore lay along the outer edge of the continental shelf of Brazil, some 200 to 250 km seaward of its present position.

The sediments of the fan, sampled by coring, prove to have been derived mostly from the land and transported by the Amazon River. They include fragments of land plants as well as fossils of shallow- and deep-water marine organisms. Also present are many graded layers, consisting of mixtures ranging from clay particles up to small pebbles. These are interpreted as turbidites.

Deep-sea fans are a major exception to the

FIGURE 5.25 Deep-sea fans. (*a*) The Amazon deep-sea fan extends beyond the continental shelf of South America to a depth of about 4700 m. (*Source:* After Damuth and Flood, 1985.) (*b*) The Indus and Bengal deep-sea fans have been built by the Indus and Ganges-Bramaphutra rivers, respectively, on the sea floor adjoining the Indian subcontinent. (*Source:* After Emmel and Curray, 1985; Kolla and Coumes, 1985.) Red lines depict channels cutting the fan. The channels probably joined as a result of turbidity currents.

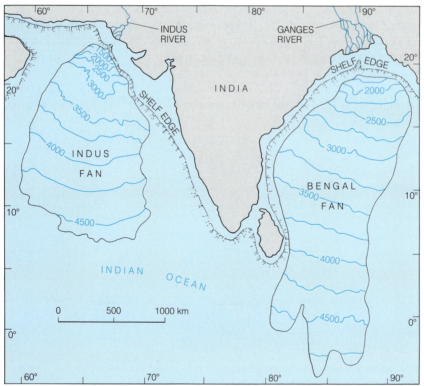

b

generalization that final deposition of land-derived sediment in the ocean is largely confined to the continental shelves. When shelves are emerged at times of lowered sea level and rivers extend across them nearly to the continental slope, the stage is set for the rapid building of deep-sea fans.

Contourites. Water overlying the continental rise is often stratified due to vertical differences in temperature and salinity. As a result, bottom currents tend to flow horizontally, following bathymetric contours on the seafloor. Such *contour currents* are responsible for moving and redistributing considerable amounts of sediment. Measured current velocities are high enough (up to 18 cm/s, or 650 m/h) to move all sediment sizes found on the rise. The *sediments deposited by contour currents* are known as **contourites.** Typically they form persistent laminae 1 mm to 10 cm thick, consist of well-sorted sands or silts, are cross-bedded, and have sharp upper and lower boundaries. Contourites often have rippled surfaces and have built dune-like features tens of meters high. They also form large depositional ridges aligned parallel with the current that are hundreds of kilometers long and tens of kilometers wide. Such large-scale constructional features show that subsea erosion and deposition is not confined to the continental shelves and rises, but is also occurring at great depths on the ocean floors.

Sediments of the Deep-Sea Floor

Sampling Techniques. The deep-sea floor is mantled with loose, fine-grained sediment. Although this sediment blanket covers more than half the Earth's surface, its systematic study dates only from the 1870s, when the British research ship H. M. S. *Challenger,* on a history-making cruise, collected a vast number of samples at carefully chosen positions. So long and thorough was the resulting study that the published reports were not completed until 18 years after the cruise ended. Most of the samples were obtained by scoops dragged along the bottom, or by small "clamshell" buckets that take a bite of sediment and then snap shut.

Since the pioneering cruise of H. M. S. *Challenger,* other expeditions of various nationalities have added greatly to the take of samples. Then, in 1947, a tremendous improvement in sampling was made possible through the invention of the *piston corer.* The device consists of a tight-fitting piston in a long metal tube which is let down to the seafloor

by an attached cable. The piston remains at the top of the sediment as the tube penetrates the bottom, thereby creating a suction that holds the sediment column in place. Cores 7 to 20 m long typically are obtained. The *hydraulic piston corer* is a different version that permits sampling of up to 200 m of sediment by repeatedly coring 4.5-m-long sections from the same hole. The resulting samples are of high quality and essentially undisturbed.

Another device is a rotary drill capable of cutting into the seafloor at water depths of up to 6000 m. Specially fitted research ships such as *Glomar Challenger* (a second and more modern *Challenger*) are used as floating drilling platforms from which a great number of cores have been obtained, each commonly hundreds of meters long (Fig. 5.26).

Sources of Sediments. Sediments that blanket the floor of the deep sea are derived from several sources. **Terrigenous sediment** (Latin for "earth-

FIGURE 5.26 A sediment core retrieved by the Deep Sea Drilling Project is brought on the deck of the *Glomar Challenger* during a cruise in the North Pacific.

FIGURE 5.27 Skeletons of foraminifera (smooth globular forms), radiolaria (coarse meshed objects), and rod-shaped sponge spicules photographed by a scanning electron microscope. Fossils are from the deep-sea core collected in the western Indian Ocean during a Deep Sea Drilling Project cruise.

born") is *sediment derived from sources on land.* Such sediment is carried to the sea by rivers, eroded from coasts by wave action, transported by wind (fine silty and clayey dust and volcanic ash), and released from floating ice. **Pelagic sediment** (from *pelagos*, Greek for "the sea") is *sediment consisting of material of marine organic origin.* It is largely composed of microscopic shells and skeletons of marine animals and plants (Fig. 5.27). Deep-sea sediments also locally include a component of **volcanic sediment** shed *from submarine volcanoes, as well as ash produced from oceanic or nonoceanic volcanic eruptions.* Finally, marine deposits contain trace amounts of **extraterrestrial sediment** (Latin for "outside the Earth"), consisting mostly of *microscopic meteoritic particles that are mixed randomly in other types of deep-sea sediment.*

Kinds and Distribution of Sediments. Analyses of samples brought up by coring devices have made it possible to sort out the various sources from which seafloor sediment is derived. The study of great numbers of samples indicates clearly that all the sediments are mixtures; no one body of sediment comes entirely from a single source. Therefore, the sediments are classified according to their

chief constituents. Seven principal kinds are described in Table 5.2, but all are mixtures and grade from one into another. Figure 5.28 is a map on which the distribution of the main sediment types is shown. The scale of the map is so small that the sediment areas are greatly generalized; a detailed map of sediment distribution would be much more complex. Even so, it would not be as accurate as a map of a comparable land area, because the samples on which it must be based are taken from points very far apart.

The distribution of several of the major sediment types on the world map is easily explained. For example, continental-margin sediments are associated primarily with continental shelves and borderlands that trap much of the sediment leaving the land. Terrigenous sediment accumulates largely as turbidite sequences along continental rises and adjacent deep-sea plains. Glacial-marine sediments are restricted to high latitudes where large numbers of icebergs, derived from ice sheets on Greenland and Antarctica, melt and drop their load of sediment to the seafloor.

Other major sediment types appear to have a more random pattern, but their distribution is, in fact, controlled by oceanic factors. Calcareous ooze

TABLE 5.2 *Major Sediment Types of the Deep-Sea and Typical Rates of Accumulation*[a]

Sediment Type	Typical Rates of Accumulation (mm/1000 yr)
Terrigenous sediment Mainly found on abyssal plains and in trenches. Mud, sand, and gravel, varying greatly from place to place.	200–2000
Glacialmarine sediment Terrigenous sediment, including nonsorted mixtures of particles of all sizes, dropped onto the seafloor from floating glacier ice (Chapter 13).	Highly variable depending on distance from source glaciers
Sediment displaced by gravity Mainly terrigenous sediment, originally deposited on the continental shelf and slope, that has moved to the deep ocean floor under the influence of gravity, by gliding, slumping, or flowing. Confined largely to deep trenches and broad fans at base of continental slope.	Highly variable
Pelagic clay (also called red or brown clay) Confined to the deep-sea floor, mostly in high latitudes or at depths greater than 4 km. Contains less than 30 percent calcium carbonate. Chief constituents are clay minerals, quartz, and micas. Since these are components of weathered soils, volcanic ash, and fine windblown dust, they are thought to come from such sources. The clay is reddish or brownish as a result of gradual oxidation during the very slow process of deposition.	1–15
Calcareous ooze Contains more than 30 percent carbonate, most of it consisting of shells and skeletons of microorganisms. Confined to regions in which surface water is warm and surface organisms exist in myriads; the resulting shells, falling like snow, accumulate on the bottom more rapidly than do the inorganic-clay particles. Because it contains much carbon dioxide, deep-sea water dissolves calcium carbonate. As the shells drift slowly down, they are gradually dissolved, but only at depths of more than about 5 km are they completely consumed. Hence calcareous ooze rarely occurs at such depths.	10–40
Siliceous ooze Contains a large percentage of skeletons built of opaline silica. Occurs where organisms with calcareous shells are few in the surface water and in areas where such shells are destroyed by dissolution before they reach the sea floor.	2–8
Authigenic sediment *Authigenic* means sediments *formed in place*. These have not been physically transported. They consist of minerals that have crystallized from seawater. The principal authigenic deposits are nodular growths of manganese minerals forming on the sea floor (Fig. 22.15).	Irregular and discontinuous

[a]See Figure 5.28 for areas of major distribution.

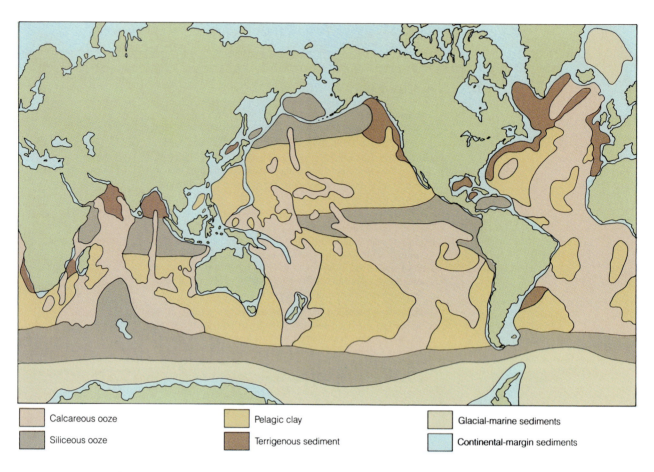

Calcareous ooze

Siliceous ooze

Pelagic clay

Terrigenous sediment

Glacial-marine sediments

Continental-margin sediments

FIGURE 5.28 Map showing generalized distribution of the principal kinds of sediment on the ocean floors. (*Source:* After Hallam, 1981.)

occurs over broad areas at low to middle latitudes where warm sea-surface temperatures favor the growth of carbonate-secreting organisms. The minute shells of these organisms settle and slowly accumulate over wide areas of the seafloor. However, calcareous oozes are not found in these same latitudes where the water is very deep. This results from the fact that cold deep-ocean waters are under high pressure and contain more dissolved carbon dioxide than shallower waters. This means that they can readily dissolve any carbonate that reaches their level. Within the oceans, *the level below which the rate of solution of calcium carbonate exceeds the rate of its deposition* is called the **carbonate compensation depth** (Fig. 5.29). In the Pacific Ocean it lies at 4 to 5 km, whereas in the Atlantic it is somewhat shallower. This explains why over much of the North Pacific and part of the central South Pacific, where the ocean floor lies at great depths, carbonate ooze is lacking. Nearly all the calcareous material settling in these areas is dissolved before reaching the bottom. Instead, the

surface is covered by reddish pelagic clay, composed of extremely fine material derived mainly from terrestrial sources. The only exceptions are some shallow platforms and ridge systems that rise above the carbonate compensation depth.

Other parts of the oceans are floored with siliceous ooze, most notably in the equatorial Pacific, in the far North Pacific, in part of the Indian Ocean, and in a belt encircling the Antarctic region. These are areas of high-biologic productivity, in part related to upwelling of waters having a high-nutrient content. In such regions, siliceous organisms predominate and become a primary component of the deep-sea sediments.

In some places the sampling device put down by exploring ships hits not sediment but bare, hard rock. Comparison with topographic records shows that some of the rocky places are cliffs and other steep slopes, from which any accumulating sediment would be expected to slide off. But still others are flat surfaces, and it is uncertain why they have no cover of sediment. Some may be re-

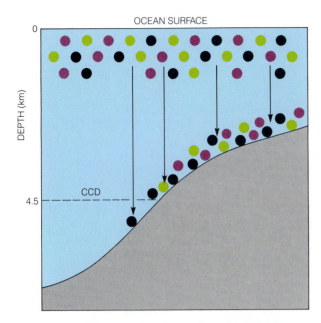

FIGURE 5.29 Diagram showing carbonate compensation depth and its effects on the remains of carbonate-secreting organisms that settle toward the bottom. Debris that is deposited before reaching that critical depth are preserved as calcareous ooze. Remains settling into deeper water are dissolved.

cent lava flows. Others, perhaps, have been scoured bare by currents crossing the floor of the deep ocean.

Because sedimentation rates in the deep sea are very low (on average less than about 3 cm/1000 yr; Table 5.2), even cores only a few tens of meters long can provide a record extending back several million years. Detailed investigations of the cores have provided information on past ocean conditions, including sea-surface temperatures, salinity, movement of icebergs, ocean currents, distribution of marine plants and animals, and changes in the volume of ice on the Earth (Chapter 20).

Although scientists once thought the oceans contained sediments from the Earth's earliest history, drilling into the deep-sea floor has penetrated no sediments older than about 200 million years, which is only the most recent 4 to 5 percent of Earth history. The drilling results are consistent with the discovery that plates of lithosphere, together with their covering of sediment, are continually destroyed when they plunge down into the deep-ocean trenches or are accreted to the margins of growing continents. Therefore, the seafloor is being continuously renewed and the older sediments continuously recycled.

ENVIRONMENTAL RECONSTRUCTIONS FROM SEDIMENTARY STRATA

Through careful study of the physical and biological features of sedimentary strata a great deal can be learned about the conditions under which the sediments were deposited. From such studies important insights are gained about past environments at the Earth's surface.

Paleogeography

Detailed regional investigations of sedimentary rocks provide the information needed to reconstruct the **paleogeography** (*the physical geography during past geologic times*) of a region at the time when a particular stratum or sequence of strata accumulated. Such a reconstruction might show the distribution of land and sea, streams and their direction of flow, mountains and lowlands, deserts, glaciers, and volcanoes. The information is often assembled in the form of a paleogeographic map or a paleogeographic diagram that depicts the environment in which each kind of sediment was deposited. Paleogeographic reconstructions are based on analogy with the environments of today's sediments, invoking the Principle of Uniformity. The physical character of sedimentary strata and their contained fossils can often lead to a rather detailed picture of landscapes in earlier geologic times.

An example of a paleogeographic diagram is shown in Figure 5.30 in which a transect across the northwestern United States has been reconstructed representing a time about 400 million years ago. It shows generalized aspects of the landscape and the marine environment, including an offshore volcanic island arc, the continental shelf and slope, and the shoreline of the North American continent—all based on the character and distribution of preserved rocks which formed at that time.

Paleogeographic maps provide us with views of the Earth's surface that may appear completely unfamiliar to us. Figure 7.22, for example, shows how the world may have looked some 80 to 90 million years ago when the continents had a very different arrangement and large areas of what is now dry land in North America, Africa, and Eurasia were vast shallow seas.

Latitude, Climate, and Relief

Ancient ice-sheet deposits in subtropical India and South America, remains of tropical plants in coal

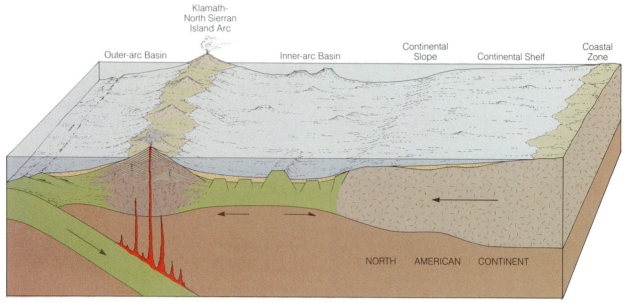

FIGURE 5.30 A paleogeographic diagram across western North America about 400 million years ago showing an island-arc system lying seaward of the continental margin. Arrows show the inferred direction of movement of crustal units. Paleogeography is reconstructed on the basis of the types and geographic distribution of rocks deposited at that time. (*Source:* After Poole et al., 1977.)

beds of frigid Antarctica, and coral reefs in Arctic Alaska all provide evidence of former surface conditions different from those in which these deposits are now found. They clearly point to environmental changes, sometimes pronounced, that cannot easily be explained in terms of the present geographic occurrence of these rocks. Instead, they imply that some sediments laid down under tropical or polar conditions now lie in climatic zones or latitudes quite different from those in which they formed. Such clues offer some of the most compelling arguments in favor of plate tectonics.

At the same time, because the latitudinal shift of land masses is an extremely slow process, the occurrence of geologically young cold-climate sediments in now-temperate environments argues strongly for a change in average temperature—just as the presence of inactive, vegetated sand dunes in a region of moderate rainfall suggests former drier conditions that permitted the sand to be blown about and piled up as dunes.

An area having high *topographic relief* (*the difference in altitude between the highest and lowest points of a landscape*) is apt to be drained by streams that have a high potential for erosion and that are actively transporting sediment down steep upland

slopes. An abundance of coarse alluvial sediments may suggest, therefore, high relief in the source area.

The uplift of mountain ranges can sometimes be inferred from fossils in sedimentary rocks. For example, ancient sediments along the Columbia River in Washington State contain a rich flora of fossil trees, including many broadleaf species that require a moist habitat. The strata now occur in an arid to semiarid steppe environment. Since the plant-bearing strata were deposited, the Cascade Range has been uplifted, blocking the moist, maritime air moving inland from the Pacific Ocean and converting the formerly tree-covered landscape into one now dominated by sagebrush.

Sedimentary rocks can also provide clues about former climate and relief from the character of the detrital particles. When a body of granite is subjected to chemical weathering it yields quartz and clay minerals, but if it were weathered mechanically, it would yield bits of the rock itself, and those bits would include feldspar. If transported and deposited quickly, the feldspar would not be destroyed by chemical weathering. The presence of relatively unweathered feldspar in sedimentary strata suggests one of two things about the origin of the sediment. Either a dry or a very cold climate with a minimum of chemical weathering prevailed, or the cutting of valleys by streams occurred on slopes so steep that the rate of chemical weathering could not keep pace with the rate of erosion by streams.

Flow Directions and Provenance

Sediments and sedimentary rocks preserve a variety of features that enable us to infer the direction of flow of the transporting agent. We have already seen that cross-strata in wind- or water-deposited sediments dip in the direction of flow. Streamlined features on bedding planes also may point in the direction of water flow, just as striations and grooves on glacially abraded rock surfaces are aligned in the direction of ice flow. A regional study of a deposit may show that the average particle size becomes finer and the degree of rounding increases in some particular direction, which can

be inferred to represent the general direction of flow. Measurements of flow-direction indicators, in sufficient numbers to provide a statistical average value for each of many sites, can enable a geologist to construct a map showing flow directions at the time a stratum accumulated (Fig. 5.31).

Through the kinds of minerals or rocks of which they are composed, sediments reflect the kind of parent rock from which they were derived. A simple example is the occurrence in a glacial deposit of boulders that were eroded from an outcrop of a distinctive rock type. It is often possible to trace such boulders back to their source area and thereby infer the direction of flow of the former glacier (Fig. 5.32). In the case of a sedimentary rock, the component mineral grains and rock fragments can help to identify the parent rocks from which the sediment was derived. Knowing where such rocks occur can then supply evidence about directions of sediment transport. Such a *determi-*

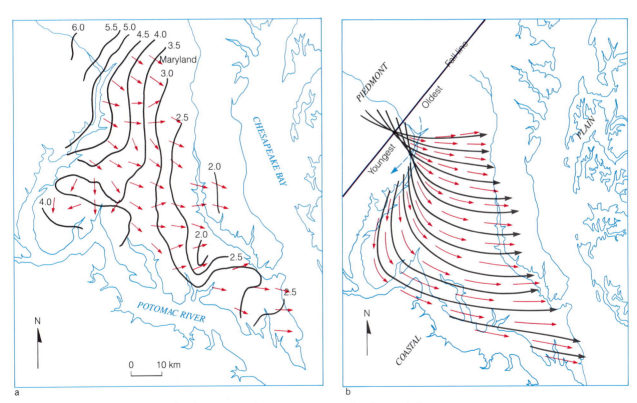

FIGURE 5.31 Reconstruction of paleocurrent directions in an upland gravel deposit in southern Maryland. (*Source:* After Schlee, 1957.) (*a*) Variations in particle size and transport directions. Contours (in millimeters) show a lateral change in the average size of gravel; arrows show the average local trend of currents, as inferred from measurements of cross-bedding. (*b*) The inferred regional current pattern during deposition of upland gravels based on particle size and local current trends.

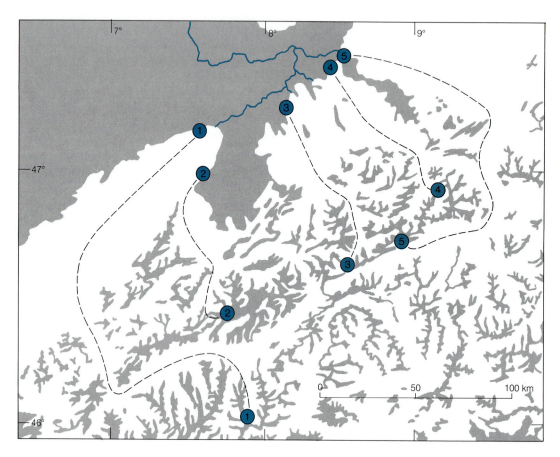

FIGURE 5.32 Map of the Swiss Alps showing positions of distinctive indicator stones transported by glaciers during the last ice age, and the source outcrops of the stones at the heads of major valleys. The dashed lines show the paths followed by each stone as it was carried from its source area to its final resting place. (*Source:* After Wick, 1981.)

nation of the place of origin of a sediment or sedimentary rock means that we have established its **provenance.**

Ocean-Water Conditions

While it may seem rather straightforward to gain information about former environmental conditions on land from information in sedimentary strata, the reconstruction of oceanic conditions in past times might appear to be more difficult. However, isotopic and paleontologic investigations of ocean-sediment cores has made it possible to define former water temperatures and salinities rather precisely. From such information, maps have been constructed that depict the global pattern of sea-surface temperatures at times in the past (Chapter 20; Fig. 20.22).

SUMMARY

1. Sediment is transported by streams, glaciers, wind, slope processes, and ocean currents. After deposition, it experiences compaction and cementation as it is transformed into sedimentary rock.

2. Clastic sediment consists of fragmental rock debris resulting from weathering, together with the broken remains of organisms. Chemical sediment forms where substances carried in solution are precipitated.

3. Various arrangements of the particles in strata are seen in parallel strata and cross-strata, uniform layers, graded layers, and nonsorted layers.

4. Particles of sediment become rounded and sorted during transport by water and air, but not during transport by glaciers. Some mass-wasting deposits display poor sorting.

5. Most sedimentary strata are built of continental detritus that is transported to the submerged continental margins. Some is trapped in basins on land where it is deposited by nonmarine agents, and a small percentage reaches the deep sea.

6. Depositional environments of nonmarine and shallow-marine sediments can be inferred from such properties as texture, degree of sorting and rounding, character of stratification, and types of contained fossils.

7. Most land-derived sediment reaching the continental shelves is deposited close to shore where it is reworked by longshore currents. Extensive areas are covered by relict sediments deposited at times of lower sea level.

8. Carbonate shelves are found on the western margins of ocean basins in low latitudes where warm waters promote growth of carbonate-secreting organisms and little or no sediment is contributed from the continents.

9. Thick evaporite deposits accumulate in restricted marine basins where evaporation is high and limited inflow provides a continuous supply of new saline water.

10. By depositing turbidites, turbidity currents have built large deep-sea fans at the base of the continental slope.

11. Chief kinds of sediments on the deep-sea floor are brownish or reddish pelagic clay, calcareous ooze, siliceous ooze, and terrigenous sediment of continental origin. Each is a mixture and grades into other kinds. Their distribution is largely related to water temperature, water depth, and surface productivity. Glacial-marine sediments are largely restricted to high latitudes near major ice sheets.

12. Sediment is constantly recycled, nearly always within the continental realm.

13. Numerous clues within ancient sedimentary strata permit depositional setting, flow directions, provenance, paleogeography, and ocean-water conditions to be reconstructed.

SELECTED REFERENCES

Blatt, H., Middleton, G. V., and Murray, R. C., 1980, Origin of sedimentary rocks, 2nd ed.: Englewood Cliffs, N.J., Prentice-Hall.

Damuth, J. E., and Kumar, N., 1975, Amazon cone: Morphology, sediments, age, and growth pattern: Geol. Soc. America Bull., v. 86, p. 863–878.

Dunbar, C. O., and Rodgers, J., 1957, Principles of stratigraphy: New York, John Wiley.

Kennett, J., 1982, Marine geology: Englewood Cliffs, N.J., Prentice-Hall.

LaPorte, L. F., 1979, Ancient environments, 2nd ed.: Englewood Cliffs, N.J., Prentice-Hall.

Larsen, G., and Chilingar, G. V. (eds.), 1979, Diagenesis in sediments and sedimentary rocks, 2 vols.: Amsterdam, Elsevier.

Reineck, H. H., and Singh, I. B., 1980, Depositional sedimentary environments, 2nd ed.: New York, Springer-Verlag.

Scholle, P. A., Bebout, D. G., and Moore, C. H. (eds.), 1983, Carbonate depositional environments: Tulsa, Amer. Assoc. Petroleum Geologists Memoir 33.

Scholle, P. A., and Spearing, D. (eds.), 1982, Sandstone depositional environments: Tulsa, Amer. Assoc. Petroleum Geologists Memoir 31.

C H A P T E R 6

Metamorphism and Metamorphic Rocks

A spray of staurolite (orange) surrounded by grains of quartz (blue), in a microscopic thin section. Staurolite is a mineral formed through metamorphism.

CAUSES OF METAMORPHISM

Igneous rock is a product of the Earth's internal processes. Sedimentary rock is a product of external processes. However, the Earth is a dynamic body—rock, once formed, may be subjected to a new set of conditions through such events as burial, heating by an igneous intrusion, or compression caused by continental collisions. As a result, new mineral assemblages may form, and new textures may develop. Changes that happen below ~ 200° C are due to diagenesis (Chapter 5). When the changes exceed those that occur during diagenesis, but happen at temperatures below melting, they are termed *metamorphism.* They include *all changes in mineral assemblage and rock texture, or both, that take place in rocks in the solid state within the Earth's crust as a result of changes in temperature and pressure.*

Changes in temperature and pressure are the two principal causes of metamorphism. However, the effects are not entirely straightforward, because their effectiveness is strongly influenced by such things as the presence or absence of fluids in rocks, the duration of heating, and how long a rock is subjected to high pressure.

Chemical Reactivity Induced by Fluids

Between the grains in a sedimentary rock there are innumerable open spaces (pores) that are filled by a watery fluid. The fluid is never pure water, and at high temperatures it is more likely to be a vapor than a liquid. Nevertheless, the intergranular fluid, for that is its best designation, plays a vital role in metamorphism. The fluid always has dissolved within it small amounts of gases, such as carbon dioxide, and salts, such as sodium chloride, plus traces of all the mineral constituents, such as quartz, that are present in the enclosing rock.

The composition of the fluid is determined by the composition of the enclosing rock and by the temperature and pressure. When the temperature and pressure change, so does the composition of the intergranular fluid. Some of the dissolved constituents move from the fluid to the new minerals forming in the metamorphic rock. Other constituents move in the other direction, from the minerals to the fluid. In this way the intergranular fluid serves as a transporting medium, or "juice," that speeds up chemical reactions in much the same way that water in a stew pot speeds up the cooking of a tough piece of meat.

When intergranular fluids are absent, or present only in tiny traces, metamorphic reactions are very slow. For example, an igneous rock, such as granite or diorite, contains few if any pores; so the amount of intergranular fluid present is miniscule. When such a "dry" rock is heated, few changes occur because the growth of new minerals means that atoms must move by diffusing through the solid minerals. Diffusion through solids is an exceedingly slow process. If, somehow, an intergranular fluid is introduced, perhaps because the rock is crushed, or otherwise deformed, diffusion of the atoms from one place to another can take place through the intergranular fluid. This is a vastly faster process, and as a result, new minerals grow rapidly and the metamorphic effects are pronounced.

As temperature and pressure increase, and metamorphism proceeds, the amount of pore space decreases and the intergranular fluid is slowly driven out of the rock. Hydrous minerals such as clays and chlorite break down to anhydrous minerals such as feldspars and pyroxenes, releasing water in the process. The released water joins the intergranular fluid and is also slowly driven out of the metamorphic rock. For this reason, rock subjected to high temperature and high pressure, for which we use the term *high grade of metamorphism,* contains few hydrous minerals. The term used for rock changed at low temperature and low pressure is *low grade of metamorphism.* Such rock contains many hydrous minerals. *The metamorphic changes that occur while temperatures and pressures are rising,* and while abundant intergranular fluid is present, are termed *prograde metamorphic effects. Changes that occur as temperature and pressure are declining,* and after much of the intergranular fluid has been expelled, are called *retrograde metamorphic effects.* Not surprisingly, prograde metamorphic effects happen more rapidly, and are more pronounced, than retrograde effects. Indeed, it is only because retrograde reactions happen so slowly that we see high-grade metamorphic rocks at all. If retrograde reactions were rapid, all metamorphic rocks would react back to clays and other low grade minerals.

Heat

When a mixture of flour, salt, sugar, yeast, and water is baked in an oven, the high temperature causes a series of chemical reactions—new compounds grow and the final result is a loaf of bread. When rocks are heated, new minerals grow and the final result is a metamorphic rock. In the case

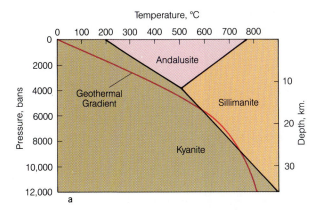
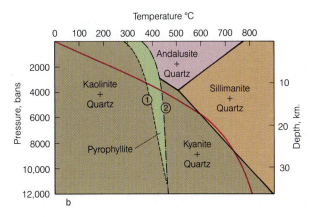

FIGURE 6.1 Regions of temperature and pressure under which different minerals form during metamorphism. (*a*) The three polymorphs of Al_2SiO_5. If a rock consisting solely of Al_2SiO_5 were buried so that it followed the pressure–temperature path of the geothermal gradient, kyanite would be replaced by sillimanite at depths where pressure ranged from 5000 to 9000 bars. (*b*) A closer approximation to an actual rock is one consisting of kaolinite and quartz. Curve (1) marks the upper limit of temperature and pressure under which kaolinite and quartz can coexist. To the right of the curve kaolinite and quartz react to form pyrophyllite. To the right of curve (2), pyrophyllite breaks down to either andalusite + quartz + H_2O or kyanite + quartz + H_2O. The sequence of minerals that would form in a pile of rock buried in such a way that pressure and temperature changed along the red geothermal gradiant, is, in order: kaolinite + quartz, pyrophyllite, kyanite + quartz, sillimanite + quartz, then kyanite + quartz again.

of the rocks, the source of heat is the Earth's internal heat. Rock can be heated simply by burial, or by a nearby igneous intrusion. But burial and the process of intrusion can also cause a change in pressure. Therefore, whatever the cause of the heating, metamorphism can rarely be considered to be entirely due to the rise in temperature. The combined effects of changing temperature and pressure must be considered together.

Pressure

We saw in Chapter 4 that the melting properties of rocks and minerals are influenced by both temperature and pressure (see Fig. 4.7, for example). Metamorphic transformations are also controlled by the dual effects of temperature and pressure. One of the clearest examples of this statement is provided by the three minerals andalusite, sillimanite, and kyanite. They are polymorphs of Al_2SiO_5, and they are only found in metamorphic rocks of appropriate composition. When a clay-rich shale is subjected to high-grade metamorphism, one of the Al_2SiO_5 polymorphs will form—

which one depends on the temperatures and pressures reached (Fig. 6.1*a*).

A shale contains aluminous minerals (such as kaolinite) that are hydrous, while the polymorphs of Al_2SiO_5 are anhydrous. Therefore, when a shale is metamorphosed, one or more chemical reactions must take place that involve water loss in order for the polymorphs of Al_2SiO_5 to form. One change involves a reaction between kaolinite with quartz (which is also present in a typical shale), to form pyrophyllite, a mineral with a sheet structure, according to the following reaction;

$$Al_4Si_4O_{10}(OH)_8 + 4SiO_2 \rightarrow 2Al_2Si_4O_{10}(OH)_2 + 2H_2O$$
kaolinite   quartz   pyrophyllite   water vapor

In turn, pyrophyllite breaks down at higher temperature to either andalusite + quartz + H_2O, or to kyanite + quartz + H_2O (Fig. 6.2) by the following reaction:

$$Al_2Si_4O_{10}(OH)_2 \rightarrow Al_2SiO_5 + 3SiO_2 + H_2O$$
pyrophyllite   kyanite   quartz   water vapor

FIGURE 6.2 Intergrowth of kyanite, quartz, and chlorite resulting from the metamorphism of an original mixture of an iron-bearing clay and quartz. The large kyanite crystal is 12 cm long. The specimen is from western Connecticut.

When Figure 6.1*a* is modified by adding the curves depicting the regions of temperature and pressure where kaolinite + quartz, and pyrophyllite occur, Figure 6.1*b* is the result. Figure 6.1*b*, which is a complicated-looking diagram, contains the kind of information used to decipher past geothermal gradients. Mineral assemblages provide a record of the temperatures and pressures reached by rocks during metamorphism. Real rocks are more complex than simply a mixture of kaolinite and quartz, so that in a real rock additional reactions occur and other minerals form besides those in Figure 6.1*b*. As a result, mineral assemblages in metamorphic rocks range widely, depending on the compositions of the rocks being metamorphosed and the temperature and pres-

sure of metamorphism. By comparing mineral assemblages seen in nature with those produced synthetically in the laboratory, it is possible to delineate the ranges of pressure and temperature conditions under which metamorphism occurs in the crust (Fig. 6.3).

So far we have discussed pressure as if it were equal in all directions, as it is in a liquid. But rock has strength, and can withstand differential pressures—by which we mean the pressure can be greater in one direction than another (Fig. 6.4). When the strength of a rock is exceeded, differential pressure will lead to effects such as shearing, flowage, and other distortions that give rise to distinctive metamorphic textures. Even if the strength of a rock is not exceeded, differential pressure can produce distinctive metamorphic textures. One commonly observed effect is a pronounced tendency of sheet-structure minerals, such as micas and chlorites, to grow so that the polymerized $(Si_4O_{10})^{-4}$ sheets are perpendicular to the direction of maximum pressure.

Time

All chemical reactions require a certain amount of time to proceed to completion. Some reactions, such as the burning of methane gas (CH_4) to yield carbon dioxide and water, happen so rapidly that they create explosions. At the other end of the scale are reactions that require millions of years to proceed to completion. Many of the chemical reactions that occur in rocks undergoing metamorphism are of the latter kind. No reliable ways have

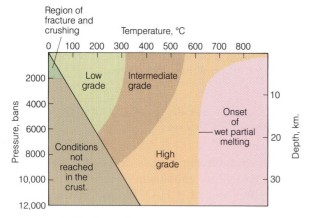

FIGURE 6.3 Regions of temperature and pressure (equivalent to depth), under which metamorphism occurs in the crust.

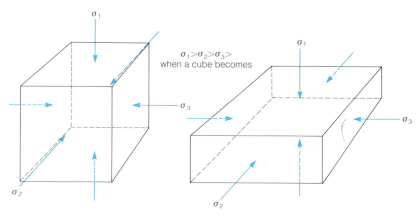

$\sigma_1 > \sigma_2 > \sigma_3 >$
when a cube becomes

FIGURE 6.4 Differential pressure. In a liquid, pressure is equal in all directions and is said to be hydrostatic. Solids, by contrast, have strength and can support pressures that are unequal in different directions. The symbol used for directed pressure or stress, as it is more commonly called, is the Greek letter σ (sigma). σ_1 is greater than σ_2; while σ_2 is greater than σ_3.

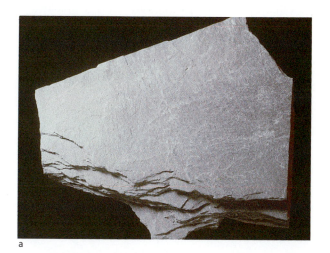

a

b

FIGURE 6.5 The grain sizes in these specimens of (*a*) slate, (*b*) phyllite, and (*c*) schist, show how continued mineral growth occurs during metamorphism. The three rocks are photographed at the same magnification and have the same chemical composition. Mineral grains in the slate are barely visible. Grains in the phyllite are large enough so they are just visible and give the specimen a pronounced sheen. Grains in the schist are large and obvious. The specimens are about 12 cm across.

c

FIGURE 6.6 A planar fabric (foliation) produced in granite by metamorphism. The foliation is caused by a parallel orientation of mica grains. The outcrop is in Namaqualand, South Africa. The pocket knife is 6 cm long.

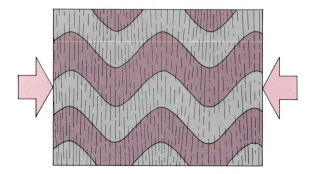

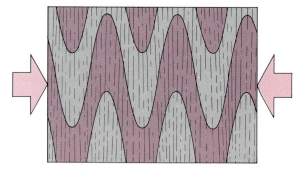

FIGURE 6.7 Two examples of slaty cleavage in folded strata. Note that even in folded strata, cleavage is sometimes approximately parallel to bedding.

FIGURE 6.8 Slaty cleavage cuts across nearly horizontal bedding. The cleavage is developed in the Martinsburg Formation, which was subjected to low-grade metamorphism. The outcrop is 70 cm across and is near Palmerton, Pennsylvania.

yet been developed to determine exactly how long a given metamorphic rock has remained at a given temperature and pressure. However, it can be readily demonstrated in the laboratory that high temperature, high pressure, and long reaction times produce large mineral grains. Thus, it is possible to draw an interesting general conclusion. The conclusion is that coarse-grained rocks are the products of long sustained metamorphic conditions (possibly millions of years) at high temperatures and pressures. Fine-grained rocks on the other hand, are products of lower temperatures, lower pressures, and shorter reaction times.

RESPONSES TO CHANGES IN TEMPERATURE AND PRESSURE

Texture

An important feature of metamorphic rocks has just been mentioned: There is a progressive increase in the grain size of minerals in a metamorphic rock as it is subjected to higher temperatures and pressures for longer times (Fig. 6.5). For a few metamorphic rocks, increased grain size is the only textural change that occurs. However, most metamorphic rocks develop additional, and conspicuous, directional textures. As metamorphism proceeds, and the sheet-structure minerals such as muscovite and chlorite start to grow, the minerals are oriented so that the sheets are perpendicular to the direction of maximum pressure. The new, parallel flakes of mica produce a texture called **foliation,** named from the Latin word, *folium,* meaning leaf. Foliation is *a plane defined by any planar set of minerals, or banding of minerals, found in a metamorphic rock* (Fig. 6.6). Foliation may be pronounced or it may be subtle, but when present it provides clear evidence of metamorphism. Foliation is not a texture that is observed in igneous and sedimentary rocks.

During the earliest stages of mineral growth, pressure is caused by the weight of the overlying rock; the new sheet-structure minerals, and therefore the foliation, tend to be parallel to the bedding planes of a sedimentary rock. But with deeper burial, or when lateral compression deforms the flat sedimentary layers into folds, the sheet-structure minerals and the foliation are no longer parallel to the bedding planes (Fig. 6.7). Regardless of the orientation of the original bedding, metamorphic rocks break readily in the direction of the foliation. When the rocks are so fine-grained that the new mineral grains can only be seen with the microscope, the breakage property is called **rock cleavage,** or **slaty cleavage,** which is defined as *the property by which a rock breaks into plate-like fragments along flat planes* (Fig. 6.8).

Slaty cleavage develops at low grades of metamorphism. Under intermediate and high grades of metamorphism, grain sizes increase and individual mineral grains can be seen with the naked eye. Foliation remains but it is no longer a flat plane. Intermediate and high grade metamorphic rocks tend to break along wavy, or slightly distorted surfaces, reflecting the presence,

and orientation, of grains of quartz, feldspar, and other minerals (Fig. 6.9). Such breakage directions arise from the **schistosity,** a term derived from the Latin, *schistos,* meaning cleaves easily, and referring to *the parallel arrangement of coarse grains of the sheet-structure minerals, like mica and chlorite, formed during metamorphism under conditions of differential pressure.*

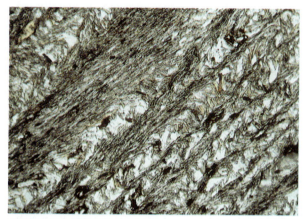

FIGURE 6.9 Thin section of a schist. The pronounced schistosity is due to a parallel arrangement of the mica grains. The photo is 1 cm wide.

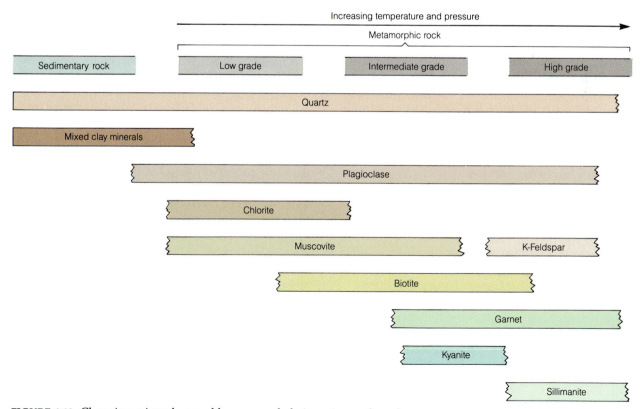

FIGURE 6.10 Changing mineral assemblages as a shale is metamorphosed from low to high grade.

Mineral Assemblages

Metamorphism produces both new textures and new mineral assemblages. As temperature and pressure rise, one new mineral assemblage follows another. For any given rock composition, each assemblage is characteristic of a given range of temperature and pressure. A few of these minerals are rarely found (or not at all) in igneous and sedimentary rocks. Their presence in a rock is usually evidence enough that the rock has been metamorphosed. The most important of these metamorphic minerals are chlorite, serpentine, epidote, pyrophyllite, talc, kyanite, sillimanite, andalusite, and staurolite. An illustration of the way mineral assemblages change with grade of metamorphism is given in Figure 6.10.

FIGURE 6.11 A coarse-grained gneiss. Minerals present are feldspar and quartz (both light colored), and biotite (dark). The streaky layers of biotite, and the parallel orientation of individual grains of biotite in the layers, are evidence that the rock is metamorphic in origin. The specimen is 8 cm across.

METAMORPHIC ROCKS

The naming of metamorphic rocks is less systematic than the naming of igneous and sedimentary rocks. This is so because certain names are based on textures, some are based on mineral assemblages, while a few are combinations of both.

The names most widely used to describe metamorphic rocks are those that are applied to the metamorphic derivatives of shales and basalts. This is so because shales and basalts are, respectively, the most abundant sedimentary and igneous rock types. Limestones, sandstones, and other rocks give rise to distinctive metamorphic rocks also, and each of the common varieties is mentioned below. More extensive details of texture, mineral assemblage, and methods of identification are given in Appendix C. Here we mention only the most important features.

Slates, Phyllites, Schists, and Gneisses

Shale is the term applied to a fine-grained, clastic, sedimentary rock. The minerals usually present in a shale include quartz, clays of various kinds, calcite, and possibly feldspar. Under conditions of low-grade metamorphism, muscovite and/or chlorite start to grow. Although the rock may still look like a shale, the tiny new mineral grains produce slaty cleavage. The low-grade metamorphic product of shale is *slate* (a word derived from Old French, *slat*, referring to the useful properties such rock has as a roofing material). Continued metamorphism, to intermediate grade, produces both larger grains of mica and a changing mineral assemblage; the rock develops a pronounced foliation and is called *phyllite* (from the Greek, *phyllon*, a leaf). Still further metamorphism produces a coarse-grained rock with pronounced schistosity, called *schist*. A comparison of the differing grain sizes in slate, phyllite, and schist, can be seen in Figure 6.5. At high grades of metamorphism, minerals start to segregate into separate bands. A high-grade rock, with coarse grains and pronounced foliation, but with the layers of micaceous minerals segregated from layers of minerals such as quartz and feldspar, is called a *gneiss* (pronounced nice, from a word in Old High German, *gneisto*, meaning to sparkle) (Fig. 6.11).

The names slate and phyllite describe textures and are commonly used without adding mineral names as adjectives. The names of the coarse-grained rocks, schists and gneisses, are also derived from textures, but in these cases mineral names are usually added as adjectives. For example, we refer to a quartz-plagioclase-biotite-garnet gneiss. The difference arises because minerals in coarse-grained rocks are large enough to be seen and readily identified.

Greenschists, Amphibolites, and Granulites

The main minerals in basalts and gabbros are olivine, pyroxene, and plagioclase, each of which is anhydrous. When a basalt is subjected to metamorphism under conditions where H_2O can enter the rock and form hydrous minerals, distinc-

FIGURE 6.12 Amphibolite resulting from metamorphism of a pillow basalt. Compare Figure 4.36. Pillow structure in this ancient basalt is deformed but can be discerned by the selvedges of pale yellow epidote formed from the original glassy rims of the pillows. The mineral assemblage of the basalt (olivine, pyroxene, anorthitic plagioclase) has been changed to hornblende, epidote, chlorite, and albitic plagioclase. The outcrop is in Namibia, near the Matchless Mine.

tive mineral assemblages develop. At low grades of metamorphism, assemblages such as chlorite + plagioclase + calcite form. The resulting rock is equivalent in metamorphic grade to a slate, but has a very different appearance. It has pronounced foliation, but it also has a very distinctive green color because of its chlorite content; it is termed *greenschist*. At intermediate grades of metamorphism, chlorite starts to break down and amphibole develops instead; the resulting rock is generally coarse-grained, and is called an *amphibolite* (Fig. 6.12). Because amphibole has a chain-structure, rather than a sheet-structure, the effect of differential pressure is to cause the amphibole to grow as elongate grains. The grains tend to line up so that their long axes are parallel, and point in the direction of least pressure. A rock that has *a parallel arrangement of elongate mineral grains* is said to possess a **lineation**. At highest grades of metamorphism, amphiboles break down to yield pyroxenes and the rock developed is called a *granulite*, or if it is very rich in pyroxene, *pyroxenite*.

The terms amphibolite and pyroxenite have mineralogical connotations, while greenschist and granulite have combined mineralogical and textural significance. Therefore, names of the minerals present are not added unless the minerals are distinctive. For example, we refer to an epidote amphibolite when epidote is present.

Marbles and Quartzites

Foliation is imparted to metamorphic rock by minerals with sheet structures, such as mica, chlorite, or serpentine. Lineation is most commonly imparted by minerals with chain structures, such as amphibole and pyroxene. Neither limestone nor quartz sandstone (when pure) contain the necessary ingredients to form sheet- or chain-structure minerals. As a result, the metamorphic derivatives of limestone and sandstone, which are *marble* and *quartzite*, respectively, commonly lack foliation and lineation.

Marble consists of a coarsely crystalline, interlocking network of calcite grains. During recrystallization of a limestone, bedding planes, fossils,

a

b

FIGURE 6.13 Textures of nonfoliated metamorphic rocks. Each specimen is 4 cm across. (*a*) Marble. Composed entirely of grains of calcite, all vestiges of sedimentary structure have disappeared during metamorphism. Calcite grains vary little in size. (*b*) Quartzite. Faint traces of the original quartz grains can barely be discerned.

and other features of sedimentary rocks are largely obliterated. The end result, as shown in Figure 6.13*a*, is an even-grained rock with a distinctive, somewhat sugary texture. Pure marble is snow white in color and consists entirely of pure grains of calcite. Such marbles are rare, even though one may not think so looking at the marble gravestones and statues in cemeteries. Most marble contains impurities such as organic matter, pyrite, limonite, and small quantities of silicate minerals, that impart various colors.

Quartzite is derived from sandstone by the filling-in of the spaces between the original grains with silica, and by recrystallization of the entire mass. Commonly, the ghost-like outlines of the original sedimentary grains can still be seen, even though recrystallization may have rearranged the original grain structure completely.

Hornfels and Tactites

Two nonfoliated, metamorphic rocks, *hornfels* and *tactites*, arise through metamorphism involving increase in temperature but in the absence of pronounced differential pressure. The places where these conditions occur are in country rocks adjacent to intrusive igneous rocks.

Hornfels is a hard, fine-grained rock, composed of an interlocked mass of uniformly sized mineral grains. Most commonly, hornfels is derived by metamorphism of shale. Because hornfels is fine-grained, it apparently forms rapidly, and under circumstances where insufficient heating time is available for large mineral grains to grow. For this reason, it is generally presumed that development of hornfels only happens adjacent to shallow intrusive bodies.

Tactite (sometimes also called *skarn*) is analogous to hornfels in that it is formed by metamorphic heating adjacent to an igneous intrusion. Unlike hornfels, tactite tends to be relatively coarse-grained. This is because tactite develops from a very reactive rock—an impure limestone that contains minerals such as quartz, clays, and chlorite as well as calcite. Upon metamorphism, tactite develops assemblages that contain minerals such as calcium-rich amphiboles, calcium-rich pyroxenes, epidotes, and calcite.

KINDS OF METAMORPHISM

The processes that result from changing temperature and pressure, and that cause the metamorphic changes observed in rocks, can be grouped under the terms *mechanical deformation* and *chemical recrystallization*. Mechanical deformation includes grinding, crushing, and the development of new textures such as rock cleavage and foliation. Chemical recrystallization includes all the changes in mineral composition, in growth of new minerals, and the loss of H_2O and CO_2 that occur as rock is heated. Different kinds of metamorphism reflect the different levels of importance of the two groups of processes.

Cataclastic Metamorphism

Purely mechanical effects do sometimes occur without any changes in mineral chemistry. But they are rare and usually localized. For example, adjacent to fractures in massive bodies of coarse-grained rocks such as granite, individual mineral grains may be shattered and pulverised. This sort of deformation occurs in brittle rocks and is called *cataclastic metamorphism*. Deep in the crust, confining pressures are high, and brittle properties of rocks are suppressed. Under such circumstances, flattening and elongation can occur without associated chemical recrystallization (Fig. 6.14). But chemical recrystallization is so much more common than simple mechanical deformation that our discussion should more profitably be centered on chemical effects. These are found in the next three kinds of metamorphism.

FIGURE 6.14 Purely mechanical deformation leads to distinctive textures. Deformation under high confining pressure has caused the quartzite pebbles in this conglomerate in Namibia, originally round, to become flattened and elongate. The pocketknife is 6 cm long.

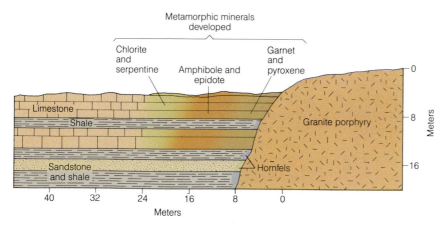

Metamorphic minerals developed

Chlorite and serpentine — Amphibole and epidote — Garnet and pyroxene

Limestone
Shale
Granite porphyry
Sandstone and shale
Hornfels

Meters: 0, 8, 16

40 32 24 16 8 0
Meters

FIGURE 6.15 Metamorphism around an intrusion of granite porphyry near Breckenridge, Colorado. Sandstones and shales have been baked to fine-grained hornfels immediately adjacent to the contact. However, in an impure limestone, a much more reactive rock, new metamorphic minerals such as garnet have been developed as far as 20 m from the contact. Note that the new minerals form a series of zones, or aureoles, around the intrusive.

Contact or Thermal Metamorphism

Contact metamorphism, which is also known as thermal metamorphism, occurs adjacent to bodies of hot igneous rock that are intruded into cooler rock of the crust. It happens in response to a pronounced increase in temperature but without extensive mechanical deformation. The temperature of the igneous rock may be as high as 1000° C, that of the intruded rocks only 200° C to 300° C. Rock adjacent to the intrusive becomes heated and metamorphosed, developing a well-defined shell, or *metamorphic aureole* of altered rock (Fig. 6.15). The width of the aureole depends on the size of the intrusive body. With a small intrusive, such as a dike or sill a few meters thick, the width of the metamorphic aureole may only be a few centimeters. But when the intrusion is large, perhaps a kilometer or more in diameter, the aureole may reach a hundred meters or more in width. The size of an aureole also depends on how susceptible the rocks are to change.

Within a metamorphic aureole it is usually found that several different and roughly concentric zones of mineral assemblages can be identified. Each zone is characteristic of a certain temperature range. Immediately adjacent to the intrusion where temperatures are high, we find anhydrous minerals such as garnet, pyroxene, and andalusite. Beyond them are found hydrous minerals such as epidote and amphibole, and beyond them, in turn, micas and chlorites. The exact assemblage of min-

erals in each zone depends, of course, on the chemical composition of the intruded rock as well as the temperature reached during metamorphism.

When heated, all kinds of rocks react in some fashion, but some are more reactive than others. Those consisting of a single mineral such as a sandstone made of pure quartz, or limestone made of pure calcite, will simply recrystallize to quartzite or marble, respectively. More commonly, the intruded rocks contain mixtures of different minerals which react to form hornfels if quartz-rich or tactite if calcite-rich.

Both cataclastic and contact metamorphism are localized phenomena and affect relatively small volumes of rock. On the other hand, the following two kinds of metamorphism, burial and regional, affect huge rock volumes.

Burial Metamorphism

Sediments, together with interlayered pyroclastic and volcanic rocks, may attain temperatures of 300° C or more when buried deeply in a sedimentary basin. There is usually an abundance of pore water and intergranular fluid in buried sediment. The fluid helps new minerals to grow, but the fabric of the metamorphic rock may look like that of an essentially unaltered sedimentary rock because there is little mechanical deformation involved. The family of minerals that particularly characterize the conditions of burial metamorphism are the zeolites. The zeolites are a family of silicate minerals with fully polymerized crystal structures containing the same chemical elements as feldspars, but also containing water. Burial metamorphism is often associated with sedimentary basins formed on the margins of plates. As temperatures and pressures increase, burial metamorphism grades into regional metamorphism.

Regional Metamorphism

The most common metamorphic rock of the continental crust occurs through areas of tens of thousands of square kilometers and the process that forms it is therefore called regional metamorphism. Unlike contact on burial metamorphism, regional metamorphism involves a considerable amount of mechanical deformation in addition to chemical recrystallization. As a result, regionally metamorphosed rocks tend to have pronounced textures; they are usually distinctly foliated and commonly strongly layered. Slate, phyllite, schist, and gneiss are the most common

varieties of regionally metamorphosed rocks, and they are usually found in mountain ranges, or eroded mountain ranges. Greenschists and amphibolites are also products of regional metamorphism, but they tend to be found in exceedingly old rocks where large segments of ancient oceanic crust of basaltic composition have been metamorphosed and incorporated into the continental crust.

Consider what happens when a pile of strata, formerly at or near the surface, becomes deeply buried and subjected to horizontal compressive forces. The strata fracture, and they become folded and buckled. In short, mechanical deformation commences. As depth of burial increases, the strata are subjected to increasing pressure and temperature. New minerals start to grow. However, rocks are poor conductors of heat; so the heating-up process is very slow. The temperature reached by a buried pile of strata depends on both depth and duration of burial. If burial is very slow, heating of the pile keeps pace with the temperature of adjacent parts of the crust (that is, a normal

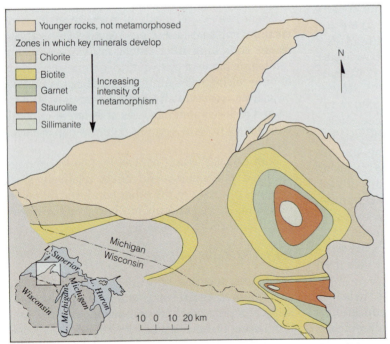

FIGURE 6.16 Zonal sequences of minerals and textures developed in regionally metamorphosed rocks reflect gradients of temperature and pressure during metamorphism. Metamorphic zones in an area of Michigan are reflected by the first appearance of index minerals. Metamorphism occurred about 1.5 billion years ago. Rocks that were deposited later than the metamorphism are unchanged. Two clearly defined centers of intense metamorphism are identified by the presence of sillimanite. (*Source:* After James, 1955.)

continental geothermal gradient is maintained). However, if burial is very fast, as it is with sediment dragged down in a subduction zone, the pile has insufficient time to heat up and conditions of high pressure, but rather low temperatures, prevail. The minerals that grow are controlled both by temperature and pressure; therefore, the mineral assemblages we observe depend on rates of burial. In a buried pile of strata both temperature and pressure vary. The larger the pile, the greater the variations. Regional-metamorphic rocks, like their contact-metamorphic equivalents, develop zonal sequences of minerals and textures in response to variations of temperature and pressure (Fig. 6.16). Unlike contact-metamorphic aureoles, however, zones of regional metamorphism tend to be broad and undulating, showing both horizontal and vertical changes. Whereas aureoles resemble a series of concentric cylinders surrounding an intrusive rock, zones of regional metamorphism are analogous to a domed pile of blankets, each blanket representing a specific region of temperature and pressure.

INDEX MINERALS AND ISOGRADS

The first geologists to make a systematic study of the distribution of minerals in a regionally metamorphosed terrain did so in the Scottish Highlands. One of them, George Barrow, observed that rocks having the same overall chemical composition (that of a shale), could be subdivided into a sequence of zones, each zone having a distinctive mineral assemblage. Each assemblage, in turn, was characterized by the appearance of new minerals. Barrow and his coworkers selected characteristic, or index minerals, which, proceeding from low-grade to high-grade, marked the appearance of each new mineral assemblage. Their index minerals were, in order of appearance, chlorite, biotite, garnet, staurolite, kyanite, and sillimanite. By plotting on maps the places where each of the index minerals first appeared in rocks having the chemical composition of shale, the workers in the Scottish Highlands defined a series of isograds. An *isograd* is *a line on a map connecting points of first occurrence of a given mineral in metamorphic rocks.* The concept of isograds is now widely used by those who study metamorphic rocks of all kinds; it is just as applicable to burial and contact metamorphism as it is to regional metamorphism. *The regions on a map between isograds* are known as **metamorphic zones**—we speak of the chlorite zone, the biotite zone, and so forth (Fig. 6.17).

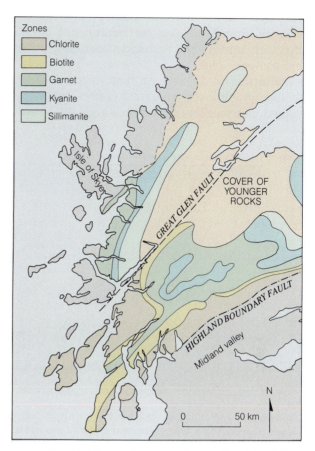

FIGURE 6.17 Metamorphic zones in the Scottish Highlands.

METAMORPHIC FACIES

Careful study of metamorphic rocks around the world has demonstrated a very important fact—the chemical compositions of rocks are little changed by metamorphism. The main changes that do occur are the addition or loss of volatiles such as H_2O and CO_2. However, the principal constituents of rocks remain fixed. The changes brought about during metamorphism, then, are changes in the mineral assemblage, not changes in the overall chemical composition of the rocks. The conclusion to be drawn from this fact is that the mineral assemblages observed in the metamorphic derivatives of common rocks are controlled by the temperature and pressure of metamorphism to which the rocks are subjected. Based on this conclusion, in 1915 a famous Finnish geologist named Pennti Eskola proposed the concept of **metamorphic facies.** The concept is that *contrasting assemblages of minerals that reach equilibrium during metamorphism within a specific range of physical condi-*

tions belong to the same facies. Eskola drew his conclusions from studies of metamorphosed basalts that were interlayered with rocks of entirely different composition.

Returning to an analogy that we used earlier (cooking), think of a large roast of beef. When it is carved, one sees that the center is rare, the outside well-done, and in between there is a region of medium-rare meat. The differences occur because the temperature was not uniform throughout. The center, or "rare-meat" facies, is a low-temperature zone; the outside, or "well-done" facies, is a high-temperature zone.

Metamorphic facies were originally described in

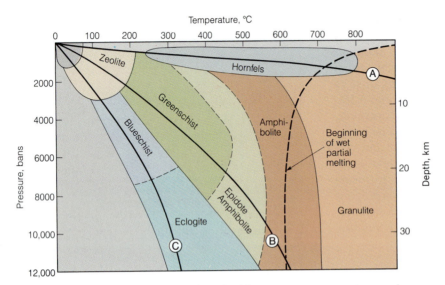

FIGURE 6.18 Metamorphic facies plotted with respect to temperature and depth. Curve A is a typical thermal gradient around an intrusive igneous rock that is causing contact metamorphism. Curve B is a normal continental geothermal gradient, and it indicates the rock types developed during slow burial of a pile of strata. Curve C is the geothermal gradient developed in a sequence of strata that are buried so rapidly the rocks cannot maintain thermal equilibrium. Rapid burial is characteristically observed in subduction zones.

TABLE 6.1 *Characteristic Minerals of Differing Metamorphic Facies for Selected Rocks*[a]

	Precursor Rock Type	
Facies Name	Basalt	Shale
Granulite	Pyroxene, Plagioclase, Garnet	Biotite, K-Feldspar, Quartz, Andalusite
Amphibolite	Amphibole, Plagioclase, Garnet, Quartz	Garnet, Biotite, Muscovite, Sillimanite, Quartz
Epidote-Amphibolite	Amphibole, Epidote, Plagioclase, Garnet, Quartz	Garnet, Chlorite, Muscovite, Biotite, Quartz
Greenschist	Chlorite, Amphibole, Plagioclase	Chlorite, Muscovite, Plagioclase, Quartz
Blueschist	Blue-Amphibole, Chlorite, Ca-Rich Silicates	Blue-Amphibole, Chlorite, Quartz, Muscovite, Lawsonite
Eclogite	Pyroxene (variety Jadeite), Garnet, Kyanite	Not Observed
Hornfels	Pyroxene, Plagioclase	Andalusite, Biotite, K-Feldspar, Quartz
Zeolite	Calcite, Chlorite, Zeolite (variety Laumontite)	Zeolites, Pyrophyllite, Na-Mica

[a]For temperature and pressure conditions of each facies, refer to Figure 6.18.

terms of recurring mineral assemblages, to each of which there was assumed to be a specific set of temperature and pressure conditions. The realization that temperature, pressure, and rock composition each play a role in determining the mineral assemblage provided the link needed to allow conditions of metamorphism to be determined through laboratory experiments, and eventually to prove Eskola's suggestion. The concept is now applied to a very wide range of temperatures and pressures. The principal metamorphic facies, together with geothermal gradients to be expected under three differing geological conditions, are shown in Figure 6.18. Because Eskola was studying metamorphosed basalts when he proposed the metamorphic facies concept, most of the names he gave to metamorphic facies reflect the mineral assemblages developed in rocks of basaltic composition. It is important to remember, however, as shown in Table 6.1, that mineral assemblages are just as much a result of rock composition as they are of the temperature and pressure of metamorphism. When comparing mineral assemblages of rocks subjected to differing grades of metamorphism, therefore, one must be certain that they have the same overall chemical composition.

MIGMATITES

The upper limits of temperature and pressure under which metamorphism occurs are the lower limits of magma generation. In the presence of an H_2O-rich intergranular fluid, the upper limit of metamorphism is marked by the onset of wet partial melting. When H_2O is present, wet partial melting of schists and gneisses starts in a region of temperature and pressure defined by the granitic magma curve (Fig. 4.16 and 6.18). Because H_2O depresses the melting temperature, the amount of H_2O present as intergranular fluid controls the amount of magma that can form by wet partial melting. The curve marking the onset of melting in Figure 6.18 is the curve for magma that is saturated with H_2O. That is, it is a curve along which the H_2O content of the magma rises as the pressure increases. Such a condition could only be obtained in the presence of an abundant intergranular fluid. More commonly, the amount of intergranular fluid is small. When there is not enough water available to produce the maximum depression of the melting curve, melting must start at a higher temperature. For this reason, the temperature and

pressure limits of metamorphism and magma generation overlap. When H_2O is abundant, magma development starts at lower temperatures; when it is limited, magma generation starts at higher temperatures. In many places, it is possible to find evidence demonstrating that while schists, gneisses, and amphibolites were forming in one part of a rock pile, melting was commencing elsewhere. It is also commonly observed that in an interlayered sequence of gneisses and schists, some layers (the H_2O-rich layers) started to melt, while adjacent but drier layers do not show any sign of melting. The result is *a composite rock containing both igneous and metamorphic portions,* called a *migmatite* (Fig. 6.19). Still further melting causes large volumes of magma to develop and to rise up through the metamorphic rock above. Eventually, a magma formed by wet partial melting will solidify as an intrusive igneous rock of granitic or granodioritic composition. As a result, we observe that batholiths and stocks of granite and granodiorite, and large volumes of regionally metamorphosed rocks, are closely associated. The geological setting in which the two rock types form is the setting of regional metamorphism, which is along subduction edges of plates, and edges of plates that carry colliding pieces of continental crust (Fig. 6.20). In subsequent chapters, we will return to the association of certain rock types and large scale tectonic features, because they are intimately associated and provide much of the evidence of ancient plate movements.

FIGURE 6.19 Example of migmatite, a partial stage in the progression from metamorphic to igneous rock. Complexly folded veins (white) have the composition of granite. They are enclosed by high-grade metamorphic rock that is biotite-rich. The granite veins represent the fraction of rock that melted at the climax of metamorphism. The specimen is 20cm wide.

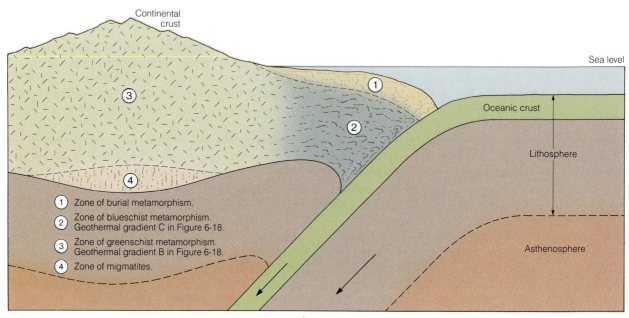

FIGURE 6.20 Diagram of a convergent plate boundary showing the approximate regions of burial metamorphism, of greenschist and blueschist styles of regional metamorphism developed along normal and subduction geothermal gradients, respectively (see Figure 6.18), and the formation of migmatites.

WHAT HOLDS ROCK TOGETHER?

We need not have had much experience in order to realize that some kinds of rock hold together with great tenacity, whereas other kinds are easily broken apart. The most tenacious rocks are igneous and metamorphic, because these possess intricately interlocked mineral grains. The growing minerals crowd against each other, filling all spaces and forming an intricate, three-dimensional jigsaw puzzle. A similar interlocking of grains holds together steel, ceramics, and bricks.

The forces that hold together the grains of sedimentary rocks are less obvious. Sediment is transformed into sedimentary rock in four ways.

1. By pressure from overlying sediment or through vibrations of the ground arising from earthquakes, the irregular-shaped grains in a sediment can be packed into a tight, coherent mass. The interlocking of grains that results from this kind of packing is not strong, as in igneous rock, but under some circumstances it can hold sediment together.

2. Water that circulates slowly through the open spaces between grains deposits new materials such as calcite, quartz, and iron oxide, which cement the grains together.

3. The weight of overlying deposits can squeeze water out of deeply buried sediment, compacting the sediment and reducing the pore space. Compaction forces small grains close together and makes more effective the capillary forces exerted by remaining films of intergranular water.

4. As sediment becomes deeply buried, its mineral grains begin to be recrystallized. Recrystallization marks the onset of diagenesis, and resembles very low-grade metamorphism in that the newly growing minerals interlock and form strong aggregates, such as those of metamorphic rocks.

WHAT BREAKS ROCK APART?

Studying an exposure of bedrock, we can see that the rock may be massive and difficult to break because no easy plane of fracture is present. Or we may observe that one or more easy directions of breakage are present. The breakage directions arise in three principal ways.

1. By original layering in a sedimentary rock, and in metamorphic rock derived from a sedimentary rock.

2. By development of rock cleavage and foliation in metamorphic rocks.

3. By development of *joints,* which are *fractures on which movement has not occurred in a direction parallel to the plane of the fracture.* Rarely do joints occur singly. Most commonly they form a widespread group of parallel joints, called a joint set (Fig. 6.21).

When rock that has been deeply buried is slowly uncovered by erosion, it is relieved of the confining pressure exerted by overlying material, and so expands, and in the process develops the fractures we call joints. Joints do not let us infer much about the origins of rocks—they are found in all of the rock families—but as we shall see in later chapters, they are extremely important in the control of weathering of rock, because they are passageways by which rainwater can enter and speed up erosion.

One special class of joints is restricted to certain igneous rocks, and in this one case they do afford a clue to origin. When a body of igneous rock cools, it contracts and sometimes fractures into small pieces in the same way that a very hot glass bottle, plunged into cold water, contracts and shatters. Cooling joints are found in igneous rock that cooled rapidly. They are also most common in tabular masses of igneous rock—dikes, sills, lava flows, welded tuffs—because the friction exerted by the enclosing country rock prevents the igneous rock masses from contracting as a coherent mass. When we see cooling joints, therefore, we infer that the rock mass is probably tabular, and that it has cooled rapidly, close to the surface. Just as a hot bottle does not fracture if allowed to cool slowly, neither does an igneous rock. Hence, if a deeply buried body of igneous rock, such as a batholith, does not develop cooling joints, we conclude that batholiths must cool slowly. Unlike shattered glass, cooling fractures in igneous rock form regular patterns. For *joints that split igneous rocks into long prisms or columns* we use the special term **columnar joints** (Fig. 6.22).

CONCLUSION

The three rock families contain the keys for understanding the internal and external activities of the Earth. Igneous rock results from internal processes, sedimentary rock from external. Metamorphic rock records the intermediate steps through which both sedimentary and igneous rock are changed and influenced by internal processes. Studies of internal and external activities reveal that new rock is being made continuously. Extrusive igneous rock is being made in all active volcanic areas and along the mid-ocean ridges,

FIGURE 6.21 This exposure of granite in Howe Sound, British Columbia, displays three directions of breakage. When the granite cooled it was a solid mass without any fractures. Subsequently three sets of joints developed, the most prominent being the vertical surfaces that break up this roadside outcrop. The field of view is 5 m wide.

FIGURE 6.22 Cooling igneous rock contracts and in many instances develops shrinkage fractures. In some fine-grained igneous rock such as this basalt, the cooling fractures occur in a system of joints that divide the rock into long, thin columns. Such joints are columnar joints. The elongate columns in the Roza flow in Washington are up to a meter in diameter and up to 6 m long.

sedimentary rock along the margins of the ocean basins and at many places on the land, and metamorphic rock in the cores of high mountains (as in the Himalaya) and adjacent to recently intruded masses of igneous rock (as beneath the Imperial Valley in southern California).

There is, seemingly, a pattern to the ways in which igneous rock is weathered, and the weathered particles transported, deposited, and converted into sedimentary rock, then into metamorphic rock, and finally again melted to form new igneous rock. Rock is the best evidence we have of the continuous motions and cyclic events taking place in and on the Earth. Rock also provides the best evidence that the rock cycle has been in operation for an exceedingly long time. We turn next, therefore, to an examination of how geological time is determined.

SUMMARY

1. Mechanical deformation, recrystallization, and chemical reactions are the processes that affect rock during metamorphism.

2. Mineral assemblages and rock textures change continually as temperature and pressure change. Metamorphic temperatures may eventually become so high that rock begins to melt. Rock that is partly metamorphic and partly igneous is called migmatite.

3. Heat given off by bodies of intrusive igneous rock causes contact metamorphism and creates contact metamorphic aureoles.

4. Regional metamorphism is produced by the Earth's internal heat. Regionally, metamorphosed rocks are produced along subduction and collision edges of plates.

5. Rocks of the same chemical composition (and subjected to identical metamorphic environments) react to form the same mineral assemblages.

SELECTED REFERENCES

Winkler, H. G. F., 1979, Petrogenesis of metamorphic rocks: 5th ed., New York, Springer-Verlag.

Best, M. G., 1982, Igneous and metamorphic petrology: San Francisco, W. H. Freeman and Co.

Ehlers, E. G., and Blatt, H., 1982, Petrology. Igneous, sedimentary and metamorphic: San Francisco, W. H. Freeman and Co.

Dietrich, R. V., and Skinner, B. J., 1979, Rocks and rock minerals: New York, John Wiley & Sons.

Turner, F. J., 1981, Metamorphic petrology: Mineralogical, field and tectonic aspects: 2nd ed., New York, McGraw-Hill.

Hyndman, D. W., 1985, Petrology of igneous and metamorphic rocks: 2nd ed., New York, John Wiley & Sons.

Ernst, W. G., ed., 1975, Metamorphism and plate tectonic regimes: New York, Dowden, Hutchinson, and Ross.

Surface geologic processes, working slowly over millions of years, have exposed rocks in the Grand Canyon of Arizona that provide a record of more than a billion years of Earth history.

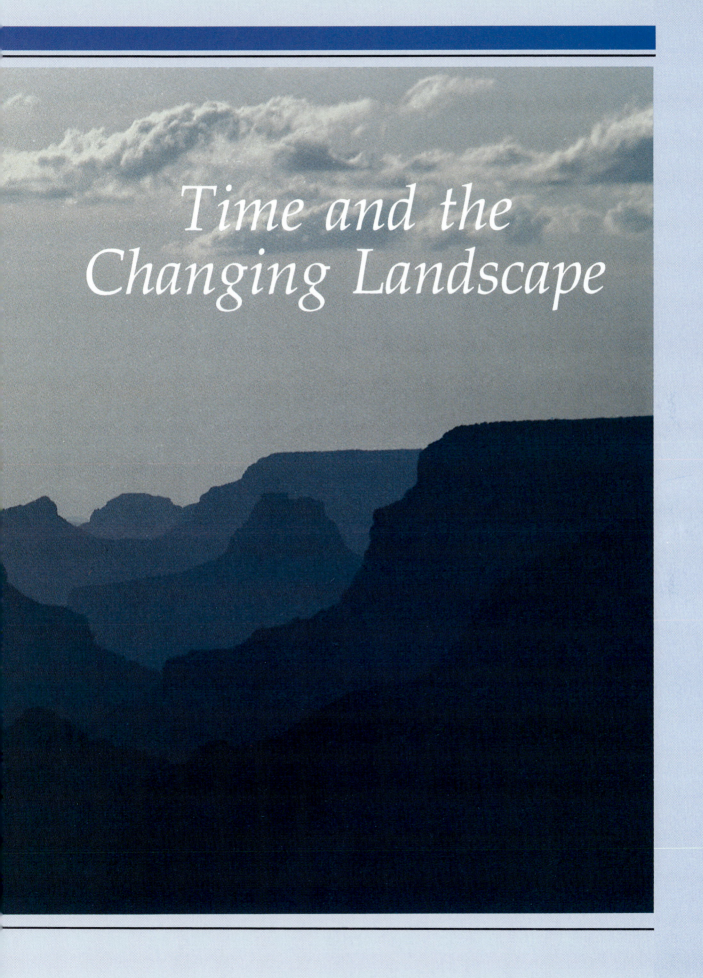

Time and the Changing Landscape

C H A P T E R 7

Stratigraphy
and the
Geologic Column

Ancient sedimentary strata of the McDonnell Ranges in central
Australia have been broadly folded and differentially eroded into an
aligned series of ridges.

STRATA AND STRATIGRAPHY

Like scholars who study the history of ancient civilizations, geologists too are historians, but they work with a much broader time scale, ranging from years to billions of years. And like historians who find the story of human events increasingly difficult to reconstruct as they trace their story farther back in time, geologists must try to piece together the history of the Earth from geologic records of past events that become more incomplete and disturbed with increasing age. Unraveling the details of Earth history is one of the major challenges in geology and one that tests both the skills and the imagination of geologists. At times it can be like assembling a complicated 5000-piece picture puzzle, only to find that more than half the pieces have been lost. How accurately can the entire picture be visualized from the randomly placed pieces that survive?

The historical information that geologists have to work with is largely in the form of stratified rocks—not just sedimentary, but also igneous and metamorphic—that crop out at the Earth's surface or can be penetrated by drilling into the ground. If we examine the rocks that are exposed in the upper walls of the Grand Canyon (Fig. 7.1) where the Colorado River has cut nearly 2 km into the Earth's crust, we can see many nearly horizontal layers. These strata formed one atop the other as sediment accumulated on the floor of a shallow sea. Such rocks contain important clues about past environments at and near the Earth's surface. If their sequence and age can be determined, they provide a basis for reconstructing much of Earth history. *The study of strata,* called **stratigraphy,** involves many different approaches, including: traditional field mapping and description of exposed rocks, study of their contained fossils, dating them by radiometric methods, and investigations of their textural characteristics and their mineralogical and isotopic composition. The regional distribution and relationships of strata can be studied using aerial photographs and images obtained by orbiting spacecraft. Strata beneath the land surface or the seafloor can be investigated by drilling, and by seismic and other geophysical surveys. The stratigrapher, therefore, has many tools at his disposal for attempting to understand and reconstruct the ever-changing environments that have led to the world as we know it.

Stratigraphic and structural studies of crustal rocks have led to an understanding of important lithospheric processes. Without a knowledge of stratigraphic principles and relationships, and of the relative ages of rock sequences, it would be impossible to work out many of the fundamental principles of physical geology. The development of plate tectonics as a viable theory came after decades of careful stratigraphic investigations that demonstrated the disposition and interrelationships of strata throughout the world. Stratigraphy, therefore, has constituted a starting point for learning about the complex details of the Earth's structure and about large-scale crustal processes.

STRATIGRAPHIC PRINCIPLES

The well-ordered and perfectly layered series of strata in the upper walls of the Grand Canyon appear to offer few serious obstacles to understanding their stratigraphic relationships. But in many mountain ranges where similar sedimentary rocks have been intensely deformed, displaced, metamorphosed, and intruded by igneous rocks the stratigraphy may appear almost undecipherable. And well it might be if we did not have some basic stratigraphic principles that enable us to unscramble even the most complex rock assemblages. These basic principles can be used to interpret not only straightforward sequences of strata like those in the Grand Canyon, but also intricately deformed rocks in tectonically disturbed regions.

Original Horizontality

A majority of preserved sedimentary rocks are of marine origin; that is, they were laid down beneath the sea, generally in relatively shallow water. Under such conditions, loose sedimentary particles tend to be deposited rather evenly over the seafloor, with each new sedimentary layer being laid down almost horizontally over older ones. This observation is consistent with *the law of original horizontality* which states that *waterlaid sediments are deposited in strata that are horizontal or nearly horizontal, and parallel or nearly parallel to the Earth's surface.* From this generalization we can infer that rock layers now inclined, or even overturned, must have been disturbed since they were deposited in a horizontal position. Of course we can visualize certain exceptions to this general rule, as in the case of sediments that accumulate in deltas, where initial inclinations of up to several tens of degrees are common. However, even when preserved as part of the rock record, such sediments are identifiable by distinctive sedimen-

FIGURE 7.1 Nearly flat-lying sedimentary formations in the upper walls of the Grand Canyon of the Colorado River resting on tilted older strata. The angular unconformity between the two sequences of strata is a surface of low relief traceable for many tens of kilometers.

tary features, so we realize they do not invalidate the law of original horizontality.

Stratigraphic Superposition

Toward the end of winter it is often possible to see a layer of old snow that is compact and perhaps also dirty, overlain by fresh, looser clean snow deposited during the latest snowstorm. Here are two layers, or strata, that were deposited in sequence, one above the other. The dirty layer underneath was deposited first and, therefore, must be the older of the two. The very simple principle involved here also applies to a whole succession of many snow layers. It applies equally well to layers of sediment and sedimentary rock. Known as *the principle of stratigraphic superposition,* it says that *in any sequence of strata, not later overturned, the order in which they were deposited is from bottom to top.*

This principle implies a scale of relative time, by which the *relative ages* of two strata can be fixed, according to whether one of the layers lies above or below the other. It does not allow us to deter-

mine the age of any stratum in years, for the stratigraphic relationships only tell us the age of one relative to the other.

Although Nicolaus Steno, a Danish scientist of the seventeenth century, appears to have been the first to grasp the significance of the principle of stratigraphic superposition, it was first forcefully presented and widely introduced to science by William Smith, an English civil engineer and land surveyor, shortly before the beginning of the nineteenth century. His profession gave him an ideal opportunity to observe not only the landscape but the rocks that underlie it. While surveying for the construction of new canals in western England, he observed many sedimentary strata and soon realized that they lie, as he put it, "like slices of bread and butter" in a definite, unvarying sequence (Fig. 7.2). He became familiar with the physical characteristics of each layer and, using the principle of stratigraphic superposition, with the sequence of the layers. By looking at a specimen of sedimentary rock collected from anywhere within a wide region, he could name the layer from which it had come and, of course, the position of the layer in the sequence.

In areas subject to tectonic deformation, overturned bedding may be present. In such cases, criteria must be sought to determine whether a succession of beds forms a normal sequence of deposits or, instead, is in reverse order (Fig. 7.3). The evidence to look for may include such stratigraphic features as graded bedding, cross-stratification, and marine fossils lying in growth position (Chapter 5), or certain features produced by deformation, such as slaty cleavage (Chapter 6).

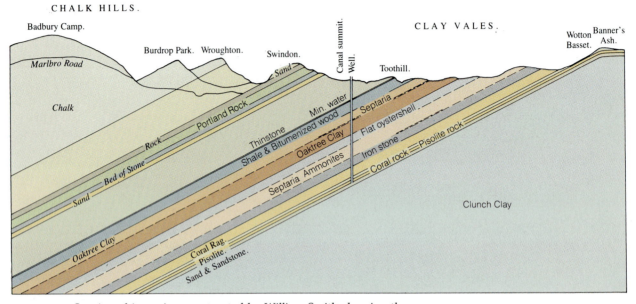

FIGURE 7.2 Stratigraphic section constructed by William Smith showing the succession of strata in north Wiltshire, England.

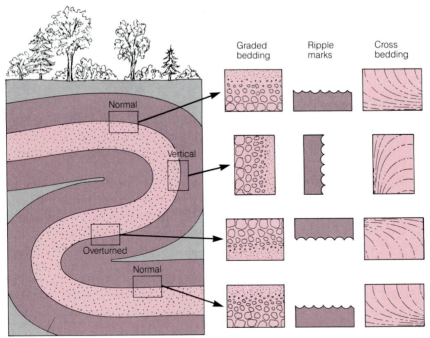

FIGURE 7.3 Sketch of an exposure showing in cross section a number of folded strata and some sedimentary features that are useful in determining whether the strata are normal (right-side-up), vertical, or overturned.

TABLE 7.1 *Percentages of Rock Types Exposed on the Continents*

| Continent | Crystalline Rocks | | | | Sedimentary Rocks |
	Extrusive	Intrusive	Metamorphic	Total	
Asia	9	12	5	26	74
Africa	4	16	22	42	58
North America	11	6	31	48	52
South America	11	2	25	38	62
Europe	3	7	3	13	87
Australia	8	11	11	30	70
Earth's surface	8	9	17	34	66

Source: Data from Blatt and Jones (1975).

Commonly, more than one criterion is required to be certain whether or not strata have been overturned.

Other Layered Rocks

Sedimentary rocks are not the only layered rocks exposed at the Earth's surface. Sequences of extrusive igneous rocks, such as lavas or tuffs, are also progressively younger from bottom to top. Many metamorphic rocks derived from layered sedimentary and igneous parent rocks also display obvious layering, and their stratigraphy can be studied and described in much the same way. For most such rocks, the principle of stratigraphic superposition applies equally well, and their relative ages can be inferred from their sequence in a stratigraphic succession. However, certain kinds of intrusive igneous rocks and metamorphic rocks do not obey the rule. For example, although a sill intruded into sedimentary strata is younger than the rocks beneath it, it is also younger than rocks immediately overlying it.

DISTRIBUTION AND AGE OF SEDIMENTARY STRATA

Sediments and sedimentary rocks mantle most of the Earth's surface. Except in areas of active volcanism, sediments cover nearly all of the seafloor (Fig. 5.28). Perhaps 95 percent of the volume of the continental crust consists of igneous and metamorphic rocks, but they are exposed over only a third of the land surface (Table 7.1). About a quarter of these are extrusive rocks. Sedimentary rocks lie at the surface over the remainder of the continents where they form a thin cover, averaging no more than about 2 km thick, resting upon the older crystalline rocks beneath (Fig. 7.4). Their thickness is highly variable, ranging from only a single thin layer up to vast sedimentary piles many kilometers thick. Many of these sequences consist of sediments that accumulated in shallow seas along continental margins. Their great thickness, together with evidence for rapid accumulation, implies that the seafloor must have been subsiding while deposition occurred.

Recent sediments can be seen nearly everywhere at the Earth's surface. Often they are thin and patchy but in some places, as at the mouths of large rivers, they may be very thick and accumulating at a rate of a meter or more per year. Along some continental margins rapid uplift and vigorous erosion of resulting highlands have resulted in deposition of thick piles of sediment. In southern Alaska, for example, coastal mountains rising to altitudes of 2000 m are composed of deformed and uplifted marine sedimentary rocks that have an aggregate thickness of about 10,000 m, yet they were deposited during the geologically brief span of only 2 to 3 million years.

The area covered by major groups of rocks is related in a general way to their age. The oldest strata yet discovered, which date back more than 3 billion years, are preserved in very few places, whereas rocks of lesser age are exposed over ever-greater areas the younger they are. This does not imply that recent rates of sedimentation have been higher than in ancient times, but merely that erosion, deep burial, metamorphism, and other processes have steadily decreased both the area and volume of strata the longer they have been subjected to such activities.

FOSSILS AND THEIR TIME SIGNIFICANCE

A *fossil* is *the naturally preserved remains or traces of an animal or a plant*. Preserved in accumulating sediment and later converted to rock, fossils form an

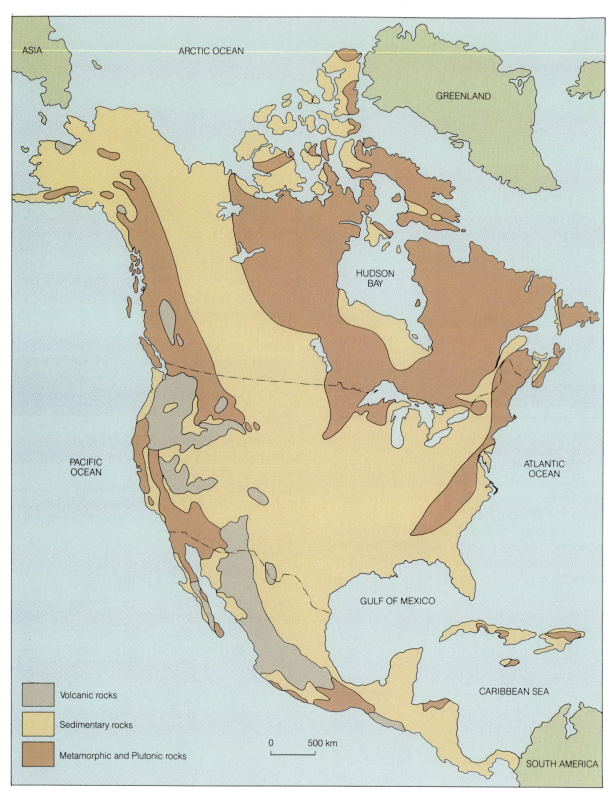

FIGURE 7.4 Map of North America showing generalized distribution of sedimentary rocks relative to igneous and metamorphic rocks. A large percentage of the metamorphic rocks consists of ancient sedimentary rocks that have been metamorphosed.

FIGURE 7.5 A portion of William Smith's geologic map (*b*) of the region about London, England, published in 1815. The map shows the distribution of London Clay, Brickearth, and Chalk—three widespread formations. Distinctive fossils in the London Clay (*a*) and Chalk (*c*) helped him to identify the two formations in the field.

ever-growing record of the kinds of plants and animals that lived when the sediments enclosing them accumulated.

As William Smith was making the study of the stratified rocks in his native England, that was to lead to the formulation of the principle of stratigraphic superposition, he also observed the fossils they contained. He soon realized that within each layer there were distinctive, characteristic fossils which enabled him to identify it, regardless of the lithology of the unit (Fig. 7.5). In other words, he recognized that each assemblage of fossils was peculiar to the stratum in which it occurred and thus constituted, in effect, an identification tag for that layer. In so doing, Smith discovered what we now call the **law of faunal succession,** which says that *fossil faunas and floras succeed one another in a definite, recognizable order.*

Today we know that this relationship between a stratum and its fossils is the result of the evolution of life through geologic time. As new species arise, they are carried from one part of the world to another through migration or shifts in the ranges of populations. The rates at which plants and animals spread or shift their living areas seem generally to have been comparable with the rates at which organisms evolve. As a consequence, the major evolutionary changes spread through large parts of the world within the generous time intervals represented by the major groups of strata. All this, however, was unknown in Smith's day and did not become clear until after 1859, when Charles Darwin published his now-famous ideas on evolution.

SEDIMENTARY FACIES

If one examines a sequence of exposed sedimentary rocks, most likely changes will be seen in their character as we move up from one layer to the next above. These differences reflect changes in depositional environments at a particular place through time. However, if any single unit is traced away from the initial outcrop, it might also be observed to change laterally. Most sedimentary strata, in fact, change character from one area to another as a result of changes in the type or intensity of transporting agencies or in the conditions under which sediments accumulate. A diversity of environments is readily seen if a traverse is made across the edge of a continent and into the adjacent ocean basin (Fig. 7.6). Within each natural zone that is crossed, distinctive sediments and associated organisms are found which serve to identify that depositional environment. *A distinctive group of characteristics within a sedimentary unit that differs, as a group, from those elsewhere in the same unit* is a **sedimentary facies.** Each facies may be represented, for example, by distinctive grain size, grain shape, stratification, color, depositional structures, or fossils. Adjacent facies merge into one another either gradually or abruptly, depending on the relationships between the two former depositional environments (Fig. 7.7).

If a sedimentary unit were exposed in section across its entire extent, it could be identified as a single unit despite changes in facies. But typically it is exposed only discontinuously. If each exposed part represents a different facies, it might be necessary to use the contained fossils to show that it is the same unit throughout. But a difficulty arises here because the assemblages of fossils in two different facies may not be exactly the same, even though the organisms they represent lived at the same time. This happens because the environ-ments in which the organisms lived were different. In the same sea, deep-water shellfish are unlike shallow-water kinds. On land, animals living in deserts are unlike those living simultaneously in moist, forested regions. However, such variations do not make it impossible to show that different

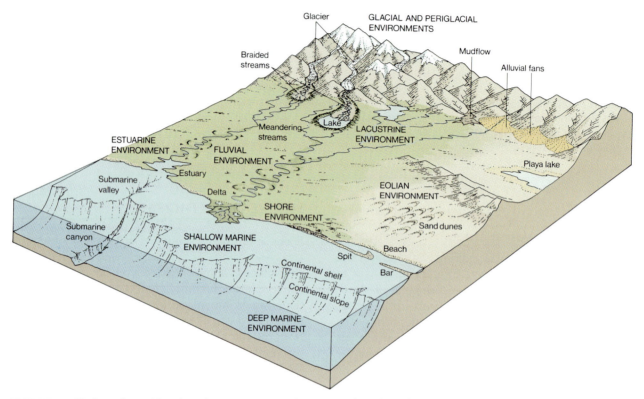

FIGURE 7.6 Various depositional environments occurring across the edge of a continent and the adjacent margin of an ocean basin.

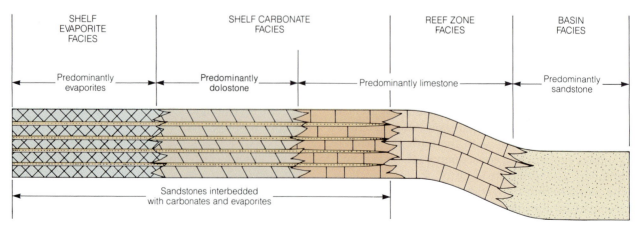

FIGURE 7.7 Geologic section across a shallow marine shelf and reef into an adjacent deep basin showing the gradational relationship among sedimentary facies. (*Source:* After Motts, 1968.)

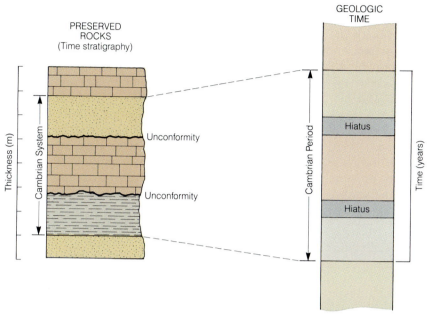

FIGURE 7.8 Relationship between a preserved section of rocks and the corresponding geologic time interval. Unconformities mark hiatuses, times for which no depositional record remains.

rock sections are equivalent; they only make it more difficult.

BREAKS IN THE STRATIGRAPHIC RECORD

The dynamic nature of the Earth's crust dictates that processes of sedimentation and erosion change both in their intensity and location through time. Rates of change can also differ greatly from place to place. For example, in the deep sea, slow and continuous sedimentation may prevail for millions or tens of millions of years with little apparent change. On the continents, by contrast, sedimentation is disrupted periodically by environmental changes that lead to intervals of erosion or nondeposition. Such changes affect not only the emergent lands but submerged continental margins as well where most of the detrital waste from the continents accumulates. Because erosion destroys some of the stratigraphic record, that part preserved is often incomplete and marked by discontinuities where intervals of geologic time, some brief and others very long, are not represented by deposits.

Unconformities and Hiatuses

An **unconformity** is *a substantial break or gap in a stratigraphic sequence that marks the absence of part of the rock record.* It records a change in environmental conditions that either caused deposition to cease for a considerable time, or erosion that resulted in loss of part of an earlier-formed depositional record, or a combination of both. While an unconformity is a physical feature we can identify in a rock sequence, we normally refer to *the lapse in time recorded by an unconformity* as a **hiatus** (Fig. 7.8).

One can easily visualize a number of ways in which unconformities might form. They include local or regional uplift of land masses, fluctuations of sea level, and changes in climate that affect the behavior of streams, glaciers, and other depositional systems. In the deep ocean, strong bottom currents can erode sediments exposed at the seafloor, and submarine landslides can disrupt and displace large bodies of sediment. The possible variations in crustal movement, erosion, and sedimentation are numerous, so there are a number of different kinds of unconformities found in crustal rocks (Table 7.2).

Among the most obvious is the **angular unconformity,** which is *an unconformity marked by angular discordance between older and younger rocks* (Fig. 2.21). It normally implies that older strata were deformed and then truncated by erosion before the younger layers were deposited across them. The degree of discordance may change laterally, from relatively low angularity to very high angularity, depending on the degree to which the underlying rocks are deformed. A well-known example is seen in the eastern part of the Grand Canyon of the Colorado River where tilted beds of ancient sedimentary rocks are overlain unconformably by marine sandstone that records the gradual submergence of the land by the sea (Figs. 7.1 and 7.9). Such *a spread of the sea over the land* is called a **transgression;** the opposite effect involving *a retreat of the sea from the land* is termed a **regression.**

A study of unconformities brings out the close relationship between crustal movements, erosion, and sedimentation. All of the Earth's land surface is a potential surface of unconformity. Some of today's surface will be destroyed by erosion, but some will be covered by sediment and preserved as a record of the present landscape. Vigorous erosion is now taking place where there has been recent uplift of the land. Erosion by streams is laying bare the records of Earth history in old rocks, and in doing so it is destroying some of those records. Meanwhile, the eroded material is being carried away and deposited elsewhere. Thus, in a sense, accumulation in one place compensates for destruction in another. The many surfaces of unconformity exposed in rocks of the Earth's crust are

TABLE 7.2 *Types of Unconformities*

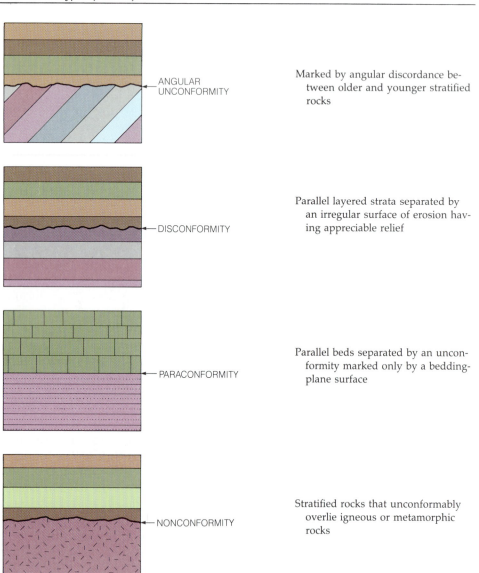

ANGULAR UNCONFORMITY — Marked by angular discordance between older and younger stratified rocks

DISCONFORMITY — Parallel layered strata separated by an irregular surface of erosion having appreciable relief

PARACONFORMITY — Parallel beds separated by an unconformity marked only by a bedding-plane surface

NONCONFORMITY — Stratified rocks that unconformably overlie igneous or metamorphic rocks

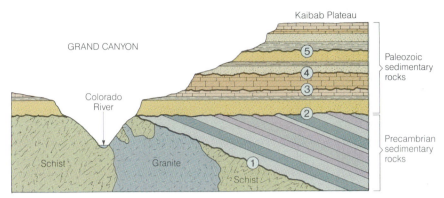

FIGURE 7.9 Geologic section through rocks exposed in the Grand Canyon. The lowest unconformity (1), separating tilted sedimentary strata from older crystalline rocks, is a nonconformity. An angular unconformity (2) separates the tilted strata from horizontally layered strata above, while three disconformities (3, 4, and 5) are seen still higher in the section.

records of former land surfaces and testify that the interactions between the internal and external processes have been going on throughout the Earth's long history.

Diastems

Unconformities record major disturbances of a depositional system and often represent lengthy hiatuses. But what of shorter breaks? We can easily see the marked changes in sedimentation that take place when a large storm causes a gently flowing stream to become a raging, eroding torrent, or the lapping waves along a beach to be transformed into crashing surf. *A short break in sedimentation resulting from normal variations about an average condition, but without a major change in the regular sedimentary pattern,* is called a **diastem.** The distinction is similar to the difference between weather (the actual conditions of the atmosphere at a particular time and place) and climate (the average weather conditions at a place over a period of years). While any particular diastem may represent only a very brief time interval, a sedimentary section may include a vast number of them, and the total missing time they represent can therefore be very great (Fig. 7.10).

STRATIGRAPHIC CLASSIFICATION

We can think of every rock stratum present at or beneath the Earth's surface as being like a chapter in a book. Each can tell us something about the

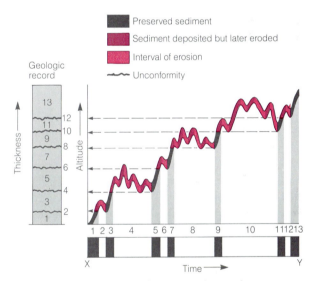

FIGURE 7.10 Incomplete depositional record in a sedimentary column reflects alternating intervals of deposition and erosion due to fluctuations of a generally rising sea level. Deposition occurs when the sea level is rising. When the sea level falls, erosion removes sediment previously deposited. In the example shown, the preserved record is punctuated by major hiatuses (4, 8, 10) represented by unconformities, as well as by shorter breaks (2, 6, 12). Only 30 percent of the total time (x–y) is represented by preserved sediment. Unconformities represent the remaining 70 percent. (*Source:* After Barrell, 1917.)

physical and biological character of a part of the Earth at some time in the geologic past. If we were to start counting the number of chapters and books represented by the rock record, we would quickly realize that available to us is a vast library of knowledge consisting of thousands upon thousands of volumes.

Just as a large research library is of little use if its collection lacks systematic organization, the rock record, if it is to prove comprehensible, must be organized in some rational manner. Because strata constitute the fundamental units in our stratigraphic library, geologists have devised a logical system to classify them as material (rock) units. At the same time, because strata also provide us with important information about geologic time, they can be classified so as to represent discrete intervals of geologic time. Such classification schemes permit us to investigate the interrelationships of different rock layers, not just locally, but between widely separated regions, and to develop an understanding of the evolution of our planet through geologic time.

Rock Stratigraphy

It is easy to identify the sedimentary rock directly above the major unconformity in the western part of the Grand Canyon as a sandstone (Fig. 7.1), but a thorough study must distinguish it from other sandstones. One respect in which any stratum differs from all others is its position in the vertical sequence of strata. Hence, we give it a designation by which its position is fixed and by which it can be catalogued and referred to. Such a unit is called a **rock-stratigraphic unit,** defined as *a body of rock having a high degree of overall lithologic homogeneity.* The basic rock-stratigraphic unit is the **formation,** which is *a stratum or collection of strata distinctive enough on the basis of physical properties to constitute a basic unit for geologic mapping.* A formation, therefore, must be thick and extensive enough to be shown to scale on a geologic map. It also must be distinguishable from the strata immediately above and below, not just at one exposure but generally wherever the unit is exposed. Within these requirements a formation can be thin or thick, to suit the geologist's convenience. Its thickness is likely to depend on the degree of detail of the field study and on the scale of the map to be made. Each formation is given a name. In North America it typically is the name of a geographic locality near which the unit is best exposed (for example, Lexington Sandstone, Fox Hills Sandstone, Green

TABLE 7.3 *Hierarchy of Rock-Stratigraphic Units*

Group
Formation
Member
Bed

River Formation). Not only sedimentary rocks but also igneous and metamorphic rocks are designated as formations (for example, San Juan Tuff, Conway Schist). Several formations appear on the geologic map prepared by William Smith (Fig. 7.5).

Formations can be subdivided into successively smaller rock-stratigraphic units (Table 7.3). *Members,* which are subdivisions of formations, can be further subdivided into *beds.* Several formations can be assembled into larger units called *groups.* Such a classification scheme is similar to that used by a librarian who classifies books into a succession of categories that are progressively more specific.

Rock-stratigraphic units in the stratigraphic classification are the basic units of general geologic field work (Appendix D) and serve as the foundation for describing strata, local and regional rock structure, economic resources, and geologic history. All such units are defined only on the basis of their observable physical characteristics. Their inferred depositional environment or geologic history play no part in their definition. Although most formations are rocks, the degree of induration is not a necessary criterion for definition. Sediments such as sand, till, and gravel, as well as other unconsolidated deposits, can also be designated as formations.

Time Stratigraphy and Geologic-Time Units

It is difficult for many of us to comprehend the immensity of geologic time, measured in thousands, millions, and billions of years, and yet we must deal with it as we seek to unravel the story of the Earth from its preserved strata. An interval of geologic time can have meaning for us only in the context of rocks that were deposited during that interval. Because geologic time is an abstract concept (whereas rocks are material objects that we can handle and study), we classify parcels of time separately from parcels of rock representing those times.

A **time-stratigraphic unit** is defined as *a unit representing all the rocks, and only those rocks, that formed during a specific interval of geologic time.* Each of its boundaries, upper and lower, is everywhere the same age. Whereas a formation is defined only on the basis of its material characteristics, and its boundaries lie where a recognizable change in physical properties occurs, a time-stratigraphic unit may include more than one rock type and its boundaries may not coincide with a formational boundary.

Time-stratigraphic units are traditionally based on the fossil assemblages they contain and are ranked so as to represent progressively shorter time intervals. Rocks comprising a *system*, the primary unit, represent a time interval sufficiently great that such units can be used all over the world. Most systems encompass time intervals of tens of millions of years (Table 7.4). Names for systems arose from the early studies of rock strata, mainly in Europe, and often derive from geographic localities (see below). Larger groupings of two or more systems are called *erathems.* Rocks representing successively smaller intervals of time within a system are referred to as *series* and *stages* (Table 7.4); while such units have the potential for worldwide application, more commonly they are used within and between geographic provinces and continents.

Intervals of geologic time are based on the rock record as expressed by time-stratigraphic units. Units of geologic time are not stratigraphic units, however, for they are nonmaterial. Each merely corresponds to the time represented by a particular time-stratigraphic unit. Like such units, they are defined to include various intervals of time. For example, a geologic *period* embraces the time during which a geologic system accumulated, while an *epoch* equates with a series, and an *age* is equivalent to a stage (Table 7.4).

Because the geologic record is incomplete and punctuated by numerous unconformities, time-stratigraphic units in a certain area may not contain a complete depositional record of the corresponding geologic time interval (Fig. 7.8). However, by piecing together many sequences of strata from different geographic areas, many of the gaps can be bridged, thereby providing us with a more complete picture of Earth history.

Type Sections

If we wish to find the meaning of a word, we consult a dictionary in which the word is carefully defined. Dictionary definitions are the primary reference standard to which people can turn to ensure that they are using words in the proper sense. In much the same way, we have reference standards for stratigraphic units. Commonly they are particular natural outcrops, called **type sections** (or **stratotypes**), chosen because they are *sections that display the primary characteristics of a stratigraphic unit in a typical manner.* Type sections should provide us, for example, with information about the thickness, composition, fossil content, and boundaries of a unit, as well as that unit's relationship to other adjacent stratigraphic units. A type section cannot be expected to display all characteristics of the unit, however, especially the degree of regional variability that it displays. The geographic area that includes the type section is called the *type area* of the unit.

Diachronous Boundaries

Depositional environments tend to change character laterally, as well as through time. A particular sedimentary facies, on the basis of which a formation is defined, may accumulate over different time intervals in different places. This means that the age of its upper and lower boundaries may differ from one place to another. In other words, the unit is said to possess **diachronous boundaries,** defined as *boundaries that vary in age in different areas.*

Figure 7.11 shows an example in which a delta is building progressively seaward. If we look at the various depositional environments, we see that marsh sediments are accumulating above a coarse, sandy facies that passes down the front slope of the delta into silty, clayey beds. These in turn grade laterally at the base of the delta front into thin, clayey facies offshore. The successive *time lines* (lines of constant age), which match the present surface, slope downward through the delta and become progressively older away from the depositional front. Facies boundaries, on the other hand, cross time lines, reflecting the progressive dis-

TABLE 7.4 *Primary Time-Stratigraphic Units, Equivalent Geologic-Time Units, and Their Average Length During the Last Half Billion Years of Earth History*

Time-Stratigraphic Units	Geologic-Time Units	Average Length (million yr)
Erathem	Era	190
System	Period	52
Series	Epoch	18
Stage	Age	7

placement of the various depositional environments in the seaward direction. If the delta sediments are preserved in the geologic record and converted to sedimentary rock, the boundaries of the formations that might then be recognized on the basis of the several sedimentary facies would not be the same age everywhere. Instead, they would be progressively younger in the direction that the delta was built.

Rock units similar to this example can be seen in a famous ancient reef complex in New Mexico (Fig. 7.12). The reef grew progressively seaward over a long period. Successive time lines pass laterally through five distinct facies representing sedimentation on the surface of a shelf, a reef front, and a marine basin. Each formation, defined by rock type, is progressively younger in the direction of the basin.

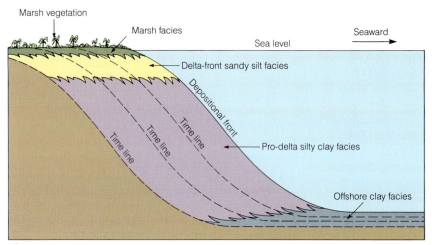

FIGURE 7.11 Diagrammatic section through a growing delta showing time lines, isochronous surfaces marking successive positions of the delta surface, that pass through different sedimentary facies. If eventually converted to rock, the resulting formations will be younger in the direction of delta growth.

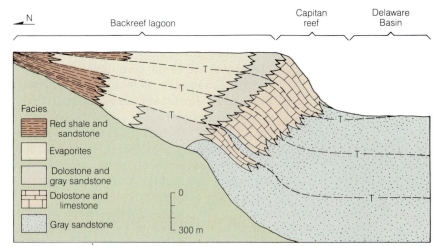

FIGURE 7.12 Cross section through marine strata in the Delaware Basin of west Texas demonstrating the relationship between different facies and time lines (T). In this example, the carbonate reef facies and the overlying dolostone and sandstone facies become younger toward the basin. (*Source:* After Wilson, 1975.)

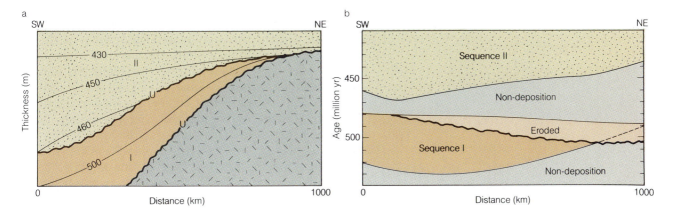

FIGURE 7.13 Unconformity-bounded rock sequences. (*Source:* After Sloss, 1984.) (*a*) Stratigraphic diagram showing a sedimentary sequence (I) separated from eroded igneous rocks by an unconformity (U) and overlain unconformably by a second sequence of strata (II). Time lines passing through the rocks are in millions of years. (*b*) Time-space diagram of the same region showing intervals of deposition and of nondeposition and erosion. Deposition was nearly continuous near the southwestern edge of the area, whereas to the northeast nondeposition and erosion prevailed over much of the time interval.

Sequences and Seismic Stratigraphy

While rocks can be classified on the basis of both physical character and time, they also can be grouped into large sequences of strata that are separated by major unconformities. Such an ***unconformity-bounded sequence*** is *a grouping of strata that is bounded at its base and top by unconformities of regional or interregional extent.* Regional unconformities are mappable across large sedimentary basins but are restricted to areas of subcontinental size. Often they are related to episodes of tectonic uplift. Interregional unconformities may be traceable across major portions of continents and into sedimentary strata of the continental shelves. Some may be of global extent. These natural sequence-bounding surfaces, therefore, reflect events in Earth history that have regional or worldwide significance.

Figure 7.13*a* is a stratigraphic cross section of a sequence of strata bounded by regional unconformities. Time-stratigraphic boundaries are seen to be truncated by the unconformities. The adjacent diagram (Fig. 7.13*b*) shows the same sequences and unconformities in relation to time and space. It is evident that the unconformities represent both an interval of nondeposition, of various length at different places, and of erosion. The unconformity at the base of the lower sequence of strata (sequence I) is diachronous, for the overlying strata were deposited progressively farther northward as the sea rose against the land. The unconformity separating this sequence from a younger one (sequence II) represents a time during which part of sequence I was eroded, as well as a subsequent interval of nondeposition when the land presumably lay above sea level. The base of sequence II is also diachronous. The lowest rocks become successively younger northward, reflecting renewed transgression of the sea across the land.

Major unconformities representing marine transgressions and regressions have been identified both in sedimentary rocks on the continents and under the continental shelves and slopes by means of *seismic stratigraphy*. This involves the use of high-resolution seismic exploration techniques to obtain a cross-sectional view of crustal rocks and sediments like that shown in Figure 7.14. In such a profile, the lines depict the arrangement of subsurface rock units that act as reflectors of seismic waves generated by exploration teams. Seismic reflection profiles, with their patterns of smooth and broken lines, make it possible to interpret the regional interrelations of strata, their structure, their thickness and probable depositional environment, and the presence, relief, and topography of unconformities. By tracing major subsurface unconformities in this manner, unconformity-

FIGURE 7.14 Seismic reflection profile, located in southern Texas near the Gulf of Mexico. The numbered shot points at top are located at the Earth's surface. To the right the reflections dip toward the Gulf, where subsidence has been taking place since Mesozoic times. The colors represent the velocity of sound in the section, facilitating recognition of lithology. The yellowish feature at mid-depth, about 6000 m, is believed to be a Jurassic reef.

bounded sequences have been identified in sedimentary strata of the continental margins that are thought to be related to marine transgressions and regressions, both of local and global character.

CORRELATION OF ROCK UNITS

William Smith's discovery that strata containing similar assemblages of fossils are broadly similar in age, no matter where they occur, was not considered by him to reflect any particular scientific principle; it was purely practical. Nevertheless, it opened the door to the correlation of sedimentary strata over increasingly wide areas. By *correlation* we mean *determination of equivalence, in geologic age and position, of the succession of strata found in two or more different areas.* Smith correlated strata, on the dual basis of physical similarity and fossil content, initially over distances of several kilometers and later over tens of kilometers. By means of fossils alone it ultimately became possible to correlate through hundreds and then thousands of kilometers.

Correlation involves two main tasks. One is to determine the ages, relative to one another, of units exposed in local sections within an area being studied. Then the ages of the units relative to a standard scale of geologic time must be found. To accomplish these goals, a geologist employs various physical and biological criteria, no one of which is necessarily more dependable or precise than the others.

Physical Criteria

Physical Continuity

Where sedimentary rocks are exposed nearly continuously over a broad area, individual beds often can be traced for considerable distances. Such correlation based on continuity of strata is generally straightforward and reliable. But eventually beds thin and die out, or merge with adjacent beds, thereby complicating and severely limiting this method of correlation. If, instead of tracing beds, we use formations, which by definition are larger units and mappable over broad areas, regional correlations can generally be made more reliably. In doing so we must make a basic assumption, namely that a formation is essentially the same age throughout. However, as we have seen, formation boundaries may be diachronous. Correlation based on physical tracing of rock units must therefore be done cautiously, keeping in mind the potential pitfalls involved.

Lithologic Similarity

Continuous or nearly continuous exposures are not always present, so commonly one is faced with correlating between widely spaced outcrops. During the time since a sediment was converted to sedimentary rock, a stratum may be deformed and eroded so that only parts of it remain. The physical matching of remnants of a formation over a broad region generally involves the use of rock characteristics that permit the unit to be distinguished from others with which it might be confused. Obvious criteria include gross lithology, mineral content,

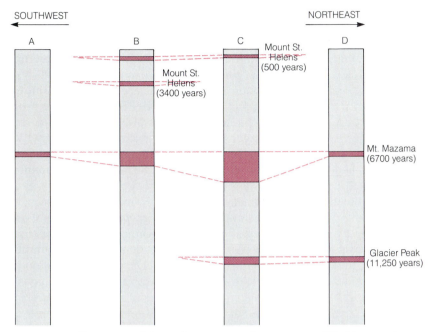

SOUTHWEST →

NORTHEAST →

A B C D

Mount St. Helens (500 years)

Mount St. Helens (3400 years)

Mt. Mazama (6700 years)

Glacier Peak (11,250 years)

FIGURE 7.15 Quaternary lake-sediment cores from western Washington demonstrate the use of distinctive ash layers from Mount St. Helens, Mount Mazama, and Glacier Peak volcanoes for regional correlation. Because the age of organic matter associated with the layers has been obtained by radiocarbon dating, the layers can be used to assess the ages of sediments associated with them.

grain size, grain shape and orientation, sedimentary structures, color, and response of the rock or sediment to weathering. Correlation on this basis is likely to be reliable over short distances, but it generally becomes less reliable through longer ones because the physical characteristics tend to change laterally (a change of facies).

Key beds can be useful in correlating major rock units. A ***key bed*** is *a thin and generally widespread bed with characteristics so distinctive that it can be easily recognized.* A correlation may be greatly strengthened if several key beds can be identified and traced from one outcrop to the next. In areas of volcanic activity, ash layers can serve as distinctive key beds for purposes of regional correlation (Fig. 7.15).

Topographic expression of a rock unit may allow it to be differentiated from adjacent beds and to be traced across the landscape (Fig. 7.16). Resistant sandstone beds, for example, may rise above adjacent lowlands underlain by more erodible shale, or a limestone unit in a moist temperate or tropical environment may display distinctive landforms caused by dissolution of the rock.

Vegetation can also be an important aid in correlation. The chemistry of one rock unit may favor

FIGURE 7.16 Folded sedimentary rocks in northeastern Bangladesh respond differently to erosion and produce distinctive patterns of small ridges and valleys that can be traced for tens of kilometers across the landscape. Such topographic features aid in mapping and correlating rock units in densely vegetated terrain and in areas with few outcrops. Distance across view is 130 km.

the growth of deciduous trees, while an adjacent unit may support mainly herbs or grasses. In using aerial photographs or satellite images for mapping surface rocks, geologists typically rely on subtle differences in tone that are related to specific vegetation types growing on different formations.

Stratigraphic Position, Intertonguing, and Sequence

If several stratigraphic sections can be linked through firm correlation of one layer, then it is reasonable to infer that the strata directly above and below that layer may likewise be correlative. One can easily visualize exceptions to this generalization, such as the presence of an unrecognized break between the layers in one section that is not present in another. If sections can be linked by correlating two different layers, then one can also suppose that the intervening strata are broadly correlative, again assuming that no sig-

nificant break in deposition has taken place. In some cases it is possible to show that different kinds of rock lying between two key beds represent different facies. If such units intertongue, then their time-equivalence is demonstrated.

In situations where strata are so similar that they cannot be distinguished by differences in composition or character, it may still be possible to correlate between sections by noting a distinctive grouping or succession of layers. An excellent example is provided by varved lake sediments. While any given layer may appear nearly identical to those above or below, minor differences in thickness and, therefore, in spacing of several adjacent layers may provide a unique signature that can be recognized in different sections, sometimes over broad regions (Fig. 7.17). Cross-correlation of varved lake deposits has provided Scandinavian geologists with a detailed chronology of ice retreat at the end of the last glaciation. The varves provide a record of year-by-year accumulation of fine-

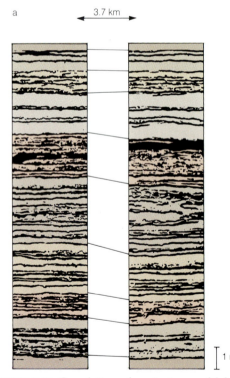

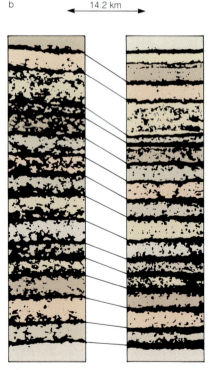

FIGURE 7.17 Correlation of ancient varved lake sediment through the recognition of distinctive patterns of successive layers that can be identified from outcrop to outcrop. Inferred correlations are indicated by lines between columns. (*Source:* After Anderson and Kirkland, 1966.) (*a*) Varved layers in sections of Toldito Limestone, New Mexico from localities 3.7 km aprt. (*b*) Varved layers in the Castile Formation from localities in New Mexico and Texas that are 14.2 km apart.

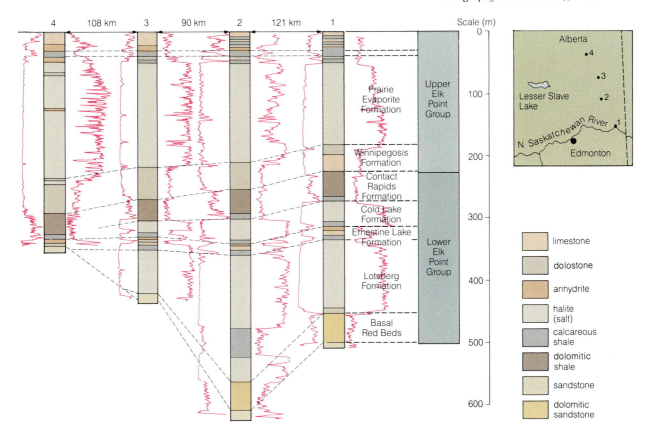

FIGURE 7.18 Correlation of formations in eastern Alberta, Canada through use of electrical properties of sedimentary strata in boreholes. Curves to the left of holes record self-potential, whereas those to the right record electrical resistivity. (*Source:* After Smith, 1981.)

grained sediments over an aggregate span of more than 3000 years.

Well Logs

Most sedimentary rocks of the continents are not exposed to view, but lie beneath the surface where information about them can be obtained directly only by drilling or indirectly by geophysical methods. Energy companies attempting to assess potential petroleum resources beneath the land surface employ various techniques to determine the nature and distribution of subsurface rock units. Numerous test wells drilled in sedimentary basins give access to buried strata. Instruments lowered into the open wells provide continuous measurements of electrical properties of the sub-surface units. The resulting signals can be compared from well to well and a correlation of strata made based on the similarity of the records (Fig. 7.18).

Radiometric Ages

Radiometric dating methods have been developed that make it possible to obtain the actual, rather than the relative, age of a rock unit. Such dates can provide the best, and sometimes the only, means of obtaining ages for strata that lack fossils. Rocks having the same or closely similar radiometric ages are considered to be correlative. Most of the methods involve a careful laboratory measurement of the amount of decay of a radioactive isotope incorporated in a rock at the time of its formation (Chapter 8). No single method is universally applicable; in fact, direct dating of very old sediments or sedimentary rocks is often not possible by such methods. In many cases ages are obtained for interstratified or cross-cutting igneous rocks. These can then provide a minimum or maximum age for the associated sedimentary unit, or can be used to bracket its age. In other cases, volcanic ash layers, which may also constitute key beds, can be dated. Because they are deposited very rapidly,

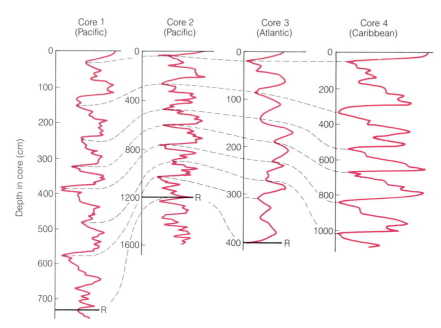

Core 1 (Pacific) Core 2 (Pacific) Core 3 (Atlantic) Core 4 (Caribbean)

FIGURE 7.19 Curves showing variations in the ratio of oxygen isotopes in marine cores from the Pacific and Atlantic oceans, and the Caribbean Sea. The close similarity in the downcore patterns makes it possible to correlate between cores. Dashed lines are spaced at 100,000-year intervals. Because the sedimentation rate of core 1 is 1 m/100,000 years, the depth scale for this core can also be read as a time scale in thousands of years. A major reversal in the magnetic signal about 730,000 years ago (labeled R), identified in three of the cores, provides a further basis for a correlation and also serves as a time-stratigraphic marker (*Source:* After Peltier and Hyde, 1984.)

the ash layers can be viewed as time lines within larger assemblages of strata.

All radiometric dates include a standard error, which is a measure of uncertainty in the precision of the age obtained. For some methods this may be about 5 percent of the measured age, whereas in others the error may be 20 percent or even more. Such uncertainties reduce the effectiveness of radiometric dates as a correlation tool, especially in the case of very old rocks. For example, a rock for which an age of 200,000,000 ± 20,000,000 years has been obtained has a dating uncertainty as great as the average length of a geologic epoch (Table 7.4).

Isotopic Variations

Isotopic studies of deep-sea sediments have disclosed that consistent variations in the ratio of two stable isotopes of oxygen, ^{18}O and ^{16}O, occur downward through the sediment column wherever a core can be obtained (Chapter 20). The variations reflect changes in the composition of the ocean waters through time and have proved invaluable in the detailed correlation of sediments less than about a million years old throughout the world oceans (Fig. 7.19).

Magnetic Properties

The magnetic properties of strata provide yet another means of correlation, especially in the case of

lavas and certain fine-grained, iron-rich sediments and sedimentary rocks. If a rock has not been significantly reheated since it formed, its magnetism records the magnetic field of the Earth at the time the unit accumulated (Chapter 8). Changes in the magnetic field lead to differences in the magnetic signal that can be measured in rocks of different age. Because the changes occur globally, and affect nearly all rocks forming at any given time, a knowledge of the changes in Earth magnetism through time provides a ready means of correlating rocks and sediments on a worldwide basis (Figs. 7.19 and 8.20).

Biological Criteria

Index Fossils and Fossil Assemblages

William Smith was the first to recognize the usefulness of fossils in identifying different strata. We now call *a fossil that can be used to identify and date the strata in which it is found, and is useful for local correlation of rock units, an* **index fossil.** To be most useful, an index fossil should have common occurrence, a wide geographic distribution, and a very restricted range in age. The best examples are swimming or floating organisms that evolved rapidly and quickly became widely distributed. Because a distinctive index fossil is instantly recognizable at an outcrop, it can provide a rapid and reliable means of correlation (Fig. 7.20). While some genera and individual species permit long-range correlation of rocks in different sedimentary basins or even on different continents, more often close dating and correlation involves using assemblages of fossils of as many different types as possible.

Stage of Evolution

Some fossil forms have undergone very rapid and complex changes in structure or ornamentation as they evolved. If their evolutionary history is well known, and the stages of evolution are reasonably well dated, the approximate age of a formation can be obtained by recognizing the stage of evolution of its contained fossils.

Climatic Inferences

Fossil plants and animals can often give us important clues about the climate when the sediments containing them were deposited. For example, plant remains or fossil pollen grains in one stratum may provide clear evidence of a former warm climate, whereas remains in a different layer may point to cold conditions. Such climatic indicators have been used to correlate otherwise undated sediments within broad geographic regions.

GEOLOGIC COLUMN AND TIME SCALE

Detailed stratigraphic studies that have been made throughout the world during the past 150 years have made it possible to assemble an increasingly long and complex *geologic column,* *a composite diagram combining in chronological order the succession of known strata, fitted together on the basis of their fossils or other evidence of relative or actual age.* This worldwide standard is being constantly added to and refined as previously unstudied rock units are described, mapped, and dated.

Standard names have evolved for the subdivision of geologic time (and also the rocks on which these time units are based). Those which can be used worldwide are eons, eras, periods, and epochs (Table 7.5).

The *geologic time scale,* *a sequential arrangement of geologic-time units, as currently understood,* is divided into four major *eons,* principally on the basis of the observable absence or presence and type of life-forms that characterized each. The term *Hadean* (Greek for "beneath the earth") is given to the earliest part of the Earth's history, an interval for which there is no known rock record. However, rocks of this age are present on other planetary bodies of the solar system whose earliest crustal rocks have been little modified since they ac-

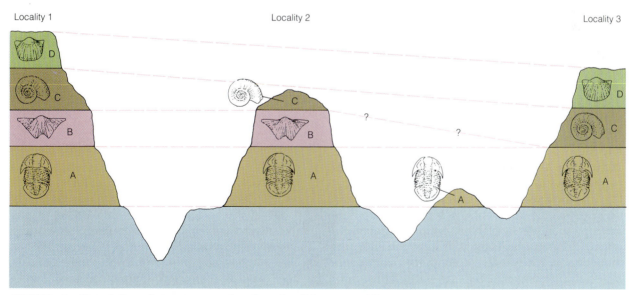

FIGURE 7.20 Correlation of strata exposed at three localities, many kilometers apart, on the basis of similarity of the fossils they contain. The fossils show that at Locality 3 stratum B is missing because C directly overlies A. Either B was never deposited there, or it was deposited and later removed by erosion before the deposition of layer C.

cumulated (Chapter 23). *Archean* (Greek for "ancient") rocks, the oldest we know of on the Earth, contain microscopic life-forms of bacterial character. *Proterozoic* (Greek for "earlier life") rocks include evidence of multicelled organisms that lacked preservable hard parts. Understandably, their record is not well known because many of these ancient rocks have been intensely deformed, metamorphosed, and eroded. *Phanerozoic* (Greek for "visible life") rocks often contain plentiful evidence of past life in the form of well-preserved hard parts. Most traditional examples of fossils that we see displayed in museums or illustrated in books are of Phanerozoic age.

Geologic *eras* encompass major spans of time that also are defined on the basis of the life forms found in the corresponding rocks. No formal eras are yet widely recognized for Archean and Proterozoic rocks, but Phanerozoic time is divided into the *Paleozoic* (Greek for "old life"), *Mesozoic* (Greek for "middle life"), and *Cenozoic* (Greek for "recent life") Eras, each name reflecting the relative stage of development of the life of these intervals (Table 7.5). Paleozoic forms of life progress from marine invertebrates to fishes, amphibians, and reptiles. Early land plants also appeared, expanded, and evolved. The Mesozoic saw the rise of the dinosaurs which became the dominant vertebrates on land. Toward the end of that era, primitive mammals appeared and later dominated the Cenozoic Era. The Mesozoic also witnessed the evolution of flowering plants, while during the Cenozoic Era grasses appeared and became an important food for grazing mammals.

The geologic *periods* have a more haphazard nomenclature. They were defined over an interval of nearly 100 years on the basis of strata that crop out in Britain, Germany, Switzerland, Russia, and the United States. The names of the periods and corresponding systems within the Paleozoic and Mesozoic Eras are partly geographic in origin, but in some cases they are based on characteristics of the strata in the place of original study (Table 7.5). Well-exposed rocks of these ages are found on nearly all the continents. In some cases the type areas provide less satisfactory examples than some distant occurrences. Triassic strata of the type area in Germany, for example, contain few marine fossils. Sections better suited for stratigraphic subdivision and correlation are found in North America and in the East Indies.

The epochs of the Tertiary Period were also defined in a piecemeal fashion. Studies of marine strata in sedimentary basins of France and Italy led

English geologist Charles Lyell to subdivide the rocks into groupings based on the percentage of their fossils that are represented by still-living species (Table 7.5). Each of the various periods of the Paleozoic and Mesozoic Eras are also subdivided into epochs, the names of which are primarily geographic in origin. They are used mainly by specialists concerned with detailed studies of these strata and their contained fossils.

The names of the geologic time scale constitute the standard time language of geologists the world over. By committing them and their relative order to memory, one can begin to comprehend numerous details of Earth history that have led to the discovery of many of the important principles of physical geology which are discussed in this book.

Significance of Major Stratigraphic Boundaries

Time-stratigraphic subdivision of crustal rocks is based primarily on important changes in the assemblages of life-forms present in the rocks as fossils. The changes reflect both the appearance of new species and also the extinction of old forms. Although marine fossils play a principal role in time-stratigraphic correlation, terrestrial (land-dwelling) forms experienced comparable changes once the Earth's landscapes began supporting plants and air-breathing animals. Major changes that form the basis for defining the boundaries of geologic eras and periods were global in scope. Therefore, their cause or causes must be sought in phenomena that affected not only the world's oceans but the lands as well.

Traditional Explanations

Two hypotheses that gained early favor suggested that the global changes responsible for major stratigraphic boundaries were caused by changes of sea level or by major tectonic events. Some geologists concluded that the geologic systems reflect major cycles of change. Each was marked by a rise of the sea against the continental margins, the attainment of a maximum level of flooding, and then the recession of marine waters. If the sea-level fluctuations were worldwide, then all coastlines would be affected. During times of low sea level, the lands presumably would be eroded. During the next cyclic rise of the sea, sediment would be deposited across the eroded landscape, thereby producing a regional unconformity marking a major stratigraphic boundary.

TABLE 7.5 *Geologic Column, Major Worldwide Subdivisions, Ages of Boundaries, and Origin of Names*

Eonothem/*Eon*	Erathem/*Era*	System/*Period*	Series/*Epoch*	Radiometric Dates[a] (millions of years ago)	Origin of Names of Periods of Paleozoic and Mesozoic and Epochs of Cenozoic
				— 0 —	
		Quaternary[b]	Holocene	— 0.01 —	Greek for "wholly recent"
			Pleistocene	— 1.6 —	Greek for "most recent"
			Pliocene	— 5.3 —	Greek for "more recent"
	Cenozoic		Miocene	— 23.7 —	Greek for "less recent"
		Tertiary[b]	Oligocene	— 36.6 —	Greek for "slightly recent"
			Eocene	— 57.8 —	Greek for "dawn of the recent"
			Paleocene	— 66.4 —	Greek for "early dawn of the recent"
		Cretaceous			Chalk (Latin = *creta*) in southern England and northern France
	Mesozoic			— 144 —	
		Jurassic		— 208 —	Jura Mountains, Switzerland and France
Phanerozoic		Triassic		— 245 —	Threefold division of rocks in Germany
		Permian	Numerous units	— 286 —	Province of Perm, Russia
		Pennsylvanian[c]		— 320 —	State of Pennsylvania
		Mississippian[c]		— 360 —	Mississippi River
	Paleozoic	Devonian		— 408 —	Devonshire, county of southwest England
		Silurian		— 438 —	Silures, ancient Celtic tribe of Wales
		Ordovician		— 505 —	Ordovices, ancient Celtic tribe of Wales
		Cambrian		— 570 —	Cambria, Roman name for Wales
Proterozoic[d]		No subdivisions in wide use		— 2500 —	
Archean[d]				— 3800(?) —	
Hadean[e]				~4650	

Note: "Carbon-bear-ing Period" is written beside Pennsylvanian and Mississippian with a bracket.

Source: Based largely on data from Palmer, 1983.

[a] Time divisions are not drawn to uniform scale.

[b] Derived from eighteenth- and nineteenth-century geologic time scale that separated crustal rocks into a four-fold division of Primary, Secondary, Tertiary, and Quaternary, based largely on relative degree of induration and deformation.

[c] Mississippian and Pennsylvanian are equivalent to Lower and Upper Carboniferous Period of Europe (named for abundance of coal in these rocks).

[d] Proterozoic plus Archean are equivalent to Precambrian.

[e] No rocks of this eon are known on Earth, but they exist on other planetary bodies in the solar system.

An alternative hypothesis rested on the assumption that there have been pulses of mountain building, perhaps cyclic in character. Such episodes would not only have disturbed crustal rocks but could have caused great disruption of biotic systems. Periods between orogenic events were times when erosion reduced the level of the land, producing widespread erosion surfaces. Some of these were subsequently buried by sediment. In areas where the rocks have been tilted or folded, angular unconformities resulted.

Each of these hypotheses had its detractors. It was pointed out that the concept of cyclically rising and falling sea level did not in itself satisfactorily explain widespread and geologically abrupt extinctions of land animals and plants, nor the appearance of new species. Arguments were also leveled against the idea of episodic mountain building by those who contended that mountain building was a more-or-less continual process, although taking place in different places at different times. At the very least, it was considered an unreliable basis for global correlation.

Sea Level and Tectonics

From detailed studies of marine strata on the continents, a curve of major sea-level fluctuations during the Phanerozoic Eon has been developed (Fig. 7.21). The evidence suggests that over much of the last 600 million years the sea has stood higher against the land than it now does. In other words, major portions of the continents were submerged by shallow seas in which marine sediments accumulated. However, the record also suggests repeated transgressions and regressions of the sea against the continental margins, although of un-

equal amplitude and period. The times of greatest marine invasion took place during the early Paleozoic Era and during the Cretaceous Period (Fig. 7.22). These intervals are represented by extensive marine sedimentary rocks on nearly all the continents. The stratigraphic record indicates that erosion of the lands and the continental margins occurred during times of marine regression, with the resulting sediment being carried to the deep sea. Interregional erosion surfaces produced at such times were subsequently buried by sediment during marine transgressions. Some of the regressions coincide with boundaries between periods and between epochs, suggesting that global sea-level changes may account, in fact, for at least some primary stratigraphic boundaries. Boundaries between unconformity-bounded sequences appear to have been times of significant change in tectonic conditions.

Plate tectonics provides a possible explanation for changes in the shapes and volumes of the ocean basins through time, changes which might cause global sea level to rise or fall. At times when the volume of rocks composing the midocean ridge systems is large, the capacity of the ocean basins is correspondingly reduced. This forces the ocean water to spill over and rise against the adjoining lands. The resulting sea-level fluctuation could be as much as 500 m.

Changes in ocean-basin volume can also occur due to crustal thickening along zones where continents collide. Such thickening effectively reduces the *area* of continental crust and causes a corresponding increase in the area of oceanic crust. The result would be a world fall of sea level, perhaps by tens of meters. In either of these cases, the rate at which such changes occur is likely to be quite slow.

On the other hand, if sea-level changes coincident with stratigraphic boundaries took place rapidly, as some have suggested, then the geologically rapid buildup and decay of continental ice sheets might provide a possible mechanism (Chapter 13). The amplitude of many of the recorded sea-level changes is far too large, however, to be explained satisfactorily in this way, for glacier-related effects are unlikely to involve more than 150–200 m of sea-level fluctuation. It has further been pointed out that the beginning of a number of sea-level cycles apparently coincide with important plate-tectonic events related to breakup of continents. As land masses separate and move apart, the newly formed margins cool and contract, allowing the sea to rise inland across them.

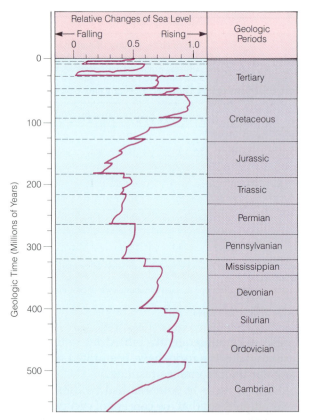

FIGURE 7.21 Reconstructed record of sea-level fluctuations during the last 600 million years. (*Source:* After Vail et al., 1977.)

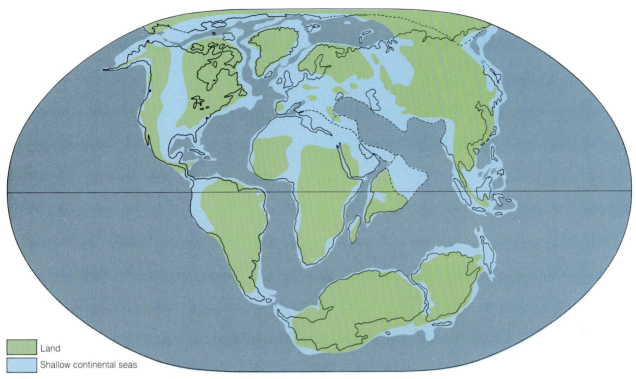

Land

Shallow continental seas

FIGURE 7.22 Map of the world showing distribution of land and water about 80 to 90 million years ago. Many parts of the continents that are now dry land were then shallow seas where marine sediments accumulated. (*Source: After Howarth, 1981.*)

At least some of the major sea-level transgressions might be explainable in this way.

Global Catastrophes

The boundary between the Cretaceous and Tertiary Systems is marked by evidence of large-scale extinctions, both on land and in sea. As many as half the genera living at the end of the Cretaceous Period did not survive into the Tertiary. An explanation for this biologic disaster has been an intriguing and perplexing geologic problem for many years. A possible explanation has recently been suggested.

In many parts of the world, the boundary is marked by a thin layer of clay, the so-called "boundary clay." Geochemical studies of the clay have disclosed an unusually high concentration of iridium and related platinum-group elements (Fig. 7.23). These elements typically are present in low concentration in crustal rocks of the Earth relative to their cosmic abundances, although very high concentrations appear in the products of some volcanic eruptions. Some geologists have suggested that the anomalous concentrations in the bound-

ary clay could have resulted from the impact of an extraterrestrial object, most likely an asteroid or comet. Consistent with this hypothesis, quartz grains found in the clay bear structures indicating high-velocity impact, and glassy spherules are found that are thought to have condensed from vaporized rock.

An impact sufficiently powerful to produce the observed features in the boundary clay would likely have formed a crater of exceptional dimensions when it hit and thrown great volumes of rock and dust into the atmosphere. Furthermore, the intense heat resulting from the impact might have incinerated forests over wide areas, sending large quantities of dark sooty smoke into the air. Sooty carbon found concentrated in the boundary clay is calculated to have been deposited at a rate some 10,000 times as great as carbon in layers directly above and below that layer. The result of such extraordinary atmospheric pollution would be to block sunlight from reaching the surface, possibly for many months. Modeling studies of the potential atmospheric effects indicate that surface temperatures would have plummeted and winterlike conditions would have prevailed for an extended

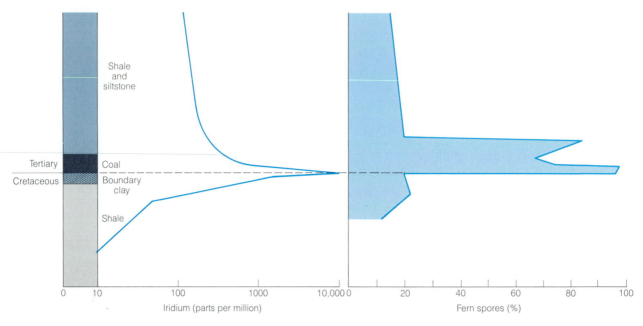

FIGURE 7.23 Stratigraphic section near Trinidad, Colorado showing the anomalously high content of the element iridium in the Cretaceous/Tertiary boundary clay (note the logarithmic scale). The percentage of fern spores increased dramatically at the same stratigraphic level, indicating that ferns replaced the previous forest flora. Following the boundary event, both iridium and fern spores returned to near-normal levels. (*Source:* After Tschudy et al., 1984.)

time. Under such conditions, photosynthesis would sharply decrease, thereby killing many forms of vegetation and disrupting major food-chains. Large-scale mortality would be expected. Those organisms that did survive would have numerous new living spaces open to them and could spread to repopulate the post-Cretaceous Earth.

Paleontological evidence from the western United States is consistent with this hypothesis, for plant populations living there were strongly disturbed at the end of the Cretaceous Period and some forms became extinct exactly at the level of the iridium-rich layer (Fig. 7.23). Immediately above the boundary clay, the vegetation was dominated by ferns, the first type of plant to colonize after a major ecological disturbance.

Calculations suggest that if an extraterrestrial object did impact the Earth at the time of the Cretaceous/Tertiary boundary, it was about 10 km in diameter. The probability that objects of this size will hit our planet is calculated to be about once in every 100 million years. Interestingly, there have been five major extinction events during the Phanerozoic Eon (the last 570 million years), or an average of nearly one per 100 million years.

As is often the case when a new idea is proposed, the asteroid/comet-impact hypothesis has provoked a lively debate. Some geologists have countered that the anomalous iridium concentration in the boundary clay is better explained by a series of major volcanic eruptions during which large volumes of magma reached the surface from the mantle, thereby leading to the release of unusually large amounts of iridium and related elements over some 10,000 to 100,000 years. They cite voluminous lavas of appropriate age that underlie the Deccan Plateau in India as possible major contributors. Although the largest known single volcanic event on the Earth, which occurred in Java about 75,000 years ago, apparently caused no worldwide extinctions, it is uncertain what the cumulative effects of numerous such eruptions would be over an extended interval.

The Cretaceous/Tertiary boundary question nicely illustrates the complexity of many fundamental problems in the Earth sciences and the varied kinds of evidence needed to solve them. Like any new idea, the impact hypothesis will either stand or fall as new evidence is sought that will permit it to be rigorously tested.

SUMMARY

1. Strata provide a basis for reconstructing Earth history and past surface environments. Most strata were horizontal when deposited and accumulated in sequence from bottom to top.

2. Sedimentary rocks cover about two-thirds of the area of the continents, but average only about 2 km thick. Many formed in shallow marine waters along the subsiding margins of continents.

3. Fossil faunas and floras preserved in sedimentary strata succeed one another in a definite sequence, providing evidence of evolutionary changes through time.

4. An extensive unit of strata may possess several facies, each determined by a different depositional environment. The boundaries between facies may be abrupt or gradational.

5. Unconformities are physical breaks in a stratigraphic sequence marking a hiatus. Angular discordance implies disturbance of rocks prior to deposition of overlying strata.

6. A formation is a fundamental rock unit for field mapping distinguished on the basis of its distinctive physical characteristics and named for a geographic locality. Formations may be assembled into groups or subdivided into members and beds.

7. Systems are rock sequences that accumulated during a specified time interval. They can be grouped into erathems or subdivided into series and stages.

8. Geologic-time units are based on time-stratigraphic units and represent the time intervals during which the corresponding rock units accumulated.

9. Although formational boundaries may coincide with boundaries of time-stratigraphic units, often their boundaries are diachronous.

10. Large groupings of strata bounded by major unconformities are recognized both on the continents and in strata of the continental shelf and slope. They record major marine transgressions and regressions related to tectonic movements and global fluctuations of sea level.

11. Correlation of strata is based on physical and biological criteria that permit demonstration of time equivalence. Reliability of correlation is greatest if several criteria are used.

12. The geologic column, pieced together over more than a century and a half, is a composite section of all known strata, arranged on the basis of their contained fossils or other age criteria.

13. The geologic time scale is a hierarchy of time units established on the basis of corresponding time-stratigraphic units. Systems and periods are based on type sections or type areas in Europe and North America. The geologic time scale constitutes the global standard to which geologists correlate local sequences of strata.

14. Major geologic boundaries often coincide with significant biological changes that include mass extinction and the appearance of new species. Some may reflect changes of world sea level. Others have been attributed to global catastrophes caused by impact of extraterrestrial bodies or to episodes of intense volcanism.

SELECTED REFERENCES

Alvarez, L. W., Alvarez, W., Asaro, F., and Michel, H. V., 1980, Extraterrestrial cause for Cretaceous-Tertiary extinction: Science, v. 208, p. 1095–1108.

Dunbar, C. O., and Rodgers, J., 1957, Principles of stratigraphy: New York, John Wiley.

Matthews, R. K., 1974, Dynamic stratigraphy: Englewood Cliffs, N.J., Prentice-Hall.

North American Commission on Stratigraphic Nomenclature, 1983, North American stratigraphic code: Amer. Assoc. Petroleum Geologists Bull., v. 67, p. 841–875.

Officer, C. B., and Drake, C. L., 1985, Terminal Cretaceous environmental events: Science, v. 227, p. 1161–1167.

Palmer, A. R., 1983, The Decade of North American Geology 1983 geologic time scale: Geology, v. 11, p. 503–504.

Phillips, J., 1844, Memoirs of William Smith: London, John Murray (reprinted 1978: New York, Arno Press).

Schlee, J. S. (ed.), 1984, Interregional unconformities and hydrocarbon accumulations: Tulsa, Amer. Assoc. Petroleum Geologists Memoir 36.

CHAPTER 8

Geological Time and Its Determination

Olduvai Gorge, Tanzania, is the resting place of some of the most ancient hominid remains ever discovered. The age of the remains has been determined by radiometric dating of pyroclastic strata.

INDIRECT ESTIMATES OF TIME

The scientists who worked out the geologic column (Chapter 7) were challenged by the question of time. They could not provide an answer. They knew the order in which the different systems had formed but they could not say whether the sediments in each system had accumulated during the same length of time. They could not provide answers to questions such as, "How much time elapsed between the end of the Cambrian and the beginning of the Permian?" or "How long was the Tertiary?" Yet the question of time is as important as the geologic column itself. An answer must be forthcoming if we are to attempt questions such as the age of the Earth, the time during which the rock cycle has been operating, the age of the ocean, how fast mountain ranges rise, or how long humans have inhabited the Earth.

Many attempts, by indirect methods, were made during the nineteenth century to subdivide the geologic column by a scale of years. One of the earliest concerned the fossil record. Charles Lyell became convinced that evolution among living species is a time-governed process. In 1867, Lyell reasoned that about 20 million years must be necessary for a complete turnover and change in a family of marine molluscs. He drew his evidence from the Tertiary Period. He then pointed out that from the end of the Ordovician to the present there had been 12 such turnovers. From that he concluded that 240 million years had elapsed since the end of the Ordovician. The actual time, we now know, is much longer—about double. Lyell's estimates were ingenious, but they contained many points of uncertainty and, thus, they were educated guesses. Two problems, in particular, plagued his efforts. He had first to guess the length of time needed for one family of molluscs to evolve. He then had to assume that all families of molluscs evolved at the same rate. In neither case could Lyell find reliable evidence, so he made a reasoned guess. Lyell was wrong, but his efforts spurred others to think of additional ways to estimate geologic time.

One of the most widely used methods consisted of rough estimates of the time during which the rock cycle has been at work. If it is assumed that sediment has always been deposited at rates equal to today's rate, it is possible, at least in theory, to say how much time was needed to accumulate all the sediment now preserved in sedimentary strata. There are several difficulties. First, rates of deposition are not everywhere constant. Second, in every pile of sediment we find gaps representing intervals where deposition ceased—diastems—and there is no way to estimate the duration of these gaps. Third, in rock older than Cambrian there are very few fossils to help put strata in their proper sequence. Because of these difficulties, early estimates of the duration of the rock cycle were unreliable and tended to be much too short. They were also highly variable. Estimates of the age of the Earth, based on the total thickness of sedimentary layers, ranged from 3 million years to 1500 million years!

A clever suggestion for estimating the age of the ocean concerned the saltiness of seawater. Since sea salts come from the erosion of common rock and reach the sea dissolved in river water, why not measure the salts in modern river water and calculate the time needed to have carried all the salts now in the sea? The answer gives an age for the ocean that is less than 100 million years—only a fraction of the age we calculate today for the ocean. The reason is that sea salts, like all chemical constituents, are cyclic, and the composition of the ocean varies very little. Salts are added both by rivers and by submarine volcanism. But salts are also removed from solution. Some are precipitated as evaporite minerals (Chapter 5), while others are removed by chemical reactions. Seawater reacts with hot volcanic rocks on the ocean floor and with clays and other mineral grains in sediments, forming new minerals in the process. Through reaction some of the dissolved salt constituents become incorporated in the new minerals. The net effect is that the composition of the ocean is essentially constant. What is actually measured is not the age of the ocean, but rather the residence time of salts in the ocean (Chapter 2).

Perhaps the most interesting estimates were those made by physicists of the time that the Earth has been a solid body. The Earth started as a very hot object they argued. Once it had cooled sufficiently to form a solid outer crust, it could only continue to cool by the conduction of heat through the solid rock. By measuring the thermal properties of rock and estimating the present temperature of the Earth's interior, they calculated a time for the Earth to cool to its present state. The physicist who used this method most effectively was the famous English scientist, Lord Kelvin. The logic used by Kelvin was faultless and his mathematical calculations were correct. However, his estimate of about 100 million years for the maximum age of the solid Earth was much too short because the data he used were incomplete. When he first made his calculation, the existence of radioactivity was not known. Radioactivity continually supplies heat to the Earth's interior. Instead of cooling rapidly, the Earth is cooling so slowly it has a nearly constant temperature over time periods as long as a few hundred million years.

Indirect methods of measuring geologic events yielded ages that were too young and uncertain. Young ages could not be reconciled with the thick pile of sedimentary strata of the geologic column, nor did it seem possible for the vast evolutionary changes seen in the fossil record to have occurred within the few hundred million years estimated for the Earth's age. To get around this dilemma what was needed was a way to measure geologic time by some process that runs continuously, that is not reversible, that is not influenced by other processes and other cycles, and that leaves a continuous record without gaps in it. At the end of the nineteenth century (1896) the discovery of radioactivity provided the needed method. By 1904, the

American scientist, Benjamin Boltwood, had demonstrated that the radioactivity of uranium could be used to determine the time of formation of uranium minerals. That discovery opened the door to radiometric dating, a new and reliable means of measuring geologic time.

RADIOMETRIC DATING

Natural Radioactivity

As we saw in Chapter 3, most of the chemical elements found in the Earth are stable and not subject to change. However, a few are radioactive and, therefore, unstable. The instability arises because there are limits within which the mass numbers of the isotopes of any element can vary (Appendix A). If the limit is exceeded, the unstable nucleus will spontaneously transform to a nucleus of either a more stable isotope of the same chemical element or to an isotope of a different chemical element. Even though the process is one of transformation—from an unstable nucleus to a more stable

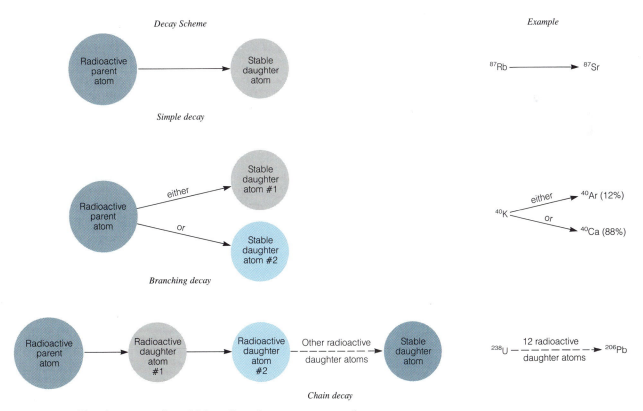

FIGURE 8.1 The three ways by which radioactive parent atoms decay to produce stable daughter atoms. Each is important in one or more of the decay systems used for radiometric dating.

one—it has become common practice to call the process radioactive decay. *An atomic nucleus undergoing radioactive decay is said to be a **parent;** the product arising from radioactive decay is called a **daughter product.***

Many of the unstable isotopes that were once in the Earth have decayed away and are no longer present. This is so because their rates of spontaneous decay were fast. A few isotopes that transform very slowly are still present, however, and it is the slow steady decay of these remaining radioactive isotopes that can be used for radiometric dating. It is the same slow radioactive decay process that creates the heat that keeps the Earth's interior hot.

The Radioactive Decay Process

A radioactive nucleus can decay in several different ways, each of which involves the emission or capture of certain atomic particles, or electromagnetic radiation, or both. The particles involved are single electrons, called beta particles (β-particles) or heavy particles consisting of two protons plus two neutrons, called alpha particles (α-particles). The electromagnetic radiation emitted is identical to a short wavelength X ray and is called a gamma ray (γ-ray). Some nuclei decay by emitting α-particles, others by emitting β-particles. Gamma rays can be emitted with either an α- or a β-particle, but in no instance do all three modes of decay occur together. A fourth and less common mode of radioactive decay occurs when a proton in a nucleus is transformed to a neutron by capture of an electron from among those in orbit around the nucleus. Loss of an α-particle results in a decrease in atomic number by 2 and a decrease in atomic mass by 4. Loss of a β-particle means that a neutron in the nucleus is changed to a proton, producing an increase in atomic number by 1 without an appreciable change in the atomic mass; the opposite effect occurs when a proton in the nucleus is transformed to a neutron through capture of an orbiting electron.

The transformation from parent to daughter can follow one of three different paths (Fig. 8.1). The first path is called **simple decay:** *a radioactive parent decays to produce a stable daughter product.* An example of simple decay is the transformation of ^{87}Rb to give ^{87}Sr. The second process is called **branching decay:** *a radioactive parent decays to produce either of two different daughter products.* An example is the decay of ^{40}K to form either ^{40}Ca by β-emission, or ^{40}Ar by electron capture. The third method of transformation is called **chain decay:** *a radioactive parent decays to form a radioactive daughter, which in turn forms another radioactive daughter, and so forth until a stable daughter product is reached.* ^{238}U provides an example of chain decay. Step by step, ^{238}U produces 13 radioactive daughter products until it finally forms the stable daughter ^{206}Pb (Appendix A: Fig. A.3).

Rates of Decay

Careful study of radioactive elements in the laboratory has shown that decay rates are unaffected by changes in the chemical and physical environment. Radioactive decay is a property of the atomic nucleus, whereas the chemical and physical environments affect the orbiting electrons. This is a particularly important point because it allows us to conclude that rates of radioactive decay are not influenced by geologic processes. Each radioactive element decays according to a distinct and measurable timetable, but all decay timetables follow the same basic law. Consider a radioactive isotope that transforms to a daughter product through simple decay. The number of decaying parent atoms continually decreases while the number of daughter atoms continually increases. The proportion, or percentage, of atoms that decay during one unit of time is constant. However, although the proportion is constant, the actual number keeps decreasing because the parent atoms are being used up.

In a mineral sample that contains atoms of a radiometric element, the overall radioactivity (the sum of the radioactivity of all the parent atoms remaining in the sample) decreases continually (Fig. 8.2). Because the rate of decrease is measured as a percentage of the number of atoms left undecayed, it is usual to designate a decay rate in terms of the **half-life** of the parent, meaning *the time required to reduce the number of parent atoms by one half.* The time units marked in Figure 8.2 are half-lives. Of course they are of equal length, just as years are. But at the end of each one, the number of atoms that decay (and, therefore, the combined radioactivity of the sample) has decreased by exactly half.

While the proportion of parent atoms declines, the proportion of daughter atoms increases. Figure 8.2 shows that the growth of daughter atoms just matches the decline of parent atoms. When the number of remaining parent atoms is added to the number of daughter atoms, the sum is the number of parent atoms that a mineral sample started with. That fact is the key to the use of radioactivity as a means of measuring time and determining ages.

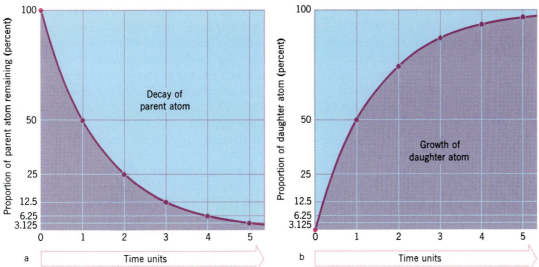

FIGURE 8.2 Curves showing decay of radioactive atoms and growth of daughter products. (*a*) At time 0, a sample consists of 100 percent radioactive parent atoms. During each time unit, half the atoms remaining decay to daughter atoms. (*b*) At time 0, no daughter atoms are present. After 1 time unit corresponding to a half-life of the parent atoms, 50 percent of the sample has been converted to daughter atoms. After 2 time units, 75 percent of the sample is daughter atoms, 25 percent parent atoms. After 3 time units, the percentages are 87.5 and 12.5, respectively.

Potassium-Argon ($^{40}K/^{40}Ar$) Dating

Several elements possess naturally radioactive isotopes. We will select one of these isotopes, potassium-40 (^{40}K), to illustrate how minerals are dated. From Table A.3 it can be seen that potassium has three natural isotopes: ^{39}K, ^{40}K, and ^{41}K. Only one, ^{40}K, is radioactive and its half-life is 1.3 billion years. ^{40}K disintegrates by branching decay. Twelve percent of the ^{40}K atoms decay by electron capture, forming ^{40}Ar, an isotope of the gas argon. The formula is:

$$^{40}K + \beta \rightarrow {}^{40}Ar$$

The remaining 88 percent of the ^{40}K atoms decay by emission of a β-particle to produce ^{40}Ca:

$$^{40}K \rightarrow {}^{40}Ca + \beta$$

Careful measurements have shown that the ratio of ^{40}Ar to ^{40}Ca daughter atoms is always the same. When a potassium-bearing mineral crystallizes from a magma, or grows within a metamorphic rock, it traps a sample of ^{40}K in its crystal structure. ^{40}Ca and ^{40}Ar are trapped in the structure also. Atoms of argon and calcium can be trapped in many ways, such as by atomic substitution for other atoms (Chapter 3) or in submicroscopic frac-

tures in the minerals. Atoms of calcium are tightly bound in the crystal because they form chemical bonds with other atoms present. However, argon is an element that has unusual atomic properties. Because its orbiting electron shells are filled, atoms of argon do not readily form chemical bonds. Thus, the daughter atoms are not chemically bound in minerals. This in turn means that at high temperatures, argon, unlike calcium, rapidly diffuses out of a mineral and does not stay trapped. When the argon content of a mineral is measured, therefore, what is determined is the ^{40}Ar accumulated during the time since a mineral started trapping and retaining argon. Although ^{40}Ar is present in a magma when a mineral crystallizes, a mineral rarely contains any initial argon because magmatic temperatures are above trapping temperatures. All the ^{40}Ar atoms in a potassium-bearing mineral, therefore, must come from decay of ^{40}K and must have accumulated since the temperature fell below the trapping temperature. All that now has to be done is to measure the amount of parent ^{40}K that remains and the amount of ^{40}Ar that has formed. The half-life of ^{40}K being known, it is a straightforward matter to calculate the ***radiometric age***—*the length of time a mineral has contained its built-in radioactivity clock.*

Dating by ^{40}K is not limited to minerals such as muscovite that contain potassium as a major element. Even minerals that contain small amounts of potassium substituting for other elements by atomic substitution will serve the purpose. Thus, hornblende, a calcium-iron-magnesium silicate, can be used because it generally contains a small quantity of potassium. Some rocks contain several different minerals that can be used for dating, and then it is possible to use the "whole rock" for dating.

K/Ar dating is most successfully applied to volcanic and pyroclastic rocks because they crystallize and cool rapidly. As a result, they have formation ages that are essentially coincident with their trapping ages. Because argon analyses can be performed with great accuracy, and because contamination by initial argon at the time of crystallization is generally not a problem, the method can be used, under ideal circumstances, for volcanic rocks as young as 100,000 years. For this reason, K/Ar dating has proved very useful in studies of paleoanthropology as well as geology (Fig. 8.3).

The moment a mineral starts accumulating ^{40}Ar is referred to as the ***setting time.*** If a mineral is subsequently reheated, perhaps by metamorphism, some or all of the contained argon may diffuse out and the time clock of the mineral is reset. While partial resetting due to loss of some, but not all, of the original argon complicates age determinations, it also creates an opportunity to determine the time of metamorphism.

Even though argon may remain locked in the center of a mineral, and be unaffected by metamorphism, partial resetting may occur through loss of argon from the marginal portion of the grain (Fig. 8.4a). Thus, the core of the mineral grain has a record of the original setting time, while the rim has a record of the time of resetting due to metamorphism. Through a clever technique, both ages can be measured. When potassium is irradiated with neutrons in a reactor, one of the stable isotopes, ^{39}K, is converted to ^{39}Ar. The isotopes of potassium are observed to be everywhere the same. Through measurement of ^{39}Ar, scientists have a means of analysing the ^{39}K content of the sample and, by calculation, the ^{40}K content. The new ^{39}Ar atoms are locked in a min-

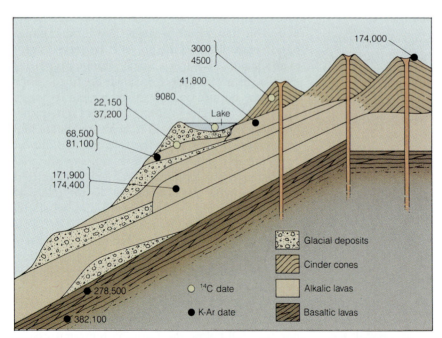

FIGURE 8.3 An example of radiometric dating. Glacial deposits on Mauna Kea volcano, Hawaii, are interlayered with basaltic lava flows and cinder cones that have been dated by the K/Ar method. The youngest glacial deposit is dated by radiocarbon and is younger than organic matter associated with cinders that are 22,150 years old, but older than sediments at the bottom of a lake that are 9080 years old. Numbers are in years before the present and are given as ranges that indicate the measurement uncertainty.

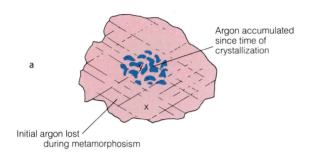

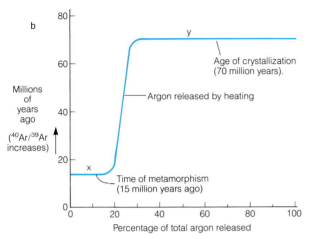

FIGURE 8.4 Resetting age of a potassium feldspar. (*a*) Grain of potassium feldspar. Argon retained in the core of the grain (*Y*), allows determination of the time of crystallization. Argon in rim of the grain (*X*) has accumulated since the time metamorphism caused earlier formed argon to leak away. (*b*) Plot of $^{40}Ar/^{39}Ar$ against age. Argon released by heating the feldspar starts with a low $^{40}Ar/^{39}Ar$ ratio, equivalent to an age of 15 million years. With higher temperatures and continued heating, argon from the interior of the grain is released; the higher $^{40}Ar/^{39}Ar$ ratio is equivalent to an age of 70 million years, the time of crystallization.

eral in the same manner as the earlier-produced ^{40}Ar daughter atoms. If, now, the mineral grain is heated and the argon collected for analysis with a mass spectrometer, it is observed that the $^{40}Ar/^{39}Ar$ ratio starts out being low, as the rims release the argon. The ratio of the earliest released argon provides a measure of the age of metamorphism—the ^{39}Ar is a measure of the potassium content and thus of the parent atoms, the ^{40}Ar is a measure of the argon daughter product produced since resetting. As heating continues, gas is eventually released from the centers of the mineral grains. The $^{40}Ar/^{39}Ar$ ratio of the released gas rises until it

reaches the value of the unaffected grain interiors (Fig. 8.4*b*). The final ratio provides a measure of the original setting time of the mineral grain. If the mineral grain came from a volcanic rock, the original setting time is essentially the same as the time of crystallization.

Rubidium-Strontium ($^{87}Rb/^{87}Sr$) Dating

Rubidium is not a very common element in the crust, but it is widespread because rubidium atoms commonly substitute for potassium and other large atoms in minerals such as micas, feldspars, amphiboles, pyroxenes, and even olivine. Approximately 28 percent of all rubidium atoms are ^{87}Rb, a radioactive isotope that decays by β-emission to an isotope of strontium (^{87}Sr):

$$^{87}Rb \rightarrow {}^{87}Sr + \beta$$

Application of $^{87}Rb/^{87}Sr$ dating is a little different from $^{40}K/^{40}Ar$ because a crystallizing mineral will usually incorporate and retain some ^{87}Sr in its structure along with the ^{87}Rb. Thus, a correction must be made for the amount of ^{87}Sr at the setting time. Use is made of another isotope of strontium, ^{86}Sr, which is not produced by radioactive decay. The initial $^{87}Sr/^{86}Sr$ ratio will be the same in all minerals in an igneous rock because the setting time is the time of crystallization and all the minerals crystallize from the same magma. By measuring $^{87}Sr/^{86}Sr$ in a rubidium-free mineral, we can apply a correction to the measured $^{87}Sr/^{86}Sr$ ratio in a mineral that contains rubidium. In the latter case, $^{87}Sr/^{86}Sr$ will increase steadily with age as ^{87}Rb decays and produces more ^{87}Sr.

A useful way to minimize error in the $^{87}Rb/^{87}Sr$ method is to employ the *isochron method*. In order to do so, a rock must contain several minerals with different rubidium contents, or it is necessary to find a differentiated igneous rock such as a layered intrusion (Chapter 4) in which several layers contain different rubidium contents. Each mineral or rock layer starts with the same $^{87}Sr/^{86}Sr$ ratio, but as time passes and ^{87}Rb decays to ^{87}Sr, the $^{87}Sr/^{86}Sr$ ratios in the different rock or mineral samples start to diverge. When the ratios are plotted as shown in Figure 8.5, a line called an *isochron* is produced. All minerals plotted on the line have the same age. The age is determined by the slope of the line.

Rb/Sr dating is most effective for very old igneous and metamorphic rocks. Because ^{87}Sr does not readily diffuse out of a mineral, except at very high temperatures, the question of resetting is not so severe as it is with K/Ar dating.

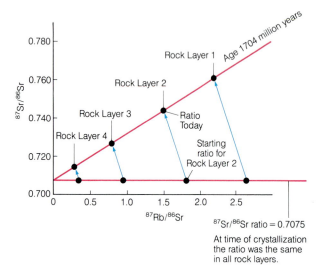

FIGURE 8.5 Rubidium-strontium whole-rock isochron dating of the layered intrusion at Sudbury, Ontario. The Sudbury intrusion contains layers of igneous rock of differing composition that crystallized over a short time interval. At the time of crystallization, the *ratio* of strontium-87 to strontium-86 was the same in each rock. The rubidium-87 *contents* of the rocks differed, however. As time passed and the rubidium-87 produced more strontium-87 by radioactive decay, the rocks changed composition along the lines marked by arrows. When they are measured today, the isotopic ratios of the four rock layers plot on a straight line called an isochron. The slope of the line is determined by the age of the rocks. The Sudbury intrusive formed 1704 million years ago. (*Source:* Data from H. W. Fairbairn et al., 1968.)

Uranium-Lead Dating

All naturally occurring uranium contains two radioactive isotopes, ^{238}U and ^{235}U, and they occur in a ratio of 138:1. Both isotopes transform through chain decays to isotopes of lead:

$$^{235}U \rightarrow {}^{207}Pb + 7\alpha + 4\beta$$

and

$$^{238}U \rightarrow {}^{206}Pb + 8\alpha + 3\beta$$

Thorium is always present in at least small amounts because thorium readily replaces uranium by atomic substitution. An isotope of thorium, ^{232}Th, also transforms by chain decay to an isotope of lead, ^{208}Pb:

$$^{232}Th \rightarrow {}^{208}Pb + 6\alpha + 4\beta$$

The fact that three radioactive isotopes—^{235}U, ^{238}U, ^{232}Th—occur together and produce different

isotopes of lead provides a convenient way to cross-check ages: in effect, a single sample provides three estimates of age. Natural lead in the crust of the earth contains the isotope ^{204}Pb in addition to isotopes 206, 207, and 208. ^{204}Pb is not produced by radioactive decay. As with ^{86}Sr, ^{204}Pb can be used to provide a correction for any ^{206}Pb, ^{207}Pb, or ^{208}Pb that may have been incorporated in a mineral at setting time. Uranium decay is particularly useful for dating very old rocks because the radioactive isotopes of uranium and thorium have very long half-lives. By making use of some of the longer-lived daughter products produced during the chain decay, uranium dating can also be used to date certain relatively young carbonate rocks, like coral reefs and dripstones in caves. However, it is necessary to be alert to possible recrystallization which can reset the radiometric clock.

Radiocarbon (^{14}C) Dating

Among the radiometric dating methods listed in Table 8.1, the one based on ^{14}C (also known as radiocarbon) is unique for two reasons. The first is that the half-life of ^{14}C is short by comparison with the half-lives of ^{40}K, ^{87}Rb, and the isotopes of uranium. The second reason is that the amount of daughter product cannot be measured.

Radiocarbon is continuously created in the atmosphere through bombardment of nitrogen-14 (^{14}N) by neutrons created by cosmic radiation. ^{14}C, with a half-life of 5730 years, decays back to ^{14}N through β-emission:

$$^{14}C \rightarrow {}^{14}N + \beta$$

The ^{14}C mixes with ordinary carbon ^{12}C and ^{13}C and diffuses rapidly through the atmosphere, hydrosphere, and biosphere. Because the rates of mixing and exchange are rapid compared with the half-life, the proportion of ^{14}C is nearly constant throughout the atmosphere. As long as the production rate remains constant, the radioactivity of natural carbon remains constant because rate of production balances the rate of decay.

While an organism is alive and is taking in carbon from the atmosphere, it contains this balanced proportion of ^{14}C. However, at death the balance is upset, because replenishment by life processes such as feeding, breathing, and photosynthesis ceases. The ^{14}C in dead tissues continually decreases by radioactive decay. The analysis for the radiocarbon date of a sample involves only a determination of the radioactivity level of the ^{14}C it contains. The daughter product, ^{14}N, cannot be measured because it leaks away and because of atmospheric contamination. In order to measure an age with ^{14}C, it is necessary to make two assumptions: (1) that the rate of production of ^{14}C has been constant throughout the last 50,000 years or so (the range of time to which the short half-life of ^{14}C limits the usefulness of the method); and (2) that all samples start with the same $^{14}C/^{12}C$ ratio.

Because the ^{14}C method is based partly on the assumption that ^{14}C in the atmosphere has been constant, the accuracy of radiocarbon dates has been checked against samples that include wooden beams, prehistoric clothing, and furniture from ancient Egyptian tombs whose dates are known independently. It has also been carefully calibrated with the age of annual growth rings of long-lived trees (Figs. 8.6 and 20.9a). Some tree-ring chronologies extend back more than 8,000 years, thereby providing calibration of radiocarbon dates over most of the Holocene Epoch. Many dates as old as 50,000 years have been calculated without the benefit of these independent checks. Although less accurate than the dates of younger samples, they are still very useful. It is interesting to note that there is a limit to ^{14}C dating of young samples due to explosions of atom bombs. Radiocarbon is one of the radioactive isotopes produced by atomic explosions in the atmosphere.

The ^{14}C content of the atmosphere has not been constant since the first atom bombs were exploded in the 1940s; it is not possible to date samples formed after 1945.

Because of its application to organisms (by dating fossil wood, charcoal, peat, bone, and shell material) and its short half-life, radiocarbon has proved to be enormously valuable in establishing dates for prehistoric human remains and for recently extinct animals. In this way it is of extreme importance in archeology. It is also of great value in dating the most recent part of geologic history, particularly the latest of the glacial ages. For example, the dates of many samples of wood taken from trees overrun by the advance of the latest of the great ice sheets and buried in the rock debris thus deposited show that the ice reached its greatest extent in the Ohio-Indiana-Illinois region not more than about 18,000–21,000 years ago. It is even possible to date glacier ice directly, for as the ice forms, bubbles of air are trapped in it. The carbon dioxide in the air bubbles can be liberated in the laboratory and dated, providing an age for the time of ice formation.

Similarly, radiocarbon dates afford the means for determining rates of geologic processes, such as: the rate of advance of the last ice sheet across Ohio; the rates of rise of the sea against the land

TABLE 8.1 *Some of the Principal Isotopes Used in Radiometric Dating*

Isotopes		Half-Life of Parent (yr)	Effective Dating Range (yr)	Minerals and Other Materials That Can Be Dated
Parent	Daughter			
Uranium-238	Lead-206	4.5 billion	10 million–4.6 billion	Zircon Uraninite and pitchblende
Uranium-235	Lead-207	710 million		
Thorium-232	Lead-208	14 billion		
Potassium-40	Argon-40 Calcium-40	1.3 billion	100,000–4.6 billion	Muscovite Biotite Hornblende Whole volcanic rock
Rubidium-87	Strontium-87	47 billion	10 million–4.6 billion	Muscovite Biotite Potassium-feldspar Whole metamorphic or igneous rock
Carbon-14	Nitrogen-14	5,730 ± 30	100–50,000	Wood, charcoal, peat, grain, and other plant material Bone, tissue, and other animal material Cloth Shell Stalactites Groundwater Ocean water Glacier ice

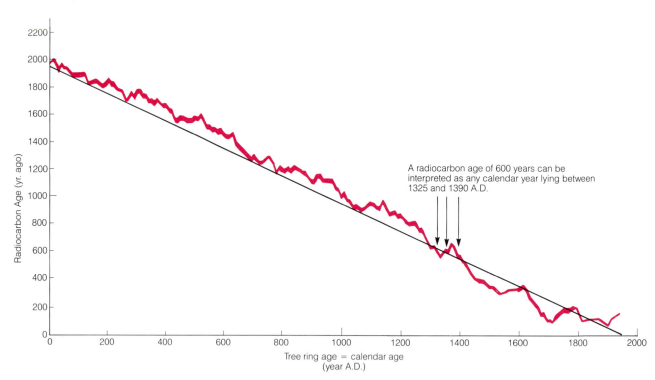

A radiocarbon age of 600 years can be interpreted as any calendar year lying between 1325 and 1390 A.D.

Radiocarbon Age (yr. ago)

Tree ring age = calendar age
(year A.D.)

FIGURE 8.6 Calibration of radiocarbon ages against tree-ring ages for the last 2000 years. Tree-ring ages are calendar ages. If radiocarbon ages were equivalent to calendar ages, they would plot along the straight line. Irregularities in the calibration curve mean that some radiocarbon ages could be equivalent to two or more calendar ages. For example, a radiocarbon age of 600 years could be any year between about 1325 A.D. and 1390 A.D. (*Source:* After Stuiver, 1982.)

while glaciers melted throughout the world; rates of soil erosion; the rates of local uplift of the crust that raised beaches above sea level; and even the frequency of volcanism.

Dating by Fission Tracks

When a radioactive isotope in a crystal structure decays by release of an α-particle, the heavy α-particle will damage a small part of the structure as it travels through the crystal. If there were a way to count the number of damaged areas in a crystal, it should then be possible to measure the number of radioactive disintegrations that had occurred. Thus, one could obtain an independent measure of the number of daughter atoms created because each area of damage is equivalent to one daughter atom.

Sensitive ways have been discovered to etch out the damaged areas—called *fission tracks*—so they can be seen under high-powered microscopes (Fig. 8.7). Then, if the concentration of the remaining radioactive isotope (such as ^{238}U) that causes the

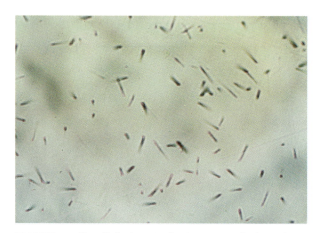

FIGURE 8.7 Fossil fission tracks in a crystal of apatite. Each track records the radioactive decay of an atom of uranium. The tracks have been etched out to make them visible. They are approximately 16 μm in length. The field of view is about 15 mm across. The apatite crystal came from one of the radial dikes at Shiprock, New Mexico (Figure 4.26), and indicates a setting age of 27 million years.

damage can be measured, it is a simple matter to calculate the age of the mineral from the abundance of tracks.

The damaged areas in crystal structures are repaired by annealing. Each mineral has a specific annealing temperature. Above the annealing temperature, fission tracks are annealed out. Below the annealing temperature, fission tracks remain unaffected. Annealing temperatures can be used to measure rates of tectonic uplift.

The annealing temperature of the mineral apatite $[Ca_5(PO_4)_3(OH,F)]$ is 100° C, of zircon (Zr_2SiO_4), 200° C, and of sphene $(CaTiSiO_5)$, 250° C. Apatite, zircon, and sphene are common accessory minerals in many granites and gneisses, and each of them contains small amounts of uranium that is present in atomic substitution. As the crust is uplifted tectonically—for example, during the formation of a mountain range—rock will be moved from a deep, hot region to a shallower, cooler region. Rocks near the top of the Alps, for example, were once buried at depths where they experienced temperatures in excess of 400° C. As a rising mass of crustal rock cools, the annealing temperatures of sphene, zircon, and apatite will be reached at specific depths, and each mineral will start recording the event through retention of fission tracks caused by radioactive decay of uranium atoms. Rocks near the top of a mountain may have a zircon fission track age of 4 million years, while those from 2 km below the summit may have a zircon fission age of 2 million years. This means that the summit rocks cooled below 200° C 4 million years ago, while those 2 km below the summit only cooled to that temperature 2 million years ago. Because the cooling is due to uplift, the rate of uplift can be calculated to be 1 mm/yr.

OTHER DATING SCHEMES

Radiometric dates are reliable and widely used but there are other dating methods that can also be used under special circumstances, and which do not rely on radioactive decay.

Annual Layering

The annual growth rings in trees have already been mentioned as a means of checking ^{14}C dates. Annual layering can also be found in the strata of certain lakes—particularly those in arctic and alpine regions. The strata are successive varves

(Chapter 5), each pair of layers comprising the deposit of a single year (Fig. 5.7). The coarse-grained unit is deposited in the summer, when the lake is open and thaw waters produce rapidly flowing streams. The fine-grained layer forms during the winter when streams flow at very low rates or, more commonly, when the lake is frozen and new sediment cannot get through the ice. Most of the materials that precipitate during winter months are the very finest-grained particles remaining in suspension from the summer months. In carefully collected samples, scientists have been able to count varved layers back as far as 10,000 years and so determine when a given layer was deposited.

Annual layering, corresponding to a season's snow accumulation, can also be found in glaciers. By drilling into the ice of large glaciers in polar and alpine regions it is possible to collect and date samples of snow, now compacted into ice, that formed a thousand or more years ago. The ages of layers determined by counting can be checked with ages determined from the ^{14}C content of ancient air bubbles trapped in the ice. The two methods usually agree quite closely.

Annual banding is also associated with the remains of some marine animals. Bands in corals represent layers of annual deposition (Fig. 8.8) that can tell us the age of a colony and permit determination of sea-surface temperatures for specific

FIGURE 8.8 Growth banding in a coral from San Cristobal, Galapagos Islands is revealed by an x-ray photograph. Each band represents a year's growth and arises as a result of rapid growth during the summer (lighter portions of the band) and slower growth during the cooler winter months (darker portion of the band).

a

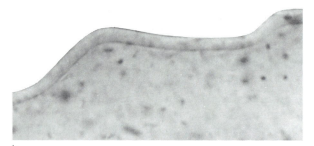

b

FIGURE 8.9 Hydration of obsidian used as a dating method. (*a*) A ceremonial obsidian blade from California. (*b*) Thin section of the obsidian in (*a*). The thin hydrated rim is visible on the edge of the obsidian. The width of the rim increases with time. The width indicates the age.

years through isotopic analysis of individual layers.

Relative-Dating Methods

A variety of methods has been developed to evaluate the *relative* ages of young geologic deposits. Most are based on the degree to which the deposits have been weathered, mass-wasted (Chapter 9), or eroded. For example, three sedimentary units may contain pebbles of fine-grained basalt that have weathered to produce a distinctive rind (Fig. 9.6). In the youngest deposit, the rinds may average only 0.5 mm thick, whereas in the preceding, older unit they are 1.5 mm thick, and in the oldest 2.5 mm thick. Statistical analysis of information on rind thickness obtained in the field may permit separation of the different units and aid in recognizing them over a large area. Weathering rinds may be useful in local regions, but they cannot be used on a global or continental scale.

Soils also provide a useful basis for separating deposits of different ages. Relatively young units are likely to have weakly developed soil profiles, perhaps characterized by a light-brownish color. Surface soils formed on older deposits will be more strongly developed, will possess a darker brown-

FIGURE 8.10 *Rhizocarpon geographicum*, a long-lived lichen growing on a boulder in a prehistoric rockfall on the south flank of the Mont Blanc Massif, Italy. The lichen is 205 mm in diameter. Based on the measured growth rate of the species in this area, the lichen, and hence the rock fall, is at least 875 years old.

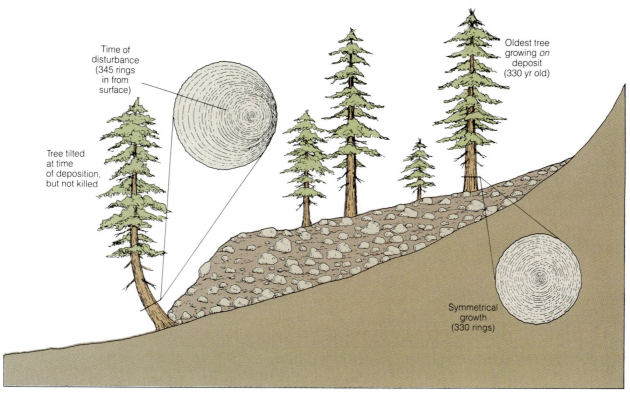

FIGURE 8.11 Use of tree-ring ages to determine the date of a rockfall deposit at the base of a steep slope. Oldest tree found growing on the deposit has 330 growth rings, indicating the deposit is at least 330 years old. The tree tilted, but not killed, at the time of the rockfall shows a change in rings from symmetrical to asymmetrical growth 345 rings in from surface. The rockfall therefore occurred 345 years ago.

ish or reddish color, and will display a noticeable buildup of clay (Chapter 9).

An interesting method of dating the obsidian tools of Stone Age man deserves mention. When glass, such as that of a freshly chipped surface of obsidian, is buried in moist soil, water starts to diffuse into the glass. The distance through which it diffuses is proportional to the square root of time ($\sqrt{t}$) and the diffusion changes the properties of the glass. By measuring the thickness of the changed, water-rich layer of the obsidian tool, it is possible to estimate the length of time the sample has been buried (Fig. 8.9). The method also has been used to date obsidian stones in glacial sediments of Yellowstone National Park, thereby providing information about glacial episodes as much as 150,000 years old.

Botanical Methods

In alpine regions, lichens soon establish themselves on newly exposed rock surfaces that may be produced by a landslide, rockfall, or a glacial retreat. Lichens are primitive plants and some species grow outward from the starting point, to produce almost perfect circular colonies. Growth rates are very slow but they can be measured. A good place to determine their growth rate is in cemeteries. By measuring the largest lichens on gravestones that have a date carved on them, a growth curve can be plotted that shows lichen size as a function of age. Some fast-growing species increase in size at a rate of at least 50 mm per century. Many lichens are long-lived, so large individuals are indications of very old colonies. In fact, some lichens growing in arctic regions are believed to be at least 8000 years old and may be the oldest continuously living organisms on our planet. If the growth rate of a species is known, then by measuring the largest lichen on a rock surface or deposit it is possible to determine a *minimum age* for exposure of the surface (Fig. 8.10).

The age of some young geologic units can be determined approximately by tree-ring dating.

Counts of rings on the oldest tree that can be found growing on a geological deposit can indicate the time the tree began to grow. If the elapsed time between formation of the deposit and establishment of the seedling was short, the tree ring age may provide a close minimum date for the unit (Fig. 8.11). If a tree were damaged or tilted (but not killed) when the deposit formed, recognition of the time of disturbance in the ring series can give an exact age.

A number of other possible dating methods are now being tested by scientists. Perhaps within the lifetimes of those who read this book, it will be possible to date accurately most of the materials we find in the Earth.

MEASUREMENT OF THE GEOLOGIC TIME SCALE

Through the various methods of radiometric dating, the dates of solidification of many bodies of igneous rock have been determined. Many such bodies have identifiable positions in the geologic column, and because of this it becomes possible to date, approximately, a number of the sedimentary layers in the column.

The standard units of the geologic column consist of sedimentary strata containing characteristic fossils, but the typical rocks from which radiometric dates (other than radiocarbon dates) are determined are igneous rocks. It is necessary, therefore, to be sure of the relative time relations between an igneous body that is datable and a sedimentary layer whose fossils closely indicate its position in the column.

Figure 8.12 shows in an idealized manner how apparent ages of sedimentary strata are approximated from the apparent ages of igneous bodies. The age of a stratum is bracketed between bodies of igneous rock, the apparent ages of which are known. In Figure 8.12 four sedimentary strata, whose geologic ages are known from their fossils, are separated by surfaces of erosion. Related to the strata are two intrusive bodies of igneous rock (*a,b*) and two sheets of extrusive igneous rock (*c,d*). From the apparent dates of the igneous bodies and the geologic relations shown, we can draw these inferences as to the ages of the sedimentary strata:

Stratum	Age (Millions of Years)	Interpretation
4	<34<30>20	Age lies between 20 and 30 million years
3	<60>34>30	Age lies between 34 and 60 million years
2	>60>34	Age of both is more than
1	>60>34	60 million years

To separate 1 from 2, dates from other localities are needed. Dates from igneous rocks elsewhere could also narrow the possible ages of 3 and 4. Through this combination of geologic relations and radiometric dating, we are able to fit a scale of time to the geologic column. The scale is being continually refined.

It is a great tribute to the work of geologists during the first half of the nineteenth century that the geologic column they established by the ordering of strata into relative ages has been fully confirmed by radiometric dating. Comparisons between the numbers column and the names column in Table

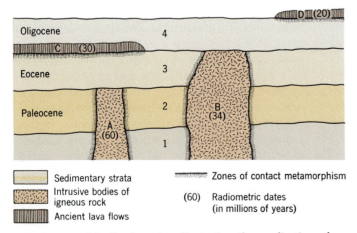

▨ Sedimentary strata	▨ Zones of contact metamorphism
▨ Intrusive bodies of igneous rock	(60) Radiometric dates (in millions of years)
▨ Ancient lava flows	

FIGURE 8.12 Idealized section illustrating the application of radiometric dating to the geologic column. For method, see the text discussion.

7.5 show this. They show also that the grouping of strata into the successively smaller subdivisions called systems, series, and stages is matched by the corresponding time units called periods, epochs, and ages. We can speak of the time units, of course, whether or not we know their dates. We could speak of events that occurred in the Devonian Period (or simply in Devonian time) based on the fossil record, even if we did not know that the dates of that period fall between 415 and 360 million years ago.

AGE OF PLANET EARTH

Table 7.5 shows us that the oldest rocks are the great assemblage of metamorphic and igneous kinds from the Archean and Proterozoic Eons, collectively known as Precambrian rocks. Of the many radiometric dates obtained from them, the youngest are around 600 million years, the oldest about 4 billion years. The Precambrian unit of the geologic column, then, existed during a *minimum* time equal to 4 billion minus 600 million years, or 3.4 billion years—a span nearly six times as long as the time elapsed since the Precambrian unit ended.

Given that some Precambrian rocks are 4 billion years old, the beginning of Planet Earth's history must be still farther back in time. The oldest radiometric dates have been obtained on individual zircon grains from rocks in Australia. Dates that are almost as old—3.8 billion years—have been obtained on a boulder of granite from a layer of conglomerate in Greenland. The existence of granite proves both that continental crust was present and that the rock cycle was operating 3.8 billion years ago. Further confirmation comes from another of the very ancient Precambrian rocks, a body of granite in South Africa. Although itself an igneous rock, this ancient granite contains great chunks of quartzite, much as a pudding contains raisins. At an earlier time, before it became enveloped by the granite magma, the quartzite must have been part of a layer of sandstone. Before that, it must have been part of a layer of loose sand. And even earlier still, an igneous rock must have been subjected to weathering and erosion to produce the grains of sand. Clearly, therefore, the rock cycle must have been operating in its present manner well before the granite magma solidified. Hence, as far back as we can see through the geologic column, we find evidence of the rock cycle and, because we see ancient sediment that

must have been transported by water, we know that when that sediment was deposited there must have been a hydrosphere.

We have been speaking of the oldest rock we can find. How much older might Planet Earth be? As we shall see in Chapter 23, strong evidence suggests that the Earth formed at the same time as the Moon, the other planets, and meteorites (small independent bodies that have "fallen" onto the Earth). Through various methods of radiometric dating and, in particular, the Rb/Sr and U/Pb systems, it has been possible to determine the ages of meteorites and of "Moon dust" (brought back by astronauts) as 4.6 billion years. By inference, the time of formation of the Earth, and indeed of all the other planets and meteorites in the solar system, is 4.6 billion years ago. Lead isotopes have been used to check this conclusion in the following manner.

Iron meteorites that are devoid of stony matter contain trace amounts of lead but are free of uranium. This is so because uranium does not replace metallic iron by atomic substitution, but lead does. When iron meteorites were formed, therefore, they incorporated a little lead but no uranium. Accordingly, the amount of ^{206}Pb and ^{207}Pb in iron meteorites must have remained unchanged from the time the meteorites formed. The only way a change could occur would be for ^{238}U and ^{235}U (the parents of ^{206}Pb and ^{207}Pb, respectively) to be present in the iron meteorite. The $^{206}Pb/^{207}Pb$ ratio in an iron meteorite is 0.903. Therefore, this must have been the $^{206}Pb/^{207}Pb$ ratio of the cosmic dust cloud from which meteorites and the planets formed.

Stony meteorites and stony planets do contain uranium. The earth is a stony planet, and its $^{206}Pb/^{207}Pb$ ratio is no longer 0.903. By measuring the ratio of $^{206}Pb/^{207}Pb$ in large samples of deep-sea clays, in seawater, and other well-mixed terrestrial materials, it can be estimated that the $^{206}Pb/^{207}Pb$ ratio for the whole Earth is presently 1.186. The ratio is slowly changing because the half-lives of ^{238}U and ^{235}U are, respectively, 4.5 billion and 710 million years. ^{235}U is disappearing faster than ^{238}U.

This means that the further back we go in the Earth's history, the smaller must have been the ratio of $^{238}U/^{235}U$. It follows then, that the further we go back the smaller must have been the ratio of the daughter products, $^{206}Pb/^{207}Pb$. It is possible to calculate that 2 billion years ago the ratio was 1, and that at 4.6 billion years ago it was 0.903, identical with the ratio in iron meteorites. While this is not absolute proof that the Earth formed 4.6 billion

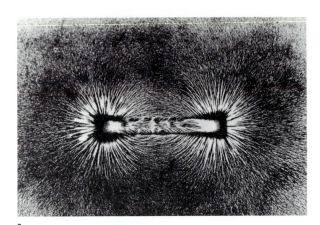
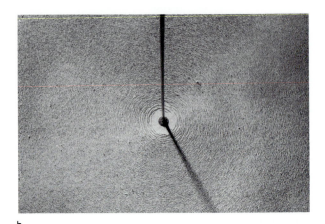

a   b

FIGURE 8.13 Lines of force in a magnetic field indicated by iron filings. (*a*) Lines of force around a bar magnet. The iron filings, which are on a sheet of paper covering the magnet, move freely and line up parallel to the lines of force. (*b*) Magnetic field around a wire carrying an electric current. The wire is perpendicular to the paper on which the iron filings sit.

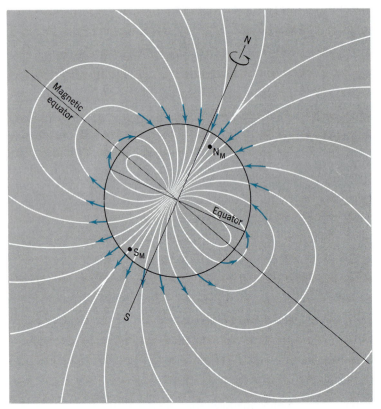

FIGURE 8.14 Magnetic lines of force surrounding the Earth. A free-swinging magnetic needle would point along the nearest line of force, with the north-seeking pole in the direction of the arrows. The axis of the magnetic field does not coincide exactly with the axis of the Earth's rotation. Where the axis of the magnetic field intersects the Earth's surface, a magnetic needle stands vertical and we define the points as the north and south magnetic poles. The north magnetic pole (N_M) lies in the arctic region of Canada; the south magnetic pole (S_M) lies in Antarctica, south of Tasmania.

years ago, it is very strong evidence in favor of the idea.

MAGNETISM AND THE POLARITY TIME SCALE

The Earth is a giant magnet. It creates an invisible force field that permeates everything in and around the Earth. The exact cause of the Earth's magnetism is not known, but it is apparently controlled by motions in the molten iron of the outer core and produced by electric currents flowing there. When an electric current flows, it does so by movement of electrons or other charged particles such as protons and ions. If an electric current flows in a wire, an invisible magnetic field surrounds the wire (Fig. 8.13). Similarly, if a wire moves through a magnetic field, an electric current flows in the wire. Electricity and magnetism are two aspects of the same phenomenon.

Magnetic lines of force surrounding the Earth define the **magnetic field** (Fig. 8.14). If a simple bar magnet is suspended on a thread so that it swings freely in any direction, the needle will come to rest along a line of force. The north-seeking pole of the bar magnet will point toward the Earth's north magnetic pole and the south-seeking pole toward the south magnetic pole. The actual orientation of the bar magnet will only be parallel to the ground surface at the magnetic equator. Everywhere else, the magnet is tilted at an angle to the horizontal, following the line of force. *The angle with the horizontal assumed by a freely swinging bar magnet is called the* **magnetic inclination** (Fig. 8.15). When the magnetic inclination is plotted all around the Earth, it is observed that there is a magnetic equator (where the inclination is zero) and that there are two magnetic poles (north and south) where the inclination is 90° and where the magnet stands vertical. There are only two magnetic poles so the Earth's magnetic field, like that of a simple bar magnet, is said to be a *dipolar field*. The north and south magnetic poles do not coincide exactly with the north and south poles of the Earth's axis of rotation (Fig. 8.14). A free-swinging bar magnet, therefore, does not point exactly to the true north pole—it points to the north magnetic pole. *The clockwise angle from true north assumed by a magnetic needle is the* **magnetic declination.**

The strength of the magnetic field fluctuates as much as 10 percent over a century, and the inclination and declination measured at any given place also vary because the positions of the magnetic poles wobble slowly and irregularly around the poles of rotation (Fig. 8.16). Neither of these properties is consistent with a magnetic field created by a solid-bar magnet buried deep in the Earth. To the contrary, they are properties consistent only with magnetic fields generated by a fluid. But how generated? The molten iron outer core is buried so deeply it cannot be seen. We can only hypothesize how the Earth's dipolar field is generated. Because the liquid core is hot, it probably contains convection cells which cause flow. The convective flow will interact with flow caused by the Earth's rotation. If a weak magnetic field is present, as it can be for any number of reasons, both from within the Earth or from outside it due to the Sun, the flowing molten metal will pass through the lines of force and an electric current will flow. The electric current in turn will create a new and stronger magnetic field, and this field in turn will help keep the electric current flowing. The process just described is called a *self-exciting dynamo*. The energy to keep it running comes from the kinetic energy of the Earth's rotation about its axis, and from the Earth's internal heat. Once started, the net electric current will flow roughly parallel to the equator. The magnetic field associated with the electric current will

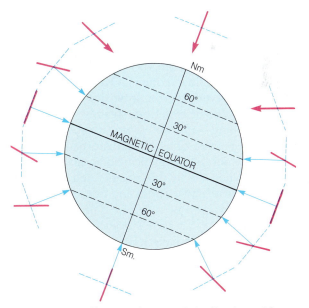

FIGURE 8.15 Change of magnetic inclination with latitude. Solid line is inclination taken up by a free-swinging magnet. Dashed line indicates a horizontal surface at each point. The magnetic poles move slowly and irregularly around the poles of rotation. By averaging a number of estimates of magnetic paleolatitudes, a good estimate can be reached for the true paleolatitude of a rock unit.

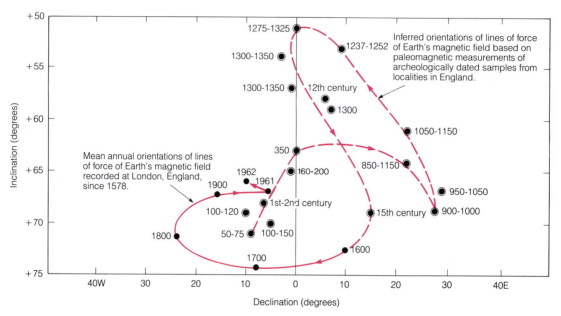

FIGURE 8.16 Variations in the declination and inclination of the Earth's magnetic field in the vicinity of London. Direct measurements made in London are shown on the solid curve; inferred positions from historical and archeological data are shown on the dashed curve. (*Source:* After Runcorn, 1964.)

have an axis perpendicular to the direction of flow of the electric current. As a result, the magnetic axis will be aligned roughly parallel to the Earth's axis of rotation.

Because the magnetic field is generated by fluid motions in the core, it can vary in strength due to fluctuations in the flow motions. Over hundreds of years it is observed that the intensity of the field can vary greatly—it is getting steadily weaker at the present time—and this is presumably due to turbulence or some other flow disturbance. It is even possible for the magnetic field to die down to near zero. When the field is regenerated, as it invariably is, the electric current and hence the magnetic poles may even be reversed. *Polarity reversals* are *changes of the Earth's magnetic field to the opposite polarity.* They have left unambiguous records in certain rocks. The length of time it takes for the field to die down and to reverse is geologically short—no more than a few thousand years—so it is possible to use polarity-reversal events as geological timemarkers.

Thermoremanent Magnetism

Certain minerals, the most important of which is magnetite, can become permanently magnetized. This property arises because orbital electrons spin-

ning around a nucleus are equivalent to an electric current and create a tiny magnetic field (Fig. 8.17). Above a temperature called the *Curie point, a temperature above which all permanent magnetism is destroyed,* the thermal agitation of atoms is such that permanent magnetism is impossible. The Curie point for magnetite is about 500° C. Above that temperature, the magnetic fields of all the tiny atoms are randomly oriented and cancel each other

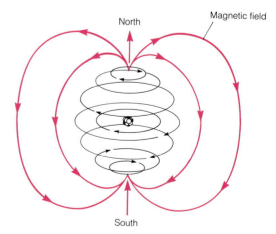

FIGURE 8.17 Movement of electrons around an atomic nucleus has the same effect as movement of electrons in a wire—they create a magnetic field.

out. Below the Curie point, the magnetic fields of adjacent atoms influence each other. Within small regions, or domains, of the solid, the magnetic fields reinforce each other (Fig. 8.18). When an external magnetic field is present, all magnetic domains in a solid that are parallel to the magnetic field become larger and expand at the expense of adjacent, nonparallel domains. Quickly, the parallel domains become predominant and a permanent magnet is the result.

Consider what happens when a lava cools. All the minerals crystallize at temperatures above about 700° C—well above the Curie points of any magnetic minerals present. As the crystallized lava continues to cool, the temperature will drop below 500° C, the Curie point for magnetite. When it does so, all the magnetite grains in the rock become tiny permanent magnets due to the Earth's magnetic field. Statistically, the magnetic poles of the magnetite grains in the lava will have the same declination and inclination as the Earth's field does. If a fragment of the lava is collected and tested it will have a distinct magnetic polarity imparted by all the tiny magnetite grains. So long as that lava lasts (until it is destroyed by weathering or metamorphism), or loses its magnetization, it will carry a record of the Earth's magnetic field at the moment it passed through the Curie point. *Permanent magnetism that is a result of thermal cooling* is called **thermoremanent magnetism.** It is the most important kind of paleomagnetism because it is the one that led to elucidation of the polarity time scale.

No magnet is actually permanent; eventually, when a magnet loses its magnetization, it is said to have relaxed. Permanent magnets have very long relaxation times. Many factors control the relaxation time; examples are mineral composition, grain size, temperature, and the strength of the original magnetization. For a rock to retain a record of the magnetic field that created the primary magnetization, at least a few magnetic grains must have relaxation times that exceed the age of magnetization. Relaxation times of a rock specimen can be measured in the laboratory as a function of temperature. Fortunately, a great many rock samples have relaxation times that are much in excess of the magnetization age. Such specimens can be used to obtain information about the Earth's magnetic field through geological ages.

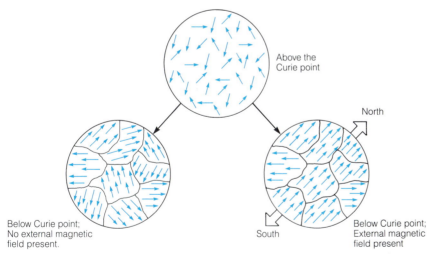

FIGURE 8.18 Magnetization of a magnetic material such as magnetite. Above the Curie point (about 500° C for magnetite), the thermal motion of atoms is so great that the magnetic poles of individual atoms, shown as arrows, point in random directions. Below the Curie point, atoms in small domains of a solid influence each other and form tiny magnets. As shown in the lower left, in the absence of an external magnetic field the domains have randomly oriented poles and the material does not produce an external magnetic field. In the presence of magnetic field (lower right), the pole directions of a majority of the domains tend to become parallel to that of the external field and the material becomes permanently magnetized. The case in the lower right is that for a magnetite grain in a crystallizing lava.

The Polarity-Reversal Time Scale

From a study of thermoremanent magnetism in lavas it was discovered that some rocks contain a record of reversed polarity. That is, when their paleomagnetism was measured, some lavas indicated a south magnetic pole where the north magnetic pole is today, and vice versa (Fig. 8.19). The ages of lavas can be accurately determined using radiometric dating techniques, especially the $^{40}K/$ ^{40}Ar method. Through combined radiometric dating and magnetic polarity measurements in thick piles of lava extruded over several million years, it has been possible to determine when magnetic polarity reversals occurred (Fig. 8.20).

Careful worldwide studies of thermoremanent magnetism in lavas have established many important facts. The first fact concerns the speed with which a reversal occurs. During a 1000- to 5000-year interval, the magnetic field slowly dies down to a very low intensity, the poles move erratically and fluctuate widely around the globe, then the magnetic field rapidly builds up again with the poles reversed. The second fact proves that polarity reversals are global effects, not local phenomena. A record of a reversal occurs on all continents and in all ocean basins at the same time. The most important fact concerns the actual record of polarity reversals. A detailed record of all changes back to the Jurassic Period has now been assembled, and still-earlier reversals are the topic of active ongoing research.

The polarity record for the past 20 million years is shown in Figure 8.20. *Periods of predominantly normal polarity* (as at present), *or predominantly reversed polarity*, *are called* **magnetic epochs** *or* **chrons.** The four most recent chrons have been named for scientists who made great contributions to studies of magnetism: Brunhes, Matuyama, Gauss, and Gilbert, respectively. It is apparent from Figure 8.20 that many of the magnetic reversals are short-lived, and that there is no obvious regularity in the reversal pattern. It is also apparent that short-term reversals sometimes occur during an epoch. *Short-term magnetic reversals are called* **magnetic events** *or* **subchrons.**

Use of magnetic reversals for geological dating differs from other dating methods. Magnetic reversal records all look the same. When the record of a reversal is found in a sequence of rocks, the problem is to know which of the many reversals it actually is. Additional information is needed. When a continuous record of reversals can be found, starting with the present, it is simply a matter of counting backward. This is the technique used in the dating of oceanic crust (Chapter 17). When the record is incomplete, it is necessary to have information such as an approximate radiometric date in order to identify a specific reversal or sequence of reversals. One of the most successful applications of polarity reversals has been with sediments rather than lavas.

Depositional- and Chemical-Remanent Magnetism

Sedimentary rocks acquire weak but permanent magnetism through the orientation of magnetic grains during or after sedimentation. As clastic sedimentary grains settle through an ocean or lake water, or even as loess particles settle through the air, many of the magnetic particles present will tend to orient themselves parallel to the magnetic lines of force. *Remanent magnetism acquired through processes of sedimentation is called* **depositional-remanent magnetism.** During diagenesis and cementation of sedimentary rocks, iron-oxide and iron-sulfide minerals, especially hematite (Fe_2O_3) and pyrrhotite (FeS), may grow in the sediment. As they grow, the magnetic domains become

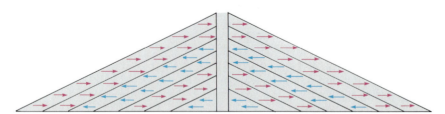

FIGURE 8.19 Lavas retain a record of the polarity of the Earth's magnetic field at the instant they cool through the Curie point. A pile of lava flows, like those in the volcanoes of the Hawaiian Islands, may record several field reversals, each of which can be dated using potassium-argon dating principles.

oriented parallel to the magnetic lines of force. *Remanent magnetism acquired through chemical precipitation and growth of magnetic minerals in a sediment is called **chemical-remanent magnetism.***

Depositional-remanent magnetism has proven to be a very sensitive and important dating technique. When fossils are present, an approximate age can be given to a sedimentary rock. Knowing the approximate age of a sediment, the exact age can be determined from the magnetic reversals.

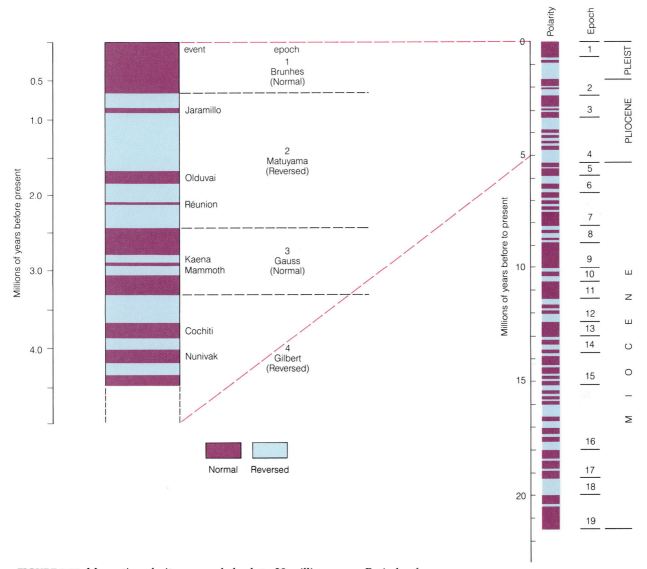

FIGURE 8.20 Magnetic polarity reversals back to 20 million years. Periods of primarily normal polarity, as today, and periods of primarily reversed polarity, are called magnetic epochs or chrons. The first four epochs have been named for people who made important contributions to our understanding of magnetism. To the base of the Miocene, 19 epochs have been identified. Within each epoch, one or more short-term polarity reversals may occur; these are termed magnetic events. During a normal epoch, events are periods of reversed polarity. The youngest events have been given names based on the location of samples in which they were discovered. Polarity reversals have now been dated back to the mid-Jurassic, approximately 162 million years ago. Polarity reversals are found in older rocks, but their exact durations have not yet been deduced.

Sediment cores recovered from the deep-sea floor (Chapters 5 and 20) can be dated very accurately using a combination of fossils and magnetic reversals.

With so many dating methods now available, geologists are starting to be able to provide quantitative answers to many questions which only a few years ago could only be approached in a descriptive way (Fig. 8.21). For example, magnetic reversals provide a way to measure rates of sedimentation in the world ocean. The problem of the time it takes for a given family of marine molluscs to evolve and disappear—the problem that defeated Lyell—can now be tackled through a variety of dating techniques. Indeed, the rates of most geological processes can now be either measured or estimated. Among the rates that can now be measured are the changes induced in geological processes by human activities.

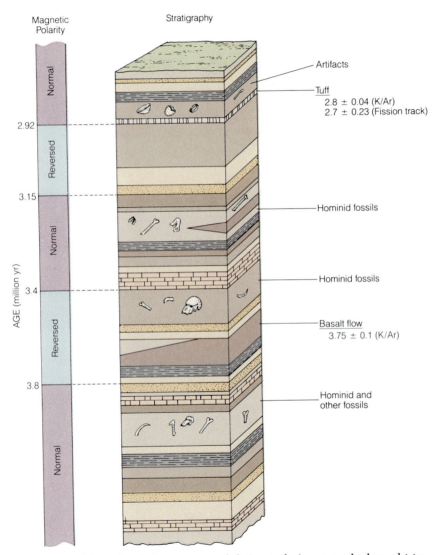

FIGURE 8.21 Example of the way several dating techniques can be brought to bear on a complex geological problem. Dates for sediments from the Haddar area of northern Ethiopia containing fossils of ancestral human beings (hominids) have been obtained by K/Ar and fission-track dating of basalt and tuff, as well as by paleomagnetism. The dates indicate that the early hominids lived in this region during the Pliocene Epoch (Table 7.5). (*Source:* After Johanson and Edey, 1981.)

SUMMARY

1. Decay of radioactive isotopes of various chemical elements is the basis of radiometric dating.

2. Potassium-argon ($^{40}K/^{40}Ar$) dating can be used both to determine the formation age of a mineral or rock, and the time of later metamorphism.

3. Rubidium-strontium ($^{87}Rb/^{87}Sr$) and uranium-lead dating techniques are most useful for dating very old rocks.

4. The age of the Earth, determined by uranium-lead dating, is 4.6 billion years.

5. Radiocarbon dating is only effective in relatively young materials (less than 50,000 years).

6. Remanent magnetism and the polarity-reversal time scale are particularly useful for dating oceanic crust, lavas, and young sedimentary rocks.

7. A sedimentary rock layer can only be dated radiometrically by being bracketed between two bodies of igneous rock to which the radiometric method can be applied.

SELECTED REFERENCES

Dalrymple, G. B., and Lanphere, M. A., 1969, Potassium-argon dating. Principles, techniques and applications to geochronology: San Francisco, W. H. Freeman and Company.

Eicher, D. C., 1976, Geologic time, 2nd ed.: Englewood Cliffs, N.J., Prentice-Hall.

Faure, G., 1986, Principles of isotope geology, 2nd ed.: New York, John Wiley & Sons.

Palmer, A. R., 1984, Decade of North American Geology geologic time scale: Geol. Soc. of America, Map and Chart Series MC-50.

Tarling, D.H., 1983, Paleomagnetism: London, Chapman and Hall.

CHAPTER 9

Weathering, Soils, and Mass-Wasting

Earth pillar produced by differential erosion of mudflow sediments at the base of the Straight Cliffs in southeastern Utah. Large sandstone boulder caps the 30 m-high column and protects the underlying unconsolidated sediment from erosion by sporadic rain storms.

EXTERNAL EARTH PROCESSES

We have now had a sweeping look at the Earth's general character—its solid surface, its gaseous, liquid, icy, and biologic envelopes. We have examined its crust (composed of minerals and rocks), the energy that drives its internal and external activities, the matter that moves gradually through the rock cycle, and the steady march of time, which is so long that even very slow processes have accomplished enormous results.

We will next examine the external processes in more detail, for they are the most easily visible. Together they form a chain in which rock is broken up, transported as sediment, and deposited to form strata. The logical processes to start with are weathering and mass-wasting. Through these processes, rock is decomposed and disintegrated, and the resulting pieces begin to move on their journey downslope.

WEATHERING

People have long sought stone that would be durable for buildings, tombstones, and other structures, but their success has been mixed. The durability of rock varies with climate, composition, texture, and degree of exposure to weather. If gravestones made of firm rock begin to crumble within only a few centuries (Fig. 9.1), what would happen to rock exposed to the atmosphere through thousands or millions of years?

Fast or slow, mechanical and chemical alteration occurs wherever the lithosphere and atmosphere meet. Their contact, however, is not easily drawn. It is a zone rather than a surface. It extends downward into the ground to whatever depth air and water penetrate. In this critical zone both hydrosphere and biosphere are also involved. Within it the rock constitutes a porous framework, full of fractures, cracks, and other openings, some of which are very small but all of which make the rock vulnerable. This open framework is continually being attacked, both chemically and physically, by water solutions. The result, given sufficient time, is conspicuous alteration of the rock.

When exposed to the atmosphere, no rock (whether bedrock or man-made structures of stone) escapes the effects of *weathering, the chemical alteration and mechanical breakdown of rock materials during exposure to air, moisture, and organic mat-*ter. The results of such alteration are often seen in highway cuts and other large excavations that expose the bedrock. In Figure 9.2, fresh, unaltered bedrock (1) grades imperceptibly upward through rock that has been altered but still retains its organized appearance (2), into loose, unorganized, earthy regolith (3) in which the texture of the fresh rock is no longer apparent. It is evident from such exposures that alteration of the fresh rock progresses from the surface downward.

The regolith seen in Figure 9.2 was formed *in situ* (in place) by alteration of the bedrock, and so we say it is *residual*. In many places, however, regolith is so different from the bedrock below that it cannot have resulted from chemical alteration of underlying rock. Instead, former regolith has been removed, and sediment has been transported from elsewhere to be deposited in its place. Both the removal of the former residual material and deposition of the sediment could have been performed by a single agent such as a river, surf along a coast, a glacier, or by two or more agents working together.

Processes of Weathering

If we could look closely at the bedrock in Figure 9.2, we would see that near the bottom of the exposure (1) the cleavage surfaces of feldspar grains in the gneiss flash brightly between grains of quartz. Higher up (2) such surfaces are lusterless and stained. Near the top (3) the grains of quartz, although still distinguishable, are separated by soft, earthy material that no longer resembles the former feldspar which has largely decomposed. Evidently, the changes that have occurred are mainly chemical and result from **chemical weathering** which is *decomposition of rocks*.

In some places, however, regolith consists of fragments identical to the adjacent bedrock. The mineral grains are fresh or only slightly altered. This relationship is commonly seen in the piles of loose rock fragments that mantle the lower parts of bedrock cliffs from which the debris obviously has been derived. When compared with the bedrock, the coarse rock fragments show little or no chemical change, implying that bedrock can be broken down not only chemically but also mechanically. Although we consider **mechanical weathering**, the *disintegration of rocks*, as being distinct from chemical weathering, the two processes generally work hand in hand and their effects are inseparably blended.

a

b

c

FIGURE 9.1 Three dated gravestones in a New England cemetery show the strong influence of rock composition on rate of chemical weathering in a humid climate. (*a*) Marble, consisting of very soluble calcite and exposed for 171 years, is greatly corroded. The entire surface is rough and the inscription is hard to read. (*b*) Medium-grained sandstone exposed for 196 years contains feldspar and micas. This rock is much less soluble than marble, but is roughened overall. (*c*) Very insoluble fine-grained slate, exposed for 275 years, is almost unaltered. Incised lettering is sharp and clear.

Mechanical Weathering

In many places regolith consists wholly of rock debris that is identical in every way with the local bedrock. Chemical alteration may be virtually undetectable, leading us to infer that the weathering processes responsible for such regolith must have been predominantly mechanical rather than chemical. Mechanical breakdown of rock is common in nature and is brought about by expansion due to removal of overlying load, growth of ice or salt crystals within fractures, the heat of fires, and activities of plants and animals.

Effects of Unloading

Rock masses buried deeply beneath the ground surface are subjected to enormous confining pressures due to the weight of overlying rock. As erosion wears down the surface, the weight, and

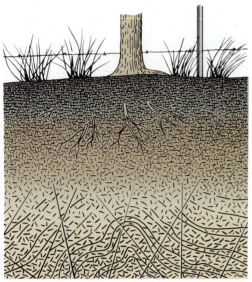

③ Loose, earthy regolith; texture and structures disappear as rock particles are slowly churned by roots, worms, and other agents.

② Bedrock weakened by chemical alteration.

① Fresh, unaltered granite gneiss with crystalline texture, wavy foliation, and joints.

FIGURE 9.2 Weathering profile showing gradation upward from fresh bedrock (granite gneiss) to earthy regolith.

therefore the pressure, are reduced. The rock may adjust to this unloading by upward expansion. As it does so, fractures can develop which appear at the surface as closely spaced joints. Rarely do joints occur singly. Most commonly they form a joint set. Joints that develop parallel to the land surface can produce extensive sheets of rock that resemble a stacked deck of cards (Fig. 9.3). Such sheets may be no more than 10 cm thick near the ground surface, but they become thicker with depth as the distance between joints increases. Generally they disappear below a depth of about 50 m.

FIGURE 9.3 Well-developed sheet jointing in massive granite forming stair-step surface on a mountainside in the Sierra Nevada, California.

Crystal Growth

Groundwater, percolating through fractured rocks, contains ions that may precipitate out of solution to form salts. The force exerted by salt crystals growing within rock cavities or along grain boundaries can be enormous and result in disaggregation or rupturing of rocks. Such effects are rather common in desert regions where precipitation of salts results from the evaporation of rising groundwater. Acid rain falling on industrial cities also can leach ions from masonry, bricks, and mortar. As they precipitate, crystallizing salts can disfigure and weaken buildings and important cultural monuments (Fig. 9.4).

In climatic regions where temperatures fluctuate about the freezing point for part of the year, water in the ground is subjected to periodic freezing and thawing. When water freezes to form ice, its volume increases by about 9 percent. In addition, as freezing occurs in the pore spaces of a rock, water is strongly attracted to the growing ice, thereby increasing the stresses against the rock. This leads to a very effective physical weathering process known as ***frost wedging,*** *the formation of ice in a confined opening within rock, thereby causing the rock to be forced apart.* The high stresses resulting from the increase in volume as ice crystallizes lead to disruptive effects. These effects are strong enough to force apart not only tiny particles but huge blocks of rock, some weighing many tons (Fig. 9.5). The disruption and breakdown is likely the result of the slow enlargement and extension of cracks as

FIGURE 9.4 Carved stone head on building in Florence, Italy shows effects of industrial pollution. Acid rainwater has caused disfiguring of the face by discoloration and disintegration of the rock.

ice crystals grow in them. For most rocks, frost wedging is probably most effective at temperatures that range from −5° to −15° C. At higher temperatures ice pressures are too low to be very effective, and at lower temperatures the rate of ice growth drops because the water necessary for crack growth is less mobile. Frost wedging is responsible for the production of most rock debris seen on the slopes of high mountains. At lower altitudes it is likely to be most important in places where the number of annual freeze–thaw cycles reaches a maximum.

Effects of Heat

Some geologists have speculated that daily heating of rock in bright sunlight followed by a comparable cooling each night should have a disruptive effect because the common rock-forming minerals expand by different amounts when heated. Surface temperatures as high as 80° C have been measured on desert rocks, and daily temperature variations of more than 40° C have been recorded on rock surfaces. Dark-colored rocks, like basalt, and rocks that do not easily conduct heat inward achieve the highest surface temperatures. Nevertheless, despite a number of careful laboratory experiments, no one has yet demonstrated that such heating and cooling have noticeable physical effects on rocks. These experiments, however, have been carried out only over relatively brief time intervals. Possibly thermal fracturing takes place only after

FIGURE 9.5 Blocks of granitic rock on the side of Taylor Valley, Antarctica that have been disrupted by frost-wedging and moved by creep.

repeated extreme temperature fluctuations over long periods of time.

Fire, on the other hand, can be very effective, as anyone knows who has witnessed the explosive shattering of a rock beside a campfire when it becomes overheated. The heat of forest and brush fires can lead to the spalling off of large rock flakes from exposed bedrock or boulders. Because rock is a relatively poor conductor of heat, an intense fire heats only the thin outer shell, which expands and breaks away. Studies of fire history in forested regions show that large natural fires, most set by lightning, may recur every several hundred years. Over long intervals of geologic time, fires may therefore have contributed significantly to the mechanical breakdown of surface rocks.

Plants and Animals

Seeds germinate in cracks in rocks to produce plants that extend their roots farther into the cracks. As trees grow, their roots wedge apart adjoining blocks of bedrock. In much the same way

FIGURE 9.6 Tree roots force apart the walls of ruined buildings of Ta Prohm at Ankor in the jungle of Cambodia. In the same manner, roots penetrating cracks in bedrock help force the rock apart.

they also disrupt sidewalks, garden walls, and even buildings (Fig. 9.6). Large trees swaying in the wind can cause cracks to widen, and if blown over can pry rock apart. Although it would be difficult to measure, the total amount of rock breakage done by plants must be very large. Much of it is obscured by chemical decay, which takes advantage of the new openings as soon as they are created.

Large and small burrowing animals (for example, rodents and ants) bring partly decayed rock particles to the surface where they are exposed more fully to chemical action. More than 100 years ago, Charles Darwin made careful observations in his English garden and calculated that every year earthworms bring particles to the surface at the rate of more than 10 tons per acre (2.5 kg/m²). After a study in the basin of the Amazon River, geologist J. C. Branner wrote that the soil there "looks as if it had been literally turned inside out by the burrowing of ants and termites." Although burrowing animals do not break down rock directly, the amount of disaggregated rock moved by them during many millions of years must be enormous. It illustrates again the cumulative effect of small forces acting over vast intervals of geologic time.

Chemical Weathering

Weathering brings about a good deal of readjustment of minerals to environments at the Earth's surface. Minerals that formed at high temperatures as components of igneous rock, and under high pressures and high temperatures in various metamorphic rocks, become unstable when exposed at the Earth's surface, where both temperature and pressure are much lower. Such minerals break down and their components form new, more stable minerals.

The active agents of rock decomposition consist of chemically active water solutions (weak acids) and water vapor. The effects of chemical weathering are therefore most pronounced in regions where precipitation and average temperatures are high enough to promote chemical reactions.

Effects on Rock-Forming Minerals

As rainwater falls through the atmosphere it dissolves small quantities of carbon dioxide, producing weak carbonic acid. Moving downward and laterally through the soil, this acid solution is strengthened by the addition of more carbon dioxide released from decaying vegetation. The carbonic acid ionizes to form hydrogen ions, which are extremely effective in decomposing minerals, and bicarbonate ions:

$$H_2O + CO_2 \rightleftharpoons H_2CO_3 \rightleftharpoons H^{+1} + HCO_3^{-1}$$

Water   Carbon   Carbonic   Hydrogen   Bicarbonate
        dioxide   acid   ion   ion

These hydrogen ions are so small that they can enter a crystal and replace other ions, thereby changing the composition.

The effectiveness of the H^{+1} ion is illustrated by the way in which potassium feldspar, a common rock-forming mineral, is decomposed by hydrogen ions in water:

$$4KAlSi_3O_8 + 4H^{+1} + 2H_2O$$

Potassium   Hydrogen   Water
feldspar   ions

$$\rightarrow 4K^{+1} + Al_4Si_4O_{10}(OH)_8 + 8SiO_2$$

Potassium   Kaolinite   Silica
ions

In this case, the H^{+1} ions enter the potassium feldspar and replace the potassium ions, which then leave the crystal and go into solution. Water combines with the remaining aluminum-silicate molecule to create the clay mineral kaolinite. This *chemical reaction, in which the H^{+1} or OH^{-1} ions of water replace ions of a mineral,* is called **hydrolysis.** It is one of the chief processes involved in the chemical breakdown of common rocks. The resulting kaolinite is a *secondary mineral,* because it was not present in the original rock. Kaolinite is the most conspicuous of the three products of the reaction. It is a common member of the group of very insoluble minerals that constitute clay, and as clay, it accumulates and forms a substantial part of the regolith. Many of the potassium ions released during the decomposition of potassium feldspar are eventually taken up by plants.

Silica, more soluble than clay minerals, remains partly in the clay-rich regolith or moves away in solution. Many of the potassium ions, likewise, escape in solution, and some, together with dissolved silica, find their way through streams to the sea. This matter carried away in solution is said to have been *leached* from the parent rock. **Leaching** is *the continued removal, by water solutions, of soluble matter from bedrock or regolith.*

The susceptibility of common rock-forming minerals to weathering is in reverse order to their crystallization from a magma (Fig. 4.23). In other words, the silicate minerals that crystallize at the highest temperatures (temperatures most different from those at the Earth's surface) tend to be the ones which weather most readily. They include olivine as well as calcium-rich feldspar, pyroxene, and amphibole. Biotite and sodium-rich feldspar are less easily weathered because they crystallize at lower temperatures. Quartz, which crystallizes at a still lower temperature, is among the most stable rock-forming minerals and experiences little obvious chemical decay. Nevertheless, over time, even quartz can be slowly taken into solution.

Iron is a common constituent of many common rock-forming minerals, including biotite, augite, and hornblende. When it is released during weathering it is rapidly changed from the ferrous form (Fe^{+2}) to the ferric form (Fe^{+3}) if oxygen is present. The result is a new red mineral, hematite:

$$4FeO + O_2 \rightarrow 2Fe_2O_3$$

Ferrous   Oxygen   Ferric
oxide   oxide

If water is present, a hydrous mineral may form through **hydration,** *the absorption of water into a crystal structure.* In the case of hematite, hydration leads to a yellowish, hydrated mineral called goethite:

$$2Fe_2O_3 + 3H_2O \rightarrow 2Fe_2O_3 \cdot 3H_2O$$

Hematite   Water   Goethite

The intensity of these colors in weathered rocks and soils can provide a clue as to the time elapsed since weathering began and to the degree or intensity of weathering.

Effects on Common Rocks

What happens in the weathering of potassium feldspar is a key to understanding the weathering of silicate rocks, such as granite, that contain this mineral. Table 9.1 contrasts the chemical weathering of granite and basalt, showing resistant minerals that persist, secondary minerals that form, and cations that are carried away in solution.

Carbonate rocks, such as limestone, are weathered in a different way. Limestone consists mainly of calcium carbonate which is only slightly soluble in pure water but undergoes the following reaction in the presence of weak carbonic acid:

$$CaCO_3 + H_2CO_3 \rightarrow Ca^{+2} + 2(HCO_3)^{-1}$$

Calcium   Carbonic   Calcium   Bicarbonate
carbonate   acid   ion   ions

The calcium and bicarbonate ions are removed in solution, leaving behind only the nearly insoluble impurities (chiefly clay and quartz) that are always present in small amounts in limestone. As lime-

Table 9.1 *Chemical Weathering of Two Great Groups of Igneous Rocks, Represented by Granite and Basalt*

	Primary Constituents		Weathering Products			
	Minerals	Cations	Colloids	Secondary minerals that form from colloids and ions	Primary minerals that persist	Soluble cations removed in solution
GRANITE	FELDSPARS	K^{+1} Na^{+1}	Silica, alumina	Clay minerals		Na^{+1} K^{+1}
	QUARTZ				Quartz	
	MICAS	K^{+1} Fe^{+2} Mg^{+2}	Silica, alumina	Clay minerals	Some mica	
	FERRO-MAGNESIAN MINERALS	Mg^{+2} Fe^{+2}	Silica, alumina	Clay minerals		Mg^{+2}
			Iron oxides	Hematite, goethite		
BASALT	FELDSPARS	Ca^{+1} Na^{+1}	Silica, alumina	Clay minerals		Na^{+1} Ca^{+2}
	FERRO-MAGNESIAN MINERALS	Mg^{+2} Fe^{+2}	Silica, alumina	Clay minerals		Mg^{+2}
	MAGNETITE	Fe^{+2}	Iron oxides	Hematite, goethite		

Primary minerals are shown in capital letters; minerals constituting the product of complete weathering are in boxes.

stone weathers, then, the residual regolith that develops from it consists mainly of clay and quartz particles.

Concentration of Stable Minerals

Not only quartz but other minerals as well are relatively stable at the Earth's surface, and so resist destruction by chemical weathering. Minerals such as gold, platinum, and diamond persist in weathered regolith, are eroded, and become sediment. Because some of these minerals are unusually dense, they settle to the beds of streams where they may collect to form a ***placer***, *a deposit of heavy minerals concentrated mechanically.* Those of economic value may be sufficiently concentrated to form valuable mineral deposits (Chapter 22).

Weathering Rinds

If a cobble of weathered basalt is cracked open, one commonly will see a discolored rim, or *rind*, surrounding a darker core of fresh unaltered rock (Fig. 9.7). Examination under the microscope reveals that the rind consists of residues resulting from chemical weathering. Similar discolored rinds form on all but the most chemically stable rock types. They gradually increase in thickness as weathering slowly attacks the solid core. As a result, geologists have found that rind thickness is a useful measure of the relative age of sediments that contain rocks of the same type and occur in comparable climatic settings.

Rinds also form in obsidian and other natural glasses when water vapor diffuses slowly through a freshly broken surface. The glass is thereby hy-

FIGURE 9.7 Basaltic stone from the eastern Cascade Range displays a well-developed weathering rind about 2 mm thick.

drated, producing a *hydration rind.* As we have seen (Chapter 8), such rinds have proved useful in dating ancient obsidian tools and sedimentary deposits.

Exfoliation and Spheroidal Weathering

As some jointed rocks weather they experience *exfoliation, the spalling off of successive shells, like the "skins" of an onion, around a solid rock core* (Fig. 9.8). It is caused by physical or chemical forces that produce differential stresses within the rock. The conversion of feldspars to clay through chemical weathering is accompanied by an increased volume of the weathered rock. This likely generates stress within the rock that causes shells to separate from the main body of the rock. In addition, rocks may experience a decrease in pressure as they are brought nearer to the Earth's surface by erosion. In some cases only a single exfoliation shell is present, but there may be 10 or more. The outermost shells tend to be parallel to joint planes and are relatively flat, whereas the innermost are progressively more spheroidal as corners become more and more rounded.

While exfoliation can take place at the surface, it also occurs below ground, for its results are often seen in newly made road cuts. The process is not

FIGURE 9.8 Spheroidally weathered volcanic rock in roadcut in the northern Cascade Range.

restricted to a particular climate, although its effects are especially noticeable in dry climates where spheroidally weathered boulders may cover the landscape (Fig. 9.9). The spheroidal forms created by exfoliation often form distinct rows running in several directions. This is because the spheroids are controlled by joints that existed in the rock before weathering began and that controlled the slow movement of water solutions through it.

At this point two important relationships should be noted. First, the effectiveness of chemical reactions increases with increased surface area available for reaction. Secondly, increased surface area results simply from subdivision of large blocks into smaller blocks. By merely subdividing a cube, while adding nothing to its volume, the surface area is greatly increased (Fig. 9.10). Repeated subdivision leads to a remarkable result. One cubic centimeter of rock subdivided into particles the size of the smallest clay minerals results in an aggregate surface area of nearly 4000 m². Weathering, itself, causes subdivision and so promotes further weathering. The way in which weathering decreases particle size is generally visible whenever an exposure displays a zone of solid rock passing upward to weathered regolith, as in Figure 9.2.

Factors That Influence Weathering

Rock Type and Structure

If different minerals react differently to weathering processes, then rock type clearly must influence decomposition. Quartz is so resistant to chemical breakdown that rocks rich in quartz are also resistant. In many places, hills and mountains that consist of granite or quartzite stand distinctly higher than surrounding terrain underlain by less-resistant rocks that contain less quartz.

The rate of weathering of a rock is influenced not only by mineral composition but also by its texture and structure. Even if a rock consists entirely of quartz (quartz sandstone, quartzite) but contains closely spaced joints or other partings, it may break down rapidly, especially if attacked by frost processes.

Contrasts in local topography often result from *differential weathering, weathering that occurs at different rates as a result of variations in the composition and structure of rocks or in intensity of weathering* (Fig. 9.11). In a sequence of alternating shale and quartz sandstone, the shale is likely to weather more easily, leaving the sandstone beds standing out in relief. If the beds are horizontal, the result is likely to be a stepped topography, with the sandstone forming abrupt cliffs between more gentle

FIGURE 9.9 Spheroidally weathered granite boulders near Iferouane in the Aïr Mountains of Niger. Thin sheetlike spalls are seen on the surfaces of the boulders.

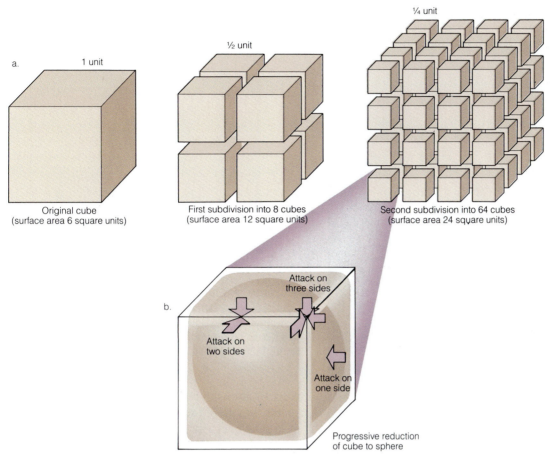

FIGURE 9.10 Subdivision and weathering of rock cubes. (*a*) Subdivision of a cube into smaller cubes. Each time a cube is subdivided by slicing it through the center of each of its edges, the aggregate surface area doubles. This greatly increases the speed of chemical reaction. (*b*) Geometry of spheroidal weathering. Solutions that occupy joints separating nearly cubic blocks of rock attack corners, edges, and sides at rates that decline in that order, because the numbers of corresponding surfaces under attack are 3, 2, and 1. Corners become rounded; eventually the blocks are reduced to spheres. Once a spherical form is achieved, energy of attack becomes uniformly distributed over the whole surface, so that no further changes in form occur.

slopes of shale. If the bedding is inclined, the sandstone will stand as ridges separated by linear depressions underlain by shale. Such topographic differences can be of considerable help to field geologists engaged in mapping the distribution of rock types in areas of limited outcrop.

Slope

When a mineral grain is loosened by weathering on a steep slope, it may be washed downhill by the next rain. With the solid products of weathering moving quickly away, fresh bedrock is continually exposed to renewed attack, so that weathered rock extends only to a slight depth beneath the surface. On gentle slopes, however, weathering products are not readily washed away, and in places accumulate to depths of 50 m or more.

Climate

Moisture and heat promote chemical reactions. Not surprisingly, therefore, weathering is more intense and generally extends to greater depths in a warm moist climate than in a dry cold one (Fig. 9.12). Rocks such as limestone and marble, which consist almost entirely of soluble calcite, are very susceptible to chemical weathering in a moist cli-

FIGURE 9.11 Differential weathering of jointed sandstone in Arches National Park Utah results in a remarkable topography of rock pinnacles and walls separated by deep clefts.

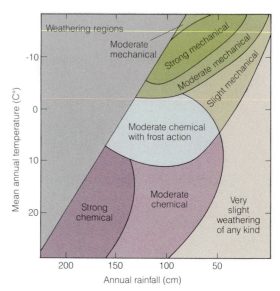

FIGURE 9.12 Climatic control of weathering process. Mechanical weathering is dominant where rainfall and temperature are both low. High temperature and precipitation favor chemical weathering. (*Source:* After L. C. Peltier, 1950.)

mate and commonly underlie subdued landscapes. In a dry climate, however, the same rocks may form bold cliffs, because with little rainfall and only patchy vegetation, the rocks infrequently come into contact with carbonic acid. In cold climates, chemical weathering proceeds very slowly. In such regions the effects of mechanical weathering are generally more obvious, with broad areas of bedrock being littered with frost-wedged rubble.

Time

Studies of the decomposition of stone in ancient buildings and monuments show that hundreds or even thousands of years are required for hard rock to decompose to depths of only a few millimeters. Granite and other hard bedrock in New England, Scandinavia, the Alps, and elsewhere still display polish and fine grooves made by glaciers during the latest glacial age, ~ 25,000–10,000 yr ago. In such cool-temperate climates it must take many

tens of thousands of years, at the very least, to create weathered regolith like that shown in Figure 9.2. However, in regions which have been continuously exposed to weathering processes for many millions of years, the zone of weathering often extends much deeper. In some tropical areas, mining operations have exposed bedrock that has been thoroughly decomposed to depths of 100 m or more.

The rates at which rocks weather have been obtained in several ways. First, experimental studies have been designed in which the length of the experiment provides time control, while the processes were speeded up by increasing temperature and available water and by decreasing particle size. Secondly, studies have been made of the degree of weathering of man-made structures, the ages of which are known historically. Thirdly, studies of radiometrically dated rock or sediments that have been exposed to weathering for thousands or millions of years can provide estimates of average rates over much longer intervals. Such studies suggest that the rates of most weathering processes decrease with time (Fig. 9.13).

The residues resulting from rock weathering, such as weathering rinds on basaltic stones, tend to be chemically stable. Their removal from the weathering zone is therefore likely to be minimal. As they build up, the rate of weathering tends to decrease. However, it may take a very long time,

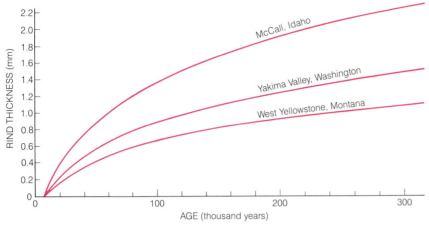

FIGURE 9.13 Graph showing change in weathering rates through time for three localities in the northwestern United States. Thickness of rinds on basaltic stones is plotted as a function estimated or known age. Differences between curves probably reflect differences in the weathering environment among the sites. All show initial rapid weathering followed by a steady decrease in rate. (*Source:* After Colman and Pierce, 1981.)

perhaps a half million years or more, before the rate slows to where weathering proceeds at a nearly constant rate.

SOILS

Origin

The physical and chemical breakdown of solid rock by weathering processes is the initial step in the formation of soil. However, soil also contains at least a little, and commonly much, organic matter mixed in with the mineral components. This organic fraction is an essential part of the usual definition of *soil: that part of the regolith which can support rooted plants.*

The organic matter in soil is derived from the decay of plant material, partly through the activity of bacteria. Living plants are nourished by decayed plant material in the soil as well as by decomposed mineral matter (derived from the regolith by chemical weathering) which is drawn upward, in water solution, through their roots. Therefore, plants are involved in the manufacture of their own fertilizer. These activities represent continual cycling of nutrients between the regolith and biosphere. With its partly mineral, partly organic composition, soil forms an important bridge between the Earth's lithosphere and its teeming biosphere. To people, it means food, and, thus, is a fundamental natural resource of every nation.

Soil Profile

As weathering of bedrock and regolith proceeds, soil gradually evolves. Normally it develops characteristic *horizons* that together constitute a *soil profile, the succession of distinctive horizons in a soil from the surface down to the unaltered parent material beneath it* (Fig. 9.14).

The uppermost horizon, called the A horizon,

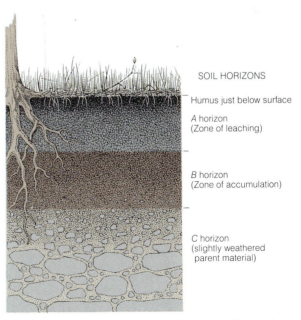

SOIL HORIZONS

Humus just below surface

A horizon
(Zone of leaching)

B horizon
(Zone of accumulation)

C horizon
(slightly weathered
parent material)

FIGURE 9.14 Horizons of a typical soil profile. Each horizon grades downward into the one below it.

typically is grayish or blackish (at least near its top) because of the addition of **humus,** *the decomposed residue of plant and animal tissues.* The A horizon has lost some of its original substance through the downward transport of clay particles and, more importantly, through the chemical leaching of soluble minerals.

The B horizon is commonly brownish or reddish, and enriched in clay and iron oxides produced by weathering of minerals within the horizon and also transported downward from the overlying A horizon. The B horizon is often characterized by structure: it breaks into blocks or prisms, each of which may be coated with clay. Although the B horizon generally is penetrated by plant roots, it contains less organic matter than the humus-rich A horizon.

The underlying C horizon does not constitute a part of the soil proper. It consists merely of slightly weathered parent material, either bedrock or regolith, in which oxidation has produced a detectable change in color.

Young or immature soils may lack a B horizon and display only an A horizon overlying a thin oxidized C horizon. As the soil develops, a B horizon appears, initially distinguishable by its color. As clay accumulates, the B horizon develops structure and the soil assumes a mature character. With the further passage of time, the B horizon slowly increases in thickness. Contrasts in the degree of soil development can often be seen as one travels across successively older sedimentary deposits exposed at the surface. In temperate latitudes, immature soils typically are found on sediments laid down during the last 10,000 years, whereas older units are capped by mature soils that display greater development with increasing age of deposits.

Soil-Forming Factors

Differences among soils, commonly reflected by differences in soil profile characteristics, result from the influence of several important soil-forming factors, namely climate, vegetation cover and soil organisms, parent material, topography,

TABLE 9.2 *Orders of the Soil Classification System Used in the United States*

Soil Order (Meaning of Name)	Main Characteristics
Alfisol (Pedalfer Soil)	Thin A horizon over a clay-rich B horizon, in places separated by a light-grey (A2) horizon. Typical of humid middle latitudes
Aridosol (Arid Soil)	Thin A horizon above a relatively thin B horizon and typically with calcium-carbonate accumulation in the C horizon. Typical of dry climates
Entisol (Recent Soil)	Very young soil with minimal development. Thin A horizon may be present
Histosol (Organic Soil)	Peaty soils rich in organic matter. Typical of cool, moist climates
Inceptisol (Young Soil)	Weakly developed soil with A horizon, and a B horizon lacking clay enrichment. May have slight calcium carbonate enrichment in the C horizon
Mollisol (Soft Soil)	Grassland soil with thick dark A horizon, rich in organic matter. B horizon may be enriched in clay. Carbonate horizon may be present
Oxisol (Oxide Soil)	Relatively infertile soil with A horizon over extremely weathered and often thick B horizon
Spodosol (Ashy Soil)	Acid soil of cool forest zones marked by highly organic surface horizon and an iron-rich B horizon
Ultisol (Ultimate Soil)	Strongly weathered soil of tropical and subtropical climates characterized by A horizon over highly weathered B horizon. Accumulation of residual silicates
Vertisol (Inverted Soil)	Old weathered soil having very high content of clays that shrink and expand as moisture varies seasonally

Source: After U.S. Department of Agriculture, Soil Conservation Service, 1975.

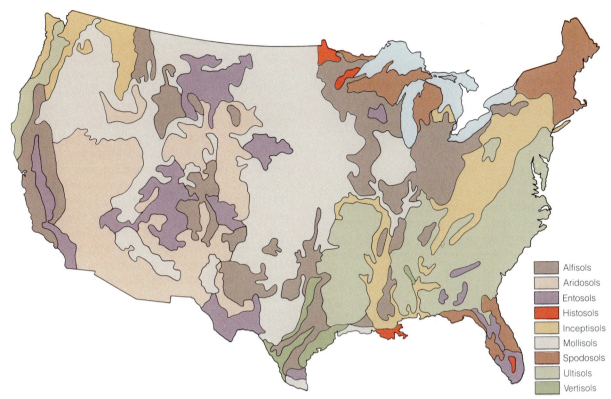

FIGURE 9.15 Major distribution of soils in the United States according to soil order. (*Source:* After USDA, Soil Conservation Service.)

and time. We have already seen that these factors influence weathering. They are also important in the evolution of soils.

Parent materials and topography differ widely and strongly influence the character of soils, especially during the early part of soil development. Climate, which in turn affects soil organisms and vegetation cover, may be an even stronger influence than bedrock in ultimately determining soil character. Under similar climatic conditions the profiles of mature soils developed on different kinds of rock become remarkably alike.

Soil Orders

Soil scientists classify soils according to their physical properties in much the same way that geologists classify rocks into rock-stratigraphic units (Chapter 7). They do so to make it easier to study and understand the great variety of soil types that exist, reflecting the interaction of the various soil-forming factors. Unfortunately, no worldwide standard system has been agreed upon. In the United States, a complex soil classification scheme has developed that makes it possible to classify soils at several levels using a hierarchy of names.

The scheme is quantitative, somewhat complicated, and designed for use by specialists. Nevertheless the largest units—the ten soil orders—provide a general framework that can be used to classify and map soils over broad regions (Table 9.2 and Fig. 9.15).

On the simplified soil map of the United States (Fig. 9.15), the pattern of soil types east of the Rocky Mountains is rather regular and, in general, reflects the regional climatic gradients of decreasing temperature from south to north and decreasing moisture toward the continental interior. In the west, the pattern is more complex because of the complicated arrangement of mountain ranges and intervening basins and plateaus. Soils, like the climate and vegetation that help determine their properties, show a relationship to topography and change character both upslope and laterally as average temperatures and precipitation change.

Caliche

In dry climates, where lack of moisture inhibits the leaching and removal of carbonate minerals, carbonates may accumulate in the upper part of the C horizon. This makes the soil strongly alkaline, in contrast to the acidic soils of humid regions. An important part of the carbonate accumulation results from evaporation of water that rises in the ground, bringing dissolved salts from below. In extensive arid areas of the southwestern United States, carbonates have in this way built up in the profile of aridosols *a solid, almost impervious layer of whitish calcium carbonate* generally known as *caliche* (Fig. 9.16).

Laterites

Oxisols differ from the other soil orders in having an *oxic horizon, a soil horizon characterized by extreme chemical alteration of the parent material.* Some oxic horizons have the interesting property of hardening after wetting and drying. The result is a material called *laterite, a hardened soil horizon characterized by extreme weathering that has led to concentration of secondary oxides of iron and aluminum.* Laterites constitute a primary source of bauxite, an ore of aluminum, from which the silica has been leached away. Because they are so hard, some lateritic soils can be cut into durable bricks and used for construction (Fig. 9.17). Where tropical forests have been cut down and the land reclaimed for agricultural purposes, the results have generally been disappointing, for the lateritic soils lack many of the important elements necessary to crops, and they often become rapidly hardened and impossible to cultivate.

Rate of Soil Formation

Although the development of soil constitutes part of the complex process of weathering, soil formation and weathering are not the same thing. In weathering, the time factor chiefly concerns the decomposition of bedrock, which is very slow and is essentially a geologic process. The time required to form a soil profile in regolith is much shorter.

A soil profile can form quickly in some environments. A study in the Glacier Bay area of southern Alaska (where because of moderate temperatures and high rainfall rapid leaching of parent material occurs) showed that within a few years after retreat of glaciers an A horizon develops on the newly revegetated landscape (Fig. 9.18). As the plant cover becomes denser, the soil becomes more acid and leaching is more effective. After about 50 years a B horizon appears and the combined thickness of the A and B horizons reaches about 10 cm. Over the next 165 years, as a mature

FIGURE 9.17 Photograph of laterite blocks used in the construction of the temples of Ankor Wat in the Cambodian jungle.

FIGURE 9.16 Profile of aridosol in central New Mexico showing whitish caliche at the top of C horizon.

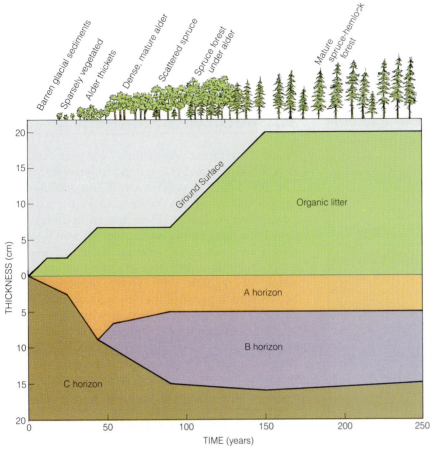

FIGURE 9.18 Progressive soil development in Glacier Bay, Alaska, over the past 250 years. With time, organic litter increases in thickness and A and B horizons develop. For first 40 years, A horizon increases in thickness to more than 5 cm, but directly overlies C horizon. B horizon then begins to develop and after another 60 years has reached a thickness of 10 cm. As vegetation changes, organic litter at the forest floor increases in thickness to about 20 cm after 150 years. (*Source:* Adapted from data from F. C. Ugolini.)

forest develops on the landscape, the A and B horizons increase in thickness to 15 cm and iron oxides accumulate in the developing B horizon.

In less humid climates, rates of soil formation are slower and it may take thousands of years for a detectable B horizon to appear. B horizons of soils in the mid-continental United States that have developed during the last 10,000 years contain little clay, whereas those dating back about 100,000 years generally display considerable clay enrichment and associated structure. By contrast, ice-free polar deserts of Antarctica are so dry and cold that sediments more than a million years old have only very weakly developed soils (entisols). Deep reddish ultisols of temperate and subtropical regions probably date to the Tertiary Period and took many millions of years to form.

The lengthy time involved in developing a mature, productive soil emphasizes the potentially disastrous results of soil erosion in agricultural regions. Agricultural soils are a prime natural resource, but once destroyed, they can only be replaced over geologically long intervals of time through weathering activity. Although they might be viewed as renewable resources over the very long term, over the lifetime of individuals, or even nations, they must be considered nonrenewable resources that should be carefully utilized and preserved.

Paleosols

If a landscape is covered by sediment or lava, the surface soil is buried and becomes part of the

FIGURE 9.19 Thick reddish-brown paleosol developed on a pumice layer near Guatemala City, Guatemala is overlain by a layer of airfall pumice and pyroclastic flows separated by thinner brownish paleosols. Fault has displaced the layers about 2 m vertically.

geologic record. The soil thereby becomes a **paleosol,** defined as *a soil that formed at the ground surface and subsequently is buried and preserved* (Fig. 9.19). It may have characteristic features that enable it to be traced over great distances. Paleosols have been identified in rocks and sediments of many different ages, but they are especially common in unconsolidated deposits of the Quaternary

Period. Distinctive and widespread paleosols have been used as time lines to subdivide, correlate, and date sedimentary sequences. They also can provide important clues about former landscapes, vegetation cover, and climate.

MASS-WASTING

Weathering forms one of the two major links between bedrock and the sediment derived from it. The other link is **mass-wasting,** *the movement of regolith downslope by gravity without the aid of a transporting medium.* This definition excludes sediment movement by water, ice, or wind, all of which are transporting media. Nevertheless, water plays an important role in mass-wasting. Saturation of regolith with water reduces friction between rock particles thereby making movement easier. This is the main reason why some mass-wasting activities are especially common and effective after long or intense rains. It is not always easy to separate weathering from mass-wasting or mass-wasting from erosion, for they constitute a continuum of processes that interact and overlap. Their end result is the gradual breakdown of solid bedrock and the redistribution of its weathered components.

Gravity

A smooth vegetated slope may appear outwardly stable and show little obvious evidence of geologic activity. Yet if we examine the regolith beneath the surface, we quite likely will find rock particles derived from bedrock that is exposed only in areas that lie farther upslope. We can deduce, therefore, that the particles have moved downslope.

The force that makes the rock particles move is *gravity,* as it pulls persistently on rock debris at the

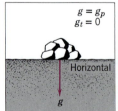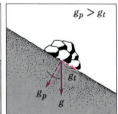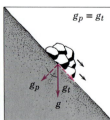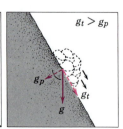

FIGURE 9.20 Effects of gravity on a rock lying on a hillslope. Gravity can be resolved into two components, one perpendicular (g_p) and one parallel (g_t) to the surface. The perpendicular component creates resistance to sliding. When g_t exceeds g_p the object will move.

TABLE 9.3 *Classification of Rapid Mass-Wasting Processes*

Dominant Material Involved

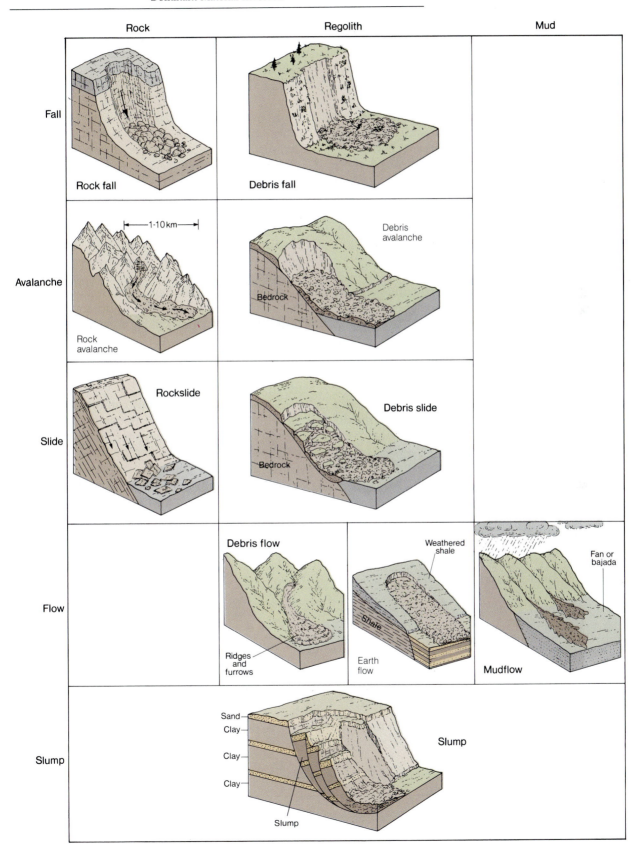

Earth's surface. On a horizontal surface, gravity holds objects in place by pulling on them in a direction perpendicular to the surface. On any slope, gravity can be resolved into two component forces. The *perpendicular component of gravity* (g_p in Fig. 9.20) acts at right angles to the slope and holds objects in place. The *tangential component of gravity* (g_t in Fig. 9.20) acts along and down the slope. When g_t exceeds g_p, objects move downhill, and we say that the slope has exceeded the **angle of repose** which is *the steepest angle, measured from the horizontal, at which rock material remains stable.*

A rock on a hilltop that reached its position because it was lifted, against the pull of gravity, as the hill was created possesses **potential energy.** This is defined as *stored energy.* The potential energy (E) of a rock of mass (m), raised to a height (h), where g is the acceleration due to gravity is

$$E = m\,g\,h$$

Similarly, moisture lifted to its position in a cloud by solar energy also has potential energy. When the moisture falls as rain, its potential energy is converted to **kinetic energy,** which is *energy that results from the motion of an object.* The kinetic energy (E) of a mass (m), moving at a velocity (v) is

$$E = \tfrac{1}{2}\,mv^2$$

In the same way, as a rock moves downslope, its potential energy is transformed into kinetic energy.

Downslope Movement

A particle loosened by weathering ceases to be bedrock and becomes regolith. A loose particle possessing potential energy will at some time move downslope. Sooner or later it reaches a stream or some other agent of transport, which will carry it farther. The beginning of its journey, the trip down the nearest slope, can be very slow or very fast, but in either case it is controlled primarily by gravity.

One way to visualize mass-wasting on a hillslope is to think of it in terms of the addition of mass (input) and the loss of mass (output). In other words, we can view it as an **open system,** defined as *any system, involving a given process or set of processes, in which material is either added or removed.* The input consists of the solid products of weathering, contributed all the way down each slope. The output consists of sediment discharged at the base of a slope where it can be carried away by a transporting agent such as a stream or a glacier. As soon as the regolith begins to move downslope it becomes sediment by definition. If input equals output, then the system is said to be in a *steady state,* or balanced, condition. Under this condition, the slope is likely to achieve an angle which at any point will permit just the quantity of regolith moving from upslope to be balanced by the quantity of material that is moving on downslope.

Mass movement is not only confined to the land, for regolith, in the form of transported sediment, covers vast areas of the seafloor. Some of this material reaches deeper water by mass movement down submarine slopes. For example, ocean-bottom surveys have disclosed topography along the margins of continents and some islands that indicate large submarine landslides. Because mass movements are so effective in moving sediment,

TABLE 9.4 *Characteristics of Some Representative Large Rock Avalanches*

Locality	Date	Volume (million m³)	Vertical Movement (m)	Horizontal Movement (km)	Calculated Velocity (km/h)
Huascaran, Peru	1971	10	4000	14.5	400
Sherman Glacier, Alaska	1964	30	600	5.0	185
Mt. Rainier, Washington	1963	11	1890	6.9	150
Madison, Wyoming	1959	30	400	1.6	175
Elm, Switzerland	1881	10	560	2.0	160
Triolet Glacier, Italy	1717	20	1860	7.2	≥125
Blackhawk, California	pre-historic	280	1220	8.0	120
Saidmarreh, Iran	pre-historic	2000	1650	14.5	340

even on relatively gentle slopes, they are very likely a major factor contributing to the development of the continental slope and rise. As on land, such sediment movement in the oceans is controlled mainly by gravity. Therefore, we can think of mass-wasting as universal, and active wherever slopes exist.

Classification of Processes

A mass-wasting event is often referred to as a **landslide,** *a general term covering a variety of mass-movement processes.* However, this oversimplifies the distinctive character and resulting landforms involved in mass movements. Although they dif-

a

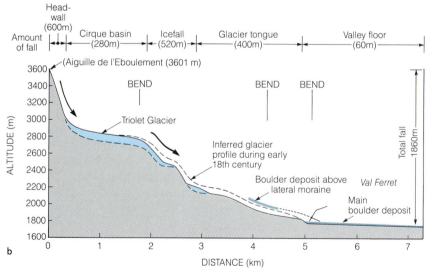

b

FIGURE 9.21 Large rockfall in upper Val d'Aosta, Italy that traveled 7 km from its source high on a mountain spur. Within only a few minutes the debris buried two communities on the valley floor, killing all the inhabitants and livestock. (*a*) Arcuate bouldery front of rockfall deposit. (*b*) Cross section diagram showing the trajectory of the rockfall, based on deposits left along the valley sides. (*Source:* After Porter and Orombelli, 1980.)

FIGURE 9.22 New apartment building at base of a steep mountain slope in the Italian Alps that was struck by large boulder falling from the cliffs above. The rockfall, which occurred one day before the owners were to move in, demolished the bedroom and most of the living room.

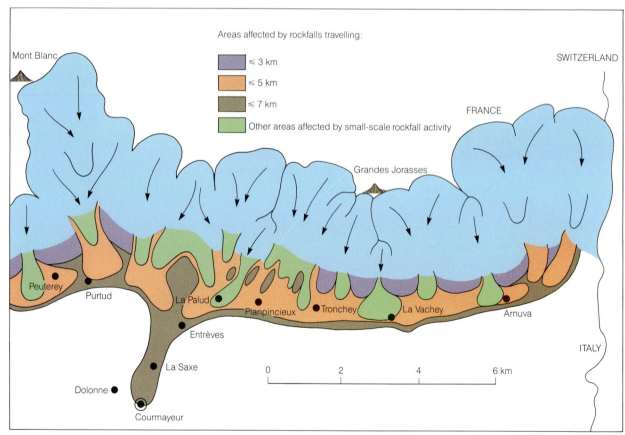

FIGURE 9.23 Map of part of the Mont Blanc massif along the French-Italian border showing zones that could be reached by giant rockfalls like that of Figure 9.21 traveling up to 3, 5, or 7 km from potential source areas high on the mountain slopes. Distribution is based on known pattern of rockfall deposits and on potential trajectories. Several small communities are built on or near large prehistoric rockfall deposits and lie within zones of potential future rockfall activity. (*Source:* After Porter and Orombelli, 1981.)

fer in other respects, these processes share one characteristic: They all take place on slopes. It would be satisfying to be able to categorize the various processes according to the type of motion each displays, but this is not strictly possible because some involve two or more distinct kinds of motion. Alternatively, it would be useful to separate them according to their velocities, but again we cannot do so strictly, for the velocity of any single process at one locality may vary from one time to another. In Table 9.3, the most rapid processes are differentiated on the basis of the dominant type of movement and the materials involved. Their separation is somewhat arbitrary, for both movement and materials tend to be gradational among processes. So are the rates of processes, which in some cases overlap. Where regolith is involved, velocity is often influenced by the relative fluidity of the moving debris or sediment which, in turn, is influenced by its water content. In the following paragraphs, we will first examine the more rapid processes, and then the slower ones.

Falls, Slides, and Avalanches

Rockfall, *the relatively free-falling of detached bodies of bedrock from a cliff or steep slope,* is generally a small-scale event that tends to occur when freezing and thawing take place frequently. A rockfall may involve the dislodgment and fall of a single rock particle, or it may involve the sudden collapse of a hugh mass of rock that breaks on impact into a vast number of smaller pieces which continue to bounce, roll, and slide downslope before friction and decreasing gradient brings them to a halt.

Large rockfalls and ***rockslides,*** *the sudden and rapid downslope movement of newly detached masses of bedrock across an inclined surface,* are much less common and occur mainly in high glaciated mountains where steep slopes abound. ***Rock avalanches*** *are large masses of falling rock that break up, pulverize on impact, and may then continue to travel downslope, often for great distances.* They move at high velocity, may involve extremely large volumes of broken rock, and can be extremely destructive (Table 9.4).

Large rockfalls and rock avalanches have had the greatest human impact in populated mountain regions like the Alps. For example, in September 1717 a large mass of rock and glacier ice fell from the crest of the Mont Blanc massif along the French-Italian border onto Triolet Glacier (Fig. 9.21). Pulverizing on impact, the fragmented debris moved rapidly downvalley where it overwhelmed two settlements before its front came to rest some 7 km from, and 1860 m lower than, the site of detachment. An estimate of the velocity can

FIGURE 9.24 Slump on a large scale in coastal cliffs at Point Firmin, California. Principal surface of slip is curved; it cuts cliff and displaces highway at two places. Erosion by surf along the base of the cliff may have been the chief factor in removing support and causing failure.

be obtained by equating the kinetic and potential energy of the rock mass

$$\tfrac{1}{2}\,mv^2 = m\,g\,h$$

and then solving for velocity

$$v = \sqrt{2\,g\,h}$$

By inserting values for gravitational acceleration (9.8 m/s) and initial distance of fall (400 m), we arrive at a velocity on impact of close to 320 km/h. As the sheet of debris reached the floor of the main valley, its momentum carried it up the opposite valley wall to a height of at least 60 m. Using the same equation, we can calculate that its velocity

a

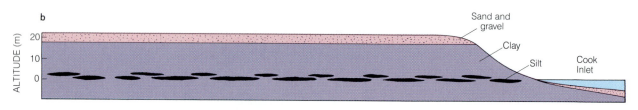

b

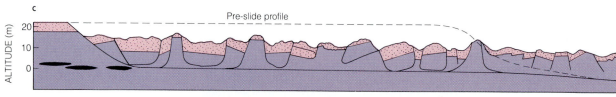

c

FIGURE 9.25 Chaotic terrain in Anchorage, Alaska caused by slope failure during the great earthquake of March 27, 1964. (*a*) Slump blocks at Turnagain Heights that have partially destroyed a suburban housing area. (*b*) Cross section through bluffs prior to earthquake. Sandy glacial gravel overlies clay containing lenses of silt. (*Source:* After NAS/NRC, 1984.) (*c*) Cross section after earthquake. Failure has occurred along the weak silt zone. Slumps have produced tilted, chaotic landscape as surface was lowered and blocks of sediment were moved laterally into the bay. (*Source:* After NAS/NRC, 1984.)

must have then been at least 125 km/hr. The total travel time, from start to finish, over the entire 7 km distance must have been between 2 and 4 min.

Large rock avalanches such as the one at Triolet Glacier can occur without warning and travel at such high velocities that they constitute, together with frequent smaller rock falls, an ever-present hazard in many mountainous areas (Fig. 9.22). The dangers can be evaluated by studying the distribution and age of previous events and constructing maps showing areas of potential rockfall hazards (Fig. 9.23).

Because rock avalanches are infrequent and extremely difficult to study, we have little observational data about the process. It has been proposed that the extreme mobility of the debris is due to its riding upon a layer of compressed air, somewhat like the manner in which a commercial hovercraft moves across a gentle surface, also on a layer of compressed air. Alternatively, it may be due to air trapped within the debris mass that causes it to behave in a fluid-like manner.

The processes described thus far primarily involve the detachment and downslope movement of rock. Although in most cases some regolith is involved, it generally constitutes a minor component of the material in motion. Equally common are comparable mass movements involving mainly regolith. The corresponding processes are referred to as *debris fall*, *debris slide*, and *debris avalanche* (Table 9.3). The debris involved may include bedrock, but the bulk of the sediment in motion is regolith, together with organic remains. It may also include such cultural debris as houses, bridges, and automobiles! Each of these processes is gradational into the analogous process that involves mainly rock.

Slumps

A **slump** is *a type of slope failure in which a downward and outward rotational movement of rock or regolith occurs along a concave-up slip surface* (Fig. 9.24). The movement occurs around a horizontal axis that lies parallel to the ground surface. The top of the displaced block usually is tilted backward, producing a reversed slope. A series of adjacent slump blocks forms a characteristic hummocky topography consisting of more-or-less concentrically aligned depressions. Slumps are common where strong, massive rock or sediment units overlie weak, deformable units. They occur most frequently in association with heavy rains or sudden shocks, such as earthquakes. Much of the structural damage to buildings in Anchorage during the great Alaska earthquake of 1964 resulted from slumping caused by failure of weak lake sediments that underlie part of the city (Fig. 9.25).

Slumps are one of the types of mass movement that we are most likely to see, for many result from human modification of the land. They are numerous along roads and highways where bordering slopes have been oversteepened by construction activity. They also are seen along river banks or seacoasts where currents or waves undercut the base of a slope.

Flows

Debris flows are a conspicuous form of mass movement involving *the downslope movement of a mass of unconsolidated regolith more than half of which is coarser than sand*. In some cases they begin with a slump, the lower part of which then continues to flow downslope (Fig. 9.26). Very slow-moving debris flows travel at a rate of no more than 1 m/yr, whereas fast ones may travel at velocities of sev-

FIGURE 9.26 Slump and debris flow near Mangaweka on North Island, New Zealand. Movement occurred following heavy rain on slope that recently had been cleared of forest cover.

FIGURE 9.27 Deposits of a large mudflow that passed along the floor of the valley leading from the upper slopes of Huascaran, a high glacier-clad summit in the Peruvian Andes. A large fall of rock, snow, and ice was converted into a mass of rapidly flowing debris that swept down the valley and buried a large town, killing many of its inhabitants.

eral km or more per hour. Commonly debris flows have an apron-like or tongue-like front. They also possess a very irregular surface marked by concentric ridges and transverse depressions that resemble the deposits of mountain glaciers. Debris flows are frequently associated with intervals of extremely heavy rainfall that lead to oversaturation of the ground.

Earthflows which *are transitional between debris flows and mudflows, are predominantly fine-grained, and have a higher water content than debris flows.* Like debris flows, they often involve weak or weathered surficial deposits, moderately steep slopes, and excessive rainfall. One type of earthflow occurs in highly porous clays that contain a great deal of interstitial water. These "quick clays" will weaken if they are stressed suddenly and strongly, as by an earthquake. A sudden shock

mobilizes the pore water thereby reducing frictional contact between grains. The clays are fluidized and will then fail abruptly. Any structure built on them or in their path may be quickly demolished.

A *mudflow* is *a flowing mass of predominantly fine-grained rock debris that generally has a sufficient water content to make it highly fluid.* As a result, mudflows tend to travel along valleys, just as streams do (Fig. 9.27).

Mudflow sediment ranges in consistency from mud as stiff as freshly poured concrete to a soup-like mixture nearly equal to that of very muddy water. In fact, after heavy rains in mountain canyons, a mudflow can start as a muddy stream that continues to pick up loose sediment until its front portion becomes a moving dam of mud and rubble, extending to each steep wall of the canyon and

sist in large measure of superposed sheets of mudflow sediment that are interstratified with alluvium (Fig. 9.28).

Because of their high density, which enables them to move large, heavy objects, mudflows can be very destructive. Buildings in the path of some mudflows have been torn from their foundations, while automobiles and even locomotives have been carried for substantial distances. Large boulders, some having diameters of 10 m or more, have been transported far out onto gentle slopes beyond mountain valleys. The occurrence of huge isolated boulders in such positions has sometimes led to the mistaken inference that they were carried by

FIGURE 9.28 Three superposed mudflow deposits, separated by buried soils, exposed in an alluvial fan in Arroyo Grande, Argentine Andes.

urged along by the force of the flowing water behind it. On reaching open country at the mountain front, the moving dam collapses, floodwater pours around and over it, and mud mixed with boulders is spread as a wide thin sheet. Sediment fans at the foot of mountain slopes in many arid regions, as in the American southwest, the central Argentine Andes, and the Hindu Kush of central Asia, con-

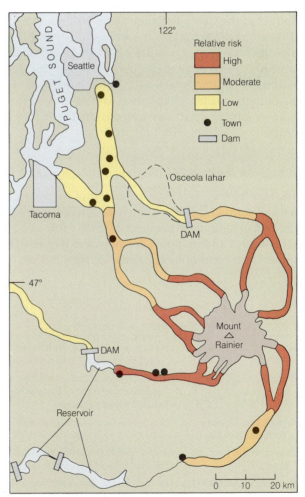

FIGURE 9.30 Map of southeastern Puget Lowland, Washington, showing areas of low, moderate, and high risk from mudflows and floods originating at Mt. Ranier volcano. The extent of the 5000-year-old Osceola lahar is shown by a dashed line. (*Source:* After Crandell and Mullineaux, 1975.)

FIGURE 9.29 Lahar from Nevado del Ruiz, a large Colombian volcano, that moved rapidly downslope during an eruption in 1985, burying most of the town of Armero and killing more than 20,000 people.

former glaciers or by huge floods. Certainly some and probably many are the result of mudflows, the muddy matrix having been largely or entirely stripped away by later erosion.

In regions of explosive volcanic activity, layers of volcanic ejecta commonly mantle the surface and are especially susceptible to mobilization as mudflows. A sudden intense rain, or an eruption that melts a large mass of snow or ice, can mobilize loose unstable sediment and can bring it into and down valleys as a *lahar,* an Indonesian term for *a mudflow that consists chiefly of volcanic debris and originates on the flank of a volcano.* Highly mobile lahars can travel great distances from a volcano and at such high velocities (100 km/hr or more) that they constitute one of the major hazards associated with some volcanic eruptions (Fig. 9.29). A lahar that originated on the slopes of Mount Rainier in the Cascade Range about 5000 years ago traveled as far as 72 km. It spread out beyond the mountain front as a broad lobe about 12 km wide where it averages about 25 m thick (Fig. 9.30). Its estimated volume is well over 1 billion m³. A similar though smaller lahar buried and destroyed the Roman city of Herculaneum during the famous eruption of Mount Vesuvius in 79 A.D.

Very Slow Processes

The slow viscous downslope movement of waterlogged soil and surficial debris is known as **solifluction.** Although such flowage is especially noticeable in polar latitudes and in alpine zones where it operates under frost conditions (Chapter 13; Fig. 13.35d), it

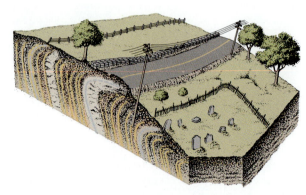

FIGURE 9.31 Effects of creep on surface features and on bedrock. Steeply dipping layered rocks have been dragged over by creep, so that they appear folded. Poles and posts affected by creep are tilted, stone fences are deformed, gravestones are tilted or fallen. Plants rooted in the zone of creep may be tilted or deformed due to slow downslope movement.

also occurs in temperate and tropical latitudes. The movement is so slow, generally only a few centimeters a year, that it can be detected only by field measurements made over several seasons. The slow flowage results in distinctive surface features, including lobes and sheets of debris that often override one another. Surface vegetation carried along with moving debris may be deformed and folded so that it resembles a crumpled carpet.

In contrast to the processes described thus far, some mass-wasting proceeds at rates that are barely detectable. *Creep, the imperceptibly slow downslope movement of regolith,* is evident in the con-

TABLE 9.5 *Factors Contributing to Creep of Regolith*

Frost Heaving	Freezing and thawing, without necessarily saturating the regolith, causing lifting and subsidence of particles
Wetting and Drying	Causes expansion and contraction of clay minerals; creation and disappearance of films of water on mineral particles causes volume changes
Heating and Cooling Without Freezing	Causes volume changes in mineral particles
Growth and Decay of Plants	Causes wedging, moving particles downslope; cavities formed when roots decay are filled from upslope
Activities of Animals	Worms, insects, and other burrowing animals, also animals trampling the surface, displace particles
Dissolution	Dissolution of mineral matter creates voids which tend to be filled from upslope
Activity of Snow	Where a seasonal snow cover is present, it tends to creep downward and drag with it particles from the underlying surface

sistent leaning of old fences, poles, and grave-stones, and in the fracture and displacement of road surfaces (Fig. 9.31). Natural and artificial exposures of bedrock often show steeply inclined layers of rock bent over in the downslope direction just below the ground surface, a result of slow differential creep.

On sloping ground a cover of dense grass or other vegetation will form a protective armor against the cutting of gullies by running water and so keep the slopes smooth. Therefore, one might suppose that ground so protected loses nothing to erosion except for mineral matter dissolved and carried off underground. However, close observation shows clearly that the regolith is creeping downslope, carrying the vegetation with it.

Several factors are responsible for creep (Table 9.5). Most are common phenomena and their contribution to creep is rather obvious. One factor, common in regions with cold winters, involves the freezing of water in the regolith. The increase in volume pushes the ground surface up. This *lifting of regolith by freezing of contained water* is called **frost heaving.** On a hillside the ground surface is lifted essentially at right angles to the slope. When thawing occurs, each particle tends to drop vertically, pulled downward by gravity. Its net motion is, therefore, a small distance downslope (Fig. 9.32). Long-term movement consists of a complex series of zigzags that can lead to significant downslope displacement.

Although creep occurs at a rate too slow to be seen, careful measurements of the downslope displacement of objects at the surface show the rates involved. As might be expected, rates tend to be higher on steep slopes than on gentle slopes. Measurements in Colorado, for example, document a rate of 1.5 mm/yr on a 19° slope but indicate a rate of 9.5 mm/yr on a slope of 39°. Rates also tend to increase as soil moisture increases. However, under wet climates vegetation cover also increases, and the roots of plants, which bind the soil together, may inhibit creep. Creep rates measured on a grassy hillside in England that has a slope of 33° were only 0.02 mm/yr.

Sediment Deposited by Mass-Wasting

Colluvium

The term **colluvium** is usually applied as a general term to *loose, incoherent deposits on or at the base of slopes and moving mainly by creep.* The particles in a

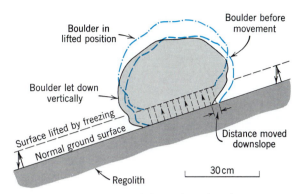

FIGURE 9.32 Stone moved downslope by alternate freezing and thawing of the ground. As freezing occurs, the stone is raised perpendicular to the ground surface which also rises; when the ground thaws, gravity pulls the stone down vertically, giving it a small but significant component of movement downslope.

body of colluvium tend to be angular and to lie in a chaotic jumble. These characteristics generally make it possible to distinguish colluvium from sediment deposited after transport in moving fluids such as water (streams, surf) and air (wind), for such sediment tends to consist of rounded particles, sorted and deposited in layers.

Sliderock and Taluses

In areas where steep cliffs prevail and mechanical weathering is active, accumulations of weathered debris usually mantle the bases of the cliffs. The particles, commonly angular and ranging in diameter from sand grains to large boulders, are loosened from the bedrock of a cliff by mechanical weathering and fall to its base where they accumulate. The resulting *apron of rock waste sloping outward from the cliff that supplies it* is a **talus** (Fig. 9.33). *The sediment composing a talus* is called **sliderock.** From cliff to talus the movement is chiefly by falling, sliding, and rolling. As they come to rest, the rock fragments typically form a steep slope that stands at the angle of repose of the coarsest particles (Fig. 9.34). Few fine particles may be visible, for they tend to settle into the large open voids between the coarser rocks. Because large falling rocks have more momentum than small particles, they tend to move farther down the slope and some may bound beyond the toe of the talus where they form a scattered array of isolated boulders (Fig. 9.33).

Rates of talus growth may undergo marked vari-

FIGURE 9.33 Taluses at the base of steep cliffs in the Argentine Andes. Large boulders roll and bound beyond the toe of the active talus where they are scattered across the surface of a river terrace.

FIGURE 9.34 Coarse, angular limestone blocks stand at the angle of repose in a talus below steep cliffs in the central Brooks Range, Alaska.

ations if the climate changes. In the Columbia Plateau region of Washington State, large taluses have accumulated along the steep walls of dry valleys (coulees) that were swept by voluminous floodwaters shortly before 13,000 years ago (Chapter 19). Just below the surface of the taluses lies a distinctive layer of whitish volcanic ash that fell about 6700 years ago. These relationships indicate that talus growth began after the floods ceased, but was largely completed prior to the ashfall. Only a small percentage of the volume of the taluses has accumulated since then. Clearly the average rate of sliderock production was far greater during the first 6700 years after the taluses began to form than in the subsequent, equally long period. The contrast in accumulation rates must reflect differences in the rate of mechanical weathering, which in turn points to significant differences in average climatic conditions.

Although new debris may be continually added to a talus, large variations in accumulation rates means that talus growth may proceed in a highly irregular or fluctuating manner. At times when sliderock production is low, the surface debris may have time to weather chemically. Simultaneously, creep transfers the debris slowly downslope. The weathering converts the coarse angular sliderock into fine-grained regolith which, with its pores of extremely small diameter, can hold much more moisture than coarse sliderock and thus can acquire both vegetation and soil.

Triggering of Mass-Wasting Events

Mass-wasting events sometimes seem to occur at random, with no apparent reason. However, the largest, most disastrous, and most numerous events commonly are related to some extraordinary activity or occurrence.

Sudden shocks, such as an earthquake, may release so much energy that slope failures of many types and sizes are triggered simultaneously. In 1929 a major earthquake (magnitude 7.7; Chapter 16) in northwestern South Island, New Zealand triggered at least 1850 landslides larger than 2500 m^2 within an area of 1200 km^2 near the quake's center. An estimated 210,000 m^3 of debris was displaced, on average, in each km^2 of land. Landslides were reported to be most numerous on well-bedded and well-jointed mudstones and fine sandstones. The Alaska earthquake of 1964 triggered many rockfalls, one of which became a huge rock avalanche that swept across the surface of Sherman Glacier, burying it with up to several meters of coarse, angular debris.

Landslides often result through *modification of slope* or *removal of support.* They typically occur where roads or highways have been cut into regolith creating an artificial slope which exceeds the angle of repose. Retaining walls may inhibit landslides, but unless they are very strong, creep of the regolith and subsequent failure may render them ineffective.

Slumps and other types of landslides may be triggered by the *undercutting* action of a stream along its bank or by surf action along a coast.

Coastal landslides are especially common during large storms which may direct their energy against rocky headlands or along the bases of cliffs of unconsolidated sediments. Seacliffs on Hawaii and many other oceanic islands retreat as wave action quarries away thin, well-jointed lava flows, thereby causing overlying rocks to collapse. The resulting debris is then quickly reworked by surf and transported away.

Landslides are frequently associated with *heavy or prolonged rains* that saturate the ground and make it unstable. Such was the case in 1925 when prolonged rains, coupled with melting snow, started a large debris flow in the Gros Ventre River basin of western Wyoming. The water saturated a permeable sandstone that overlies impermeable shale and dips toward the valley floor, creating conditions that were ideal for slope failure. An estimated 37 million m^3 of rock, regolith, and organic debris moved rapidly downslope and created a natural dam that ponded the river. Sixty years later, the scar at the head of the slide is still quite obvious, as is the hummocky topography downslope.

Volcanic eruptions are still another means of initiating mass-wasting events. Stratovolcanoes consist of an inherently unstable pile of interstratified lava flows, rubble, and pyroclastic layers. Unconsolidated deposits generally lie at the angle of repose. On glaciated volcanoes, slopes may be oversteepened by glacial erosion. During eruptions, slope failure is common and often widespread. If a volcano supports glaciers or extensive snowfields, melting of snow and ice can release large quantities of water which combine with unconsolidated deposits on the slopes and move rapidly downvalley as lahars (Fig. 9.29). Lahars are so widespread in valleys surrounding some volcanoes in the Cascade Range that a substantial percentage of the total eruptive products from a vent actually lie beyond the volcanic cone itself. Mount St. Helens, an especially active member of the chain, has produced lahars throughout much of its 35,000-year history, most recently during the huge eruption of 1980.

As the human population grows and cities and road systems expand across the landscape, the likelihood that mass-wasting processes will affect people increases. Although it may not always be possible to predict accurately the occurrence of events that will trigger mass movements, a knowledge of the processes and their relationship to local geology can lead to intelligent planning which can help reduce the loss of lives and property.

SUMMARY

Weathering

1. The zone of weathering extends to whatever depth air and water penetrate. Water solutions, which enter the bedrock along joints and other openings, attack the rock chemically and physically, causing breakdown and decay.

2. Mechanical and chemical weathering, although involving very different processes, generally work together.

3. Subdivision of large blocks into smaller particles increases surface area and thereby accelerates weathering.

4. Growth of crystals, especially ice and salt, along fractures and other openings in bedrock is a major process of mechanical weathering.

5. Daily heating of rocks by the sun followed by nocturnal cooling may cause little or no breakdown, but intense fires can lead to spalling of rock surfaces.

6. The wedging action of plant roots and the churning of rock debris by burrowing animals can have large cumulative effects over time.

7. Carbonic acid is the prime agent of chemical weathering; heat and moisture speed chemical reactions.

8. Chemical weathering converts feldspars into clay minerals. Quartz is resistant to chemical decay and commonly is deposited as sand grains.

9. The effectiveness of weathering depends on such factors as rock type and structure, surface slope, local climate, and the time over which weathering processes operate.

Soils

10. Soils consist of weathered regolith capable of supporting plants. Soil profiles display distinctive horizons, the character of which depends on such factors as climate, vegetation cover and soil organisms, composition of parent material, topography, and time.

11. The A horizon is rich in organic matter and has lost soluble minerals through leaching. Clay accumulates in the B horizon together with substances leached from the A horizon. Both overlie the C horizon, which is slightly weathered parent material.

12. Soils are classified into soil orders based on their physical characteristics. Caliche is a common component of many arid-region soils and forms in the upper part of the C horizon. Laterites are typical of tropical climates, display extreme weathering, and have concentrations of iron and aluminum oxides.

13. Paleosols are buried soils that can provide clues about former topography, plant cover, and climate.

Mass-Wasting

14. The pull of gravity causes unstable rock particles to move downslope without the aid of a transporting medium. It affects rock debris on land as well as beneath the sea.

15. Falling rock masses that break up on impact can travel downslope at high velocity and be extremely destructive. Slumps are much more common and may be triggered on unstable slopes by heavy rains or sudden shocks.

16. The mobility of mudflows increases as water content rises. Lahars constitute one of the most hazardous phenomena associated with some active volcanoes.

17. Although solifluction and creep are extremely slow processes, they are widespread on hillslopes and therefore are very effective over long periods in moving rock and regolith downslope.

18. Colluvium is generally angular nonsorted sediment deposited mostly by creep. Rockfall leads to accumulation of sliderock on taluses at the base of cliffs.

19. Mass-wasting events are most frequently triggered by sudden shocks, modification of natural slopes, undercutting of slopes by streams or waves, heavy or prolonged rains, and volcanic eruptions.

SELECTED REFERENCES

Weathering and Soils

Birkeland, P. W., 1984, Soils and geomorphology: New York, Oxford University Press.

Carroll, Dorothy, 1970, Rock weathering: New York, Plenum Press.

Colman, S. M., and Dethier, D. P., 1986, Rates of chemical weathering of rocks and minerals: New York, Academic Press.

Gauri, K. L., 1978, The preservation of stone: Sci. American, v. 238, p. 126–136.

Hunt, C. B., 1972, Geology of soils. Their evolution, classification, and uses: San Francisco, W. H. Freeman.

Ollier, C. D., 1969, Weathering: New York, Elsevier.

Mass-Wasting

Crandell, D. R., 1971, Postglacial lahars from Mount Rainier volcano, Washington: U.S. Geological Survey Professional Paper 677.

Hsu, K. J., 1975, Catastrophic debris streams (sturzstroms) generated by rockfalls: Geol. Soc. America Bull., v. 86, p. 129–140.

Porter, S. C., and Orombelli, G., 1981, Alpine rockfall hazards: Am. Scientist, v. 69, p. 67–75.

Selby, M. J., 1982, Hillslope materials and processes: Oxford, Oxford University Press.

Voight, B., ed., 1978, Rockslides and avalanches, 1. Natural phenomena: New York, Elsevier.

C H A P T E R 1 0

Groundwater

Rugged and scenic karst landscape eroded in carbonate rocks at
Guilin, China.

THE HYDROLOGIC CYCLE AND GROUNDWATER

Food and water are both necessary for us to live, although we can survive longer without food than water because our bodily biochemical activities are based on water solutions. Ready access to water, whether from streams, lakes, springs, or direct rainfall, is therefore a primary human need. Most early cities and towns were founded along or close to streams that provided a reliable source of water. With growth of population, the streams often became insufficient for human needs. People then resorted to bringing water from a more distant source through canals, or obtained water from underground supplies.

As society has become increasingly industrialized, communities have generated ever-larger amounts of human and industrial wastes, a good deal of which has inevitably found its way into the very water that people must rely on for their existence. In nearly every city, water has become a critical resource. In many places it is dwindling both in quantity and quality, creating important questions for the communities involved: Will there be enough water to sustain our future needs? Is its quality adequate for the uses to which we put it? Is the water being used efficiently, and with a minimum of waste?

The answers to these questions demand that we know some basic things about the water supply: the quantity of the Earth's water; where it is located; how it is cycled; the techniques for using it efficiently; and ways of keeping its quality from declining.

Distribution of Water on the Earth

The Earth is unique among the planets and moons of our solar system in having an abundance of liquid water at and near its surface. This water resides in several primary reservoirs, not only as liquid water but also as vapor and ice. In Table 10.1, we see an estimate of the water volume in each of these reservoirs, as well as its percentage of the total. The bulk of the Earth's free water, more than 97 percent, resides in the world's oceans. Next in importance are the myriad bodies of snow and ice that occupy the high mountains and polar latitudes of our planet. All the remaining water, including that in the atmosphere, lakes and streams, and the ground, amounts to less than one percent of the total. Yet this is the water we are most conscious of and rely on in our daily life.

The volume of water in the different natural reservoirs is approximately constant over short time periods, but water in each reservoir is always in a state of movement and is continually being cycled from one reservoir to another (Fig. 10.1). Over long periods of time, the percentages of water in the different reservoirs can change dramatically. For example, during a glacial age vast quantities of water are transferred from the oceans to land where, as snow and ice, water accumulates to form numerous mountain glaciers, as well as great continental ice sheets that reach thicknesses of thousands of meters.

The Hydrologic Cycle

Both the day-to-day and long-term cyclic changes that we can observe in the Earth's hydrosphere are powered by heat from the Sun which evaporates water from the ocean and land surface. The water vapor thus produced enters the atmosphere and moves with the flowing air. Some of it condenses and is precipitated as rain or snow back into the ocean or onto the land. Rain falling on the ground surface may either drain off in streams, percolate into the ground, or be evaporated back into the air where it is further recycled. Snow may remain on the ground for one or more seasons until it melts and the meltwater flows away. Snow that nourishes glaciers remains locked up much longer, through many years or even thousands of years, but eventually it too melts or evaporates and ultimately returns to the oceans. Part of the water in the ground is taken up by plants which return water to the atmosphere through *transpiration*.

The hydrologic cycle provides a good example of the conversion of energy from one form to another. The Earth's land surface stands, on average,

TABLE 10.1 *Estimated Global Water Inventory*

Reservoir	Volume (thousand km³)	Volume (%)
Rivers	1.25	0.0001
Atmosphere	13	0.001
Soil moisture	67	0.005
Freshwater lakes	125	0.009
Saline lakes and inland seas	104	0.008
Groundwater (to 4 km depth)	8350	0.615
Glacier ice	29,200	2.150
Oceans	1,320,000	97.212
Total	1,357,860	100.0001

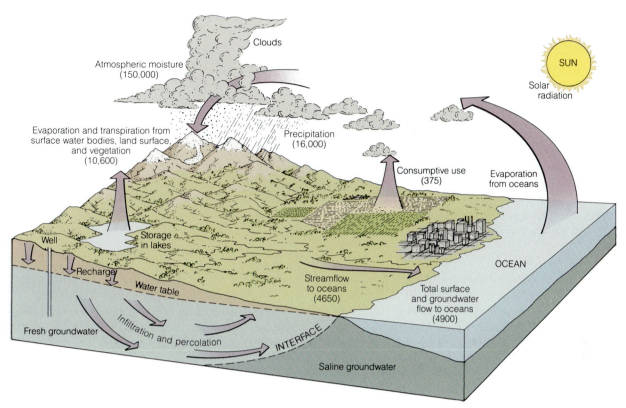

FIGURE 10.1 Hydrologic cycle, showing the gross daily water budget of the conterminus United States in millions of cubic meters of water a day. (*Source: After U. S. Geological Survey, 1984.*)

some 700 to 800 m above the surface of the ocean (Fig. 2.7). The Sun's radiant energy evaporates water from the ocean and the resulting water vapor rises and becomes part of the atmosphere. When the water vapor condenses, then precipitates as rain onto the land, the radiant energy has been converted into potential energy. The potential energy is transformed into kinetic energy as the water erodes the land surface and transports sediment downhill. We make use of this same principle when we build a dam across a stream to form a reservoir. The dammed-up water conserves and concentrates the potential stream energy for use in generating electrical power.

The quantity of water that participates each year in the hydrologic cycle has been estimated for the United States by comparing measurements of the amounts evaporated and precipitated at many places on land and at sea, as well as the amounts carried by streams, glaciers, and through the ground. If we make similar estimates for the entire world, the total quantity of water involved is about 410,000 km^3, or enough water to cover an area the size of England or the state of Illinois to a depth of about 3 km. From such estimates we can calculate that an amount of water equal to the combined volumes of all the oceans goes through the global hydrologic cycle once in every 3200 years.

Groundwater in the Earth's Water Inventory

As seen from Table 10.1, less than 1 percent of the water on the Earth is *groundwater,* defined simply as *all the water contained in spaces within bedrock and regolith.* Although the total volume of groundwater is small, it is about 35 times greater than the volume of water lying in fresh-water lakes or flowing in streams on the Earth's surface.

Nearly all the Earth's groundwater (except for a very small amount that is released from magma during volcanic eruptions) has its origin in rainfall. It is always slowly moving on its way back to the ocean, either directly through the ground or by flowing out onto the surface and joining streams (Fig. 10.1).

In ancient Greece, some 2500 years ago, it was thought that the water in the ground originated as seawater, driven into the rocks by the winds and somehow desalted, or that it was created in some manner from rocks and air deep below the surface.

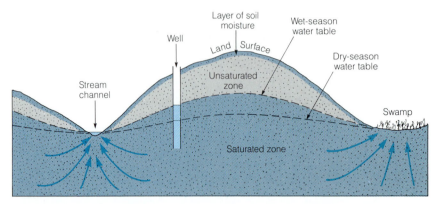

FIGURE 10.2 Diagram showing the position of the saturated zone, the water table, and the unsaturated zone in a typical groundwater system.

Later it was recognized that rivers are fed, at least in part, by springs emerging from the ground. It was also recognized that the discharge of rivers does not raise the surface of the sea appreciably. That groundwater is derived mainly from precipitation was first recognized by Marcus Vitruvius, a Roman architect of the time of Christ, who wrote a treatise on aqueducts and water supply, a matter of great practical importance to the Romans.

Although valid, Vitruvius's statement that groundwater comes from rain was not established on a quantitative basis until the seventeenth century. Then Pierre Perrault, a French physicist, measured the mean annual rainfall for a part of the drainage basin of the Seine River in eastern France and the mean annual runoff from it in rivers. After allowing for loss by evaporation, he concluded that the difference beween the amounts of rainfall and runoff was ample enough, over a period of years, to account for the amount of water in the ground. Today we accept rainfall as being the source of most groundwater.

Depth of Groundwater

People have been able to establish permanent settlements, not only in well-watered country but also in desert lands, by exploiting groundwater resources. Few areas exist in which holes, intelligently located and sunk far enough into the ground, do not find at least some water. In a moist country the depth of an adequate well may have to be only a few meters. In a desert it may have to be hundreds. Water is present beneath the land nearly everywhere, but whether it is present in usable quantity depends on depth of occurrence, kinds of rock present, and kinds and amounts of substances dissolved in the water. For this reason, some places are much more favorable than others for obtaining useful supplies of groundwater.

More than half of all groundwater, including most of the water that is usable, occurs within about 750 m of the Earth's surface. The volume of water in this zone is estimated to be equivalent to a layer of water approximately 55 m thick spread over the world's land area. Below a depth of about 750 m, water decreases in amount, gradually though irregularly. Holes drilled for oil have found water as deep as 9.4 km. A deep hole drilled by Soviet scientists on the Kola Peninsula encountered water solutions at depths of more than 11 km. Even though water may be present in crustal rocks at such depths, the pressure exerted by overlying rocks is so high and openings in rocks are so small that it is unlikely water can move freely through them.

Water Table

Much of our knowledge of where groundwater occurs has been learned from the accumulated experience of generations of people who have dug or drilled millions of wells. This experience tells us that a hole penetrating the ground ordinarily passes first into a **zone of aeration** (or **unsaturated zone**), *the zone in which open spaces in regolith or bedrock are filled mainly with air.* The hole then enters the **saturated zone,** *the zone in which all openings are filled with water. The upper surface of the saturated zone* is the **water table,** which normally slopes toward the nearest stream or lake (Fig. 10.2). In moist climatic regions the water table ordinarily lies within a few meters of the surface. Whatever its depth, the water table is a very significant surface, because it represents the upper limit of all readily usable groundwater.

MOVEMENT OF GROUNDWATER

Most of the groundwater within a few hundred meters of the surface is in motion. Unlike the swift flow of rivers, which is measureable in kilometers per hour, groundwater moves so slowly that velocities are expressed in centimeters per day or meters per year. To understand why the movement is so slow, we must know something about the porosity and permeability of rocks.

Porosity and Permeability

The limiting amount of water that can be contained within a given volume of rock or sediment depends on the *porosity* of the material, which is *the proportion (in percent) of the total volume of a given body of bedrock or regolith that consists of pore spaces* (open spaces). A very porous rock is one containing a comparatively large proportion of open pores, regardless of their size. Sediment is ordinarily very porous, ranging from 20 percent in some sands and gravels to as much as 50 percent in some clays. The sizes and shapes of the constituent particles and the compactness of their arrangement affect the porosity (Fig. 10.3*a* and *b*). In a sedimentary rock, the degree to which pores have become filled with cementing substances (Fig. 10.3*c*) also affects porosity. The porosity of igneous and metamorphic rocks generally is low, except where joints and fractures have developed in them.

Permeability is *the capacity for transmitting fluids.* A rock of very low porosity is also likely to have low permeability. However, high-porosity values do not necessarily mean high-permeability values, because size and continuity of the openings influence permeability in an important way. The relationship between the size of openings and the molecular attraction of rock surfaces plays a large part. Molecular attraction is the force that makes a thin film of water adhere to a rock surface despite the force of gravity. An example is the wet film on a pebble that has been dipped in water. If the open space between two adjacent particles in a rock is small enough, the films of water that adhere to the two particles will come into contact with each other. The force of molecular attraction therefore extends right across the open space (Fig. 10.4). At ordinary pressures, the water is held firmly in

a

b

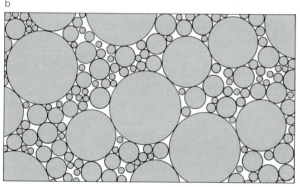

c

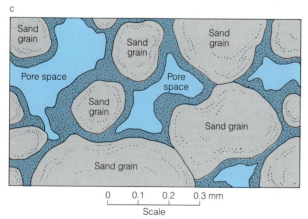

FIGURE 10.3 Contrasts in porosity of sediments. (*a*) A porosity of 32 percent in a reasonably well-sorted sediment. (*b*) A porosity of 17 percent in a poorly sorted sediment in which fine grains fill spaces between larger ones. (*c*) Reduction in porosity of an otherwise porous sediment due to the presence of a cementing agent.

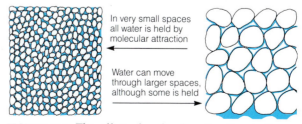

FIGURE 10.4 The effect of molecular attraction in the intergranular spaces of a fine sediment (left) and a coarser sediment (right). The scale is larger than natural size.

place and permeability is low. That is what happens in a wet sponge before it is squeezed. The same thing happens in clay, the particles of which have diameters of less than 0.005 mm (Table 5.1).

By contrast, in a sediment with grains at least as large as sand (0.06–2 mm) the open pores commonly are wider than the films of water adhering to the grains. Therefore, the force of molecular attraction does not extend across them effectively, and the water in the centers of the openings is free to move in response to gravity or other forces (Fig. 10.4). Such sediment is permeable. As the diameters of the openings increase, permeability increases. Gravel with its very large openings is more permeable than sand and can yield large volumes of water to wells.

Movement in the Unsaturated Zone

Water from a rain shower soaks into the soil, which usually contains clay resulting from the chemical weathering of bedrock. Due to its content of extremely fine clay particles, the soil is generally less permeable than underlying coarser regolith or rock. The low permeability and the fine clay cause part of the water to be retained in the soil by forces of molecular attraction. This is the layer of soil moisture shown in Figure 10.2. Some of this moisture evaporates directly, but much is taken up by plants which later return it to the atmosphere through transpiration (Fig. 10.1).

Water that molecular attraction cannot hold in the soil seeps downward until it reaches the water table. In fine-grained sediment a narrow

fringe as much as 60 cm thick immediately above the water table is kept wet by *capillary attraction,* the same force that draws ink through blotting paper and kerosene through the wick of a lamp. With every rainfall, more water is supplied from above, but apart from soil moisture and the capillary fringe, the unsaturated zone is likely to be nearly dry between rains (Fig. 10.2).

Movement in the Saturated Zone

The movement of groundwater in the saturated zone, called **percolation,** is similar to the flow of water when a saturated sponge is squeezed gently. Water moves slowly by percolation through very small open spaces along parallel, threadlike paths. Movement is easiest through the central parts of the spaces but diminishes to zero immediately adjacent to the sides of each space because there molecular attraction holds the water in place.

The force of gravity supplies the energy for percolation of groundwater. Responding to that force, water percolates from areas where the water table is high toward areas where it is lowest. In other words, it flows toward surface streams or lakes (Fig. 10.5). Only part of the water travels directly down the slope of the water table by the shortest route. Much of it flows along innumerable long curving paths that go deeper through the ground. Some of the deeper paths turn upward against the force of gravity and enter the stream or lake from beneath. This happens because, in the saturated zone, the water at any given height (such as h_1 in Fig. 10.5) is under greater pressure beneath a hill

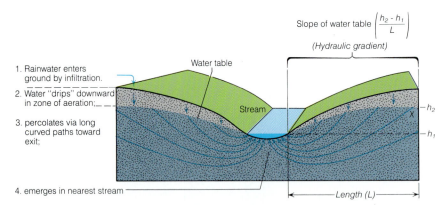

FIGURE 10.5 Movement of groundwater in uniformly permeable rock. Long curved arrows represent only a few of many possible paths. At any point, such as X, the slope of the water table (the *hydraulic gradient*) is determined by $(h_2 - h_1)/L$ where $h_2 - h_1$ is the height of X above the point of emergence in the surface stream, and L is the distance from X to the point of emergence.

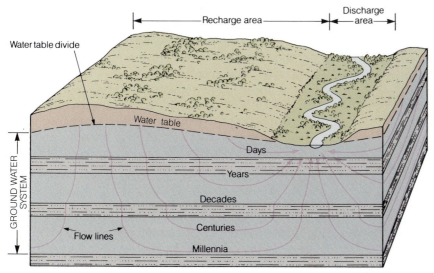

FIGURE 10.6 Recharge and discharge areas in a humid landscape. The time required for groundwater to reach the discharge area from the recharge area depends on the route and distance of travel. (*Source:* After Heath, 1983.)

than beneath a stream. The water therefore tends to move toward points where pressure is least.

Laboratory models have been made in which dye is injected into the percolating groundwater at various depths. Paths followed by the dye resemble those in Figure 10.5, where they turn upward beneath a model stream at the base of a hill. The rate at which the water moves along these paths decreases sharply with increasing depth, indicating that most of the groundwater entering a stream travels along shallow paths not far beneath the water table. Groundwater that moves downward to the bottom of deep basins may reside there for long periods of time and may be very old.

NATURE OF THE GROUNDWATER SYSTEM

The shallower part of the Earth's groundwater system operates continually as a small but integral part of the hydrologic cycle. This system both stores water and transmits it, and therefore acts both as reservoir and conduit. Water entering the system moves through the ground and escapes into stream valleys on its way toward the ocean.

Recharge and Discharge Areas

Water enters the groundwater system as precipitation falling on **recharge areas,** which are *areas where water is added to the saturated zone* (Fig. 10.6). It

moves through the system to **discharge areas,** which are *areas where subsurface water is discharged to streams or to bodies of surface water*. The areal extent of recharge areas is invariably larger than that of discharge areas. In humid regions, recharge areas encompass nearly all areas except streams and their adjacent floodplains. In more arid regions, recharge occurs mainly in mountains and in the alluvial fans that border them, as well as along channels of major streams that are underlain by permeable alluvium. The Platte River in Nebraska and the Nile River in Egypt are examples of streams that flow from mountains having substantial rainfall into much drier regions in which the water table lies deep beneath the surface. Water from these rivers leaks downward and recharges the groundwater below (Fig. 10.7).

In the United States, annual rates of recharge range from zero in arid desert regions to as much as 600 mm in humid regions underlain by very permeable soils and rocks. The time it takes for water to move through the ground from the recharge area to the nearest discharge area depends on rates of flow and the distance between them. It may take only a few days, or possibly thousands of years in cases where water moves through the deeper parts of the groundwater system (Fig. 10.6)

Fluctuations of the Water Table

The water table is a subdued imitation of the ground surface above it. It is high beneath hills

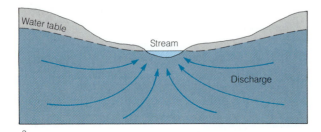

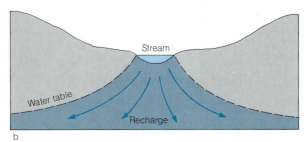

FIGURE 10.7 Recharge of streams in a humid region and a dry region. (*Source:* After Dunne and Leopold, 1978.) (*a*) In a humid region, where water is added throughout large recharge areas, the water table slopes toward the rivers and water flows into the streams. (*b*) In arid regions, direct recharge is minimal, and the water table lies below the riverbed. During runoff events, water percolates downward from the bed and banks of the channel to resupply the groundwater at depth.

and low at valleys because water tends to move toward low points in the topography where the pressure on it is least. If all rainfall were to cease, the water table would slowly flatten and gradually approach the levels of the valleys. Percolation would diminish and cease, and the streams in the valleys would dry up. In times of drought, when rain may not fall for several weeks or even months, we can sense the flattening of the water table in the drying up of wells. When that occurs we know that the water table has fallen to a level below the bottoms of the wells. It is repeated rainfall, dousing the ground with fresh supplies of water, that maintains the water table at a normal level.

Discharge and Velocity

What, then, determines the steepness of the slopes of a water table and the rate of flow of the percolating groundwater? In Figure 10.5, we can see that the slope of a water table between any point (X) at some height (h_2) and the height where it emerges (h_1) is measured by the difference in height ($h_2 - h_1$) divided by the horizontal distance l. This is *the*

slope of the water table and is commonly referred to as the **hydraulic gradient.**

It was discovered by experiment that the rate of flow of groundwater increases as the slope of the water table increases, as long as the permeability of the ground remains uniform. In 1856, Henri Darcy, a French engineer, proposed a formula that can be used to express the flow rate of groundwater through rock, an expression now known as Darcy's Law. This law can be expressed as a simple equation

$$Q = KAh/l$$

where Q is the rate of flow (in cm/day), K is a coefficient that reflects the permeability of the rock or sediment, A is the cross-sectional area through which flow can occur, and h/l is the hydraulic gradient. The law states that the rate of groundwater flow through permeable material is directly proportional to the product of the cross-sectional area through which flow can occur, the permeability, and the hydraulic gradient.

Because of the large amount of friction involved in percolation, flow rates are slow. Normally velocities range between half a meter a day and several meters a year. The largest rate yet measured in the United States, in exceptionally permeable material, is only about 250 m/yr.

The average velocity of percolating water is measured between pairs of wells by various techniques. In one method, two wells with metal casings are connected to form an electric circuit. A chemical compound that is an efficient conductor and is soluble in water is poured into the upslope well and percolates downslope. On its arrival at the lower well it creates a short circuit between the well casing and the electrode which is recorded on an ammeter. The distance between the wells divided by the elapsed time gives the velocity.

Economy of the Groundwater System

We can speak of the *economy* of a groundwater system as *a measure of the input of water to and outflow of water from the entire system.* If the inflow (recharge) just balances the outflow (discharge), then the economy is balanced and the water table will maintain a given level. If either input or outflow changes, then the economy will change and the water table will adjust. Apart from permeability and rate of flow, which are likely to change as water flows from one kind of rock or sediment to another, the important factor in the groundwater system is the slope of the water table. This will change as the climate changes, thereby affecting the overall hydrologic balance of the land.

AQUIFERS

Water-Table Aquifers

When we look for a good supply of groundwater we search for an *aquifer* (from the Latin, "water carrier"), *a body of permeable rock or regolith saturated with water and through which groundwater moves.* Bodies of gravel and sand are commonly good aquifers, and so are many sandstones. However, the presence of a cementing agent between grains of a sandstone reduces the diameter of the openings and so reduces the effectiveness of these rocks as aquifers.

It might seem that claystones, igneous rocks, and metamorphic rocks would not be aquifers, because the spaces between their mineral grains are extremely small. Furthermore, samples measured in the laboratory are impermeable. However, what is true for laboratory samples does not necessarily apply to large bodies of the same rock. Many such rock bodies contain fissures, spaces between layers, and other openings such as joints that permit free circulation of groundwater. The Hawaiian Islands, for example, are built of massive piles of thin-bedded, well-jointed, and porous lava flows that hold large natural reservoirs of groundwater.

More than half the land area of the United States is underlain by one or more aquifers (Fig. 10.8). About 30 percent of the groundwater used for irrigation is obtained from an extensive aquifer that lies at shallow depths beneath the High Plains (Fig. 10.9). The aquifer is tapped by about 170,000 wells and supplies water for more than 20 percent of the irrigated land of the country. The aquifer consists of a number of young sandy and gravelly rock units that lie at depths of less than 350 m, and it averages about 65 m thick. The water table slopes gently from west to east and water flows through the aquifer at an average rate of about 30 cm/day.

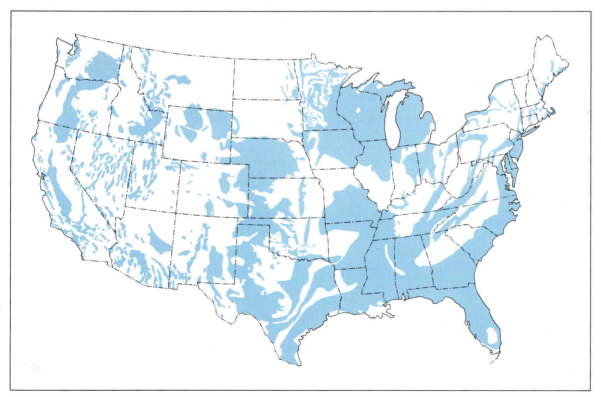

FIGURE 10.8 Map of the United States showing areas (shaded) underlain by one or more aquifers that can yield, in individual wells, at least 1.9 cubic meters (50 gallons) per minute of water containing no more than 0.2 percent dissolved solids. (*Source:* After Thomas, U. S. Department of Agriculture.)

Development of groundwater irrigation in the High Plains was spurred by severe regional drought in the 1930s and again in the 1950s (Chapter 12). Annual recharge of the High Plains aquifer from precipitation is much less than the amount of water being withdrawn, so the inevitable result is a long-term fall of the water table. A dramatic increase in pumping rates has led to serious water-level declines. In parts of Kansas, New Mexico, and Texas, the saturated thickness has declined by more than 50 percent. The resulting decreased water yield and increased pumping costs have led to major concern about the future of irrigated farming on the High Plains.

Springs

A **spring** is a *flow of groundwater emerging naturally at the ground surface*. The simplest spring is an ordinary or *gravity spring* which issues from a place where the land surface intersects the water table. An example of such a spring is illustrated in Figure 10.10. A vertical or horizontal change in permeability is a common reason for the localization of springs (Fig. 10.11). If a porous sand overlies a relatively impermeable clay, water percolating downward through the sand will flow laterally when it reaches the underlying clay and may emerge as spring where the stratigraphic contact

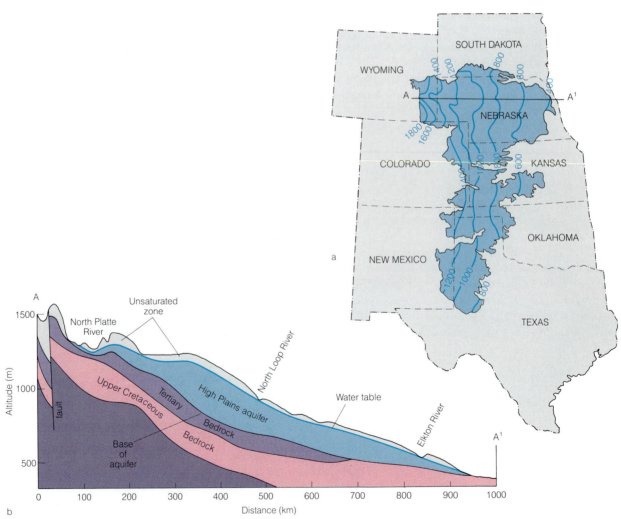

FIGURE 10.9 High Plains aquifer. (*a*) Map showing the distribution of aquifer and contours on the water table. Water flow is generally east, perpendicular to the contour lines. (*Source:* After Gutentag et al., 1984.) (*b*) Cross section along profile *A–A'* showing the relation of High Plains aquifer to underlying bedrock units and the position of the water table.

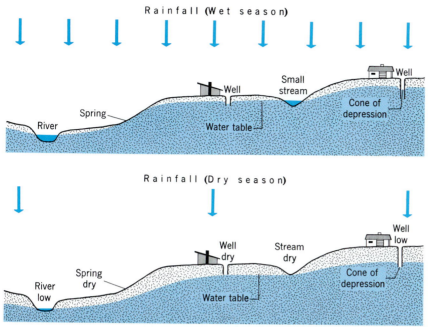

Rainfall (Wet season)

Rainfall (Dry season)

FIGURE 10.10 Wells and a spring in homogeneous rock, showing cones of depression and the effect of seasonal fluctuation of the water table. The slopes of the water table are steeper in the wet season, when the input of water into the system is greatest, than in the dry season, when the input is least.

crops out, as along the side of a valley or a coastal cliff. Springs may also be localized along structural features, such as faults. In fact, one way in which a fault can be identified at the land surface is by a series of springs aligned along its trace.

Small springs are found in all kinds of rocks. Almost all large springs issue from lava, limestone, or gravel aquifers. Many of the largest flow from volcanic rocks and are localized by porous lavas that overlie layers of impermeable volcanic ash or other impermeable units such as lahars.

Most springs display variations in rates of flow. Some show seasonal fluctuations related to seasonal variations in precipitation. Others may show daily fluctuations, related to use of water by plants. Discharge will be greatest between midnight and sunrise, and decrease steadily as vegetation withdraws water from the aquifer during daylight hours.

Ordinary Wells

An ordinary well fills with water simply because it intersects the water table (Fig. 10.10). If lateral flow cannot replenish water being lifted from the well, the water level lowers. This creates a *cone of de-*

pression, a conical depression in the water table immediately surrounding a well (Fig. 10.10). In most small domestic wells the cone of depression may be hardly discernable. Wells pumped for irrigation and industrial uses, however, withdraw so much water that the cone can become very wide and steep, and can lower the water table in all wells of a district. Figure 10.10 shows that a shallow well can become dry at times, whereas a deeper well in the vicinity may yield water throughout the year.

If rocks are not homogeneous, the yields of wells may vary considerably within short distances. Massive igneous and metamorphic rocks (Fig. 10.12a), for example, are not likely to be very permeable except where they are cut by fractures. A hole that does not intersect fractures is therefore likely to be dry. Because fractures generally die out downward, the yield of water to a shallow well can be greater than to a deep one. Discontinuous bodies of permeable and impermeable rock or sediment (Fig. 10.12b) result in very different yields to wells. They also create *perched water bodies* (*water bodies that occupy basins in impermeable sediments or rocks, perched in positions higher than the main water table*). The impermeable layer catches and holds the water reaching it from above.

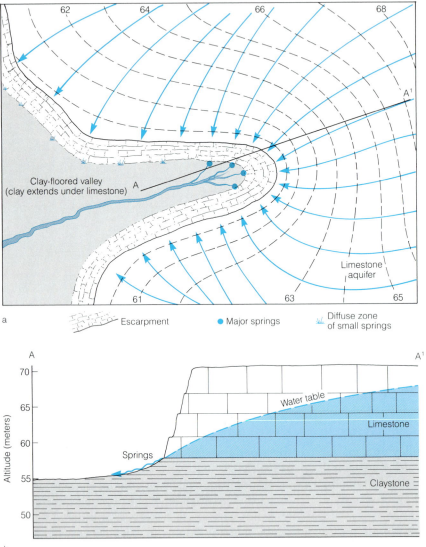

a

b

FIGURE 10.11 Map (*a*) and cross section (*b*) showing an unconfined aquifer overlying an impermeable bed. Dashed contours represent the altitude of the water table in meters and are called equipotential lines. Solid arrows show the direction of flow and are everywhere perpendicular to equipotential lines. At the foot of the scarp is a zone of seeps or small springs. (*Source:* After Dunne and Leopold, 1978.)

Confined Aquifers

In some regions the geometry of the inclined rock layers makes possible a special pattern of ground-water circulation. Three essentials of the pattern are: a series of inclined strata that include a permeable layer sandwiched between impermeable ones, rainfall to feed water into the permeable layer where that layer intersects the land surface, and a fissure or well so situated that water from the sandstone can escape upward through the impermeable rock above (Fig. 10.13). When these es-

sentials are present, we have an *artesian system.* The name comes from the French province of Artois (Artesium during the Roman era), in which, near Calais, the first well of this type in Europe was developed. The input consists of precipitation, which enters the permeable layer (now an aquifer by definition) and percolates through it. The ouput consists of water forced upward by hydrostatic pressure through fissures or wells that penetrate the capping rock.

Figure 10.13 illustrates such a system. Except for a thin zone close to the surface that lies above the

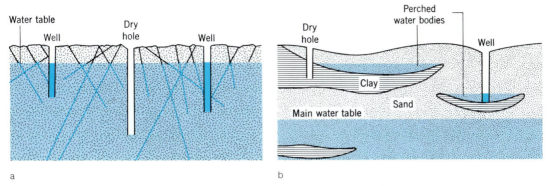

a   b

FIGURE 10.12 Ordinary wells and adjacent dry holes in rock that is not homogeneous. (*a*) In fractured massive rocks such as granite. (*b*) In bodies of permeable sand containing discontinuous bodies of impermeable clay. Two perched water bodies are shown.

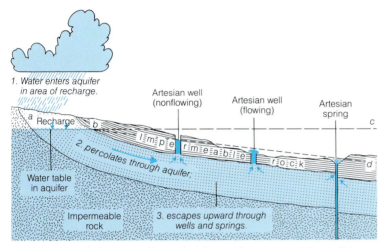

FIGURE 10.13 Artesian wells and springs. Three essential conditions are an aquifer, an impermeable roof, and water pressure sufficient to make the water in any well rise above the aquifer. The water rises, in any well, to the height (*bc*) of the water table in the recharge area (*ab*), minus an amount determined by the loss of energy in friction of percolation. Thus the water rises only to the line *bd*, which slopes downward away from the recharge area.

water table, the whole series of strata is saturated with water. In the impermeable roof rock the water is motionless, because it is held in place by capillary attraction in tiny spaces between mineral grains. But in the aquifer it moves, provided only that water can escape from the system through fissures or wells. Percolation in the aquifer, however, is confined between the impermeable strata above and below. It moves past the water that is held motionless in those strata. The aquifer is like a broad, flat, sand-filled pipe or conduit, holding its groundwater confined under pressure of the column of water that extends up to the water table at its upper end.

If, in the area of recharge shown in the illustration, rainfall reaches the ground in greater volume

FIGURE 10.14 Linear series of vertical shafts surrounded by mounds of excavated dirt define the course of ancient artesian water systems (called *kanats*) on an alluvial plain west of Kashan, Iran.

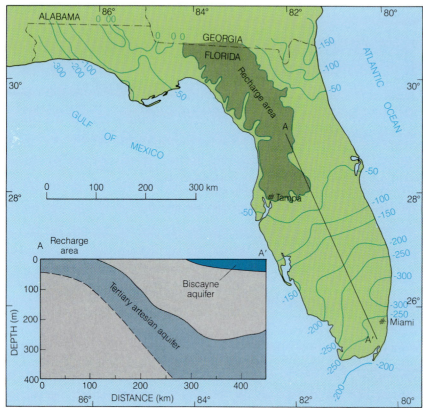

FIGURE 10.15   Map of the Florida peninsula showing depth to top of the Tertiary limestone artesian aquifer (in meters) and the area of recharge. Inset diagram shows a stratigraphic section along line A-A'. In the southern part of the state, the Tertiary aquifer lies below a surface aquifer in younger rocks. (*Source:* After Cederstrom et al., 1979.)

than that of water discharged through fissures or wells, only enough water to balance output can enter the system. The excess flows away over the surface. On the other hand, if wells draw out of the system more water than can enter it from the available rainfall on the area of recharge, the yield of the wells will diminish to a quantity small enough to be balanced by the recharge.

A well of this kind is an *artesian well, a well in which water rises above the aquifer. Natural springs that draw their supply of water from a confined aquifer* in the same manner are called *artesian springs.* Under unusually favorable conditions, hydraulic pressure can be great enough to lift water above ground level, creating fountains as much as 60 m high.

Tapping an artesian system is a very ancient art. Some 4000 years ago many artesian wells, some as much as 100 m deep, were in existence (Fig. 10.14). The well near Calais, France, was made in 1126 A.D. and is still flowing today, so in that area, at least, the rate of withdrawal has not seriously exceeded supply.

Tertiary Limestone Artesian System

A major artesian aquifer confined to Tertiary limestone strata in Florida is full of caves and smaller openings, intricately interconnected, that have been dissolved in the rock. In the central and northwestern parts of the peninsula this unit of rocks is exposed at the surface, but to the east and west it is covered by younger overlying strata as it slopes downward toward both the Atlantic and Gulf coasts (Fig. 10.15). The aquifer contains impermeable clayey beds interstratified with the limestone which are responsible for creating a series of aquifers, one atop another.

The age of water at various places within the system has been determined by measuring the age of radiocarbon (^{14}C) in HCO_3^{-1} dissolved in the water. Samples were collected in a series of wells along a line 133 km long, and approximately parallel to the dip of the aquifer. Most of the radiocarbon enters the ground as precipitation falling on the recharge area and moves through the aquifer with the groundwater. The age was found to increase systematically away from the recharge area. From the sum of the differences in age between samples from pairs of wells in the series an average percolation rate of 7 m/yr was calculated. Based on this average velocity, water in the well farthest from the recharge area has been in the ground for nearly 19,000 years.

A small error enters into these calculations, for "young" groundwater has been percolating down into the aquifer, and is still doing so, along the entire 133 km-long transect. This "young" water dilutes the older water already in the aquifer, so the average age of the water sampled in each well is a little younger than it would be if no water had entered the system from above. Therefore, the travel time calculated for water in the most distant well is a little too small.

GEOLOGIC ACTIVITY OF GROUNDWATER

Dissolution

As soon as rainwater infiltrates the ground, it begins to react with minerals in regolith and bedrock and weathers them chemically (Chapter 9). This *chemical weathering process, whereby minerals and rock material pass directly into solution,* is known as **dissolution.** The dissolved matter contained in the

groundwater reappears in streams as dissolved load and is carried to the sea, where it joins other substances in solution and eventually enters into the building of limestone and other marine sedimentary rocks.

Carbonate rocks, such as limestone, are especially susceptible to dissolution. By measuring over a period of time the amount of dissolution observed on small, precisely weighed limestone tablets placed at different sites in various areas, it is possible to calculate the average rate at which limestone landscapes are being lowered by chemical processes alone. In temperate regions, with high rainfall, a high water table, and a nearly continuous cover of vegetation, landscapes are being lowered by dissolution at rates of up to about 10 mm/1000 yr. In dry regions, with scanty rainfall, low water tables, and discontinuous vegetation, rates are much lower.

More than a century ago, a measurement of the dissolution rate in England was made in a churchyard by noting that fossil crinoids (sea lilies) projected 2.5 mm above the surface of limestone gravestones that had been dressed by stonemasons 50 years before. Therefore, the minimum average rate of surface lowering was 50 mm/1000 yr. A similar value was obtained by noting that blocks of sandstone deposited on a limestone surface in Yorkshire about 12,000 years ago by the last continental ice sheet to cover Britain now stand on

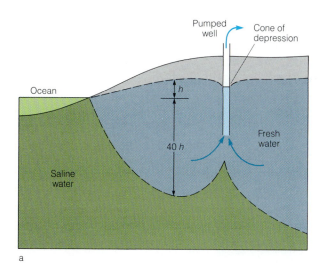 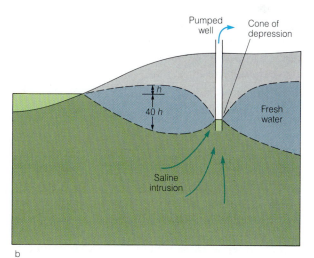

a   b

FIGURE 10.16 Seawater contamination of wells. (*Source:* After T. Dunne and L. B. Leopold, 1978.) (*a*) Near the coast, groundwater occurs as a lens over salty marine water. The height of the lens above sea level is equal to 1/40 of its depth below sea level. (*b*) Heavy pumping of groundwater forms a large cone of depression both at the top and base of the groundwater lens, eventually permitting saline water to enter the well.

pedestals of limestone as much as 50 cm high. After the ice melted away, the limestone was progressively dissolved, except for the rock protected beneath the sandstone boulders. In recent years, precise measurements of dissolution rates have been obtained with a microerosion meter, an instrument that can measure reliably the lowering of a rock surface to about the nearest 0.005 mm. Such measurements made on limestone terrain in New South Wales, Australia, gave dissolution rates averaging 16–29 mm/1000 yr. Measured rates of dissolution by groundwater in carbonate terrains of the United States make it apparent that the erosion rate can be greater than the average erosional reduction of the surface by mass-wasting, sheet erosion, and streams.

Chemical Content of Groundwater

Analyses of many wells and springs show that the elements and compounds dissolved in groundwater consist mainly of chlorides, sulfates, and bicarbonates of calcium, magnesium, sodium, potassium, and iron. We can trace these substances to the common minerals in the rocks from which they were derived by weathering. As might be expected, the composition of groundwater varies from place to place according to the kind of rock in which it occurs. In much of the central United States the water is "hard," that is, rich in calcium and magnesium bicarbonates, because the bedrock includes abundant limestones and dolostones that consist of those carbonates. In some places within arid regions the concentration of dissolved sub-

FIGURE 10.17 Logs of petrified wood weathering out of mudstone layers in Petrified Forest National Park, Arizona.

stances, notably sulfates and chlorides, is so great that the groundwater is unfit for human consumption. In particularly dry regions, evaporation of water in the unsaturated zone leads to deposition not only of calcium carbonate but of sodium sulfate, sodium carbonate, and sodium chloride. Soils containing these precipitates are loosely termed "alkali soils." They are unsuitable for agriculture because crops will not grow in them. Such conditions may result from slow circulation of groundwater through sedimentary rock, from which salts are dissolved.

Along coasts, fresh groundwater is separated from seawater along a narrow transition zone. Any pumping from a coastal aquifer can reduce the seaward discharge of fresh groundwater and may cause a landward and upward shift in the saline interface. Excessive pumping that exceeds the natural regional flow to the sea may cause saline water to move inland until it reaches major pumping centers. Such seawater intrusion can then contaminate basic water supplies (Fig. 10.16).

Chemical Cementation and Replacement

The conversion of sediment into sedimentary rock is primarily the work of groundwater. A body of sediment lying beneath the sea is generally saturated with water, as is sediment lying in the saturated zone beneath the land. Substances in solution in the water are precipitated as cement in the spaces between rock particles that form the sediment. This activity transforms the loose sediment into firm rock. Calcite, silica, and iron compounds (mainly oxides) are, in that order, the chief cementing substances.

Less common than the deposition of cement between the grains of a sediment is *replacement, the process by which a fluid dissolves matter already present and at the same time deposits from solution an equal volume of a different substance.* Evidently replacement takes place on a volume-for-volume basis because the new material preserves the most minute textures of the material replaced. Petrified wood is a common example (Fig. 10.17). Replacement is not confined to wood and other organic matter, for it can affect mineral matter as well.

Caves and Cave Deposits

Limestone, dolostone, and marble are carbonate rocks that consist of the minerals calcite and dolomite in various proportions. These rocks underlie millions of square kilometers of the Earth's sur-

FIGURE 10.18 Dripstone formations adorn a large limestone cavern in Carlsbad Caverns National Park, New Mexico. In places, stalactites have merged with stalagmites to form columns, the sides of which are sometimes ornately fluted.

face. Although carbonate minerals are nearly insoluble in pure water, they are readily dissolved by the carbonic acid formed by the interaction of carbon dioxide and rainwater, which percolates into the groundwater reservoir. As a result, the groundwater becomes charged with calcium and bicarbonate ions, as shown by the following reactions, the first of which we encountered in the discussion of chemical weathering (Chapter 9):

$$H_2CO_3 = H^{+1} + HCO_3^{-1}$$

Carbonic   Hydrogen   Bicarbonate
acid   ion   ion

The hydrogen ions attack the calcite and dissolve it:

$$CaCO_3 + H^{+1} = HCO_3^{-1} + Ca^{+2}$$

Calcite   Hydrogen   Bicarbon-   Calcium
       ion   ate ion   ion

Such dissolution is a form of chemical weathering just as much as is the decomposition of igneous rock that contains feldspar and ferromagnesian

FIGURE 10.19 Drop of water collects at the end of a growing stalactite in Carlsbad Caverns. As the water loses carbon dioxide, a tiny amount of calcium carbonate precipitates from the solution and is added to the end of the dripstone formation.

FIGURE 10.21 Sacred well at Chichen Itza, a ruined Mayan city on the Yucatan Peninsula. This cenote, formed in flat-lying limestone beds, contained a rich store of archeological treasures that were cast into the water with human sacrifices.

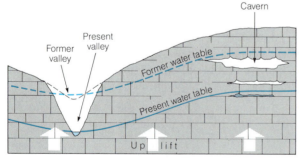

FIGURE 10.20 Possible history of a cavern containing dripstone. The cavern was excavated below the water table. When streams deepened their valleys, the water table was lowered as it adjusted to the deepened valleys. This left the cavern above the water table.

minerals. In both cases the weathering attack occurs along joints and other partings in the bedrock. Whereas in granite the quartz and other resistant minerals remain, nearly all the substance of a body of pure limestone can be carried away in solution in the slowly moving groundwater.

The process of dissolution creates cavities of many sizes and shapes. *A natural underground opening, generally connected to the surface and large enough for a person to enter* is called a **cave.** A very large cave or system of interconnected cave chambers is often called a *cavern.* Although most caves are small, some are of exceptional size. The Carlsbad Caverns in southeastern New Mexico include one chamber 1200 m long, 190 m wide, and 100 m high. Mammoth Cave, Kentucky, consists of interconnected caverns with an aggregate length of at least 48 km. The recently discovered Good Luck Cave on the tropical island of Borneo has one chamber

FIGURE 10.22 Sink, nearly 130 m in diameter and 45 m deep, near Montevallo, Alabama, formed at 2 p.m. on December 2, 1973. Debris of the roof forms a talus that conceals the bedrock. The bottom contains a pond.

so large that into it could be fitted not only the world's largest previously known chamber (in Carlsbad Cavern, New Mexico), but also the largest chamber in Europe (in Gouffre St. Pierre Martin, France), and the largest chamber in Britain (Gipping Hill Hole, England).

Some caves have been partly filled with insoluble clay and silt, originally present as impurities in the limestone and gradually concentrated by dissolution. Others contain partial fillings of *dripstone, a deposit chemically precipitated from dripping water in an air-filled cavity,* and *flowstone, a deposit chemically precipitated from flowing water in the open air or in an air-filled cavity.* Both are commonly composed of calcium carbonate. The precipitates take on many curious forms, which are among the chief attractions to cave visitors. The most common shapes are *stalactites* (*icicle-like forms of dripstone and flowstone, hanging from ceilings*), *stalagmites* (*blunt "ici-*

FIGURE 10.23 Typical karst landscape near Bowling Green, Kentucky, with numerous closed basins. Those intersecting the water table contain small lakes.

FIGURE 10.24 Distinctive karst landscape depicted in seventeenth-century Chinese scroll painting.

cles'' of flowstone projecting upward from cave floors), and **columns** (stalactites joined with stalagmites, forming connections between the floor and roof of a cave) (Fig. 10.18).

As its name implies, dripstone is deposited by successive drops of water. As each drop forms on the ceiling of a cave, it loses a tiny amount of carbon dioxide gas and precipitates a particle of calcium carbonate (Fig. 10.19). This chemical reaction is simply the reverse of the one by which calcium carbonate is dissolved by carbonic acid.

Dripstone can be deposited only in caves that are filled with air and therefore lie above the water table. Yet many, perhaps most, caves are believed to have formed below the water table, as is suggested by their shapes and by the fact that some caverns are lined with crystals, which can only form in an aqueous environment. How can we reconcile these apparently conflicting observations? The answer most probably lies in a fall of the water table due either to uplift of the land and accompanying downcutting of streams, or to a change of climate which causes a lowering of the regional water table. Caves excavated during a period of high water table would emerge into the zone of aeration as the water table falls, and dissolution

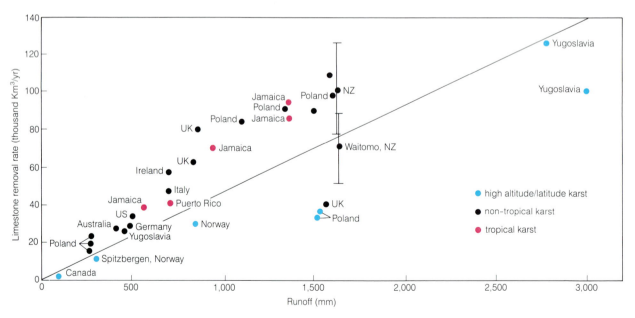

FIGURE 10.25 Rates of limestone erosion plotted as a function of runoff. The principal factor determining limestone removal is the amount of available water. The line separates predominantly soil-covered karst landscapes of tropical and nontropical regions from karst in high latitudes or at high altitudes where bare rock prevails. (*Source:* After Jennings, 1983.)

could then give way to deposition of dripstone (Fig. 10.20).

Sinkholes

In contrast to a cave, a *sinkhole* (also commonly called a *doline*) is *a large solution cavity open to the sky*. Some sinkholes are caves whose roofs have collapsed. Others are formed at the surface, where rainwater is freshly charged with carbon dioxide and is most effective as a solvent. Many sinkholes that are located at the intersections of joints, where downward movement of water is most rapid, have funnel-like shapes.

Sinkholes of the Yucatan Peninsula in Mexico, which are locally called *cenotes*, have high, vertical sides and contain water because they extend below the water table (Fig. 10.21). The cenotes were the primary source of water for the ancient Maya, and formerly supported a considerable population in Yucatan. A large cenote at the ruined city of Chichen Itza was sacred and dedicated to the rain gods. Remains of more than 40 human sacrifices, mostly young children, have been recovered from the cenote, together with huge quantities of jade, gold, and copper offerings.

In the widespread carbonate landscape of the Florida Peninsula new sinkholes are forming constantly. In one small area of about 25 km², more than 1000 collapses have occurred in recent years (Fig. 10.22). In this case the cause may be lowering of the water table by drought and excessive pumping of local wells. Although these sinkholes are relatively small, a far older sinkhole near Mammoth Cave, Kentucky, has an area of about 13 km².

Karst Topography

In some regions of exceptionally soluble rocks, sinks and caverns are so numerous that they combine to form a peculiar topography characterized by many small closed basins (Fig. 10.23). In this kind of landscape the drainage pattern is irregular. Streams disappear abruptly into the ground, leaving their valleys dry, and then reappear elsewhere as large springs. Such terrain is called *karst topography* after the Karst region of Yugoslavia where it is strikingly developed. It is defined as *an assemblage of topographic forms resulting from dissolution of the bedrock and consisting primarily of closely spaced sinkholes*. While most typical of carbonate rock terrain, karst can also occur in areas underlain by gypsum. In the United States, karst topography

is developed over wide areas of Kentucky, Tennessee, southern Indiana, northern Florida, and northern Puerto Rico. One of the most famous and distinctive karst regions lies near Guilin in southeastern China, the dramatic landscape of which has inspired both classical Chinese painters and present-day photographers (Fig. 10.24 and chapter frontispiece).

Sinkholes and caves record the destruction of a very large volume of carbonate rock. The rates at which karst landscapes develop appear to depend on several factors. In general, high rates of limestone removal are associated with high *runoff, the fraction of precipitation that flows over the land surface* (Fig. 10.25). At the same time, bare-rock karsts

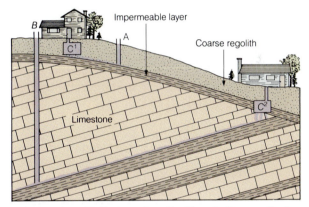

FIGURE 10.26 Pollution of wells. The shallow well, *A*, was unwisely located a short distance downslope from a septic tank *C*¹, and received polluted drainage (black) from it. The owner then drilled a deeper well, *B*. This well tapped layers of cavernous limestone dipping toward it from the lower septic tank, *C*². The water flowed through openings in the limestone, and reached the bottom of well *B* unpurified by percolation. The well owner must relocate his septic tank or else dig a shallow well located upslope from *C*¹.

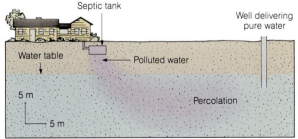

FIGURE 10.27 Purification of contaminated groundwater in sand and gravel during percolation through a short distance.

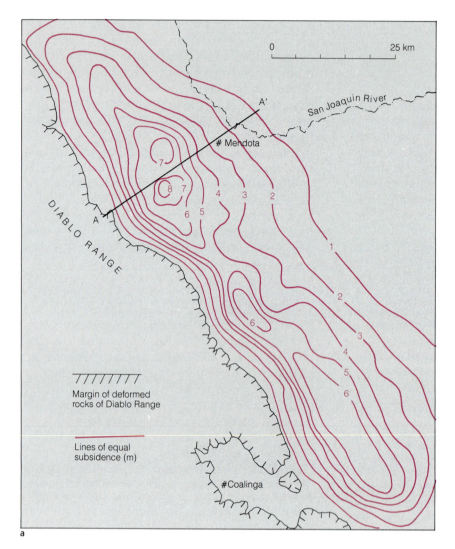

a

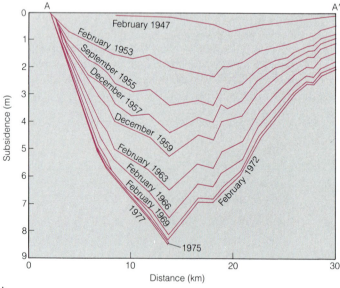

b

FIGURE 10.28 Ground subsidence due to withdrawal of groundwater in the San Joaquin Valley, California. (*Source:* After Ireland et al., 1984.) (*a*) Map showing amount of land subsidence, in meters between 1926 and 1972. (*b*) Cross section near Mendota, California along line *A–A'* showing progressive subsidence measured between 1947 and 1977. (*c*) Markings on this power pole show approximate positions of the land surface in 1955, 1963, and 1975, over which time the land subsided nearly 9 m due to pumpage of groundwater.

form more slowly than those mantled by soil, demonstrating the role that carbon dioxide in the soil plays in generating carbonic acid. Another important factor is rock structure, not only because it affects rates of erosion but also because it results in different forms of karst relief.

Long-term rates of karst evolution are often difficult to evaluate. In middle and high latitudes, which have been repeatedly affected by glaciation, environmental conditions seldom have been stable for very long. Even in tropical regions, which have the most impressive and varied examples of karst terrain, the rate of karst development is likely to have fluctuated through time. We must be cautious about extrapolating measured recent rates of limestone dissolution into the geologic past. The history of changing climates (Chapter 20) makes it evident that erosion rates must have varied as precipitation and river discharge rates changed, and as water tables rose and fell due to major fluctuations of climate.

c

1955

1963

1975

SAN JOAQUIN VALLEY
CALIFORNIA
Bench Mark S661
MAXIMUM SUBSIDENCE
29.3 feet (8.93 m.)
1925-1975

WATER AND PEOPLE

Human beings require reliable supplies of usable water for survival. As cities grow in size the problems of water supply may become acute. The location of new supplies, the balancing of need with supply, and the problem of maintaining water quality become of increasing importance in any area that is experiencing growth in population and industry.

Finding Groundwater

In the days when the population of North America was mostly rural, in any region of fairly abundant rainfall a well could be dug to a few meters' depth with a good chance that it would yield enough water for the use of a family. Sometimes the sites for such wells were located by persons who used forked sticks and other kinds of "divining rods" and who claimed to possess supernatural powers. The search for water by this means, often called "dowsing," dates back at least to Old Testament times. Although no scientific basis for this kind of claim is known to exist, use of the divining rod persists, partly because in many areas shallow supplies of groundwater are so widespread that successful results would be numerous even though sites were located at random. If the diviner were asked to indicate where water is *not* present below the ground surface and if his predictions were then tested by drilling holes, the statistical results would soon reveal how sound his claims are. However, because little money is spent on drilling holes in attempts to avoid water, this test has never been made.

Today the average depth of wells is deeper, and the rock drill has largely replaced the spade and pickax. Groundwater is being found in aquifers that are well hidden. Some are buried deeply beneath regions which, at the surface, are dry. Before a well is drilled, the exposed rock or regolith is examined in detail in an effort to determine conditions far beneath the surface.

Artificial Recharge

Although most recharge is supplied directly by rainfall, the intense demand for water in some areas has led to artificial recharging of the ground. One example is the practice of *water spreading* in dry parts of the American West. A common way to spread water for recharge is to build a low dam across a stream valley. This holds water back that would otherwise flow away and allows it to seep downward and recharge aquifers beneath the stream bed. The water thereby, stored underground is withdrawn through wells as needed.

In some regions an aquifer may be recharged with used water. This practice has increased as air conditioning, which requires a large volume of water, has become commonplace in hot, dry regions. Some cities have laws requiring that water used for air conditioning be returned to the ground, where it successfully builds up the water table. This illustrates the basic principle of groundwater conservation: that withdrawal of water must, over the long run, be balanced by recharge. Otherwise either recharge must be increased or withdrawal curtailed.

Water Quality

The *quality* of a body of water refers to its temperature and the amount and character of its content of mineral particles, dissolved substances, and organic matter (chiefly bacteria) in relation to its intended use. The most common source of water pollution in wells and springs is sewage. The infection most often communicated by polluted waters is typhoid. Drainage from septic tanks, broken sewers, privies, and barnyards contaminates groundwater. If water contaminated with sewage bacteria passes through sediment or rock with large openings, such as very coarse gravel or cavernous limestone, it can travel long distances without much change (Fig. 10.26). If, on the other hand, it percolates through sand or permeable sandstone, it can become purified within short distances, in some cases less than about 30 m (Fig. 10.27). The difference lies in the aggregate internal surface area of the material through which it percolates. The force of molecular attraction holds water and promotes its purification by (1) mechanically filtering out bacteria (water gets through but most of the bacteria do not), (2) destruction of bacteria by oxidation, and (3) destruction of bacteria by other organisms, which consume them. Purification goes on in the zone of aeration as well as in the zone of saturation. Because clay particles are much smaller than sand particles, it might be thought that clay, with its much larger internal surface area, would be the ideal medium for purification. However, it is not ideal, for it is almost impermeable. Particles of sand are large enough to permit rapid percolation, yet small enough to permit purification within short distances. For this reason purification plants that treat municipal water supplies and sewage percolate

these fluids through sand. Unfortunately, dissolved chemicals (including pesticides and heavy metals) are not filtered out by this technique.

A substantial proportion of domestic sewage

FIGURE 10.29 Fissure located about 15 km east of Mesa, Arizona believed to be caused by subsidence of the ground due to removal of large quantities of underground water.

TABLE 10.2 *Examples of Subsidence of Cities Due Partly or Wholly to Groundwater Withdrawal*

City	Maximum Subsidence (m)	Area Affected (km²)
Bangkok	1.00	800
Denver	0.30	320
Houston	2.70	12,100
London	0.30	295
Long Beach/ Los Angeles	9.00	50
Nagoya	2.37	1300
New Orleans	2.00	175
Osaka	3.00	500
Shanghai	2.63	121
Tokyo	4.50	3000
Venice	0.22	150

Source: Data from Dolan and Goodell, 1985.

passes through septic tanks and then mingles with groundwater. The mixture gradually becomes purified as it percolates toward the nearest streams. In many areas, though, domestic and much industrial waste is dumped unaltered directly into surface streams. Although purification can be accomplished during stream transport, the distances involved are much greater than those required for the purification of water in the ground, and the amounts of sewage in many rivers are far too great to be dealt with by natural processes. In densely populated industrial countries, such situations constitute serious public-health problems.

Underground Storage of Hazardous Wastes

One of the leading environmental concerns of industrialized countries is the necessity of dealing with dangerous waste products, especially those which are highly toxic or radioactive. Experience has demonstrated that surface dumping quickly leads to contamination of surface and subsurface water supplies, and can result in serious health problems and even death. Countries with nuclear capacity have the special problem of disposing of high-level radioactive waste products, substances so highly toxic that even minute quantities can prove fatal if released to the surface environment.

Most studies concerning disposal of toxic and nuclear wastes have concluded that underground storage is appropriate, provided safe sites can be found. In the case of high-level nuclear wastes that can remain dangerous for tens or hundreds of millennia due to the long half-lives of the radioactive isotopes, a primary requirement is a site that will be stable over such intervals of time. The only completely safe sites would be ones where waste products and their containers would not be affected chemically by water, physically by natural deformation such as earthquakes, or accidently by people.

The placement of toxic-waste products underground, even far underground, immediately raises concerns about groundwater. Water is a nearly universal solvent, and its weakly acid character means that any toxic substance put in contact with it, including the container that holds the substance, is likely to corrode, dissolve, and be transported away from the site of storage. As we have seen, water is present in crustal rocks to depths of many kilometers and in many of those rocks it is circulating at rates of 1–50 m/yr. Over tens or hundreds of thousands of years, even such slow rates can move dissolved substances over great dis-

tances and introduce them to more rapidly flowing parts of the hydrologic system.

The safe long-term storage and eventual disposal of toxic and nuclear wastes at underground sites therefore requires considerable knowledge of local and regional groundwater systems. It also requires that we gain an understanding of how these systems are likely to change in the future as a result of crustal movements, local and global climatic change, and other natural factors that can affect the stability of a storage site.

Land Subsidence

In some places withdrawal of groundwater has led to subsidence of the ground over large areas. Such effects are particularly widespread in the southwestern United States (Fig. 10.28) where pumping of water for irrigation has led to disruptions of the ground surface (Fig. 10.29); structural damage of

FIGURE 10.30 Leaning Tower of Pisa, the tilting of which accelerated as groundwater was withdrawn from underground aquifers to supply the growing city.

buildings, roads, and bridges; damage to buried cables, pipes, and drains; and increase in areas subject to surface flooding.

Especially alarming is the subsidence taking place within many major cities as groundwater is consumed (Table 10.2). Particularly susceptible are cities built on unconsolidated sediments containing large quantities of water, such as river floodplains, deltas, coastal plains, or the floors of lakes or swamps. Withdrawal of groundwater from numerous closely spaced wells may lead to large-scale lowering of the water table. The dewatering of unconsolidated sediments reduces fluid pressures that help support the load imposed by buildings and other structures. Therefore, the load is transferred from the pore fluids to the granular particles of the aquifer, resulting in rearrangement of the grains and compaction of the sediment.

A well-known example is Mexico City, built on the site of the ancient Aztec capital of Tenochtitlan in the middle of a former shallow lake and now subsiding differentially as groundwater is withdrawn to support a growing population. Venice, already partially submerged by the Adriatic Sea and underlain by a kilometer of unconsolidated sediments, began to subside at an increasing rate as the groundwater was pumped from wells on the nearby mainland to meet industrial demands.

Pisa's famous Leaning Tower is built on unstable floodplain sediments of the Arno River (Fig. 10.30). It began tilting as construction commenced in 1174. The tilting increased in the 1960s as withdrawal of groundwater from deep aquifers accelerated subsidence. Recent stabilization of the foundation has temporarily countered the effect and should make the tower stable for many centuries, providing future groundwater withdrawal is strictly controlled.

Consumption and Future Supplies

In most parts of the world the supplies of surface water and groundwater remain fairly constant. In

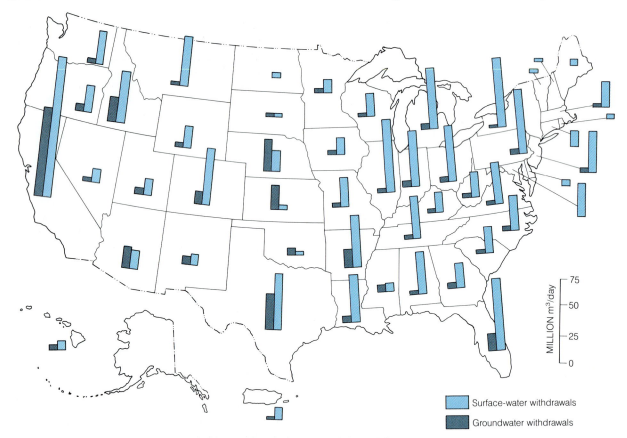

FIGURE 10.31 Water use in the United States during 1980. Withdrawal from groundwater reservoirs is greatest in those states having arid to semiarid climates and limited surface runoff. States with the largest withdrawals require huge quantities of water for irrigation of crops. (*Source:* U. S. Geological Survey, 1984.)

those places where people live in a simple agricultural economy, their withdrawal of water from streams, lakes, and the ground affects the available supply hardly at all because they use little water—perhaps less than 40 liters of water per person each day. By contrast, in industrial societies like the United States, the average per capita use of water may be 100–200 times as great, largely due to the prodigious use of water for irrigating croplands. Such a large consumption of water can substantially affect the natural state of balance, particularly in dry regions where the demand for irrigation water is greatest.

In the United States, the rate of surface reservoir construction has slowed perceptably because the best sites have already been developed, whereas the use of groundwater has accelerated. In three critical areas—the High Plains, the Colorado River basin, and California—high rates of groundwater withdrawal have occurred in recent years. Although such withdrawals have provided adequate water supplies for these areas, the high withdrawal rates cannot continue indefinitely. As the water table continues to fall, decreasing yields and increasing costs of pumping are likely to result in a gradual decline in production.

In parts of Arizona and Israel, it has been found by radiocarbon dating that irrigation water drawn from deep wells is more than 10,000 years old. Such "old" water is obviously a relict of earlier times. It represents rain that fell when climates were generally cooler and when less surface water was lost to evaporation. Under today's warmer and drier climates, rates of recharge are often much less than the present high rates of withdrawal. Therefore, the water used today in many dry areas is being "mined," and used up at a nonreplaceable rate. This means that if current withdrawal rates are maintained, the supply now in the ground will eventually be exhausted. In some irrigation areas, even if all pumping of wells were stopped, it would require more than 100 years to recharge the aquifers.

One of the most obvious characteristics of industrialization is the concentration of people in cities. Today more than two thirds of the population of the United States lives in or near cities. Not only do urban dwellers use more water for all purposes, but the buildings in which they live and work, as well as the streets and roads that give access to them, reduce local recharge from rainfall. In cities, the ground is covered largely with buildings, concrete, and asphalt, all of which send water along the surface and through sewers, rather than allow-

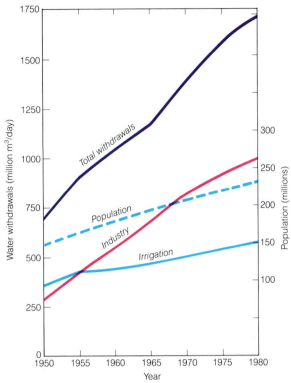

FIGURE 10.32 Trends in groundwater withdrawal in the United States (1950–1980) for industrial and irrigation purposes, compared with the population trend. Although agricultural use closely parallels population increase, industrial use has risen sharply and now accounts for the greatest part of groundwater withdrawal (*Source:* U. S. Geological Survey, 1984.)

ing it to soak naturally into the ground. It is urbanization, more than growth of population overall, that has created a demand for water which in many cities threatens to exceed supply. Furthermore, the available supply is less than it would be because of pollution (chemical, bacterial, or thermal) of many streams, some lakes, and some groundwater reservoirs by industrial wastes, agricultural poisons, and sewage.

In Figure 10.31, a comparison of the amounts of surface water and groundwater withdrawn in the United States during 1980 can be seen. While stream flow supplies much of the demand in eastern states, in many of the drier western states groundwater withdrawals constitute a high percentage of the total. The states with the greatest water consumption are those having large populations, substantial industrialization, or intensive agricultural production. Over the past 30 years the population of the United States increased about one and a half times while the use of water increased two and a half times. Although the increase in water use for irrigation has kept pace with population growth, industrial use has risen more sharply and accounts for most of the per capita increase in water consumption (Fig. 10.32).

Much of the water withdrawn from the ground and from surface sources and used for a variety of human purposes is recycled to these natural reservoirs to be reused. If groundwater is withdrawn for irrigation purposes, that part not evaporated or used by plants will either soak downward and return to the groundwater reservoir or will run off and enter surface streams from which it may once more be withdrawn for human use. This can happen repeatedly to a parcel of water during its long journey from the middle of a continent on its way toward the ocean.

Many authorities believe that the shortage of water in some large American cities, although real, is less a matter of inadequate quantity than of inefficient planning and development. They believe that except for the dry southwestern part of the country, potential supplies of water, *if properly managed*, need not be overtaxed for some time to come.

SUMMARY

1. More than 97 percent of the Earth's water resides in the oceans. Snow and ice account for about another 2 percent while less than 1 percent occurs as rivers, lakes, groundwater, and water vapor.

2. Water is constantly cycled from one natural reservoir to another. Although over short periods the amounts in each reservoir remain relatively constant, over longer spans substantial changes can occur.

3. Groundwater, derived almost entirely from rainfall, occurs nearly universally.

4. The water table is the top of the saturated zone. Its form is a subdued imitation of the ground surface above it.

5. Groundwater flows chiefly by percolation, at rates far slower than those of surface streams. With constant permeability, velocity of flow of groundwater increases as the slope of the water table increases.

6. In moist regions groundwater percolates away from hills and emerges in valleys. In dry regions it is likely to percolate away from beneath large surface streams thereby recharging the ground.

7. A groundwater system is an open system in each segment of which a steady state is approached.

8. Major supplies of groundwater are found in aquifers, among the most productive of which are porous sand, gravel, and sandstone.

9. Groundwater flows into most wells directly by gravity, but into artesian wells under hydrostatic pressure. Withdrawal of water through wells creates cones of depression in the water table.

10. Groundwater dissolves mineral matter from rock. It also deposits substances as cement between grains of sediment, thereby reducing porosity and converting the sediments to sedimentary rock.

11. In carbonate rocks, groundwater not only creates caves and sinkholes by dissolution but also, in some caves, deposits calcium carbonate as flowstone and dripstone.

12. Polluted water percolating through permeable sand or sandstone can often become purified within short distances.

13. Safe underground storage of hazardous wastes requires a thorough knowledge of groundwater systems and how they will respond to future crustal movements and changes of climate.

14. Where groundwater withdrawal exceeds the rate of recharge, the lowering of the water table can lead to subsidence of the ground.

15. In some places, groundwater is being mined more rapidly than it can be replenished by recharge. Artificial recharge and proper management can sustain groundwater supplies for future use.

SELECTED REFERENCES

Davis, S. N., and De Wiest, R. J. M., 1966, Hydrogeology: New York, John Wiley & Sons.

Dolan, R., and Goodell, H. G., 1986, Sinking cities: Am. Scientist, v. 74, p. 38–47.

Dunne, T., and Leopold, L. B., 1978, Water in environmental planning: San Francisco, W. H. Freeman.

Freeze, R. A., and Cherry J. A., 1979, Groundwater: Englewood Cliffs, N.J., Prentice-Hall.

Heath, R. C., 1983, Basic groundwater hydrology: U.S. Geological Survey Water-Supply Paper 2220.

Heath, R. C., 1984, Ground-water regions of the United States: U. S. Geological Survey Water-Supply Paper 2242.

Jennings, J. N., 1983, Karst landforms: Am. Scientist, v. 71, p. 578–586.

Leopold, L. B., 1974, Water, a primer: San Francisco, W. H. Freeman.

Monroe, W. H., 1976, The karst landforms of Puerto Rico: U. S. Geological Survey Prof. Paper 899.

Moore, G. W., and Nicholas, G., 1964, Speleology. The study of caves: New York, D. C. Heath.

Price, M., 1985, Introducing groundwater: London, Allen and Unwin.

Trudgill, S., 1985, Limestone geomorphology: White Plains, N.Y., Longman.

U. S. Geological Survey, 1984, National water summary 1983—Hydrologic events and issues: U. S. Geological Survey Prof. Paper 2250.

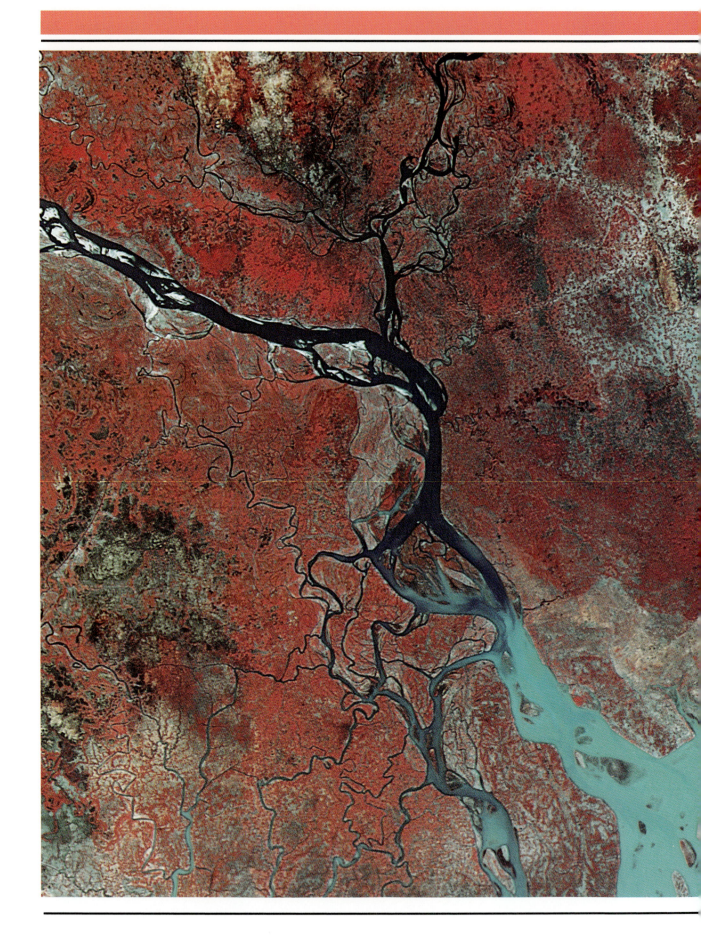

C H A P T E R   1 1

Streams and Drainage Systems

Ganges River near the head of its delta in Bagladesh. One of the
world's largest rivers, it carries an extremely large load of sediment
from the Himalaya to the Bay of Bengal. The main channel is braided
and marked by many alluvial islands, whereas smaller tributaries
have meandering patterns.

STREAMS IN THE LANDSCAPE

Almost anywhere we travel over the Earth's land surface we can see evidence of the work of running water. Even in places where no rivers flow today, we are likely to find deposits and landforms that tell us water has been instrumental in shaping the landscape. Most of these features can be related to the activity of streams that are part of complex drainage systems. We define a *stream* as *a body of water that carries rock particles and dissolved substances, and flows down a slope along a clearly defined path*. The path is the stream's *channel,* and the rock particles constitute the bulk of its **load,** *the material the stream moves or carries*.

Streams play an important role in our lives for many reasons. They are an important source of water for human and industrial consumption. They are a comparatively small but essential source of energy. Many rivers are avenues of transportation. They also have great scenic and recreational value. The floors of stream valleys are generally fertile, and building on them is easy. Therefore, they tend to invite large populations which then must face the danger of damage by floods as well as the necessity of controlling pollution from the discharge of wastes into the streams.

In addition to their immediate practical and esthetic importance, streams are vital geologic agents for the following reasons:

Streams carry most of the water that goes from land to sea, and so are an essential part of the hydrologic cycle.

Streams transport sediment to the ocean, where it is deposited and can ultimately become part of the rock record. Sampling the loads of many rivers in many countries tells us that every year streams transport from land to ocean about 18 billion metric tons of solid rock debris plus about 4 billion tons in solution. Thus, about 82 percent of the average stream load consists of visible rock particles. The rest is invisible, dissolved substances which are the product of chemical weathering.

Streams shape the surface of the continental crust. Most of the Earth's landscapes consist of stream valleys separated by higher ground, and are the result of weathering, mass-wasting, and stream erosion working in combination (Fig. 11.1). Other agents, such as glaciers and wind, have also locally shaped the land surface, but even in such regions the imprint of stream activity is commonly present.

FIGURE 11.1 Satellite view of southern flank of the Himalaya in Nepal showing effects of streams in shaping the landscape. Total relief in the picture is more than 8000 m. Mt. Everest (Chomolungma) (8850 m) lies at upper right corner of scene. Master streams flowing down the steep slope of the mountains are joined by tributaries that follow belts of weak rock and fault zones. Emerging onto the low-gradient alluvial plain, the streams flow south to join the Ganges River in northern India. Distance across the scene is about 100 km.

EROSION BY RUNNING WATER

Erosion of the land by water begins even before a distinct stream has been formed. It occurs in two ways: by impact as raindrops hit the ground, and by sheets of water that result from heavy rains (Fig. 11.2). As raindrops strike bare ground they dislodge small particles of loose soil, spattering them in all directions. On a slope the result is net displacement downhill. One raindrop has little effect, but the number of raindrops is so great that together they can accomplish a large amount of erosion.

The average annual rainfall on the area of the United States is equivalent to a layer of water 76 cm thick covering the entire land surface. Of this layer, 45 cm returns to the atmosphere by evaporation and transpiration (Fig. 10.1) and 1 cm infiltrates the ground, recharging the groundwater. The remaining 30 cm forms runoff. We can subdivide the runoff into **overland flow,** *the move-*

FIGURE 11.2 Rainsplash and sheetwash on a very gently sloping sandy soil in Amboseli National Park, Kenya. The water around each raindrop splash is more turbid than in the surrounding sheetflow, indicating that the impacting drops dislodge soil particles that are then carried away by sheetwash.

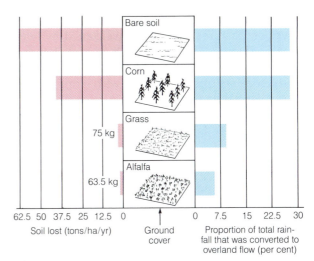

FIGURE 11.3 Effect of plant cover on rate of sheet erosion measured over four years at Bethany, Missouri. The soil is silty, slope is 4.5° and annual rainfall is 1000 mm. The measurements show that grass and alfalfa, with their continuous network of roots and stems, are nearly 300 times as effective as "row crops," such as corn, in holding the soil in place. Erosional loss from bare soil shown here occurs at a rate of about 45 cm/100 yr.

ment of runoff in broad sheets or groups of small, interconnecting rills, and **stream flow,** *the flow of surface water in a well-defined channel.* Stream flow is very obvious, but overland flow is less obvious. Usually it occurs only through short distances before it concentrates into channels and forms streams. Such flow takes place wherever rainfall is greater than the capacity of the ground to absorb it. *The erosion performed by overland flow is called* **sheet erosion.**

The effectiveness of raindrops and overland flow in eroding the land are greatly diminished by vegetation. Over a large proportion of the world's natural landscapes, excluding areas covered with ice and snow, the regolith is protected by a mantle of vegetation. The leaves and branches of trees break the force of falling raindrops and cushion their impacts upon the ground. More importantly, the intricate network of roots, especially grass roots, forms a tight mesh that holds soil in place, greatly reducing erosion. Not only does the mesh retain soil, but it also holds water. Like a sponge, it absorbs rainwater and retains it for a time, letting it percolate slowly down through the soil beneath. As a consequence, less water runs off over the surface.

In this case we are dealing with natural vegetation. But where crops are grown, during part of each year the surface is ordinarily bare. On unvegetated, sloping fields, pastures that are too closely grazed, and areas planted with widely spaced crops such as corn, rates of erosion can be high. Sheet erosion creates no obvious valleys, so the damage it does to bare soil was not fully realized until accurate measurements began. Many ex-

FIGURE 11.4 Effects of accelerated sheetwash erosion on a silty soil in the Samburu District, northern Kenya. Ground surface has been lowered about 60 cm in the last 50 years, as is indicated by the exposed tree roots below the remnants of the original ground surface.

periment stations maintained by government agencies now measure the erosion of soil. The measurements have practical value not only to farmers, but also to everyone else with a stake in the economy because they show that sheet erosion is a menace to soil left unprotected on slopes. In recognition of this fact, wise farmers reduce areas of bare soil to a minimum and prevent the grass cover on pastures from being weakened by overgrazing (Fig. 11.3). If crops such as corn, tobacco, and cotton must be planted on a slope, strips of such crops are often alternated with strips of grass or similar plants that resist sheet erosion.

Erosion of slopes does not result solely from human activities. Under certain natural conditions

the splash of raindrops and the work of sheet erosion are so effective that they combine to remove large volumes of fine rock particles. For example, in some subtropical grasslands all the rainfall is concentrated within a single rainy season. During the long dry season, evaporation so depletes soil moisture that grass becomes sparse, covering no more than 40–60 percent of each square meter of ground. Although kept bare by natural causes, the soil is as vulnerable to erosion as soil laid bare by farming (Fig. 11.4).

GEOMETRY OF STREAMS

Stream Channel

A stream's channel is designed as an efficient conduit for carrying water. The **discharge,** or *quantity of water that passes a given point in the channel per unit time,* varies both along the channel and through time. In response, most channels are self-adjusting, continually modifying their shape to changing conditions. The size of any particular channel cross section reflects the typical stream conditions at that place. However, it may not be large enough to carry exceptionally large flows, which inevitably overtop the stream banks and

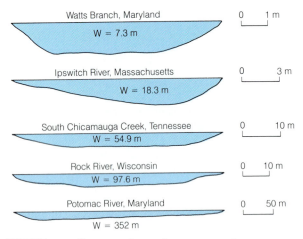

FIGURE 11.5 Cross sections of some natural streams, drawn so their widths (W) are at the same scale. In general, the wider the channel, the larger is the ratio of width to depth. (*Source:* After Leopold et al., 1964.)

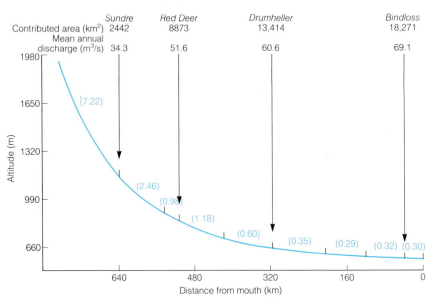

FIGURE 11.6 Long profile of Red Deer River in southern Alberta, Canada, showing typical hyperbolic shape. The size of the land area that contributes water to the stream increases downstream, as does discharge. The slope of the channel (in parentheses) is measured in m/km for nine segments along the course, and decreases steadily downstream. (*Source:* After Campbell, 1977.)

spread across the adjoining *floodplain,* which is *the part of any stream valley that is inundated during floods.*

Cross-Sectional Shape

A stream channel typically is rounded in cross section. Very small streams may be as deep as they are wide, whereas very large streams usually have widths many times greater than their depths (Fig. 11.5). It follows that along any stream channel the ratio of width to depth is likely to change downstream as the volume of water increases. It also changes as the volume and character of the debris load change.

Long Profile

The *gradient* (or *slope*), which is *a measure of the vertical drop over a given horizontal distance,* may be 60 m/km or even more for a mountain stream, whereas near the mouth of a large river it may be 0.1 m/km or even less. The average gradient of a river decreases downstream, and so its *long profile* (*a line drawn along the surface of a stream from its source to its mouth*) is generally a curve that decreases in gradient downstream (Fig. 11.6). In geometrical terms, the profile has a shape like a section of a hyperbola. In reality, though, it only approximates such a form, because irregularities along the channel of a stream usually cause the long profile to depart from perfect regularity. Such a local change in gradient may occur, for example, where a channel passes from a belt of resistant rock into one that is more erodible. It may result where a tributary stream, joining a main stream, supplies more water to the channel, thereby causing greater erosion immediately downstream. If a tributary supplies more sediment than the main stream can move, a deposit may result that causes a local change in the stream gradient. A fault that crosses the path of a stream and offsets the bedrock can also cause a change of gradient, as can a landslide or lava flow that forms a temporary dam. A man-made dam likewise introduces an irregularity in the long profile of a stream channel, usually creating an extensive reservoir upstream from the obstruction.

Velocity Distribution

The flow of water within a channel is not uniform. In a natural channel, velocity decreases toward the bed and the channel walls because of increasing

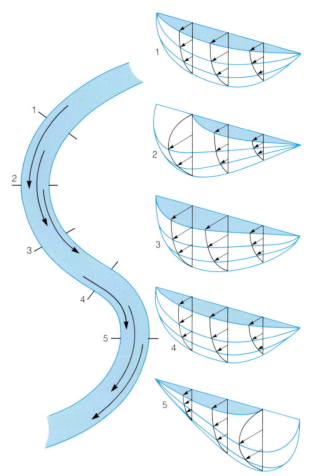

FIGURE 11.7 Velocity distribution in cross sections through a sinuous channel (length of arrows indicate relative flow velocities). The zone of highest velocity lies near the surface and toward the middle of the stream where the channel is relatively straight (sections 1 and 4). At bends, the maximum velocity swings toward the outer bank and lies below the surface (sections 2 and 5). (*Source:* After Leopold, 1964.)

frictional resistance to flow. In large rivers the maximum velocity in a straight symmetrical channel segment is at or near the surface in midchannel (Fig. 11.7). In sinuous channels, the maximum velocity zone swings toward the outside of each bend, becoming closer to the bank as the radius of curvature of the channel decreases.

Channel Patterns

Straight Channels

If we look out the window of an airplane and examine carefully the streams on the ground below, we see that almost all follow sinuous paths across

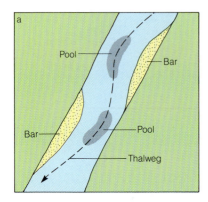
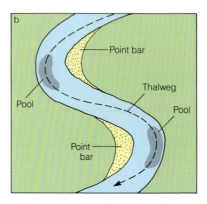
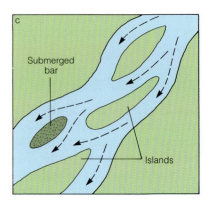

FIGURE 11.8 Features associated with (*a*) straight, (*b*) meandering, and (*c*) braided streams. The arrows indicate the direction of stream flow. (*Source:* In part after Ritter, 1978.)

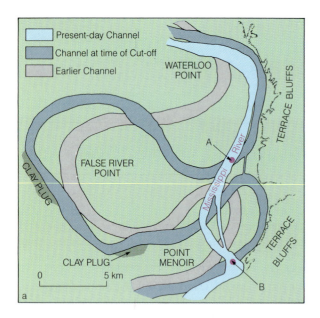

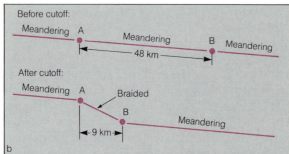

FIGURE 11.9 Cutoff of a meander loop of the Mississippi River in Louisiana. (*Source:* After Russell, 1967.) (*a*) Downvalley migration of the river was halted when the channel encountered a body of clay in the floodplain sediments. This allowed the next meander loop to advance and finally cut off the river segment surrounding False River Point. (*b*) The new, shorter channel had a steeper gradient than the abandoned course, and a braided pattern developed.

the land surface. Except where the landscape is strongly influenced by joints or faults, straight channel segments are rare. Generally, they occur for only brief stretches before they turn sinuous.

Close examination of a segment of straight channel in the field shows that it has many of the features of sinuous channels. *A line connecting the deepest parts of the channel*, called the **thalweg**, typically does not follow a straight path equidistant from the banks, but wanders back and forth across the channel (Fig. 11.8*a*). Where the thalweg swings to one side of a channel, a deposit of sediment tends to accumulate on the opposite side where velocity is lower. Such a deposit is called a **bar**, *an accumulation of alluvium formed where a decrease in stream velocity causes deposition.* Because the thalweg is sinuous, the bars alternate on opposite sides of the channel. Clearly, a straight channel implies neither a symmetrical, unchanging streambed nor a straight thalweg.

Meandering Channels

The pattern of most streams is a series of bends. Very commonly the bends are smooth, looplike, and similar in size. Such a *looplike bend of a stream channel* is called a **meander** [from the Menderes River in southwestern Turkey (Latin = Meander), noted for its winding course]. The meanders of the Mississippi River and other large streams are under continual study by engineers and geologists concerned with problems related to floods and navigability. These people know that meanders are not accidental, that they occur most commonly in channels having gentle gradients in fine-grained alluvium, and that they occur even in streams having no load at all. The meandering pattern reflects the way in which a river minimizes resistance to flow and dissipates energy most nearly uniformly along its course. Therefore, it is a pattern of equilibrium.

A meander changes position almost continually, as shown by year-to-year measurements of river channels and by artificial streams in the laboratory. The shift or migration of a meander is accomplished by erosion on the outer banks of meander bends and deposition on the inside of the bends. Along the inner side of each meander loop, where water is shallowest and velocity is lowest, coarse sediments accumulate as a distinctive *point bar* (Fig. 11.8*b*).

Meanders tend to migrate in the downvalley direction as well as across the floodplain. Slumping of the stream banks occurs most frequently along the downstream side of a meander bend. This is where the thalweg, and therefore the highest current velocity, impinges on the channel side, causing erosion and undercutting of the banks. As a result, meanders tend to migrate slowly down the valley, subtracting from, and adding to, various pieces of real estate along the banks. This constant shifting of land from one bank to the other can lead to legal disputes over property lines and even over the boundaries between countries and states.

The behavior of streams in laboratory channels shows that if the bank sediment is uniform, meanders are symmetrical and migrate downvalley at the same rate. However, the material of natural banks generally is not uniform. Wherever the downstream limb of a meander encounters less-erodible sediment, such as clay, its migration can be slowed. Meanwhile the upstream limb, migrating more rapidly, may intersect and cut into the slower-moving limb (Fig. 11.9a). Thus, the channel bypasses the loop between the two limbs and the cut-off loop is converted into a curved *oxbow lake*. The new course is shorter than the older course, so the channel gradient is steeper there and the stream length is shortened (Fig. 11.9b).

The aggregate length of the Mississippi River channel abandoned through cutoffs since 1776 amounts to nearly 600 km. However, the river has not been shortened appreciably because the segments lost through cutoffs have been balanced by lengthening resulting from enlargement of other meanders.

Braided Channels

Water in a braided stream flows through two or more adjacent but interconnected channels separated by bars or islands (Fig. 11.8*c*). If a stream

FIGURE 11.10 Intricate braided pattern of Brahmaputra River where it flows out of the Himalaya en route to the Ganges delta. Noted for its huge sediment load, the river is as much as 8 km wide during the rainy monsoon season.

is unable to transport all the available bed load, it may deposit the coarsest sediment as a bar which locally divides the flow and concentrates it in the deeper stretches to either side. As the bar builds up, it may emerge above the surface as an island and become stabilized by vegetation that anchors the sediment and inhibits erosion.

Large braided rivers, like the Brahmaputra where it emerges from the high mountains of central Asia (Fig. 11.10), are marked by numerous wide shallow channels, abundant sediment load, and constant shifting of channel courses. Over the span of an hour or two, a person standing near the side of such a river can observe channels move laterally as a result of bank erosion, or become suddenly dry as one channel captures the flow of another. Although at any moment the active channels of a large braided stream may cover no more than 10 percent of the width of the entire channel system, within a single season all or most of the surface sediment may be reworked by the laterally shifting channels.

The braided pattern tends to form in streams having highly variable discharge and easily erodible banks that can supply abundant sediment load to the channel system. If a meandering stream is unable to move all the load it is transporting through a certain channel segment, a change to a braided pattern will apparently increase the ability of the same discharge to move a greater load. This is possible because the cumulative width of the channel system becomes greater and the slope locally increases (Fig. 11.9). The braided pattern, therefore, seems to represent an adjustment by which a stream is able to transport a larger bed load more efficiently.

DYNAMICS OF STREAM FLOW

The channels of most natural streams consist partly or wholly of sediment. As streams move this material from place to place, their channels are continually being altered. Because stream and channel are closely related and are ever-changing, they should be examined together as an interrelated system.

Factors in Stream Flow

Five basic factors control the manner in which a particular stream behaves: the *channel dimensions* (width and depth), expressed in meters (m); the *gradient*, expressed in meters per kilometer (m/km); the *average velocity*, expressed in meters per second (m/s); the *discharge*, expressed in cubic meters per second (m^3/s); and the *load*, consisting of rock particles plus matter in solution, expressed in metric tons per cubic meter ($m.t./m^3$). Unlike rock particles that constitute the mechanical load, dissolved matter generally makes little difference to the behavior of the stream.

In a stream system there is a continual interplay among these factors. Discharge is calculated from measurements made systematically at selected points along large and small streams for evaluating water supply, irrigation potential, and flood control. The U.S. Geological Survey, for example, maintains nearly 6500 measurement points, called *gaging stations*, in various parts of the United States.

The measurements show that as discharge changes, velocity and channel shape also change. The relationship can be expressed by the formula:

$$Q \quad = \quad w \quad \times \quad d \quad \times \quad v$$

Discharge	Width	Depth	Velocity
(m^3/s)	(m)	(m)	(m/s)

Depth varies continuously across the stream, and velocity differs at every point in any cross section. Therefore, these values are expressed as averages,

which are difficult to obtain accurately. When discharge changes, as it does continually, one or more of the three terms must change accordingly. With increased discharge, the velocity also increases. The stream erodes and enlarges its channel, rapidly if it flows on alluvium, much more slowly if it flows on bedrock. The increased load is carried away. This continues until the increased discharge can be accommodated in a larger channel and by faster flow. In contrast, when discharge decreases, some of the load is dropped, decreasing the channel depth and width, and the velocity is reduced by increased friction. In these ways width, depth, and velocity are continually readjusted to changing discharge.

A dramatic example of changes in stream factors can be seen when floods occur. During 1956, the channel of the Colorado River at Lees Ferry, Arizona, experienced a major change in dimensions as discharge increased and then declined (Fig. 11.11). Prior to the flood, the channel averaged about 2 m deep and 100 m wide. As discharge increased in late spring, the water rose in the channel and erosion scoured the bed until at peak flow the channel was about 7 m deep and 125 m wide. Together with an increase in velocity, the enlarged channel was now able to accommodate the increased flood discharge and carry a greater load. As discharge fell, the stream was unable to transport as much sediment, and the excess load was dropped in the channel, causing its floor to rise. At the same time, the water level fell, thereby returning the cross-sectional area to its preflood dimensions.

Thus, a stream and its channel are related intimately. We can think of them as a single system. The channel is so responsive to changes in discharge that the system, at any point along the stream, is continually close to a balanced condition.

Changes Downstream

Traveling down a river from its head to its mouth, one can see that orderly adjustments occur along it. For example, (1) discharge increases (Fig. 11.6), (2) width and depth of the channel increase (Fig. 11.12), (3) velocity increases slightly (Figs. 11.12 and 11.13), and (4) gradient decreases (Figs. 11.6 and 11.12).

The demonstration that velocity increases downstream seems to contradict the common observation that water rushes turbulently down steep mountain slopes and flows smoothly over nearly

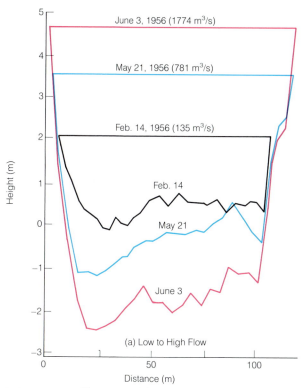
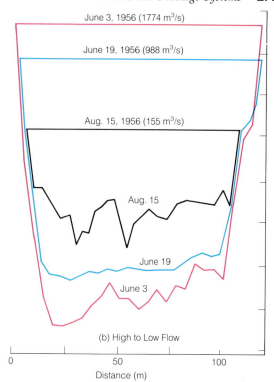

FIGURE 11.11 Changes in the cross sectional area of the Colorado River at Lees Ferry, Arizona during 1956. As discharge increased from February to June, the channel floor was scoured and deepened and the water level rose higher against the banks. During the falling-water phase, the river level fell and sediment was deposited in the channel, decreasing its depth (*Source: After Leopold et al., 1964.*)

flat lowlands. However, the physical appearance of a stream is not a true measure of its velocity, which increases downstream as tributaries introduce more water, thereby increasing the discharge.

Base Level

As a stream flows downslope and eventually enters the sea, its potential energy falls to zero, at which point it no longer has the ability to deepen its channel. This *limiting level below which a stream cannot erode the land* is called its **base level.** The *ultimate base level* for most streams is global sea level, projected inland as an imaginary surface (Fig. 11.14). Exceptions are streams that drain into closed interior basins having no outlet to the sea. Where the floor of a tectonically downfaulted basin lies below sea level (for example, Death Valley, California, or the basin of the Dead Sea in the Middle East), such streams can erode their channels below the level of the world ocean.

For a stream ending in a lake, the level of the lake acts as a *local base level,* for the stream cannot erode below it (Fig. 11.14). Such local base levels may temporarily put a halt to the stream's ability to erode downward. But if the lake is destroyed by erosion at its outlet, the stream, having acquired additional potential energy, would then be able to deepen its channel. By doing so it would adjust its long profile to the changed conditions.

The Graded Stream

A change in discharge of a stream, or in the erodibility of the rock or sediment on which it flows, leads to an adjustment in the shape of the stream's channel cross section and long profile. In adjusting to such changes, the stream modifies its channel so that irregularities are minimized and the least energy is expended in the movement of water and sediment along its course. Frequently, this involves enlargement of the channel cross section by erosion, as in the example of the Colorado River flood, or reducing its dimensions through deposition. The overall tendency, therefore, is toward a

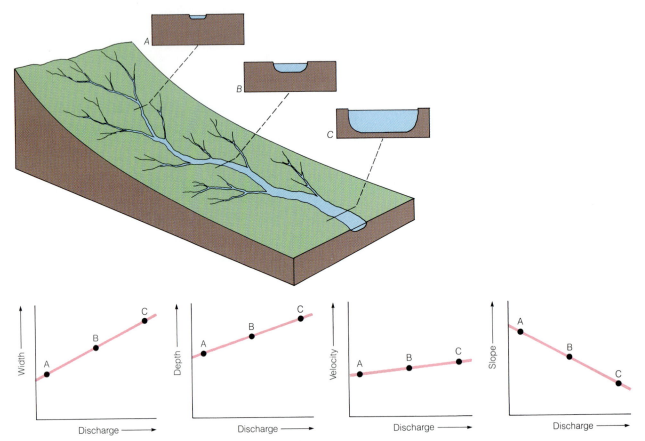

FIGURE 11.12 Changes in the downstream direction along a stream system. Discharge increases as new tributaries join the main stream. Width and depth of the channel are shown by cross sections *a*, *b*, and *c*. Relative velocity is indicated by the length of arrows. Graphs show relationship of discharge to channel width and depth, to velocity, and to slope at the same three cross sections.

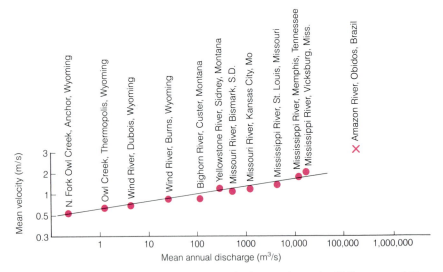

FIGURE 11.13 Relationship of velocity to discharge along the Yellowstone-Missouri-Mississippi River system. As discharge increases downstream, average velocity also increases. (*Source:* After Bloom, 1978.)

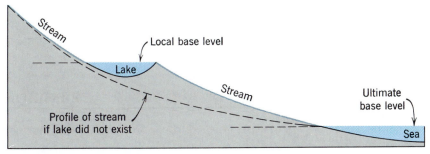

FIGURE 11.14 Relationship of a stream to its ultimate base level (the sea) and to a local base level (a lake) along its course.

smooth long profile — a profile of equilibrium in which all factors are in a state of balance. Geologists have long used the term *grade* in referring to a river that has achieved such a balance. A ***graded stream*** is conceived as *one in which the slope has become so adjusted, under conditions of available discharge and prevailing channel characteristics, that the stream is just able to transport the sediment load available to it*. If any of the controlling factors change, then the stream will adjust to absorb the change and restore an equilibrium condition.

Although the concept of a graded stream is both reasonable and useful, in fact it is unlikely that a condition of perfect equilibrium is ever achieved in natural stream systems. In the drainage basin of a typical river, changes are constantly taking place that upset the balance. A passing rain cloud may suddenly increase the discharge in one tributary, collapse of a bank may locally introduce an excess mass of sediment to the channel, or the stream may abruptly encounter a less-erodible rock or sediment along its course. Adjustments take place as the stream reacts to such events; however, each event perturbs the system and leads it away from a state of perfect balance. For this reason, it is more appropriate to think of a stream as reaching a condition of *quasi-equilibrium* (seemingly in equilibrium), in which adjustments are continually taking place.

Large-Scale Changes in Equilibrium

Significant changes in the equilibrium condition of a stream system may require much longer periods of adjustment. For example, a major storm can introduce a sudden increase in discharge requiring substantial channel adjustment through a large part of the stream system. A long-term change in climate can also have dramatic effects on a stream if it results in a change in mean annual discharge,

in the seasonal distribution of precipitation, or in the vegetation cover and related rates of runoff (Fig. 11.15). Fluctuations of world sea level may lead to major changes in river systems as streams adjust their long profiles to a rise or fall of base level. Tectonic activity can also cause streams to change course, to become ponded, or to erode downward or build up their channels in response to local or regional crustal movements. In each of these cases, both long-term adjustments of drainage systems and shorter-term adjustments related to seasonal, local, or small-scale events are likely.

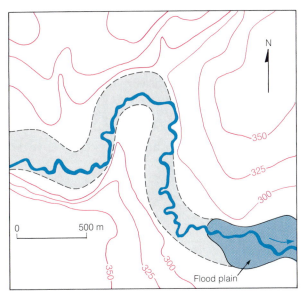

FIGURE 11.15 Map showing part of Windrush River in Oxfordshire, England. Channel width and amplitude of meander loops is very much smaller than the width and amplitude of the loops of the meandering valley, implying that the stream formerly had a much larger discharge and therefore greater width and larger meanders. (*Source:* After Dury, 1964.)

Floods

Every spring the media carry stories of minor or major disasters caused by floods. Communities are inundated, buildings and automobiles are carried away, and valuable farmland is submerged to depths of several meters. The state government may designate the flooded lands as "disaster areas," thereby qualifying them for federal assistance loans. Although the people affected by such events are frequently surprised and even outraged at what the rampaging stream has done to them, geologists tend to view floods as normal and expectable events.

The seasonal distribution of rainfall causes many streams to rise seasonally in *flood,* which is *discharge great enough to cause a stream to overflow its banks* (Fig. 11.16). As discharge increases during a flood, so does velocity. This has the double effect of enabling a stream to carry not only a greater load, but also larger particles. An extreme example of the exceptional force of floodwaters resulted from the collapse, in 1928, of the large St. Francis Dam in southern California. As the dam gave way, the water behind it rushed down the valley, moving blocks of concrete weighing as much as 9000 metric tons (9 million kg) through distances of more than 750 m. Because natural floods are also capable of moving very large objects as well as great volumes of sediment, they are able to accomplish considerable geologic work.

Flood Frequency

Major floods can be disastrous events, causing both loss of life and extensive property damage (Table 11.1), so it is highly desirable to be able to predict their occurrence. There is no certain way of knowing when a major flood will occur. However,

FIGURE 11.16 Mekong River in Kampuchea (formerly Cambodia), swollen by monsoon rains and colored brown with its load of silt. Villages, built on poles and perched on the highest parts of the levees, are surrounded by water during times of flood.

by analyzing the frequency of occurrence of past floods of different size, it is possible to establish the probable interval, in years, between floods of a given magnitude (termed the *recurrence interval*). The recurrence interval is found by plotting on a

TABLE 11.1 *Fatalities from some Disastrous Floods*

River	Date	Fatalities (est.)	Remarks
Hwang Ho, China	1887	900,000	Flood inundated 129,500 km² (50,000 mi²). Many villages swept away
Johnstown, Pennsylvania	1889	2100	South Fork Dam failed. Wall of water 10–12 m high rushed down valley
Yangtze, China	1911	100,000	Formed lake 130 km long and 50 km wide
Yangtze, China	1931	200,000	Flood extended from Shanghai to Hankow
Vaiont, Italy	1963	2000	Landslide into lake produced wave that overtopped dam and inundated villages below

Sources: NOAA, U.S. Geological Survey, and Encyclopedia Americana, 1983.

probability graph the frequency of floods of different magnitudes that a stream has experienced during a period of record. The resulting curve is then used to estimate how frequently a flood of a certain magnitude is likely to recur. In Figure 11.17, for

example, a flood having a discharge of 1750 m³/s is likely to recur once in every 10 years, whereas a larger flood of 2500 m³/s will probably recur only once in every 50 years.

Exceptional floods — well outside the stream's normal range — occur infrequently, only once in many decades or even centuries. A large flood that recurs, for example, only once in every 50 years is referred to as a 50-year flood. During such floods the geologic work accomplished may be prodigious, but they happen so rarely that in most cases their long-term effects on the landscape probably are less than the cumulative effect of normal seasonal flooding throughout the lengthy intervening periods.

Flood Control on the Mississippi River

Along the banks of the Mississippi River, as along most other large rivers, are found *natural levees, broad, low ridges of fine alluvium built along both sides of a stream channel by water that spreads out of the channel during floods.* The levees have been built up by natural events over a long period, but beginning early in the eighteenth century they also were heightened artificially by earth dikes designed to hold in ordinary floods and protect buildings and farmlands on the adjacent floodplain (Fig. 11.18). At selected points spillways were built to allow the water of the highest floods to escape harmlessly into natural channels that parallel the main chan-

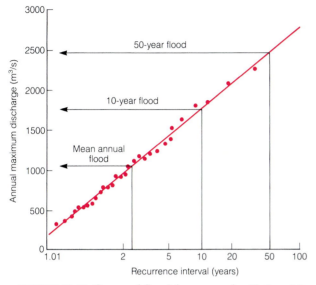

FIGURE 11.17 Curve of flood frequency for Skykomish River at Gold Bar, Washington, plotted on a probability graph. Once in every 10 years a flood of close to 1750 m³/s can be anticipated, whereas only once in every 50 years is there likely to be a flood of 2500 m³/s. (*Source:* Data from U. S. Geological Survey; after Dunne and Leopold, 1978.)

FIGURE 11.18 Artificial levee, built along margin of the Mississippi River, constrains floodwaters in this 1903 photo. A subsequent breach caused water to inundate floodplain on left to depth of 2 m.

nel of the Mississippi. The heightened levees narrowed the channel by as much as one-third. Under natural conditions, a flood would likely lead to widening and deepening of the channel to accommodate the increased discharge. However, the artificially fixed width of the channel now caused the stream not only to erode its bed but also to rise much higher than in former years in order to compensate for the greater flow. The levee-heightening program was based on the assumption that it was principally the lower river that must be controlled if destructive floods were to be prevented. Gradually, however, it became evident that all the rest of the vast watershed must also be considered.

While the levees along the Mississippi were being built ever higher, other things were happening over a large part of the river basin. Former woodland and grassland were being converted into cropland. Swamps and other wet areas on farms were drained, with the water flowing through tile drains instead of passing through the ground by percolation. Much of the land was being overgrazed, thereby impairing grassland sod and partly destroying the ability of the soil to retain water. Buildings, paved streets, highways, and other structures were being built, each with ditches and other conduits designed to lead rainwater rapidly off the surface.

All these changes heightened and narrowed the peaks of floods, and so made matters worse in the downstream areas. Since the 1930s, the diversion of runoff from farmland has been reduced somewhat by plowing along contours instead of up and down the slopes, alternating strips of grass with strips of crops along the contours of slopes, and leaving a litter of organic trash scattered over bare fields to impede surface water and allow it to percolate into the ground. This attempt to restore the natural balance has had positive effects. However, because the human impact on this and other rivers remains great, the consequences are often serious and unanticipated.

Artificial Dams

The practice of building artificial dams across rivers, thus creating a reservoir upstream from the dam, has been widespread in many countries. In the United States, examples of whole series of

FIGURE 11.19 Dams and reservoirs in the lower Colorado River region, showing existing and proposed canals (dotted lines). Dams were built between 1938 and 1964.

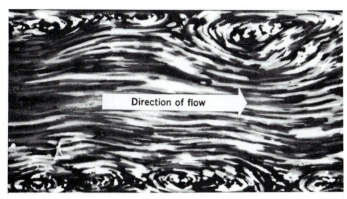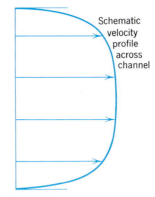

FIGURE 11.20 Turbulent flow in an open channel, as viewed from above. Because of drag, turbulence at the surface is greatest near the channel sides.

dams along a single river and its tributaries include the Tennessee, the Missouri, the lower Columbia, and the lower Colorado river systems. Some of the dams make possible the generation of hydroelectric power, and they also reduce seasonal floods by allowing lowered reservoirs to fill during times of flood. The Missouri River dam system also provides water for irrigation in the comparatively dry climate of the Great Plains, as does water from Columbia River reservoirs which irrigates the wheatfields of the otherwise dry Columbia Plateau.

The Colorado River system generates power, provides irrigation water, and, through aqueducts more than 400 km long, furnishes much of the municipal water supply of Los Angeles and San Diego. Figure 11.19 shows the essentials of the system of dams and reservoirs. Parker Dam provides the municipal water supplies. The dams downstream from it (including the Morelos Dam in Mexico) furnish irrigation water. At the three dams upstream, Colorado River water generates electric power. Thus, the water is used twice. First it falls steeply and generates power and then, via distribution systems, it supplies cities and irrigates crops. So fully is it used that downstream from the Morelos Dam the river is sometimes hardly more than a trickle. In addition, there is now increased loss of water through evaporation, promoted by the extensive surfaces of reservoirs that are exposed to dry desert air. Of course the reservoirs, which are artificial lakes, trap nearly all the sediment that the river formerly carried uninterruptedly to the ocean. The accumulating sediment will eventually fill the reservoirs, making them useless; however, the sedimentation process is slow, and filling will not be complete for at least several hundred years.

THE STREAM'S LOAD

A stream's load of solid particles consists largely of *coarse particles that move along or close to the streambed* (the **bed load**) and *fine particles suspended in the stream* (the **suspended load**). Where deposited on land, these solid particles constitute alluvium. In addition to solid particles there is also *matter dissolved in stream water* (the **dissolved load**), chiefly a product of chemical weathering.

Turbulent Flow

Most molecules of water in a stream do not move along straight or parallel paths. Instead, they move in many directions and at different speeds, a kind of motion known as *turbulent flow.* Turbulence is important in the movement of sediment in a stream, for it helps keep the smaller particles suspended within the moving water so they can be transported downstream. Turbulence in fast streams is much greater than in slow ones. Turbulence is generated along the sides and bottom of the stream channel where the flowing water experiences the greatest frictional drag (Fig. 11.20).

Bed Load

Trying to describe the difference between bed load and suspended load is not as easy as it may sound, for particles that move only along the bed at one moment may become suspended if velocity suddenly increases, or particles in suspension may settle and move along the bottom if velocity drops. Under any particular condition of flow, however, the bed load consists of the part of the total load that is not continuously in suspension or in solution.

The average rate of movement of bed-load particles is less than that of the water, for the particles are not in constant motion. Instead they move discontinuously, individually or in groups, by rolling or sliding. Where forces are sufficient to lift a particle off the bed, it may move short distances by **saltation,** a motion that is intermediate between suspension and rolling or sliding. It involves *the progressive, forward movement of a particle in a series of short intermittent jumps along arcing paths.* Saltation will continue as long as currents are turbulent enough to pull particles off the bed and permit them to travel some distance downstream as they settle under the pull of gravity.

Although the bed load generally amounts to between 5 and 50 percent of the total load of most streams, few reliable measurements have been made because of the difficulties involved.

Suspended Load

The turbid character of many streams is due to the presence of fine particles of silt and clay moving in suspension. In some streams, the suspended load is so dense and so typical that it figures in the

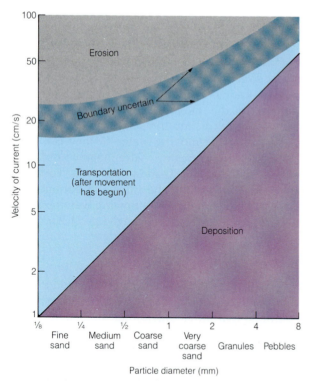

FIGURE 11.21 Average velocity at which uniformly sorted sediment particles of different sizes are eroded, transported, and deposited (*Source:* After Rubey, 1938.)

stream's name. For example, the Yellow River (Huang Ho) of China is yellow because of the great load of yellowish silt it erodes and transports seaward from widespread deposits of loess that underlie much of its basin.

Because upward-moving currents within a turbulent stream exceed the velocity at which particles of silt and clay can settle toward the bed under the pull of gravity, such particles tend to remain in suspension longer than they would in nonturbulent waters. They can settle and be deposited only where velocity decreases and turbulence ceases (Fig. 11.21), as on a floodplain, in a lake, or in the sea. The transport of particles as suspended load, therefore, differs greatly from that of bed-load particles which tend to move only intermittently.

Most of the suspended load in streams is derived from two sources. One part is fine-grained regolith (mostly soil) washed from areas unprotected by vegetation, including plowed fields. The other is sediment eroded and reworked by the stream from its own banks.

Dissolved Load

Even the clearest streams contain dissolved chemical substances that constitute a part of their load. Although in some streams the dissolved load may represent only a few percent of the total load, in others it amounts to more than half (Table 11.2). Streams that receive large contributions of groundwater generally have higher dissolved loads than those whose water comes mainly from surface runoff.

Only seven ions comprise the bulk of the dissolved content of most rivers: bicarbonate, calcium, sulfate, chloride, sodium, magnesium, and potassium. Fluorine, silica, iron, manganese, and nitrate ions are also usually present in small amounts. The relative percentages of ions in the dissolved load vary among small- to medium-sized streams because of differences in local geology (Fig. 11.22). However, among large streams variations in the percentages tend to be small, probably because the larger a drainage area, the closer the average composition of its rocks comes to that of all rocks exposed at the Earth's surface.

Competence and Capacity

When a clear, gently flowing stream is transformed into a torrent of turbid water during flood season, its ability to transport sediment obviously can increase dramatically. Normally, this ability is

TABLE 11.2 *Dissolved and Suspended Load in Streams*

River and Location	Suspended Load (billion kg/yr)	Dissolved Load (billion kg/yr)	Dissolved Load (% of Total Load[a])
Little Colorado, Woodruff, Arizona	1.6	0.02	1.2
Colorado, San Saba, Texas	3.02	0.208	6.4
Bighorn, Kane, Wyoming	1.60	1.82	12
Colorado, Cisco, Utah	15.0	19.4	23
Mississippi, Red River Landing, Louisiana	284.0	358.8	26
Delaware, Trenton, New Jersey	1.003	0.83	45
Juniata, New Port, Pennsylvania	0.322	0.566	64

Source: Data from Leopold et al., 1964.
[a]Total load = bed + suspended + dissolved load.

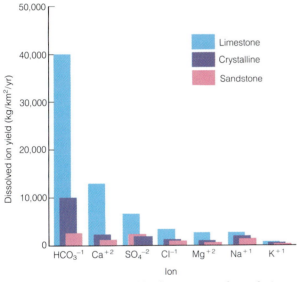

FIGURE 11.22 Annual yield of major ions from drainage basins underlain by limestone, sandstone, and crystalline rock types. HCO_3^{-1} includes some CO_2 released to groundwater by decaying vegetation and absorbed from atmosphere. (*Source:* Data from U. S. Geological Survey; after Peters, 1984.)

referred to in terms of a stream's *competence* and *capacity*.

Competence is *the size of particles a stream can transport under a given set of hydrologic conditions,* measured as the diameter of the largest sediment that can be moved as bed load. The competence depends mainly on velocity, but not in a simple 1-to-1 ratio. Rather, the largest particle moved in a stream varies as the sixth power of the velocity. In other words, if the velocity is doubled, the diameter of the largest sediment in transport will increase 2^6, or 64, times. This shows why a major flood can move very large boulders, or even objects like railroad locomotives, while under average flow conditions the same stream may be able to transport only fine gravel. Competence changes along the course of a stream as velocity changes, but it also changes through time at any place along its course. The results of such changes can often be seen in exposed sections of stream sediment in which alternating coarse and fine layers record variations in the transporting ability of the stream.

Capacity is *the potential load a stream can carry,* measured as the volume of sediment passing a given point in the channel per unit of time. It depends especially on discharge, channel gradient, and the character of the load. Controlled laboratory experiments show that increasing the slope or the discharge, while keeping other factors constant, causes an increase in a stream's load. Numerous field measurements of the relationship between discharge and load lend support (Fig. 11.23). The actual mix of coarse and fine sediment in the bed load, which influences channel roughness and therefore the ease with which sediment is moved into suspension, also affects the capacity.

Downstream Changes in Grain Size

If competence is related to velocity and discharge, then one might expect the average size of sediment to increase in the downstream direction as both discharge and velocity increase. In fact, the opposite is true; sediment normally decreases in coarseness downstream. In mountainous headwaters of

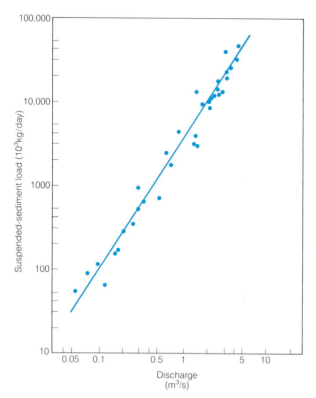

FIGURE 11.23 Relationship between suspended sediment load of Rio Grande near Bernalillo, New Mexico, and the discharge for 1952. (*Source:* After Leopold et al., 1964.)

tion in grain size. Thus, although large rivers may be capable of transporting a very coarse load, they may have only fine sediment available along their channels.

DEPOSITIONAL FEATURES OF STREAMS

Floodplains

When a stream rises in flood, it may overflow its banks and inundate the floodplain (Fig. 11.24). Many streams are bordered by natural levees that confine the stream under normal flow conditions. The fine alluvium of which levees are chiefly built becomes still finer away from the river and grades into a thin cover of silt and clay over the rest of the floodplain. Natural levees are built, and are continually added to, only during floods so high that the floodplain is converted essentially into a lake deep enough to submerge the levees. As the water flows out of the submerged channel during a flood and across the adjacent submerged floodplain, depth, velocity, and turbulence decrease abruptly at the channel margins. The decrease results in sudden, rapid deposition of the coarser part of the suspended load (usually fine sand and silt) along the margins of the channel. Farther away from the channel, finer silt and clay settle out in the quiet water. In the vicinity of Kansas City, Missouri, during an exceptional flood in 1952, Missouri River water deposited a layer of silt as much as 15 cm thick over wide areas of the floodplain. In some places fences and other obstacles caused silt and fine sand to accumulate to thicknesses as great as 1.5 m.

Under ordinary conditions flood-deposited silt is beneficial to agricultural lands because it contains organic matter, washed from soils on the watershed, which acts as fertilizer. Although flood-control dams serve to control rampaging spring floodwaters, thereby protecting property and lives, they also deprive the floodplain of its natural annual accumulation of organic-rich sediment that constantly renews its fertility.

large rivers, tributary streams mostly flow through channels floored with coarse gravel that may include boulders as large as a meter or more in diameter. Because fine sediment is easily moved, even by streams having low discharge, it is readily carried away by small mountain streams, leaving the coarser sediment behind. Through time, the coarse bed load is gradually reduced in size by abrasion and impact as it moves slowly along. When the stream eventually reaches the sea, its bed load may consist mainly of fine alluvium, often no coarser than sand. Some coarse sediment may be temporarily stored on floodplains or in terraces where it will weather, thereby leading to further reduc-

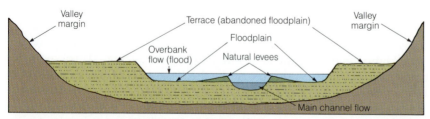

FIGURE 11.24 Main elements of an alluvial valley.

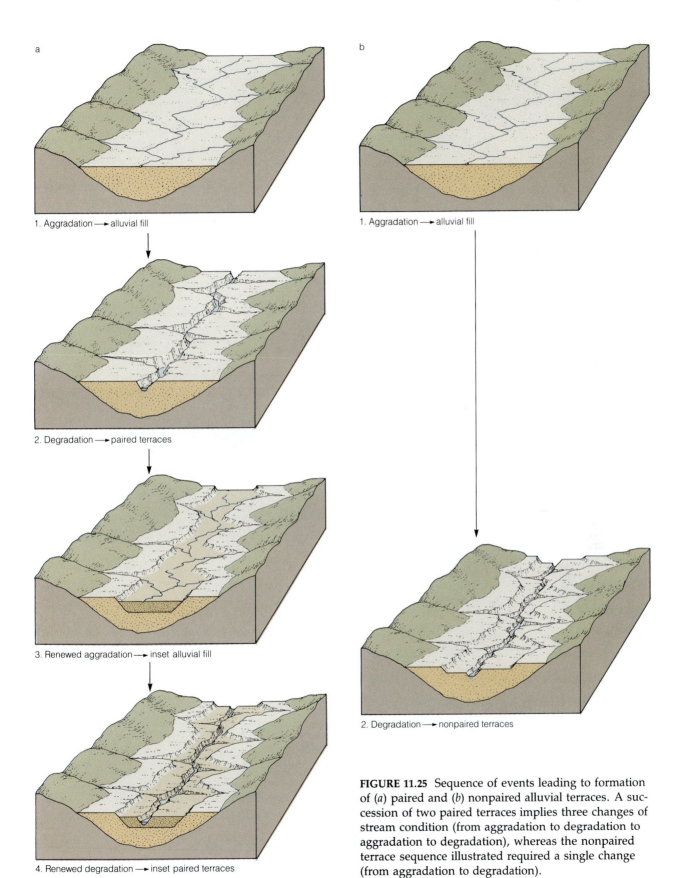

a

1. Aggradation ⟶ alluvial fill

2. Degradation ⟶ paired terraces

3. Renewed aggradation ⟶ inset alluvial fill

4. Renewed degradation ⟶ inset paired terraces

b

1. Aggradation ⟶ alluvial fill

2. Degradation ⟶ nonpaired terraces

FIGURE 11.25 Sequence of events leading to formation of (*a*) paired and (*b*) nonpaired alluvial terraces. A succession of two paired terraces implies three changes of stream condition (from aggradation to degradation to aggradation to degradation), whereas the nonpaired terrace sequence illustrated required a single change (from aggradation to degradation).

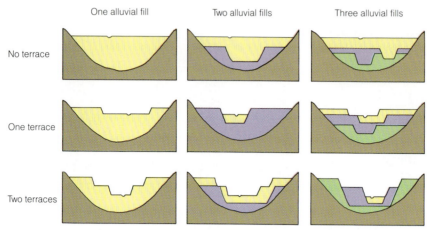

FIGURE 11.26 Valley cross sections showing examples of possible relationships of terraces to alluvial fills. A valley with no terraces may nevertheless have had a long and complicated fluvial history, as revealed by subsurface alluvial fills. Conversely, a valley with several terrace levels may have only a single alluvial fill, reflecting a relatively simple history. (*Source:* After Leopold et al., 1964.)

FIGURE 11.27 Alluvial fan built into Death Valley, California, a down-faulted desert basin with white, salt-encrusted playas on its floor.

Terraces

Most stream valleys contain **terraces,** which are *abandoned floodplains formed when a stream flowed at a level above the level of its present channel and floodplain* (Fig. 11.24). Their presence indicates that a change in the equilibrium condition of the stream has occurred. Terraces are erosional features, formed as a stream erodes downward through its deposits to a new level. A terrace may be underlain by sediments, by bedrock, or by both. In other words, a terrace is a landform and is distinct from the materials that compose it.

In some valleys, terraces occur at many levels, implying a complex history of stream events. Terrace remnants on opposite sides of a valley that lie at the same level are termed *paired terraces.* They generally are assumed to be of the same age and to have resulted from the abandonment of a former floodplain as the stream began to erode downward. Terraces on opposite sides of a stream that do not match are *nonpaired terraces* and must differ in age, for a stream cannot flow at different levels simultaneously. They are generally interpreted as having formed when a stream shifted laterally from one side of a valley to the other as it was cutting downward.

The distinction between paired and nonpaired terraces is important, for each set of paired terraces implies an interval of deposition (also called *aggradation*) to produce a sedimentary fill or of lateral erosion across alluvium or bedrock, followed by an episode of downcutting (also called *degradation*). A series of such terraces, therefore, implies a succession of aggradational and degradational episodes. By contrast, an entire series of nonpaired terraces may reflect but a single episode of downcutting (Fig. 11.25). Many terrace sequences include examples of both types, making interpretation difficult. If the terraces are incorrectly interpreted, too many or too few events in the stream's history may be inferred. A correct reconstruction of stream history generally requires not only an understanding of the terrace landforms but also of the sediments that compose them (Fig. 11.26).

Alluvial Fans

When a stream flows through a steep highland valley and comes out suddenly onto a nearly level valley floor or plain, it experiences an abrupt decrease of slope and a corresponding decrease in competence. Therefore, it deposits that part of its load which cannot be transported on the gentler slope. As sediments accumulate along its course, the stream shifts laterally toward lower ground. Through constant shifting of the channel, the deposit takes the form of an **alluvial fan,** defined as *a fan-shaped body of alluvium typically built where a stream leaves a steep mountain valley* (Fig. 11.27). The profile of the fan, from top to base in any direction, has the same curved form characteristic of the long profiles of streams. The exact form of the profile depends chiefly on discharge and on the diameters of particles in the bed load. Hence, no two fans are exactly alike. A small stream carrying a load of coarse particles builds a shorter, steeper fan than a larger stream carrying a load of finer particles. The area of a fan generally is closely related to the size of the area upstream from which its sediments are derived (Fig. 11.28).

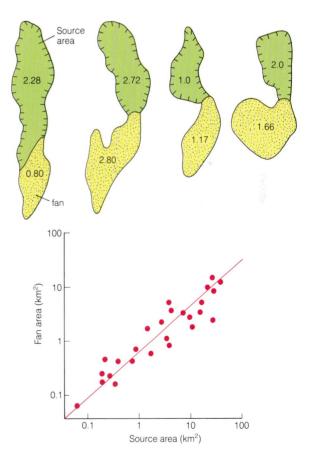

FIGURE 11.28 Relationship between the area of an alluvial fan (in km^2) and the area of its source basin illustrated by four representative fans and their source basins in the Death Valley region, California. The larger the drainage basin, the larger the area of the fan at its mouth. (*Source:* Based on data from Denny, 1965.)

A fan is originally localized by a decrease of slope. As soon as its long profile has become smooth, further deposition occurs as water is redistributed through a network of spreading channels, resulting in net reduction in discharge, velocity, and competence in each channel. Deposition may also occur as water percolates down into the underlying porous fan sediments, thereby reducing surface discharge and competence. In some cases, such percolation will cause a stream to disappear near the top of a large fan, only to reappear again near the base of the fan.

Deltas

As the water of a stream diffuses into the standing water of the sea or a lake, its speed is checked, it loses both competence and capacity, and it deposits its load to form a delta. Although deltas are of several kinds, the type easiest to recognize (and one of the most common) is one built by a stream entering a lake (Fig. 11.29a). In form it somewhat resembles an alluvial fan, but it differs from a fan in two ways: first, stream flow is checked by standing water; and, second, the level surface of the lake sets an approximate limit to the accumulating deposit, the top of which is flatter than the profile of a fan and nearly horizontal.

Particles of the bed load are deposited first, in order of decreasing weight. Then the suspended sediments settle out. A layer representing one depositional event (such as a single flood) is sorted, grading from coarse sediment at the stream mouth to finer sediment offshore. The accumulation of

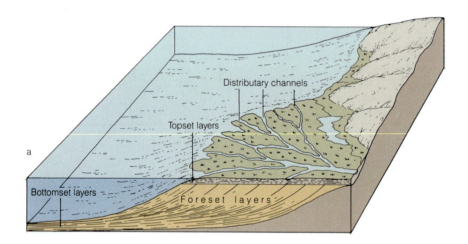

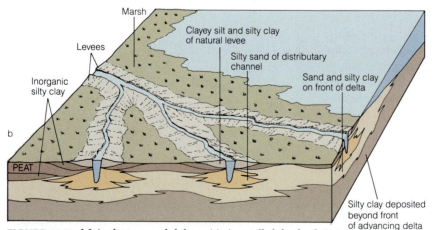

FIGURE 11.29 Main features of deltas. (*a*) A small delta built into a lake, showing topset, foreset, and bottomset layers. (*b*) Part of a large delta built into the sea, showing intertonguing relationship of coarse channel deposits, and finer sediments deposited on the front of the delta and beyond. (*Source:* After Pettijohn et al., 1972.)

many successive layers creates an embankment that grows progressively outward, like a highway fill made by dumping. *The coarse, thick, steeply sloping part of each layer in a delta* is a **foreset layer.** Traced seaward, the same layer becomes rapidly thinner and finer, covering the bottom over a wide area. This *gently sloping, fine, thin part of each layer in a delta* is a **bottomset layer.**

As successive layers are deposited, the coarse foreset layers progressively overlap the bottomset layers. The stream gradually extends outward over the growing delta, so that its course is lengthened. During floods, it erodes the tops of the foreset layers and part of the suspended load accumulates in areas beside the channel, but at other times part of the bed load is deposited within the channel and the suspended load is transported beyond the channel mouth. The channel deposits and interchannel deposits together form the **topset layers** of the delta, defined as *the layers of stream sediment that overlie the foreset layers in a delta.*

At times of flood the stream spills out of its channel and forms distributary channels, through which the water flows independently, multiplying the topset deposits. Radiating distributary channels can give the delta a crudely triangular shape like the Greek capital letter delta (Δ), from which the deposit derives its name (Fig. 11.30).

Some of the world's greatest rivers, among them the Ganges-Brahmaputra, the Huang Ho, the Amazon, and the Mississippi, have built massive deltas at their mouths. Each delta has its own peculiarities, determined by such factors as the stream's discharge, the character and volume of its load, the shape of the bedrock coastline at the delta, the offshore topography, and the intensity and direction of longshore currents and waves. Most major rivers transport large quantities of fine suspended sediment, the bulk of which is carried seaward as the fresh river water overrides denser salt water at the coast. The fine sediment then settles out to form the gently sloping front of a marine delta. Where currents and wave action are so strong that most sediment reaching the coast is reworked, delta formation may be inhibited. However, if the rate of sediment supply exceeds the rate of removal by erosion along the coast, then a delta will be built seaward. The Mississippi River delivers such a large load to the Gulf of Mexico that long fingerlike bodies of sediment are deposited along and around distributary channels to form a complex delta front (Fig. 2.15). The coarsest sediment lies along the channel, while finer sediment reaches the front of the delta and also accumulates

FIGURE 11.30 Delta of the Nile River, as seen from an orbiting spacecraft. Meandering distributary channels cross the flat, vegetated delta surface and build lobes of sediment where they enter the Mediterranean.

between distributaries during times of overbank flooding. The result is a complex intertonguing of facies (Fig. 11.29b).

The Mississippi delta, with an area of 31,000 km^2 (not counting the submarine part), is really a complex of several coalescing subdeltas built successively during the last several thousand years (Fig. 11.31). Each subdelta was begun by a flood that created a new distributary. Only the present active delta has a digitate front. The fronts of the adjacent inactive subdeltas have been extensively modified by coastal erosion and by gradual submergence as the crust slowly subsides under the weight of the accumulating sedimentary pile.

DRAINAGE SYSTEMS

Drainage Basins and Divides

Every stream or segment of a stream is surrounded by its **drainage basin,** *the total area that contributes*

FIGURE 11.31 The Mississippi River has built a seres of overlapping subdeltas (numbered 1 to 7) while occupying successive distributary channels. The ages of subdeltas are given in radiocarbon years before present. Compare fig. 2.15. (*Source:* After Morgan, 1970.)

water to the stream. The line that separates adjacent drainage basins is a **divide** (Fig. 11.32). Drainage basins range in size from less than a square kilometer to vast areas of subcontinental dimension (Fig. 11.33). In North America, the huge drainage basin of the Mississippi River encompasses an area that exceeds 40 percent of the area of the contiguous United States. Not surprisingly, the areas of drainage basins bear a close relationship to both the length and mean annual discharge of the streams that drain them (Fig. 11.34).

Stream Order

The arrangement and dimensions of streams in a drainage basin tend to be orderly. This can be verified by examining a stream system on a map and numbering the observed stream segments according to their position, or *order*, in the system. The smallest segments, without tributaries, are classified as first-order streams. Those with only first-order tributaries are second-order streams, while third-order streams have first- and second-order tributaries (Fig. 11.32). If the frequency of segments of each order is then tabulated, it can quickly be seen that for any stream system, the frequency of segments increases with decreasing stream order. In other words, a stream system is somewhat like a tree, with a trunk and numerous branches. There is only one stream segment of highest order (the main stream), but there are many tributaries, with their numbers increasing, the shorter they are. This orderliness is like that inherent in a stream's long profile, in which gradient decreases systematically from head to mouth, while discharge, velocity, and channel dimensions increase. All these relationships imply that in response to a given quantity of runoff, stream systems develop with just the size and spacing required to move the water off each part of the land with greatest efficiency.

Drainage Density

Stream channels are unequally distributed over the landscape. In some areas they are closely spaced, whereas in others they are far apart (Fig. 11.35). The differences are primarily related to geology and climate. A good way to analyze the contrasts in relative spacing of streams is to measure the **drainage density,** defined as the *ratio of the total length of all stream segments in a drainage basin to the area of the basin.* The ratio provides a way of placing numbers on the "texture" of the topography. High

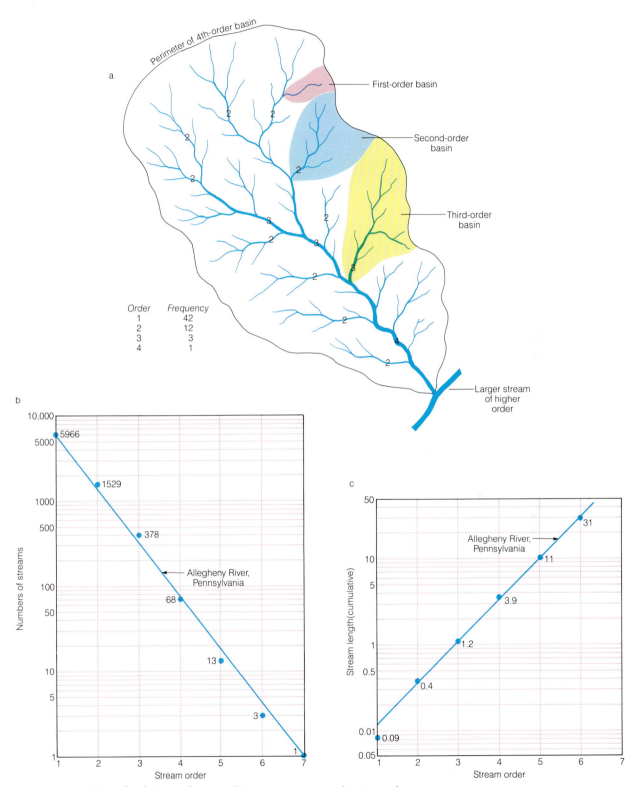

FIGURE 11.32 Map of a drainage basin and its stream system showing tributary channel segments numbered according to stream order. Both the number of tributaries and their length are related to stream order, as shown in the two graphs. (*Source:* Data for *b* and *c* from Morisawa, 1962.)

values (closely spaced streams) are found for regions underlain by very erodible or impermeable rocks. They also are common where the climate favors sparse vegetation and high rainfall intensity (the badlands topography of many desert areas be-

ing a good example). Low values are generally related to areas underlain by resistant rocks or by rocks through which rainfall quickly infiltrates, as well as areas having a continuous cover of vegetation.

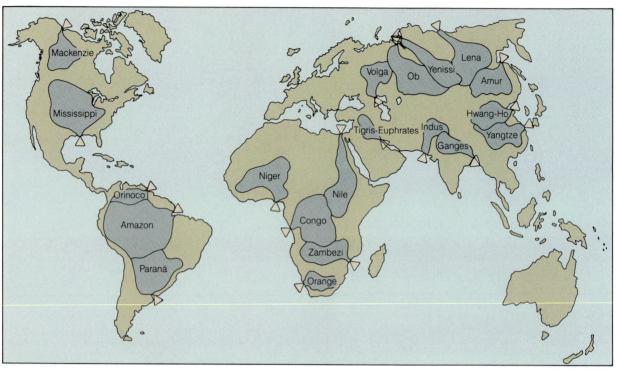

FIGURE 11.33 World's largest drainage basins, showing location of deltas at their mouths. (*Source:* After Evans, 1981.)

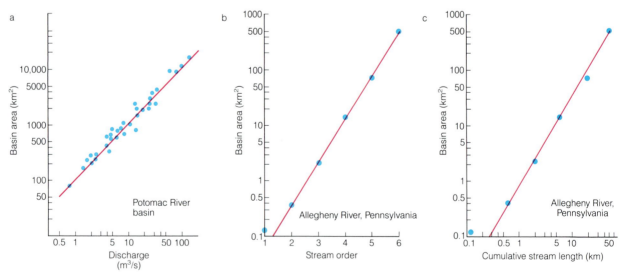

FIGURE 11.34 Relationship of drainage basin area to annual discharge, stream length, and stream order. (*Source:* Hack, 1957 and Morisawa, 1962.)

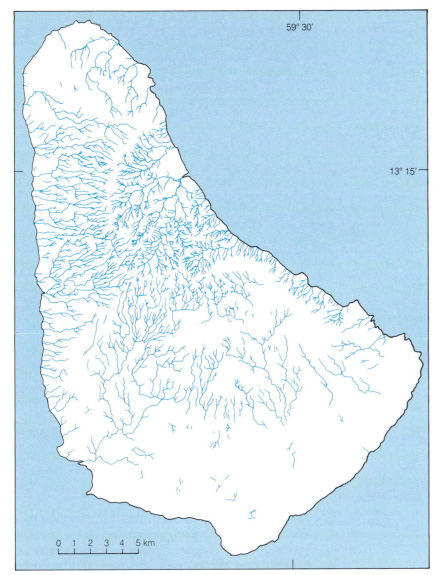

FIGURE 11.35 Map of the island of Barbados showing contrasting drainage density. The north-central part of the island is underlain by young rocks, including shales, that are relatively impervious. The extreme northern and southern parts of the island are underlain by porous coral limestone in which some streams end in sinkholes. (*Source:* After Derbyshire et al., 1979.)

Drainage Evolution

A system of streams does not necessarily require much time to develop, as indicated by the following example. In August 1959, an earthquake occurred at Hebgen Lake, near West Yellowstone, Montana. The movement tilted the country in such a way that a large area of silt and sand, formerly part of the lakebed, emerged and was subjected to runoff. Small-size drainage systems began to de-

velop immediately. Sample areas were surveyed and mapped one and two years after the earthquake occurred. The results showed the same basic geometry that characterizes much larger and older systems. The small, newly formed valleys, together with the areas between them, were disposing of the available runoff in a highly systematic way, and all within a period of two years after the surface had emerged from beneath the lake.

As a drainage system develops, details of its pat-

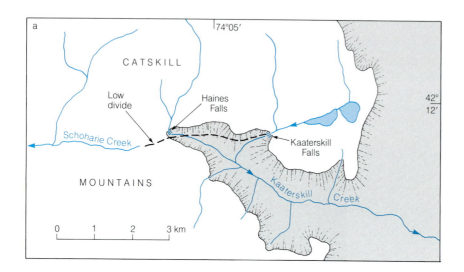

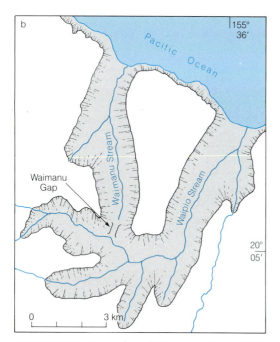

FIGURE 11.36 Examples of stream capture. (*a*) Steep-gradient Kaaterskill Creek has extended its channel headward up the high eastern scarp of the Catskill Plateau in New York where it has intercepted and captured the headwaters of low-gradient westward-flowing Schoharie Creek. Dashed line shows former course of Schoharie Creek. (*b*) Waipio Stream on the island of Hawaii has cut rapidly headward along a system of water to Waimanu Stream. It also has tapped into groundwater reservoirs confined by numerous vertical dikes. The head of Waimanu Valley, below Waipio Gap, is now filled with landslide sediments that are too massive to be removed by the reduced discharge of Waimanu Stream.

tern change. New tributaries are added, and some old tributaries are lost due to **stream capture** (or **piracy**), *the diversion of a stream by the headward growth of another stream* (Fig. 11.36). In the process some stream segments are lengthened and others are shortened. Just as the hydraulic factors within a stream are constantly adjusting to changes, so too is the drainage system constantly changing and adjusting as it grows. Like the stream channel, it is a dynamic system tending toward a condition of equilibrium.

Stream Patterns and Classification

Not only the profiles of streams, but also their patterns as seen on a map, are affected by the kinds of rock on which they are developed. Stream patterns are affected also by rock structure, and therefore by the geologic history of the areas in which they occur (Table 11.3). An experienced geologist can tell a great deal about the geology of an area simply by analyzing the pattern of surface streams.

On the basis of their patterns and other characteristics, streams can be classified into several categories, labeled consequent, subsequent, antecedent, and superposed. The streams in each group have distinctive origins and histories.

A **consequent stream** is *a stream whose pattern is determined solely by the direction of slope of the land.* Therefore, consequent streams often are found in massive or gently sloping rocks and commonly have dendritic patterns. The drainage system that developed on the freshly emerged sediments of Hebgen Lake following the earthquake of 1959 is one example. Another is the system of stream channels that have developed on the slopes of vol-

TABLE 11.3 *Stream Patterns*

	Pattern	Characteristics
Dendritic	Dendritic	*Irregular branching of channels ("treelike") in many directions.* Common in massive rock and in flat-lying strata. In such situations, differences in rock resistance are so slight that their control of the directions in which valleys grow headward is negligible.
Parallel	Parallel	*Parallel or subparallel channels that have formed on sloping surfaces underlain by homogeneous rocks.* Parallel rills, gullies, or channels are often seen on freshly exposed highway cuts or excavations having gentle slopes.
Radial	Radial	*Channels radiate out, like the spokes of a wheel, from a topographically high area,* such as a dome or a volcanic cone.
Rectangular	Rectangular	*Channel system marked by right-angle bends.* Generally results from the presence of joints and fractures in massive rocks or foliation in metamorphic rocks. Such structures, with their cross-cutting patterns, have guided the directions of valleys.
Trellised	Trellis	*Rectangular arrangement of channels in which principal tributary streams are parallel and very long,* like vines trained on a trellis. This pattern is common in areas where the outcropping edges of folded sedimentary rocks, both weak and resistant, form long, nearly parallel belts.
Annular	Annular	*Streams follow nearly circular or concentric paths along belts of weak rock* that ring a dissected dome or basin where erosion has exposed successive belts of rock of varying degrees of erodibility.
Centripetal	Centripetal	*Streams converge toward a central depression,* such as a volcanic crater or caldera, a structural basin, a breached dome, or a basin created by dissolution of carbonate rocks.
Deranged	Deranged	*Streams show complete lack of adjustment to underlying structural or lithologic control.* Characteristic of recently deglaciated terrain whose preglacial features have been remodeled by glacial processes.

canoes, the roughly conical shape of which typically leads to a distinctive radial drainage pattern (Fig. 11.37 and Table 11.3).

A *subsequent stream* is *a stream whose course has become adjusted so that it occupies belts of weak rock or other geologic structures.* When such belts are long and straight, subsequent streams constitute the long straight tributaries characteristic of trellis drainage patterns (Table 11.3). Many examples can be found in the Valley and Ridge country of the Appalachian Mountains. A further excellent example is the system of nearly parallel streams, carved in erodible belts of deformed rock, that flow southward from the highlands of central China toward the plains of southeast Asia. (Fig. 11.38).

An **antecedent stream** is *a stream that has maintained its course across an area of the crust that was raised across its path by folding or faulting.* The name comes from the fact that the stream is antecedent to (older than) the uplifting. Such streams are often recognizable because they cross topographi-

cally high ridges through deep gorges, rather than taking a more obvious path around the end of the uplift. The lower Yakima River of Washington State is such a stream, for it has managed to maintain its course across a succession of high ridges that were uplifted across its path during the last 10 million years (Fig. 11.39).

A **superposed stream** is *a stream that was let down, or superposed, from overlying strata onto buried bedrock having a lithology or structure unlike that of the covering strata.* Most superposed streams began as consequent streams on the surface of the covering strata. The streams' paths, therefore, were not controlled in any way by the rocks on which they are now flowing. The course of the Vaal River, which crosses the Vredefort Dome in southern Africa, provides an example (Fig. 11.40). As a cover of sedimentary strata on which the river was flowing was stripped away by erosion, the river became superposed on the tilted older rocks beneath.

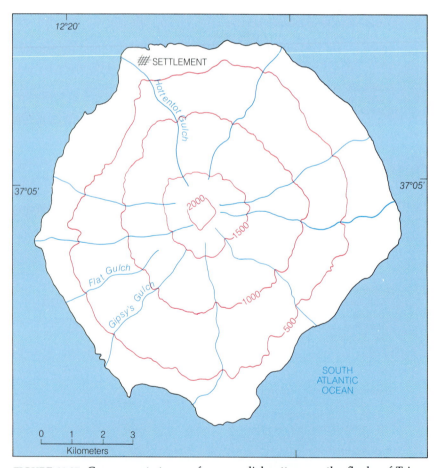

FIGURE 11.37 Consequent streams form a radial pattern on the flanks of Tristan de Cunha, a young volcanic island in the South Atlantic Ocean.

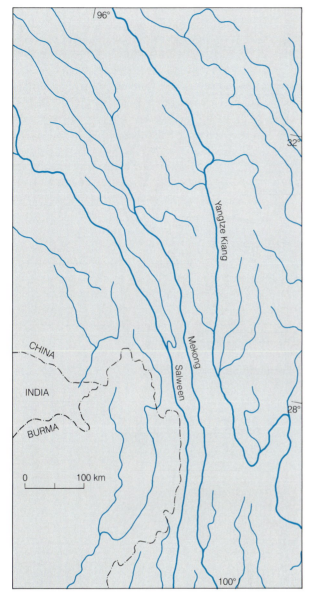

FIGURE 11.38 Nearly parallel streams, draining high regions of central China, flow through mountainous terrain where the Himalaya turns abruptly south near the Burmese border. The subsequent streams follow weak rock belts in highly deformed terrain.

SEDIMENT YIELD

Potential Energy for Erosion

The average altitude of the 48 contiguous United States is about 750 m and the average runoff for the country as a whole is 30 cm/yr. The potential energy of this water (equal to the mass of the water × the height of the land) is converted into the

kinetic energy of flow of all streams draining the country. Part of that energy is spent in the geologic work of eroding and transporting rock particles. This kinetic energy makes streams the prime movers of rock waste from lands to ocean and among the chief agencies by which the sedimentary rock of the Earth's crust has accumulated.

Factors Influencing Sediment Yield

Some streams run clear nearly all the time, whereas others are constantly muddy in appearance and are quite obviously transporting a sizeable load. Such contrasts suggest that some land areas are being eroded more rapidly than others. The differences are related to a combination of geologic, climatic, and topographic factors, including rock type and structure, local climate, and relief and slope. These factors control the sediment yield from a drainage basin, generally expressed as volume or weight per unit area per unit time.

Climate influences erosion in several ways. Intuitively, it might be expected that the greater the precipitation, the greater the erosion. However, in humid regions, the nearly continuous vegetation cover tends to inhibit erosion. In vegetated areas with high average precipitation, therefore, erosion rates may actually be less than in some dry regions that lack a continuous cover of vegetation. Field measurements suggest that the greatest sediment yields are from small basins characterized by landscapes that are transitional between full-desert conditions and grassland (Fig. 11.41). With increasing humidity and the development of forest cover, sediment yields tend to decline.

The highest measured sediment yields are from basins that drain mountainous terrain having steep slopes and high relief. As one might expect, rates are also much higher in areas underlain by erodible clastic sediments or sedimentary rocks, or by low-grade metamorphic rocks, than in areas where crystalline or highly permeable carbonate rocks crop out. Structural factors also play a role, for rocks that are more highly jointed or fractured are more susceptible to erosion than massive ones.

Yet another factor that influences sediment yield is human activity. Through human history, the character of the landscape has been changed tremendously, especially due to clearing of forests, development of cultivated land, damming of streams, and construction of cities. Each of these activities has affected erosion rates and sediment yields in the drainage basins where they have occurred. Sometimes the results are dramatic. For

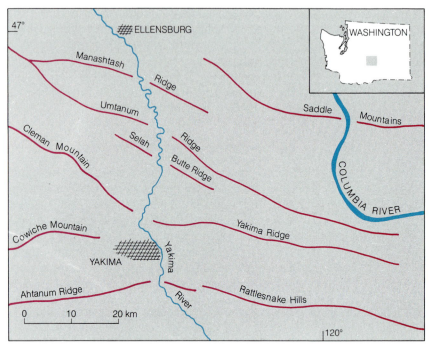

FIGURE 11.39 Antecedent segment of Yakima River in central Washington State has maintained a path through ridges of basalt that have been uplifted across its path during the last 10 million years.

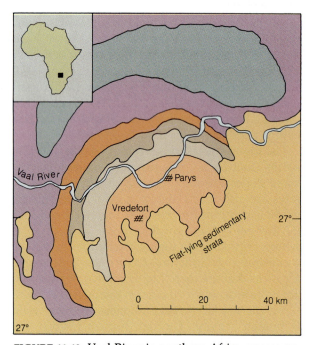

FIGURE 11.40 Vaal River in southern Africa crosses ancient deformed rocks of the Vredefort Dome. The stream, whose course formerly lay across nearly flat-lying sedimentary strata that overlies the older rocks, was superposed upon the older rocks beneath as the covering layer was stripped away by erosion.

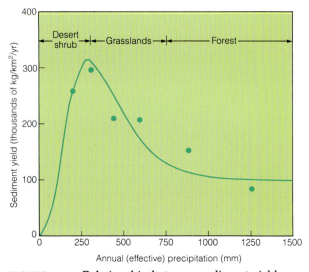

FIGURE 11.41 Relationship between sediment yield and precipitation. As precipitation increases, so does sediment yield, so long as vegetation cover remains restricted. Once the moisture level is sufficient to support continuous vegetation, erosion is reduced and sediment yield declines. (*Source:* After Langbein and Schumm, 1958.)

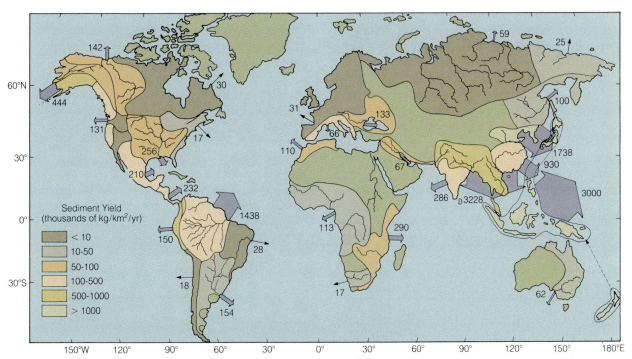

FIGURE 11.42 Average annual discharge of suspended sediment from various drainage basins of the world. Width of arrows is related to the amount of sediment entering the sea. Numbers are annual input in billions of kilograms. Annual sediment yields (thousands of kg/km²) for major basins are shown by patterns. Areas shown in greenish pattern contribute minimal sediment to the oceans. (*Source:* After Milliman and Meade, 1983.)

example, in parts of eastern United States, areas cleared for construction produce between 10 and 100 times more sediment than comparable rural areas or natural areas that are vegetated. On the other hand, in urbanized areas sediment yield tends to be low because the land is almost completely covered by buildings, sidewalks, and roads that protect the underlying rocks and sediments from erosion.

Discharge of Sediment to the Oceans

Rivers constitute the principal means by which sediment eroded from the land is transported to the oceans. If the sediment yield for all the world's rivers were known, it would be possible to calculate the rate at which the Earth's landmasses are being progressively worn down. Although estimates of global sediment yield can be made, the results are imprecise because accurate measurements are available for very few of the world's rivers. Measurements generally do not include bed

load, which for some rivers may be greater than suspended load. Neither do they include unusually large (catastrophic) floods, a single one of which may transport as much sediment as is moved by the same stream during an average year.

The continent discharging the most sediment to the ocean is Asia (Fig. 11.42). It also is the conti-

TABLE 11.4 *Discharge of Sediment to the Oceans by Streams*

Land Area	Sediment Discharge (billion kg/yr)
Asia	6349
Oceanic Islands	3000
South America	1788
North America	1020
Africa	530
Central America	442
Europe	230
Eurasian Arctic	84
Australia	62
Total	13,505

Source: Data from Milliman and Meade, 1983.

nent on which the greatest average stream sediment loads have been measured. Asian rivers entering the sea between Korea and Pakistan are believed to contribute nearly half the total world sediment input to the oceans (Table 11.4). Second to Asia is the combined area of the large western Pacific islands of Indonesia, Japan, New Guinea, New Zealand, the Philippines, and Taiwan. Taiwan is especially remarkable because it produces only slightly less sediment than the entire contiguous United States! South and North America, with their vast area, high mountains near their western margins, and large river systems draining their interiors are next in order. Most of the sediment leaving Europe is trapped in the

Black and the Mediterranean seas. Like Australia, Africa discharges relatively little sediment for its size, probably because of low average altitude and widespread aridity.

For many drainage basins the measured and estimated sediment yields reflect conditions that are probably quite different from those of only a few decades ago, because much of the sediment that formerly reached the sea is now being trapped in reservoirs behind large dams. The high Aswan Dam now intercepts most of the sediment that used to be carried by the Nile to the Mediterranean Sea. Robbed of its sediment supply, the productive Nile Delta is now being actively eroded along parts of its seaward coast.

SUMMARY

1. As an integral part of the hydrologic cycle, streams are the chief means by which water returns from the land to the sea. They shape the continental crust and transport sediment to the oceans.

2. Raindrop impact and sheet erosion are effective in dislodging and moving regolith on bare, unprotected slopes.

3. The average gradients of streams decrease from head to mouth, so their long profiles tend to be smooth hyperbolic curves.

4. Straight channels are rare. Meandering channels form on gentle slopes and where load is small to moderate. Braided patterns develop on steeper slopes and where bed load is large.

5. Discharge, velocity, and cross-sectional area of a channel are interrelated such that when discharge changes, the product of the other two factors also changes to restore equilibrium.

6. As discharge increases downstream, channel width and depth increase, and velocity increases slightly.

7. World sea level, projected inland as an imaginary surface, constitutes base level for most streams. A local base level, such as a lake, may temporarily halt downward erosion upstream.

8. Streams tend toward a graded condition in which slope is adjusted so that the available sediment load can just be transported. Changes continually occur that upset a stream's balance, requiring adjustments to be made in the channel factors.

9. Streams experiencing large floods have increased competence and capacity, and so are capable of transporting great loads of sediment as well as very large boulders. Exceptional floods, however, have a low recurrence interval.

10. Although bed load in some streams may amount to as much as 50 percent of the total load, it is very difficult to measure accurately. Most suspended load is derived from erosion of fine-grained regolith or from stream banks. Streams that receive large contributions of groundwater commonly have higher dissolved loads than those deriving their discharge principally from surface runoff.

11. During floods, streams overflow their banks and construct natural levees that grade laterally into silt and clay deposited on the floodplain. Terraces are due to the abandonment of a floodplain as a stream erodes downward.

12. Alluvial fans are constructed where a stream experiences a sudden decrease in gradient, as when it leaves a steep mountain valley. It thereby loses competence and deposits the coarser fraction of its load. The area of a fan is closely related to the size of the drainage basin upstream from which its sediments originated.

13. A delta forms where a stream enters a body of standing water and loses its ability to transport sediment. The contact between steeply sloping foreset beds and overlying topset beds marks the approximate surface of the water body into which the delta is built.

14. A drainage basin encompasses the area supplying water to the stream or stream system that drains it. Its area is closely related to the stream's length and annual discharge.

15. Streams are so arranged that the length of stream segments decreases with decreasing stream order. The density, or spacing, of streams on the landscape commonly reflects local climate and/or lithology.

16. Streams can be classified as consequent, antecedent, subsequent, or superposed, depending on the influence of lithology and structure on their development.

17. Sediment yield is influenced by lithology, structure, climate, and topography. The greatest sediment yields are recorded in small basins that are transitional from desert to grassland conditions, and in mountainous terrain with steep slopes and high relief. Under moist climates, vegetation anchors the surface, thereby reducing erosion.

18. Asia discharges more sediment to the oceans than any other continent, reflecting high sediment loads of rivers that drain its mountainous interior.

SELECTED REFERENCES

Chorley, R. J., Schumm, S. A., and Sugden, D. E., 1984, Geomorphology: London, Methuen.

Czaya, E., 1981, Rivers of the world: New York, Van Nostrand Reinhold.

Denny, C. S., 1965, Alluvial fans in the Death Valley region, California and Nevada: U. S. Geological Survey Professional Paper 466.

Dunne, T., and Leopold, L. B., 1978, Water in environmental planning: San Francisco, W. H. Freeman.

Gregory, K. J., and Walling, D. E., 1973, Drainage basin form and process: London, Edward Arnold.

Leopold, L. B., Wolman, M. G., and Miller, J. P., 1964, Fluvial processes in geomorphology: San Francisco, W. H. Freeman.

Milliman, J. D., and Meade, R. H., 1983, World-wide delivery of river sediment to the oceans: J. Geol., v. 91, p. 1–21.

Ritter, D. F., 1986, Process geomorphology, 2nd ed.: Dubuque, Iowa, W. C. Brown.

Schumm, S. A., ed., 1972, River morphology: Stroudsburg, Pa., Dowden, Hutchinson, and Ross.

Schumm, S. A., 1977, The fluvial system: New York, John Wiley & Sons.

C H A P T E R 1 2

Deserts and Wind Action

Crest of a dune complex wanders sinuously across the Namib Desert near the west coast of Africa.

GEOGRAPHIC DISTRIBUTION OF DESERTS

Although the word *desert* means literally a deserted, unoccupied, or uncultivated area, the modern development of artificial water supplies has changed the original meaning of this word by making many dry regions habitable. *Desert* has become a synonym for *arid land, whether "deserted" or not, in which annual rainfall is less than 250 mm (10 in.) or in which the evaporation rate exceeds the precipitation rate.* Aridity, then, remains the chief characteristic of any desert.

Arid lands of various kinds add up to about 25 percent of the total land area of the world outside the polar regions. In addition, there is a smaller though still large percentage of semiarid land in which the annual rainfall ranges between ~ 250 and 500 mm (10 and 20 in.). These dry and semidry areas form a distinctive pattern on the world map (Fig. 12.1). The meaning of the pattern becomes clear as soon as we grasp the general plan of circulation of the atmosphere.

Circulation of the Earth's Atmosphere

The atmosphere is continually in motion and circulates in a definite pattern. The basic reason is that more of the Sun's heat is received per unit of land surface near the equator than near the poles. The heated air near the equator expands, becomes lighter, and rises, like the rise of boiling water in a teakettle. High up, it spreads outward toward both poles. On the way it gradually cools, becomes heavier, and sinks. Meanwhile, beneath it lies still cooler, heavier air. This air, chilled in higher latitudes, forms a return flow toward the equator. The returning air replaces the warm rising air and, in turn, is heated and rises.

The Earth's rotation interferes with what would otherwise be a simple circulation pattern. The Coriolis effect, named after the nineteenth-century Frenchman who first analyzed it, causes any body that moves freely with respect to the rotating solid Earth to veer toward the right in the Northern Hemisphere and toward the left in the Southern Hemisphere, regardless of the direction in which the body may be moving. Flowing water (such as an ocean current) and flowing air (wind) respond to it. So too do small bodies such as projectiles, if they travel over long paths.

The Coriolis effect breaks up the simple general flow of air between the equator and the poles into belts (Fig. 12.2). At about latitude 30° in the Northern Hemisphere some of the high-level, north-flowing equatorial air descends toward the Earth's

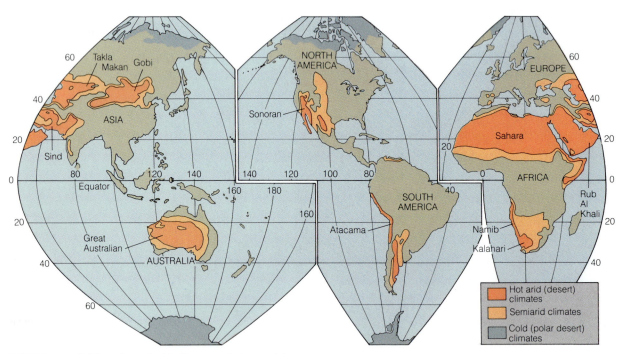

FIGURE 12.1 Arid and semiarid climates of the world and the major deserts associated with them. Very dry areas of the polar region include areas known as polar deserts.

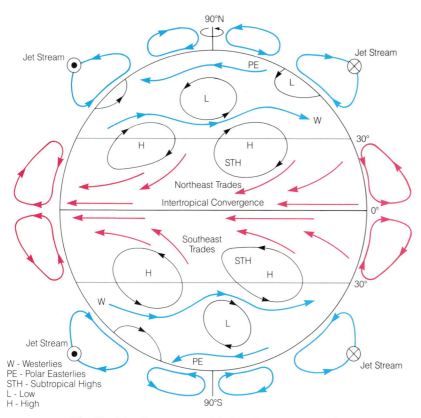

FIGURE 12.2 The Earth's planetary wind belts, shown schematically. Circulation within the six latitudinal belts of circulation are shown in cross sections, with arrows indicating the direction of air flow. Dots in circles labeled "Jet Stream" indicate flow upwards, seemingly out of the page; those having a cross in the circles represent flow into the page. The jet stream actually is not fixed, but wanders across middle to high latitudes in both hemispheres, with predominantly westerly winds. The high and low pressure zones are mainly features of the lower atmosphere.

solid surface. As it descends, this cold dry air becomes warmer. This means that near that latitude, right around the world, climates are warm and dry, skies are clear, and rain is scarce.

The descending air spreads out along the surface, toward the north and south. In the Northern Hemisphere the south-flowing air responds to the Coriolis effect by moving toward the right (west) with respect to the rotating Earth beneath. The south flow thus becomes a southwest flow. In everyday language we call such a wind a northeast wind, for the direction from which it is blowing. As Figure 12.2 shows, these winds form a belt of northeast *trade winds* extending around the world.

Returning to the northern edge of this belt, where air is descending, we now follow the air that

is spreading toward the north. This flow also moves toward the right so that northward flow becomes northeastward flow, forming winds that blow from the southwest. These winds are the belt of westerly winds or *westerlies* that encircle the world.

At higher latitudes, somewhere between 40° and 60°, the westerlies encounter the cold heavy air that flows from the polar region toward the equator. This air, of course, also moves toward the right and flows southwest as *polar easterlies*. Where warmer and lighter air of the westerlies encounters this cold polar air, some of it tends to rise and return toward the equator. The polar air itself becomes warmer, gradually rises, and returns toward the pole.

To see what happens in the Southern Hemisphere, we need only repeat this description, substituting left for right turns. The result is a belt of southeast trade winds, a belt of descending air near 30° south latitude, a belt of westerlies, and a belt of polar easterlies—thus completing a symmetrical pattern for the Earth. As a result of this pattern, much of the arid land in both hemispheres (Fig. 12.1) is centered between latitudes 15° and 35°. Not surprisingly, the world's major deserts are found in these two belts of generally clear skies and low rainfall, where dry tropical air is descending.

Kinds of Deserts

The most extensive arid lands are associated with the two circum-global belts of dry subtropical air. Examples include the Sahara and Kalahari deserts of Africa, the Rub-al-Khali Desert of Saudi Arabia, and the Great Australian Desert (Fig. 12.1). These and other subtropical deserts comprise one of five classes of deserts that are recognized.

A second type is found in continental interiors far from sources of moisture where warm summers and dry cold winters prevail. The Gobi and Takla Makan deserts of central Asia fall into this category.

Yet a third, more local kind of desert is found on the lee side of mountain ranges. The mountains create a barrier to the flow of moist air producing a *rainshadow* effect. As the air rises against the windward slope of a range it cools, thereby enabling it to retain less and less moisture. The bulk of its moisture is then lost through precipitation. Air reaching the lee side of the mountain range is deficient in moisture, resulting in a dry climate over the country beyond. The lofty Sierra Nevada in eastern California forms such a barrier and is largely responsible for the arid climate of the desert basins immediately east of it.

The fourth category constitutes coastal deserts. They occur locally along the margins of continents where upwelling cold seawater cools passing marine air, thereby decreasing its ability to hold moisture. As the air encounters the warm land, its limited moisture condenses giving rise to coastal fogs. However, the air contains too little moisture to generate much precipitation. Coastal deserts of this type can be found in Chile, Peru, and southwest Africa, and are among the driest places on Earth.

The deserts mentioned thus far are hot deserts, where rainfall is low and summer temperatures are high. Other vast deserts occur in the polar regions

where precipitation is also extremely low due to the sinking of cold dry air. However, the surfaces of these *polar deserts*, unlike those of more tropical latitudes, are often underlain by abundant water, but nearly all is in the form of ice. Even in midsummer, with the sun above the horizon for 24 h, the temperature may remain below freezing. Good examples of polar deserts are found in northern Greenland, arctic Canada, and in the ice-free valleys of Antarctica. Such deserts are considered to be the closest earthly analogues to the surface of Mars, where temperatures also remain below freezing and the rarefied atmosphere is extremely dry.

DESERT CLIMATE AND VEGETATION

The arid climate of a hot desert results from the combination of three factors:

High temperature. The highest temperature recorded in the United States, 56° C (132.8° F), was at Death Valley, a desert area in southeastern California. The world's record temperature of 57.7° C (135.9° F) was measured at a place in the Libyan Desert in northern Africa.

Low Precipitation. At Death Valley annual precipitation averages between 20 and 50 mm, while in the Atacama Desert of northern Chile intervals of a decade or more have passed without measureable rainfall.

High Evaporation Rate. The higher the temperature, the greater is the rate of evaporation. If most of the precipitated water evaporates, little is left for streams and vegetation. In parts of the southwestern United States, evaporation from lakes and reservoirs amounts to as much as 250 mm annually—some 10 to 20 times more than the annual precipitation.

Besides the aridity, resulting from these three factors, deserts are generally characterized by frequent strong winds. During daylight hours, the air over especially hot areas is heated and expands in volume causing it to rise. Strong winds are produced as surface air moves in rapidly to take the place of the rising hot air.

The vegetation in deserts is a direct reflection of dry climate. Usually the vegetation is discontinuous. Where grass is present, it is likely to be thin and grow only in clumps. More commonly, desert plants consist of low bushes growing rather far

FIGURE 12.3 Jointed sandstone at Dead Horse Point, Utah breaks into coarse blocks that litter the steep talus at base of the cliff.

apart, with bare areas between them. This pattern of vegetation promotes active movement of sediment by the wind, as well as by running water.

GEOLOGIC PROCESSES IN DESERTS

No major geologic process is restricted entirely to desert regions. Rather, the same processes operate with different intensities in moist and arid landscapes. As a result, in a desert the forms of the land, the soils, and the surface sediments show some distinctive differences from those of humid regions.

Weathering and Mass-Wasting

In a moist region, regolith covers the ground almost universally and is comparatively fine-textured because it usually contains clay, a product of chemical weathering. Its downslope motion occurs mainly by creep and it is covered with almost continuous vegetation. As a result of creep, hillslope form can usually be described as a series of curves.

By contrast, the regolith in a desert is thinner,

FIGURE 12.4 Prehistoric drawings have been etched in dark coating of rock varnish on rock outcrop in Owens Valley, California. Removal of oxide coating has exposed underlying unweathered rock of lighter color.

less continuous, and coarser in texture. Much of it is the product of mechanical weathering. Although chemical weathering takes place, its intensity is greatly diminished because of reduced soil moisture. Slope angles developed by downslope creep become adjusted to the average particle size of the regolith; the coarser the particles, the steeper the slope required to move them. As the particles created by mechanical weathering tend to be coarse, slopes are generally steeper than in a moist region.

Mechanically weathered fragments of rock tend to break off along joints, leaving steep, rugged cliffs. Hills with cliffy slopes, particularly where layers of exposed bedrock are nearly horizontal, are common in dry regions (Fig. 12.3).

In many desert areas, the light hues of recently deposited sediments contrast with darker hues of older deposits. The darker color can generally be attributed to the presence of **rock varnish,** *a thin* *dark shiny coating, consisting mainly of manganese and iron oxides, formed on the surface of stones and rock outcrops in desert regions after long exposure* (Fig. 12.4). Concentration of manganese in the varnish is thought to be due either to release of the element from desert dust (which settles on the ground and is weathered following wetting by summer rainstorms) or, alternatively, to the manganese-

FIGURE 12.5 A box canyon in Canyon de Chelly area, Arizona has steep walls and a flat floor underlain by sandy alluvium.

concentrating activity of microorganisms that live on rock surfaces. The longer a stone has been exposed to weathering, the greater is the concentration of the coating oxide. The ratio of mobile to nonmobile elements in the varnish has been used as a dating method, for the ratio changes with time. This has permitted the dating of such features as alluvial fans, taluses, old lake shorelines, and archaeological sites in several desert areas.

Streams

Most streams that originate in deserts never reach the sea, for they soon disappear as evaporation takes place and water soaks into the ground. Exceptions are long rivers like the Nile in Egypt and the Colorado in the southwestern United States, both of which originate in high mountains that receive abundant precipitation. Such rivers carry so much water that they keep flowing to the ocean despite great losses where they cross a desert.

The scarce plants in deserts are no great impediment to surface runoff, and the loose dry regolith is eroded easily. Typical violent rainstorms are likely to be accompanied by "flash" floods that transport large quantities of sediment suddenly and swiftly. The debris is deposited as alluvium that forms fans at the bases of mountain slopes and on the floors of wide valleys and basins. In many deserts such sudden and spectacular dis-

FIGURE 12.6 Playa, with salt deposits at its surface, about 240 km north of Broken Hill, Australia. Annual precipitation now is close to 200 mm, but stabilized sand dunes in foreground suggest that in the past the climate favored less vegetation cover and dune migration.

charges constitute a primary mechanism of landscape modification.

Often streams in flood effectively undercut the sideslopes of their valleys causing the slopes to cave. Then, as the flood subsides, the load is deposited rapidly, creating a flat alluvial surface. The result is a steep-sided, flat-bottomed "box canyon" that is characteristic of many dry regions (Fig. 12.5).

Playa Lakes

In an arid region, water is rarely abundant enough to flow into a basin and maintain a lake on its floor. Although streams that flow down from a highland infrequently reach the center of the nearest basin, after an exceptionally large rainstorm some of them discharge enough water to convert the basin floor into a shallow lake that may last several days or a few weeks. In the dry region of western United States *a dry lakebed in a desert basin* is called a *playa* (Spanish for "beach"); when runoff is sufficient, a temporary *playa lake* is created. Many playas are white or grayish because of precipitated salts at their surfaces (Fig. 12.6). Such deposits, formed by repeated generation and evaporation of temporary lakes, can reach a thickness of tens of meters and constitute a primary source of important industrial chemicals (Fig. 5.3). In cases where the lake water can percolate downward through the basin floor before evaporation saturates the water with salts, little or no salt is precipitated and the playa sediments consist mainly of clay.

Groundwater

In arid regions groundwater is recharged only from the scanty local rainfall plus water that enters as streams or via artesian aquifers. Deep groundwater may be very old and a relict of former times when evaporation rates were reduced by lower air temperatures. Useful supplies, nevertheless, are often small and the water table generally lies well below the surface. Under these conditions, if recharge is not kept in balance with the rate of withdrawal, further drawdown of the water table will occur, and this valuable natural resource may be depleted.

Wind

Wind is an effective geologic agent in dry regions, and landforms resulting from erosion and deposition by wind are often widespread. However, con-

FIGURE 12.7 Dry stream channels (wadis) crossing the desert landscape on Sinai Peninsula just west of the Gulf of Aqaba are evidence of former climates marked by greater surface runoff.

trary to popular belief, most deserts are not characterized mainly by sand dunes. Only a third of Arabia, the sandiest of all dry regions, and only a ninth of the Sahara are covered with sand. Much of the nonsandy area of deserts is crossed by systems of stream valleys or is covered by alluvial fans and alluvial plains (Fig. 12.7). Thus, even in deserts it is apparent that more geologic work is done by streams than by wind.

DESERT LANDFORMS

Fans and Bajadas

Alluvial fans can develop under a wide range of climatic conditions, but they are especially common in arid and semiarid lands. They also constitute a characteristic desert landform. In such areas they may be the major source of groundwater for irrigation. In some places entire cities have been built on alluvial fans or fan complexes (for example, San Bernardino, California and Teheran, Iran). Many fans in Iran, Afghanistan, and Pakistan are dotted with mounds of debris that mark the sites of deep shafts connecting horizontal tunnel systems. These were designed to collect water within the upper reaches of fans for use in surface irrigation. Some such systems are ancient and predate the Islamic era.

In desert basins of the American Southwest, the Middle East, and central Asia, alluvial fans form a prominent part of the landscape. In these regions they border highlands, with the top of each fan lying at the mouth of a mountain canyon. Where a mountain front is straight and the canyons are widely spaced, each fan will encompass an arc of about 180° (Fig. 11.7). More typically, one finds a *bajada* (Spanish for "slope"), *a broad alluvial apron composed of coalescing adjacent fans* that has an undulating surface due to the convexities of the component fans (Fig. 12.8).

Pediments

In deserts, landforms sculpted by erosion differ from those of regions receiving more rainfall. When the landscape consists of a mountain range and an adjacent basin, two geological relationships are common. In some situations alluvial fans along the mountain base merge outward into a bajada. In other cases, a sloping surface at the mountain base closely resembles a merging row of fans, but the surface is not underlain by fan sediments. Instead of being built up of thick alluvium, the surface is erosional and cut across bedrock. Scattered over it are rock fragments, some brought by running water from the adjacent mountains and some derived by weathering from the rock immediately beneath. Downslope, the rock particles gradually thicken to form a continuous cover of alluvium. The eroded bedrock surface, which may extend for many kilometers along the mountain front, passes beneath the margin of the basin fill.

The bedrock landform is called a *pediment* because the eroded surfaces on opposite sides of a mountain mass together resemble the triangular pediment or gable of a roof. A **pediment,** then, is *a sloping surface, cut across bedrock and thinly or discontinuously veneered with alluvium, that slopes away from the base of a highland in an arid or semiarid environment* (Fig. 12.9a). The kinds of bedrock on which pediments are cut are those that yield easily to erosion in such a climate. The profile of a pediment, like that of a fan, is concave-up, becoming steeper toward the mountain front. We associate such a form with the work of running water, and the pediment surface is commonly marked by faint, shallow channels. It is likely, therefore, that pediments are the work mainly of running water, flowing as definite streams, sheet runoff, or rills.

Because geologic processes work slowly in deserts, the formation of pediments is not easily observed. Eyewitness accounts of sheetfloods and of lateral erosion by stream floods (accompanying intense desert storms) has led both processes to be proposed as agents responsible for pediment formation. It is generally agreed that pediments are slopes across which sediment is transported. However, the exact way in which pediments form is

FIGURE 12.8 Bajadas, formed by coalescing alluvial fans, flank rugged desert mountain ranges in southeastern California. Contrasts in color of the alluvium reflect differences in bedrock of mountain source areas.

still not established. As more is learned about former climates (Chapter 20), it becomes increasingly evident that many now dry desert regions were wetter in the past. Therefore, could pediment formation occur mainly under more humid conditions? If true, pediments might be largely a relict of

a

b

FIGURE 12.9 Pediments truncating sedimentary strata along mountain fronts. (*a*) Pediment eroded on tilted sedimentary strata in the northern Mojave Desert of California. (*b*) Partially dissected pediment cut across steeply dipping sedimentary rocks meets steep eastern slope of Sheep Mountain, Wyoming at a distinct angle. The surface of the pediment is capped by a thin layer of alluvial gravel.

wetter times and present-day erosional activity may provide a misleading picture of how pediments form.

A pediment meets the mountain slope at its head not in a curve but at a distinct angle (Fig. 12.9*b*). This suggests that mountain slopes in the desert do not become gentler with time, as they would in a wet region where chemical weathering and creep of regolith are dominant. Instead, they seem to adopt an angle determined by the resistance of the bedrock, and maintain that angle as they gradually retreat under the attack of weathering and mass-wasting. In this way, retreat of the mountain slope should lengthen a pediment at its upslope edge. The growth of the pediment, at the expense of the mountain, may continue until the entire mountain has been consumed. In fact, pediments in various stages of development have been identified, lending credence to this deduction. Some form relatively narrow belts along mountain fronts whereas others penetrate well back into mountains along major valleys. In some cases pediments extend to divides and may meet with pediments growing headward from the other side. Given enough time, as mountain ridges are fragmented and reduced in area, a landscape could be largely worn away to a surface of low relief.

Bornhardts and Inselbergs

In many arid and semiarid regions there occur *steep-sided mountains, ridges, or isolated hills rising abruptly from adjoining monotonously flat plains.* These features, called **inselbergs** (German for "island mountains"), resemble rocky islands standing above the surface of a broad, flat sea. **Bornhardts,** *a special type of inselberg having rounded or domal form* (Fig. 12.10), are named after a German explorer, Wilhelm Bornhardt, who described such features in East Africa late in the last century. Although they have been reported from nearly all latitudes and from many environmental settings, ranging from coastal to interior and arid to humid, they are especially common and well developed in treeless grasslands in the middle of stable continents. Numerous examples can be found in southern and central Africa, northwestern Brazil, and central Australia.

Field evidence suggests that bornhardts have formed in areas of massive or resistant rock (most commonly granite or gneiss, but also sedimentary rocks) that contrasts with surrounding rocks which are more susceptible to weathering (Fig. 12.11). Differential weathering over long periods of time

FIGURE 12.10 Ayres Rock, a massive bornhardt that rises about 360 m above a surrounding flat plain in central Australia.

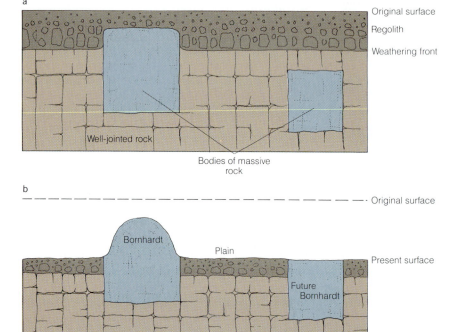

FIGURE 12.11 Evolution of bornhardts. As the landscape is slowly worn down by erosion, bodies of massive rock, less-weathered than surrounding more jointed rock, emerge as high-standing residual landforms. (*Source:* After Twidale and Bourne, 1975.)

has lowered adjacent terrain while leaving these rock masses standing high. Once formed, the residual bare rock hills tend to shed water, whereas surrounding debris-mantled plains absorb water and therefore are weathered more quickly. Their presence thus reinforces the processes that are thought to generate these unique landforms.

Because weathering tends to proceed more rapidly in warmer than in colder climates, bornhardts are especially common in low latitudes. Although low precipitation in desert areas might seem to argue against their development, it should be remembered that most of the weathering takes place below the ground surface where moisture is more plentiful. Furthermore, in many of these regions the climate was periodically wetter in the past than it is today.

Bornhardts tend to remain as stable parts of the landscape and may persist for tens of millions of years. Some are believed to date back to the Mesozoic Era, in which case they may have remained prominent landforms since the time of the dinosaurs.

WIND ACTION

Although the effects of wind action are visible in deserts, winds are not the dominant process that shapes desert lands. The foregoing section on land sculpture makes it clear that in deserts, as in moist regions, running water is the principal agent. Nevertheless, in many local areas landforms resulting from wind erosion or deposition predominate. Both wind activity and the resulting landforms are often referred to as *eolian* (after *Aeolus*, the Greek god of wind). Although most typical of arid and semiarid regions, eolian deposits are often found along seacoasts as well as in temperate zones where they provide evidence of former eolian activity under drier and windier conditions.

Transport of Sediment by Wind

We are now in a position to compare flowing air with flowing water as an agent that erodes and deposits sediment. Wind is normally turbulent. Like a stream, its velocity increases with height above the bed, in this case the ground surface. Also like a stream, it can carry coarser particles as bed load and finer ones as suspended load. The two loads can be seen moving in any desert windstorm. The bed load, a layer rarely as much as 1 m thick, consists of moving sand (Fig. 12.12). The

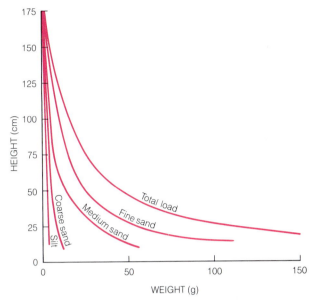

FIGURE 12.12 Graph showing percentage of different sediment sizes that were moved by wind at various heights above the ground in the Coachella Valley, California during a 146 day period in July-December, 1953. The bedload was concentrated mainly below 1 meter height. (*Source:* After Sharp, 1964.)

sand grades upward into clouds of suspended silt and clay particles that may reach great heights and travel over great distances.

Bed Load

Experiments with sand blown artificially through glass-sided wind tunnels show that sand grains move by saltation along arcing trajectories, much as they do in a stream of water (Fig. 12.13). In this mode of sediment transport, the particles follow a motion that is intermediate between suspension and rolling or sliding. The jumps involve elastic bounces that are similar to those of a ping-pong ball.

Most sand grains probably enter an airstream by bouncing or being knocked into the air by the impact of another grain. If a wind is strong enough, it can start a grain rolling along the surface where it may impact another grain and knock it into the air. When this second particle hits the ground it will impact other grains, some of which are projected upward into the air stream. Within a very short time the air close to the ground may contain a very large number of sand grains that move along with the wind in arcing trajectories as long as the wind velocity is great enough to keep them moving.

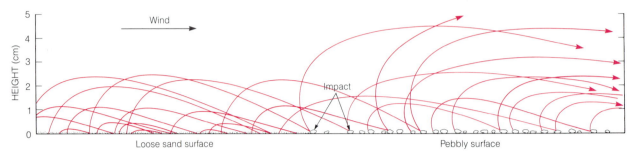

FIGURE 12.13 Movement of sand grains by saltation. Impacted sand grains bounce into the air stream and are carried along by the wind as gravity pulls them back to the land surface where they impact other grains, repeating the process. After reaching its maximum height, each follows a ballistic trajectory, like that of a bullet.

FIGURE 12.14 The famous "sliding rocks" of Racetrack Playa in Death Valley, California have moved over this normally dry lake bed through distances of up to 260 m. The stones probably are moved by strong winds following heavy rains when the playa surface becomes a layer of slippery clay.

The constant impact among moving sand grains causes their surfaces to become finely pitted. The resulting frosted appearance, similar to that of frosted glass, serves to differentiate eolian grains from sand grains shaped by other processes, such as running water or glacier ice (Fig. 5.20b).

The jumping sand grains seldom rise far off the ground. In laboratory experiments they generally travel within 10 cm of the surface. In desert country the sand often rises to about half a meter, as shown by abrasion marks on utility poles and fence posts which are sandblasted to that height but no higher. In very strong winds, such blast effects commonly extend no more than a meter above the ground. However, fine sand and coarse silt moving at high velocity within several meters of the ground can etch the window glass of vehicles, giving it a frosted appearance. Because blowing sand always moves close to the land surface it is easily stopped by obstacles in its path that can promote the formation of dunes.

A remarkable example of the movement of large particles by the wind is seen in the spectacular "sliding stones" of Racetrack Playa in Death Valley, California (Fig. 12.14). Stones weighing up to 25 kg lie at the end of long grooved tracks showing that they have moved across the smooth, normally dry lake bed over distances as great as 260 m. Field studies have shown that movement is related to wet stormy weather. The stones are probably moved at velocities of up to a meter a second by strong winds following deposition of a thin layer of fine slippery clay on the playa surface.

Suspended Load

Obstacles on the ground in the path of wind flow, whether coarse sediment, vegetation, or man-made objects, create a very thin layer of "dead," motionless air immediately above the surface (Fig. 12.15). The thickness of the dead layer equals 1/30 the height of an obstacle, regardless of velocity. Over ground that is covered with pebbles 3 cm in diameter, the dead-air layer would be 1 mm thick. Thin though it is, it plays a controlling part in the movement of all particles finer than medium-size sand grains. The reason why it does not influence the movement of coarser grains is that each such grain constitutes an obstacle that projects above the dead-air layer in the zone of turbulent air. But particles of silt and clay within that layer are so small and closely packed that they present a smooth surface to the wind which cannot lift them off the ground directly. Instead, they commonly are moved into the airstream by the impacts of saltating sand grains or through physical disruption of the smooth surface.

We can see how this happens by looking at a dusty desert road covered by dry silt on a windy day. The wind blowing across the road generates little or no dust. But a vehicle driving over the road creates a choking cloud, which is blown a short distance before settling once more to the ground. The passing wheels have broken up the crusted

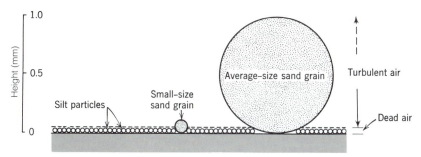

FIGURE 12.15 Silt particles form a smooth surface that lies within the dead-air layer caused by sand grains and other large obstacles to flowing air. The thickness of the dead-air layer is 1/30 the diameter of the average-size sand grain.

FIGURE 12.16 The "Dust Bowl" during the 1930s. A dust cloud approaches Springfield, Colorado at 4:47 P.M. on May 21, 1937. Total darkness lasted some 30 minutes.

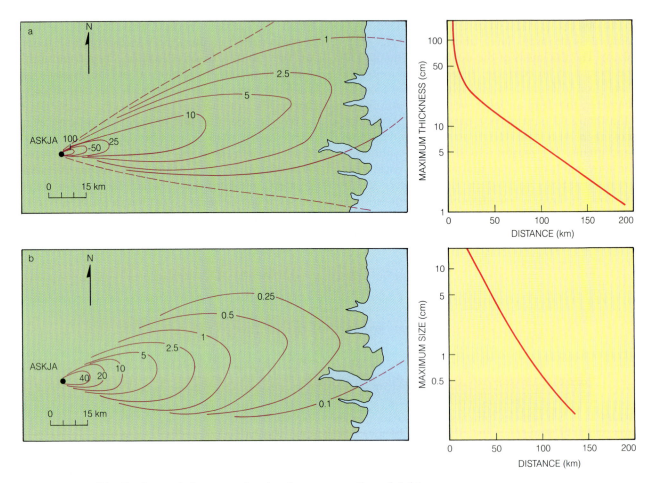

FIGURE 12.17 Distribution and character of tephra from an eruption of Askja Volcano, Iceland in 1975 (*Source:* After Sparks et al., 1981.) *a.* Isopach map and graph show thickness, in cm, of tephra deposit downwind from source. *b.* Isopleth map and graph show maximum particle size, in cm, downwind from source.

surface of powdery silt that was too smooth to be disturbed by the wind.

Once in the air, fine particles constitute the wind's suspended load. They are continually tossed about by eddies, like particles in a stream of turbulent water, while gravity tends to pull them toward the ground. Meanwhile they are carried forward. Although in most cases suspended sediment is deposited fairly near its place of origin, strong winds are known to carry fine particles thousands of kilometers before depositing them. For example, fine quartz dust blown across the Pacific Ocean from central Asia has been detected in the Hawaiian Islands where it forms an anomalous foreign component of surface soils. Reddish dust deflated from the Sahara Desert has settled on the decks of ships in the Atlantic Ocean and also is deposited on glaciers in the Alps. The amount of sediment actually moved by the atmosphere, year

in and year out, is probably only a fraction of a percent of its potential capacity, for the air is rarely if ever fully loaded.

During the great windstorms in the dry years of the 1930s, however, loads became unusually large. In a particularly great storm on March 20, 1935, when the sky looked much as in Figure 12.16, the cloud of suspended sediment extended 3.6 km (about 12,000 ft) above the ground, and the transported load in the area of Wichita, Kansas, was estimated at 35 million kg/km^3 in the lowermost 1.6 km of air. Sampling of sediment on flat roofs of buildings showed that about 280,000 kg of silt particles, or about 5 percent of the load suspended in the lowermost layer of air, were deposited on each square kilometer of land that day. Enough sediment was carried eastward on March 21 to bring temporary twilight conditions in midday over New York and New England, 3000 km beyond the principal source area in eastern Colorado. The distance and travel time imply wind velocities of about 80 km/h (50 mph).

Volcanic Ash

Not all wind-transported sediment originates by being picked up from the ground. Large quantities of pyroclastic debris can be ejected into the atmosphere during explosive volcanic eruptions. Although coarse and dense particles fall out quickly downwind from the source vent, small particles of lower density may be carried for great distances. Fine ash that reaches the stratosphere may circle the Earth many times before it finally falls to the ground. The particles that fall out during an eruption commonly form an elongate plume of sediment that decreases in particle size and thickness downwind from the source volcano (Fig. 12.17).

EROSION BY WIND

Flowing air erodes in two ways. **Deflation** (from the Latin word meaning "to blow away"), *the picking up and removal of loose rock and soil particles by wind*, provides most of the wind's load. The second process, *abrasion* of rock by wind-driven rock particles, is analogous to abrasion by running water.

Deflation

Deflation on a large scale happens only where there is little or no vegetation and where loose rock

particles are fine enough to be picked up by the wind. Areas of significant deflation are found mainly in deserts; others include ocean beaches and the shores of large lakes, floodplains of large meltwater streams, and—of greatest economic significance—bare plowed fields in farmland during times of drought, when no moisture is present to hold the soil particles together.

In most areas the results of deflation are not easily seen, inasmuch as the whole surface tends to be lowered irregularly. In some places, however, measurement is possible. In the dry 1930s, deflation in parts of the western United States amounted to 1 m or more within only a few years (Fig. 12.18), a tremendous rate compared with estimates of long-term average regional rates of erosion (only a few centimeters per thousand years).

The most conspicuous evidence of deflation consists of basins excavated by the wind. Tens of thousands occur in the semiarid Great Plains region of North America from Canada to Texas. Most are less than 2 km long and are only a meter or two deep. In wet years they are clothed with grass and some even contain shallow lakes. An observer seeing them at such times would hardly guess their origin. However, in dry years soil moisture evaporates, grass dies away, and wind deflates the bare soil.

Where sediments are particularly prone to deflation, depths of deflation basins can reach 50 m or more. The immense Qattara Depression in the Libyan Desert of western Egypt, the floor of which lies more than 100 m below sea level, has been attributed to intense deflation. In any basin, the depth to which deflation can reach is limited

FIGURE 12.18 Part of a large area deflated during the "Dust Bowl" years. A thickness of 1 m of deflation can be measured directly. The plant roots in the residual hummock, just above the dark band (an ancient soil), mark the position of the ground surface before deflation.

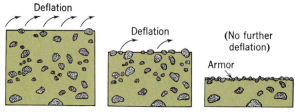

FIGURE 12.19 Three stages in the development of a deflation armor.

FIGURE 12.20 Desert pavement on the floor of Searles Valley, California. A layer of gravel, too coarse to be moved by the wind, covers finer sediments and inhibits further deflation.

FIGURE 12.21 Ventifacts litter the ground surface near Lake Vida in Victoria Valley, Antarctica. The most intensely abraded surfaces are inclined to the right, in the direction from which strong winds blow off the East Antarctic Ice Sheet.

finally only by the water table. As deflation approaches the level of the water table, the surface soil is moistened, thereby encouraging the growth of vegetation which inhibits further wind erosion.

A further inhibition to deflation is the development of a cover of rock particles too large to be removed by the wind. Deflation of alluvium, which generally consists of silt, sand, and pebbles, may create such a cover (Fig. 12.19). The sand and silt are blown away, or locally removed by sheet erosion, but the pebbles remain. When the surface has been lowered just enough to create a continuous cover of pebbles, the ground has acquired a *deflation armor, a surface layer of coarse particles concentrated chiefly by deflation.* Such armors are also called *desert pavement,* because long-continued removal of the fine particles makes the pebbles settle into such stable positions that they fit together almost like the blocks of a cobblestone pavement (Fig. 12.20).

Ventifacts

In desert areas, bedrock and loose stones are abraded by wind-driven sand and silt, which can shape and polish them to a high degree. A *ventifact* is *any bedrock surface or stone which has been abraded and shaped by wind-blown sediment.* It is recognized by its smooth, polished surfaces, which may be pitted and/or fluted, and by facets that typically are separated from each other by sharp keel-like edges (Fig. 12.21).

In the laboratory, sandblasting of pieces of plaster of Paris demonstrates that the facets always face the wind (Fig. 12.22). A stone can be worn down flush with the ground by enlargement of a single facet. If the stone is shifted or rotated, or if the wind direction varies from time to time, two or more facets may be cut on it. The dominant erosive agents near the bed are saltating sand grains. Well above the bed the coarsest fraction of the suspended load becomes important, and presumably

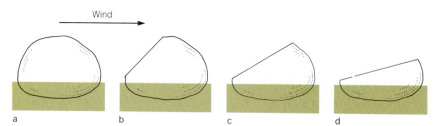

FIGURE 12.22 Four stages in the cutting of a ventifact. The stone becomes a ventifact between stages A and B.

is also responsible for cutting the remarkable flutes and grooves seen on some large ventifacts.

Yardangs

Among the most common eolian landforms in some desert regions are *elongate and streamlined, wind-eroded ridges* called **yardangs** (from the Turki word *yar*, meaning steep bank). Typically, they are carved from indurated sediments or from highly weathered crystalline rocks, and have a shape similar to an inverted ship's hull (Fig. 12.23). Individual yardangs range up to a few tens of kilometers long and up to 100 m high. Generally they occur in groups. Yardangs probably begin to form by differential erosion along irregular surface depressions that parallel the wind direction and increase in size as wind abrasion deepens and broadens the depressions, often leaving sharp intervening ridges. Although reasonably common in some desert regions, they were largely ignored by geologists until recently. Renewed interest in them has arisen because images of the surface of Mars show that yardangs are widespread on that planet. They apparently are indicators of strong, unidirectional winds and therefore can be used to infer the path of prevailing winds at the Martian surface.

WIND-TRANSPORTED SEDIMENTS

As sediment is moved by the wind, the sand grains forming the bed load move rather slowly and are deposited quickly when wind velocity subsides, whereas the finer particles of silt and clay which form the suspended load travel faster, longer, and much farther before they settle to the ground. The wind typically deposits sand in heaps or small hills, whereas the finer sediment tends to be deposited as a smooth blanket across the landscape.

Sand Ripples

Sheets of well-sorted sand that have accumulated at the surface are inherently unstable, even under gentle winds. As the wind passes over the edge of such an accumulation, saltation begins to move the smaller, most easily transported grains. Grains too large to be moved are left behind. The saltating finer grains impact the surface at some average distance downwind, setting more fine particles in motion; there, another accumulation of coarse grains develops as the fine sand moves onward.

Through this process, ripples of sand form with their long axes oriented perpendicular to the wind direction (Fig. 12.24). Such *sand ripples* are often seen on the windward sides of dunes. Under very strong winds the ripples disappear, for then all

FIGURE 12.23 Aligned yardangs, resembling inverted ship's hulls, parallel the direction of prevailing wind in the coastal desert of Peru. Cut in erodible Tertiary sedimentary strata, the largest reach lengths of up to 500 m.

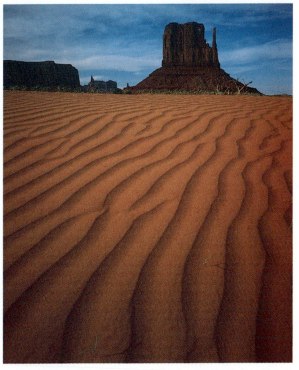

FIGURE 12.24 Sand ripples cross the surface of a desert sand sheet on the floor of Monument Valley, Arizona.

grains can be moved and sorting is less likely to take place.

Dunes

A **dune** is *a mound or ridge of sand deposited by wind.* Generally a dune forms where an obstacle distorts the flow of air. Velocity within a meter or two of the ground varies with the slightest irregularity of the surface. On encountering an obstacle, wind sweeps over and around it, but leaves a pocket of slower-moving air immediately behind the obstacle. In these pockets of low velocity, moving sand grains drop out and form mounds. The growing mounds in turn influence the flow of air. As more sand piles up, the mounds join together to form a single dune.

A dune is asymmetrical. It has a steep, straight lee (downwind) slope and a gentler windward slope (Fig. 12.25). Sand grains move up the windward slope by saltation to reach the crest of the dune. As most saltation jumps are much shorter than the length of the lee face, grains making it past the brink of the dune generally fall onto the lee surface near its top. The subtle bulge thus created through grain-by-grain accumulation eventually reaches an unstable angle, whereupon the sand avalanches (or "slips") downward, spreading the grains in the bulge down the lee face. For this reason, *the straight lee slope of an active dune* is known as the **slip face.** The avalanching returns the slip face to the angle of repose, which typically is 30–34°.

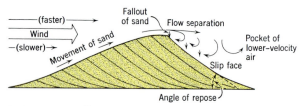

FIGURE 12.25 Cross section through a dune showing development of windward and leeward slopes, and internal stratification.

FIGURE 12.26 Huge sand dune in the Alashan Plain of western China reaches a height of several hundred meters.

The angle of slope of the windward flank varies with wind velocity and grain size, but it is always much less than that of the slip face, which it meets at a sharp angle. The asymmetry of a dune provides a means of inferring the direction of the wind that shaped it, for the slip face always lies on the side toward which the prevailing wind is blowing.

Many dunes grow to heights of 30–100 m, and some massive desert dunes in the western Alashan Plain of China are reported to reach heights of more than 500 m (Fig. 12.26). The height to which any dune can grow probably is determined by the upward increase in wind velocity, which at some level will become great enough to carry the sand grains up into suspension off the top of a dune as fast as they arrive there by saltation up the windward slope.

The accumulation and avalanching of sand grains on the slip face of a dune produces cross-strata much like the foreset layers in a delta (Fig. 12.25). As erosion of the windward slope continually removes layers already deposited, the sand grains are transported to the dune crest where they can avalanche onto the slip face to form new layers.

Transfer of sand from the windward to the lee side of an active dune causes the whole dune to migrate slowly downwind. Measurements of desert dunes of the barchan type (Table 12.1) show rates of migration as great as 25 m/yr. The migration of dunes, particularly along coasts just inland from sandy beaches, has been known to bury houses, fill in irrigation canals, and even threaten the existence of towns (Fig. 12.27). In such places, sand encroachment is countered most effectively by planting vegetation that can survive in the very dry sandy soil of the dunes. Continuous plant cover inhibits dune migration for the same reason that it inhibits deflation; if the wind cannot move sand grains across it, a dune cannot migrate.

Dune type is controlled by the degree of vegetation cover, as well as by strength of the wind and the amount of available sand (Fig. 12.28). Where sand is plentiful and lack of moisture inhibits growth of vegetation, strong winds build barchans and transverse dunes. If less sand is available, longitudinal dunes tend to form. As moisture increases and vegetation begins to encroach, parabolic dunes predominate. With a further increase in vegetation and with declining wind strength, dune formation may cease.

A dune covered with grass can be reactivated wherever the grass is killed off in patches allowing deflation to begin. This might result from drought

FIGURE 12.27 Ruins of Incan palace of Mama Cuna (1440–1530 A.D.) buried by shifting coastal sand dunes at Pachacamac, south of Lima, Peru.

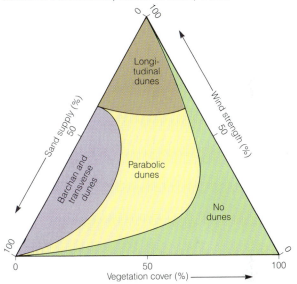

FIGURE 12.28 Relation of dune type to vegetation cover, amount of sand available, and wind strength in the Navajo Country of northeastern Arizona. (*Source:* After Hack, 1941.)

FIGURE 12.29 Blowout beside Anaktuvuk River in northern Alaska. Vegetation covering sand dune has been broken, allowing deflation to occur.

TABLE 12.1 *Principal Types of Dunes Based on Form*

Dune Type	Definition and Occurrence	
Barchan dune	*A crescent-shaped dune with horns pointing downwind.* Occurs on hard, flat floors of deserts. Constant wind and limited sand supply. Height 1 m to more than 30 m	
Barchanoid ridge	*A row of connected crescent-shaped dunes* oriented transverse to wind direction.	
Transverse dune	*A dune forming an asymmetrical ridge transverse to wind direction.* Occurs in areas with abundant sand and little vegetation. In places grades into barchans	
Parabolic dune	*A dune of U-shape with the open end of the U facing upwind.* Some form by piling of sand along leeward and lateral margins of a growing blowout in older dunes	
Linear dune	*A long, straight, ridge-shaped dune parallel with wind direction.* As much as 100 m high and 100 km long. Occurs in deserts with scanty sand supply and strong winds varying within one general direction. Slip faces vary as wind shifts direction	
Star dune	*An isolated hill of sand having a base that resembles a star in plan.* Ridges converge from basal points to central peak as high as 100 m. Tends to remain fixed in place in area where wind blows from all directions	
Reversing dune	*An asymmetrical ridge intermediate in character between a transverse dune and a star dune.* Forms where strength and duration of winds from nearly opposite directions are balanced.	

(*Source:* After McKee, 1979)

or from trampling of the vegetation by animals. A bare patch is thereby converted into a **blowout,** *a small shallow deflation basin excavated in loose sand* (Fig. 12.29).

Some dunes, of which the barchan is the best example, are built by winds blowing from one direction, whereas others are built by winds that shift direction from time to time. As long as the dune form exists, we can infer wind direction from the orientation of the slip face (Fig. 12.30). Even after erosion has destroyed the dune form, and after what is left of the dune sand has been converted into sandstone and later reexposed by erosion, we can determine the former direction of the wind from the cross-strata, which are always inclined downwind (Fig. 5.9).

Sand Seas

Some large deserts contain *vast tracts of shifting sand* known as **sand seas** (also referred to by the Hamitic word **ergs**). Some of the best examples are found in northern and western Africa and in the vast desert regions of the Arabian Peninsula (Fig. 12.31). They contain a variety of dune forms, ranging from low mounds of sand and barchans to huge dune complexes that form a seemingly endless and monotonous landscape.

Loess

Most regolith contains a small proportion of fine sediment deposited from suspension in the air, but so thoroughly mixed with other materials as to be indistinguishable from them. Over some wide areas wind-deposited sediment is so thick and uniform that it constitutes a distinctive deposit and may control the primary landscape characteristics. It is known as **loess** (German for "loose") and is defined as *wind-deposited silt, commonly accompanied by some fine sand and clay.*

Most loess is massive and lacks stratification, apparently because grains of different sizes settled progressively from the air and were deposited at random. Also, plant roots, worms, and other burrowing organisms have churned the sediment countless times during and since its deposition. Where exposed, loess commonly stands at such a

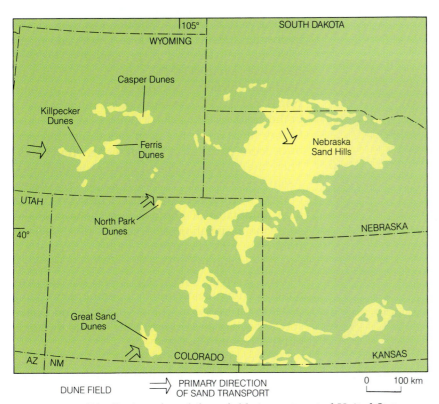

FIGURE 12.30 Distribution of sand dune fields in west-central United States formed during the last 100,000 years. Primary direction of sand transport for some fields, deduced from dune form, is shown by arrows. (*Source:* After Ahlbrandt et al., 1983.)

FIGURE 12.31 Northwestern part of large sand sea in the central Namib Desert of Nambia in western Africa. Large, complex linear dunes pass shoreward into barchanoid ridges that have been reworked along the coast to form pronounced spits.

FIGURE 12.32 Loess exposed in steep face of bluff near Xian, China. The silty loess was derived by deflation of desert regions to the northwest. Section in bluff was deposited over approximately the last 100,000 years. Dark brown zone at level of figure is a paleosol.

steep angle that it forms cliffs (Fig. 12.32), just as though it were firmly cemented rock. This is the result of the fine grain size of loess, in which molecular attraction is strong enough to make the particles very cohesive. Porosity is extremely high, commonly exceeding 50 percent. Loess, therefore, absorbs and holds water, and typically supports productive soils. The corn-growing region of Illinois, Iowa, northern Missouri, and eastern Nebraska is famous for its productivity, which partly is due to most of its soils being developed in loess.

Minerals composing loess are chiefly quartz, feldspar, micas, and calcite. The particles are generally fresh, and show little evidence of chemical weathering other than the slight oxidation of iron-bearing minerals that has occurred since deposition and which gives a yellowish-brown tinge to the deposit as a whole.

Loess possesses two characteristics that indicate it was deposited by wind. It forms a rather uniform blanket, mantling hills and valleys alike through a wide range of altitudes, and it contains fossils of land plants and animals. Air-breathing snails are especially common, but remains of small and large mammals also are found.

The distribution of loess shows that its principal sources were deserts and the floodplains of glacial meltwater streams. Desert loess covers enormous areas that lie to leeward of major deserts. The loess that covers some 800,000 km^2 in central China, and reaches a thickness of more than 300 m in some places (Fig. 12.33), was blown there from the floors of the great desert basins of central Asia. No satisfactory mechanism has been found that can produce loess particles in hot desert environments. Therefore, it is likely that the extensive and thick Asian loess deposits resulted from the breakdown of rocks by frost action and glacial processes in the high mountains of northwestern China, and the subsequent deflation of resulting fine particles from large fans of alluvium in adjacent desert basins. The beginning of loess accumulation in China about 2.5 million years ago coincided approximately with major uplift of mountains and plateaus in western China which led to widespread cooling and drying of the climate throughout this region (Chapter 19).

Loess of glacial origin is abundant in the middle part of North America (especially Nebraska, South Dakota, Iowa, Missouri, and Illinois) and in east-central Europe (especially Hungary and Czechoslovakia). It has two distinctive features. First, the shapes and mineral composition of its particles resemble ''rock flour'' produced by the grinding ac-

FIGURE 12.33 Loess deposits mantling hillslopes in Shaanxi Province, China, reach a thickness of more than 150 m. Reddish-brown bands crossing the section are paleosols. Erosion of the deposits, which were laid down over a period of about 2.5 million years, generates a vast load of silt that is responsible for the striking color of the Yellow River.

FIGURE 12.34 Active deflation of meltwater sediments downstream from Tasman Glacier in the Southern Alps of New Zealand. Loess is accumulating on the surface of vegetated glacial deposits in the foreground.

tion of glaciers (Chapter 13). Second, glacial loess is thickest immediately in the lee of rivers, such as the Mississippi and Missouri, which are known to have been swollen by meltwater during glacial times. At those times the areas just outside the margins of large glaciers were very cold and windy. The sediment-laden glacial streams deposited their load so continuously that plants could not gain a permanent foothold. The floodplains remained largely bare and, therefore, were easily deflated (Fig. 12.34). The wind-blown sediment settled out, forming deposits 8–30 m thick adjacent to the source valleys (Fig. 12.35) but which thinned downwind to thicknesses of only 1–2 m.

Why was the silt not picked up again and again and carried even farther by the wind? One reason lies in the stability of a silt surface, which results from its fine grain size (Fig. 12.15). Another reason is that the silt settled out chiefly on grassland, and to some extent on woodland, as is known from the

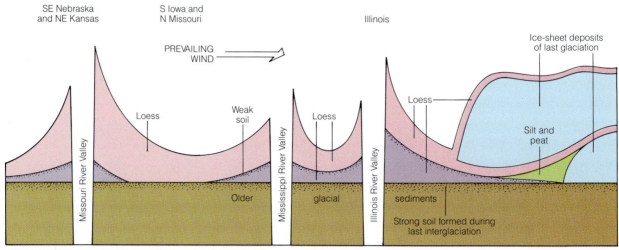

FIGURE 12.35 Cross section of loess deposits in the upper Mississippi Valley. Loess is thickest and coarsest adjacent to major river valleys, especially on their downwind sides, indicating that the valleys were the primary source areas for the sediment. (*Source:* After Willman et al., 1968.)

types of fossils found in the loess. Once on the ground in environments such as these, silt would be in no danger of further deflation.

DROUGHTS AND DESERTIFICATION

Droughts and Soil Erosion

Because climates fluctuate continually, regions ordinarily suitable for agriculture sometimes experience dry periods during which soil erosion by wind may reach disastrous proportions. During the 1930s an enormous volume of soil was blown away from parched, unprotected plowed fields in the "Dust Bowl" region of the Great Plains. Sand was piled up along fences and around farm buildings, and finer particles were blown eastward to be deposited over wide areas. A good deal of the sediment fell into the Atlantic Ocean.

What determined the location of the "Dust Bowl"? That area closely approximates the largest area in the United States in which average wind velocity is great enough both to move sand grains and to keep coarse silt grains in suspension. All that was needed further was a long succession of dry years—the drought years of the 1930s. This drought period gained much attention because of its great impact on crop production during the time of the Great Depression. Similar disastrous droughts have occurred periodically both in this region and in many others (Fig. 12.36).

Desertification

Desertification, defined as *the invasion of desert into nondesert areas,* can result from natural environmental changes as well as from human activities. The major symptoms are declining groundwater tables, increasing saltiness of water and topsoil, reduction in supplies of surface water, unnaturally high rates of soil erosion, and destruction of native vegetation. Although we can find evidence of natural desertification events in the geologic record, there is increasing concern that human activities, regardless of natural climatic trends, can in themselves help promote widespread desertification.

In the region south of the Sahara Desert lies a belt of very dry grassland known as the Sahel (Arabic for *border*). There the annual rainfall is normally only 100–300 mm and most of it falls during a single short season.

In the early 1970s the Sahel experienced the worst drought of this century (Fig. 1.3). For several years in succession the rains failed, causing adjacent desert to spread southward—according to one estimate as much as 150 km. The effects of the drought extended from the Atlantic to the Indian Ocean, a distance of 6000 km, and affected a population of at least 20 million people, many of them seminomadic herders of cattle, camels, sheep, and goats. The results of the drought were intensified by the fact that between about 1935 and 1970 the human population had doubled, and with it that of the livestock. This increase of people and animals

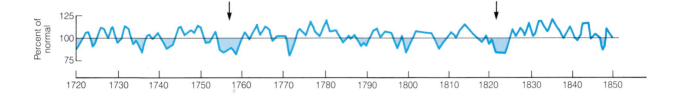

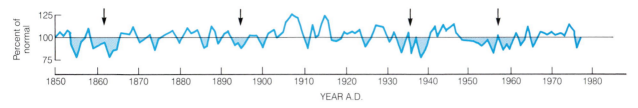

FIGURE 12.36 Drought history in Great Plains region of the United States, reconstructed from tree rings. The graph shows variations in tree-ring growth, related to precipitation, relative to normal growth (100%). Intensity, duration, and regional extent of droughts of "Dust Bowl" years of the 1930s were at least equaled by those of the 1750s, 1820s, 1860s, and 1890s (arrows). The years 1934, 1938, and 1939 were among the 10 driest years since 1700 A.D. (*Source:* After Stockton and Meko, 1983.)

led to severe overgrazing, so that with the coming of the drought the grass cover failed almost completely. Some 40 percent of the cattle—a great many millions—died. Millions of people suffered from thirst and starvation. Many succumbed as vast numbers migrated southward in search of food and water. By 1975 the rains returned briefly. Then in the 1980s, the continuing drought, felt es-

pecially in Ethiopia and the Sudan, led to widespread famine. Mass starvation was only alleviated by worldwide relief efforts.

Along the ancient Silk Road in central China, long abuse of the land has also created deserts from productive grasslands. In one area, lying inside the great northward bend of the Yellow River, incursions of herdsmen since the ninth century

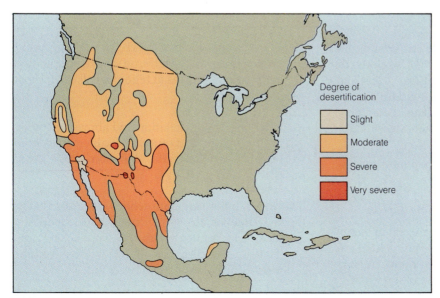

FIGURE 12.37 Status of desertification in North America. Vast areas of the American West show evidence of moderate to severe desertification. (*Source:* After Dregne, 1977.)

B.C., together with poor farming practices, destroyed the delicately balanced natural environment and allowed the desert to advance. Ruins of the capital city of the Xia Dynasty (2205–1766 B.C.) and 11 other large ancient cities are now covered with shifting sand dunes.

Although the impact of desertification on human life in North America is less severe than in more densely populated regions of the world, it nevertheless has important and far-reaching implications for the continent's food, water, and energy supplies, as well as its natural environment. A recent assessment concludes that 2,850,000 km² or nearly 37 percent of the continent's arid lands have experienced "severe" desertification, while some 27,000 km² have undergone "very severe" desertification (Fig. 12.37). In the latter regions, large shifting sand dunes have formed, erosion has largely denuded the landscape of vegetation, numerous gullies have developed, and salt crusts have accumulated on nearly impermeable irrigated soils. Within the United States, about 10 percent of the land area—an area approximately the size of the original 13 states—has been severely or very severely affected by desertification. Mostly this has been brought about by overgrazing, by excessive withdrawal of groundwater, and by unsound water-use practices, in part allied with population increase and expanded agricultural production.

How can these detrimental effects be halted and even reversed? The answer lies largely in an understanding of the geologic principles involved, and in intelligent application of measures designed to reestablish a natural balance in the affected areas. For example, in China, sand dunes have been stabilized by planting shrubs and trees to decrease the strength of high-velocity winds. Downwind from such windbreaks, the air velocity is reduced in a wake extending a distance equivalent to many times the height of the trees. Reducing the wind velocity diminishes the likelihood of serious deflation and halts the migration of dunes. In one area where intense windstorms carried sands that completely buried all houses and destroyed an entire wheat crop, the subsequent construction of a 6 km-long windbreak of deep-rooted, fast-growing poplar trees substantially reduced the wind velocity. The dunes were then leveled by hand, and the area is now productive farmland. In one sandy area of northeastern China, a newly planted forest belt measuring 500 by 800 km protects more than 90,000 km² of farmland from the effects of the wind.

In the United States reversal of the trend toward desertification is complicated by the fact that the federal government subsidizes both the exploitation and conservation of arid-land resources. Reduction of incentives to exploit arid lands beyond their natural capacity, coupled with long-range planning aimed at minimizing the negative effects of human activity, should help in reaching the desired goal. Because the arid West supplies some 20 percent of the nation's total agricultural output, the long-term benefits could be substantial.

SUMMARY

1. Hot deserts, constituting about a quarter of the world's nonpolar land area, are regions of slight rainfall, high temperature, excessive evaporation, relatively strong winds, sparse vegetation, and interior drainage. Polar deserts occur at high latitudes where descending cold, dry air creates arid conditions.

2. No major geologic process is confined to deserts, but in them mechanical weathering, flash floods, and winds are effective geologic agents. The water table generally is low.

3. Fans, bajadas, and pediments are conspicuous features of many deserts. Pediments are mainly shaped by running water and are surfaces across which sediment is transported.

4. Bornhardts form in massive or resistant rocks and may remain as persistent landforms for millions of years.

5. Wind carries a bed load of saltating sand grains close to the ground and a suspended load of fine particles at higher levels. Sorting of sediment results.

6. Wind erodes by deflation and abrasion, chiefly in dry regions and on beaches. It creates deflation basins, blowouts, deflation armor, ventifacts, and yardangs.

7. Dunes often originate where obstacles distort the flow of air. Bare dunes have steep slip faces and gentler windward slopes. They migrate in the direction of wind flow, forming cross-strata that slope downwind.

8. Dunes are classified according to form as barchan dunes, transverse dunes, longitudinal dunes, parabolic dunes, and star dunes.

9. Loess is deposited chiefly downwind from deserts and from active glacial meltwater streams. Once deposited, it is stable and is little affected by further wind action.

10. Recurring natural droughts can lower the water table, cause high rates of soil erosion, and destroy vegetation, thereby leading to the invasion of deserts into nondesert areas. Overgrazing, excessive withdrawal of groundwater, and other human activities can promote desertification. It can be halted or reversed by measures that restore the natural balance.

SELECTED REFERENCES

Bagnold, R. A., 1941 (repr. 1954), The physics of blown sand and desert dunes: New York, William Morrow.

Brookfield, M. E., and Ahlbrandt, T. S., eds., 1983, Eolian sediments and processes: New York, Elsevier.

Bull, W. B., 1977, The alluvial-fan environment: Progress in Phys. Geography, v. 1, p. 222–270.

Cooke, R. U., and Warren, A., 1973, Geomorphology in deserts: Berkeley, Univ. California Press.

Greeley, R., and Iversen, J., 1985, Wind as a geological process: Cambridge, Cambridge University Press.

Hadley, R. F., 1967, Pediments and pediment-forming processes: Jour. Geol. Education, v. 15, p. 83–89.

Mabbutt, J. A., 1977, Desert landforms: Cambridge, M.I.T. Press.

McGinnies, W. G., Goldman, B. J., and Paylore, P., eds., 1968, Deserts of the world: Tucson, Univ. Arizona Press. (Contains maps of all deserts.)

McKee, E. D., ed., 1979, A study of global sand seas: U.S. Geol. Survey Prof. Paper 1052.

Péwé, T. L., ed., 1981, Desert dust: origin, characteristics, and effect on man: Geol. Soc. America Spec. Paper 186, 303 p.

Sharp, R. P., 1963, Wind ripples: Jour. Geology, v. 71, p. 617–641.

Sheridan, D., 1981, Desertification of the United States: Washington, D.C., Council on Environmental Quality.

Twidale, C. R., 1982, The evolution of bornhardts: Am. Scientist, v. 70, p. 268–276.

Walker, A. S., 1982, Deserts of China: Am. Scientist, v. 70, p. 366–376.

CHAPTER 13

Glacial and Periglacial Landscapes

Vertical aerial photograph of a part of Malaspina Glacier, Alaska, shows intricately deformed moraines that have been compressed into tight folds. Stagnant ice zone below forested ridges (red color) is pockmarked with kettles.

GLACIERS

Glaciers as Part of the Cryosphere

A person living in New York City who wished to see a glacier firsthand would have to travel some 2500 km west to the Rocky Mountains or a nearly equal distance north into eastern Canada to find even a few small examples. Yet, about 20,000 years ago thick glacier ice covered the entire landscape between western Canada and New York City. The southern margin of the ice followed an irregular line from the Montana Rockies across the Great Plains to central Illinois, then eastward beyond New York City to the now-submerged continental shelf beyond southern New England. If people inhabited North America at that time, they could not have lived at the site of downtown New York because Manhattan Island lay buried beneath the glacier.

Similarly, Stone Age people, known to have been living in Europe since long before the culmination of the last glacial age, were driven southward by another vast ice sheet that spread across northern Europe and stretched from the British Isles in the west far eastward into the region of the Soviet Union.

Not only during the last glaciation, but also during numerous earlier ones, glaciers formed where none exist today and caused dramatic changes in the Earth's surface environments. However, glacial ages have not typified all of Earth history; instead, they seem to have been clustered in intervals that were separated by many millions of years. Such intervals are now thought to be due to the shifting of lithospheric plates that move continents from warm latitudes into colder ones. These crustal movements cause mountain ranges to be uplifted, thereby changing the paths of ocean and wind currents and permitting glaciers to form and expand in size. The succession of glacial and interglacial ages within these longer episodes seems to be controlled mainly by slight but important changes in the position of the Earth relative to its primary source of energy, the Sun. These changes affect the amount and seasonal distribution of radiation striking any point on the Earth's surface (Chapter 20). Relatively small changes in energy input apparently can lead to substantial changes in the amount and distribution of glacier ice. This in turn can affect surface conditions over much of the planet, for as ice sheets form and grow in subpolar latitudes, world sea level falls and the paths and

moisture content of wind systems change, causing a shift of the Earth's climatic zones.

Glaciers constitute an important part of the cryosphere, that portion of the planet where temperatures are so low that water exists primarily in the frozen state. Other components of the cryosphere include snow, perennially frozen ground, and sea, lake, and river ice. Most glacier ice on the Earth resides in the polar regions, above the Arctic and Antarctic circles. In these regions sea ice forms a vast sheet over the polar seas, but its extent fluctuates seasonally. Because it is so thin (generally 3 m or less), it is not as important volumetrically as the ice contained in large ice sheets, but it has an important effect on global climate because of its highly reflective surface.

Large fluctuations also occur in the areal extent and volume of glaciers, but on a longer time scale. Widespread deposits and glacially eroded terrain beyond existing glaciers point to changing climatic conditions on the Earth, for glaciers and other

FIGURE 13.1 Glaciers in eastern Greenland spill from cirques carved in mountain flanks and enter deep ice-eroded valleys that constrain them and channel their flow.

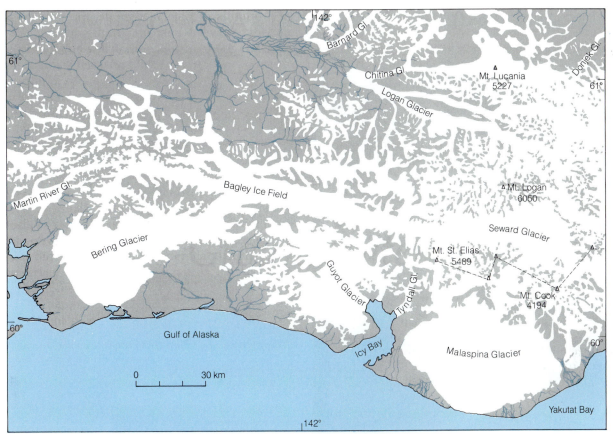

FIGURE 13.2 Glacier complex of the Icefield Ranges along the Alaska Yukon border. Malaspina Glacier and Bering Glacier are broad piedmont lobes fed by ice draining from a large intermontane icefield. The base of each lies far below sea level. If they were to recede, long deep fjords would appear and extend far inland from the present coast.

forms of ice are very sensitive to climate. Therefore, a study of their past distribution can provide important information about global changes of climate over millions of years.

Forms of Glaciers

Defined simply, a *glacier* is *a body of ice, consisting largely of recrystallized snow, that shows evidence of downslope or outward movement due to the pull of gravity.* On the basis of form and extent, several classes of glaciers can be distinguished (Table 13.1). The smaller types are confined by surrounding topography that determines their shape and direction of movement. The smallest glaciers mostly occupy protected hollows or depressions on the sides of mountains. Larger glaciers spread downward onto valley floors where their shapes are controlled mainly by the structure and erosional pattern of the bedrock landscape on which they lie (Fig.

13.1). Most of the world's glaciers are no more than about 1–2 km long and have areas of less than several square kilometers, but some large valley glaciers in the high mountains of Alaska and central Asia reach lengths of tens of kilometers.

Still larger glaciers spread beyond the confining walls of mountain valleys and onto gentle slopes where topography exerts little control on their form. Malaspina Glacier and Bering Glacier, the two largest glaciers of this type in North America, occupy lowlands along the Gulf of Alaska and are fed by extensive intermontane ice fields in the coastal ranges of Alaska and Yukon Territory (Fig. 13.2). Ice caps of various sizes cover mountain highlands or lower-lying lands at high latitude, and display generally radial outward flow. One well-studied example is the small tropical Quelccaya Ice Cap of the Peruvian Andes which covers 70 km^2 and lies at altitudes of 4950–5645 m. Vatnajökul, a far larger ice cap in Iceland, measures

TABLE 13.1 *Principal Types of Glaciers, Classified According to Form*

	Glacier Type	Characteristics
	Cirque glacier	Occupies bowl-shaped depression on the side of a mountain (*Sexton Glacier, Glacier National Park, Montana*)
	Valley glacier	Flows from cirque(s) onto and along floor of valley (*Trimble Glacier, Alaska Range*)
	Fjord glacier	Occupies a submerged coastal valley and its base lies below sea level. May have steep terminus that recedes rapidly by frontal calving. (*Muir Glacier, Glacier Bay, Alaska*)
	Piedmont glacier	Terminates on piedmont slopes beyond confining mountain valleys and is fed by one or more large valley glaciers. (*Malaspina Glacier, Alaska*)

TABLE 13.1 (Continued)

	Glacier Type	Characteristics
	Ice cap	Dome-shaped body of ice and snow that covers mountain highlands, or lower-lying lands at high latitudes, and displays generally radial outward flow. (*South Patagonian Ice Cap, Chile and Argentina*)
	Ice field	Extensive area of ice in a mountainous region that consists of many inter-connected alpine glaciers. Lacks domal shape of ice caps. Its flow is strongly controlled by underlying topography. (*Juneau Icefield, Alaska*)
	Ice sheet	Continent-sized masses of ice thick enough to flow under their own weight and which overwhelm nearly all land within their margins. (*West Antarctic Ice Sheet*)
	Ice shelf	Thick glacier ice that floats on the sea and commonly is located in coastal embayments. (*Pine Island Ice Shelf, Antarctica*)

FIGURE 13.3 Vertical satellite image of Vatnajökul ice cap near the southeast coast of Iceland. The firn limit separates new white snow of the accumulation area from darker ice of the ablation area. Moraine bands on the glacier surface are oriented in the direction of ice flow. Dark bands of volcanic ash, locally deformed by flow, cross the ablation zone in several places. Braided streams have built extensive outwash plains beyond the glacier front.

about 100 by 140 km across and terminates close to sea level (Fig. 13.3).

Ice sheets are the largest glaciers on the Earth. These continent-sized masses of ice overwhelm nearly all the land surface within their margins. Modern ice sheets are confined to Greenland and Antarctica, and collectively comprise about 95 percent of all glacier ice on our planet. During glacial ages, as we shall see later, ice sheets also covered extensive portions of North America and Eurasia. The Greenland Ice Sheet, which has an area approximately equal to that of the United States west of the Rocky Mountains, reaches such a great thickness (some 3000 m) that the crust of the Earth beneath much of it has been depressed below sea level by its weight. Were the glacier suddenly to melt away, the island of Greenland would have the unusual form of an elongated ring of land enclosing an extensive arm of the sea.

On many maps, Antarctica appears to be covered by a single vast glacier, but in reality it consists of two large ice sheets that meet along the lofty Transantarctic Mountains (Fig. 13.4). The East Antarctic Ice Sheet is the larger of the two and covers the continent of Antarctica. It is the only truly polar ice sheet on Earth, for the North Pole lies at the center of the deep Arctic Ocean which is covered only by a thin layer of sea ice. Because of its ice sheet, Antarctica has the highest average altitude and the lowest average temperature of all the continents. The smaller West Antarctic Ice Sheet overlies numerous islands of the Antarctic archipelago. Like its more massive neighbor, *terrestrial* portions rest on land that rises above sea level while *marine* portions cover land lying below sea level. Measurements of ice thickness obtained by sending radio waves through the glacier where they then bounce off rocks at its base show that Antarctic ice reaches a thickness of 3600 m or more. The combined estimated volume of the two ice sheets of close to 24×10^6 km^3 would be sufficient to raise world sea level by nearly 60 m if the ice were to waste away entirely (Fig. 13.5).

Ice caps typically have a relatively simple geometry and constitute a single broad dome. Large ice sheets, on the other hand, may be more complex and consist of several domes from which ice flows radially to the ice margin or to broad interdome saddles where the ice flow diverges downslope. The location of ice domes and ice saddles determines the flow path of ice within an ice sheet, but such features do not necessarily remain fixed. Instead, they may shift position with time as an ice sheet grows or shrinks in size.

Ice shelves occur at several places along the margins of the Greenland and Antarctic Ice Sheets, as well as locally in the Canadian Arctic islands. They are mainly located in large coastal embayments (Fig. 13.4), are attached to land on one side, and their seaward margin generally forms a steep ice cliff rising as much as 50 m above sea level. The largest ice shelves extend hundreds of kilometers seaward from the coastline and can reach a thickness of at least 1000 m. They are nourished by ice streams flowing off the land, as well as by direct snowfall on their surface.

Large tabular icebergs produced by marginal breakup of floating ice shelves can form "ice islands;" some, discovered drifting in the Arctic Ocean, have been used as remote stations by polar scientists. Recently, studies have been made regarding the feasibility of towing massive tabular icebergs from Antarctic waters to arid countries of the Middle East where they could provide large quantities of fresh water for agricultural use. Although the bergs would have to be towed very slowly northward across warm equatorial waters where substantial melting would occur, calculations suggest that sufficient ice would remain at the end of a trip to make such a venture feasible. Some day desert countries may be growing abundant food crops irrigated by meltwater that originated as snowflakes falling in central Antarctica tens of thousands of years ago!

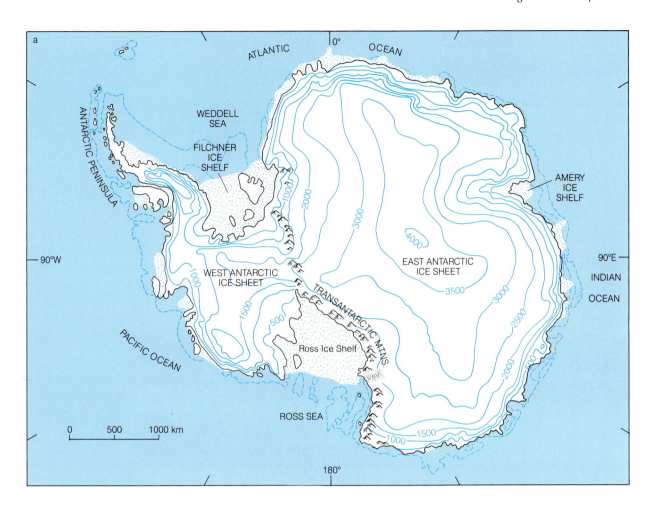

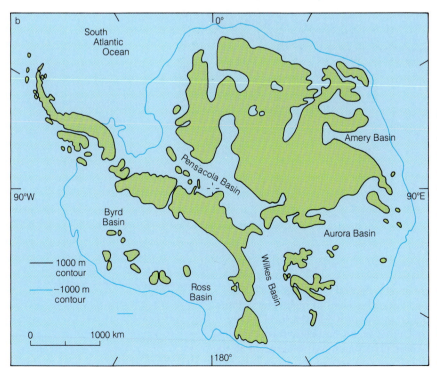

FIGURE 13.4 Ice sheets and ice shelves in the Antarctic region. (*a*) The East Antarctic Ice Sheet overlies the continent of Antarctica whereas the much smaller West Antarctic Ice Sheet overlies a volcanic island arc and adjacent sea floor. Three major ice shelves occupy large embayments. The ice sheets and ice shelves of Antarctica cover an area nearly equal to that of Canada and the conterminous United States. (*Source:* After Denton et al., 1984.) (*b*) Map showing how Antarctica would look if all ice were removed and the land then adjusted isostatically. (*Source:* After Drewry, 1983.)

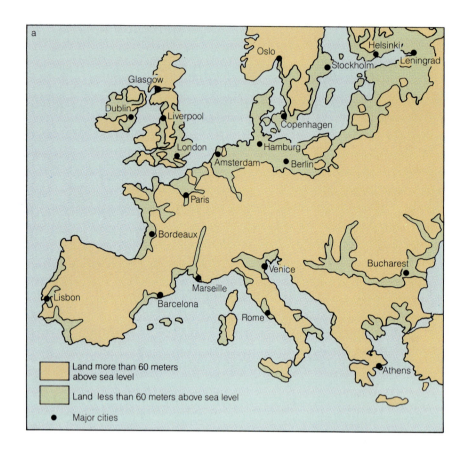

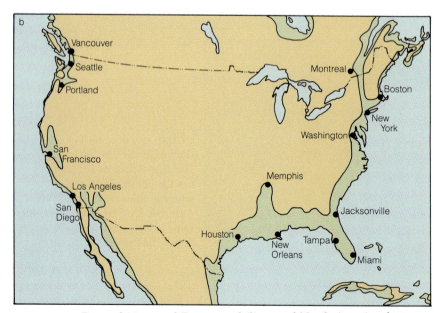

FIGURE 13.5 Parts of (*a*) coastal Europe and (*b*) central North America that would be flooded if all the glacier ice on Antartica were to melt. The volume of ice involved is nearly equivalent to a 60 m rise of world sea level. Many major cities of western Europe and the United States lie below the 60 m contour and would therefore be totally or partially submerged. Melting of all the additional ice on the earth would cause the oceans to rise about another 6 m.

Temperatures in Glaciers

Except for a thin surface layer that is chilled below freezing each winter, the ice throughout many glaciers is at the **pressure melting point,** *the temperature at which ice can melt at a particular pressure* (Figs. 3.6 and 13.6). Under such conditions meltwater and ice can exist together in equilibrium. **Temperate** or **warm glaciers,** *in which the ice is at the pressure melting point throughout,* are found mainly in low and middle latitudes. At high altitudes and latitudes, where the mean annual air temperature lies below freezing, ice temperature drops below the pressure melting point and little or no seasonal melting occurs. *Glaciers whose ice remains below the pressure melting point* are termed **polar** or **cold glaciers. Subpolar glaciers,** an intermediate type, may *have surface temperature at the freezing point in summer, but temperatures beneath the upper meter or two of ice remain below freezing.* Some large glaciers that originate in high mountains may contain polar or subpolar ice in their higher parts but have temperate ice in their lower parts. Warm basal ice may also occur in the deepest parts of large thick ice sheets where very high pressure at the glacier bed allows the ice to reach the pressure melting point.

Ice temperature is very important in controlling the way glaciers move and their rate of movement. Meltwater at the base of temperate glaciers acts as a lubricant and permits the ice to slide across its bed. By contrast, polar glaciers are so cold they are frozen to their bed. The motion they display does not involve basal sliding, and their rate of movement is greatly reduced.

Snowline

The two chief requirements for the existence of glaciers are adequate snowfall and low temperatures, both of which depend on climate. Such requirements are fulfilled at high latitudes and at high altitudes. These conditions also are more frequently met in moist coastal regions than in the dry interiors of continents. It is not surprising, therefore, that existing ice sheets lie in high latitudes and are surrounded by marine waters. Most of the world's numerous smaller glaciers are found 1) in moist coastal mountain systems, such as the high cordillera of northwestern North America; 2) on polar and subpolar islands like Spitsbergen and Iceland; and 3) in the rugged mountain ranges of central Asia, which are so high that very low temperatures offset the prevailing aridity and allow glaciers to develop.

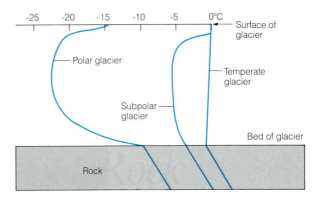

FIGURE 13.6 Temperature profiles for polar, subpolar, and temperate glaciers. Ice in temperate glaciers is at the pressure melting point from surface to bed, whereas in polar glaciers the temperature remains below freezing and the ice is frozen to its bed. In subpolar glaciers only a thin surface zone may seasonally reach the melting point. (*Source:* After Meier, 1964.)

Glaciers can only form at or above the **snowline,** *the lower limit of perennial snow.* Because its altitude is controlled mainly by temperature and precipitation, the snowline rises from near sea level in polar latitudes to altitudes of about 5000–6000 m in the tropics, and it also rises inland from moist coastal regions toward the drier interiors of large islands and continents (Fig. 13.7). Where high coastal peaks intercept moist air traveling onshore, resulting in strong climatic contrasts on opposite sides of mountain ranges, the snowline rises inland with a steep gradient.

Mass Balance

The mass of a glacier is constantly changing as the weather varies from season to season and, on longer time scales, as local and global climates change. These ongoing environmental changes cause fluctuations in the amount of snow added to the glacier surface, and in the amount of snow and ice lost by melting. These, in turn, determine the **mass balance** of the glacier, which is *a measure of the change in total mass during a year.*

Mass balance is measured in terms of **accumulation,** *the addition of mass to the glacier,* and of **ablation,** which is *the loss of mass to the glacier.* Accumulation occurs mainly as snowfall, whereas ablation takes place mainly through melting. In the case of subpolar and polar glaciers, however, ablation may result from evaporation of meltwater from the ice surface, or from direct vaporization without the ice passing through a liquid phase (a process called *sublimation*). Ablation may also involve melting at the base of ice shelves, or the breaking off of bergs into the sea or into lakes marginal to the ice.

If, during a year, more mass is added to a glacier than is lost, then the result is a positive mass balance (Fig. 13.8). By contrast, if more mass is lost than gained, the glacier experiences a negative mass balance. If the balance is mainly positive over a period of years, it means that the glacier is increasing in mass. Accordingly, the front, or *terminus*, of the glacier is likely to advance as the glacier grows. A succession of predominantly negative years normally leads to retreat of the terminus. Alternatively, if no net change in mass occurs, the glacier is in a balanced state. If this condi-

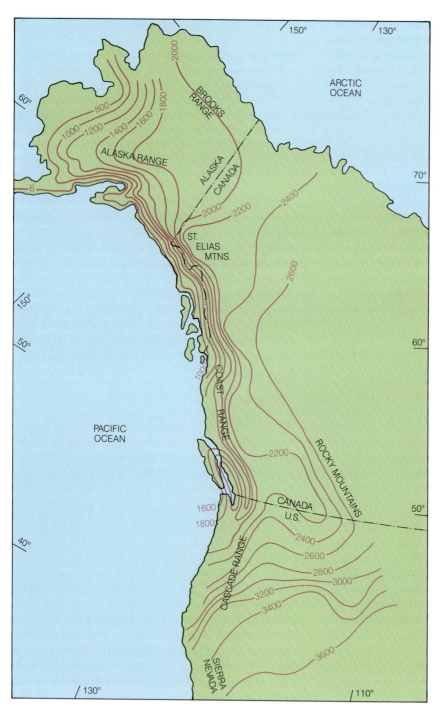

FIGURE 13.7 Contours, in hundreds of meters, show the regional pattern of glacier equilibrium-line altitudes throughout northwestern North America at the end of the 1961 balance year. The surface defined by the contours rises steeply inland from the Pacific coast in response to increasingly drier climate, and also from north to south in response to rising mean annual temperatures. (*Source:* After Meier and Post, 1962.)

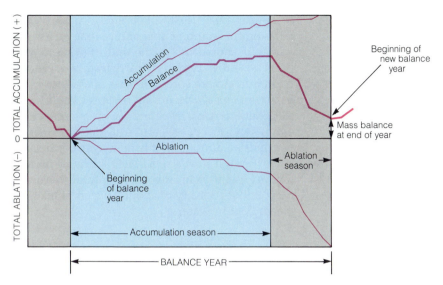

FIGURE 13.8 Diagram showing how accumulation and ablation determine glacier mass balance (heavy line) over the course of a balance year. The balance curve rises during the accumulation season as mass is added to the glacier, then falls during the ablation season as mass is lost. Mass balance at the end of the balance year reflects the difference between mass gain and mass loss. (*Source:* After International Commission of Snow and Ice, 1969.)

tion persists, the terminus is likely to remain relatively stationary.

If a mountain glacier is viewed at the end of the summer ablation season, two zones are generally visible on its surface. An upper zone, the ***accumulation area,*** is *that part of a glacier covered by remnants of the previous winter's snowfall and is an area of net gain in mass.* Below it lies the ***ablation area,*** *a region of net loss characterized by a dark-toned surface of bare ice and old snow from which the previous winter's snowcover has largely melted away* (Figs. 13.1 and 13.3). If a glacier is close to a balanced condition, then on average about two thirds of its total area lies in the accumulation area. The ***equilibrium line*** separates the accumulation area from the ablation area and, therefore, *marks the level on the glacier where net loss equals net gain.* On temperate glaciers it coincides with the lower limit of fresh snow at the end of the summer (the snowline). When a balanced condition exists, the equilibrium line lies approximately midway in altitude between the terminus and the head of the glacier.

Being very sensitive to climate, the equilibrium line fluctuates in altitude from year to year and is higher in warm dry years than in cold wet years (Fig. 13.9). Its position during a glacial age is many hundreds of meters lower than during an interglacial age, which helps explain the great contrast in the areal extent of glaciers at such times.

Response of the Glacier Terminus

Although measurement of the mass balance of a glacier may provide an excellent indication of its current "state of health," observations of its marginal fluctuations are not as good an indicator, for there normally is a lag in the time it takes for the terminus to respond to a change in climate. The lag reflects the time it takes for the effects of an increase or decrease in accumulation rate above the equilibrium line to be transferred through ice flow to the glacier terminus. The length of the response lag depends on the size and flow characteristics of a glacier, and will be longer for large glaciers than for small ones, and longer for cold glaciers than for warm ones. For glaciers of modest size in temperate latitudes (like those in the European Alps), response lags may range from several years to a decade or more. This partly explains why in any area having glaciers of different sizes, fluctuations of glacier margins may not be synchronous.

Conversion of Snow to Glacier Ice

Glacier ice is basically a metamorphic rock, for it consists of interlocking crystals of the mineral ice and has been deformed by flow due to the weight of overlying snow and ice. Newly fallen snow consists of hexagonal crystals we know as snowflakes.

It is very porous, having a density less than a tenth that of water. Air easily penetrates the pore spaces where the delicate points of snowflakes gradually disappear due to evaporation. The resulting water vapor then condenses, mainly in constricted places near the centers of ice crystals. In this way, snowflakes gradually become smaller, rounder, and thicker, and the pore spaces between them

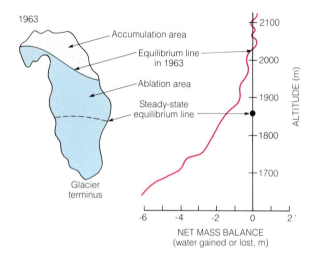

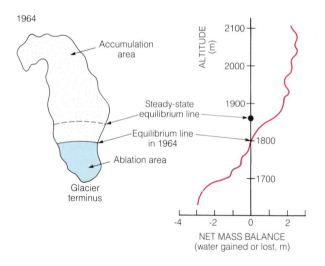

FIGURE 13.9 Maps of South Cascade Glacier in Washington State at end of 1963 and 1964 balance years, showing the position of the equilibrium line relative to the position it would have under a balanced (steady-state) condition. Curves show values of mass balance as a function of altitude. During 1963, a negative balance year, the glacier lost mass and the equilibrium line was high (2025 m). In 1964, a positive balance year, the glacier gained mass and the equilibrium line was low (1800 m). (*Source:* After Meier and Tangborn, 1965.)

disappear (Fig. 13.10). The entire mass of snow takes on the granular appearance that we associate with old snowdrifts at the end of winter. In the process, snow is transformed from a loose sediment into a more cohesive mass. *Snow that survives a year or more of ablation and achieves a density that is transitional between snow and glacier ice* is called **firn**. Ultimately, firn passes into true glacier ice when it becomes so dense that it is no longer permeable to air. Although now a rock, such ice has a far lower melting point than any other naturally occurring rock, and its density of about 0.9 g/cm^3 means that it will float in water.

Movement of Glaciers

Glacier Flow

Newly formed ice in a glacier consists of a myriad of randomly oriented small interlocking crystals having a texture like that of igneous rocks (Fig. 13.11). At some point, depending on the steepness of the surface on which it is lying and on the surrounding temperature, a mass of compacted snow and ice begins to deform and flow downslope under the pull of gravity. The flow takes place mainly through movement within individual ice crystals. Crystals in the accumulating glacier are subjected to higher and higher stress as the weight of the overlying snow and ice increases. Under this stress, deformation (*creep*) takes place along internal planes in an ice crystal in much the same way that playing cards in a deck slide past one another if the deck is pushed from one end. As movement proceeds, differential pressures between crystals cause some to grow at the expense of others, and the resulting larger crystals end up having a similar orientation. This leads to increased efficiency of flow, for the internal creep planes of all crystals now are approximately parallel (Fig. 13.12).

Importance of Water

In temperate glaciers, ice flow is enhanced by the presence of water, especially at the glacier bed. Measurement of the progressive deformation and downglacier displacement of a borehole drilled to the bed of a glacier shows that only a part of the observable surface motion is due to internal creep. An important component of movement results from sliding of the glacier across its bed (Fig. 13.13). In some glaciers basal sliding may contribute up to 90 percent of the total measured velocity.

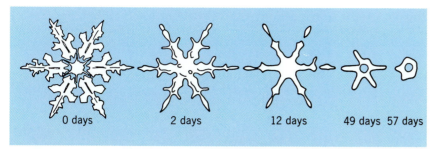

FIGURE 13.10 Conversion of a snowflake into a granule of old snow. Delicate points of a snowflake disappear through melting and evaporation. The resulting water refreezes and vapor condenses near the center of the crystal, making it denser. (*Source:* After Bader et al., 1939.)

FIGURE 13.11 Thin section of glacier ice showing interlocking structure of ice crystals. An average crystal in this view has a diameter of about 3 mm.

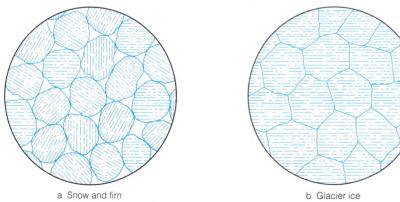

a Snow and firn   b Glacier ice

FIGURE 13.12 Arrangement of ice crystals in snow, firn, and glacier ice. (*a*) Random arrangement of crystals in snow and firn. Crystal planes along which creep could occur are not parallel. (*b*) Oriented crystals of glacier ice. Crystals are arranged with creep planes parallel, enhancing the ease of internal flow.

Crevasses

The surface portion of a glacier, having little weight upon it, is brittle. Where a glacier flows over an abruptly steepened slope, such as a bed-rock cliff, the surface ice is subjected to tension and it cracks. The cracks open up and form **crevasses,** which are *deep, gaping fissures in the upper surface of a glacier* (Fig. 13.26). Although they are often difficult and dangerous to cross, crevasses are

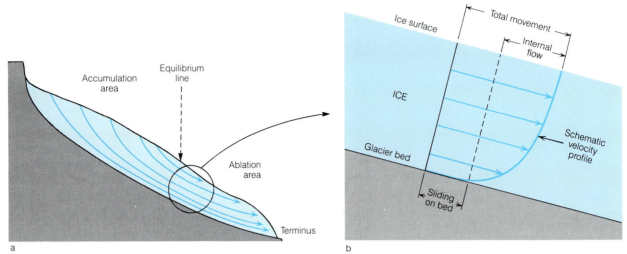

a
b

FIGURE 13.13 Flow of ice within a glacier. (*a*) Snow accumulating above the equilibrium line is compacted and flows downward and toward the terminus. The flow lines emerge at the surface below the equilibrium line in the ablation area. (*b*) In a vertical velocity profile through a temperate glacier a portion of the total observed movement is due to internal flow within the ice (with velocity increasing upward from the bed), whereas part is due to sliding of the glacier along its bed, lubricated by a film of meltwater.

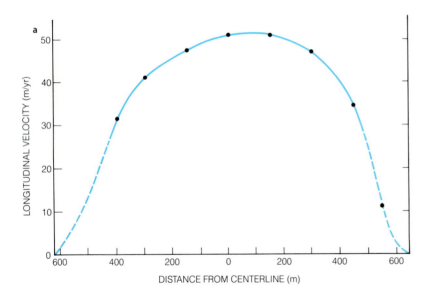

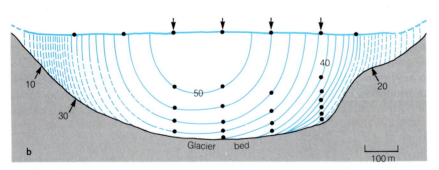

FIGURE 13.14 Velocity of flow within a valley glacier. (*Source: After Raymond, 1971.*) (*a*) Map showing down-glacier velocity profile across Athabasca Glacier in the Canadian Rocky Mountains. Velocity is highest at the center of the glacier and decreases rapidly toward margins. (*b*) Vertical section through Athabasca Glacier at right angles to the direction of flow showing velocity distribution (m/yr). Arrows mark positions of boreholes. Velocity decreases toward bed and margins.

rarely as much as 50 m deep. At greater depths internal flow prevents crevasses from forming. Because it cracks at the surface, yet flows at depth, we can liken a glacier to the upper layers of the Earth itself, which consist of a surface zone that cracks and fractures (the lithosphere) and a deeper zone (the asthenosphere) that can flow very slowly.

Rates of Flow

The surface velocity of a valley glacier can be measured by surveying from the sides of the valley, at intervals of time, a line of markers or targets extending across the glacier (Fig. 13.14*a*). Results show that surface ice in the central part of the glacier moves faster than ice at the sides, similar to the velocity distribution in a river. The reduced rates of flow toward the margins are due to frictional drag against the valley walls. A similar reduction in flow rate toward the bed is observed in a vertical profile of velocity, obtained by drilling a borehole through a glacier and measuring the angle of the hole with an inclinometer lowered into it. If measurements at different depths are repeated after a year, the annual flow as a function of depth can be determined (Fig. 13.14*b*).

If a glacier is in a balanced condition, then the ice passing through any transverse vertical section in the accumulation area must equal the amount of snow added to the surface upglacier. At the same time, ice passing through any transverse section in the ablation area must equal the amount of ice lost between that section and the terminus. Therefore, the ice flowing through any cross section must steadily increase downglacier toward the equilibrium line, then decrease downglacier away from the equilibrium line. This means that ice-flow velocity should be highest near the equilibrium line.

High rates of flow are observed where a glacier moves over an abrupt cliff to form a steep *icefall*. However, flow velocities in most glaciers range from only a few centimeters to a few meters a day, or about the same slow rate that groundwater percolates through crustal rocks. Probably hundreds of years have elapsed since ice now exposed at the terminus of a very long glacier fell as snow near the top of its accumulation area.

Directions of Flow Within a Glacier

Although snow continues to pile up in the accumulation area each year, while melting removes snow and ice from the ablation area, the surface profile of a glacier does not change much because ice is transferred from the accumulation area to the ablation area. If the altitude of the surface is to remain relatively unchanged, then ice flow cannot parallel the surface. Instead, the movement must be downward in the accumulation area, where mass is being added, and upward in the ablation area, where mass is being lost (Fig. 13.13). Crystals of ice that enter the glacier near its head, therefore, have a long path to follow before they emerge near the terminus, whereas those falling closer to the equilibrium line may travel only a short distance through the glacier before reaching the surface again.

Measurements show that marginal parts of the ice sheets in Antarctica are flowing at rates of about 50 m/yr (about 15 cm/day). Ice streams within the body of the ice sheets follow large valleys in the underlying bedrock and are not always visible at the surface. Such ice streams, somewhat analogous to well-defined currents in the oceans, are flowing as much as ten times faster than the ice around them.

Nonstable Behavior

Most glaciers slowly expand or contract in size as the climate fluctuates, but certain glaciers change dimensions rapidly and in a way that is either unrelated, or only secondarily related, to climatic change.

Glacier Surges

From time to time certain glaciers experience *surges,* which are *unusually rapid rates of movement marked by dramatic changes in glacier flow and form.* Such events appear mainly to affect temperate valley glaciers, although sectors of ice caps have been observed to surge and it is believed that portions of ice sheets may also be capable of surging. When a surge occurs, a glacier seems to go berserk. Before a surge, the lower part of the glacier often consists of stagnant ice. As the surge begins the boundary between active ice in the upper glacier and stagnant ice below moves rapidly downglacier and a chaos of crevasses and ice pinnacles forms. Although the terminus does not always advance, in some cases advances of up to several kilometers have been observed. Rates of movement as great as 100 times those of normal glaciers and averaging as much as 6 km a year have been measured during surges. Most surges run their course within a year or two, after which the lower part of the

FIGURE 13.15 Glaciers experiencing normal flow and surge flow. (*a*) Banded moraines of Barnard Glacier, Alaska, are oriented parallel to ice-flow direction, indicating stable flow. (*b*) Moraines of Tweedsmuir Glacier, Alaska, are contorted by periodic surges of tributary ice streams.

glacier will slowly revert to a stagnant condition before the next surge begins, generally a decade or more in the future.

A surging glacier can frequently be identified from the air or on satellite images because bands of rock debris on its surface tend to be intricately folded, in contrast to the generally parallel pattern of debris bands on nonsurging glaciers (Fig. 13.15). The reasons for glacier surging are not fully understood. It is generally believed that as water accumulates beneath a glacier over a period of years high pressures are generated within the water that lead to widespread separation of the ice from its bed. The resulting effect is similar to what happens when a rapidly moving automobile encounters a wet street during a rainstorm: The weight of the car places the water layer beneath the tires under so much pressure that they are floated off the wet pavement and the vehicle slides along out of con-

trol. The high velocities observed during glacier surges, therefore, probably result from greatly enhanced basal sliding rather than from any significant change in internal flow rate.

Calving Glaciers

When Captain George Vancouver sailed up the coast of southeastern Alaska in the late eighteenth century he plotted on his charts the positions of a number of large ice streams that issued from the high coastal mountains and terminated near the rocky shore. Today, in their place, one finds long open *fjords.* Such *deep glacially carved valleys submerged by the sea* extend many tens of kilometers back into the mountains. The dramatic recession of these fjord glaciers during the past century and a half at rates far in excess of typical glacier retreat rates on land is due to frontal *calving,* a process that involves *the progressive breaking off of icebergs from a glacier that terminates in deep water* (Fig. 13.16). Although the base of a fjord glacier may lie far below sea level along much of its length, its terminus can remain stable as long as it is "grounded" against a shoal. However, if the glacier's mass balance becomes negative, the front will recede into deeper water and calving can proceed. Once it commences, calving may continue rapidly and irreversibly until the glacier front once again becomes grounded, generally near the head of the fjord. Calving glaciers may undergo extensive retreat while nearby glaciers that rest on dry land fluctuate through only short distances in response to climatic variations. The final breakup and disappearance of the interior parts of the great ice sheets of eastern North America and Europe during the last glaciation very likely involved the retreat of calving margins that rapidly cut back into the ice sheets and caused them to collapse.

While most of the large fjord glaciers of coastal Alaska have retreated well back into their source regions during recent decades, an exception is Columbia Glacier which has remained relatively stationary with its terminus near the point of its greatest recent advance. During the early 1980s, signs of imminent retreat were detected along its front as calving increased and the terminus began to recede into deeper water. Calculations indicate that the rate of recession is likely to increase dramatically as the terminus calves, releasing a vast number of icebergs. Many of the bergs are likely to drift across nearby shipping lanes where large oil tankers enter and leave the port of Valdez at the southern end of the Alaska Pipeline, thereby creating potential hazards to navigation. When retreat is

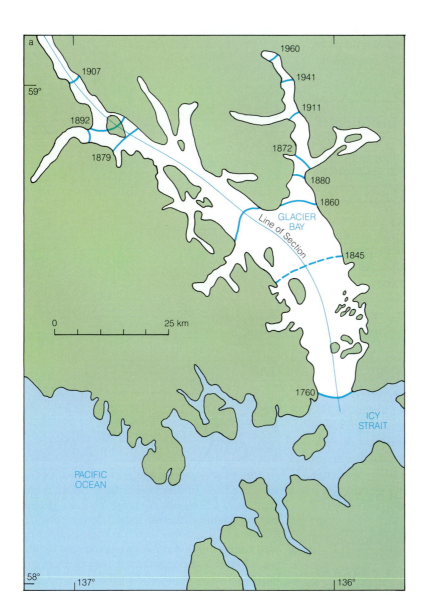

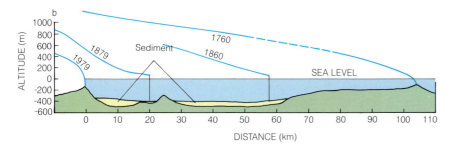

FIGURE 13.16 Rapid glacier recession in Glacier Bay, Alaska, resulting from frontal calving. (*a*) Map of Glacier Bay showing historically observed positions of the ice front between 1760 and 1960. Over these two centuries, the glacier retreated about 100 km into the upper reaches of its fjord system. (*b*) Section along length of Glacier Bay showing four successive profiles. Note steep vertical front of calving terminus and sediment deposited on fjord bottom since recession of the glacier. (*Source:* After Brown et al., 1982.)

complete, perhaps late in the next century, a newly exposed long fjord system will extend as an arm of the sea far back into the coastal mountains.

GLACIATION

Landscapes in Canada, northern United States, and northern Europe differ from those somewhat farther south, a principal reason being that these northern regions have been glaciated. *Glaciation,* defined as the *modification of the land surface by the action of glacier ice,* occurred so recently that weathering, mass-wasting, and erosion by running water have not had time to alter the landscape appreciably. Except for a cover of vegetation, the appearance of these glaciated landscapes has remained nearly unchanged since they emerged from beneath the ice. Like the geologic work of other surface processes, glaciation involves erosion, transport, and deposition of sediment.

Glacial Erosion and Sculpture

In changing the surface of the land over which it moves, a glacier acts collectively like a plow, a file, and a sled. As a plow it scrapes up weathered rock and soil and plucks out blocks of bedrock; as a file it rasps away firm rock; and as a sled it carries away the load of sediment acquired by plowing and filing, along with additional rock debris fallen onto it from adjacent slopes.

Small-Scale Erosional Features

The base of a temperate glacier is studded with rock particles of various sizes. The fragments move with the flowing ice across underlying bedrock and produce *glacial striations,* which are *long subparallel scratches inscribed on a rock surface by rock debris embedded in the base of a glacier,* as well as larger grooves that also are aligned in the direction of ice flow (Fig. 13.17). Striations are abraded as well on the moving rock fragments themselves. In places, fine particles of sand and silt in the basal ice act like sandpaper, and polish the rock until it has a smooth reflective surface. At the same time the basal ice drags at the bedrock, breaking off blocks (usually along joints or fractures) and quarrying them out. Blocks are mainly removed on the downglacier sides of hillocks, whereas on the up-glacier sides abrasion and polishing of the rock is dominant. The asymmetry of the resulting landforms clearly indicates the direction in which the glacier was moving (Fig. 13.18).

FIGURE 13.17 Recently deglaciated bedrock surface beyond Findelen Glacier, Swiss Alps. Debris carried at the base of the glacier produced grooves, striations, and polish on bedrock as the ice flowed forward in the direction of the Matterhorn. Crescentic marks on the rock surface represent fractures produced by the impact of large stones against the rock surface, which slopes gently in the up-glacier direction.

Landforms of Glaciated Mountains

Cirques and Related Forms. Most of the world's high mountains owe their scenic grandeur to sculpture by present and former glaciers. These mountains bear a distinctive suite of landforms attributable to glacial erosion that are lacking in nonglaciated uplands. Among the most characteristic is the *cirque, a bowl-shaped hollow on a mountainside, open downstream and bounded upstream by a steep slope (headwall), and excavated mainly by frost-wedging and by glacial plucking and abrasion* (Fig. 13.19). The floors of many cirques are rock basins. Some contain small lakes, called *tarns,* ponded behind a bedrock threshold at the edge of the cirque.

A cirque probably begins to form beneath a large

FIGURE 13.18 Asymmetrical glacially sculptured bedforms in front of Franz Josef Glacier in New Zealand's Southern Alps. The glacier flowed from right to left. Up-glacier slopes are smooth and polished. Scarps facing downvalley result from the plucking of bedrock blocks by flowing ice.

snowbank or snowfield just above the snowline. Meltwater infiltrating rock openings beneath the snow refreezes and expands, disrupting the rock and dislodging fragments. Small rock particles are then carried away by snowmelt runoff during periods of thaw. This activity gradually creates a depression in the land and enlarges it. As the snowbank turns into a glacier, plucking helps to enlarge the cirque still more and abrasion at the bed will further deepen it.

Most cirques owe their form to repeated episodes of glaciation. During interglacial periods, rock fragments dislodged from headwalls of ice-free cirques by frost action accumulate as taluses, which are transported away when glaciers reform during the next glaciation. Through successive glacial and interglacial ages cirques on opposite sides of mountain crests expand headward, creating a characteristic assemblage of features for which we use names given to them by Alpine mountaineers (Fig. 13.20).

Glacial Valleys. Glaciated valleys differ from ordinary stream valleys in several ways. Their chief characteristics, not all of which are present in every case, include a cross profile that is trough-like (U-shaped) and a floor that lies below the floors of tributary valleys, from which streams often descend as waterfalls or cascades (Fig. 13.20). The tributaries are referred to as *hanging valleys*. They "hang" above the floor of the main valley

FIGURE 13.19 Cirque carved in the side of the Aiguille Noire in the Mont Blanc massif of the northern Italian Alps. During glacial ages, ice spilled over the threshold of the cirque as an icefall and joined a large glacier in the main valley below.

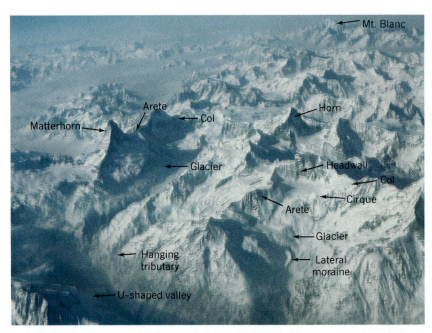

FIGURE 13.20 Photograph of central Swiss Alps near Zermatt showing alpine glacial landforms. Major U-shaped valleys head in cirques and have hanging tributary valleys. The sharp-pointed Matterhorn (middle distance) is a classic **horn,** *a sharp-pointed peak bounded by the intersecting walls of three or more cirques.* Many peaks are connected by an **arête,** *a jagged, knife edge ridge created where glaciers have eroded back into the ridge,* or by a **col,** *a gap or pass in a mountain crest where the headwalls of two cirques intersect.*

because tributary ice streams can merge with the glacier in the main valley well above its base. Unlike a river, in which water from tributaries quickly mixes with the main stream, tributary glaciers retain their identity and flow beside, or are inset within, the main valley glacier. Both the main valley and the tributary valleys are shaped by erosion at the sides as well as the base of the glacier. The long profile of a glaciated valley floor also may be marked by steplike irregularities and shallow basins. These are related to the spacing of joints in the rock, which influences the ease of glacial plucking, or to changes in rock type along the valley. Finally, the valley is likely to head in a cirque or group of cirques.

Some valleys are glaciated from head to mouth. Others are glaciated only in their headward parts; downstream their form has been shaped largely by mass-wasting processes and stream erosion. The resulting contrasts in landscape form make it relatively easy to distinguish glaciated from nonglaciated valley segments on topographic maps, on aerial photographs, or on images taken by satellites orbiting 100 km or more above the Earth's surface (Fig. 13.21).

FIGURE 13.21 Vertical satellite image of high mountains bordering the Indus River in northern Pakistan. Glaciated valley segments (G) with typical U-shaped cross profile and cirques (C) at their head contrast with nonglaciated V-shaped valleys (V) downstream.

Fjords. Long, deep fjords are common features along the mountainous west-facing coasts of Norway, Alaska, British Columbia, Chile, and New Zealand (Fig. 13.22). The form and depth of many fjords imply glacial erosion of 300 m or more. Sognefjord in Norway reaches a depth of 1300 m, yet near its seaward end it shallows to only about 150 m. The characteristic overdeepening of fjord floors far below sea level is due mainly to deep glacial erosion resulting from thick and fast-moving ice. The frequent linear geometric arrangement of fjord systems suggests that geologic structures in the bedrock exert a strong control on glacial erosion. Unlike streams, which cease to erode when they reach the sea, glaciers can erode their beds far below sea level. For example, a large coastal glacier 300 m thick, and having a specific gravity close to 0.9, can continue to erode its bed until it is in water about 270 m deep, whereupon it begins to float and basal erosion ceases.

Landforms Associated with Ice Caps and Ice Sheets

Abrasional Features. Landscapes glaciated by ice sheets display the small-scale erosional features typical of most glaciated terrain. Striations, especially, have been helpful to geologists in reconstructing the flow lines of long-vanished northern ice sheets. Erosional features indicative of basal sliding are common in the central zones of former ice sheets where the ice was between 3 and 4 km thick, indicating that basal ice was at the pressure melting point. However, in a peripheral zone, evidence of glacial erosion is often less obvious, leading to the conclusion that thinner ice there may have been very cold and largely frozen to its bed.

Where ice sheets overwhelmed mountainous terrain, as in the cordillera of northwestern North America, the upper limit of glaciation can frequently be seen where smooth, abraded mountain slopes pass abruptly upward into rugged frost-shattered peaks and mountain crests. In such terrain, some divides between adjacent drainages are broad and smooth and show evidence of glacial abrasion and plucking where they were overridden by ice (Fig. 13.23).

Streamlined Forms. In many areas that lie near the outer edge of former ice sheets, the land surface has been molded into smooth, nearly parallel ridges that range up to many kilometers in length. These forms resemble the streamlined bodies of supersonic airplanes and offer minimum resis-

tance to glacier ice flowing over and around them. The best-known variety of streamline forms is the **drumlin,** *a streamlined hill consisting of glacially deposited sediment and elongated parallel with the direction of ice flow* (Fig. 13.24). Not all such landforms consist

FIGURE 13.22 Icebergs break away from the calving front of a fjord glacier in southwest Greenland. This fjord was filled by an arm of the sea as the glacier retreated back rapidly by frontal calving.

FIGURE 13.23 High pass in northern Cascade Range, Washington that was overridden by ice sheet flowing south from Canada during the last glacial age. Ice over the adjacent valleys was at least 1500 m thick. Sharp-crested summits to the left and in the distance stood above the ice sheet as **nuntaks,** *areas of ice-free land rising above the surface of a glacier.*

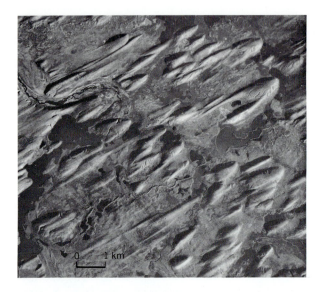

FIGURE 13.24 A field of drumlins near Snare Lake, Saskatchewan, Canada. The higher, wider, and blunter side of each drumlin points toward the direction from which the ice flowed (from the northeast). Some of the hills have tapered, streamlined tails.

of contemporaneous ice-laid sediment. Some are cored by preexisting glacial deposits, and these too are drumlins. Others are shaped by glacial erosion of bedrock; even though they have the form of a drumlin they are not true drumlins, but *rock drumlins*. These landforms also owe their streamlined shape to molding by flowing ice, and their long axes lie parallel to the direction of flow of the glacier that produced them.

Lake Basins. The margins of ice sheets typically are lobate, a result of control of ice flow by subglacial topography. In the north-central United States, for example, the irregular southern margin of the last continental ice sheet marked the terminus of lobes that flowed along former drainage courses and deepened them into basins that now contain the Great Lakes. These lakes and other large ones in central Canada lie along the boundary between resistant metamorphic rocks and more erodible sedimentary rocks that overlie them. As the spreading ice sheet rose against and crossed the low northward-facing sedimentary escarpment, glacial erosion created an arcuate series of basins that filled with meltwater as the ice receded. These basins still retain large lakes today. Analogous lake basins are found inside glacial limits within and near many mountain ranges. The famous Alpine lakes of Switzerland and northern

Italy, like those of the southern Andes and New Zealand, are glacially deepened basins surrounded by glacial deposits.

Glacial Transport

A glacier differs from a stream in the way in which it carries its load of rock particles. Part of its load can be carried at its sides and even on its surface. A glacier can carry much larger pieces of rock and it can transport large and small pieces side by side without segregating them according to size and density into a bed load and a suspended load. Because of these differences, deposits made directly from a glacier are neither sorted nor stratified.

The load of a glacier typically is concentrated at its base and sides because these are the areas where glacier and bedrock are in contact and where abrasion and plucking are effective. Much of the rock material on the surface of valley glaciers arrived there by rockfalls from adjacent cliffs.

A good deal of the load in the base of a glacier consists of very fine sand and silt. The particles are mostly fresh and unweathered. As revealed under the microscope, they have jagged, angular surfaces (Fig. 5.20*a*). These *fine rock particles, the products of glacial crushing and grinding,* are referred to as *rock flour.* They differ from the more rounded and chemically weathered particles found in sediments of nonglaciated areas.

Much of the transfer of sediment from a glacier to the ground occurs by release of particles as the surrounding ice melts. Therefore, glacial deposition takes place below the equilibrium line where melting is dominant.

Glacial Deposits

Drift, Till, and Stratified Drift

Sediment deposited directly by glaciers or indirectly by meltwater in streams, in lakes, and in the sea together constitute *glacial drift,* or simply *drift.* The term *drift* dates from the early nineteenth century when it was vaguely conjectured that all such deposits had been "drifted" to their resting places during the biblical flood of Noah or by some other ancient body of water. Included within drift are several kinds of sediment that form a gradational series ranging from nonsorted to sorted types.

At one end of the range is *till,* which is *nonsorted drift deposited directly from ice.* The name was given by Scottish farmers long before the origin of the sediment was understood. The constituent rock particles in till are not sorted according to size or density, but lie just as they were released from the ice (Fig. 13.25*a*).

Probably most till is plastered onto the ground, bit by bit, from the base of flowing ice in the ablation area (Fig. 13.26). We say "probably" because as yet the process has not been directly observed at the glacier bed. Most tills are a random mixture of rock fragments consisting of a *matrix* of fine-

FIGURE 13.25 Glacial deposits. (*a*) Bouldery till in an end moraine of an alpine glacier in the eastern Cascade Range showing wide range in grain size and lack of sorting. (*b*) Striated boulder embedded in till on a land surface that was exposed by recent retreat of Ruitor Glacier in the Italian Alps.

grained sediment surrounding larger stones of various sizes. The till matrix consists largely of sand and silt particles derived by abrasion of the glacier bed and from reworking of preexisting fine-grained sediments. Pebbles and larger rock fragments in till often have faceted surfaces, the result of abrasion, and some are striated (Fig. 13.25*b*). Both the stones and the coarser matrix grains in till tend to have their longest axis aligned in the direction of ice flow.

Glacial marine drift, which closely resembles till, is *sediment deposited in the sea from floating ice shelves or bergs.* As an iceberg or the base of an ice shelf melts, the contained sediment is released and falls to the seafloor. Unlike till, the elongate rock particles in glacial marine drift are randomly oriented, for they settle through water rather than being deposited by flowing ice. Furthermore, such drift may contain the remains of marine organisms, still in growth position.

Where many icebergs are generated and the resulting sedimentation rate is high, a continuous layer of glacial marine drift will accumulate. However, where bergs are less plentiful and fine-grained sediment is settling out continuously from suspension, stones dropped from passing bergs will plunge into marine sediments on the seafloor, deforming them. Such *dropstones* are diagnostic features of glacial marine environments, as well as ice-marginal lake environments where bergs are produced from a calving glacier front.

A number of other nonglacial sediments resemble till and are easily confused with it. This has sometimes led to misinterpretations about the former extent and age of glaciations. For example, colluvium, mudflow sediments, landslide deposits, and some very poorly sorted alluvium at first glance look very much like till. However, the presence of faceted and striated stones, rock particles of distant origin, and an underlying grooved or striated rock pavement help to identify till.

Stratified drift, by contrast, is *drift that is both sorted and stratified.* It is not deposited by glacier ice, but by meltwater emanating from the ice. Stratified drift ranges from coarse, very poorly sorted sandy gravels that are transitional into till, to fine-grained, well-sorted silts and clays deposited in quiet-water environments.

Deposits of Active Ice

Moraines. In actively flowing glaciers, sediment transported by the ice is plastered onto the ground as till or is released by melting at the glacier margin where it either accumulates as a moraine or is re-worked by meltwater and transported beyond the terminus. *Widespread drift with a relatively smooth surface topography consisting of gently undulating knolls and shallow, closed depressions* is known as **ground moraine.** Most commonly it consists of till that blankets the landscape and that may reach a thickness of 10 m or more. **End moraines,** on the other hand, are *ridge-like accumulations of drift deposited along the margin of a glacier.* The terminal part is a *terminal moraine,* while the lateral part is a *lateral moraine,* but both are normally part of a single continuous landform (Fig. 13.27). End moraines can

form by bulldozing action of the glacier front, by slumping of loose surface debris off the glacier margin as the ice melts, by repeated plastering of drift from basal ice onto the ground, or by streams of meltwater that build up deposits of stratified drift at the glacier margin. They range in height from a few to hundreds of meters. The great height and thickness of some lateral moraines are due to repeated accretion of drift upon them during successive ice advances. Buried land surfaces, marked by soils and organic remains, can sometimes be seen exposed in lateral moraines, and provide evidence of their composite character.

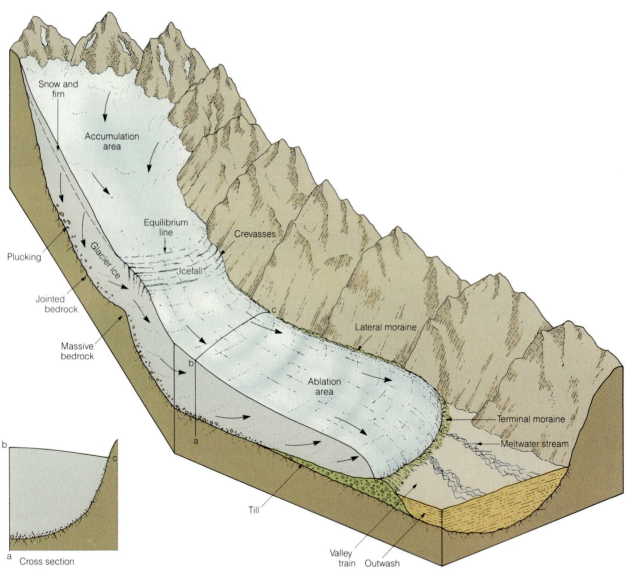

FIGURE 13.26 Main features of a valley glacier and its deposits. The glacier has been cut away along its center line; only half is shown. Crevasses form where the glacier passes over a steeper slope at its bed. Length of arrows are proportional to velocity of flow.

FIGURE 13.27 Sharp, arcuate end moraines mark the successive position of the margin of an ice-age valley glacier in the eastern Sierra Nevada, California. The glacier originated in cirque basins beneath the distant high peaks and flowed down steep tributary valleys. Several tributary ice streams joined to form a single ice tongue that terminated on gentle ground beyond the mouth of the valley.

Erratics and Boulder Trains. Some of the boulders and smaller rock fragments in till are the same kind of rock as the bedrock on which the till was deposited, but many are of other kinds, having been brought from greater distances. *A glacially deposited rock or rock fragment whose composition differs from that of the bedrock beneath it* is an **erratic** (Latin for "wanderer"). The presence of foreign stones on the land surface was one of the earliest recognized proofs of former glaciation. Many erratics form part of a body of drift, but others lie isolated on the ground. Some erratics are enormous, and have estimated weights of thousands of tons (Fig. 13.28). Such boulders are far larger than can be transported by an ordinary stream of water.

In areas that have been glaciated by ice sheets, erratics derived from some distinctive bedrock source are often so plentiful and easily identified that their distribution can be plotted on a map. The resulting plot may have a fanlike shape, spreading out from the area of outcrop and reflecting the diverging pattern of ice flow. Such *a group of erratics which are spread out fanwise* is a **boulder train,** so named in the nineteenth century when rock particles of all sizes were called boulders. In Canada, boulder trains have been used to prospect for mineral deposits in regions where the bedrock is obscured by glacial drift.

Outwash. *Stratified drift deposited by streams of meltwater as they flow away from the glacier margin is called* **outwash** ("washed out" beyond the ice). Such streams typically have a braided pattern because of the large sediment load they are moving, and they have sedimentary characteristics like those of many nonglacial braided streams. If the

FIGURE 13.28 Large erratic boulder of granite embedded in an end moraine of the last glaciation on Tierra del Fuego in southernmost Chile. The local bedrock is sedimentary. The nearest possible source area for the boulder lies in the high Cordillera Darwin to the south, on the opposite side of a deep fjord system.

streams are free to swing back and forth widely beyond the glacier terminus, they can build *a body of outwash that forms a broad plain,* an **outwash plain,** like that lying beyond the front of Vatnajökul, a large ice cap on Iceland (Fig. 13.3). By contrast, meltwater streams confined by valley walls will build a **valley train,** *a body of outwash that partly fills a valley.*

When a glacier retreats, the sediment load supplied to the meltwater stream is greatly reduced and the underloaded stream is therefore able to cut down into its valley train to produce *outwash terraces.* Series of terraces are common in valleys that have experienced repeated glaciations (Fig. 13.29). Generally, each major terrace can be traced upstream to an end moraine or former ice limit.

Deposits of Stagnant Ice

When rapid ablation greatly reduces ice thickness in the terminal zone of a large glacier, movement may virtually cease. Sediment carried by meltwater flowing over or beside the nearly motionless stagnant ice is deposited as stratified drift, which slumps and collapses as the supporting ice slowly melts away. This *stratified sediment deposited in contact with supporting ice* is called **ice-contact stratified drift.** It is recognized by abrupt changes in grain size, by distorted, offset, and irregular stratification, and by extremely uneven surface form. Bodies of ice-contact stratified drift are classified according to their shape (Fig. 13.30). Some extensive end-moraine systems of former ice sheets consist of broad belts of **kettle-and-kame** **topography,** *an extremely uneven terrain resulting from wastage of debris-mantled stagnant ice and underlain by ice-contact stratified drift* (Fig. 13.30). Many of the multitude of lakes that dot the land surface in the states of Michigan and Wisconsin occupy kettles in such terrain.

Reconstructing Former Glaciers

The landforms and sediments left by a former mountain glacier can be used to reconstruct its geometry. From the surface form, the altitude of the equilibrium line can also be estimated, thereby providing information about the climatic environment in which the glacier existed. A terminal moraine or the head of a valley train marks the downvalley limit of the glacier, while lateral moraines, erratics, and the upslope change from glacially sculptured rock surfaces to unglaciated frost-shattered slopes above permit the level of its upper surface to be approximated. The headward portions of the glacier are delimited by cirques. Measured altitudes of such features can be used to reconstruct the glacier's topography (Fig. 13.31). If one then assumes that the accumulation area occupied two thirds of the total area of the glacier, as is the case for most present glaciers in a balanced state, then the altitude of the equilibrium line can be calculated. Using such techniques, it has been possible to map the distribution of former glaciers in many parts of the world and determine the regional pattern of glacier equilibrium lines during and since the last glaciation.

FIGURE 13.29 Outwash terraces rise above meandering Cave stream on South Island, New Zealand.

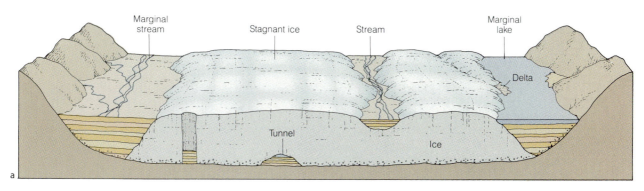

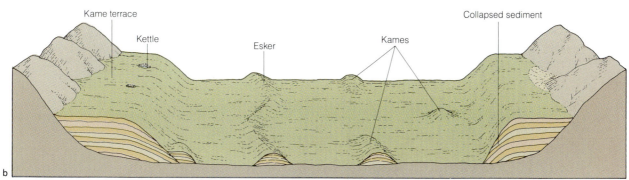

FIGURE 13.30 Origin of ice-contact stratified draft associated with stagnant-ice terrain. (*a*) Nearly motionless melting ice furnished temporary retaining walls for bodies of sediment deposited chiefly by meltwater streams and in meltwater lakes. (*b*) As ice melts, bodies of sediment slump, creating a kettle-and-kame topography. Resulting landforms include a **kame,** *a body of ice-contact stratified drift in the form of a knoll or hummock;* a **kame terrace,** *a terracelike body of ice-contact stratified drift along the side of a valley;* a **kettle,** *a basin in glacial drift, created by melting-out of a mass of underlying ice;* and an **esker,** *a long, narrow ridge, commonly sinuous, composed of stratified drift.*

Reconstruction of continental ice sheets is a more difficult exercise, for although end moraines that mark ice-sheet limits can be mapped on land, significant portions of former ice sheets terminated beyond present coastlines on the submerged continental shelves where detailed information is difficult to obtain. Furthermore, the large ice sheets covered virtually all the terrain within their margins, and little land projected above the ice surface. Consequently, within most of the region that was covered by ice there is no way to measure former ice thickness directly from geologic features. Instead, ice thickness and glacier profiles have been calculated numerically on the basis of ice-flow properties derived from laboratory and field measurements of existing glaciers; the results, however, are only approximations. Therefore, we do not yet have an accurate measure of the area,

thickness, topography, or volume of former ice sheets. Nevertheless, a good deal is known about directions of ice flow, for much of the land surface over which the glaciers moved is abundantly marked with striations, grooves, drumlins, and other aligned ice-flow indicators.

PERIGLACIAL LANDSCAPES

Land areas beyond the limit of glaciers where low temperature and frost action are important factors in determining landscape characteristics are spoken of as ***periglacial zones.*** Modern periglacial conditions are most widespread in polar and subpolar regions and at high altitudes. However, during past glacial ages such zones extended far beyond their present limits and into now-temperate latitudes. Geolo-

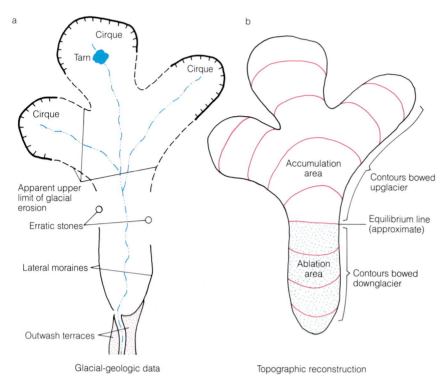

a

b

Glacial-geologic data

Topographic reconstruction

FIGURE 13.31 Method of reconstruction of a former glacier from geologic data. (*a*) Glacier-marginal features (end moraines, upper limits of erratics and ice-eroded bedrock, cirque headwalls) are used to determine the areal extent of the glacier. (*b*) Topography of the glacier is reconstructed using altitude of ice-marginal features as a basis for drawing contours. The equilibrium line is assumed to lie at the contour above which two-thirds of the total area of the glacier lies. Contours above the equilibrium line are bowed up-glacier; those below are bowed down-glacier. Contours are drawn perpendicular to the general trend of ice-flow indicators, such as striations.

gists have discovered and mapped the distribution of relict periglacial features throughout much of central Europe and across northern United States in zones that were marginal to the great Pleistocene ice sheets. Active and relict periglacial features have also been found widely in many other areas, including Alaska, Canada, Siberia, northern and western China, Patagonia, and the ice-free areas of Antarctica.

Today periglacial conditions are found over more than 25 percent of the Earth's land areas. Despite their vast areal distribution, periglacial landscapes remained little known until recently due to their remoteness and their low number of inhabitants. However, many of these regions have been receiving increasing attention because of their important energy resources and mineral wealth. Long pipeline systems in Alaska, northern Canada, and Siberia now carry newly discovered oil and natural gas across periglacial landscapes.

New settlements are springing up in areas underlain by thick frozen ground. This has created an urgent need for information about periglacial processes and environmental conditions that has spurred field and laboratory research on cold-climate phenomena.

Permafrost

A common feature of periglacial regions is perennially frozen ground, generally known as **permafrost.** It is defined as *sediment, soil, or even bedrock that remains continually at a temperature below 0° C for an extended time* (from two years to tens of thousands of years). Such conditions exist mainly in the circumpolar zones of each hemisphere, as well as at high altitudes. The largest areas of permafrost occur in northern North America, northern Asia, and in the high, cold Qinghai-Xizang Plateau of western China (Fig. 13.32). It also has been found on many high mountain ranges, even including some lofty summits in tropical and subtropical latitudes. The southern limit of continuous permafrost in the Northern Hemisphere (as opposed to discontinuous patches) generally lies where the annual air temperature is between -5 and $-10°$ C.

Most permafrost is believed to have originated during the last glacial age or earlier glacial ages. Remains of wooly mammoth and other extinct Pleistocene animals, which have been found well preserved in frozen ground, indicate that permafrost existed at the time of their death.

The depth to which permafrost reaches depends

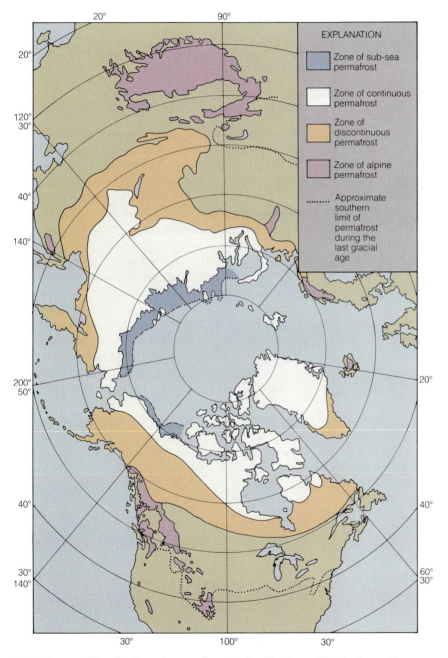

FIGURE 13.32 Distribution of permafrost in the Northern Hemisphere. Continuous permafrost lies mainly north of the 60th parallel and is most widespread in Siberia and arctic Canada. Extensive alpine permafrost underlies the high, cold plateau region of central Asia. Smaller isolated bodies occur in high mountains of the western United States and Canada. (After Péwé, 1983.)

FIGURE 13.33 Thaw lake in thermokarst terrain near Yakutsk in northern Siberia where the ground is underlain by thick permafrost.

not only on the average air temperature but also on the rate at which heat flows upward from the Earth's interior and on how long the ground has remained continuously frozen. The maximum reported depth of permafrost is about 1400–1500 m in Siberia. Thicknesses of about 1000 m have been reported in the Canadian Arctic, and of at least 600 m in northern Alaska. These areas of very thick permafrost all occur in high latitudes outside the areas of former ice sheets. The ice sheets would have insulated the ground surface and, where thick enough, actually caused ground temperatures beneath them to rise to the pressure melting point. On the other hand, nonglaciated open ground unprotected from subfreezing air temperatures could have become frozen to great depths during prolonged cold periods.

A thin surface layer of ground that thaws in summer and refreezes in winter is known as the **active layer.** Within it the thawed ground in summer tends to become very unstable and subject to movement. The permafrost beneath, however, is capable of supporting large loads without being deformed. Many of the landscape features we associate with periglacial regions reflect movement of rock particles within the active layer during freezing and thawing.

Permafrost presents unique problems for people living on it. If a building is constructed directly on the surface, the warm temperatures developed inside when the house is heated are likely to thaw the underlying permafrost making the ground unstable. Arctic inhabitants learned long ago that they must place the floors of their buildings on pilings above the surface so that cold air can circulate freely beneath, thereby keeping the ground frozen.

Wherever tundra vegetation that forms a continuous cover over some permafrost landscapes is ruptured, melting can begin. In nature, this may lead to the collapse of the ground and formation of *thaw lakes.* The resulting topography resembles karst terrain (Chapter 10) and is known as *thermokarst* (Fig. 13.33). Thawing can also commence through human activity, and the results can be environmentally disastrous. Large wheeled or tracked vehicles crossing arctic tundra can quickly rupture it. The water-filled linear depressions that result from thawing can remain as features of the landscape for many decades.

Periglacial Landforms

The seasonal freezing and thawing action in the active layer disrupts the mineral soil and promotes differential sorting of surface sediments into varied surface patterns. *More-or-less symmetrical patterned forms due to frost action* are collectively known as **patterned ground.** They include such features as circles, polygons, nets, and stripes, which are descriptive terms for the resulting patterns (Fig. 13.34). While such features are common in ground underlain by permafrost, many also occur in areas that only experience seasonal frost.

Ice-wedge polygons, which are *large polygonal features that form by contraction and cracking of frozen ground,* are diagnostic of continuous permafrost. Water that enters the cracks expands as it freezes so that after many repeated cycles a typically wedge-shaped body of ice is produced (Fig. 13.35a). Some polygons reach diameters of 150 m (Fig. 13.35b) and may extend to depths of several meters. If the climate changes, causing the ice to thaw, sediment may fill the resulting cavity.

FIGURE 13.34 Sorted circles, about 4 m in diameter, form a striking pattern on the barren landscape at Brogerhalvoya in western Spitsbergen.

Identification of such geologic features beyond the limits of former ice sheets in Europe and North America provides evidence of glacial-age permafrost in areas that now lie 1000 km or more from the nearest contemporary permafrost (Fig. 13.35*c*).

Other typical periglacial landforms, characteristic of at least discontinuous permafrost, include **pingos** which are *large, generally conical ice-cored mounds that commonly reach heights of 30–50 m and are formed from freezing of water within the permafrost* (Fig. 13.36*a*). Pingos are common features in coastal regions of arctic Alaska and Canada, and from a distance resemble small volcanoes. The crest is sometimes ruptured and collapsed due to exposure and melting of the lenslike ice core. Remains of collapsed pingos also have been found in former periglacial zones beyond the limits of Pleistocene ice sheets.

A **rock glacier,** another characteristic periglacial landform, is *a glacierlike tongue or lobe of angular rock debris containing interstitial ice or buried glacier ice that moves downslope in a manner similar to glaciers* (Fig. 13.36*b*). Active rock glaciers may reach thicknesses of 50 m or more and advance at rates of up to about 5 m/yr. They are especially common in high interior mountain ranges, such as the Swiss Alps, the Rocky Mountains, and the northern Andes of Argentina.

Gelifluction, *the slow downslope movement of saturated sediment associated with frost action in cold-climate regions,* produces lobes and sheets of debris that creep slowly down hillslopes (Fig. 13.36*c*). Although measured rates of movement are low, generally less than 10 cm/yr, gelifluction is so widespread on arctic landscapes that it constitutes a highly important agent of mass transport.

THE GLACIAL AGES

History of the Concept

As early as 1821 European scientists began to recognize features characteristic of glaciation in places far from any existing glaciers. They drew the then remarkable conclusion that glaciers must once have covered wide regions. Consciously or unconsciously, they were applying the principle of uniformity. The concept of a glacial age with widespread effects was first proposed in 1837 by Louis Agassiz, a Swiss scientist who achieved considerable fame through his hypothesis. Although at first many regarded the idea as outrageous, gradually, through the work of many geologists, the

a

b

c

FIGURE 13.35 Modern and ancient ice-wedge polygons. (*a*) Active ice wedge exposed in bank along the Aldan River in northern Siberia. Growth of the wedge has caused deformation of frozen river silts as its thickness has increased. (*b*) Ice-wedges form distinctive polygonal patterned ground enclosing shallow lakes near the shore of the Arctic Ocean. (*c*) Polygonal ice-wedge casts exposed during road construction near Rawlings, Wyoming.

concept gained widespread acceptance. Today we have a basic understanding of the nature of the glacial ages, although important questions remain unanswered.

a

b

c

FIGURE 13.36 Common periglacial features. (*a*) Ibyuk Pingo rises steeply above surface of Mackenzie Delta in northwestern Canada. (*b*) Active rock glacier, fed by rockfall from steep slopes above, advances across valley margin in the Alaska Range. (*c*) Gelifluction sheets creeping slowly downslope on mountain (to right) in Italian Alps have overridden moraines on valley floor.

Extent of Former Glaciers

During the second half of the nineteenth century geologists mapped the distribution of glacial drift and other characteristic features in order to determine the extent of former glaciers. Their studies focused on the regions of North America and Europe that were most accessible and which became the "classic" regions of glacial-geologic studies. Only after the middle of the present century were extensive surveys of more remote lands undertaken. Despite uncertainty regarding the local extent of ice in many areas, the global distribution of former glaciers is reasonably well known (Fig. 13.37). On a global scale, the areas of former glaciation add up to an impressive total of more than 44 million km², or about 29 percent of the entire land area of the Earth. Today, for comparison, only about 10 percent of the world's land area is covered with glacier ice. Of this area, 84 percent lies in the Antarctic.

Directions of Flow

In most mountainous regions, former glaciers flowed down existing valleys which determined their direction of movement. The larger ice tongues terminated in adjacent lowlands or spread out to form lobate ice fronts similar in form to the modern Bering and Malaspina glaciers of Alaska. In glaciated continental interiors, streamlined landforms and small-scale abrasion features show that the great ice sheets spread outward from their source regions in a radial pattern. Flow directions are determined also by tracing erratics of conspicuous rock types to their places of origin (Fig. 5.30). For example, native copper found as far south as Missouri has been traced to rock outcrops on the south shore of Lake Superior.

Lowering of Sea Level

Whenever large glaciers formed on the land, the moisture needed to produce and sustain them was derived primarily from the oceans. As a result, sea level was lowered in proportion to the volume of ice on land. During each glacial age, world sea level probably fell by 100 m or more (Fig. 13.38), thereby causing large expanses of the shallow continental shelves to emerge as dry land. At such times the Atlantic coast of the United States between New York and Florida lay about 150 km east of its present position. Based on fossils dredged from the seafloor on the continental shelf, the emergent coastal plain was forested with spruce and pine and its animal population included mammoths, mastodons, and other now-extinct mammals. At the same time, lowering of sea level joined Britain to France where the English Channel now lies, and North America and Asia formed a continuous landmass across what is now the Bering Strait. These and other land connections allowed plants and animals, including humans, to pass freely between land areas that now are separated by ocean waters.

Depression of the Crust

The weight of the massive ice sheets caused the crust of the Earth to subside beneath them, an effect described further in Chapter 16. The contrast between the density of crustal rocks (about 2.7 g/cm³) and glacier ice (about 0.9 g/cm³) means that an ice sheet 3 km thick might cause the crust to subside by as much as 1 km. The Hudson Bay region of central Canada, which formerly lay beneath the central part of an ice sheet at least 3 km thick, is still rising in response to removal of the ice load (Fig. 13.39). As a result, Hudson Bay is becoming progressively shallower and its area is diminishing in size as the water is slowly decanted. A similar change is affecting the floor of the Baltic Sea and surrounding lands. This region lies near the center of the former Scandinavian Ice Sheet and continues to rise as the crust adjusts to deglaciation.

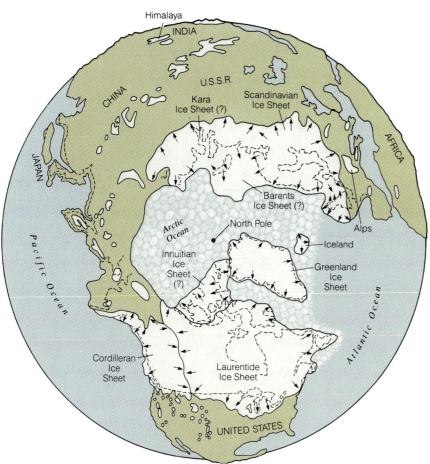

FIGURE 13.37 Areas of the Northern Hemisphere that were covered by glaciers during the last glacial age. Arrows show general directions of ice flow. Coastlines are shown as they were at that time, when world sea level was about 100 m lower than present. Sea ice is shown covering the Arctic Ocean and extending south into the North Atlantic. Some scientists postulate that thick ice shelves, rather than sea ice, covered these portions of the ocean. The extent of former glacier ice over the Barents and Kara seas, as well as in parts of northern North America, is controversial.

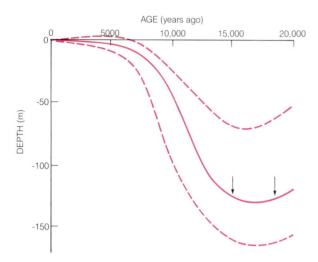

FIGURE 13.38 Curve showing estimates of the mean position of world sea level over the last 20,000 years (solid line). Estimates for maximum lowering range from about 75 to 165 m, reflecting considerable uncertainty. The time of lowest sea level (arrows) is also uncertain. The upper part of the curve, spanning the last 10,000 years, is better known. Dashed lines mark the limits within which most data lie.

FIGURE 13.39 Emerged beaches on the southwest shore of Hudson Bay, Canada. They were formed shortly after the continental ice sheet disappeared from this region and were progressively uplifted as the land rose isostatically in postglacial time. The highest beaches in this region lie 250–300 m above the present level of the bay (upper left corner).

Repeated Glaciation

Most of the glacial drift seen at the land surface is fresh and little weathered. From that observation it was realized very early that the glaciation must be geologically recent. However, in the middle nineteenth century geologists began to find exposures showing that the layer of comparatively fresh surface drift overlies another layer whose upper part is chemically weathered. This led to the realization that there had been two glaciations separated by a long enough interval of time to produce surface weathering to a depth of a meter or more. Before the beginning of the twentieth century, evidence had been found of not merely two but three or more glaciations, during each of which ice covered approximately the same geographic areas.

Radiometric dating indicates that the latest glaciation culminated between about 20,000 and 14,000 years ago. Older drifts are not as well dated, but the earliest evidence of middle-latitude ice-sheet glaciations reaches back several million years. The time that embraced the majority of these glaciations is called the Pleistocene Epoch (Table 7.3). Although evidence now shows that some high-latitude glaciers developed even earlier during the Cenozoic Period, the Pleistocene was an interval when glaciers were especially widespread on the Earth (Chapter 20).

Evidence of still older glacial ages appears repeatedly in the rock record throughout at least the last half of Earth history. The evidence includes occurrences of *tillite, till that has been converted to solid rock,* in strata of several different ages (Fig. 5.12). Such evidence tells us that the Earth's climate has fluctuated repeatedly, causing glaciers to form and later disappear and creating important environmental changes throughout the world.

SUMMARY

1. Glaciers consist of ice which has been transformed from snow by compaction, recrystallization, and flow. They form part of the cryosphere, the part of the planet where water exists primarily in the frozen state.

2. The major types of glaciers, based on their ge-

ometry, are cirque glaciers, valley glaciers, ice caps, ice sheets, and ice shelves.

3. Ice in temperate glaciers is at the pressure melting point and water exists at the glacier bed. Polar glaciers consist of ice that is below the pressure melting point and are frozen to their bed. In subpolar glaciers, a thin surface zone reaches the melting point in summer, but the ice beneath is below freezing.

4. Glaciers depend for their survival on low temperature and adequate precipitation. They bear a close relationship to the snowline which is low in polar regions and rises to high altitudes in the tropics.

5. The mass balance of a glacier is measured in terms of accumulation and ablation. The equilibrium line separates the accumulation area from the ablation area and marks the level on the glacier where net gain is balanced by net loss.

6. Glaciers flow under their own weight. Their surface zone is brittle; however, at greater depth internal flow occurs.

7. The motion of temperate glaciers includes both internal flow and sliding along the bed. In polar glaciers, which are frozen to their bed, motion is much slower and involves only internal flow.

8. Surges involve extremely rapid flow, probably related to excess water at the glacier bed. Frontal calving can lead to rapid recession of glacier margins that recede into deep water.

9. Glaciers erode rock by quarrying and abrasion. They transport the waste and deposit it as drift.

10. Mountain glaciers convert stream valleys into U-shaped troughs with hanging tributaries and with cirques at their heads. Mountain areas that project above glaciers are shaped by frost action into angular landforms that contrast with glacially smoothed terrain downslope.

11. Fjords are excavated far below sea level by thick, fast-flowing ice streams in coastal regions.

12. Flow directions of ice sheets are inferred from striations, grooves, and drumlins.

13. The load, carried chiefly in the base and sides of a glacier, includes rock fragments of all sizes, from fine rock flour to large boulders.

14. Till is deposited directly by glaciers, while glacial marine drift is deposited on the seafloor from floating glacier ice. Stratified drift is deposited by meltwater. It includes outwash deposited as outwash plains or valley trains beyond the ice margin, and ice-contact stratified drift deposited upon or against stagnant ice.

15. Ground moraine is built up beneath a glacier, whereas end moraines (both terminal and lateral) form at the glacier margins.

16. Permafrost is widespread in periglacial regions. Where continuous, it may reach a thickness of 1000 m or more, but toward its limit it becomes patchy and discontinuous. In summer, a thin active layer at the surface thaws and may become unstable.

17. Patterned ground results from repeated freezing and thawing of the ground. Distinctive relict periglacial features permit mapping the extent of former permafrost conditions.

18. During glacial ages huge ice sheets repeatedly covered northern North America and Europe, eroding bedrock and spreading drift over the outer parts of the glaciated regions. As the ice sheets grew, world sea level fell and the crustal rocks beneath them subsided due to the added weight.

SELECTED REFERENCES

Agassiz, L., 1967, Studies on glaciers (Neuchatel, 1840), Translated and edited by A. V. Carozzi: New York, Hafner.

Denton, G. H., and Hughes, T. J., 1981, The last great ice sheets: New York, John Wiley & Sons.

Flint, R. F., 1971, Glacial and Quaternary geology: New York, John Wiley & Sons.

LaChapelle, E. R., and Post, A. S., 1971, Glacier ice: Seattle, University of Washington Press.

Prest, V. K., 1983, Canada's heritage of glacial features: Geological Survey of Canada Misc. Rept. 28, 119 p.

Sugden, D. E., and John, S., 1976, Glaciers and landscape: London, Edward Arnold.

Swiss National Tourist Office, 1981, Switzerland and her glaciers: Berne, Kummerly and Frey.

Washburn, A. L., 1980, Geocryology: London, Edward Arnold.

CHAPTER 14

The Oceans and Their Margins

Bora Bora, the eroded remnant of a large mid-Pacific shield volcano in the Society Islands, is encircled by a barrier reef that has grown progressively upward as the island slowly subsides.

THE WORLD OCEAN

If, by some mysterious means, we could briefly remove all the water from the oceans and then view the dry Earth from space, we would see that continents, standing high above the seafloor, end where their bordering continental slopes meet the deep ocean floor. If we could then examine the rock in such a place, we would find that the foot of the continental slope is the place where continental crust meets oceanic crust. Exploring further, we would observe that the world ocean actually occupies several great basins, each floored with oceanic crust and rimmed with continental crust.

Within the ocean, beyond the continental slope, lies the remote world of the deep-ocean floor. With devices for sounding the sea bottom and for sampling its sediment, teams of oceanographers and marine geologists have explored the ocean floor and greatly expanded our geologic knowledge of submarine regions. Scuba-diving geologists have visited, photographed, and mapped areas of seafloor at depths as great as 70 m, and observers in specially designed submersible craft have visited the deepest-known places in the ocean.

Because of this intensive research, involving many nations, the oceans are gradually giving up their secrets. The romanticist in each of us can not help regretting that beliefs and legends built up through more than 3000 years of human history have vanished: the singing mermaids, the strange and threatening gods of the sea, the fabled cities and castles believed to have sunk into watery deeps, the monsters of seafarers' tales. These and other poetic visions have faded away as scientific knowledge has steadily increased and fables have been replaced with facts. The newly gained knowledge can help us understand this fragile environment that covers more than two thirds of the Earth's surface and is responsible for a large part of the Earth's biological heritage.

Dimensions

Seawater covers 71 percent of the Earth's surface. The remaining 29 percent that is land is not evenly distributed. About 40 percent of the Northern Hemisphere is land, while in the Southern Hemisphere only 20 percent is land. This uneven distribution of land and water plays an important part in determining the paths along which water circulates in the ocean.

The greatest ocean depth yet measured, about 11 km, lies near the island of Guam in the western Pacific. This is more than 2 km farther below sea level than Mount Everest rises above sea level. The average depth of the sea, however, is about 3.8 km, compared to an average height of the land of only 0.75 km.

Knowing the area of the sea and its average depth, we can calculate the present volume of seawater to be about 1.35 billion km^3. We say *present* volume because the volume fluctuates somewhat over thousands of years, with the growth and melting of continental glaciers (Chapters 13 and 20).

Composition

About 3.5 percent of average seawater, by weight, consists of dissolved salts (Fig. 14.1), enough to make the water undrinkable. It is enough also, if it were precipitated, to form a layer of solid salts about 56 m thick over the entire seafloor.

The measure of the sea's saltiness is termed **salinity.** We commonly express salinity in parts per thousand, rather than in percent (parts per hundred). Average seawater, therefore, has a salinity of 35 parts per thousand. The principal elements that contribute to the salinity of the sea are sodium and chlorine. When seawater is evaporated, more than three quarters of the dissolved matter is precipitated as common salt (NaCl). However, seawater contains most of the other natural elements as well, many of them in such low concentrations that they can be detected only by extremely sensitive analytical instruments. Nevertheless, as can be seen in Figure 14.1, more than 99.9 percent of the salinity is accounted for by only nine ions.

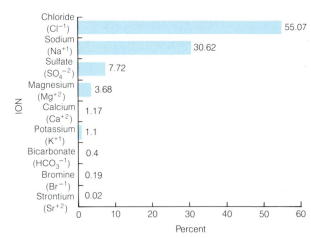

FIGURE 14.1 Graph showing weight percent of major constituents of sea water, listed as principle ions in solution.

Where do these ions come from? Each year streams carry 2.5 billion tons of dissolved substances to the sea. These are soluble substances leached from rock during chemical weathering, together with a small amount of soluble material carried up from the mantle and erupted in volcanic gases. The quantity of dissolved ions added by rivers over the millions of years of Earth history far exceeds all those now dissolved in the sea. Why, then, doesn't the sea have a higher salinity? The reason is that chemical substances are being removed at the same time they are being added. Some elements, such as silicon, calcium, and phosphorus, are withdrawn from seawater by aquatic plants and animals to build their shells or skeletons. Some, such as potassium and sodium, are absorbed and removed by clay particles and other minerals as they slowly settle to the seafloor. Still others, such as copper and lead, are precipitated as sulfide minerals in claystones and mudstones which are rich in organic matter. The net result of these and other processes of extraction, taken together with what is being added, is that the composition of seawater remains virtually unchanged.

Age and Origin

The Earth's oldest rocks include water-laid sedimentary strata similar to those we see being deposited today. Therefore, we are sure that as far back in history as we can see, which is 3.8 billion years, the Earth has had liquid water on its surface. Indeed, water has probably been present almost as long as the planet has existed as a solid body.

Where did the water come from? We can be sure that the ocean was created between 4.6 billion years ago, when the Earth formed, and 3.8 billion years ago when the oldest known rock was made, but we cannot be sure *how* it formed. Most probably, water condensed from steam produced during primordial volcanic eruptions. Because volcanic activity has characterized all of Earth history, the world ocean must have increased in volume through time. Water is, of course, slowly removed from the hydrosphere by being buried along with sediments as they are slowly transformed into sedimentary rocks. Water also can react with rocks and become incorporated into the chemical composition of their minerals. At the same time, water continues to be added to the hydrosphere through volcanic activity. It is difficult to say whether the subtraction and addition of water are exactly equal, but the difference is likely to be so small that

we can assume the hydrosphere has essentially reached a state of balance.

Has the ocean always been salty? The best evidence of the sea's saltiness in the past is the presence, in marine strata, of salts precipitated by evaporation of seawater. Strata containing marine evaporite deposits are common in young sedimentary basins, but are not known from rocks older than about a billion years. Possibly this is because ancient deposits of soluble evaporite minerals have been completely removed from the geologic record through dissolution by groundwater. Thus, although we can be certain that the sea has been salty for a billion years, perhaps it has been salty for much of its history.

TOPOGRAPHY BENEATH THE OCEANS

As we sail over or fly above the broad expanses of ocean, we receive few hints of the complex seafloor topography beneath. Yet the topography is just as varied, irregular, and fascinating as the familiar land topography we see around us. Present on the seafloor are long mountain chains, valleys and canyons, featureless plains, great escarpments, and steep-sided volcanoes. If we compare a topographic profile across the Atlantic Ocean basin with one across North America (Fig. 14.2), submarine terrain is seen to be just as rugged as that on land and the relief is even greater.

Continental Shelves

The continental shelves, which form the rims of the deep and vast ocean basins, are overlapped by the world ocean (Fig. 2.9). Their flat upper surfaces add about 10 percent to the areas of the continents, and the continental slopes beyond add at least half as much more. The shelves average about 60 km wide, but the local variation is large, from 1300 km off the Arctic coast of Siberia down to almost zero. Over great distances, the shelf off the Pacific coasts of North and South America averages only a few kilometers wide.

The seaward edges of the shelves are covered everywhere by at least 100 m of water, and in places by as much as 600 m. Thus, the edge of a shelf is defined not by depth of water but by a distinct steepening of slope.

Many decades ago, when little was known about the seafloor, it was supposed that the shelves were enormous shallow platforms cut by ocean waves. However today, with many more facts in hand, we realize that wave erosion has played only a minor part in creating the shelves. Mostly, they have been built up by deposition of sedimentary strata instead of being carved out of pre-existing rock.

The shelf and slope off the state of Maryland are good examples of how these features gradually evolve (Fig. 14.3). Subsurface information obtained from seismic exploration (Chapter 16) and drill-hole records show that the shelf consists of a wedge of sedimentary strata, as much as 13 km thick near the present shoreline, that thins both landward and seaward. Offshore these strata overlie a sloping floor of very old igneous and metamorphic rock. The kinds of sediment and fossils in the drill cuttings show that this great wedge consists mainly of sediment that was deposited in

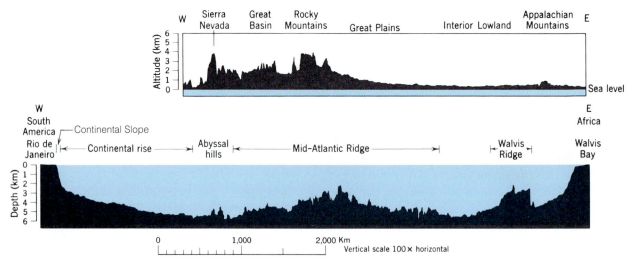

FIGURE 14.2 Topographic profile across the Atlantic Ocean basin shows that the topography is fully as rugged as along a profile across North America. On these comparative profiles, the vertical scale is exaggerated to emphasize the relief.

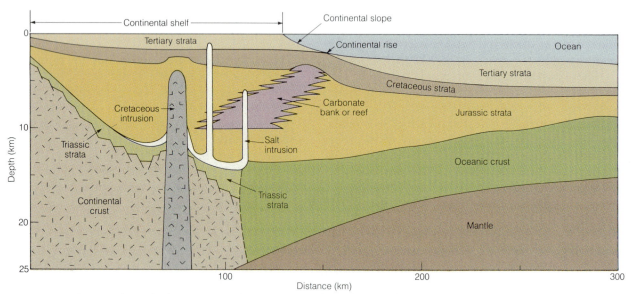

FIGURE 14.3 Section through the continental margin along the Baltimore Canyon Trough off the state of Maryland in the eastern United States. Subsurface features, interpreted from seismic records, include probable carbonate reefs and intrusive bodies of salt and volcanic rocks that penetrate thick piles of sedimentary strata. (*Source:* After J. A. Grow, 1981.)

shallow seawater. Because the oldest sedimentary layers now lie far below sea level, they must have sunk below the positions at which they were deposited. The layers are in various stages of transformation into rock, some of which are between 100 and 180 million years old. This and similar sections across eastern North America tell us that over long periods the margins of continents have been receiving sediment from rivers and have been gradually subsiding.

The age of the shelf off eastern North America cannot be much more than about 180 million years. Before a shelf could be built, there had to be an Atlantic Ocean basin into which to build it. The evidence suggests that the basin began to form at about that time, when the gigantic American lithospheric plate began to move slowly away from the lands that once joined it on the east (Chapters 2 and 18). When the break occurred, the severed continental margins—those of the Americas on one side, and of Europe and northern Africa on the other—were thinned and bent slightly downward to form a long, narrow basin. Seawater flowed into the basin, creating the infant Atlantic Ocean. In this new ocean, sediment brought by rivers that flowed out of the adjacent continents began to accumulate.

As the drifting movement continued at a rate of a few centimeters per year, North American rivers built deltas into the sea and formed a growing continental shelf. Younger layers extended outward over older ones, and were draped over the shelf and continental slope as the margin of the continent slowly subsided isostatically under the weight of the accumulating sediment.

Other shelves, like that of southern California, are narrow, irregular in surface form, and apparently consist mainly of hard bedrock. They seem to have resulted from partial drowning and faulting of the western margin of the North American continent, accompanied by prolonged erosion by wave action that in places has cut broad, flat benches near sea level.

Continental Slope and Rise

Continental slopes form the outer edge of the continental crust where it abuts against the oceanic crust. Commonly, they grade downward and outward into the continental rise (Figs. 14.2 and 14.3). This region is underlain by a vast pile of sediment, consisting of waste derived from the adjacent continent and from sedimentary debris that slumps down from the shelf and upper slope above. Much

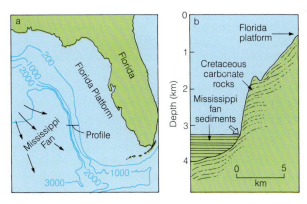

FIGURE 14.4 Continental margin of eastern Gulf of Mexico. (*Source:* After C. K. Paull et al., 1984.) (*a*) Western edge of the continental shelf off Florida forms a steep scarp of eroded carbonate rocks. Contours in meters. (*b*) Profile across the Florida Escarpment shows the abrupt topographic boundary (arrow) where even-bedded sediments of the Mississippi deep-sea fan bury the base of the steep undersea scarp. Note the vertical exaggeration.

of it consists of turbidites, submarine debris flows, and contourites (Chapter 5). Typically, the continental slope and rise pass seaward with decreasing gradients into deep gentle plains. Locally, however, the continental slope ends abruptly against flat-lying sediments of the deep-sea floor. For example, west of Florida the base of the slope has been buried by the outer part of the encroaching Mississippi deep-sea fan (Fig. 14.4).

Submarine Canyons

The continental slope and rise, like the shelf, generally have a rather smooth surface. However, cut into them are many remarkable valleys, some so deep that they have been called *submarine canyons* (Fig. 14.5). These valleys, which are found around the world, are as much as 1 km deeper than the adjacent seafloor and have steep sideslopes. Some are as long as 370 km, but the average probably is closer to 50 km. The shortest canyons tend to be steep, with gradients of up to 60 m/km, whereas long ones have gradients averaging close to 10 m/km.

The origin of submarine canyons has been a puzzle. The valleys extend down to water depths of 3 km or more—far too deep for the canyons to have been excavated by ordinary rivers, even at times of glacially lowered sea level. Some, such as those off the mouths of the Hudson, Congo, and Ganges rivers, line up with valleys that cross the continental shelves, while others apparently have no such associated valleys. Some join at the base of

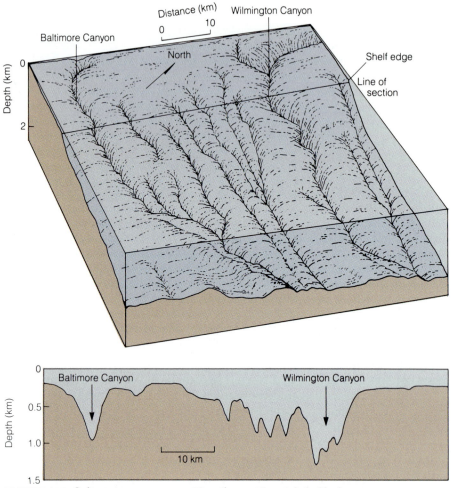

FIGURE 14.5 Submarine canyons cutting the continental shelf of eastern United States off Maryland. The canyons, which head at or near the shelf edge, are as much as 1 km deep and have tributaries like those of large streams on the land. (*Source:* After B. A. McGregor, 1984.)

the continental slope and continue across the rise as a single channel to the edge of an adjacent abyssal plain at depths of 5000 m or more.

Observers in submersible research vessels have noted features within the canyons, such as ripple and scour marks on surface sediments and polished bedrock slopes, that indicate downslope movement of sediment. Cores of sediment from the canyon floors display features typical of turbidity currents. Very likely, therefore, turbidity currents play an important role not only in moving sediment downslope, but in actually eroding the canyons.

It is apparent that many active canyons serve as conduits for transporting sediment from the continental shelf to the deep sea. Such canyons typically head close to the shoreline or near the mouth of a river where longshore sediment drift or direct influx of stream load can supply sediment to canyon heads where turbidity currents are generated. Canyons heading near the edge of the submerged continental shelf apparently were active during glacial times when sea level was lower and streams flowed across the emerged continental shelf to the edge of the slope. Canyons are especially numerous along parts of the continental margins that lie adjacent to areas formerly covered by large glaciers, as along the southern edge of the Bering Sea off Alaska and along eastern North America between Labrador and Virginia, where at least 190 canyons have been discovered. Beyond such regions, fewer canyons are found.

Active erosion of a submarine canyon apparently requires a continuous and abundant supply of sediment to its head which can form dense sediment-laden turbidity currents that scour the can-

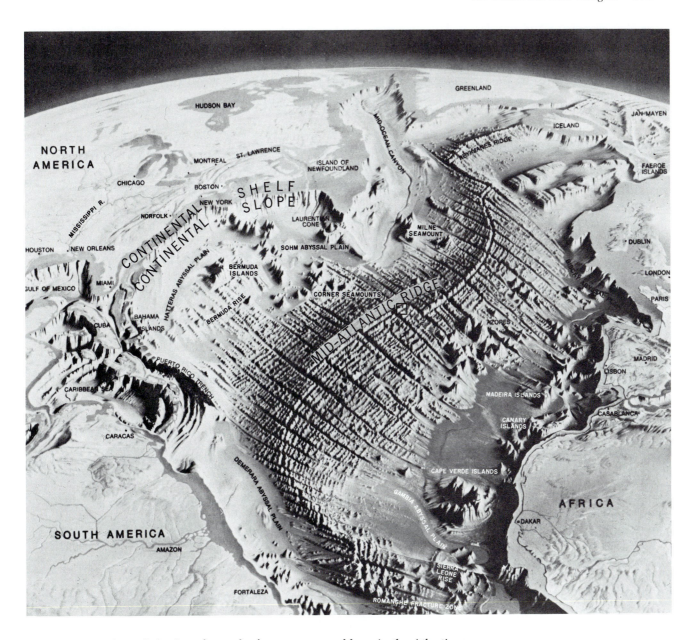

FIGURE 14.6 An artist's view shows the features we would see in the Atlantic Ocean basin if all water were removed. Slopes appear steeper than they actually are because the vertical scale has been exaggerated. Along the center of the basin is the Mid-Atlantic Ridge, a great volcanic mountain chain, in places broken and offset by huge fractures that form steep scarps at sharp angles to the ridge. Away from the ridge lie seamounts, both singly and in groups. Farther away are abyssal plains, smooth-floored parts of the ocean floors that lie adjacent to continental rises. Above, the continental slopes rise to meet the gently sloping continental shelves.

yon sides as they move rapidly downslope. Active canyons clearly are ones that now receive such a supply of sediment, whereas inactive ones seem to have been cut mainly during times when sea level was lower and sediment was able to reach their heads.

Oceanic Ridge System

The most striking feature of the submarine profile in Figure 14.2 is the Mid-Atlantic Ridge. It is part of the oceanic ridge system, a chain of mountains

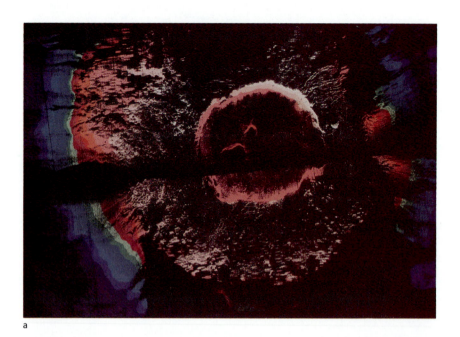

a

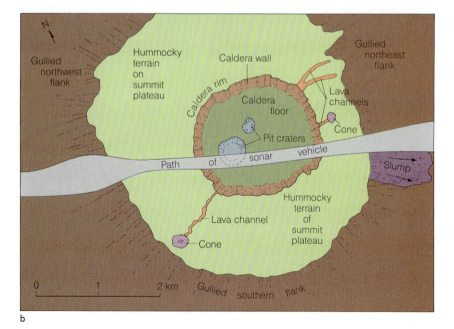

b

FIGURE 14.7 Submarine volcano on the crest of the East Pacific Rise. (*a*) Side-looking sonar image of a young seamount. Although lying far below sea level, the seamount has a form which is very similar to that of volcanic cones built on land. Topographic relief is shown by different colors, with depths ranging from 2650 m (purple) to 1640 m (white). (*b*) Map showing major geologic features identifiable on the sonar image.

some 84,000 km long that twists and branches in a complex pattern through the ocean basins. It marks the spreading edges of moving plates of lithosphere (Fig. 2.12). This great mountain chain would be one of the most impressive features we would see if we could view a dry Earth from out in space. The ridge dominates the artistic view in Figure 14.6 where it is some 1500 km wide. The ridge crest rises 1000–3000 m above the surrounding ocean floor and has an average depth of 2500 m. It is highest along its axis, from which its flanks slope away rather symmetrically on either side. The topography is often as rugged as that on land, and relief locally is as much as 600 m.

A narrow valley, or rift, that in places runs down the center of the Mid-Atlantic Ridge has been the object of intense international study in recent years. It has many features in common with the great rift valley system of East Africa. A small portion of the rift, examined by French and American scientists from deep-diving submarines, was found to be bounded by steep walls as much as 1200 m high and to be broken by long, subparallel fractures in the seafloor. An inner rift valley lying within the larger central rift is the principal site of active volcanism. The distinctive central rift is continually renewed because it is the gap between two plates of lithosphere that move continually apart from each other.

The central rift is characterized by unusually high heat flow and intense submarine volcanism, due to magma welling up periodically from the mantle. In addition to the sheeted lava flows and pillow basalts described in Chapter 4, undersea images show submerged volcanoes similar in form to those erupted on land (Fig. 14.7).

At several places around the world the oceanic ridge reaches sea level and forms oceanic islands. The largest of these is Iceland, which lies on the center of the Mid-Atlantic Ridge; there we can examine part of the central rift valley on land (Fig. 14.8). As along the submarine extension of the ridge, the rift valley on Iceland is the principal area of volcanic activity, and the floor of the valley has many active fissures that parallel the steep bounding walls of the rift. The volcanic rocks of Iceland, just as elsewhere along the ridge system, tend to be progressively older away from the central rift valley in either direction.

Seamounts and Oceanic Islands

Looking again at Figure 14.6, we see that the ocean floor is dotted with many steep-sided mountains,

FIGURE 14.8 A fault scarp 40 m high borders the western margin of the slowly widening central rift valley of Iceland where new oceanic crust is being generated on this emergent portion of the Mid-Atlantic Ridge.

some alone, some in groups. Nearly all such mountains are volcanic in origin. Where they reach the surface of the sea, as in the Azores or Hawaii, they form volcanic islands. Where a submarine volcano does not reach the surface, but is *an isolated submerged volcanic mountain standing more than 1000 m above the seafloor,* it is called a *seamount.*

A seamount with a more-or-less flat top well below sea level is called a *guyot* (Fig. 14.9). Seismic surveys (Chapter 16) of guyots disclose a capping of coral reef and sediment. The coral caps of some guyots that have been drilled or dredged have yielded ancient fossils. Because the fossils are of types that live only in shallow water, the reefs and the underlying volcano must have experienced progressive submergence.

Support for the idea that guyots are sunken islands comes from three different types of **coral reef,** *a ridge of limestone built by colonial marine organisms.* Coral reefs abound in tropical seas. They are built by vast numbers of tiny colonial organisms that secrete calcium carbonate. Such organisms require shallow, clear water of near-normal salinity in which the temperature remains above about 18° C. Reefs, therefore, are built only at or close to sea level and are characteristic of tropical latitudes.

Three principal reef types are commonly recognized (Fig. 14.10). A **fringing reef** is *a reef directly attached to or bordering the adjacent land.* It therefore lacks a **lagoon,** which is *a bay inshore from an enclosing reef or island paralleling a coast.* Typically, a fringing reef has a table-like upper surface as much as 1 km wide, and its seaward edge plunges steeply into deeper water. A **barrier reef** is *a reef separated from the land by a lagoon* that may be of considerable length and width. Such a reef may lie far off the

FIGURE 14.9 (*a*) Part of the ocean floor in the mid-Pacific as it would appear if the water were drained away. In the foreground is a valley cut into the flank of a seamount, the top of which is out of view. In the distance are flat-topped guyots. (*b*) Profile through Horizon Guyot in the mid-Pacific based on seismic evidence. The top of the sunken volcanic island is capped by a cover of coral reef rocks and pelagic sediment. (*Source:* After Winterer and Metzler, 1984.)

a

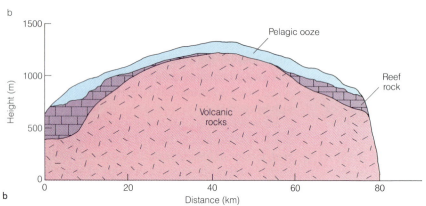

b

coast of a continent, as is the case with the Great Barrier Reef off Queensland, Australia. An **atoll** is *a coral reef, often roughly circular in plan, that encloses a shallow lagoon.* Atolls generally are surrounded by deep water of the open ocean and range in diameter from as little as 1 to as much as 130 km.

It is not difficult to see that if a volcanic island slowly subsided, a fringing reef would have to grow upward if its organisms were to survive near sea level. With continued subsidence, the approximately conical volcanic island would become smaller in area and the fringing reef would become an offshore barrier reef (Fig. 14.10*b*). Eventually,

the volcanic island would disappear, to be replaced by an atoll. This theory of atoll formation, first proposed by Charles Darwin after his epic voyage on H. M. S. *Beagle*, has been substantiated by borings that have penetrated thick coral formations on atoll islets and eventually encountered volcanic rock at depth.

More than 2000 seamounts and guyots have been discovered, mostly in the Pacific, and some oceanographers believe the total will eventually reach well over 10,000. Where oceanic islands and seamounts have been dated by radiometric methods, they often turn out to be younger than the surrounding ocean floor. This means that such

a

b

c

FIGURE 14.10 Chief kinds of tropical coral reefs. (*a*) Fringing reef on the island of Molokai in the Hawaiian Islands. (*b*) Barrier reef enclosing island of Moorea in the Society Islands. A narrow lagoon separates the high island, the remnant of a formerly active volcano, from a shallow reef flat. (*c*) Rongelap atoll in the Marshall Islands of the central Pacific Ocean. Volcanic rocks now lie far below the rim of coral that defines the modern atoll.

islands could not have formed at oceanic ridges and then been rafted on a moving plate of lithosphere to their present sites. Instead, it must indicate that volcanism occurs at spots on the seafloor away from the spreading edges of the plates. The discovery that some lines of oceanic islands and seamounts get progressively older in one direction (Fig. 17.23) suggests that local, more-or-less fixed "hot spots" in the mantle generate magma that moves upward toward the seafloor. As a plate of lithosphere moves over the hot spot, a succession of oceanic volcanoes forms that slowly evolve into atolls and ultimately into guyots (Chapter 19).

Seafloor Trenches

The greatest depths in the ocean (7000 to nearly 11,000 m) occur in seafloor trenches. As much as 200 km wide and 25,000 km long, trenches mark places where moving plates of lithosphere plunge down into the mantle. Most trenches lie in the Pacific Ocean (Fig. 14.11). Some, such as the Aleutian Trench and the Tonga-Kermadec Trench, are far from a continental shore. Others, such as the Peru-Chile Trench, lie immediately adjacent to a continent and in places are filled with sediment derived from the nearby landmass.

Abyssal Plains

The final major topographic feature we must mention is the **abyssal plain,** *a large flat area of deep-sea floor having slopes of less than about 1 m/km.* Abyssal plains lie adjacent to the continental rise, which is a primary source of sediment (Fig. 14.6). They generally are found at depths of 3000–6000 m and range in width from about 200–2000 km. Such features are most common in the Atlantic and Indian oceans which have large sediment-laden rivers entering them and lack bounding deep-sea trenches that can act as sediment traps. Where they occur, the original seafloor topography has been completely buried beneath a blanket of fine sediment, much of which has been transported downslope by turbidity currents.

OCEAN CIRCULATION

The restless sea is always in motion, at generally slow rates to be sure, although in places at velocities comparable with swift rivers. There are several immediate causes of motion, but the chief motive power is solar energy, with some help from the

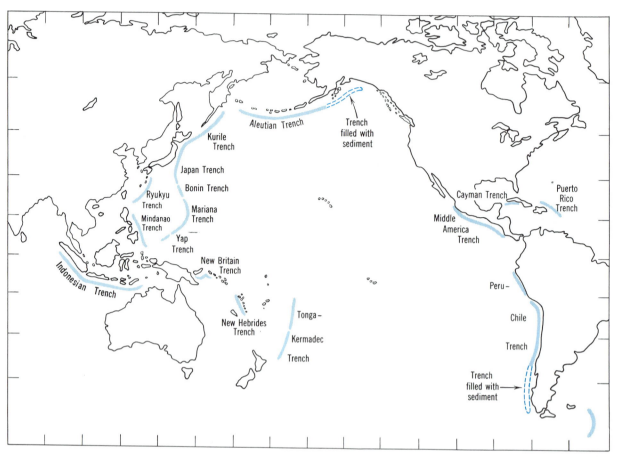

FIGURE 14.11 Sea-floor trenches occur mainly in the Pacific Ocean. They mark places where plates of lithosphere plunge down into the mantle. Trenches off southern Alaska and southern South America are locally filled with sediment eroded and dumped into the adjacent oceans by glaciers.

energy of the Earth's rotation and from the gravitative pull of the Sun and Moon. Surface ocean currents and density currents are geologically important in the deep oceans, whereas tidal currents, waves, and longshore currents are effective geologic agents along coasts.

Surface Ocean Currents

Surface ocean currents are *broad, slow drifts of surface water.* They are set in motion by the prevailing surface winds. Air that flows across a water surface causes waves, but it also drags the water slowly forward, creating a current of water as broad as the current of air but rarely more than 50–100 m in depth. The marked effect of winds on the ocean is evident if a map of surface ocean currents (Fig. 14.12) is compared with the positions of the belts of prevailing winds (Fig. 12.2). In low latitudes surface seawater moves westward with the trade

winds. The general westerly direction of the North and South Equatorial Currents (Fig. 14.12) is reinforced by the Earth's rotation. Both north and south of the equator, the westerly moving currents eventually are deflected where a current encounters a coast, and by the Coriolis effect (Chapter 12).

In the North Atlantic, the North Equatorial Current flows west. Deflected water that flows north is represented by the Florida Current, the Gulf Stream, and the North Atlantic Current. The currents transfer warm equatorial water to higher latitudes and, therefore, have considerable effect on climates. Part of the North Atlantic Current eventually moves into the Arctic Ocean, taking heat with it, and part is deflected south along the European and African coasts as the Canary Current. By this time the southward-flowing water has lost so much heat that it is cooler than the surrounding tropical water. Approaching the equator, it has completed its circulation and once more flows westward, dragged by the trade winds.

We can follow a similar pattern in the northern Pacific Ocean, where we have this sequence: North Equatorial Current, Kuroshio Current, North Pacific Current, and the cool California Current. In the Southern Hemisphere we also find great circular movements of surface seawater, but the direction of rotation is counterclockwise, unlike the clockwise pattern north of the equator. However, there is another major difference between circulations north and south of the equator: the far-southern oceans are not impeded by continents. This makes possible a major globe-circling movement of water, the West-Wind Drift. This moves water from one ocean to another, gradually mixing their waters, a process that is completed about every 1800 years.

Although rates of movement are generally slow, in confined areas surface ocean currents can flow rapidly. In the narrow strait between Florida and Cuba, for example, the rate approaches 5 km/h.

Circulation in the Deep Ocean

Density Currents

Throughout the ocean, beneath the great currents set up by winds in the shallow surface zone,

deeper and much slower circulation of water is also occurring. It is caused by differences in density. A **density current** is *a localized current within a body of water caused by dense water sinking through less-dense water.* Turbidity currents, as we have seen (Chapter 5), are a type of density current resulting from flow of sediment-laden water. However, when seawater gets colder or becomes more saline, its density also increases. Dense water tends to sink, displacing less-dense water below.

In the polar regions, surface water becomes chilled by the cold atmosphere. Also, it acquires increased salinity by freezing to form sea ice. As the ice (nearly pure H_2O) forms, the dissolved ions remain in solution in the residual seawater, which becomes more saline and therefore denser. As its density increases by cooling and the formation of sea ice, the polar water sinks, then flows slowly along the seafloor toward the tropics. Deep, cold polar water from the Antarctic even crosses the equator into the Northern Hemisphere.

Bottom-Water Circulation

In the Atlantic Ocean several clearly defined sources of cold, dense bottom water have been identified:

(1) Water in the Gulf Stream and North Atlantic

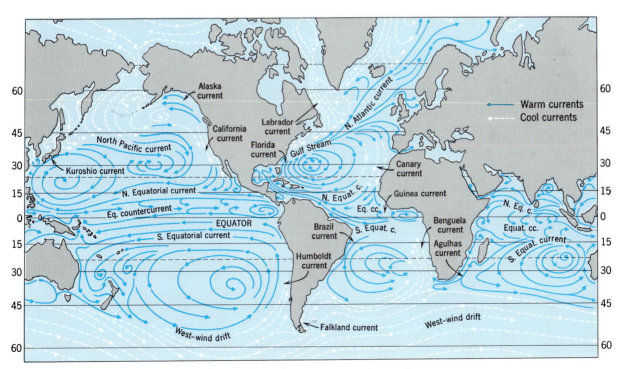

FIGURE 14.12 Surface ocean currents form a distinctive pattern, curving to the right in the Northern Hemisphere and to the left in the Southern.

Current is highly saline as a result of evaporation in low latitudes. Having a higher salinity than most seawater, when it reaches the Arctic region and becomes chilled, it sinks to the bottom. This cold North Atlantic Deep Water (Figs. 14.13 and 14.14*a*) flows south, almost into the Antarctic region, before it is obscured by mixing.

(2) In the Antarctic region, water is chilled during the winter, and at the same time becomes more saline through the formation of sea ice. The major input of this water to deep-ocean circulation comes from the large embayment of the Weddell Sea (Fig. 13.4*a*). The cold, dense Weddell Sea Bottom Water sinks to the ocean floor where it travels around the Antarctic continent in a clockwise direction and also flows northward, crossing the equator and reaching intermediate northern latitudes (Figs. 14.13 and 14.14*a*).

(3) Antarctic Intermediate Water forms by chilling of the highly saline waters of the Brazil Current (Figs. 14.12 and 14.14*a*).

(4) The Mediterranean Sea is an enclosed basin with strong evaporation, especially in its eastern part. Mediterranean surface water, although warm, becomes very saline due to evaporation and sinks. The denser water eventually flows as a deep current through the Strait of Gibraltar and into the Atlantic. To counterbalance the flow of deep water out of the Mediterranean, a fast-moving surface current flows in through the strait. Thus, because of differences in density between two great water bodies, the Strait of Gibraltar carries two currents, one above the other, flowing in opposite directions. During the Second World War, German submarine captains, wishing to escape detection by the British base at Gibraltar, submerged their boats, stopped the engines, and were carried into the Mediterranean in the higher, inflowing current. When wishing to exit, they dived deeper and were carried westward into the Atlantic in the denser return flow.

Deep circulation in the Pacific and Indian oceans differs from Atlantic circulation in that all the deep water comes from the Antarctic region (Fig. 14.13). In the northern Pacific, there is no large source of deep, cold water because a shallow barrier at the Bering Strait prevents deep Arctic water from breaking through. It is possible for deep water

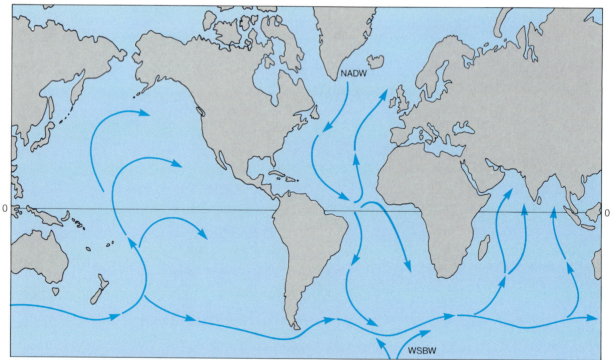

FIGURE 14.13 Generalized flow pattern of bottom waters in the world's oceans at a depth of about 4000 m. The major inputs are North Atlantic Deep Water (NADW), which enters from the vicinity of Greenland, and Weddell Sea Bottom Water (WSBW), which enters from the margin of Antarctica near the South Atlantic.

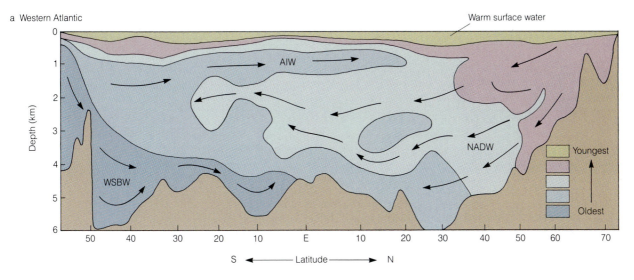

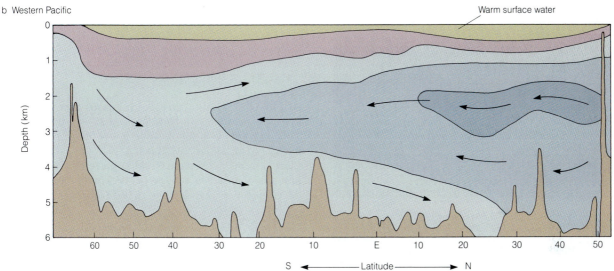

FIGURE 14.14 Cross sections through the western Atlantic Ocean (*a*) and the western Pacific Ocean (*b*) showing the primary subsurface circulation pattern. Warm water near the surface in both oceans overlies colder water flowing from the polar regions. Primary water masses in the Atlantic are North Atlantic Deep Water (NADW), Antarctic Intermediate Water (AIW), and Weddell Sea Bottom Water (WSBW). In the western Pacific cold Antarctic bottom water extends well north of the Equator. From data collected by Geochemical Ocean Section Study (GEOSECS). (*Source:* After M. Stuiver and H. G. Ostlund, 1980; H. G. Ostlund and M. Stuiver, 1980.)

originating in Antarctica to flow as far north as California and Japan (Fig. 14.14*b*).

The residence time of deep water in the world's oceans has been calculated by measuring the distribution of radiocarbon in the waters. An average parcel of deep water will remain in the Atlantic or the Indian Ocean for about 275 yr, and for 510 yr in the Pacific. The oldest ocean water is found in the North Pacific, as one might surmise from studying Figure 14.13. The deep waters of the entire world ocean are replaced about every 500 yr.

Upwelling and Sinking

Prevailing winds, blowing offshore or onshore, cause vertical movements of seawater (Fig. 14.15). *Upwelling* is caused by offshore winds. It is the principal way in which cold, deep water is brought to the surface. *Sinking* is the opposite effect and is caused by onshore winds. Because deep water is colder than surface water, when it wells up and reaches the surface it cools the air and creates fog.

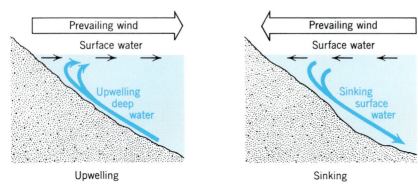

FIGURE 14.15 Cross section of a coast showing the effect of wind on the vertical movement of sea water. The prevailing wind blows surface water away from the coast, causing deep water to well up. The opposite effect—sinking—happens when the prevailing wind blows toward the shore.

a

b

c

FIGURE 14.16 Tidal fluctuations in coastal zones. (*a*) Coastline near Nelson, New Zealand at high tide. (*b*) Same coast about 12 hours later at low tide. (*c*) Inrushing tide in an estuary at Moncton, New Brunswick, Canada forms a turbulent wall-like wave of water (a *tidal bore*) that moves rapidly (10–15 km/hr) against the river flowing gently from right to left.

This is the source of the great summer fog banks along parts of the Pacific coasts of North and South America. Deep water also tends to be rich in nutrients such as phosphorus and nitrogen. When it reaches the sunny surface where organisms can thrive, microscopic plant life blooms in abundance and fish populations that feed on the plants expand too. For this reason, some of the world's most productive fishing grounds are associated with areas of upwelling.

Tidal Currents

Tidal currents are caused mainly by the twice-daily tidal bulges that pass around the Earth (Chapter 3). In the open sea, tidal effects are small; however, in bays, straits, estuaries, and other narrow places, tidal fluctuations can generate rapid currents. Tidal flows can approach 25 km/h in places, and tidal heights of more than 16 m are known (Fig. 14.16). Such fast-moving currents, although restricted in extent, readily move sediment around. Large linear sand ridges can be built paralleling such currents, as well as sand waves (which are large sand ripples, oriented perpendicular to the current direction).

Waves

Wave Motion

Ocean waves are generated by winds that blow across the surface. Figure 14.17 shows the significant dimensions of a wave traveling in deep water where it is unaffected by the bottom far below. The motion of a wave is very different from the motion of any parcel of water within it. As wind sweeps across a field of grain or tall grass, the individual stalks bend forward and return to their positions, creating a wavelike effect. In similar fashion the *form* of a wave in water moves continuously forward, but each water parcel revolves in a loop, returning, as the wave passes, very nearly to its former position. This looplike, or *oscil-*

lating, motion of the water, first determined theoretically, was later proved by injecting droplets of colored water into waves in a glass tank and then photographing their paths with a movie camera.

Waves receive their energy from wind, and so can receive it only at the surface of the water. Because the wave form is created by a looplike motion of water parcels, the diameters of the loops at the water surface exactly equal wave height (*H* in Fig. 14.17). Downward from the surface there is progressive loss of energy, expressed in diminished diameters of the loops. At a depth equal to only half the **wavelength,** *the distance between successive wave crests or troughs* (*L* in Fig. 14.17), the diameters of the loops have become so small that motion of the water is negligible.

Erosion by Waves

The depth *L*/2 is the effective lower limit of wave motion; therefore, it must also be the lower limit of erosion of the bottom by waves. This depth is generally referred to as the *wave base.* In the Pacific Ocean, wavelengths as great as 600 m have been measured. For them, *L*/2 equals 300 m, a depth half again as great as the outer edge of the average continental shelf. Although the wavelengths of most ocean waves are far less than 600 m, it is

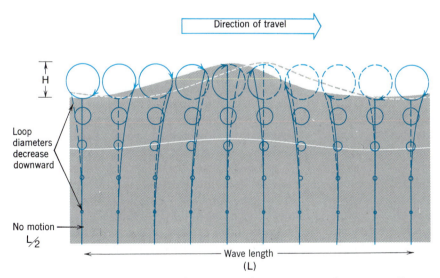

FIGURE 14.17 Looplike motion of water parcels in a wave in deep water. To follow the successive position of a water parcel at the surface, follow the arrows in the largest loops from right to left. This is the same as watching the wave crest travel from left to right. Parcels in smaller loops underneath have corresponding positions, marked by continuous, nearly vertical lines. Dashed lines represent wave form and parcel positions one-eighth period later.

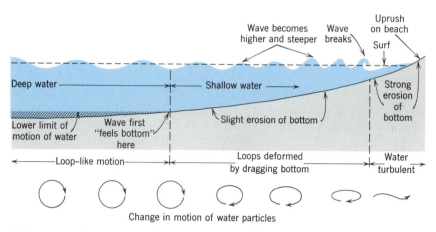

FIGURE 14.18 Waves change form as they travel from deep water through shallow water to shore. Circles and ellipses are not drawn to scale with the waves shown above. Compare with Figure 14.17.

nevertheless possible for very large waves approaching these dimensions to affect even the outer parts of continental shelves, which average 200 m in depth. Landward of depth $L/2$ the wave motion continuously lifts and drops fine particles of bottom sediment, very slowly moving them seaward along the gently sloping bottom. Such erosion is very slow and not at all spectacular, but over a million years or more the cumulative result is great.

What happens in the shallow water at the shore is more rapid, sometimes spectacular, and always different in style. When a wave moving toward shore reaches depth $L/2$ its base meets the bottom and the form of the wave begins to change. The looplike paths of water parcels gradually become elliptical, and velocities of the parcels increase. Interference of the bottom with wave motion distorts the wave by increasing its height and shortening wavelength. Often the height is doubled. This means the wave grows steeper. Because the front part of the wave is in shallower water than the rear part, it is steeper than the rear. Eventually, the steep front becomes unable to support the wave, the rear part continues forward, and the wave collapses or *breaks* (Fig. 14.18).

When a wave breaks, the motion of its water instantly becomes turbulent, like that of a swift river. Such "broken water" is called *surf*, defined as *wave activity between the line of breakers and the shore.*

In turbulent surf each wave finally dashes against rock or rushes up a sloping beach until its energy is expended; then it flows back. Water piled against the shore returns seaward in an irregular and complex way, partly as a broad sheet along the bottom and partly in localized narrow channels. The returning water is mainly responsible for the currents known to swimmers as "undertow."

Surf possesses most of the original energy of each wave that created it. This energy is quickly consumed in turbulence, in friction at the bottom, and in moving the sediment that is thrown violently into suspension from the bottom. Although fine sediment is transported seaward from the surf zone, most of the geologic work of waves is accomplished by surf shoreward of the line of breakers.

How deep below sea level can surf erode rock and move sediment? The answer depends on the depth at which waves break. Most ocean waves break at depths that range between wave height and 1.5 times wave height. Because waves are seldom more than 6 m high, the depth of vigorous erosion by surf should be limited to 6 m times 1.5, or 9 m below sea level. This theoretical limit is confirmed by observation of breakwaters and other structures, which are found to be only rarely affected by surf at depths of more than about 7 m. The surf zone, the place where high-energy turbulent water cuts into the land, is limited then to the narrow vertical range extending from sea level down to about 7 m below sea level.

During great storms surf can strike effective blows well above sea level. The west coast of Scotland is exposed to the full force of Atlantic waves. During a great storm on that coast a solid mass of stone, iron, and concrete weighing 1200 metric tons (1,200,000 kg) was ripped from the end of a breakwater and moved inshore. The damage was repaired with a block weighing more than 2300 tons, but five years later storm waves broke off and

FIGURE 14.19 A nearly horizontal wave-cut bench has formed along the coast at Bolinas Point, California as the surf, acting like an erosional saw, has cut into the exposed tilted sedimentary rocks.

moved that one too. The pressures involved in such erosion were about 27 tons/m². Even waves having much smaller force break loose and move blocks of bedrock from sea cliffs, partly by compressing the air in fissures, which then pushes out blocks of rock.

The vertical distance through which water can be flung against the shore would surprise anyone whose experience of coasts is limited to periods of

calm weather. During a winter storm in 1952, again on the west coast of Scotland, the bow half of a small steamship was thrown against a cliff and left there, wedged in a big crevice, 45 m above sea level.

Another important kind of erosion in the surf zone is the wearing down of rock by wave-carried rock particles. By continuous rubbing and grinding with these tools, the surf wears down and deepens the bottom and eats into the land, at the same time smoothing, rounding, and making smaller the tools themselves. As we have seen, this activity is limited to a depth of only a few meters below sea level. The surf, therefore, is like an erosional knife-edge or saw, cutting horizontally into the land (Fig. 14.19).

Transport of Sediment

The rock particles worn from the coast by surf, as well as those brought to the coast by rivers, are intermittently in motion. They are dragged or rolled along the bottom, lifted in irregular jumps, or carried in suspension, according to their size and to the varying energy of waves and currents. In the surf, as can be seen on almost any beach, sediment is moved to and fro, shoreward and seaward.

Seaward of the surf zone, in deeper water, bot-

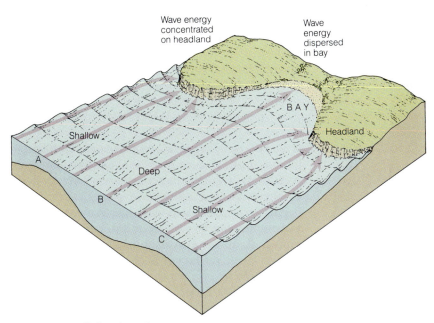

FIGURE 14.20 Refraction of waves concentrates wave energy on headlands, and disperses it along shores of bays. Oblique view shows eight waves that become more distorted as they approach the shore over a bottom that is deepest opposite the bay.

a

b

FIGURE 14.21 Longshore currents and beach drift. (*a*) Waves arriving obliquely onshore along a coast near Oceanside, California generate longshore current that moves sediment down the coast. (*b*) Surf swashes obliquely onto a Brazilian beach and forms a series of arcuate cusps as the water loses momentum and flows back down the sandy slope.

tom sediment is shifted by unusually large waves during storms and by currents, with net movement seaward. Each particle is picked up again and again, whenever the energy of waves or currents is great enough to move it. As the particle gets into ever-deeper water it is picked up less and less frequently. With increasing depth, energy related to wave motion decreases, so only finer-sized grains can be moved. As a result, the sediment becomes sorted according to diameter, from coarse in the surf zone to finer offshore.

In the gradual building of a continental shelf by sedimentation, the deposited sediments normally grade seaward from sand into mud. This gradation is true not only of the particles eroded from the shore by surf but also of the particles contributed by rivers, whose currents carry suspended loads into the sea. Calculations made along some coasts indicate that the volume of sediment contributed by rivers is much greater than that contributed by surf erosion. The actual proportions, of course, differ from place to place depending on several varying factors.

Off some middle- to high-latitude coasts a consistent offshore gradation from coarse to fine sediments is not always observed. Instead, belts of coarse gravel are found in ridgelike banks far beyond the surf zone. In many cases such sediment coincides with the crests of end moraines of former glaciers that terminated on the emergent continental shelves when sea level was lower and the shore lay far beyond its present position. Submergence of the moraines by rising sea level has subjected their crests to attack by waves and currents, producing concentrates of coarse particles as the fine sediment was swept away.

Wave Refraction

A wave approaching a coast over an undulating bottom cannot encounter the bottom along all parts of its crest simultaneously. As each part does so, the wave slows down. Wavelength at that part begins to decrease, and wave height increases. The wave gradually swings around to parallel the bottom contours and is said to be refracted (Fig. 14.20). **Wave refraction** is *the process by which the direction of a series of waves, moving in shallow water at an angle to the shoreline, is changed.* Thus, waves approaching the shore in deep water at an angle of 40° or 50° may, after refraction, reach the shore at an angle of 5° or less. Waves passing over a submerged ridge off a headland will converge on the headland. Convergence, plus the increased wave height that accompanies it, concentrates wave energy on the headland. Conversely, refraction of waves approaching a bay will make them diverge, diffusing their energy at the shore. Because of refraction, headlands are eroded more vigorously than are bays, so that in the course of time irregular coasts tend to become smoother and less indented.

Longshore Currents and Beach Drift

Despite refraction, most waves reach the shore at an angle, however small. The oblique approach of waves sets up longshore movement of two distinct kinds, both of which move sediment *along* the coast. The first kind consists of a **longshore current,** *a current, within the surf zone, that flows parallel to the shore.* It is generated as waves that arrive obliquely to the shore are refracted (Fig. 14.21a). Because waves are generated by winds, the direction of longshore currents may change seasonally if the prevailing wind directions change, thereby causing changes in the direction of arriving waves. Such currents easily move fine sand suspended in the turbulent surf.

Meanwhile, on the exposed beach itself, a second kind of movement alongshore is occurring. Because waves generally strike the beach at an angle, the swash of each wave travels obliquely up the beach, but the backwash flows straight down the slope of the beach. The result is *beach drift,* a zigzag movement of sand and pebbles, with net progress along the shore (Fig. 14.21b). The greater the angle of waves to shore, the greater the longshore movement. Pebbles tagged and timed have been observed to drift along a beach at a rate of more than 800 m/day.

When the volume of sand moved by longshore currents is added to that moved by beach drift, the total can be very large. The amounts of sediment transported by these two shore processes can result in significant changes to beaches used by people, as will be seen later in this chapter.

THE SHAPING OF COASTS

The coast of a landmass is a great boundary between two realms, land and water. At a coast,

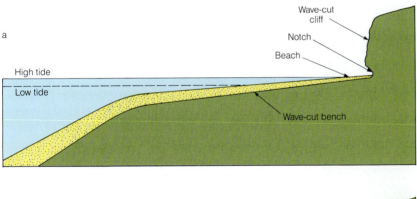

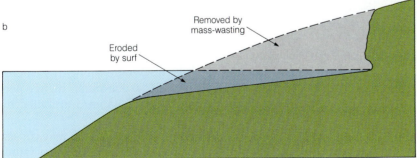

FIGURE 14.22 (*a*) Principal features of a shore profile along a cliffed coast. (*b*) The comparatively large proportion of material removed by mass-wasting relative to that eroded by surf. Notching of the cliff by surf action undermines the rock or sediment which collapses and is reworked by surf.

ocean waves that may have traveled unimpeded through thousands of kilometers encounter an obstacle to their further progress. They dash against firm rock, erode it, and move the eroded rock particles. Over the long term the net effect is substantial.

Waves and the currents created by waves are the agents responsible for most of the erosion of coasts, as well as for most of the transport and deposition of the sediment created by wave erosion or washed into the sea by rivers. As sediment is moved outward from the coast and is deposited offshore, it contributes to the progressive growth of continental shelves, and thus to the growth of the continents.

The Shore Profile

To understand the changes made by the sea along a coast, we must first look at what happens at the

FIGURE 14.23 Enlargement of wave-cut notch at the base of a steep bluff beside Pololu Valley on the island of Hawaii ultimately leads to the collapse of the cliff composed of jointed, thin-bedded lava flows.

surface along the **shore profile,** *a vertical section along a line perpendicular to the shore.* If we combine this with what we know about the forces that act along (parallel to) the shore, we will then have a three-dimensional picture of coastal activities.

Elements of the Profile

Seen in the profile, the usual elements of a cliffed coast (Fig. 14.22) are a wave-cut cliff and wave-cut bench, both the work of erosion, and a beach, the result of deposition. On noncliffed coasts, the beach constitutes the primary shore environment.

A **wave-cut cliff** is *a coastal cliff cut by surf.* Acting like a horizontal saw, the surf cuts most actively at the base of the cliff. As the upper part of the cliff is undermined, it collapses and the resulting debris furnishes rock particles that are reworked by the surf. An undercut cliff which has not yet collapsed may have a well-developed *notch* at its base (Fig. 14.23). The notch is a concave part of the shore profile, overhung by the part above. Other erosional features associated with cliffed coasts include *sea caves, sea arches,* and *stacks* (Fig. 14.24). Each is the result of differential erosion as surf attacks the cliff, resulting in its gradual retreat.

A **wave-cut bench** is *a bench or platform cut across bedrock by surf.* It slopes gently seaward and is extended progressively landward as the cliff retreats. Some benches are bare or partly bare, but most are covered with sediment that is in transit from shore to deeper water. The shoreward parts of some benches are exposed at low tide (Fig. 14.19). If the coast has been raised by recent faulting, as has occurred along parts of the Gulf of Alaska during large earthquakes, a wave-cut bench can be wholly exposed (Fig. 14.25).

The *beach* is regarded by most people as the sandy surface above water along a shore. Actually it is more than this. We define a **beach** as *wave-washed sediment along a coast, extending throughout the surf zone.* In this zone, as we have seen, sediment is in very active movement. The sediment of a beach is derived in part from erosion of adjacent cliffs or from cliffs elsewhere along the shore. However, along most coasts a much higher percentage of it comes from alluvium contributed by rivers.

On low, open shores a typical beach may consist of several distinct elements. The first is a rather gently sloping **foreshore,** *a zone extending from the level of lowest tide to the average high-tide level.* Next is a **berm,** which is *a nearly horizontal or landward-sloping bench formed of sediment deposited by waves.*

FIGURE 14.24 Stack and sea arch along the French shore of the English Channel near Étretat carved in horizontally bedded white chalk. The surf hollows out sea caves in the most erodible part of the bedrock. A cave cut through a headland is transformed into an arch. Isolated remnants of the cliff stand as stacks on a wave-cut bench.

Beyond this lies the **backshore,** *a zone extending inland from a berm to the farthest point reached by waves* (Fig. 14.26). On some beaches the backshore ends in a wave-cut scarp or in a line of sand dunes.

Steady State Along a Coast

As noted at the beginning of this chapter, the line along which water meets land is a scene of conflict that causes erosion and the creation, transport, and deposition of sediment. Through these activities, the form of the land slowly changes, and the water in motion moves and shapes the sediment derived from the land. The forces that fashion the shore profile—cliff, bench, and beach—tend to reach and maintain a condition of equilibrium or *steady state*, a compromise in the water/land conflict.

The compromise is reached in several different ways. On a beach, for instance, the *swash* of a wave running up the beach as a thin sheet of water moves sediment upslope, while gravity pulls it back again. More energy is needed to move pebbles than to move sand grains downslope. Therefore, the pebbles moved by the swash remain until the slope becomes steep enough for them to be carried back again. This is partly why gravel beaches are generally steeper than beaches built of sand. Another factor is permeability. Beaches composed of coarse gravel are highly permeable compared to sandy beaches. Much of the water swash-

FIGURE 14.25 Marine platform at Cape Yakataga, Alaska raised above sea level by recent uplift. The wave-cut surface truncates steeply dipping sedimentary strata.

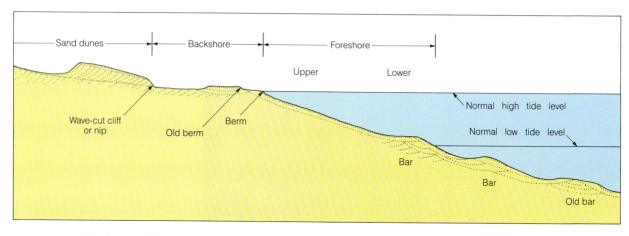

FIGURE 14.26 Typical profile across a beach showing foreshore, berm, and backshore elements. Length of profile about 100 m. Vertical scale exaggerated about twice.

FIGURE 14.27 Mouth of Yangste River at Shanghai, China. Light-colored silty sediment is carried seaward, settles to seafloor, and is added to prograding delta.

narrower. In calm weather, the exposed beach is likely to receive more sediment than it loses and consequently becomes wider. Storminess may be seasonal, resulting in seasonal changes in beach profiles. Along parts of the Pacific coast of the United States, for example, winter storms tend to carry away fine sediment, and the remaining coarse fraction assumes a steep profile. In calm summer weather, fine sediment drifts in and the beach assumes a more gentle profile. At any time, however, the beach profile represents an average steady-state condition among the forces that are shaping it.

Depositional Features Along Coasts

Up to this point we have been describing the erosional effects of surf and the shaping of the shore profile. The deposits made by surf and by the currents it sets up are equally important, for they result in large part from longshore movement of sediment and occur as recognizable landforms.

Deltas

The position of the outer limit of a delta—the extent to which it projects seaward from the land—is also a compromise between the rate at which the river delivers sediment at its mouth and the ability of currents and waves to erode the sediment and move it elsewhere along the coast (Fig. 14.27). The great size of the Mississippi delta (Figs. 2.15 and 11.31) testifies to the huge volume of sediment carried by its parent river, and to the relative ineffectiveness of waves in destroying it. The Columbia

ing upon them quickly moves downward into the beach sediment, thereby reducing the volume and the transporting capability of the backwash.

During storms the increased energy in the surf erodes the exposed part of the beach and makes it

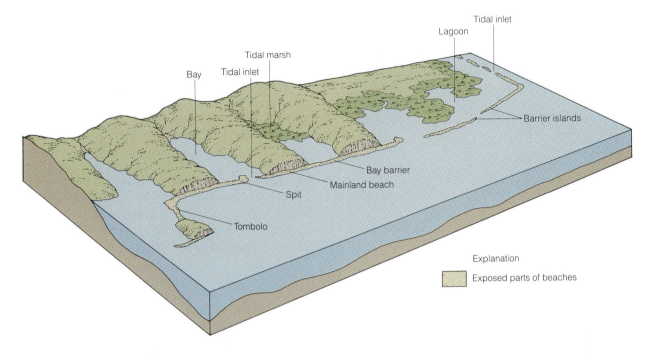

FIGURE 14.28 Common depositional shore features along a stretch of coast. Local direction of beach drift is toward the free end of spits.

River of the northwestern United States formerly transported a large load of sediment to the Pacific coast (much of the sediment is now trapped behind numerous hydroelectric and irrigation dams built along its course). Yet there is no large delta at its mouth. Winter storms and the persistent action of waves moving shoreward from the North Pacific erode sediment as quickly as it arrives. The sediment is moved laterally along the shore and is added to extensive spits built across major coastal embayments.

Spits, Bay Barriers, and Other Forms

Common among other conspicuous forms on many coasts is the **spit,** *an elongate ridge of sand or gravel that projects from land and ends in open water* (Fig. 14.28). Well-known large examples are Sandy Hook on the southern side of the entrance to New York Harbor, and Cape Cod, Massachusetts (Fig. 14.29). Most spits are merely continuations of beaches, built by beach-drifted sediment dumped at a place where the water deepens, as at the mouth of a bay. When the spit has been built up to sea level, waves act on it just as they would act on a beach. Much of a spit, therefore, is likely to be above sea level, although the tip of it cannot be. The free end curves landward in response to currents created by surf.

A **tombolo** is *a ridge of sand or gravel that connects an island to the mainland or to another island* (Fig. 14.28). It forms in much the same way as a spit does.

Along an embayed coast with abundant sedi-

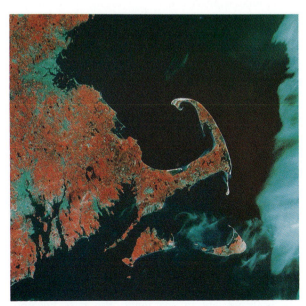

FIGURE 14.29 Spit forming on Cape Cod, Massachusetts. Sediment derived by the reworking of glacial sediments is transported mainly northward by longshore drift. Provincetown lies inside recurving end of spit where an eddy carries sediment around the point and into Cape Cod Bay.

FIGURE 14.30 View north along trend of barrier island off coast of North Carolina from top of lighthouse at Cape Hatteras. Pamlico Sound, about 45 km wide, separates the island from the mainland.

ment supply, a **bay barrier** may form as *a ridge of sand or gravel that completely blocks the mouth of a bay* (Fig. 14.28). It develops as beach drift lengthens a spit across a bay in which tidal or river currents are too weak to scour away the spit as fast as it is built.

Barrier Islands

A **barrier island** (Figs. 14.28 and 14.30) is *a long island built of sand, lying offshore and parallel to the coast*. Such islands are found along most of the world's lowland coasts. Large-sized examples are Coney Island and Jones Beach (New York City's coastal playground areas), the long chain of islands centered at Cape Hatteras on the North Carolina coast, and Padre Island, Texas, some 130 km long.

Although barrier islands originate in various ways, many were probably built as the rapid rise of sea level at the end of the last glacial age began to slow and stabilize about 5000 to 6000 years ago (Fig. 13.38). Shells collected at depths of 5–10 m from the basal deposits of some barriers have radiocarbon ages of about 5000 years, supporting this hypothesis. As the sea slowly rose across the very gentle continental shelf off the southeastern United States, very likely waves breaking in shallow water some distance offshore eroded the bottom and piled up sand to form long bars. Gradually built up above sea level, the bars became barrier beaches, and ultimately barrier islands. With continued slow rise of sea level over the last several thousand years, the barrier islands moved progressively landward as the edge of the continent was slowly submerged.

A barrier island generally consists of one or more ridges of dune sand related to successive shorelines occupied as the island formed. During great storms, surf washes across low places in the barrier and erodes it, cutting inlets that may remain open permanently. At the same time, fine sediment is washed into the lagoon between barrier and mainland. Sediment is thereby transferred landward by overwash, as well as laterally by longshore drift, as an island evolves. It is apparent that the development of barrier islands must be closely related to such factors as sediment supply, direction and intensity of waves and nearshore currents, the shape of the offshore profile, and the relative stability of sea level.

Sediments and Their Stratification

Beaches, barriers, and related features consist of the coarser fraction of whatever range of rock particles is contributed by erosion of sea cliffs or by rivers. Quartz is the most durable of common minerals in continental rocks, and generally occurs as crystals or crystal aggregates having the diameter of sand grains. It is not surprising, therefore, that most continental beaches consist chiefly of quartz sand. Bedrock that breaks down into larger pieces, along joints and other surfaces, makes beaches of gravel.

Dragged back and forth by the surf and turned over and over, particles of beach sediment become rounded by abrasion, much as do comparable particles in streams. Although beach gravel and gravelly alluvium may be very similar in appear-

FIGURE 14.31 Cross-stratified sands and interbedded gravel layers in a beach on the Olympic coast of Washington State.

ance, on many beaches pebbles and cobbles assume a distinctive flattened or discoid shape (Fig. 5.31). Beach sediments also tend to be better sorted than stream sediments of comparable coarseness.

If spits, bay barriers, and the exposed parts of beaches are examined where natural erosion or man-made cuts expose them in section, generally they are seen to be well stratified (Fig. 14.31). Their seaward parts consist of thin layers, gently inclined at various angles. In the lower foreshore zone of beaches, the sediments display cross-stratification, with beds dipping both seaward and landward. Sediments of the backshore tend to be more irregular. Some beds display gentle cross-stratification, and even lenses of silt and clay, whereas others give evidence, through foreset layering and poorer sorting, of high waves that washed entirely across the beach, spit, or barrier and deposited sediment on the far side. In such instances the result is somewhat similar to the stratification of a sand dune. However, the foreset layers are less variable in direction than those in a dune because the angle that waves make with the beach commonly varies

less than the angle between wind direction and the crest of a dune.

Constructional Features Along Coasts

While all coasts are subjected to the erosional effects of waves and currents, in some places constructional processes may *prograde* (build out) the coastline more rapidly than it can be destroyed by surf. We have already seen that the seaward expansion of a delta depends largely on such a relationship. However, other constructional activities can be observed along some coasts.

Organic Reefs

Many of the world's tropical coastlines consist of reefs constructed by marine organisms and composed largely of coral and coralline algae. The seaward edge is subject to strong wave action and is where the hardiest, most wave-resistant species live. Those less able to withstand the violent impact of waves are confined to the lee sides of the reef.

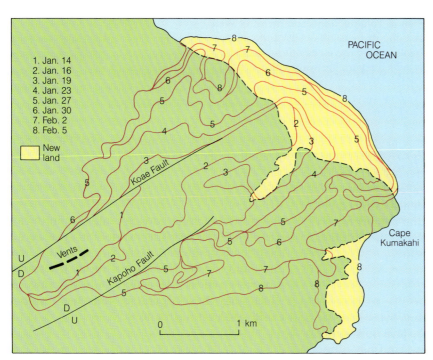

FIGURE 14.32 Construction of a lava delta where lava enters the sea from an eruption along the east rift zone of Kilauea volcano, Hawaii. As the fresh lava meets the water, it is quickly quenched and breaks up, building a delta of lava debris over which the flow continues to move, thereby extending the new landform progressively seaward. (*Source:* After Richter et al., 1970.)

FIGURE 14.33 Emerged wave-cut bench at Tongue Point, southwest of Wellington, New Zealand. Crustal uplift along this coast has raised the former sea floor to form a broad emergent platform. A wave-cut cliff marks the seaward edge of the platform, below which a younger wave-cut bench is forming.

Coral reefs will continue to grow as long as sea level remains constant or changes very slowly. However, growth will cease if a reef is exposed by rapid uplift or an abrupt fall of sea level. A reef will also stop growing if rapid subsidence carries it below the shallow depth in which the organisms can survive, or if the temperature of the surrounding water falls below some critical value.

Lava Flows

Coastal areas subject to volcanic activity may expand seaward and change form as lava flows enter the sea. Substantial tracts of new land have thus been created along the coasts of Iceland and Hawaii, for example (Fig. 14.32). When hot lava enters the sea, it is quickly quenched and breaks apart. The resulting fragments accumulate to build a lava delta. The shoreline will continue to advance seaward as fluid lava moves across the top of the expanding delta surface.

Coastal Evolution

The world's coasts do not all fall into easily identifiable classes. Their variety is great because their configurations depend largely on the structure and erodibility of coastal rocks, the geologic processes which are active and the time over which they have operated, and the history of world sea-level fluctuations.

Structural Control

Some coasts, like most of the Pacific coast of North America, are steep and rocky and consist of mountains or hills separated by deep valleys. Others, like the Atlantic and Gulf coasts from New York

City to Florida and onward into northern Mexico, cut across a broad coastal plain that slopes gently seaward, and are festooned with barrier islands. These coasts represent two extremes, between which are many intermediate kinds. Each owes its general character to its structural setting. The rugged and mountainous Pacific coast lies along the margin of the American lithospheric plate which is continually being deformed where it interacts with adjacent plates to the west. Uplifted and faulted marine terraces are common features along parts of this coast and similar ones that are emerging from the sea (Fig. 14.33). By contrast, the eastern continental margin lies within the same lithospheric plate, but in a region which is tectonically passive. There the old bedrock has low relief and much of the coastal zone borders young sedimentary deposits of the Atlantic and Gulf coastal plains.

Where rocks of contrasting erodibility are exposed along a coast, marine erosion will commonly produce a shoreline that is strongly controlled by rock type and structure. Such control is especially impressive in folded sedimentary or metamorphic belts that have been partially submerged (Fig. 14.34). It also is responsible for some of the world's deeply embayed fjord coasts, the pattern of which reflects deep glacial erosion along regional fracture systems and subsequent drowning as glaciers retreated to the heads of their fjords (Fig. 14.35).

Coastal Processes

The effectiveness of erosional and depositional processes is not equal along all coasts. Coasts lying at latitudes between about 45° and 60° are subject to higher-than-average storm waves, whereas subtropical east-facing coasts are subjected to infrequent but often disastrous hurricanes (called *typhoons* west of the 180th meridian). In the polar regions sea ice becomes an effective agent of coastal erosion. These and other factors influence the amount of energy expended in erosion along the shore and, together with structural and compositional properties of the exposed rocks, contribute to the variety of coastal landforms.

Changing Sea Level

Whatever their nature, nearly all coasts have experienced *submergence* due to the worldwide rise of sea level that has occurred during the last 15,000

FIGURE 14.34 Structurally controlled coast near Coos Bay, Oregon. Embayed shoreline results from drowning of eroded belt of highly deformed sedimentary rocks.

FIGURE 14.35 Satellite view of embayed coast of southern Alaska showing deep fjords carved by glaciers flowing along major structural zones.

years as the Earth emerged from a glacial age (Fig. 13.38). The magnitude of the sea-level fluctuations caused by the buildup and decay of ice-age glaciers has far exceeded the vertical range of observable shore processes. As a result, evidence of lower, as well as somewhat higher, stands of the sea can be found along many coasts.

Because of the recent submergence, evidence of lower, glacial-age sea levels is almost universally to be found only beyond the present coastlines and to depths of 100 m or more. Former beaches, coastal sand dunes, and similar features on the inner continental shelves mark shorelines built by the rising sea at the end of the glacial age, and later drowned.

Evidence of higher sea levels is related mainly to past interglacial ages. Inland from the Atlantic coast of the United States from Virginia to Florida are many marine beaches, spits, and barriers, the highest of which reach an altitude of more than 50 m. It is thought that as a group, these landforms owe their present altitude to a combination both of broad upward arching of the crust as well as to submergence during times when climates were warmer than now, glaciers were smaller, and sea level was therefore higher. The position of such features above present sea level points to *emergence* of the land following their formation.

Varied evidence demonstrates that many coastal and offshore features are relicts of times when sea level was either higher or lower than now. The youngest deposits along a coast often form a thin blanket over older, similar units that date to earlier times. Repeated emergence and submergence over many glacial–interglacial cycles, each accompanied by erosion and redeposition of shoreline deposits, has resulted in complex coastal landform assemblages that test the ability and experience of geologists to interpret them correctly.

Rise and fall of sea level are universal movements, affecting all parts of the world ocean at the same time. Uplift and subsidence of the land, also causing emergence or submergence along a coast, are piecemeal movements, generally involving only parts of landmasses. Geologically rapid *relative* changes of sea level may characterize such regions; for example, some coastal stretches of North America and Europe that were covered by large ice sheets as recently as 10,000 years ago are still rising, as they adjust isostatically, at rates of nearly 1 cm/yr. This uplift has been in progress ever since the ice sheets began to melt away, and has raised some former ocean beaches hundreds of meters above present sea level (Fig. 13.39). Similarly, vertical tectonic movements at the boundary of converging plates of lithosphere have elevated beaches and tropical reefs to positions far above sea level (Figs. 14.33 and 14.36). Because changes of land and sea may occur simultaneously, either in the same or opposite directions, unraveling the history of sea-level fluctuations along a coast can be a difficult and challenging exercise.

Protection Against Shoreline Erosion

The approximate equilibrium among the forces that operate on coasts is occasionally interrupted by exceptional storms that erode cliffs and beaches at rates far greater than the long-term average. During a single storm in 1944, cliffs of compact sediment on Cape Cod in eastern Massachusetts retreated up to 5 m, or more than 50 times the normal annual rate of retreat. Such infrequent bursts of rapid erosion not only can be quantitatively important in the natural evolution of a coast, but can also have a significant impact on coastal inhabitants. More than 75 percent of the population of the United States now lives in coastal belts that include only 5 percent of the land area of the nation. The concentration of such large numbers of people in coastal areas means that infrequent large storms not only can be hazardous to life, but can cause extraordinary damage to property.

A strip of shore that consists of comparatively erodible rock or sediment, such as that on Cape Cod, can be protected from erosion in several

FIGURE 14.36 Uplifted coral reefs along the margin of Huon Peninsula in eastern Papua, New Guinea. The reefs, formed at sea level during the last several hundred thousand years, have been progressively uplifted to altitudes of as much as several hundred meters.

ways. A cliff can be clad with an armor consisting of tightly packed boulders so large that they can withstand the onslaught of storm waves. It can also be defended by a strong *seawall* built parallel to the shore on foundations deep enough to prevent undermining by surf during storms. Both structures protect cliffs against ordinary storms, at least, but both are expensive.

Because of their great recreational value, beaches in densely populated regions justify greater expense for maintenance than most headlands do. A beach, however, presents a special sort of problem. As a result of beach drift (Fig. 14.21), what happens on one part of a beach affects all other parts that lie in the downdrift direction. For example, a seawall, dock, or other structure built at the updrift end of a beach reduces the amount of sand available for beach drift. The surf becomes underloaded and makes good the loss by eroding sand from along the beach. Small beaches have been completely destroyed by this process in only a few years.

Such erosion can be checked, at least to some extent, by building groins at short intervals along the beach. A *groin* is *a low wall, built on a beach, that crosses the shoreline at a right angle* (Fig. 14.37). Many groins contain openings that permit some water to pass through them. Groins act as a check on the rate of beach drift and so cause sand to accumulate against their updrift sides. Some erosion, however, occurs beyond them on their downdrift sides.

Another way of protecting a beach that is being eroded is to bring in sand artificially and pile it on the beach at the updrift end. Surf then erodes the pile and drifts the new sand down the length of the beach. Using this method, however, the sand that artificially nourishes a beach must be continuously replenished. As can be imagined, both the feeding of a beach and the construction and maintenance of groins are expensive.

Beaches in southern California are deteriorating for another reason, likewise the result of human interference. Most of the sand on those beaches is supplied, not by erosion of wave-cut cliffs, but by alluvium dumped into the sea by streams at times of flood. The floods themselves cause damage to man-made structures along the stream courses, so dams have been built across the streamways to control flooding. Of course, the dams also trap the sand and gravel carried by the streams, thus preventing the sediment from reaching the sea. This in turn has affected the balance among the factors involved in longshore currents and beach drift,

FIGURE 14.37 Groins built along the shoreline of Miami Beach, Florida to prevent excessive loss of sand by longshore drift at this popular resort area. Sand piles up on the upcurrent side of each groin.

and has resulted in significant erosion of some beaches.

A similar situation has developed along the Black Sea coast of the Soviet Union. Of the sand and pebbles that form the natural beaches there, 90 percent was supplied by rivers as they entered the sea. During the 1940s and 1950s, three things occurred: large resort developments including high-rise hotels were built at the beaches; by construction of breakwaters, two major harbors were extended into the sea; and dams were built across some rivers inland from the coast. All this construction interfered with the steady state that had existed among the supply of sediment to the coast, longshore currents and beach drift, and deposition of sediment on beaches. By 1960, it was estimated, the combined area of all beaches along the coast had decreased by 50 percent. Then beachfront buildings began to sag or collapse as the surf undermined and ate away at their foundations. An ironic twist to the chain of events lies in the fact that large volumes of sand and gravel were removed from beaches for use as concrete aggregate, not only to construct buildings but also the dams that cut off the supply of sediment to the coast.

SUMMARY

1. Oceans cover 71 percent of the Earth's surface. Beneath them lies a topography as rugged and diverse as the topography of the continents.

2. The volume of the Earth's seawater is nearly constant, but fluctuates slightly as the amount of glacer ice on land changes. The total amount of water in the hydrosphere is close to a steady state.

3. Various substances in ionic form have been in the sea as long as seas have existed on the Earth. They are added continually to the sea by streams, which derive them as products of chemical weathering. Because ions are extracted by a variety of processes, the composition of the sea remains virtually unchanged.

4. Continental shelves, with their accompanying continental slopes, add about 15 percent to the combined area of the continents. Most continental shelves consist of thick aprons of sediment washed out from the continents and deposited mostly in relatively shallow water.

5. Topographic features of the deep sea include submarine valleys and canyons, mid-ocean ridges, seafloor trenches, abyssal plains, seamounts, and guyots.

6. Probably most seamounts are volcanic cones. Atolls and guyots point to subsidence of oceanic crust as it moves away from mid-ocean ridges and mid-plate hot spots and slowly cools.

7. Oceanic ridges occupy all the ocean basins as a world-circling chain of seafloor mountains. Oceanic ridges are the lines along which magma from the mantle adds new rock to the edges of growing plates of lithosphere. Trenches are places where older parts of the lithosphere plunge back into the mantle.

8. Surface seawater circulates as currents in a number of huge subcircular cells that rotate clockwise in the Northern Hemisphere and counterclockwise in the Southern Hemisphere. Surface ocean currents are driven by winds and move warm equatorial water toward the polar regions.

9. Deep ocean circulation is controlled by density currents caused by evaporation and by chilling of water in high latitudes. Cold water sinks in the polar regions and moves slowly toward, or even across, the equator.

10. Most of the geologic work of waves is performed by surf at depths of 9 m or less.

11. In deep water waves have little or no effect on the bottom. At a depth equal to half the wavelength (usually less than 300 m) waves can begin to stir the bottom.

12. Wave refraction tends to concentrate wave erosion on headlands and to diminish it along the shores of bays.

13. Longshore currents and beach drift transport great quantities of sand along coasts.

14. On rocky coasts the shore profile includes a wave-cut cliff, a wave-cut bench, and a beach. On gentle, sandy coasts the beach typically consists of a foreshore, berm, and backshore.

15. Depositional shore features include beaches, spits, bay barriers, barrier islands, and tombolos. Most barrier islands form offshore, in areas where a rising sea advances over a gently sloping coastal plain.

16. Constructional landforms along coasts include tropical organic reefs that may be separated from the shore by a lagoon, and lava flows that build lava deltas where they enter the sea.

17. Nearly all coasts have experienced recent submergence due to postglacial rise of sea level. Some have experienced more complicated histories of emergence and submergence due to tectonic and isostatic movements on which are superimposed the worldwide sea-level rise.

18. The shape of coasts partly reflects the amount of energy available to erode and deposit sediment. Rock structure and degree of erodibility help dictate the form of rocky coasts.

19. A shore cliff can be protected for a time, at least, by a seawall or an armor of boulders. A beach can be protected, at least temporarily, by a series of groins or by importation of sand.

SELECTED REFERENCES

Anikouchine, W. A., and Sternberg, R. W., 1973, The world ocean: Englewood Cliffs, N.J., Prentice-Hall.

Bascomb, Willard, 1964, Waves and beaches. The dynamics of the ocean surface: Garden City, N.Y., Anchor Books, Doubleday.

Dolan, R. B., Godfrey, P. J., and Odum, W. E., 1973, Man's impact on the barrier islands of North Carolina: Amer. Sci., v. 61, p. 152–162.

Gross, M. G., 1986, Oceanography, 3rd ed.: Englewood Cliffs, N.J., Prentice-Hall.

Heezen, B. C., and Hollister, C. D., 1971, The face of the deep: New York, Oxford University Press.

Inman, D. L., 1954, Beach and nearshore processes along the southern California coast: California Division of Mines Bulletin 170, Chap. 5, p. 29–34.

Kennett, J. P., 1982, Marine geology: Englewood Cliffs, N.J., Prentice-Hall.

Scientific American, 1983, The Ocean: San Francisco, W. H. Freeman.

Shepard, F. P., 1973, Submarine geology, 3rd ed.: New York, Harper and Row.

Shepard, F. P., and Dill, R. F., 1966, Submarine canyons and other sea valleys: Chicago, Rand-McNally.

Shepard, F. P., and Wanless, H. R., 1971, Our changing coastlines: New York, McGraw-Hill.

Van Andel, Tjeerd, 1977, Tales of an old ocean: New York, W. W. Norton.

The eastern side of the Sierra Nevada is a normal fault. The Sierra, comprised largely of intrusive igneous rocks, has been recently subjected to glaciation.

The Dynamic Earth

Deformation of Rock

Normal faults displace strata in Utah.

HOW IS ROCK DEFORMED?

Preceding chapters have dealt largely with external activities that can be seen and studied as they happen—activities that are driven by the Sun's heat energy. The next four chapters discuss internal activities that cannot be seen in progress, activities such as bending of rock, earthquakes, uplifting of mountains, and movement of tectonic plates. Such activities are driven by the Earth's internal heat energy.

Internal and external activities just offset each other, and so are in a steady state. The rock cycle brings new rock to the surface as fast as old rock is removed by erosion. The balance between internal and external activities provides indirect evidence that materials inside the Earth must be capable of movement; for without internal movement to counteract erosion, how could continents remain above sea level? Yet the crust and mantle are vast solid masses. How can movement occur within them by means other than formation and rise of magma? To answer the question we must consider how rock can be deformed.

Solids can be deformed in three basically different ways. The first is by *elastic deformation,* which is *the reversible or nonpermanent deformation that occurs when an elastic solid is stretched or squeezed and the force is then removed.* All solids are elastic, and rocks are no exception. The second way the shape of a solid can be changed is by fracture. When the limits of elastic deformation are exceeded, a solid may fracture. Fractures produce permanent or irreversible deformation. The third way solids are deformed is by *ductile deformation,* which is also irreversible. A solid exhibiting ductile deformation behaves elastically under low pressure, but a point is reached where *elastic properties cease and ductile flow occurs.* Examples of ductile deformation can be seen in folded strata and distorted foliation in gneisses and schists. Ductile deformation is common in metamorphic rocks.

WHAT CONTROLS DEFORMATION?

To understand elastic deformation, as well as bending, flowage, and fracture in rocks, it is helpful to review some of the elementary properties of solids. Knowledge of rock deformation comes largely from laboratory experiments in which cylinders or cubes of rock are squeezed and twisted under controlled conditions. The terms used to describe deformation in the laboratory are convenient ones for us to employ also.

When a solid is subjected to a squeezing, stretching, or twisting force, we use the term *stress,* rather than pressure, for the deforming force. Stress and pressure are measured in the same units, force per unit area. One commonly used measure of stress is the *bar,* a unit defined in Chapter 6. A bar is approximately equal to an atmosphere, which is 101,325 Pa (15 lb/in^2). Why use stress instead of pressure? Stress has the connotation of direction, whereas pressure does not. When we discuss deformation of solids, therefore, we identify the directions of maximum and minimum stress. Pressure, on the other hand, is commonly thought of as being equal in all directions, as in a fluid. The term directed pressure means the same thing as stress, and is often used when development of metamorphic textures is discussed.

When a substance is stressed, it responds by changing size or shape, or both. The term used to describe change in shape or volume is *strain.* If the length of a stressed rod is reduced by 10 percent, we say the longitudinal strain is 10 percent. Similarly, if the volume of a solid body is decreased by 10 percent when it is squeezed, we again say it has suffered a 10 percent volumetric strain. **Stress,** therefore, is a measure of *the magnitude and direction of a deforming force.* **Strain** is a *measure of the changes in length, volume, and shape in a stressed material.*

Elastic Deformation

The famous British scientist, Sir Robert Hooke (1635–1703), was the first to discover that for elastic materials, provided the strain is not large, a plot of stress against strain yields a straight line. Hooke proved his point by using a spring (Fig. 15.1); however, his finding is equally true for rocks or for any other solid elastic body. The important point to remember is that within the elastic range, a stressed solid returns to its original size and shape when the stress is removed. This is the reason that earthquake vibrations do not generally leave any mark of their passage through rocks—the vibrations do not exceed the elastic limit of the rocks they pass through.

There is a limit beyond which an elastic solid can no longer be deformed and at which rupture will occur. A cylinder of marble, for example, when stressed parallel to its axis, will rupture at about 750 bars (Fig. 15.2a). We say that the marble has been deformed, or has ruptured, by *brittle fracture.*

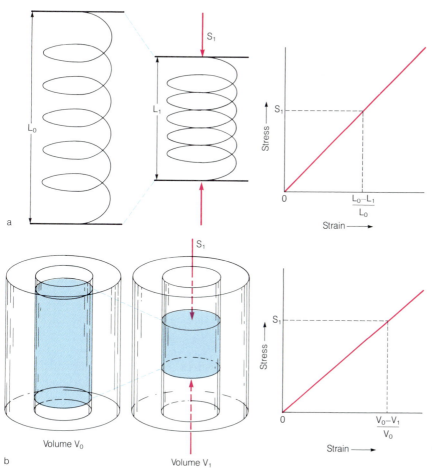

FIGURE 15.1 Hooke's law states that for elastic solids, strain is proportional to stress. (*a*) A spring, compressed by stress S_1, has its length reduced from L_0 to L_1. A plot of the strain ($L_0 - L_1/L_0$) against stress produces a straight line. (*b*) A cylinder of rock constrained by a tight metal jacket and compressed by stress S_1, has its volume reduced from V_0 to V_1. A plot of strain ($V_0 - V_1/V_0$) against stress produces a straight line.

Ductile Deformation

If another cylinder of marble is jacketed and subjected to a confining pressure while it is being stressed parallel to its long axis, a very different result is obtained. As shown in Figure 15.2*b*, the stress–strain curve rises through the elastic region; then at point *Z*, known as the *yield point*, the curve starts to flatten out. If, at point X^1, the stress is removed, the marble will change its shape along the curve X^1Y. A permanent strain, equal to *XY*, has been induced in the marble, but fracture has not occurred. The permanent change in shape is due to *ductile deformation*. If instead of relieving the stress at point X^1 in Figure 15.2*b* the marble had been stressed still further, rupture would even-

tually occur. The marble would then have been changed both by ductile deformation and rupture.

Influence of Temperature and Confining Pressure

To evaluate deformation in rocks fully we must attempt to estimate the relative importance of brittle fracture versus ductile deformation in solids. The essential conditions controlling the relative importance of the two kinds of deformation are: (1) confining pressure, (2) temperature, (3) time, and (4) composition. The higher the temperature, the weaker and less brittle a solid becomes. A rod of iron or glass is difficult to bend at room temperature. If we try too hard, both will break. However,

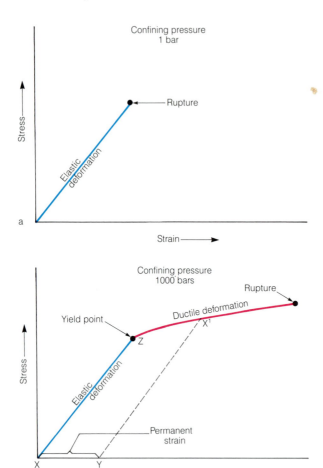

a

Confining pressure
1 bar

Stress ——→

Elastic deformation

•— Rupture

Strain ——→

b

Confining pressure
1000 bars

Stress ——→

Yield point

Z

Elastic deformation

Ductile deformation

Rupture

X¹

Permanent strain

X   Y

Strain ——→

FIGURE 15.2 Typical stress-strain curves for two kinds of irreversible deformations. (*a*) Brittle fracture occurs after a solid has been elastically deformed then the strength of the rock exceeded. (*b*) Following elastic deformation (X to Z), the yield point Z marks the onset of ductile deformation. If, at point X^1, the stress is removed, the solid will return to an unstressed state along path X^1Y. The distance XY is a measure of the permanent, irreversible, strain produced by ductile deformation. If the stress is not released at X^1, but is increased and maintained, the strength of the solid is exceeded when rupture occurs.

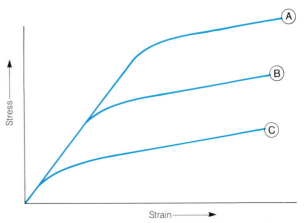

Stress ——→

Strain ——→

A

B

C

FIGURE 15.3 Typical stress-strain curves for a rock such as granite, held under a high confining pressure, but with differing temperatures and strain rates. Curve A: low temperature and high strain rate. Curve B: high temperature, high strain rate. Curve C: high temperature, low strain rate.

confining pressures, are therefore part of the reason why solid rock can be bent and folded.

Influence of Time

The effect of time on deformation of rock is vitally important, but as with confining pressure, it is not obvious from common experience. Stress applied to a solid is transmitted by all the constituent atoms of the solid. If the stress exceeds the strength of the bonds between atoms, either the atoms must move to another place in the crystal lattice in order to relieve the stress, or the bonds must break. But atoms in solids cannot move rapidly. Nevertheless, if the stress builds up slowly and gradually, and is maintained for a long period, the atoms have time to move, and the solid can slowly readjust and change shape by folding and flowing. The term used for time-dependent properties of rocks is ***strain rate,*** by which we mean the *rate at which a rock is forced to change its shape or volume.* Strain rates are measured in terms of change of volume per unit volume per second. For example, a strain rate that is sometimes used in laboratory experiments is 10^{-6}/s, by which is meant a change in volume of one millionth of a unit volume per unit volume per second. Strain rates in the Earth are much slower than this— about 10^{-14} to 10^{-15}/s. The lower the strain rate, the greater the tendency for ductile behavior to occur.

both can be readily bent if they are heated to redness over a flame. The effect of confining pressure is less familiar in common experience. Confining pressure tends to keep a solid mass in the form of a single body and hinders the formation of fractures. At high confining pressures, therefore, it is easier for a solid to bend and flow than to break. Weakening of rock by the high temperatures associated with deep burial, and at the same time loss of brittleness caused by high

A comparison of the influences of temperature, confining pressure, and strain rate can be seen in Figure 15.3. Low temperatures, low pressures, and high strain rates enhance brittle fracture. These are characteristic of the crust (especially the upper crust), and as a result fractures are common in upper-crustal rocks. High temperature, high pressure, and low strain rates, which are characteristic of the deeper crust and mantle, reduce the likelihood of brittle fracture and enhance the ductile properties of rock.

Influence of Rock Composition

The composition of a rock has a pronounced effect on its strength properties. There are two aspects to the composition. First, the films of water that fill the tiny spaces between mineral grains reduce the strength of individual mineral grains and of the whole rock. Thus, wet rocks have a greater tendency to be deformed by ductile effects than do dry rocks. Second, the minerals in a rock have a strong influence on strength properties because some minerals (such as quartz) are very strong, while others (such as calcite) are weak.

Minerals susceptible to ductile deformation are those which deform readily by creep, as ice does in a glacier (Chapter 13). Minerals with pronounced

ductile properties are halite, the carbonate minerals calcite and dolomite, and the sheet-structure silicate minerals such as clay, chlorite, mica, serpentine, and talc. Minerals which have less-pronounced ductile properties are those with isolated silicate tetrahedra—garnet and olivine—and the minerals polymerized in three dimensions, such as quartz and feldspar (Chapter 3). Chain-structure minerals, such as amphiboles and pyroxenes, have

FIGURE 15.4 Boudinage structure in a sequence of metamorphosed Jurassic tuffs, Ritter Range, Sierra Nevada, California. The grade of metamorphism is greenschist facies. The broken unit is rich in epidote and was deformed by brittle fracture, while the enclosing units, which contain quartz, feldspar, and mica, displayed ductile deformation.

FIGURE 15.5 Large-scale deformation is displayed by the folded strata in the Macdonnell Ranges, west of Alice Springs, Australia. The strata, which are Late Proterozoic and Early Paleozoic in age, were deformed about 300 million years ago. The duration of the deforming event is not known. The photograph was taken from a space shuttle in September, 1984, and covers an area approximately 60 × 100 km.

FIGURE 15.6 On December 25, 1965, a fault abruptly cut one of the main roads near Kilauea Volcano, Hawaii. The difference in height of the two sides of the fault is 1 m, but the damage was increased greatly by slump along the fault interface. Faulting occurred when large volumes of magma, previously held in deep-lying magma chambers, were rapidly erupted.

ductile properties that are intermediate between those of quartz and the micas.

Rocks that readily deform by ductile deformation are limestone, marble, shale, slate, phyllite, and schist. Rocks that tend to fail by brittle fracture rather than ductile deformation are sandstone and quartzite, granite, granodiorite, and gneiss.

When a sequence of interlayered rocks, some ductile and some strong and brittle, is stretched during deformation, an interesting difference in deformation style can be observed. The brittle layers fracture into elongate blocks called *boudins* (after a French word for sausage). The ductile layers flow into the fracture producing a structure called *boudinage* (Fig. 15.4).

DEFORMATION IN PROGRESS

So far we have discussed deformation in terms of small-scale features, such as laboratory experiments and individual minerals. It is rare to observe deformation taking place on such a local scale in the Earth. The deformation is usually too slow and too deeply buried to be observed. It is possible,

however, to find abundant evidence of large-scale movements (Fig. 15.5). Large-scale deformation is the sum of innumerable small, local deformations. Small-scale deformation can be studied in the laboratory, so that laboratory experiments are the key to understanding the large-scale deformation that can be seen in mountain ranges.

Most large-scale movements happen so slowly, which means at such low strain rates, they cannot be measured over a few tens or even a few hundreds of years. A few movements can be detected, however, and for convenience we divide large-scale movement of the crust into two groups: *abrupt movement*, involving brittle fracture, in which blocks of the crust suddenly move a few centimeters or a few meters in a matter of minutes or hours; and *gradual movement*, involving ductile deformation in which slow, steady motions occur without any abrupt jarring.

Abrupt Movement

Abrupt movement involves fracture and movement along the fracture. We call *a fracture along which the opposite sides have been displaced relative to*

FIGURE 15.7 An orange grove in southern California planted across the San Andreas Fault. Movement on the fault displaced the originally straight rows of trees. The direction of motion is such that trees in the background moved from left to right relative to the trees in the foreground.

FIGURE 15.8 Scarp formed by a fault that crosses 9th Avenue, Anchorage, Alaska. Appearing during the earthquake of March 27, 1964, the fault shows vertical displacement of approximately 2 m and horizontal displacement of about 1 m. The position of the roadway shows that the upper block has moved from left to right relative to the lower block. The small fault in the foreground likewise shows horizontal movement of about 0.5 m. Fault displacement is more recent than the snow.

each other a *fault.* Stress builds up slowly as elastic deformation occurs; then, when the strength of the rock is exceeded, fracturing occurs. Once fracturing has started, friction prevents continual steady slippage. Instead, stress again builds up slowly until friction between the two sides of the fault is overcome. Then abrupt slippage occurs again. If the stresses persist, the whole cycle of slow buildup, culminating in an abrupt movement, repeats itself many times. Although the extent of movement on a large fault may eventually total many kilometers, it is the sum of numerous small, sudden slips. Each sudden movement may cause an earthquake and, if the movement occurs near the Earth's surface, may disrupt and displace surface features. In doing so, it may leave clear evidence of the amount of movement (Fig. 15.6).

During the San Francisco earthquake of 1906, abrupt horizontal movement occurred along the San Andreas Fault. Roads and fences that crossed the fault were offset by as much as 7 m. In 1940 another earthquake occurred, again with horizontal movement along the same fault—this time in the Imperial Valley, nearly 800 km southeast of San Francisco. The displacement, as much as 5.5 m, was registered accurately by offset rows of fruit trees as well as by broken fences. Horizontal movement, therefore, seems to be a habit of the San Andreas Fault (Fig. 15.7). As we shall see in Chapter 17, this is because that fault marks a boundary along which two plates of the lithosphere are sliding past each other. In this case the surface movement is a record of deformation caused by the forces that move the plates—forces that operate hundreds of kilometers down within the Earth.

Abrupt vertical movements are more obvious than horizontal movements, as is obvious in Fig-

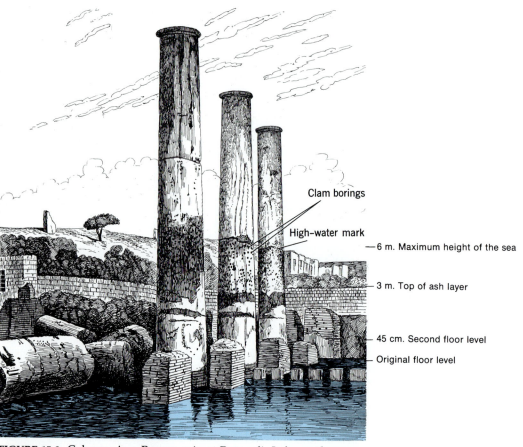

FIGURE 15.9 Columns in a Roman ruin at Pozzuoli, Italy, as they appeared in 1828. Borings made by marine clams indicate former submergence.

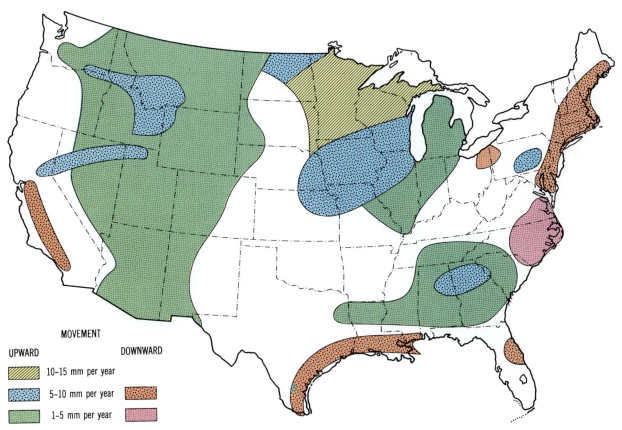

MOVEMENT

UPWARD   DOWNWARD

▨ 10–15 mm per year

▨ 5–10 mm per year   ▨

▨ 1–5 mm per year   ▨

FIGURE 15.10 Accurate measurements over a 100-year period show that in large areas of the United States the surface is slowly moving up or down. Subsidence along the coasts of California and the Gulf of Mexico is believed to have been caused in part by withdrawal of gas, oil, and water, which allows subsurface reservoirs to collapse. Uplift near the Great Lakes is a rebound effect following melting of the last ice sheet. The causes of movements in other areas are not known with certainty. Those areas in which no movement is shown are not necessarily stationary. They are simply areas in which measurements are very few. (*Source:* After Hand, 1972.)

ure 15.6, where fault cliffs as much as 1 m high cut a main road in Hawaii. Most faults, however, display a combination of horizontal and vertical motion (Fig. 15.8).

The largest abrupt displacement actually observed occurred in 1899 at Yakutat Bay, Alaska, during an earthquake. A stretch of the Alaskan shore (including the beach, barnacle-covered rocks, and other telltale features) was suddenly lifted as much as 15 m above sea level. This visible displacement may actually be less than the total amount, because the fault is hidden offshore and the block of crust on the other side of it, entirely beneath the sea, possibly moved downward, thus adding to the total displacement.

Gradual Movement

Movement along faults is not always abrupt, nor is it always accompanied by earthquakes. Measurements along the San Andreas Fault reveal places where slow, steady slipping occurs, sometimes reaching a rate as high as 5 cm a year. This seems to be a case in which continuing ductile deformation is happening at depths of 100 km or more, and brittle fracture is occurring near the surface.

A classic example of slow, gradual changes in the level of the land can be seen in the ruins of an ancient Roman marketplace known as the Temple of Serapis, west of Naples. Three columns left standing have been bored into by a distinctive

FIGURE 15.11 Wave-cut benches far above sea level on San Clemente Island off the coast of Southern California. In this air photo taken in 1960, ten benches are visible. The height of the cliff behind each bench varies, indicating that some upward movements were greater than others.

marine clam at a height of about 6 m above the floor (Fig. 15.9); the shells of the clams still line some of the borings. Along the shore near the ruin is sediment that contains abundant shells of ordinary clams like those now living in the adjoining bay; these deposits are exposed in bluffs as much as 7 m above present sea level. At first thought, one might try to explain these observations by a worldwide rise and later falling of sea level while the land remained stationary. However, such fluctuation would have left a record on all the world's coasts. As the evidence cited above is found only within a limited area near the old ruin, it is necessary to conclude that at some time after the Romans built the temple, this part of the Italian coast slowly sank, and then within more recent time was reelevated. The probable reason for the rising and sinking is the movement of magma beneath the volcanoes that surround Naples. A reconstruction of the likely sequence of events is as follows: The temple was constructed about the second century B.C., and apparently it started to subside almost immediately because a new floor was soon constructed 45 cm above the original mosaic floor. Some time later—though how much later is still

unknown—an eruption of volcanic ash from one of the nearby volcanoes buried the court area of the temple under 3 m of ash. Continuing subsidence eventually allowed the sea to invade the entire structure, so the top of the ash became the floor of the sea. The portions of the columns marked by the clam borings range from about 3–6 m from the floor of the temple; the top of the bands mark the maximum height of the sea, and the bottom of the bands mark the top of the ash.

The time when the land started to rise again is not known exactly, but according to local records it was well under way by 1500 A.D., and was nearly complete by 1638 A.D. The Temple of Serapis was offered by Charles Lyell, one of the great geologists of the nineteenth century, as an example of the slow movements that continually shift, distort, and deform the crust. It may be that across the entire surface of the Earth, no spot is completely stationary. Measurements by U.S. Government surveyors over the past 100 years, for example, reveal great areas of the United States where the land is slowly sinking and other places where it is slowly rising (Fig. 15.10). The causes of these vast, slow movements are not all well-understood, but the movements do prove that the solid Earth is not as rigid as it seems at first sight, and that great internal forces are continually deforming its crust.

EVIDENCE OF FORMER DEFORMATION

With such convincing evidence of present-day deformation of the Earth's crust, we might reasonably expect to find a great deal of evidence of former deformation—indeed, we do. Studies of land and sea-bottom topography provide abundant evidence of vertical movements, and in some areas the distribution of various kinds of rock provide clear evidence that horizontal movements through distances as great as several hundred kilometers have occurred. We will discuss a few examples of such evidence, beginning with topographic features.

Topographic Features

In many parts of the world, well-developed, wave-cut benches stand one above another like stairsteps (Fig. 15.11). Some of the lowest steps, still decorated with barnacles, terminate inland against typical wave-cut cliffs. Along the southern California coast and nearby islands, the highest recognizable terraces are now more than 450 m above the sea. Because these terraces are found along only a part of that coastal area, we reason that a segment

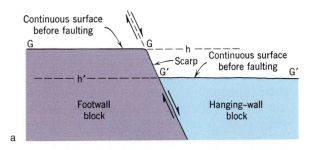

a

b

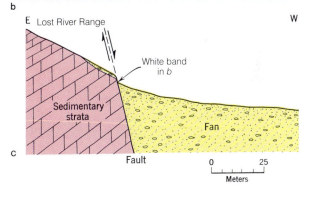

c

FIGURE 15.13 Fault that displaces surface of ground (*a*). Scarp made by displacement of a formerly level surface G-G' may mean that the hanging-wall block moved down from position h to h'; that the footwall block moved up from position h'; or that both blocks moved to some degree, to create the net displacement shown. (*b*) Scarp at west base of Lemhi Range, Idaho, formed in 1984 by abrupt movement along a fault that generated an earthquake. The whitish band of newly exposed rock marks displacement at the top of the valley fill. (*c*) Vertical section across lower part of Lost River Range, approximately in the direction of the photographer's line of sight in (*b*). Half arrows indicate relative movement of crustal blocks.

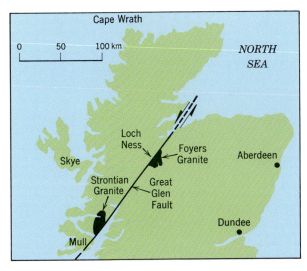

FIGURE 15.12 The Great Glen Fault in Scotland cuts a granite mass into two separate segments, the Strontian Granite and the Foyers Granite. Horizontal movement along the fault has separated the masses by approximately 100 km. (*Source:* After Kennedy, 1946.)

of the Coast Ranges has risen in a succession of pulses, separated by pauses long enough to allow surf to create cliffs and beaches. The altitude of individual terraces varies appreciably from place to place, indicating that the amount of uplift was irregular. Apparently, the uplift of coastal California is caused by pressures developed between two moving plates of lithosphere. The suture between the Pacific and American Plates passes very close to the region of uplift (Fig. 2.11).

A variety of topographic evidence indicates large-scale subsidence. Accurate surveys of the seafloor north of the Aleutian Islands reveal a sub-

marine topography of high ridges and hills separated by valleys that unite in what appears to be a well-developed drainage system. The best explanation seems to be the submergence of a broad land area that was shaped by mass-wasting and stream erosion. Was this drowning of a former landscape caused by the rise of sea level or by the sinking of the land? We know the level of the sea was raised by the return of water that had been locked up in ice sheets during the glacial ages. The total rise from this cause is estimated to have been about 100 m. However, the drowned hills and valleys near the Aleutian Islands lie at depths greater

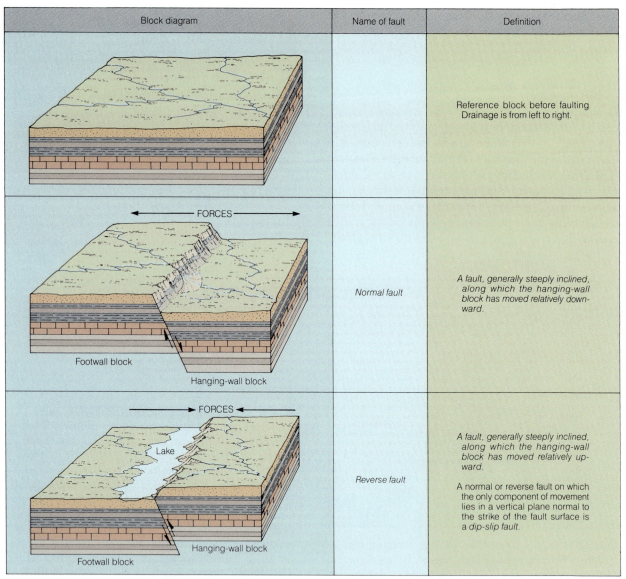

Block diagram	Name of fault	Definition
		Reference block before faulting Drainage is from left to right.
	Normal fault	*A fault, generally steeply inclined, along which the hanging-wall block has moved relatively downward.*
	Reverse fault	*A fault, generally steeply inclined, along which the hanging-wall block has moved relatively upward.* A normal or reverse fault on which the only component of movement lies in a vertical plane normal to the strike of the fault surface is a *dip-slip fault.*

FIGURE 15.14 Principal kinds of faults, the directions of forces that cause them, and some of the topographic changes they cause.

than 400 m. We infer, therefore, that a large part of the submergence was caused by the sinking of land.

Bedrock Features

When we examine the distribution of various kinds of rocks, we find a great deal of evidence of former movement, yet one striking example should suffice. A remarkable fault, the Great Glen Fault, crosses Scotland from southwest to northeast. The trace of the fault is a line of easy erosion, and a valley containing a string of lakes now marks its path. One lake is Loch Ness, home of the famous Lock Ness Monster. When, during the Paleozoic Era, the Great Glen Fault was active, it severed and displaced the metamorphic zones, so that the grade of metamorphism can be seen in places to change abruptly from the biotite zone south of the fault to the sillimanite zone north of it (Fig. 6.17). Even more striking evidence is presented by the two fragments of a large granite mass severed by the fault. The fragments now lie on opposite sides of the fault, approximately 100 km apart, and thus give striking evidence of large-scale horizontal movement (Fig. 15.12).

Block diagram	Name of fault	Definition
	Strike-slip fault	*A fault on which displacement has been horizontal. Movement of a strike-slip fault is described by looking directly across the fault and by noting which way the block on the opposite side has moved. The example shown is a left-lateral fault because the opposite block has moved to the left. If the opposite block has moved to the rithe it is a right-lateral fault. Notice that horizontal strata show no vertical displacement.*
	Oblique-slip fault	*A fault on which movement includes both horizontal and vertical components. See also Fig 15.8. Forces are a combination of forces causing strike-slip and normal faulting.*
	Hinge fault	*A fault on which displacement dies out (perceptibly) along strike and ends at a definite point. Figure 15.8 shows a small example located in the foreground of the photograph, between the viewer and the man walking away from the camera. Forces are the same as those causing normal faulting.*

Not all evidence of movement and deformation observed in bedrock is as obvious as the Great Glen Fault. But once we learn to recognize it, evidence of deformation is seen to be very widespread—so much so that a special branch of geology, *structural geology,* has the study of rock deformation as its primary focus. In order to evaluate the Earth's internal activities, we must be able to recognize and evaluate evidence of rock deformation. First, we will examine deformation by fracture.

DEFORMATION BY FRACTURE

The brittle properties of rock are the properties that most commonly lead to fracture. Rock in the crust, especially rock close to the surface, tends to be brittle. As a result rock near the surface tends to be cut by innumerable fractures, most of them joints (Chapter 6). The fractures speed erosion, serve as channels for the circulation of groundwater, provide entryways along which magma is intruded, and in many places serve as the openings in which veins of valuable minerals are deposited.

Most fractures are small, and little or no slippage has occurred along them. They are like small cracks in a pane of window glass. Along a few fractures, however, visible movement has occurred, and the fractures are therefore faults.

Generally, there is no way of telling how much movement has occurred along a fault, nor which side of the fault has moved. In an ideal case, for example, if a single mineral grain or a pebble in a conglomerate has been cut through by the fault and the halves carried apart a measurable distance, the amount of movement can be determined. Yet even with this case it is not possible to say whether one block stood still while the other moved past it, or whether both sides shared in the movement. The only way movement can be determined is to observe it happening. Precise surveys of points on the ground before and after movement on an active fault can indicate which side is moving. Such a check is made continuously along the San Andreas Fault. However, most faults are old features whose former expression at the Earth's surface was destroyed by erosion long ago. Under such circumstances we can never know which side moved. In classifying fault movements, therefore, it is only possible to speak of *apparent* and *relative* displacements.

Most faulting occurs along fractures that are inclined. To describe this inclination geologists have adopted two old mining terms. From a miner's viewpoint, one *wall* of an inclined vein overhangs him, while the other is beneath his feet. Because veins of ore commonly occupy openings created by faults, we use the old miner's terms in the following way: The **hanging wall** is *the surface of the block of rock above an inclined fault; the surface of the block of rock below an inclined fault is the* **footwall** (Fig. 15.13). These terms, of course, do not apply to vertical faults.

Faults are grouped into classes according to (1) the inclination of the surface along which fracture has occurred and (2) the direction of relative movement of the rock on its two sides. The common classes of faults, together with the changes in local topography they sometimes create, are listed in Figure 15.14. The standard planes of reference in classifying faults are the vertical and the horizontal. Along some faults, movement is confined to one of these two reference planes, although (as shown in Fig. 15.8) along other faults both vertical and horizontal movements occur. Not all movement on faults is in straight lines. Sometimes fault blocks are rotated; we classify such faults as hinge faults (Fig. 15.14). Many normal and reverse faults become hinge faults as they approach the point where they dissipate.

Normal Faults

Normal faults are caused by tensional forces that tend to pull the crust apart, and also by forces tending to expand the crust by pushing it upward from below. There are many zones in the crust that have been repeatedly deformed by normal faulting. Commonly, two or more similarly trending normal faults enclose an upthrust or down-dropped segment of the crust. As shown in Figure 15.15, a down-dropped block is a *graben* if it is bounded by two normal faults, and a *half-graben* if subsidence occurs along a single fault. An upthrust block is a *horst*. The central, steep-walled valley that runs down the center of the oceanic ridge and cuts through Iceland (Figs. 2.11 and 14.8) is a graben. Perhaps the world's most famous system of grabens and half-grabens is the African Rift Valley (Fig. 15.16), which runs north–south through more than 6000 km. Within parts of it are volcanoes where magma has followed channelways that lead upward along the fault surfaces.

Normal faults are innumerable. Horsts and grabens are also very common, although none is as spectacular as the African Rift Valley. The north–south valley of the Rio Grande in New Mexico is a

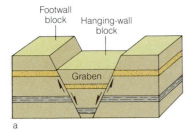

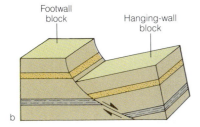

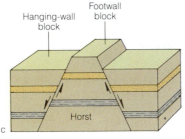

d

FIGURE 15.15 Horsts and grabens formed when tensional forces produce normal faults. (*a*) ***Graben,*** *a trench-like structure bounded by parallel normal faults,* formed when hanging-wall block that forms the trench floor moves downward relative to the footwall blocks. (*b*) ***Half-graben,*** *a trench-like structure formed when the hanging-wall block moved downward on a curved fault surface.* (*c*) ***Horst,*** *an elevated elongate block bounded by parallel normal faults,* formed when the elevated footwall block moves upward relative to the hanging-wall block. (*d*) The African Rift Valley in central Kenya, seen here in a LANDSAT image, is a series of horsts and grabens. The image is about 70 km wide and 140 km long. To the east (right-hand side) is a high plateau bounded by a series of normal faults. Within the valley, normal faults run due north. Several volcanic cones are visible; magma is presumed to rise up the faults bounding the grabens.

graben. The valley in which the Rhine River flows through western Europe follows a series of grabens, and Lake Baikal in central Asia, the Earth's deepest lake, is located in a very deep graben (Fig. 15.17). A spectacular example of normal faulting is found in the Basin and Range Province in Utah and Nevada. There, movement on a series of parallel and subparallel, north–south, normal faults

has formed horsts and half-grabens that are now mountain ranges and sedimentary basins. The province is bounded in the east by the western edge of the Wasatch Range and continues westward to the eastern edge of the Sierra Nevada (Fig. 18.8).

Reverse Faults

Reverse faults arise from compressive forces. Movement on reverse faults pushes older rocks over younger ones, thereby shortening and thickening the crust.

A special class of reverse faults, called *thrust faults* and generally known as *thrusts,* are *low-angle reverse faults with dips less than 45°.* Such faults, common in great mountain chains, are noteworthy because along some of them the hanging-wall block has moved many kilometers over the foot-wall block. In most cases the hanging-wall block, thousands of meters thick, consists of rocks much older than those adjacent to the thrust on the foot-wall block (Fig. 15.18). The strata above some

FIGURE 15.16 African Rift Valley in Tanzania. The eastern wall of one of the many rifts that comprise the African Rift Valley, a giant system of grabens and half-grabens several thousand kilometers in length. The valley floor in which Lake Manyara now lies was originally at the same height as the plateau above, but has been lowered by movement on a normal fault. The fault surface has been modified by erosion, so the valley walls are no longer straight.

FIGURE 15.17 Lake Baikal, in central Asia, the world's deepest lake, is in a graben. The view is of the southern end of the lake, looking to the north.

a

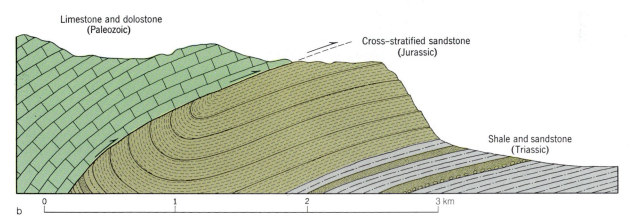

Limestone and dolostone
(Paleozoic)

Cross–stratified sandstone
(Jurassic)

Shale and sandstone
(Triassic)

0   1   2   3 km

b

FIGURE 15.18 Keystone Thrust, west of Las Vegas, Nevada. (*a*) Air view northward shows the fault clearly defined by a color contrast in the strata adjacent to it. Light-colored Jurassic sandstone, forming a cliff nearly 600 m high (right), lies below the fault; dark-colored Paleozoic limestones and dolostones (left) lie above it. (*b*) Section drawn across photograph above and extending somewhat farther east and west. Canyons crossing the fault reveal that it steepens downward toward the west and crosses overturned layers of the sandstone. Farther east the fault becomes essentially parallel to the sedimentary layers, both below and above.

thrusts lie nearly parallel to those beneath, and so may appear, deceptively, to represent an unbroken sequence.

Strike-Slip Faults

Two famous strike-slip faults have already been mentioned—the San Andreas Fault and the Great Glen Fault (Fig. 15.12). Strike-slip faults are those in which all the fault movement is horizontal (Fig. 15.14). The San Andreas is a right-lateral strike-slip fault (Fig. 17.19). Apparently, movement may have been occurring along it from at least as long ago as Cretaceous time to the present day—in other words, through at least 65 million years. The total movement is not known, but some evidence

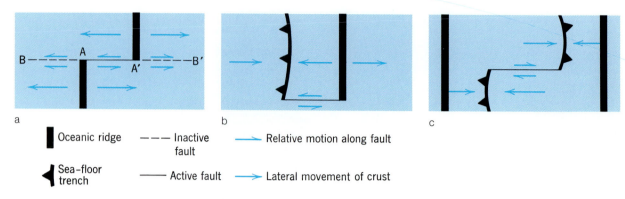

a

b

c

| | Oceanic ridge | - - - | Inactive fault | → | Relative motion along fault |
| | Sea-floor trench | —— | Active fault | → | Lateral movement of crust |

FIGURE 15.19 Transform faults, a special class of strike-slip faults forming a globe-encircling network with oceanic ridges and sea-floor trenches. (*a*) An oceanic ridge is offset by a transform fault. Crust on both sides of the two ridge segments is moving laterally away from the ridge. Between segments of the ridge, along A-A', movement on the two sides of the fault is in opposite directions. Beyond the ridge, however, along segments A-B and A'-B', movement on both sides of the fault is in the same direction. A transform fault does not, therefore, cause the ridge segments to move continuously apart. (*b*) A transform fault joins an oceanic ridge with a sea-floor trench. New crust formed at a spreading edge (the oceanic ridge) moves laterally away from the ridge and plunges back into the mantle at a place marked by the sea-floor trench. The triangles on the trench point in the direction of movement of the downward-plunging crust. (*c*) Transform fault joining two sea-floor trenches.

suggests that it now amounts to more than 600 km. The Great Glen Fault is a left-lateral strike-slip fault; it is not presently active.

Transform Faults

Strike-slip faults arise most commonly as a result of movement between adjacent plates of lithosphere. A special kind of strike-slip fault is widespread in the oceanic crust. For example, many of the great fracture zones that cut the oceanic-ridge system are strike-slip faults. In Figure 14.6, each of the numerous fracture zones that offset the Mid-Atlantic Ridge is a strike-slip fault. Indeed, they are so common that it has been suggested that the three major structural forms marking sites of deformation of the Earth's crust are seafloor trenches, oceanic ridges, and strike-slip faults. These features link together to form continuous networks encircling the earth. When one feature terminates, another commences; their junction point is called a *transform*. J. T. Wilson, a Canadian scientist who first recognized the network relation, proposed that *the special class of strike-slip faults that links major structural features* be called **transform faults.** Close study of the strike-slip faults that off-

set the oceanic ridges proved Wilson's suggestion correct. As seen in Figure 15.19, movement along transform faults is a consequence of the continuous addition of new crustal material along oceanic ridges, the lateral movement of older crust away from the ridge, and its consumption beneath sea-floor trenches.

Evidence of Movement Along Faults

Often we find fractures in rock but cannot tell at first glance whether or not movement has occurred along them. For example, in uniform, even-grained rock such as granite, or in a pile of thin-bedded strata, no one of which is unique or distinctive, we would not see displacement of any obvious features. However, examination of the fault surface, or of rock immediately adjacent to it, commonly reveals signs of local deformation, indicating that movement has occurred. Under special circumstances, even the direction of movement can be deciphered.

Adjacent to some faults, the bending of strata or other internal features can be seen. Large and small *structures created by bending adjacent to faults* are known collectively as **fault drag** (Fig. 15.20).

Movement of one mass of rock past another can cause the fault surfaces to be smoothed, striated, and grooved. *Striated or highly polished surfaces on hard rocks, abraded by movement along a fault,* are **slickensides.** Parallel grooves and striations on such surfaces record the direction of most recent movement (Fig. 15.21*a*).

Not all fault surfaces are slickensides. In many instances fault movement crushes rock adjacent to the fault into a mass of irregular pieces, forming *fault breccia* (Fig. 15.21*b*). Most intense grinding breaks the fragments into such tiny pieces that they may not be individually visible even under a microscope.

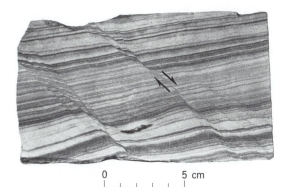

```
0                    5 cm
```

FIGURE 15.20 Fault drag. Near small faults that cut sandstone, thin layers have been bent during movement. Direction of bending indicates that direction of last movement was shown by arrows. The drag in this case indicates these are normal faults.

DEFORMATION BY BENDING

Bending may consist of broad, gentle warping that extends over hundreds of kilometers, or it might be close, tight flexing of microscopic size, or anything in between. Regardless of the volume of rock involved or the degree of warping, we refer to the bending of rocks as *folding*. Before we discuss folds and folding, it is necessary to become familiar with the terms used to describe them.

The simplest fold is a **monocline,** *a one-limbed flexure, on both sides of which the strata either are horizontal or dip uniformly at low angles* (Fig. 15.22). An easy way to visualize a monocline is to lay a book on a table. Then drape a handkerchief over one side of the book and out onto the table. So draped, the handkerchief forms a monocline.

Most folds are more complicated than monoclines. *An upfold in the form of an arch* is an **anticline** (Fig. 15.23). *A downfold with a troughlike form* is a **syncline.** As we see in Figure 15.23, *the sides of a fold* are the **limbs,** and *the median line between the limbs, along the crest of an anticline or the trough of a syncline,* is the **axis** of the fold. *A fold with an inclined axis is* said to be a **plunging fold,** and *the angle between a fold axis and the horizontal* is the **plunge** of a fold. *An imaginary plane that divides a fold as symmetrically as possible, and that passes through the axis,* is the **axial plane.**

Many folds, such as those in Figure 15.23, are nearly symmetrical. Others, however, are not symmetrical; strong deformation may create complex shapes. The common forms of folds are shown in Figure 15.24. If only fragmentary expo-

a   b

FIGURE 15.21 Slickensides are the polished and striated surfaces produced by faulting. (*a*) Specimen of a slickenside developed on basalt. (*b*) Fault breccia. Angular gneiss fragments (dark) broken by faulting set in a matrix of rock flour and calcite. Titus Canyon, Death Valley.

sures of bedrock are available, it is apparent that there might be difficulties in deciding whether a given fold is overturned or not. We must know whether a layer is right-side up or upside down in order to decide which limb of a fold it is in. This is not always possible, but in some cases sedimentary structures, such as mud cracks and graded layers, do record this (Fig. 7.3). In other examples

FIGURE 15.22 A monocline in southern Utah that interrupts the generally flat-lying sedimentary strata of the wide Colorado Plateau. On both sides of the monocline the exposed layers are nearly horizontal. View looking south.

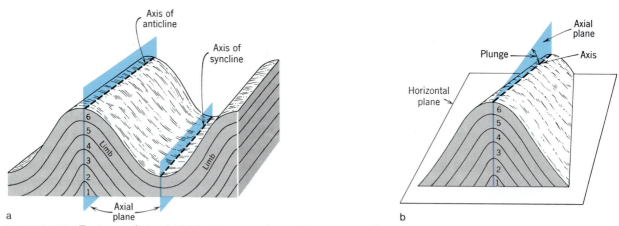

FIGURE 15.23 Features of simple folds. Upper surface of the youngest layer (6) slopes toward the axis of the syncline but away from the axis of the anticline. (*a*) Fold axis horizontal. (*b*) Fold axis plunging.

only careful, thorough mapping of all bedrock exposures can provide the answer.

Folds are often so large that when we examine a single exposure we are not even aware we are seeing folded rock. Nevertheless, when all the exposures of a particular rock body are plotted and a geologic map is prepared (Appendix D), folds can be recognized from the distribution of the various

Name	Description	
Symmetrical	Both limbs dip equally away from the axial plane.	
Asymmetrical	One limb of the fold dips more steeply than the other.	
Overturned	Strata in one limb have been tilted beyond the vertical. Both limbs dip in the same direction, though not necessarily at the same angle.	
Recumbent	Axial planes are horizontal. Strata on the lower limb of anticline and upper limb of syncline are upside down.	
Isoclinal	Both limbs are essentially parallel, regardless whether the fold is upright, overturned, or recumbent.	

a

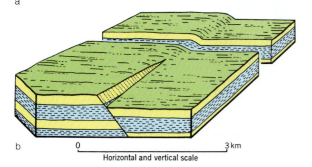

b

0   3 km
Horizontal and vertical scale

FIGURE 15.24 (*a*) Five kinds of folds. (*b*) Hinge fault (front block) passes laterally into monocline (rear block).

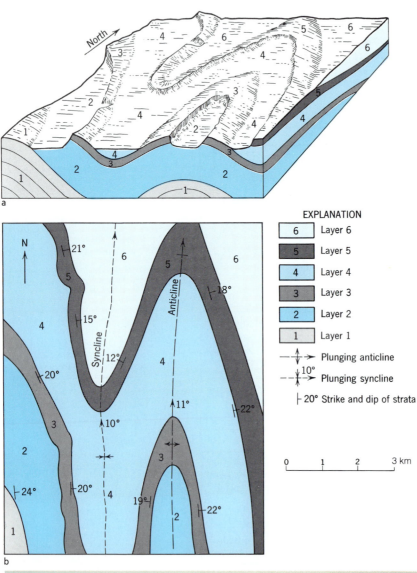

EXPLANATION

6	Layer 6
5	Layer 5
4	Layer 4
3	Layer 3
2	Layer 2
1	Layer 1

Plunging anticline

Plunging syncline

⊢20° Strike and dip of strata

0   1   2   3 km

FIGURE 15.25 Distinctive topographic forms and distinctive patterns in the distribution of various kinds of rock reveal the presence of plunging folds. (*a*) Block diagram showing topographic effects. (*b*) Geologic map of area shown in (*a*). Appendix D gives further details on the preparation and reading of geologic maps.) (*c*) Air view of the plunging anticline at the northern end of Sheep Mountain, Wyoming. The view is towards the northeast. Resistant sandstone layers make jagged, low ridges; shale layers erode readily and form the valleys. The curve of the sandstone ridges points in the direction of the plunge.

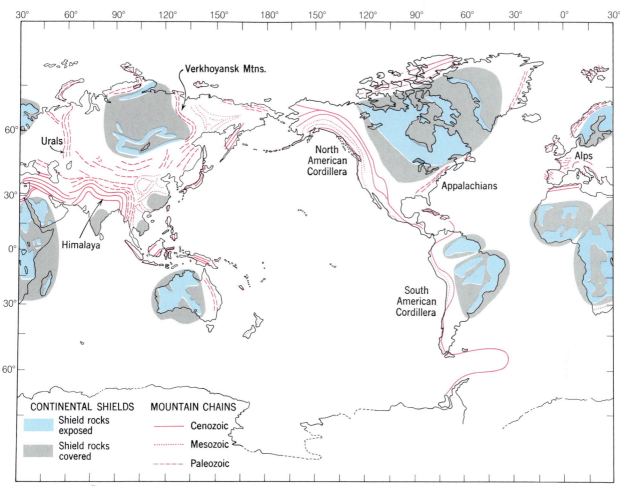

FIGURE 15.26 Continental shields and the mountain chains formed since the beginning of the Paleozoic Era. Each shield is itself a patchwork of remnants of cratons and older mountain chains.

rock types. Differences of erodibility in adjacent strata can also lead to distinctive topographic forms by which we can recognize the presence of folds (Fig. 15.25).

Gentle upward or downward warping of the crust, which is really gentle folding on a very large scale, is common. It is likely to be most evident in arid regions where, because of the absence of concealing regolith and the presence of deep dissection by streams, wide areas of rock layers are exposed. Most commonly, however, vegetation and regolith prevent direct observation, and the presence of warping may not even be apparent from examination of several exposures. Inclinations of strata may be only 1 or 2°, and such small slopes are not conspicuous. Yet if a stratum has a persistent dip of 2°, the altitude of the layer changes by about 38 m/km. Therefore, careful mapping reveals even the most gentle warping. Upwarping

forms a dome, and down-warping creates a basin. Domes are particularly important because some of them are the sites of large accumulations of oil and gas (Chapter 21).

Relations Between Faults and Folds

Folds and faults can be related to each other in various ways. As we have just seen (Fig. 15.20), strata near active faults may be folded by the effects of frictional drag along the planes of movement. A steeply inclined fault may pass, either upward or laterally, into a fold and thus die out. The hinge fault shown in Figure 15.24*b* is an example of how a high-angle fault can become a monocline. Typically, the projected continuation of the fault surface coincides with the axial plane of the related fold.

Thrusts sometimes form in sedimentary strata

that have been compressed and folded into overturned folds. Compression can break a fold, generally near the axial plane, and a combined fold and thrust fault results. Apparently the Keystone Thrust (Fig. 15.18) was formed in this fashion.

Not only have strata been deformed as the result of active forces of compression, but some thrusts have themselves been folded as if they were strata. A few thrusts have even been overturned. Deciphering the sequence of events in such a case can be a challenging and difficult task.

REGIONAL STRUCTURES

When the crust is viewed on the scale of a continent, two distinctly different kinds of structural units can be distinguished. The first unit is a kind of core or nucleus of very ancient rock and is called a *craton.* The term is applied to *a portion of the Earth's crust which has attained tectonic stability and has been little deformed for a prolonged period.* Rocks within cratons may be deformed, but the deformation is invariably ancient. Cratons are the cores around which continents seem to have grown (Fig. 15.26).

Draped around cratons are the second kind of crustal building unit, *orogens* or *orogenic belts,* which are *elongate regions of the crust that have been intensely folded and faulted during mountain-building processes.* Orogens differ in age, history, size, and origin; however, all were once mountainous terrains. Only the youngest orogenic belts are mountainous today; ancient orogenic belts that are now deeply eroded reveal their history through the kinds of rock they contain and the way they are deformed. Within an orogenic belt it is sometimes observed that several different geological units are present. *An elongate series of mountains belonging to a single geologic unit* is called a *mountain range.* Excellent examples are the Sierra Nevada in eastern California and the Front Range in Colorado. *A group of ranges similar in general form, structure, and alignment, and presumably owing their origin to the same general causes,* constitute a *mountain system.* The Rocky Mountain system is a great assemblage of ranges, all formed within a few million years of each other, that extend northward beginning near the Mexican border, through the United States, and up to western Canada. The term *mountain chain* is used somewhat more loosely to designate *an elongate unit consisting of numerous ranges or systems, regardless of similarity in form or equivalence in age.* An example is the gigantic mountain chain that runs along the western edge of the Americas, from the tip of South America to northwestern Alaska, and that includes all the systems and ranges in between. This broad belt of ranges is also called the *American Cordillera.*

All the major mountain chains that formed during the last 600 million years (that is, since the beginning of the Paleozoic Era) are depicted in Figure 15.26. There is abundant evidence that much older mountain chains once existed on the Earth, since even in the cratons we find belts of intensely deformed and folded rocks which are now deeply eroded. They remain as mute evidence that great mountain ranges once towered above. Indeed, it is probable that most fragments of continental crust were once part of mountain ranges, and that continents have grown to their present sizes by the welding of younger and younger mountain chains onto the growing cratons.

Cratons and Continental Shields

By careful mapping of structures and dating of rocks, geologists have been able to divide North America and adjacent islands into several major tectonic units (Fig. 15.27). The ancient craton, within which most rocks are older than 2.5 billion years and all are older than 1 billion, can be seen over much of eastern Canada. Within the United States the cratonic rocks only crop out in a small region around Lake Superior. Nevertheless, by drilling through the cover of sedimentary rocks we know that the craton lies below much of the central United States and part of western Canada. Within this covered cratonic region, labeled the Interior Lowland on Figure 15.27, the covering sedimentary rocks are nearly flat-lying and little deformed. To distinguish *the portion of a craton where rocks are exposed at the surface* we employ the term *continental shield. That portion of a craton that is covered by a thin layer of little deformed sediments* is called a *stable platform.*

Rocks within cratons are of all kinds—sedimentary, metamorphic, and igneous—but two groups of igneous rocks form an especially distinctive unit. The first are ancient basalts, similar in composition to modern seafloor basalts, that have been changed by low-grade metamorphism into chlorite schists. Within these ancient, altered, and metamorphosed basalts (known as *greenstone*), it is still possible to find evidence of pillow structure (Fig. 15.28), indicating that the basalts were extruded under water. The second kind of igneous rock forms small, granitic plutons. The granites in-

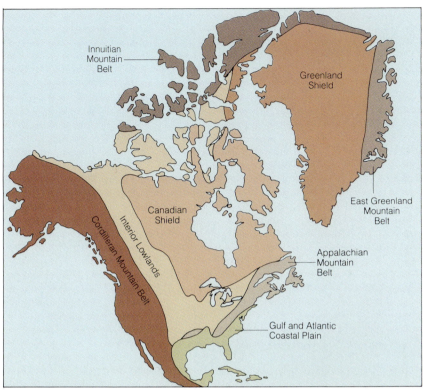

FIGURE 15.27 Major tectonic features of North America. The shields contain rocks older than 2 billion years. The Interior Lowland is underlain by rocks 2 billion years or older, but is covered by flat-lying Paleozoic sedimentary rocks. The edges of the Interior Lowland mark the northern, western, and southern margins of the North American craton. The mountain belts surrounding the craton are younger than 600 million years.

trude the greenstone and produce a distinctive pattern known as a *granite-greenstone terrane* (Figs. 15.29 and 19.23).

Most rocks within cratons are older than 2.5 billion years, which means that cratons formed during the Archean Eon. Recent studies of the North American craton have revealed that the craton itself consists of smaller fragments with differing geological histories. There are actually two large cratons separated by an ancient (1.9-billion-year-old) orogen (Fig. 15.30). Within each large craton unit, still smaller bodies of cratonic rock can be discerned. The small cratonic fragments were probably minicontinents during the Archean Eon. By about 2 billion years ago, the smaller fragments had become welded together to form the larger cratons we see today. As the larger cratonic fragments moved, orogenic belts were formed around their margins. The existence of ancient orogenic belts, formed as a result of collision between cratons, is the best evidence available to support the idea that plate tectonics operated at least as far back in time as 2 billion years ago.

Surrounding the ancient rocks in the core of the North American craton is a collar of deeply eroded, complexly deformed orogens ranging in age from 1.0 to 1.9 billion years. Most of the rock in

FIGURE 15.28 Pillow basalt older than 3 billion years in the Barbeton Mountainland, South Africa. The basalt has been metamorphosed to the greenschist facies, but the outlines of pillows, up to 1 m across, are beautifully preserved despite their extreme age.

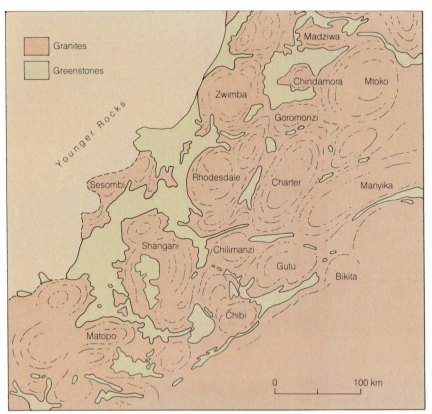

FIGURE 15.29 A striking pattern of small, rounded granite batholiths surrounded by greenstones is observed in the Zimbabwe craton. Each batholith has a local name. Compare Figure 19.23. (*Source:* After Macgregor, 1951.)

the collar now lies beneath the sedimentary rocks of the stable platform. Nevertheless, samples obtained through drilling, and observations made in those places where the rocks crop out, indicate that the 1.0- to 1.9-billion-year-old rocks are the eroded remnants of former mountain belts.

Surrounding the stable platform are even younger belts of highly deformed rocks, forming what resembles a showy necklace of young mountain systems that have not yet been worn down. These are our familiar mountain systems such as the Appalachians, the Rockies, and the Sierra Nevada. All these have been formed during the last 0.6 billion years. Therefore, the structure of North America suggests that an ancient craton had formed by the welding of smaller fragments of crust. The continent then grew slowly larger as successive mountain systems were formed along its margins.

There are two schools of thought concerning the growth of continents and the amount of continental crust. One suggests that the total volume of continental crust grew rapidly between 4.0 and 2.5

billion years ago, but has remained constant through the last 2.5 billion years and is being continually recycled. The second school of thought, to which the authors of this text are attracted, maintains that the continental crust is still slowly increasing by addition of magma derived from partial melting of subducted oceanic crust. As an example, the Andes in South America are thought to be now growing larger by addition of magma from the mantle. To counter this argument, the "constant volume of crust" school maintains that a small amount of continental crust must somehow be subducted into the mantle. The question of growth or constancy is still open and is a fascinating area for research.

Mountains

Although we all know that mountains are rocky masses standing above the surrounding terrain, it is not easy to classify mountains on the basis of their geology. Mountains display such a great variety of rocks and structures that no two are identical. If we concentrate on the details, we are in danger of seeing only the foliage and missing the forest. The most helpful way to organize thoughts about mountains, and to see through the foliage, is to identify the single, most characteristic feature, and use it for classification. On this basis it is possible to identify three principal kinds of mountains.

1. Volcanic mountains.
2. Fault-block mountains.
3. Fold-and-thrust mountains.

Volcanic Mountains

Some of the world's most beautiful and scenic mountains are volcanoes. Mount Fuji, Mount Etna, Mount Rainier, Mount Mayon, and Mount Kilimanjaro are examples. Volcanic mountains differ in a fundamental way from other mountains in that they are formed by deposition of volcanic rock and not by deformation of preexisting crust. Although volcanic mountains are found on land, they are far more abundant on the seafloor. In some chains of seafloor mountains, such as the chain of volcanoes that forms the Hawaiian Islands, the higher peaks protrude above sea level. In other chains (for example, in the oceanic ridges) the entire chain is submerged.

A special class of volcanic mountain chain, an island arc, is defined in Chapter 17. It is a great arcuate belt of andesitic and basaltic volcanic islands formed over a subduction zone in oceanic crust. Some island arcs are 2000 km or more in length. The Aleutian Islands are a conspicuous island arc; another arc runs from Kamchatka through the Kurile Islands and down through Japan; yet another consists of the islands of Sumatra, Java, Sumba, and Timor.

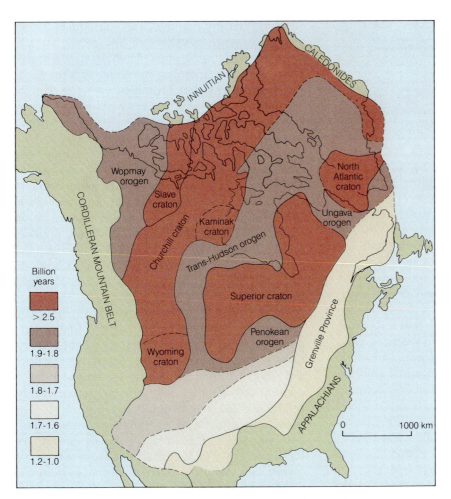

FIGURE 15.30 Assemblage of small cratons in the North American craton. Accretion to the cratons of the Wopmay and Trans-Hudson Orogens was apparently complete by about 1.9 billion years ago. (*Source:* After Kerr, 1985.)

FIGURE 15.31 Volcanoes of the Cascade Range. Each volcano has been active during the last 2 million years. (*Source:* After Tabor and Crowder, 1969.)

Where a subduction zone occurs beneath an edge of a continent, the volcanoes that form are located on continental crust. Instead of an island arc, therefore, the result is a chain of stratovolcanoes on land. An example of a volcanic mountain chain formed in this manner is the Cascade Range in the northwestern United States (Fig. 15.31). This is a range of huge young, andesitic volcanoes, running from Lassen Peak, California (at the south end) to Mount Baker, Washington (more than 900 km farther north) and also includes several volcanic peaks in southern British Columbia. These snow-covered giants (all active during the past few million years, and some, such as Mount St. Helens, still active today) were erupted onto a platform of older, folded, and deeply eroded rocks. The trend of the line of volcanoes cuts directly across many other geologic features, and for this reason scientists believe that the build-

ing of the mountains is controlled by wet partial melting of subducted oceanic crust (Chapter 4). The piece of oceanic crust that is being subducted and is responsible for forming the Cascade Range is the Juan de Fuca Plate.

Fault-Block Mountains

In many parts of the world isolated mountain ranges stand abruptly above surrounding plains. Study reveals that these ranges are separated from the intervening lowland areas by faults of great displacement. The ranges seem to be giant pieces of crust punched upward from below. These are fault-block mountains. Rock within the mountains commonly contains evidence that former fold-and-thrust mountains once occupied the same sites, but that erosion had worn them down before the fault blocks formed. An example of mountains of this kind is the Adirondack Mountains of northern New York. Originally a fold-and-thrust mountain range of Precambrian age, the Adirondacks were eroded down to a surface of low relief that was stable for millions of years. Then, in the Cenozoic Era, the region again became active and the modern Adirondacks were lifted up as a series of great fault blocks. *How* and *why* are a puzzle. One suggestion is that the eroded remnants of the ancient mountain range still had a root of low-density rocks beneath them, so that an upward-pressing force existed. The force is called an isostatic restoring force, and it is discussed in more detail in Chapter 16. If the force were not great enough to overcome the friction that opposes movement on faults, nothing would happen. However, if movement of the lithosphere caused a slight warping or twisting of the crust, the friction might be overcome and the fault blocks could then rise to form the mountains.

One of the most extensive fault-block mountain systems in the world lies in parts of Idaho, Oregon, Nevada, Utah, California, Arizona, New Mexico, and northern Mexico. Known as the *Basin and Range Province*, it contains a spectacular development of fault-blocks (horsts) separated by sediment-filled valleys (grabens) (Fig. 15.32). The province is underlain by sedimentary strata deposited on older Precambrian rocks during the Paleozoic Era. Following a period of folding, the region was deeply eroded during Mesozoic and Cenozoic times when it supplied sediment to form strata that later developed into the Coast Ranges of California and the Rocky Mountains. Starting about 25 million years ago, and accompanied by

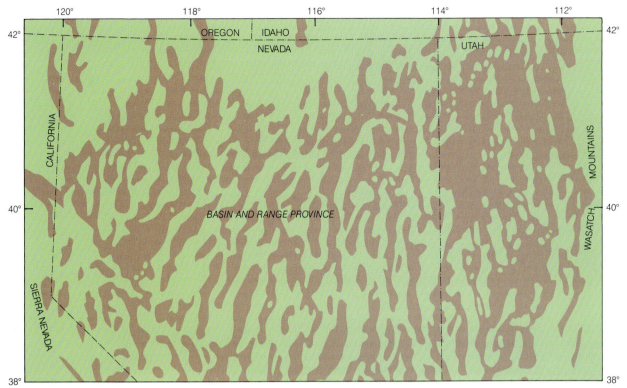

FIGURE 15.32 Map of portion of the Basin and Range Province in Nevada and Utah. The Wasatch Mountains form the eastern border of the province; the Sierra Nevada, the western border. Green areas are mountains (horsts), brown areas, basins (grabens). Boundaries of basins are faults. (Simplified from Geologic Map of the United States, U.S. Geological Survey, 1974.)

extensive volcanic activity, the region broke up into a series of blocks 30 to 40 km in width and as much as 150 km long, bounded by steeply inclined faults. The tilting of the blocks is pronounced (Fig. 15.33).

The geology of the Basin and Range Province is very complex and has led to long-term controversies over the origin of the ranges. More than one process has apparently been involved because both thrust faults and normal faults are present. In some instances it is possible to show that a given fault has, at different times, acted both as a thrust and as a normal fault. The faults that bound the blocks are long and straight, and inclined at angles of 50 to 70°. There is some evidence to suggest that at depth, many of the steep faults are curved and flatten to inclinations of ≤ 30° to the horizontal, and that extensive movement has occurred on the nearly flat faults. Overall, it appears that the Basin and Range has undergone extension due to tensional or pull-apart forces. Just how much extension has occurred remains in question, but some authorities suggest the width of the province has

FIGURE 15.33 Faulting in the Basin and Range Province. Exhumed fault surface, Dixie Valley, Nevada. The hanging-wall rocks have been removed by erosion, thus exposing the normal fault on which the horst (left) was raised relative to the grabens. Note the slickensides on the surface.

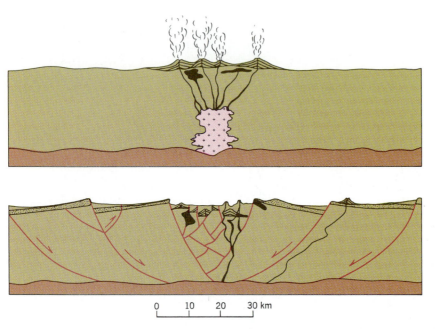

FIGURE 15.34 Possible model for formation of structures in the Basin and Range Province. Warping of the crust and extrusion of large quantities of lava and volcanic ash were followed by collapse along a series of steeply inclined normal faults that become flatter at depth. (*Source:* After Mackin, 1968.)

increased by 10 percent or more. The cause of the extension is also a matter of open debate. Many authorities point to the volcanism that accompanied the faulting as evidence that rising magma may have heated and distended the crust, thereby forming the kind of structures shown in the simplified diagram in Figure 15.34. The source of the magma, as shown in Figure 18.8, was probably a subduction zone overridden by North America.

Fold-and-Thrust Mountains

Fold-and-thrust mountains are spectacular, complex structures. Occurring in great arc-shaped systems a few hundred kilometers wide, they commonly reach several thousand kilometers in length. The words *fold* and *thrust* indicate their most characteristic features. Strata have been compressed, faulted, folded, and crumpled, commonly in an exceedingly complex manner. Although folding and thrust faulting are the key features, other kinds of mountain-building processes participate in the making of fold-mountain systems also; metamorphism and igneous activity are always present. Examples are widespread: the Appalachians, the Alps, the Urals, the Himalaya, and the Carpathians are all fold-and-thrust mountain systems. Indeed, the Alps, Carpathians, and

the Himalaya belong to a gigantic fold-and-thrust mountain chain formed during the Mesozoic and Cenozoic Eras. The term Alpine mountains is used by many geologists as being synonymous with fold-and-thrust mountains.

All fold-and-thrust mountain systems share another feature related to their folded strata. They develop from exceptionally thick piles of sedimentary strata, commonly 15,000 m or more in thickness. Early in the study of mountain ranges, it was realized by the American scientist James Hall that so huge a thickness called for an unusual sort of basin in which the sediments could accumulate. Another famous nineteenth-century American scientist, J. D. Dana, discussed this fact in his analysis of the history of the Appalachians. He pointed out that both subsidence of the crust to form an unusually deep basin and filling of the basin with sediment must have preceded the final deformation and uplift that created the mountains. Dana coined the name *geosyncline* for the basin in which the Appalachian sediments accumulated. Studies of other fold mountains show that thick sediment piles preceded all of them. The term **geosyncline**, therefore, has wide application and describes *a great trough that has received thick deposits of sediment during its slow subsidence through long geologic periods.*

The strata of fold-mountain systems are predominantly marine; we can draw this conclusion from the presence in them of marine fossils. In systems such as the Alps, the marine strata are mostly of deep-water origin. In others, such as the Appalachians, the sediments apparently accumulated in shallow water. Regardless of water depth, the kinds and thicknesses of sediments found in geosynclines lead to two important conclusions. First, geosynclines are predominantly oceanic features. Second, the great thicknesses of sediments, which commonly exceed the greatest depths of the ocean, indicate that the catchment area for the sediment must have been sinking while it was being filled.

Some geosynclines occur in pairs. An example can be observed in the Appalachians. Two elongate geosynclines, roughly parallel, once occupied the region from Newfoundland southwest to Alabama. One geosyncline lay directly adjacent to the continent and became filled with shallow-water sediment (some marine, some nonmarine) which we now see as the limestone, sandstone, coal, and other strata that are common in most of the Appalachians. The other geosyncline lay farther east and farther offshore. It became filled with deeper-water sediment, all of which is marine. The sediment is of the sort we see today at the foot of the continental slope off the east coast of North America. The deep-water geosyncline also contained some volcanic rock. We do not want to give the impression that the two geosynclines were completely separate. They were not. They merged, one into the other, and overlapped along their margins. Nowhere is there a break in sedimentary strata.

Two similar and modern sediment piles occur side-by-side along the present-day Atlantic coast of North America. They provide evidence of how geosynclines form. One, the shallow-water geosyncline, is a wedge of sediment forming the continental shelf. It is underlain by continental crust. The other geosyncline is a great pile of deep-water sediment that largely underlies the continental rise. It is partly underlain by oceanic crust. Geosynclines, therefore, are not really synclines or even basins, but are continental margins where continental crust joins oceanic crust. The places where oceanic and continental crust are joined are in the centers of tectonic plates. The joins are sometimes called passive continental margins. It is here, in the interior of plates, along passive continental margins, that sediment derived by weathering of the adjacent continent is accumulated.

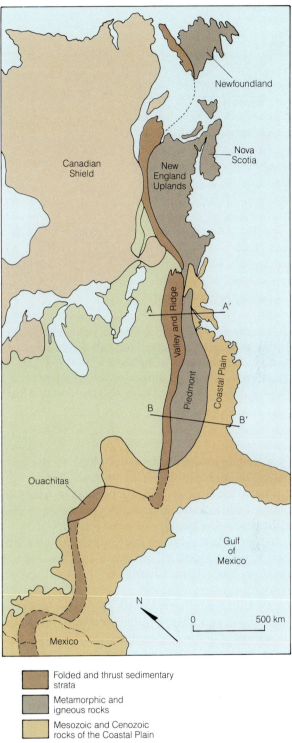

FIGURE 15.35 The Appalachian Mountain System runs from Newfoundland to the Mexican border. Part of the eastern and southern margins of the system is covered by younger sediments of the coastal plain. Figures 15.36 and 15.37 are cross sections drawn approximately along the lines A-A' and B-B', respectively.

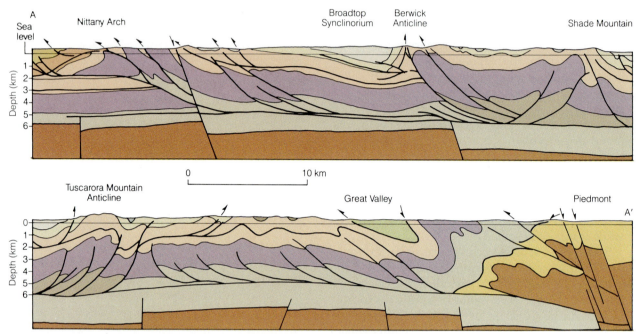

FIGURE 15.36 Section through the Valley and Ridge Province of the Appalachians in Pennsylvania, along the line A-A′ in Figure 15.35. The structure includes both folding and thrusting. The prominent stratum is a limestone of middle Cambrian to early Ordovician age. (*Source:* After Woodward, 1985.)

Geosynclines are an essential phase in the formation of fold-and-thrust mountains. Before we approach the final step in the process of forming fold-and-thrust mountains (Chapter 18), let us describe briefly the geology of some typical fold-and-thrust mountain systems and compare them with that of other kinds of mountain systems. First, consider the Appalachians, a fold-and-thrust mountain system 2500 km in length that borders the east and southeast coasts of North America (Fig. 15.35) and that continues offshore beneath the sediment of the continental shelf. The shallow-water sediment in the old western geosyncline contains mud cracks, ripple marks, fossils of shallow-water organisms, and in places fresh-water materials such as coal. The sediment was deposited on a basement of metamorphic and igneous rock, and becomes markedly thicker away from the former western shore (that is, it thickens from west to east).

Most, but not all, of the sediment in the western geosyncline has now been deformed. Today, if we approach the central Appalachians from western New York and western Pennsylvania, we see, first, the former sediment occurring as essentially flat-lying, undisturbed strata. Continuing east-

ward, we notice the same strata thicken and become gently folded and thrust-faulted. In eastern Pennsylvania, in the region known as the Valley and Ridge Province, the strata have been bent into broad anticlines and synclines. The province gets its name because valleys have been developed by erosion in the most erodible strata, composed of limestone, dolostone, and claystone. The valleys alternate with prominent ridges formed by very resistant strata, chiefly sandstone (Fig. 15.36).

In the mountains further south, in Tennessee and the Carolinas, a different style of deformation is apparent. Here, strata have been deformed both by thrust faulting and by folding; however, in most places, thrust faults predominate. An example of deformation can be observed by approaching the southern Appalachians from the northwest as shown in Figure 15.37, a section drawn along the line *B–B′* in Figure 15.35. Huge, thin slices of sedimentary strata were pushed westward, each successive slice riding upward and over earlier slices. The surface or layer along which movement occurred is known as a *detachment surface*, and the slice that moved is commonly referred to by the French name, *décollement*. A significant and puzzling feature of a décollement is that the style of deformation above the detachment surface is usually different from that below; that is, the weaker sedimentary strata above the detachment surface have been fractured, moved along thrust faults, and stacked like a series of thin cards, while the older basement rocks below tend to have resisted faulting and large-scale translation and to have been deformed by ductile deformation.

Proceeding farther east, toward the region from which the thrust slices came, we see the core of the Appalachians. Here the ancient basement rocks can be examined. The deep-water sediments that were deposited in the eastern geosyncline also can be seen. These strata are increasingly metamorphosed, and deformation becomes increasingly intense, the farther east we go. Folds become isoclinal and then overturned, and faulting is prevalent. In places, fragments of the old basement can be seen to have been thrust up over younger sedimentary strata. Finally, we reach a region where intense metamorphism has occurred and where granite batholiths have been emplaced.

We naturally ask how well the Appalachian picture can be applied to other fold-and-thrust mountains. The answer is that similar features are found in all of them. The Alps and associated mountain ranges in southern Europe (Fig. 15.38) were formed later than the Appalachians, during the Mesozoic and Cenozoic Eras. Nevertheless, the two systems have many features in common. For instance, the Jura Mountains, which form the northwestern edge of the Alps, have the same folded form and origin as the Valley and Ridge Province (Fig. 15.39a). Also, the Jura were formed from shallow-water sediments. In the high Alps, which correspond to the now deeply eroded Appalachians that can be seen in Connecticut, Vermont, Virginia, and Maryland, thrusting appears to have developed on a much grander scale than in the Appalachians (Fig. 15.39b). The high Alps are composed of deeper-water marine sedimentary strata.

The Canadian Rocky Mountains, a magnificent mountain system much less eroded than the Appalachians, can also be compared. A section through the Canadian Rockies at about the latitude of Calgary, Alberta (Fig. 15.40) reveals all the features we have described for the Appalachians. A central or core zone has been intensely metamorphosed. In it, parts of the older basement rocks have been thrust upward, and in the marginal region folding and extensive thrust faulting are evident. The thrust sheets have moved eastward, away from the core zone. It is apparent that each sedimentary unit becomes thinner as it is followed from west to east, indicating that the core zone coincides with the thickest part of the old geosyncline.

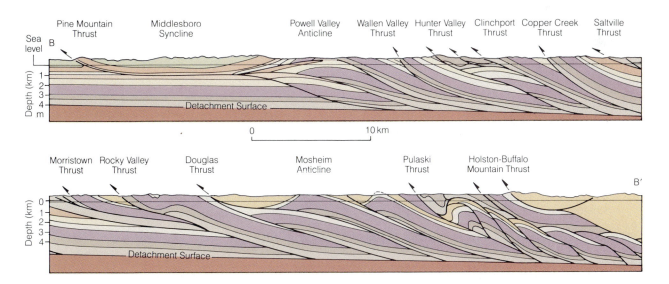

FIGURE 15.37 Section through the Valley and Ridge Province of the Appalachians in Tennessee and North Carolina, along the line B-B′ in Figure 15.35. Compare with Figure 15.36. Development of décollement by thrust faulting is predominant and folding less important in the southern Appalachians. (*Source:* After Woodward, 1985.)

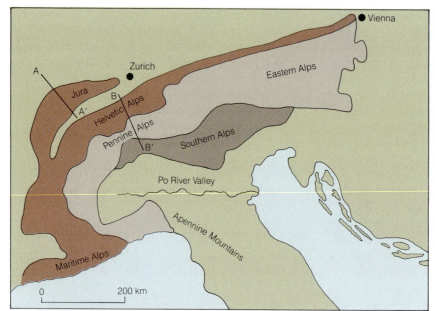

FIGURE 15.38 Map of the major units of the Alps in Switzerland and Austria. The mountains were formed as a result of compressive forces operating in a southeast to northwest direction. Figures 15.39(*a*) and (*b*) are along the lines A-A' and B-B', respectively.

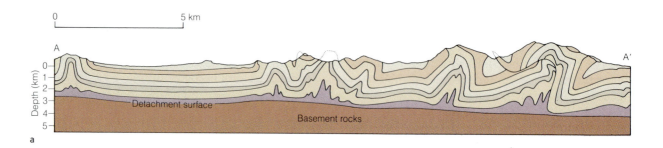

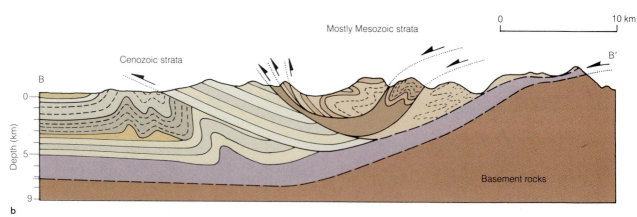

FIGURE 15.39 Two sections through portions of the Alps. (*a*) Section through the Jura Mountains along line A-A' in Figure 15.38. The cover of weak sedimentary rocks has been deformed by folding and by slippage along the detachment surface. The term décollement was first introduced to describe the large-scale slippage of cover rocks in the Jura. (*b*) Section through the Alps in central Switzerland along the line B-B' in Figure 15.38. Strata have moved northward along great thrust faults which later were themselves folded. A major thrust fault separates overlying strata from basement rocks. (*Source:* After Heim, 1922.)

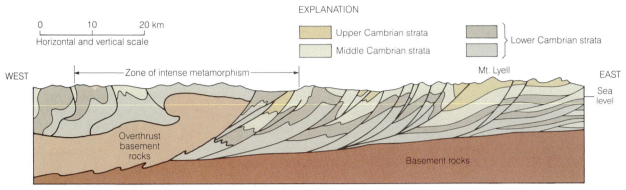

EXPLANATION

Upper Cambrian strata   Lower Cambrian strata

Middle Cambrian strata

0   10   20 km
Horizontal and vertical scale

WEST   EAST

Zone of intense metamorphism

Mt. Lyell

Sea level

Overthrust basement rocks

Basement rocks

FIGURE 15.40 Section through the Canadian Rocky Mountains at about the latitude of Calgary, Alberta. The zone of intense metamorphism coincides with the region of maximum uplift and maximum deformation. Farther east, where strata become progressively thinner, the pile has been greatly thickened by movement along thrust faults. The sense of movement is such that each fault block has moved toward the east, riding over the block beside it. (*Source:* After Prince and Mountjoy, 1970.)

MOUNTAINS AND PLATE TECTONICS

Throughout the nineteenth and most of the twentieth century, geologists struggled to provide an adequate explanation for the tremendous shortening of strata indicated by folding and thrusting in mountains such as the Alps. One popular idea was that the Earth is cooling and contracting, and that mountains are surface wrinkles formed by contraction. Early in the twentieth century the cooling hypothesis was proved wrong when radioactive decay was shown to release heat (Chapter 3). Even if the Earth is cooling, contraction should cause the surface to be wrinkled more or less uniformly, and it is obvious that mountains are highly localized.

The theory of plate tectonics was initially proposed to explain data derived from the seafloor and the oceanic crust. When the consequences of lateral movement of plates of lithosphere were examined, it soon became apparent that many puzzling features of the continental crust might be also explained. One of the first triumphs of the theory was the explanation it provided for the localization of fold-and-thrust mountain ranges along converging plate boundaries.

The theory of plate tectonics was proposed in 1967. By 1970 details of a plate tectonic origin for fold-and-thrust mountains were being elucidated. The piles of sediment along passive continental margins—the old geosynclines—become contorted, compressed, and thickened due to the subduction process, to collision between two continental masses, or to a combination of subduction

and collision (Figs. 15.41 and 15.42). Overturned folds and thrusts are the results of collisions. The sediment shed from a continental land mass and accumulated along a passive margin, is deformed, metamorphosed, and welded back on to the edge of the continent by the collision process. The fact that successive belts of deformed rock are roughly parallel can also be explained by plate tectonics. When two continents collide they sweep up the sediment from the seafloor and form a new mountain range that separates them. The weakest part of the new supercontinent is apparently the collision zone. When the supercontinent is broken up again at a later time, the break will occur along the approximate line of collision. J. T. Wilson, the Canadian scientist who first recognized the importance of transform faults, also pointed out that the sequence of orogens represented by the Grenville and Appalachian provinces (Fig. 15.30) was evidence for successive openings and closings of the Atlantic. Presumably, he argued, a new subduction zone will one day develop along the eastern margin of North America, the present-day Atlantic Ocean will start to close, and eventually another continental collision will fold the sediment along the continental margin into a new mountain range. The new range will lie to the east of the Appalachians and will be in the interior of a large continent, much as the Ural Mountains are today in the center of the Eurasian continent. *The cycle of opening and closing of oceans, accompanied by successive fragmentations and collisions of continents,* is called a **Wilson Cycle.**

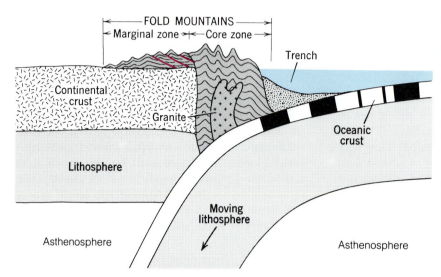

FIGURE 15.41 Schematic section showing how fold mountains are formed when a plate of lithosphere plunges downwards beneath the edge of a continent. Strata originally accumulated in a deep-water geosyncline are crumpled, metamorphosed, and intruded by granite, which forms the core of mountains. Strata of shallow-water geosyncline are pushed and slid sideways to form marginal zone of mountains. On seaward side of the mountains, a sea-floor trench forms. (*Source:* After Dewey and Bird, 1970.)

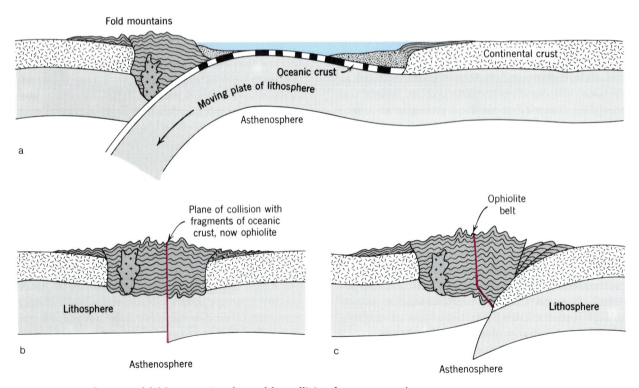

FIGURE 15.42 System of fold mountains formed by collision between continents. (*a*) Continental margin, of the type now found on the west coast of South America, forms a chain of fold mountains. Moving plate of lithosphere carries a second continent on a collision course. (*b*) Collision between continents crumples geosynclines and increases size of mountain system. (*c*) The downward-moving plate of lithosphere becomes detached, but the edge of the remaining segment of plate is partly thrust under the edge of the stationary plate, causing further elevation of the fold mountains. (*Source:* After Dewey and Bird, 1970.)

The concept of a Wilson Cycle seems to be correct in principle, but there are many details of local geology that the cycle does not explain, for example, the lateral movement of large slices of crust by transform faults. That portion of North America that lies west of the San Andreas Fault is moving northwesterly relative to the rest of the continent (Fig. 17.20). In effect, the continental margin is being rearranged by a process that is neither subduction nor continental collision. Recent studies have shown that most of the western margin of North America is made up of translated and rearranged blocks of crust. Individual blocks are called *allochthonous terranes*. Most recently, it has been shown that the eastern margin of North America also contains allochthonous terranes, but in that case the motion occurred during the Paleozoic Era. Before attempting a more detailed inquiry into such topics as plate tectonics, allochthonous terranes, and the history of the crust, it is helpful to consider the Earth's internal properties in order to see why plates move.

SUMMARY

1. Solids can be deformed in three different ways: by elastic deformation, by ductile deformation, and by fracturing.

2. High confining pressure and high temperatures enhance ductile properties. Low temperatures and low-confining pressure enhance elastic properties and failure by fracture if the elastic strength is exceeded.

3. The rate at which a solid is deformed (strained) also controls style of deformation. High strain rates lead to fractures; low strain rates cause ductile deformation.

4. Fractures along which slippage occurs are called faults. Normal faults are caused by tensional forces, reverse and thrust faults by compressional forces. Strike-slip faults are vertical fractures which have horizontal motion.

5. Folds are formed by ductile deformation.

6. Two major structural units can be discerned in the continental crust. Cratons are ancient portions of the crust that are tectonically stable. Surrounding the cratons are orogenic belts of highly deformed rock, marking the site of ancient mountain ranges.

7. Mountains can be divided into three kinds based on their principal geological features; they are: volcanic mountains, fault-block mountains, and fold-and-thrust mountains.

8. Fold-and-thrust mountains are formed along subduction edges of plates by the compression and thickening of sediments accumulated along passive continental margins.

SELECTED REFERENCES

Billings, M. P., 1972, Structural geology, 3rd ed.: Englewood Cliffs, N.J., Prentice-Hall.

Burchfiel, B. C., 1983, The continental crust: Sci. American, v. 249, no. 3, p. 130–145.

Cox, A. and Hart, R. B., 1986, Plate tectonics; How it works: Palo Alto, CA., Blackwell.

Hills, E. S., 1972, Elements of structural geology, 2nd ed.: New York, John Wiley & Sons.

CHAPTER 16

Earthquakes, Isostasy, and the Earth's Internal Properties

Gaping fissures opened in a residential area of Anchorage, Alaska, during the earthquake of March 27, 1964.

EARTHQUAKES

Every year the Earth experiences many hundreds of thousands of earthquakes. Fortunately, only one or two are large enough, or close enough to major centers of population, to cause loss of life. Certain areas are known to be earthquake prone, and special building standards are required to make structures as resistant to damage as possible. However, all too often an unexpected earthquake will devastate an area where buildings are not adequately constructed. One such example is the earthquake that destroyed parts of old Mexico City in 1985 (Fig. 16.1).

Sixteen earthquake disasters are known to have caused 50,000 or more deaths (Table 16.1). The most disastrous earthquake on record occurred in 1556, in Shen-Shu Province, China, where an esti-

FIGURE 16.1 The Hotel DeCarlo was one of the buildings that collapsed during the great earthquake that struck Mexico City in 1985. Proper building design can minimize damage. Nearby buildings of sturdier construction withstood the shaking.

mated 830,000 people died. Those people lived in cave dwellings excavated in loess (Chapter 12), which collapsed as a result of the quake. The second most disastrous earthquake also occurred in China, at T'ang-Shan, in 1976. The town was completely destroyed and approximately 700,000 people are believed to have died. The third most disastrous earthquake hit Calcutta, India in 1737, where an estimated 300,000 people perished. In 1920 a quake in Kansu Province, China killed 180,000 people. During the quake of 1908 at Messina, Italy, 160,000 people lost their lives, and the Japanese earthquake of 1923 killed 143,000 in Tokyo and Yokohama. Since 1900 there have been 39 earthquakes, worldwide, in each of which 500 or more people have died.

No locality on the Earth's surface is free from earthquakes, but in some regions the quakes that do occur are weak and, consequently, not very dangerous to people or dwellings. For example, scientists believe that in southern Florida, southern Texas, and parts of Alabama and Mississippi, the probability of damaging earthquakes is almost zero. All other parts of the United States have experienced damaging quakes in the past, and more can be expected to occur in the future (Fig. 16.2).

Most Americans think immediately of California when earthquakes are mentioned. It is probable, however, that the most intense earthquake to jolt the continent in the past 200 years was centered near New Madrid, Missouri. Three earthquakes of great size occurred on December 16, 1811, January 23, and February 7, 1812. The actual sizes are un-

TABLE 16.1 *Earthquakes During the Past 800 Years That Have Caused 50,000 or More Deaths*

Place	Year	Estimated Number of Deaths
Shen-shu, China	1556	830,000
T'ang-shan, China	1976	700,000
Calcutta, India	1737	300,000
Kansu, China	1920	180,000
Messina, Italy	1908	160,000
Tokyo and Yokohama, Japan	1923	143,000
Chihli, China	1290	100,000
Peking, China	1731	100,000
Naples, Italy	1693	93,000
Shemaka, U.S.S.R.	1667	80,000
Kansu, China	1932	70,000
Silicia, Turkey	1268	60,000
Catania, Italy	1693	60,000
Lisbon, Portugal	1755	60,000
Quetta, Pakistan	1935	60,000
Calabria, Italy	1783	50,000

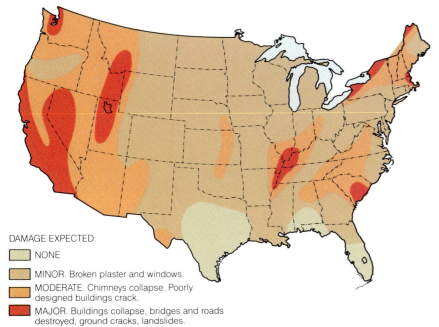

DAMAGE EXPECTED:

NONE

MINOR. Broken plaster and windows.

MODERATE. Chimneys collapse. Poorly
designed buildings crack.

MAJOR. Buildings collapse, bridges and roads
destroyed, ground cracks, landslides.

FIGURE 16.2 Seismic-risk map of the United States. Zones refer to maximum earthquake intensity and, therefore, to maximum destruction that can occur. The map does not indicate frequency of earthquakes. For example, frequency in southern California is high, but in eastern Massachusetts it is low. Nevertheless, when earthquakes occur in eastern Massachusetts, they can be as severe as the more frequent quakes in southern California. (*Source:* After Algermissen, 1969.)

known because instruments to record them did not exist at the time, but from the local damage caused, plus the fact that tremors were felt and minor damage occurred as far away as New York and Charleston, South Carolina, it is possible to estimate that the largest of the quakes was as large or larger than that which leveled San Francisco in 1906. But it is an ill wind that blows no one any good. Although earthquakes are sometimes dangerous and destructive to humans, they are also the most powerful tools available to scientists for studying the Earth's interior.

In marked contrast to the accessibility of the Earth's surface, the interior regions are hidden from view. It is possible, of course, to sample unseen regions by drilling, but the deepest hole ever drilled, in northern Russia, is less than 15 km deep. So far, no one has drilled to the base of the crust, let alone into the mantle, so there is no way by which it is possible to examine directly the Earth's internal structure. To discover what lies between the surface and the unseen center of the Earth, it is necessary to use indirect means. How this is accomplished and the conclusions that can

be drawn provide one of the most exciting success stories of science.

The circumference of the Earth was calculated long ago by Eratosthenes of Alexandria (Fig. 16.3). Calculation of the volume of the Earth is a straightforward matter once the circumference is known. The Earth's mass was first calculated by comparing the measurements of the gravitational pull between the Earth and the Moon with the gravitational pull between two metal spheres of known mass. Knowing the volume and mass, it is simple to calculate that the Earth's density is 5.5 g/cm^3. Yet rocks in the crust have densities that range from 2.6 to 3.3 g/cm^3. It is apparent that at least some part of the interior of the Earth must consist of material more dense than 5.5 g/cm^3. It would not be possible to proceed easily beyond this inference without some way of "seeing" inside the Earth. Fortunately there is a way. The vibrations, or waves, caused by earthquakes have the ability to travel completely through the Earth. By making careful measurements it is possible to use those waves to obtain a sort of giant X-ray picture of what is inside.

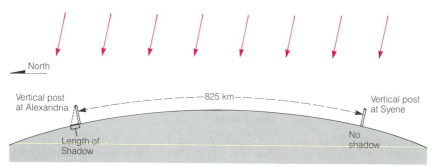

FIGURE 16.3 Method used by Eratosthenes to estimate the circumference of the Earth. The distance between Syene and Alexandria in Egypt is 825 km. Alexandria is north of Syene. When the Sun is directly above Syene at midday, a vertical post does not cast a shadow. At the same moment, a vertical post in Alexandria does cast a shadow. Erathostenes reasoned that the effect occurred because the Earth is curved. From the length of the shadow at Alexandria he calculated that the distance from Syene to Alexandria is 1/50 of the circumference of the Earth.

Origin

Earthquakes and the vibrations they cause happen when the Earth is suddenly jolted, as if struck by a giant hammer. Make an experiment yourself. Have a friend hit one end of a wooden plank or the top of a wooden table with a hammer while you press your hand on the other end. You will feel vibrations set up in the plank or tabletop by the energy of the hammer blow. The harder the blow, the stronger the vibrations. The reason you can feel the vibrations is that some of the energy imparted by the hammer blow is transferred by elastic deformations through the solid wood. Fortunately, giant hammers don't hit the Earth, but a bomb blast or a violent volcanic explosion will serve as an energy source just as well. So too will the sudden slipping of rock masses along a fault, causing two hard, rocky surfaces to slide suddenly past each other.

Sudden movement along faults is the cause of most earthquakes. But it cannot be that simple sliding occurs every time pressure is applied to a fault; some earthquakes are millions of times stronger than others. The same energy that in one case will be released by thousands of tiny slips and earthquakes will in another case be stored and released in a single giant earthquake. The answer seems to be that provided fault surfaces do not slip easily, energy can be stored in elastically deformed bodies of rock, just as in a steel spring that is compressed. As discussed in Chapter 15, an elastically deformed solid returns to its original shape when the deforming force is removed.

Evidence supporting the idea of energy being stored in elastically deformed rocks came first from studies of the San Andreas Fault. During long-term field observations in central California, beginning in 1874, scientists from the U.S. Coast and Geodetic Survey determined the precise positions of many points both adjacent to and distant from the fault. As time passed, movement of the points revealed that the crust was slowly being bent. For some reason, in the area of measurement near San Francisco the fault was locked and did not slip. On April 18, 1906, the two sides of the fault shifted abruptly. The stored energy was released as the bent crust snapped back to its former position, thereby creating a violent earthquake. Repetition of the survey then revealed that the bending had disappeared (Fig. 16.4).

Most earthquakes occur in the brittle rock of the lithosphere. As discussed in Chapter 15, brittleness is the tendency for a solid to fracture when the deforming force exceeds the limits of elasticity. At great depth, temperatures and pressures are too high for brittle fracture to happen. Under such conditions, bodies neither fracture nor store large amounts of elastic energy, but instead undergo ductile deformation and permanent changes of shape, even after the deforming forces have been removed. Earthquakes, then, are a phenomenon of the outer, cooler portion of the Earth.

How Earthquakes Are Studied

The name given to *the study of earthquakes* is **seismology,** a word that comes directly from the ancient Greek term for earthquakes, *seismos.* When an earthquake occurs, an observer at the surface of the Earth can detect shocks and vibrations. The *device used to study the shocks and vibrations caused by earthquakes* is a **seismograph.** The ideal way to study the vibrations and motions of the Earth's surface would be from a stable platform that sits above the surface, but is not connected to the Earth. An observer on the platform would not be influenced by the vibrations and so could make accurate measurements of the shaking surface below. An observer who must stand on the vibrating surface will move with the surface, making the act of measurement much more difficult. The observer standing on the vibrating surface faces the same difficulty a sailor in a small boat faces in trying to make an accurate measurement of the sea's surface—there is no stable, fixed reference point from which to make that measurement.

To overcome the frame-of-reference problem, most seismographs make use of **inertia,** which is *the resistance a large mass has to sudden movement.* If a heavy mass, such as a block of iron, is suspended from a light spring (Fig. 16.5), the iron block has so much inertia it will tend to remain almost stationary when the spring is suddenly extended. If the spring is connected to the ground, and the ground vibrates, the spring will expand and contract but the iron block will stay almost stationary. Then the distance between the ground and the iron mass can be used to sense vertical displacement of the ground surface. Horizontal displacement can be similarly measured by suspending a large mass from a string to make a pendulum (Fig. 16.5). Because of its inertia, the mass does not keep up with the horizontal-ground motion, and the difference between the pendulum and ground movement records the horizontal-ground motions. Seismographs with inertial masses are commonly used in groups so that the vertical plus horizontal motions in three directions can be measured simultaneously.

Another kind of device, called the Benioff strain seismograph, employs two concrete piers in the ground spaced at a distance of about 35 m. Attached to one pier is a long, rigid, silica-glass tube (Fig. 16.5). The other pier carries a very sensitive detector to measure even the slightest movement in the end of the silica-glass rod. Strain seismographs are commonly installed in mines, tunnels, and other places where a constant temperature can be maintained, and where wind and other disturbances are minimal.

Modern seismographs are incredibly sensitive. Vibrational movements as tiny as one hundred millionth (10^{-8}) of a centimeter can be detected. Indeed, many instruments are so sensitive they can detect motions due to wind, moving automobiles blocks away, and even ocean waves and tides several kilometers from the seashore.

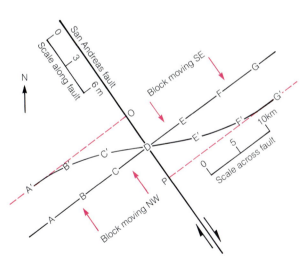

FIGURE 16.4 An earthquake caused by sudden release of stored elastic energy. Sketch based on detailed surveys near the San Andreas Fault, California, before and after the abrupt movement that caused the earthquake of 1906. The seven survey points, *A* to *G*, were originally aligned. Slowly, movement of the two fault blocks bent the crust and displaced the points to new positions, *A'* to *G'*. Friction between the two sides of the fault prevented steady slippage. Suddenly, the frictional lock was broken and the rocks on either side of the fault rebounded. The surveyed points lay along the lines *A'O* and *PG'*. The sudden offset along the fault, distance *OP*, was 7 m.

Seismic Waves

The point of the first release of energy that causes an earthquake is called the **earthquake focus.** The focus generally lies at some depth below the surface; so for convenience we define *that point on the Earth's surface that lies vertically above the focus of an earthquake* as the **epicenter** (Fig. 16.6). A good way to describe the location of an earthquake focus is to state the location of its epicenter and its depth.

How is the energy of an earthquake transmitted from the focus to other parts of the Earth? As with

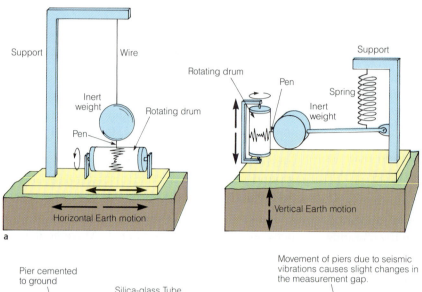

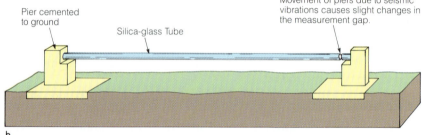

FIGURE 16.5 Seismographs measure vibrations sent out by earthquakes. (*a*) Two kinds of inertial seismographs. The pendulum device measures horizontal motions, the spring supported device measures vertical motions. Both seismographs are based on the principle that the inertia, or resistance to motion, tends to keep a heavy mass suspended on a string or spring, motionless, while the supporting mechanism, anchored in the Earth, moves with the seismic vibration. (*b*) A strain seismograph. A rigid silica-glass tube, as much as 35 m long, is supported on a solid concrete pier anchored to the ground. A second pier carries a sensitive electronic measuring device to record any movement in the end of the rod. The distance between the two piers changes when a seismic vibration disturbs the surface.

any vibrating body, waves (vibrations) spread outward from the focus. The waves, called *seismic waves*, spread out in all directions from the focus, just as sound waves spread in all directions when a gun is fired. Seismic waves are elastic disturbances, so unless the elastic limit is exceeded, the rocks through which they pass return to their original shapes after passage of the waves. Very weak rocks, such as poorly cemented sediments, are sometimes permanently deformed if they are close to the epicenter of a very strong earthquake. In most instances, however, the elastic limits of rocks are not exceeded and it is not possible to tell, by examining a rock, whether seismic waves have passed through at some time in the past. Seismic

waves must be measured and recorded while the rock is still vibrating. For this reason, many continuously recording seismograph stations are installed around the world.

Seismic waves are of two kinds. *Body waves* travel outward from the focus, passing entirely through the Earth. *Surface waves*, on the other hand, are guided by the Earth's surface, with only a loose constraint imposed by the atmosphere and the ocean. Body waves are analogous to rays of light, or sound waves, which travel outward in all directions from their points of origin. Surface waves are analogous to the ringing of a bell, in that they arise from the movement of the surface due to vibration of the entire mass of the Earth.

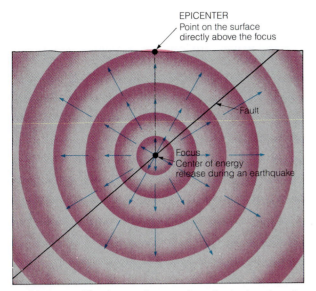

FIGURE 16.6 The focus of an earthquake is the site of initial movement on a fault and the center of energy release. The epicenter is the point on the Earth's surface directly above the focus.

Body Waves

Seismic body waves are of two kinds. *Compressional waves* deform materials by change of volume in the same way that sound waves do, and consist of alternating pulses of compression and expansion acting in the direction of travel (Fig. 16.7). Compression and expansion produce changes in the volume and density of a medium. Compressional waves can pass through solids, liquids, or gases because each can sustain changes in density. When a compressional wave passes through a medium, the compression pushes atoms closer together. Expansion, on the other hand, is an elastic response to compression and it causes the distance between atoms to be increased. Movement in a solid, subjected to compressional waves, is back and forth in the line of the wave motion. Compressional waves have the greatest velocity of all seismic waves—6 km/s is a typical value for the uppermost portion of the crust. The fastest travelers, compressional waves, are the first ones to be

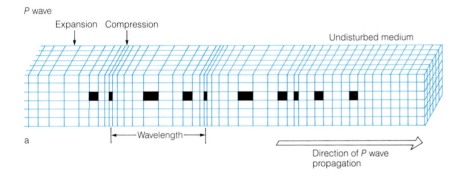

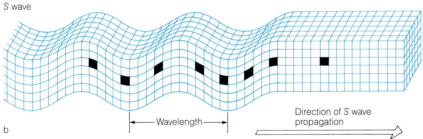

FIGURE 16.7 Difference between seismic body waves of the *P* and *S* types. (*a*) *P* waves cause alternate compressions and expansions passing through rock. An individual point in the rock will move back and forth parallel to the direction of *P*-wave propagation. As wave after wave passes through, a square will repeatedly change its shape to a rectangle then back to a square. (*b*) *S* waves cause a shearing motion as they pass through the rock. An individual point in the rock will move up and down, perpendicular to the direction of *S*-wave propagation. A square will repeatedly change to a parallelogram then back to a square again.

recorded by a seismograph and they are therefore called *P* (for *Primary*) waves.

The second kind of body waves are *shear waves*. They deform materials by change of shape but not change of volume. Because gases and liquids cannot change shape, shear waves can only be transmitted by solids. Shear waves consist of an alternating series of sidewise movements, each particle in the deformed solid being displaced perpendicular to the direction of wave travel. A typical velocity for a shear wave in the upper crust is 3.5 km/s. Shear waves, slower than *P* waves, reach a seismograph some time after a *P* wave arrives, so they are called *S* (for *Secondary*) waves (Fig. 16.8).

Surface Waves

Surface waves are caused, as mentioned previously, by the Earth ringing like a bell. Such a motion causes the shape and/or size of the Earth to change, just as a ringing bell changes in shape and size as it vibrates. Such motions are whole-body oscillations, but to an observer on the surface of the Earth, these oscillations are detected as movements of the surface. For this reason they are called surface waves.

Surface waves are due to oscillations of two kinds. *Spheroidal* oscillations involve mainly a change in the Earth's volume, whereas *torsional*

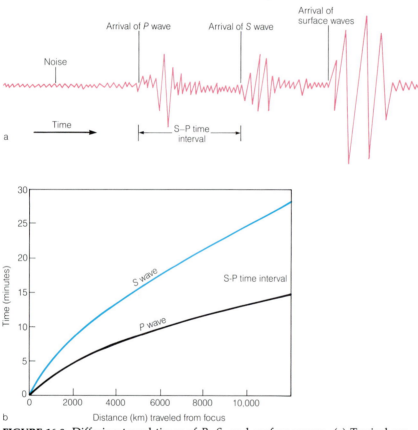

FIGURE 16.8 Differing travel times of *P*, *S*, and surface waves. (*a*) Typical record made by a *seismograph*. The *P* and *S* waves are body waves. They leave the earthquake focus at the same instant and travel outward in all directions. The fast-moving *P* waves reach the seismograph first, and some time later, the slower-moving *S* waves arrive. The delay in arrival times is proportional to the distance traveled by the waves. The surface waves are due to the Earth vibrating like a bell (Fig. 16.9). (*b*) Average travel–time curves for *P* and *S* waves in the Earth. (*Source:* After Bullen, 1954.) When the arrival times of *P* and *S* waves are recorded by seismographs at three different locations, the exact location of the earthquake focus can be calculated (Fig. 16.10). Note how the curves bend, indicating increased speed with distance, due to deeper penetration in the Earth.

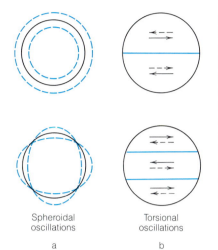

Spheroidal
oscillations

Torsional
oscillations

a

b

FIGURE 16.9 Surface waves are caused by free oscillation of the vibrating Earth. They are of two kinds: spheroidal oscillations (called Rayleigh waves), in which both the volume and shape may vary; and torsional oscillations (called Love waves), in which the volume is unchanged but the shape varies. (*a*) The simplest spheroidal oscillations occur when the Earth simply expands and contracts radially, and the sphere becomes deformed to a football shape. To an observer on the surface of the Earth, spheroidal oscillations cause up–down, rotary motions, rather like *S* waves. (*b*) The simplest torsional oscillations occur when the northern hemisphere is twisted in the opposite direction from the southern hemisphere. The next-simplest oscillation divides the Earth into three twisting slices. To an observer on the surface of the Earth, torsional oscillations cause back–forth rocking motions, rather like *P* waves.

oscillations involve only a change in the Earth's shape (Fig. 16.9). The properties of free oscillations are therefore similar to those of *P* and *S* waves. To an observer at the surface, the oscillations indeed appear very similar to ordinary *P* and *S* waves. However, they travel more slowly than *P* and *S* waves, and in addition they must pass around the Earth rather than through it. Thus, surface waves are the last to be detected by a seismograph (Fig. 16.8). Surface waves are not readily separated, but those caused by spheroidal oscillations are called Rayleigh waves, and those by torsional oscillations, Love waves, after the two English scientists who first recognized them.

Location of Epicenter

The location of an earthquake's epicenter can be determined from the arrival times of the *P* and *S* waves. *P* waves travel faster than *S* waves, and arrive first, so the further a seismograph is away from an epicenter, the greater the time difference between the arrival of the *P* and *S* waves (Fig. 16.10). When it has been determined how far an epicenter lies from a seismograph, the seismologist takes a map and draws a circle with a radius equal to the calculated distance. The exact position of the epicenter can be determined when data from three or more seismographs are available—it lies where the circles intersect.

Direction of Motions on a Fault

When one block of a fault slips past another and causes an earthquake, it is possible, in some instances, to determine the direction of motion on

the fault from the seismograph record. This is exceedingly important information about those parts of the world where direct observations of the geology are not possible or at best exceedingly difficult. One example of this is the great transform faults that cross the ocean floor.

The information used to determine the direction of movement on a fault is contained in the arrival of body waves. Consider the *P*-wave record. If the first arrival is a compressive pulse, the release of elastic energy, and the fault motion, must be *toward the* seismograph (Fig. 16.11*a*). If it is an expansion, the fault motion must be away from the seismograph. In Figure 16.11*b*, the effect of an earthquake caused by movement on a strike-slip fault is shown. It is apparent that the first *P*-wave motion observed depends on the location of the seismograph. By plotting first motions from several seismographs, the fault movement can be determined. The actual radiation pattern of body waves is in three dimensions, not two as suggested by Figure 16.9. It is not possible to distinguish up–down movement from back–forth movement using *P* waves. There is a 90° ambiguity inherent in the *P*-wave radiation pattern. The first arrival of the *S* wave is used to resolve the ambiguity, because the *S*-wave oscillations also carry the signature of the direction of the first motion. Determination of several movements on the same fault provides still further information—the variety of fault, normal, reverse, or strike slip. As we shall see in Chapter 17, this is essential information for determining the dynamics and geometry of moving plates of lithosphere. Indeed, the evidence from first motions that spreading edges are associated with normal faults, subduction edges with reverse faults, and transform fault edges with

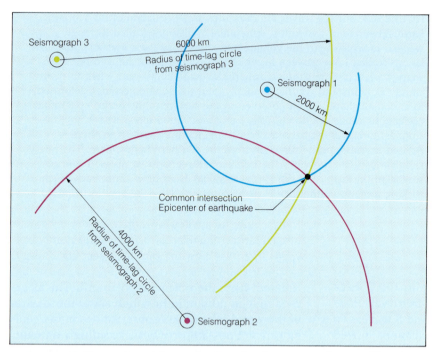

FIGURE 16.10 Illustration of the method used to locate an epicenter. The effects of an earthquake are felt at three different seismograph stations. The time differences between the first arrival of the *P* and *S* waves depends on the distance of a station from the epicenter. Using the curves in Figure 16.8, the following distances are calculated:

	Time Difference	Calculated Distance
Seismograph 1	1.0 min	2000 km
Seismograph 2	5.6 min	4000 km
Seismograph 3	7.5 min	6000 km

On a map, a circle of appropriate radius is drawn around each of the stations. The epicenter is where the three circles intersect.

strike-slip faults, is a powerful confirmation that the theory of plate tectonics is correct.

Magnitudes and Numbers of Earthquakes

Very large earthquakes (of the kind that destroyed San Francisco in 1906, Tokyo in 1923, T'ang-Shan, China in 1976, and parts of Mexico City in 1985) are, unfortunately, relatively frequent. In earthquake prone regions, such as San Francisco and the surrounding area, very large earthquakes occur about once a century. In some areas the time is shorter, in others, longer—a century is an approximate average. This means that the time needed to build up elastic strain energy to a point where the frictional locking of a fault is overcome, is about one hundred years. Many small earthquakes may occur along a fault during this time, but even so, elastic energy is accumulating because the fault is

locked. When the lock is broken and an earthquake occurs, the elastic energy is released during a few terrible minutes. By careful measurement of elastically strained rocks along the San Andreas Fault, seismologists have found that about a billion (10^9) ergs of elastic-strain energy can be accumulated in 1 m³ of strained rock. This is not very much—it is only equivalent to 100 J or about 25 cal of heat energy—but when billions or trillions of cubic meters of rock are strained, the total amount of stored energy can be enormous. The elastic energy stored in 10^{10} m³ is 10^{19} erg. The energy released by the atom bomb that destroyed Hiroshima during World War II was 4.0×10^{19} erg!

Measurements of elastically strained rocks before an earthquake, and of unstrained rocks after an earthquake, can provide an accurate measure of the amount of energy released. The task is very time consuming, and all too frequently measurements of the length of the fault, the amount of slip on the fault, and other necessary data are simply not available. Therefore, seismologists have developed a way to estimate the energy released by measuring the amplitudes of the seismic waves recorded on a seismograph. The *Richter magnitude scale*, named after the seismologist who developed it, is the most widely used system and is defined by the amplitudes of the *P* and *S* waves recorded on a special kind of seismograph. The convention is to measure, or by suitable calculation adjust the measurement to, a signal recorded at a dis-

tance of 100 km from the epicenter of an earthquake (Fig. 16.12). Because wave signals vary in strength by factors of a hundred million or more, it is impractical to use a scale divided into equal increments. Therefore, the Richter scale is logarithmic, which means it is divided into steps called *magnitudes*, starting with magnitude 1 and increasing upward. Each unit increase in magnitude corresponds to a tenfold increase in the amplitude of the wave signal. Thus, a magnitude 2 signal has an amplitude that is ten times larger than a magnitude 1 signal, and a magnitude 3 is a hundred times larger.

The energy in a wave is a function of both amplitude (A) and frequency (ω) which is the number of waves that pass a given point each second. Energy is proportional to $\omega^2 A^2$. Thus, if one Richter scale corresponds to a tenfold increase in A, the increase in *energy* should be proportional to A^2, which is to say, a hundredfold. However, the range of frequencies differs from one earthquake to another (in particular, the most energetic earthquakes have higher proportions of low-frequency waves). As a result, the energy increase corresponding to one Richter scale increase, when summed over the whole range of frequencies in a wave record, is only a thirtyfold increase. Thus, the difference in energy released between an earthquake of magnitude 4 and one of magnitude 7 is $30 \times 30 \times 30 = 2700$ times!

How big can earthquakes get? The largest recorded to date have Richter magnitudes of about 8.5, which means they release about as much energy as ten thousand atom bombs of the kind that decimated Hiroshima at the end of World War II. Possibly earthquakes do not get any larger than this because rocks cannot store more elastic-strain energy. If they are strained further, they fracture and so release the energy.

The dangers of earthquakes are profound and the havoc they can cause is often catastrophic. Their effects are of six principal kinds: the first two, ground motion and faulting, cause damage directly; the other four effects cause damage indirectly as a result of processes set in motion by the earthquake.

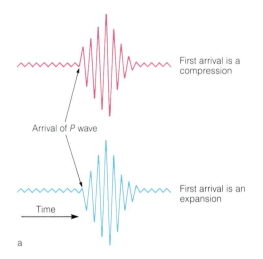

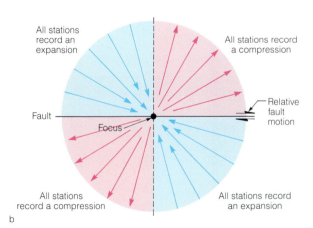

FIGURE 16.11 Initial motion of seismic body waves used to determine direction of movement on a fault. (*a*) Initial motion of a *P* wave detected by a seismograph is either a push away from the focus (that is, arrival of a compression), or a pull toward the epicenter (arrival of an expansion). (*b*) By plotting the first motions detected at a number of seismograph stations, the direction of movement on a fault can be uniquely determined. The example shown is for right-lateral motion on a strike-slip fault (Chapter 15).

1. *Ground motion* results from the passage of seismic waves, and especially surface waves, through surface-rock layers and regolith. The motions can damage and sometimes completely destroy buildings. Proper design of buildings (including such features as steel framework and foundations tied to bedrock) can do much

to prevent such damage, but in a very strong earthquake even the best buildings may suffer some damage.

2. Where fault offsets break the ground surface, buildings can be split, roads disrupted (Fig.

15.6), and any feature that crosses or sits on the fault can be broken apart.

3. A secondary effect, but one that is sometimes a greater hazard than moving ground, is *fire*. Ground movement displaces stoves, breaks

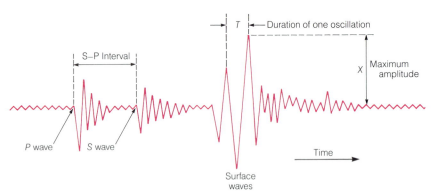

FIGURE 16.12 Determination of Richter magnitude from a seismograph record. Divide the maximum amplitude, X, measured in steps of 10^{-4} cm on a suitably adjusted seismograph, by the duration of one oscillation, T, in s. Then add a correction factor, Y, determined from the S–P wave arrival time interval. X/T is a measure of the maximum energy reaching the seismograph and the epicenter. The formula for the magnitude, M, is $M = \log x/T + Y$.

FIGURE 16.13 Landslides scar a hillside in Yenan, China. One of the worst earthquake disasters occurred when landslides started by an earthquake buried the people living in caves dug in the soft loess that mantles a large portion of northern China.

gaslines, and loosens electrical wires, thereby starting fires. In the disastrous earthquakes that struck San Francisco in 1906, and Tokyo and Yokohama in 1923, probably more than 90 percent of the damage to buildings was caused by fire.

4. In regions of hills and steep slopes, earthquake vibrations may cause regolith to slip, cliffs to collapse (Fig. 16.13), and other rapid mass-wasting movements to start (Chapter 9). This is particularly true in Alaska, parts of southern California, China, and hilly places such as Iran and Turkey. Houses, roads, and other structures are destroyed by rapidly moving regolith.

5. The sudden shaking and disturbance of water-saturated sediment and regolith can turn seemingly solid material to a liquidlike mass such as quicksand. This is called liquefaction and it was one of the major causes of damage during the earthquake that destroyed much of Anchorage, Alaska on March 27, 1964 (Fig. 9.25), and that caused apartment houses to sink and collapse in Niigata, Japan that same year (Fig. 16.14).

6. Finally, there are *seismic sea waves* (commonly called by their Japanese name, *tsunami*) that occur following violent movement of the seafloor. Seismic sea waves, often incorrectly called tidal waves, have been particularly destructive in the Pacific Ocean. About 4.5 h after a severe submarine earthquake near Unimak Island, Alaska in 1946, such a wave struck Hawaii. The wave traveled at a velocity of 800 km/h. Although the height of the wave in the open ocean was less than 1 m, the height increased dramatically as the wave approached land. When it hit Hawaii the wave had a crest 18 m higher than normal

FIGURE 16.14 Ground motion during a 1968 earthquake, Amori Prefecture, Japan, caused the sediment underneath this fertile farmland to lose coherence and develop fluidlike properties. Once weakened, the ridge and embankment slumped and flowed outward. Compare Figure 9.25; the same phenomenon created much of the damage in Anchorage, Alaska, during a 1964 earthquake.

high tide. This destructive wave demolished nearly 500 houses, damaged a thousand more, and killed 159 people.

Because damage to the land surface and to human property is so important, a scale of earthquake-damage intensity (called the *Modified Mercalli Scale*) has been developed based on observed damage. The correspondence among damage caused, Richter magnitudes, and the estimated number of earthquakes is listed in Table 16.2.

THE EARTH'S INTERNAL STRUCTURE

We have seen that *P* and *S* waves are body waves that travel through rock at differing velocities. These waves respond to changing elastic properties of rock in differing degrees. The arrival times of *P* and *S* waves at seismographs stationed around the world provide records of waves that have traveled through the Earth along many different paths. From such records it is possible to calculate how the elastic properties change and where there are distinct boundaries between layers having sharply different properties.

Layers of Differing Composition

If the Earth's composition were uniform, and if there were no polymorphic changes of the minerals present, the velocities of *P* and *S* waves would

increase smoothly with depth. This is so because higher pressure leads to an increase in the density and the rigidity of a solid and this, in turn, leads to a change in the elastic properties. With an Earth of uniform composition it would be possible to calculate the pressure effects on velocities and to predict how long it would take seismic waves to pass through. Observed travel times differ greatly from such predictions. The only way the differences can be accounted for is to suppose that velocities do not change smoothly with depth, and that composition is not constant throughout.

To find out where the supposed composition changes happen, it is necessary to make use of additional properties of waves. Seismic body waves behave like light waves and sound waves, which is to say they can be both transmitted through a medium and also *reflected* and *refracted*. Reflection is the familiar phenomenon that can be seen when light bounces off the surface of a mirror, or a glass of water. Seismic body waves are reflected by numerous surfaces in the Earth; examples are the boundary between the core and the mantle, and that between the mantle and the crust. Refraction is a less familiar phenomenon. It occurs whenever a wave velocity changes. The velocity change can be either gradual or abrupt. An abrupt change is seen when a ray of light strikes a surface of water, some of the light is reflected, but some also crosses the surface and travels through the water. The velocity of light is different in water than in air, and the ray path is sharply bent at the

TABLE 16.2 *Earthquake Magnitudes and Frequencies for the Entire Earth and Damaging Effects*

Richter Magnitude	Number Per Year	Modified Mercalli Intensity Scale[a]	Characteristic Effects of Shocks in Populated Areas
<3.4	800,000	I	Recorded only by seismographs
3.5–4.2	30,000	II and III	Felt by some people
4.3–4.8	4800	IV	Felt by many people
4.9–5.4	1400	V	Felt by everyone
5.5–6.1	500	VI and VII	Slight building damage
6.2–6.9	100	VIII and IX	Much building damage
7.0–7.3	15	X	Serious damage, bridges twisted, walls fractured
7.4–7.9	4	XI	Great damage, buildings collapse
>8.0	One every 5–10 yr	XII	Total damage, waves seen on ground surface, objects thrown in the air

Source: After B. Gutenberg, 1950.
[a]Mercalli numbers are determined by the amount of damage to structures and the degree to which ground motions are felt. These depend on the magnitude of the earthquake, the distance of the observer from the epicenter, and whether an observer is in or out of doors.

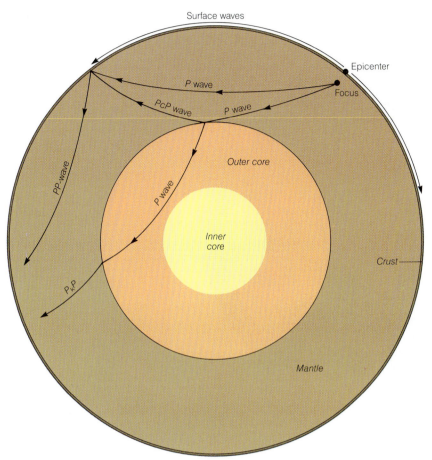

FIGURE 16.15 Schematic illustration of *P* waves radiating from an earthquake focus in the upper mantle. Seismographs at some places will receive both direct *P* waves as well as reflected and refracted *P* waves. A *P* wave reflected off the surface is called a *PP* wave; one reflected off the core–mantle boundary is a *PcP* wave; one refracted through the liquid outer core is a *PkP* wave. Note that refraction causes the wave paths to be curved. *S* waves also show reflection and refraction effects except that *S* waves cannot pass through the outer core because it is a liquid.

surface. The ray is said to have been refracted. Similarly, a seismic wave can be both reflected and refracted by a surface in the solid Earth (Fig. 16.15). Refraction occurs whenever there is a change in wave velocity. Because seismic-wave velocities increase with depth, wave paths are curved by refraction (Fig. 16.15).

Both *P* and *S* waves are strongly influenced by a pronounced boundary at a depth of 2900 km. When *P* waves reach that boundary, they are reflected and refracted so strongly that the boundary actually casts a *P*-wave shadow over part of the Earth (Fig. 16.16). Because the boundary is so pronounced, it can be inferred that it is the place where the comparatively light silicate material of the mantle meets the dense metallic iron of the

core. The same boundary casts an even more pronounced *S*-wave shadow, but the reason is not reflection or refraction. Shear waves cannot traverse liquids. Therefore, the huge *S*-wave shadow lets us infer that the outer core is liquid.

Refraction and reflection of seismic waves define three very pronounced boundaries separating four fundamental zones within the Earth. The boundaries correspond to those separating the crust from the mantle, the mantle from the outer core, and the outer core from the inner core.

The Crust

Early in the twentieth century the boundary between the Earth's crust and mantle was

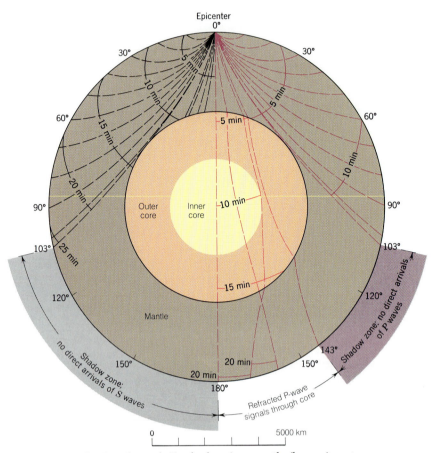

FIGURE 16.16 Section through Earth showing mantle (brown), outer core (orange), and inner core (yellow). Paths of *P* waves from an earthquake focus with epicenter at 0° (top) shown in right half only. Paths of *S* waves shown in left half. Distances reached by waves at five-minute intervals are indicated. Reflection and refraction of *P* waves at the mantle–core boundary create a *P*-wave shadow zone from 103 to 143°. Because *S* waves cannot pass through a liquid, an *S*-wave shadow exists between 103 and 180°. (*Source:* After B. Gutenberg, *Internal Constitution of the Earth*, 1950. By permission Dover Publications, Inc., New York.)

demonstrated by a scientist named Mohorovicic (Mo-ho-ro-vitch-ick) who lived in what today is Yugoslavia. He noticed that in measurements of seismic waves arriving from an earthquake whose focus lay within 40 km of the surface, seismographs within 800 km of the epicenter recorded *two* distinct sets of *P* and *S* waves. He concluded that one pair of waves must have traveled from the focus to the station by a direct path through the crust, whereas the other pair represented waves that had arrived slightly earlier because they had been refracted by a boundary at some depth in the Earth. Evidently, the refracted waves had penetrated a deeper zone of higher velocity below the crust, had traveled within that zone, and then had been again refracted upward

to the surface (Fig. 16.17). From his conclusions Mohorovicic hypothesized that a distinct compositional boundary, strongly influencing seismic

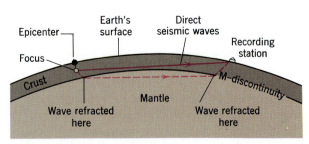

FIGURE 16.17 Travel paths of direct and refracted seismic waves from shallow-focus earthquake to nearby seismograph station.

waves, separates the crust from an underlying zone of differing composition. Scientists now refer to this boundary as the **Mohorovicic discontinuity** and recognize it as *the seismic discontinuity that marks the base of the crust.* Since the name is a tongue twister, the feature is commonly called the **M-discontinuity,** and in conversation is shortened still further to **moho.**

By seismic methods it is possible to determine the thickness of the crust. Seismic-wave velocities can be measured on different rock types in the laboratory and in the field. When the velocities of waves received at a number of seismographs are

calculated, laboratory measurements can be used to determine the depth of the moho, and to estimate the probable composition of the crust. Beneath ocean basins the crust is thin, in most localities averaging 10 km. Elastic properties of the oceanic crust are those characteristic of basalt and gabbro. But in the continental crust both thickness and composition are very different. The continental crust ranges in thickness from 20 to nearly 60 km, and tends to be thickest beneath major mountain masses (Fig. 16.18), a fact discussed further on in this chapter. Velocities in the continental crust are distinctly different from those in the oceanic

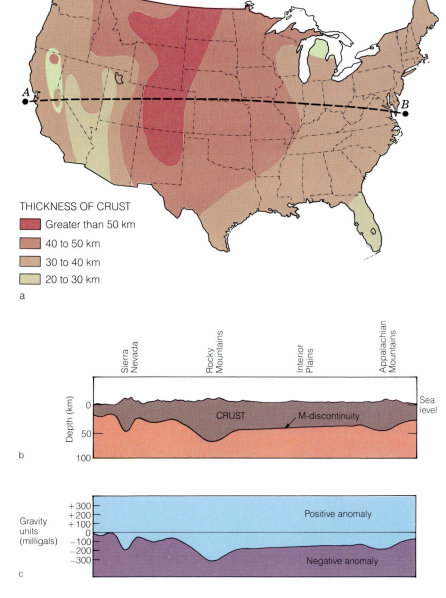

THICKNESS OF CRUST

■ Greater than 50 km
■ 40 to 50 km
■ 30 to 40 km
■ 20 to 30 km

a

b

c

FIGURE 16.18 Crust beneath the United States. (*a*) Thickness of crust beneath United States, determined from measurements of seismic waves. (*b*) Section through the crust along the line *A-B* (above). The crust tends to thicken beneath major mountain masses such as the Sierra Nevada, the Rocky Mountains, and the Appalachians. (*Source:* After Pakiser and Zietz, 1965.) (*c*) Profile of gravity traverse, adjusted for latitude, free air, and Bouguer corrections. The negative gravity anomalies over the Sierra, the Rockies, and the Appalachians are due to the roots of low-density rocks beneath these topographic highs.

crust. They indicate elastic properties like those of rock such as granite and diorite, although at some places just above the M-discontinuity, velocities close to those of oceanic crust are often observed. These conclusions agree well with what is known about the composition of the crust from other lines of evidence such as geological mapping and deep drilling. The agreement gives us confidence in drawing conclusions about the mantle, where these other lines of evidence are scarce.

Seismic Exploration of the Crust

Artificial earthquakes provide a powerful way of mapping buried structures in the crust. In the common method an artificial earthquake is created by an explosion, or by striking the ground surface sharply and repeatedly with a powerful air hammer. The waves generated by the explosion travel down through the rocks, they are reflected from the upper surfaces of buried strata, from faults and other discontinuities, and travel back to seismographs located on the surface (Fig. 16.19). The exact focus of the artificial earthquake is known, as are the exact positions of the seismographs, so a very accurate and informative picture can be prepared of the unseen structures below.

Seismic exploration using man-made earthquakes can be used equally well at sea. At sea, recording seismographs are towed behind a ship, and the seismic waves must pass through water as well as rock (Fig. 16.20). In many instances, seismic records are even better at sea than they are on land.

The Mantle

The mantle is something of an enigma. Although it is huge and seems to control much of what happens in the crust, it cannot be seen. *P*-wave velocities in the crust vary between 6 and 7 km/s. Beneath the M-discontinuity, velocities are greater than 8 km/s. Laboratory tests show that rocks common in the crust, such as granite, gabbro, and basalt, all have *P*-wave velocities of 6 to 7 km/s. But rocks that are rich in the dense minerals olivine, pyroxene, and garnet, such as peridotite, have velocities greater than 8 km/s. We therefore suppose, these minerals must be among the principal materials of the mantle. This evidence is one of the principal reasons we infer that the upper region of the mantle has the composition of a peridotite. This inference is consistent with what little is known about the composition of the upper part of

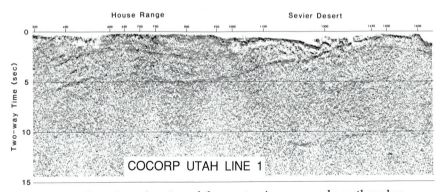

FIGURE 16.19 Seismic exploration of the crust using manmade earthquakes. P waves generated at the surface by striking the ground with air hammers are reflected at depth and recorded by strings of geophones. The section shown is in western Utah. The numbers at the top indicate distance; 100 units equals 10 km. The vertical numbers refer to travel times in seconds. Because velocities differ at different places in the crust, the section is somewhat distorted but, as a general rule, a two-way travel time of 10 s is approximately equivalent to a depth of 30 km.

The seismic line is in the Basin and Range Province. The prominent structure starting near the Sevier Desert and traceable toward the west for nearly 100 km is a fault known as the Sevier Desert detachment. (*Source:* Allmendinger, et al., 1983.)

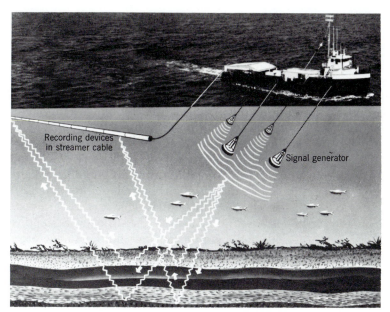

FIGURE 16.20 Seismic exploration at sea. Signal generators send out waves that travel through seawater but are reflected by layers of sediment beneath the seafloor. Reflected waves are recorded by detectors housed in the streamer cable, which may be as long as 2.5 km. (*Source:* Courtesy Continental Oil Company and American Petroleum Institute.)

the mantle—for example, with the ophiolite complex shown in Figure 4.37, and with rare samples of mantle rocks from **kimberlite pipes**, *narrow pipelike masses of igneous rock, sometimes containing diamonds, that intrude the crust but originate deep in the mantle* (Fig. 16.21).

The Low-Velocity Layer. When *P*- and *S*-wave velocities are calculated for different regions of the mantle—using the arrival times of direct, refracted, and reflected waves—the manner in which velocities are found to change with depth is far from regular. In general, both the *P* and *S* waves increase with depth throughout the crust and mantle, but some rapid and striking changes interrupt the velocity increase (Fig. 16.22).

The first of the major changes in velocity starts at a depth of about 100 km below the surface. The *P*-wave velocity at the top of the mantle is about 8 km/s. The overall *P*-wave velocity increases to 14 km/s at the core/mantle boundary. From the base of the crust to a depth of about 100 km, the velocity rises slowly to about 8.3 km/s. Then, the velocity starts to drop slowly to a value just below 8 km/s and it remains low, to a depth of about 350 km. The zone of reduced, or low velocity, is not sharply defined. It is better developed beneath the oceans than beneath the continents. Although the

FIGURE 16.21 Kimberlite, Monarch Pipe, South Africa. Fragments of rock from deep in the mantle are carried upward by the forceful intrusion of kimberlite magma. Rounded fragments are the transported blocks; fragmental, grayish background material is the kimberlite.

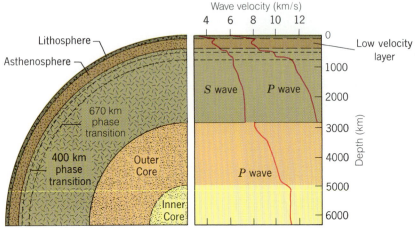

FIGURE 16.22 Variation of seismic-wave velocity within the Earth. Abrupt changes occur, at the boundaries between the crust and mantle and between the mantle and core, due to change in composition. Other abrupt changes occur at depths of 400 and 700 km below the surface because of phase transitions.

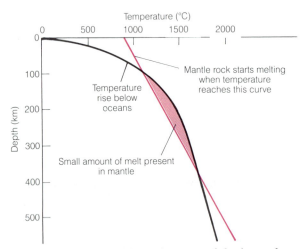

FIGURE 16.23 A possible explanation of the low-velocity zone. When the curve marking the onset of melting of mantle rock crosses the curve (black) defining the geothermal gradient, a small amount of liquid develops. As little as 1 percent of the rock could melt. A similar state of partial melting, determined experimentally for peridotite, is shown in Figure 4.15. The small amount of melt created would be enough to account for the plastic properties of the low-velocity zone.

upper boundary of the layer is reasonably sharp and well defined, the lower boundary is rather diffuse. The low-velocity layer can be seen as a small trough, or blip, in both the *P*- and *S*-wave velocity curves in Figure 16.22. There is no evidence that suggests the low-velocity layer is a zone where density decreases or, for that matter, where the composition changes. To account for the velocity changes, therefore, we infer that the zone that starts at a depth of about 100 km and continues to a depth of 300 to 350 km, has the same composition as the mantle immediately above and below, but is, by some means, less rigid, less elastic, and more plastic than the regions above and below.

A possible explanation of this low-velocity layer is that at depths between 100 and 350 km the geothermal gradient reaches temperatures close to the onset of partial melting of mantle rock (Fig. 16.23). If the explanation is correct, either the rigidity drops sharply close to the melting curve, or melting actually starts and a small amount of liquid develops and forms very thin films around the mineral grains, thus, serving as a lubricant. The amount of melting, if it occurs at all, must be very small, because the low-velocity zone does transmit *S* waves, and *S* waves cannot pass through liquids. Thus, rock in the low-velocity zone remains solid, but very close to melting. Any liquid, like a thick film of oil, merely serves to lubricate the grains, and at the same time reduce wave velocities by reducing the elastic properties.

An integral part of the theory of plate tectonics is the notion that plates of lithosphere can slide over a somewhat plastic zone in the mantle. The importance of the low-velocity layer for the theory is readily apparent. The top of the low-velocity layer coincides with the base of the lithosphere. Furthermore, as stated in Chapter 4, basaltic magma apparently originates by partial melting of rock in the mantle at depths of 100 to 350 km. Thus, the low-velocity zone, at least beneath the oceans, coincides with the asthenosphere.

We have discussed the asthenosphere as if it were a region of uniform thickness, lying everywhere at a constant depth of 100 km. But the asthenosphere cannot be constant, either in depth or in thickness. Beneath the oceans the top of the asthenosphere rises, in places to depths as shallow as 20 km, being closest to the surface near an oceanic ridge but progressively deeper away from the ridge. Beneath the continents the top of the asthenosphere is closer to 100 km, but it sinks as deep as 150 km beneath the thickest parts of the crust and beneath seafloor trenches. The schematic section through the upper portions of the Earth, shown in Figure 16.24, illustrates the present understanding of the variability of the asthenosphere.

Phase Transitions in the Mantle. From the *P*-wave curve in Figure 16.22 it is apparent that there are two places in the mantle where the velocity increases sharply. One is about 400 km deep, the other 670 km deep. It is also apparent that the *S*-wave velocity is not so strongly affected as the *P*-wave. The sudden increase in velocity seems to be too gradual to be accounted for by a change in composition; the cause must be something else. A probable explanation is suggested by laboratory experiments. When olivine is squeezed at a pressure equal to that which exists at a depth of 400 km the atoms rearrange themselves into a more dense polymorph (Chapter 3). This process of *atomic repacking caused by changes in pressure or temperature* is called a **phase transition.** In the case of olivine, the

repacking involves a change to a structure resembling that found in a family of minerals called the spinels, of which magnetite is a well-known example. The structural repacking involves a 10 percent increase in density. It is likely that the increase in seismic-wave velocities at 400 km is caused by the olivine–spinel phase transition rather than by compositional changes. The density increase determined from seismic-wave velocities is almost exactly 10 percent.

The velocity increase at 670 km is more difficult to explain. The observed increase in density is again about 10 percent, but the boundary is diffuse. It is not possible to determine from seismic evidence whether it is solely due to a phase transition, a compositional change, or both. Some scientists suggest that the increase at 670 km results solely from a phase transition involving the pyroxenes that are presumed to be present in the mantle. Others have suggested alternative phase changes, such as the rearrangement of silicate tetrahedra to create more dense structures, in which case each silicon atom is surrounded by six oxygen atoms rather than four. Still other suggestions center on compositional changes. One of the intriguing pieces of evidence comes from earthquakes. The deepest earthquakes have foci of 670 km. Because deep earthquakes are associated with subducted slabs of cool, oceanic lithosphere, it has been suggested that a compositional boundary may exist at 670 km, and that lithosphere can sink no deeper. Opposing this point of view is new evidence derived from a research technique called

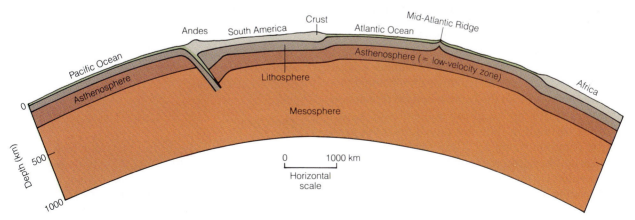

FIGURE 16.24 Asthenosphere varies in thickness and depth. Section through crust and upper mantle shows that the low-velocity layer corresponding to the asthenosphere is deeper beneath continents than beneath ocean basins and dips sharply down beneath the Andes. The section appears distorted because the vertical scale is twice the horizontal scale.

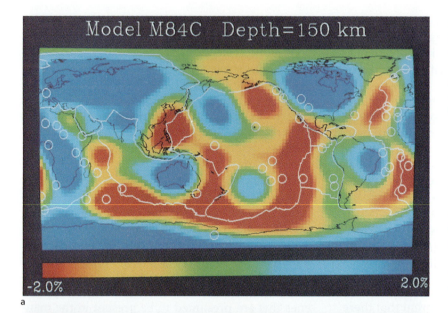

a

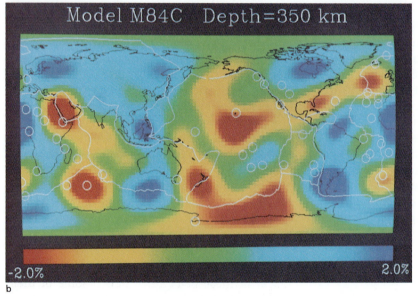

b

FIGURE 16.25 Lateral heterogeneity in the upper mantle is revealed through seismic tomography. Seismic waves travel faster through cooler, more rigid material (shown in blue), slower in hotter, less rigid material (red). The two tomographic images show the mantle at depths of 150 and 350 km below the surface. White lines show plate boundaries; white circles are centers of volcanic activity. Note that the red-colored, low velocity zones lie beneath spreading edges at 150 km depth but show much less correspondence at 350 km depth. (*Source:* Dziewonski and Anderson.)

seismic tomography. The method is similar to that used in medicine in which a three-dimensional picture of the interior of the human body is developed from slight differences in the intensities of X rays passing through in different directions. CAT scan is the common name for X-ray tomography. In a similar manner, inhomogeneities in the mantle can be revealed by measuring slight differences in the frequencies and velocities of seismic waves (Fig. 16.25).

The various possibilities for the 670 km seismic discontinuity are still being tested by experiments in laboratories around the world. Below 670 km, compression causes seismic-wave velocities to increase slowly and regularly until the abrupt boundary between the mantle and the core is reached at 2900 km.

The Core

Seismic waves indicate the way in which density increases with depth. Aided by both increasing pressure and phase transitions, density increases slowly from about 3.3 g/cm³ at the top of the mantle to about 5.5 g/cm³ at the base of the mantle. We already know that the mean density of the whole Earth is 5.5 g/cm³. Therefore, to balance the less-dense crust and mantle, the core must be composed of material with a density of at least 10 to 11 g/cm³. The only common substance that comes close to fitting this requirement is iron. Iron meteorites are samples of material believed to have come from the core of an ancient, now disintegrated, tiny planet. All iron meteorites contain a little nickel and the Earth's core presumably does too. Both pure iron and iron-nickel mixtures (even in the molten state) seem to be more dense than seismic-wave velocities suggest for the density of the core. But small amounts of silicon, sulfur, and other chemical elements with low atomic weights could dissolve in molten iron and produce the required velocity (Fig. 16.26). *P*-wave reflections indicate the presence of a solid inner core enclosed within the molten outer core. The two cores (outer and inner) appear to be identical in composition. The reason for the change from a liquid to a solid lies in the effect of pressure on the melting temperature of iron. As the center of the Earth is approached, pressure rises to values millions of times greater than atmospheric pressure. Temperature rises also, but not steeply enough to offset the effect of pressure. From the base of the mantle (at a depth of 2900 km) to a depth of 5350 km, temperature and pressure are so balanced that iron is molten. But at a depth of 5350 km another strong reflecting and refracting boundary occurs and the boundary has properties consistent with a change from a liquid to a solid. Apparently, from 5350 km to the center of the Earth, rising pressure overcomes rising temperature, and iron is solid, creating the solid core.

WORLD DISTRIBUTION OF EARTHQUAKES

Now that a picture has been developed of the way in which seismic waves are used to build a picture of the Earth's interior, let us turn to the pattern of occurrence of earthquakes. This pattern suggests a great deal about the shape and motions of the plates of lithosphere.

Although no part of the Earth's surface is exempt from earthquakes, several *seismic belts,* or *large tracts,* are *subject to frequent earthquake shocks* (Fig. 16.27). Of these the most obvious is the *Circum-Pacific belt,* for it is here that about 80 percent of all recorded earthquakes originate. The belt follows the mountain chains in the western Americas from Cape Horn to Alaska, crosses to Asia where it extends southward down the coast, through Japan, the Philippines, New Guinea, and Fiji, where it finally loops far southward to New Zealand. Next in prominence, giving rise to 15 percent of all earthquakes, is the *Mediterranean-Asiatic belt,* extending from Gibraltar to Southeast Asia. Lesser belts follow the mid-ocean ridges.

Seismic belts are places where a lot of the Earth's internal energy is released. Therefore, we might expect other manifestations of internal energy to appear in these belts also. Indeed some of them do. Mid-ocean ridges, deep-sea trenches, andesitic volcanoes, and many other features that outline the margins of plates of lithosphere, coincide with, or closely parallel, these margins. Compare Figure 16.27 with Figures 2.11 and 4.19, to see that earthquake belts outline the plate boundaries.

The depths of earthquake foci around the edges of the plates also have a story to tell. Most foci are no deeper than 100 km, because, as already mentioned, earthquakes occur in brittle rocks and the brittle lithosphere is only 100 km thick. However, a few earthquakes do originate at greater depths. The epicenters of deep earthquakes, with foci deeper than 100 km, are plotted in Figure 16.27*b*. It is noteworthy that the deep earthquakes are not associated with oceanic ridges. When Figure 16.27*b* is compared with Figure 14.11, however, it

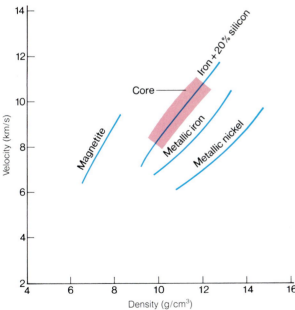

FIGURE 16.26 Density in the fluid outer core plotted against *P*-wave velocity. Compressive wave velocities determined in the laboratory show that neither molten iron nor molten nickel have the correct range of density and velocity. Addition of a light element, such as silicon, to molten iron or to a mixture of nickel and iron produces the correct density and velocity.

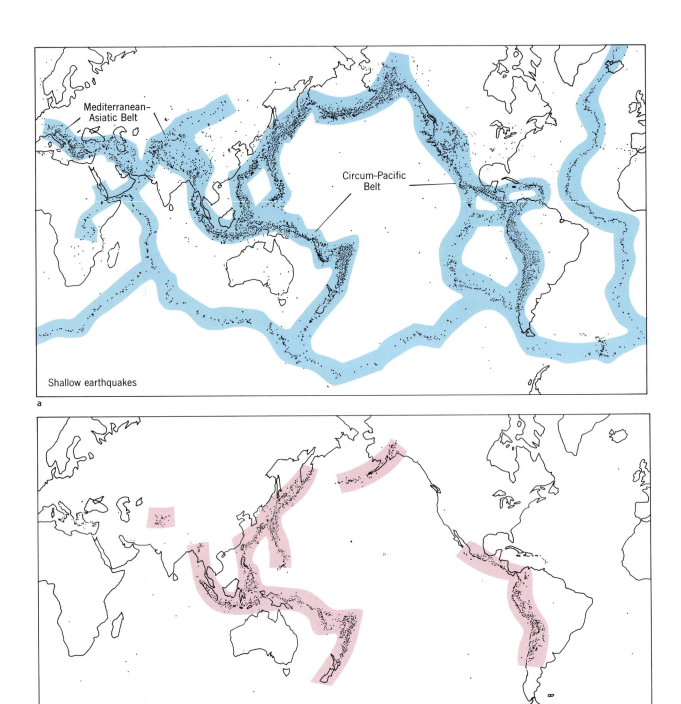

Mediterranean–
Asiatic Belt

Circum–Pacific
Belt

Shallow earthquakes

a

Deep earthquakes

b

FIGURE 16.27 Seismic belts and epicenters of earthquakes recorded by the U.S. Coast and Geodetic Survey between 1961 and 1967. Each dot represents a single earthquake. (*a*) Earthquakes of all depths are plotted. Most are shallow, with foci within 100 km of the surface. The epicenters fall into well-defined seismic belts (dark blue shading) that coincide closely with the margins of plates of the lithosphere. (*b*) Epicenters of earthquakes having foci deeper than 100 km. They form belts that coincide closely with the sea-floor trenches. (*Source:* After Barazangi and Dorman, 1969.)

is immediately apparent that all deep earthquakes are closely related to seafloor trenches. Those trenches mark the places where lithosphere sinks down into the mantle.

Detailed study of deep-earthquake foci beneath a seafloor trench (Fig. 16.28) shows that the foci follow a well-defined zone, sometimes called a *Benioff zone*. This important observation is strongly suggestive that deep earthquakes originate within the relatively cold, downward-moving plate of lithosphere. Because some earthquake foci can be as deep as 670 km, it must be concluded that rapidly descending lithosphere can retain at least some brittle properties to that depth. The reason that earthquakes do not seem to originate at depths below 670 km remains an unsolved problem, but as previously mentioned, may result from a physical barrier presented by the 670-km seismic discontinuity. It may also be that even a rapidly sinking slab of lithosphere is sufficiently hot by the time it reaches a depth of 670 km that it has become ductile rather than brittle.

Locations of earthquakes reveal a great deal about the structure of the moving lithosphere. As remarked at the beginning of this chapter, seismic waves generated by the earthquakes serve as a kind of giant X ray that let us infer the Earth's internal structure. But they provide a static picture, a sort of snapshot of the way things are at the moment. In order to discover how the Earth changes over time as it responds to forces that make materials move and flow, it is necessary to include other observations besides those from seismology. The most revealing is information derived from measurements of the Earth's gravitational attraction.

GRAVITY ANOMALIES AND ISOSTASY

The fact that the Earth is a deformable body and that it is not a perfect sphere were discussed in

Chapter 2. The shape of the Earth is only approximately spherical; careful measurement reveals that it is actually an ellipsoid that is slightly flattened at the poles and bulged at the equator. The approximately spherical shape of the Earth is due to gravity which pulls every particle on and in the Earth toward the center of its mass (Fig. 2.5). The equatorial bulge and polar flattening are caused by the centrifugal force arising from the Earth's rotation around its axis (Fig. 2.6).

Because the Earth's radius at the equator is 21 km larger than it is at the poles, the pull exerted by the Earth's gravitational attraction is slightly greater at the poles than it is at the equator. Thus, a man who weighs 90.5 kg (199 lb) at the North Pole would observe his weight decreasing slowly and steadily to 90 kg (198 lb) by simply traveling to the equator. If the weight-conscious traveler made very exact measurements as he traveled, he would observe that his weight changed irregularly, rather than smoothly. From this he could conclude that the pull of gravity must change irregularly. If the traveler went one step further, and carried *a sensitive device for measuring the pull of gravity at any locality* called a **gravimeter** (or **gravity meter**), he would indeed find an irregular variation.

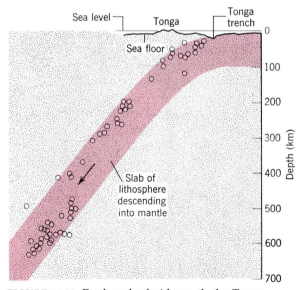

FIGURE 16.28 Earthquake foci beneath the Tonga Trench, Pacific Ocean, during several months in 1965. Each circle represents a single earthquake. The earthquakes are believed to be generated by downward movement of a comparatively cold slab of lithosphere that is plunging slowly back into the mantle. (*Source:* After Isacks et al., 1968.)

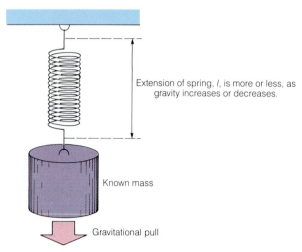

FIGURE 16.29 A gravimeter (= gravity meter) is basically a known mass of metal suspended on a sensitive spring. The mass exerts a greater or lesser pull on the spring as gravity changes from place to place, extending the spring more or less. The mass of metal and the spring are contained in a vacuum together with exceedingly sensitive measuring devices.

The simplest form of a gravimeter is a pendulum. When a pendulum is displaced from its rest position, gravity exerts a restoring pull. The stronger the pull, the faster the pendulum swings back. From the speed of the swing (the period), the force of gravity can be calculated at any place. Modern gravimeters consist of a heavy mass suspended by a sensitive spring (Fig. 16.29). When the ground is stable and free from vibrations due to earthquakes, the pull exerted on the spring by the heavy mass provides an accurate measure of the gravitational pull. Modern gravity meters are incredibly sensitive. The most accurate and sensitive devices in operation can measure variations in the force of gravity as tiny as one part in a hundred million (10^{-8}).

In order to compare the pull of gravity from point to point on the Earth, three corrections must be applied to gravimeter measurements.

1. A correction must be applied for the latitude of the place of measurement. This correction takes care of the departure of the Earth's shape from a perfect sphere.

2. Topographic variations on the Earth's surface mean that measurements on mountains are made further from the center of the Earth than are measurements made in valleys. A gravimeter can detect the increase in gravitational pull when it is moved as little as a meter from the

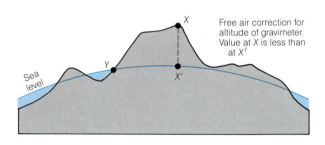

Free air correction for altitude of gravimeter. Value at *X* is less than at *X*'

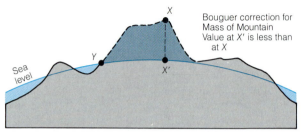

Bouguer correction for Mass of Mountain Value at *X*' is less than at *X*

FIGURE 16.30 In order to compare gravitational force measured on a mountain top, *X*, with that at sea level, *Y*, the value at *X* must be corrected for altitude, increasing it to the value it would have at *X*'. The gravitational attraction of the mountain mass between *X* and *X*' must then be corrected by subtracting the attraction corresponding to the amount of rock between *X* and *X*'.

top of a table to the floor. Therefore, all measurements must be adjusted for elevation. In order to do so, gravimeter readings are adjusted as if they were made at the surface of a reference ellipsoid that lacks topographic variations. The Earth's reference ellipsoid corresponds approximately with sea level. The topographic correction is called the *free-air correction* (Fig. 16.30).

3. The final correction that must be made is an adjustment to account for the free-air correction. When a gravimeter reading is adjusted to the reference ellipsoid value, account must be taken of the gravitational attraction exerted by the rock lying between the gravimeter and the ellipsoid. The correction that completes the adjustment for topography is called the *Bouguer correction*; it has an effect opposite to that of the free-air correction.

After corrections are made for latitude, free air, and Bouguer, the adjusted figures for the force of gravity might be expected to be the same everywhere on the Earth. In fact the adjusted figures reveal large and significant variations called *gravity anomalies*. The anomalies are due to bodies of rock having differing densities. A simple example of an anomaly is shown in Figure 16.31. From the anomalies, a great deal of important information can be derived.

The thickness of the crust beneath the United States, as determined from seismic measurements of the M-discontinuity, is shown in Figure 16.18. Beneath the three major mountain systems (the Appalachians, the Rockies, and the Sierra Nevada) the crust is thickened. In profile (Fig. 16.18*b*), the crust beneath the mountains resembles icebergs with high peaks, but with massive roots below the waterline. The accuracy of this analogy is demonstrated by the gravity profile across the United States, shown in Figure 16.18*c.* Negative gravity anomalies are observed where the crust is thickest. The anomalies are caused by the roots of low-density rock beneath the mountains, just as the basin of low-density sediments produced the gravity anomaly shown in Figure 16.31. The reason a root of low-density rock forms in the first place provides some interesting insights into the Earth's physical properties. Mountains stand high and have roots beneath them because they are comprised of low-density rocks and are supported by the buoyancy of weak, easily deformed but more dense rocks below. Mountains are, in a sense, floating. Rock can flow and be deformed under very slow strain rates, so that, strange as it may seem, the topographic variations observed at the surface of the Earth do not arise from the strength of the crust, but rather from its buoyancy. The flotational property is embodied in the property known as isostasy that was defined in Chapter 2.

The great ice sheets of the last glaciation provide an impressive demonstration of isostasy. The weight of a large continental ice sheet will depress the crust, but when the ice melts, the land surface slowly rises again. A spectacular example of glacial depression and rebound is shown in Figure 13.39. The effect is very much like pushing a block of wood into a bucket of thick, viscous oil. When the wood is released, it slowly rises again to an equilibrium position determined by its density. The speed of its rising is controlled by the viscosity of the oil. Just like the block of wood, glacial depression and rebound means that somewhere in the mantle, rock must flow laterally when the ice depresses the crust, and then must flow back again when the deforming force is removed (Fig. 16.32). The flow must be slow because parts of northern Canada and Scandinavia are still rising even though the ice sheets melted away about 5,000 years ago.

Continents and mountains are composed of low-density rock, and they stand high because they are thick and light; ocean basins are topographically low because the thin oceanic crust is composed of dense rock. Isostasy and differences in the density of rocks beneath the continents and the oceans, therefore, are the reasons that the Earth has two pronounced topographic levels, as shown in Figure

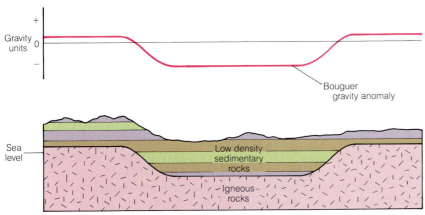

FIGURE 16.31 Example of a gravity anomaly. The cross section through part of the crust reveals a basin filled with low-density sedimentary rocks sitting on a basement of more dense igneous rocks. Gravity measurements, corrected for latitude, free-air, and Bouguer effects, reveal a pronounced gravity low over the basin. The magnitude of the Bouguer anomaly can be used to calculate the thickness of rocks of the basin.

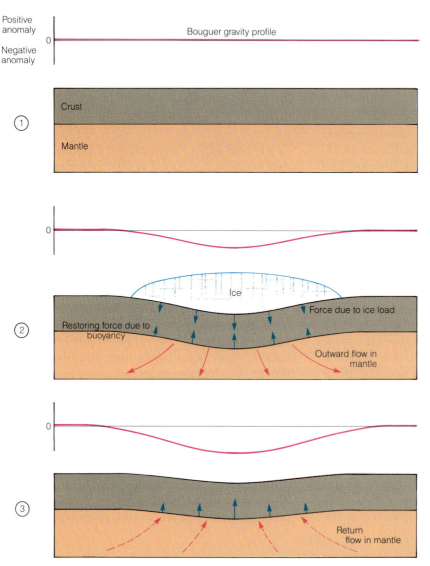

FIGURE 16.32 Schematic diagram illustrating the depression of the crust by a continental ice sheet. (1) Prior to formation of the ice sheet there is no gravity anomaly. (2) When the ice sheet forms it depresses the crust. At some depth in the mantle, material must slowly flow outward to accommodate the sagging crust. (3) When the ice melts, buoyancy slowly restores the crust to its original level. A negative Bouguer gravity anomaly continues until the depression is removed. The viscosity of the mantle controls the rate of flow and, therefore, the slowness of recovery.

2.7. The depth at which rock is weak enough to flow like a viscous fluid in order to produce the buoyancy effects of isostasy is called the ***depth of compensation.*** It is defined as *that depth above which segments of crust and upper mantle act as blocks that rise or sink depending on the mass and density of the individual blocks.* The depth of compensation corresponds approximately to the bottom of the lithosphere where rock is hot enough to be weak and easily deformed. The depth of compensation does not occur at the same depth everywhere. Beneath the oceans it seems to be nearer the surface than beneath the continents. Local effects can also influence the depth of compensation. For example, when a mass of hot and, therefore, less-dense rock rises up in the mantle, the beginning of fluid-like

properties happen at a shallower depth. When a region of the upper mantle is occupied by low-density hot rock, the overall density of that block of lithosphere is reduced and a topographic high is the result. An example can be seen in Figure 16.18, where the Basin and Range Province, a topographically high area, is underlain by hot, low-density mantle rock rather than a root of light, continental crust. Similarly, there are broad swells and rises in the oceanic crust that are underlain by shallow bodies of hot, low-density mantle rock.

The important points to be drawn from this discussion of isostasy are that the crust and indeed all of the lithosphere act as if they are floating on the asthenosphere. Floating is not exactly the correct word, because the Earth is solid. But the system is buoyant and acts as if it were floating. Sometimes it is observed that a mountain has too little root for its mass; sometimes, as in the seafloor trenches, it is observed that low-density crust has been dragged down to form a root without a mountain mass above it. These and many other situations lead to local gravity anomalies. The anomalies do not seem to become very large. This suggests that the Earth is always moving toward an isostatic balance. Indeed, isostasy is the principal explanation for vertical motions of the Earth's surface, as plate tectonics is the principle explanation for lateral mo-

tions. An important consequence of isostatic balance is that if the weight of a mountain mass is removed, the depressed crust will slowly "bob" back up. Thus, as erosion removes material from a mountain, thereby reducing the mountain's mass, isostatic adjustment will cause the entire mountain to rise slowly. The root below the ancient, deeply eroded Appalachians is smaller than the root beneath the younger and higher Sierra. As we shall see in the next three chapters, such isostatic rebounding is a vital feature in the development and history of the Earth's crust.

SUMMARY

1. Abrupt movement on faults is responsible for most earthquakes, many of which cause destructive damage to dwellings and other man-made structures.

2. Ninety-five percent of all earthquakes originate in the Circum-Pacific belt and the Mediterranean-Asiatic belt. The remaining 5 percent are widely distributed.

3. Energy released at an earthquake's focus radiates outward as P (compressional) waves, and as S (shear) waves. Earthquake energy also makes the Earth vibrate like a giant bell. The vibrations are sensed as surface waves.

4. From the study of seismic waves, scientists infer the internal structure of the Earth by locating boundaries or discontinuities in its composition. Pronounced compositional boundaries occur between the crust and mantle and between the mantle and the outer core.

5. Within the mantle there are two zones, at depths of 400 and 670 km, where sudden density changes produce seismic-wave discontinuities. The 400 km change is produced by a

phase transition of olivine. The 670 km change might be due to either a phase transition, a compositional change, or a combination of both.

6. The lithosphere is approximately 100 km thick and rigid; it overlies a plastic zone within which seismic waves have low velocities. The low-velocity zone coincides with the asthenosphere.

7. The base of the crust is a pronounced seismic discontinuity called the M-discontinuity. Thickness of the crust varies from 20 to 60 km in continental regions, but is only about 10 km thick beneath the oceans.

8. The two forces that give the Earth its spherical shape are gravity, acting inward, and centrifugal force caused by rotation, that acts outward. The Earth is not a perfect sphere. It bulges at the equator and is flattened at the poles.

9. The outer portions of the Earth are in approximate isostatic balance; in other words, they act as though they were huge icebergs floating in fluid substrata.

SELECTED REFERENCES

Board of Earth Sciences, National Academy of Sciences, 1983, The lithosphere: Washington, D.C., National Academy of Sciences.

Bolt, B. A., 1978, Earthquakes: A primer: San Francisco, W. H. Freeman and Co.

McKenzie, D. P., 1983, The Earth's mantle: Sci. American, v. 249, no. 3, p. 114–129.

Press, F., 1975, Earthquake prediction: Sci. American, v. 232, no. 5, p. 14–23.

Walker, B., 1982, Planet Earth. Earthquake: New York, Time-Life Books.

Wyllie, P. J., 1975, The Earth's mantle: Sci. American, v. 232, no. 3, p. 50–63.

CHAPTER 17

Dynamics of
the Lithosphere

An active strike-slip fault, locally known as the Great Fault, is slicing the
South Island of New Zealand into two pieces.

A NEW PARADIGM

A revolution makes people reexamine old ideas. The scientific revolution that was initiated by the hypothesis of plate tectonics has been no exception. First enunciated in the late 1960s, this revolution shows no signs of abating. Every corner of geology has had to be looked at through new eyes, every deduction reexamined, every conclusion rethought, and every question asked again. Throughout this book we have tried to integrate the abundant fruits of the revolution; they appear in almost every chapter. Plate tectonics is no longer just a hypothesis; in mid 1984, scientists working for NASA announced the first measurements of plate motions using satellites and lasers. Geological deductions were proved correct—plates really are moving with measured velocities between 1.5 and 7 cm/yr. As with all successful concepts in science, plate tectonics has provided simplified explanations for many seemingly complex problems. It has provided testable answers for long-standing problems (such as the origin of ocean basins), and it has shown that seemingly unrelated features (such as mid-ocean ridges, and grabens on continents) are actually closely related. In short, plate tectonics is so widely accepted that we are going to devote this chapter entirely to the topic.

SEARCH FOR A SOLUTION

One cannot but wonder why continents have their peculiar shapes, why ocean basins are where they are, why mountain ranges, earthquake belts, volcanoes, and many other major features occur where they do. Such wonderings prompted many scientists to think that there might be a single, underlying cause for the whole array of the Earth's major features. Scientists speculated for more than a hundred years about possible causes, before the hypothesis of plate tectonics was suggested.

During the nineteenth century people favored the idea that the Earth was originally hot—a molten mass—and that it has been gradually cooling, contracting, and compressing the crust. They pointed to fold-mountain ranges and seismic belts as the places where most of the contraction now occurs and has occurred in the past. Contraction did explain some features, but it did not help with questions about the shapes and the distribution of continents. Nor did it help explain the great rift valleys and other features where the crust was clearly in a state of tension rather than compression. Also, as we now know, the theory of contraction has a fundamental flaw: The decay of naturally radioactive atoms keeps the Earth's interior hot so that the rate of cooling is vastly slower than is suggested by the evidence for contraction.

When it was realized that the Earth's internal heat was caused by radioactive decay, some scientists suggested that the Earth might actually be heating up. A much smaller Earth, they suggested, could once have been covered largely by continental crust. Heating would cause the Earth to expand and the continental crust would then crack and break into fragments. As expansion continued, the cracks would grow into ocean basins, and through the cracks basaltic magma would rise up from the mantle to build new oceanic crust. Although the theory of an expanding Earth did not easily account for fold-mountain ranges, it did offer a plausible explanation for the approximately parallel coastlines of adjacent continents, such as Africa and South America. Nevertheless, the expansion theory has other fundamental problems. There is no evidence (such as an increase in the rate of production of magma) to suggest that the Earth is heating. To get around the flaws in both the expansion theory and the contraction theory, the effects of other forces on the crust were examined. Such forces include the Earth's rotation and the Moon's gravitational pull. The centrifugal force due to rotation and the gravitational pull were both suggested to be forces that could possibly deform continents, just as the Earth's rotation and the Moon's gravity influence the motions of ocean currents. Calculations showed, however, that both forces are far too weak to cause large-scale deformation, so the ideas were abandoned. By the middle of the twentieth century all the reasonable suggestions concerning the shapes and positions of continents seemed to have been exhausted. The time was ripe for a totally new approach.

Wegener and Continental Drift

One key suggestion was made early in the twentieth century, soon after the contraction theory had collapsed. Alfred Wegener, a German meteorologist, suggested in 1912 that continents drift slowly across the surface of the Earth, sometimes breaking into pieces and sometimes colliding with each other. Lateral motion could produce both compressive and tensional forces at the same time. The front edge of a moving continent would be in a

state of compression, the interior and tail ends would be in a state of tension. Similar suggestions had, in fact, been made in earlier times by others, but it was Wegener who presented the most persuasive arguments and who carefully gathered evidence. His *theory of continental drift* originated when he attempted to explain the striking parallelism of the edges of the shorelines on the two sides of the Atlantic Ocean, especially the shores of Africa and South America. Other bits of favorable evidence were quickly found. These bits of evidence supported the hypothesis that the world's landmasses had once been joined together in a single great supercontinent which was dubbed *Pangaea* (pronounced pan-jee-ah, meaning "all lands") (Fig. 17.1). According to the hypothesis, Pangaea was somehow disrupted during the Triassic, and its fragments (the continents of today) slowly drifted to their present positions. Proponents of the theory likened the process to the breaking-up of a sheet of ice that floats in a pond. The broken pieces, they argued, should all fit back together again, like pieces of a giant jigsaw puzzle. Figure 17.1a shows that a jigsaw reconstruction indeed works well.

Some of the more impressive types of evidence presented by Wegener and his colleagues are that Triassic and earlier fossils of identical land-dwelling animals are found in South America and in southern Africa, but nowhere else. Fossils of identical trees are found in South America, India, and Australia. Fragments of what is apparently a single mountain system are observed on both sides of the North Atlantic Ocean (Fig. 2.13). However, the most impressive clue consists of unmistakable evidence that about 300 million years ago a continental ice sheet covered parts of South America, southern Africa, India, and southern Australia (Fig. 17.1b). The ice sheet resembled the one that covers Antarctica today; evidence of its existence is so well preserved that thousands of glacial striations (Chapter 13) reveal the directions in which the ice flowed. However, if 300 million years ago continents were in the positions they occupy today, the ice sheet would have had to cover all the southern oceans, and in places would even have had to cross the equator! A glacier of such huge size could only mean that the world climate was exceedingly cold. Yet if the climate had been cold, why had no evidence of glaciation at that time been found in the Northern Hemisphere? The dilemma would be explained neatly by continental drift. Three hundred million years ago the regions covered by ice lay in high, cold latitudes. Indeed

those regions were adjacent to the South Pole (Fig. 17.1a). At that time, therefore, the Earth's climates need not have been greatly different from those of today.

Among those who supported the ideas of Alfred Wegener was a distinguished South African geologist, Alexander du Toit. As some of the most convincing evidence concerning continental drift is to be found in the Southern Hemisphere, du Toit set to work to gather and verify all data. Du Toit carefully noted that evidence for Pangaea was not completely convincing. Equally likely was the suggestion that a great southern supercontinent, Gondwanaland, coexisted with a northern supercontinent, Laurasia, which were separated by a narrow sea called Tethys. Today the sediments laid down in Tethys are seen squeezed, thrust, folded, and metamorphosed in the great Alpine–Himalayan mountain chain. Present-day workers agree that Gondwanaland (Australia, Africa, South America, India, New Zealand, and Madagascar) and Laurasia (North America, Europe, and Asia) did have separate identities, but that for a time (at least during the Permian and Triassic) they were indeed joined together to form Pangaea and that the Tethys sea formed a great bay between Africa and Eurasia, like an ancient Mediterranean Sea that opened to the east (Fig. 17.1a).

Despite the impressive evidence that favored the drifting of continents, many scientists remained unconvinced, largely because the fluid-like properties of the asthenosphere had not been discovered. Indeed, the concept of a lithosphere and an asthenosphere having different physical properties without parallel changes in composition was hardly suspected. Wegener struggled with the problem of how solid, brittle rocks could move through or over other solid, brittle rocks. The fact that continental rocks differed from oceanic rocks was known, and of course the fact that the continental land surface stood high above the seafloor was well known. Putting these facts together, Wegener suggested that, despite problems of strength, continental crust must somehow slide over oceanic crust. Opponents, many of whom were geophysicists, quickly argued that because of the great frictional resistance to such sliding, rigid continental crust simply could not slide over rigid oceanic crust without both crusts disintegrating. The process is like trying to slide two sheets of coarse sandpaper past each other.

Wegener died in 1930. Although debate continued, its pace slowed down. True, the geological evidence of glaciers and fossils suggested that con-

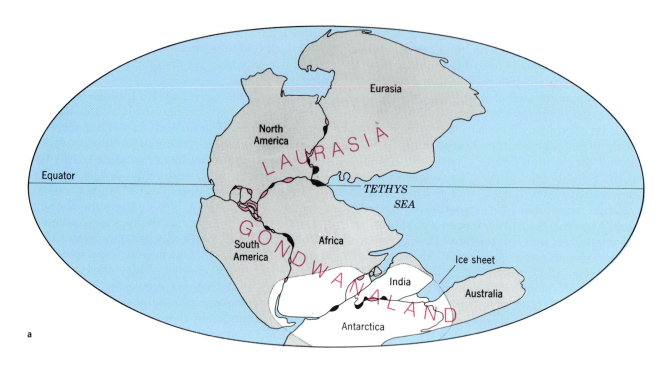

a

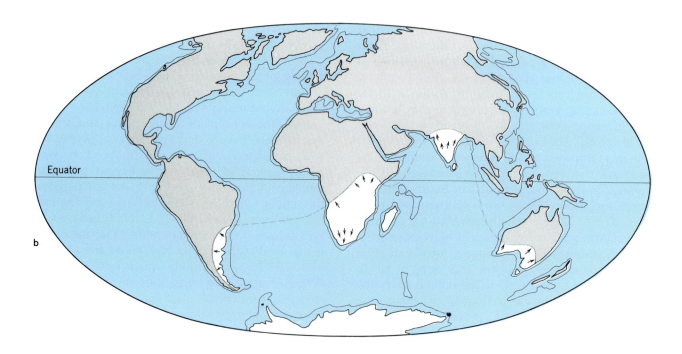

b

tinents might have moved; however, geophysicists, studying the Earth's physical properties, became convinced that movement could not have occurred, and elaborate explanations were erected to explain the geological evidence. The situation became a stalemate. Despite the work of du Toit, more and more scientists discarded the hypothesis that continents had drifted because an acceptable mechanism could not be found to explain the movement.

FIGURE 17.1 The continents attained their present shapes when Pangaea broke apart 200 million years ago. (*a*) Shape of Pangaea, determined by fitting together pieces of continental crust along a contour line 2000 m below sea level. This line coincides with the foot of the continental slope. It is the line along which continental crust meets oceanic crust. In a few places some overlap (black) occurs: in others, small gaps (red) are found. These are places where existing maps are poor or where later events have modified the shapes of continental margins. White area is the region affected by continental glaciation 300 million years ago. (*b*) Present continents and the 2000 m contour below sea level. White area is where evidence of the old ice sheets exists. Arrows show directions of movement of the former ice. The dashed line joining the glaciated regions indicates how large the ice sheet would have to be if the continents were in their present positions at the time of glaciation (*Source:* After Dietz and Holden, 1971 and du Toit, 1937.)

Apparent Polar Wandering

A turning point in the debate came during the 1950s. From the mid 1950s to the mid 1960s geophysicists made a number of remarkable discoveries. The first of the key discoveries arose through studies of paleomagnetism. As previously discussed, when certain igneous and sedimentary rocks form, they become weakly magnetized (Chapter 8), and the direction of magnetization preserves a "fossil" record of the Earth's magnetic field at the *time* and *place* of formation. Three essential bits of information are contained in the fossil magnetic record. The first is the polarity— whether the magnetic field was normal or reversed at the time of formation. The second piece of information is the direction of the magnetic pole at the time the rock formed. Just as a free-swinging magnet today will point toward the magnetic poles, so does paleomagnetism record the direction of the magnetic poles from the point of rock formation. The third piece of information, and the one that provides the needed information to say how far away the magnetic pole lay, is the magnetic inclination. As discussed in Chapter 8, the magnetic inclination changes from zero at the magnetic equator to 90° at the magnetic pole (Fig. 8.15). The paleomagnetic inclination is therefore a record of the place between the pole and the equator (that is, the magnetic latitude) where the rock was formed. Once the magnetic latitude of a rock and the direction of the magnetic poles are known, the position of the magnetic pole at the time of formation is determined.

Geophysicists who were studying paleomagnetic pole positions during the 1950s found evidence suggesting that the poles wandered all over the globe. They referred to the strange plots of paleo-pole positions as *paths of apparent polar wandering*. The geophysicists were puzzled by evidence for polar wandering. The Earth's magnetic field, they knew, was caused by motions in the liquid outer core, and the motions in turn were caused by the Earth's rotation.

Although the magnetic poles might wobble a little, they should always remain close to the poles of rotation. Determination of magnetic latitude should therefore be a good approximation to the geographic latitude. When it was discovered that the path of apparent polar wandering measured in North America differed from that in Europe (Fig. 17.2), the geophysicists were even more puzzled. Somewhat reluctantly, they concluded that because it is unlikely that the magnetic poles have moved, it is more likely that the continents and the magnetized rocks had moved instead. In this way the hypothesis of continental drift was revived, but a mechanism to explain how the movement occurred was still lacking.

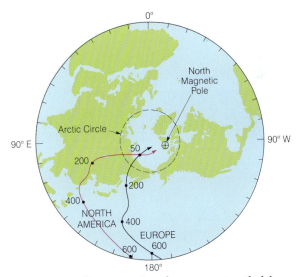

FIGURE 17.2 Curves tracing the apparent path followed by the north magnetic pole through the past 600 million years. Numbers are millions of years before the present. The curve determined from paleomagnetic measurements in North America (red curve) differs from that determined by measurements made in Europe (black curve). Wide-ranging movement of the pole is unlikely; therefore, it is concluded that it is not the pole, but the continents that have moved. (*Source:* After Northrop and Meyerhoff, 1963.)

Seafloor Spreading

Help came from an unexpected quarter. All the early debate about continental drift, and even the data about apparent polar wandering, had centered on evidence drawn from the continental crust. Before 1950, the 70 percent of the Earth's surface that is covered by oceanic crust was still largely unknown and unexplored. During the 1950s the mysteries of the ocean floor started to be revealed. Features such as the topography of the mid-ocean ridge, the fact that a graben runs down

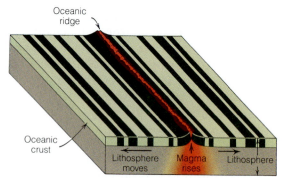

FIGURE 17.3 Schematic diagram of oceanic crust. Lava extruded along an oceanic ridge forms new oceanic crust. As lava cools, it becomes magnetized with the polarity of the Earth's field. Successive strips of oceanic crust have alternate normal polarity (black) and reversed polarity (gray).

the center of the ridge, indicating tensional forces, and the fact that sediments thicken away from the ridge were remarkable revelations. In 1962, the revelations led Harry Hess of Princeton University to hypothesize that the seafloor moves sideways, away from the oceanic ridges. This hypothesis, which came to be called *the theory of seafloor spreading*, was soon proved correct, and once again it was geophysicists who used paleomagnetism to provide the proof.

The theory of seafloor spreading postulated that magma rose from the interior of the Earth and formed new oceanic crust along the mid-ocean ridge. Hess could not explain what made the crust move away from the ridge, but he nevertheless proposed that it did and that as a consequence the oceanic crust became older the further it moved. Two tests for the Hess theory were soon proposed. One test was suggested by J. Tuzo Wilson. He argued that if there are long-lived magma sources deep in the mantle there should be lines of volcanic islands formed as the seafloor moves over the magma source, and there should be a steady progression in the age of volcanism on the islands. As discussed later in this chapter and in Chapter 19, the Hawaiian chain of islands provides a striking confirmation of Wilson's suggestion. A second and more powerful test of the Hess theory was proposed by three geophysicists: Vine (who was a

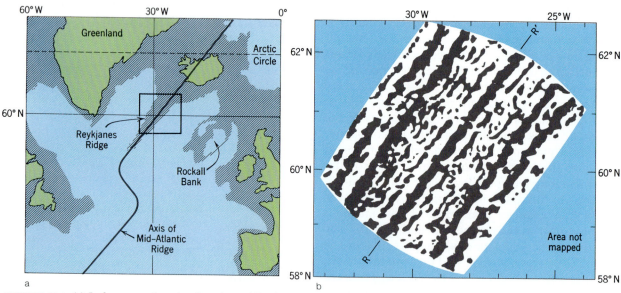

FIGURE 17.4 (*a*) Index map showing location of Reykjanes Ridge, a portion of the Mid-Atlantic Ridge southwest of Iceland. (*b*) Map of the magnetic striping of rock on the sea floor. *R-R'* is the center line of Reykjanes Ridge. Strips of rock with normal polarization (black) alternate with reversely polarized rock (white). (*Source:* After Heirtzler, Le Pichon and Baron, 1966.)

student at the time), Matthews, and Morley. The Vine-Matthews-Morley suggestion concerned thermoremanent magnetism of the oceanic crust.

When lava is extruded at the oceanic ridge, or when gabbro is intruded, the rocks become magnetized and acquire the magnetic polarity that existed at the time they cooled through the Curie point. If new lava and new gabbro are forming and are continually moving away from the oceanic ridge, the oceanic crust should contain a continuous record of the Earth's magnetic polarity. Magnetic data gathered as a result of antisubmarine defense research proved this point. The crust is, in effect, a very slowly moving magnetic tape recorder in which successive strips of oceanic crust are magnetized with normal and reversed polarity (Fig. 17.3). Seafloor magnetism can easily be measured with instruments carried in ships or airplanes. An example of the results is given in Figure 17.4. It was a simple matter to match the sort of pattern observed in Figure 17.4*b* with the record of magnetic polarity, such as that shown in Figure 8.20. The distinctive magnetic striping allowed the age of any place on the seafloor to be determined. Because the ages of magnetic polarity reversals had been so carefully determined, magnetic striping also provided a means to estimate the speed with which the seafloor moved. In places it was found to be remarkably fast, reaching values as high as 10 cm/yr.

PLATE TECTONICS: A POWERFUL THEORY

Proof that the seafloor moves, and acceptance by geophysicists of the new evidence that continents also move, were the two spurs needed for a new, all-embracing theory to emerge. It was quickly forthcoming. Although a lot of the evidence to support the theory came from geophysicists, all branches of geology and paleontology combined to provide the needed evidence. The essential points in formulating a theory of plate tectonics were, first, that the low-velocity zone (soon identified with the asthenosphere) is exceedingly weak and has viscous, fluid-like properties. The rigid, stronger lithosphere, on the other hand, seems to form coherent sheets which can slide sideways over the weak, underlying asthenosphere. This answered the objections that the geophysicists had directed at Wegener—movement could occur without massive resistance from friction. The lithosphere is much thicker than the crust, however, so one con-

sequence of the theory was that as the lithosphere moved the crust was simply rafted along as a passenger. Continents move, to be sure, but they only do so as portions of larger plates, not as discrete entities.

The second essential point answered by the theory of plate tectonics concerned the destruction of old oceanic crust. If, as the theory of seafloor spreading required, new oceanic crust is continually created along the mid-ocean ridges, either the Earth must be expanding and the oceans must be getting larger, or an equal amount of old crust must necessarily be destroyed in order to maintain an Earth of constant size. The answer was provided by the previously unexplained Benioff zones (Chapter 16). These slanting zones of deep earthquake foci are the places where old, cold lithosphere is sinking back into the asthenosphere and mesosphere. In this way, cool and still brittle lithosphere can sink to great depths. In a simplified form, the basic elements of plate tectonics are shown in Figure 17.5.

Structure of a Plate

The hypothesis of plate tectonics maintains, as has already been pointed out, that the surface of the Earth is covered by six large and many small plates of lithosphere, each about 100 km thick, sliding over the fluid-like asthenosphere (Fig. 2.11). The plates are believed to be rigid, or nearly so, moving as single coherent units; that is, the plates do not crumple and fold like wet paper, but act more like semirigid sheets of plywood floating on water. The plates may flex slightly, causing gentle up- or down-warping of the crust, but the only places where intense deformation occurs is at edges along which plates impinge on each other. Such plate margins are *active zones*; plate interiors are *stable regions*.

Plates have three kinds of margins: *divergent* margins along which two plates move apart from each other; *convergent* margins along which two plates move toward each other; and *transform fault* margins along which two plates simply slide past each other (Fig. 17.6). Each margin creates distinctive topography in its vicinity, and is associated with a distinctive kind of earthquake activity and volcanism. These features are summarized in Table 17.1 and are briefly discussed below.

Divergent Margins

As defined in Chapter 2, a **spreading edge** or **divergent margin** is *the new growing edge of a plate*. It is a line along which two adjacent plates move apart from each other, and along which new lithosphere is created. Divergent margins are places where crust is being stretched by tensional forces. The kinds of faults associated with tensional forces are invariably normal faults (Fig. 15.12). Therefore, normal faults and grabens (Fig. 15.13) are associated with divergent margins.

Earthquakes along spreading edges tend to have low magnitudes and shallow foci. This is so because the ductile asthenosphere comes close to the surface beneath a spreading edge. The kind of volcanic activity along a spreading edge is almost always basaltic. Where the spreading edge lies below the sea, the lavas either form prominent pillow structures, or they form as thin sheets that spread

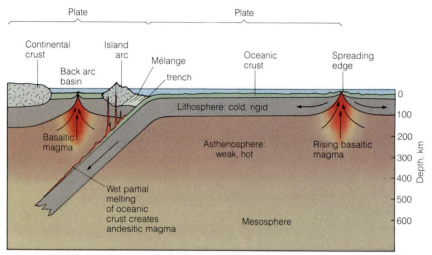

FIGURE 17.5 Cross section through crust and upper mantle showing details of a tectonic plate. Lithosphere is capped by oceanic crust formed by basaltic magma rising from the asthenosphere. Moving laterally, the lithosphere accumulates a thin layer of marine sediment and eventually starts sinking into the asthenosphere. At the point of sinking, an oceanic trench is formed and sediment deposited in the trench, plus sediment from the moving plate, is compressed and deformed to create a mélange. The sinking oceanic crust eventually reaches the temperature where wet partial melting commences and forms andesitic magma that rises to form an island arc of stratovolcanoes. Behind the island arc, on an adjacent plate, tensional forces lead to the development of a back-arc basin, as discussed in the text.

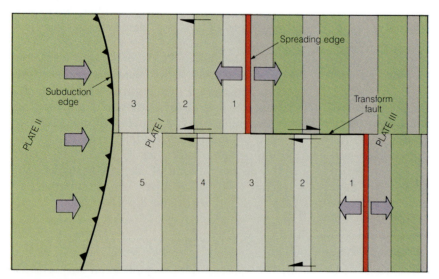

FIGURE 17.6 Simplified structure of a plate. The spreading edge is offset by the transform fault that divides the plate. Bars are magnetic time lines of oceanic crust. Broad arrows show the direction of plate motion.

TABLE 17.1 *Kinds of Plate Margins and Characteristic Features*

Crust on Each Plate	Feature	Kind of Margin		
		Divergent	Convergent	Transform Fault
Oceanic–Oceanic	Topography	Oceanic ridge with central rift valley	Seafloor trench	Ridges and valleys created by oceanic crust
	Earthquake	All foci less than 100 km deep	Foci from 0–700 km deep	Foci as deep as 100 km
	Volcanism	Basaltic pillow lavas	Andesitic volcanoes in an arc of islands parallel to trench	Volcanism rare; basaltic along "leaky" faults
	Example	Mid-Atlantic Ridge	Tonga-Kermadec Trench; Aleutian Trench	Kane Fracture
Oceanic–Continental	Topography	—	Seafloor trench	—
	Earthquake	—	Foci from 0–700 km deep	—
	Volcanism	—	Andesitic volcanoes in mountain range parallel to trench	—
	Example	(No examples)	West coast of South America	(No examples)
Continental–Continental	Topography	Rift valley	Young mountain range with folded crust	Fault zone that offsets surface features
	Earthquake	All foci less than 100 km deep	Foci as deep as 300 km over a broad region	Foci as deep as 100 km throughout a broad region
	Volcanism	Basaltic and rhyolitic volcanoes	No volcanism. Intense metamorphism and intrusion of granitic plutons	No volcanism
	Example	African Rift Valley	Himalaya, Alps	San Andreas Fault

outward from the normal faults in the central graben. A graben is formed where a spreading edge breaks continental crust. The graben will tend to become filled with sediment and some of that sediment may be assimilated by the rising basaltic magma to produce magmas of differing composition (Chapter 4). At times it is even possible that some of the sediment near the base of a graben may melt and produce small quantities of magma with very unusual compositions; such magmas may erupt onto the floor of the graben as unusual lavas.

Magnetic Records and Plate Velocities

The magnetic polarity record implanted in oceanic crust at a spreading edge provides some vital clues. Working outward from an active ridge, the crust becomes progressively older. The first magnetic reversal recorded away from a ridge crest is that at 730,000 years (when the reversed polarity of the Matuyama Epoch changed to today's normal polarity) (Fig. 8.20). Subsequent reversals are located in succession away from the ridge. The oldest reversals recorded in oceanic crust date back to the middle Jurassic, about 165 million years ago. Jurassic-aged crust occurs in the western Pacific, and is the most ancient part of the present seafloor. When the positions and ages of the magnetic time lines have been located (as shown in Figure 17.7), it is apparent that plate velocities can be calculated. From the symmetrical spacing of magnetic time lines on both sides of the ridge it appears that both plates move away from a ridge at equal rates. Appearances can be deceiving, however. The same pattern of magnetic time lines in Figure 17.7 would be observed if the African Plate were stationary and both the Mid-Atlantic Ridge and the North American Plate were moving westward. Later in this chapter evidence will be presented to substantiate the suggestion that mid-ocean ridges do indeed move. All that can be deduced from magnetic time lines, therefore, is the velocity of one plate *relative* to another. An answer to the question of *absolute* velocities requires more information.

Relative plate velocities vary greatly from place to place on the globe. They do so for two reasons.

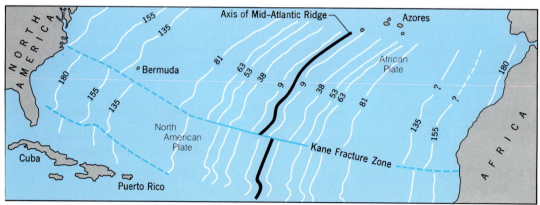

FIGURE 17.7 Age of the ocean floor in the central North Atlantic, deduced from magnetic striping, increases regularly away from the axis of the Mid-Atlantic Ridge. Numbers give ages in millions of years before the present. The Kane Fracture Zone, observed near the center of the oceanic ridge, continues across the Atlantic and causes consistent offsetting of the age contours. (*Source:* After Pitman and Talwani, 1972.)

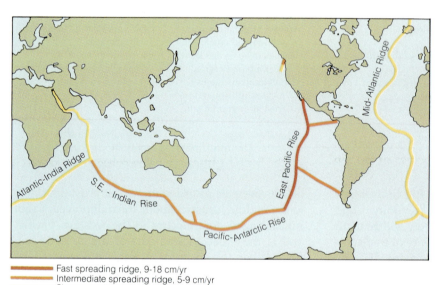

Fast spreading ridge, 9-18 cm/yr
Intermediate spreading ridge, 5-9 cm/yr
Slow spreading ridge, 1-5 cm/yr

FIGURE 17.8 Spreading rates of principal mid-oceanic ridges. Fast spreading rates mean plates move away from each other between 9 and 18 cm a year. Intermediate rates are 5 to 9 cm a year; slow rates are 1 to 5 cm a year. Spreading rates influence the topography of the mid-ocean ridge (Fig. 17.11).

First, the relative motions across some mid-ocean ridges are much greater than they are across others (Fig. 17.8). The reasons for high relative velocities are not known with certainty, but they appear to be related to the amount of continental crust sitting on a plate. Plates that do not carry a large load of continental crust, as is the case for the Pacific and Nazca Plates, have high velocities. Plates with large loads of continental crust, such as the North American and Eurasian plates, have low velocities.

The second reason that plate velocities vary from place to place has to do with the geometry of motion on a sphere. One might think, intuitively, that all points on a plate move with the same velocity. That is incorrect. Our intuitions would only be cor-

rect if plates of lithosphere were flat (like sheets of plywood) and moved over a flat asthenosphere (like plywood floating on water); then, all points on the plate *would* move with the same velocity. However, plates of lithosphere are pieces of a shell on a spherical Earth, so they are curved, not flat. In the geometry of a sphere, any movement on the surface can be described as a rotation about an axis of the sphere. A consequence of rotation and, therefore, of a curved plate moving over the surface of a sphere is that different parts of a plate move with different velocities.

To picture how points on a plate move with different velocities, imagine a plate so large that it forms a hemispherical cap covering half the Earth

(Fig. 17.9). The cap moves independently of the Earth's rotation and rotates instead about an axis of its own, colloquially called a *spreading axis.* In the figure, point *P,* where the spreading axis reaches the surface, is a *spreading pole.* Point *P* has no velocity of movement because it is the fixed point around which the hemispherical cap rotates. However, point *E,* at the edge of the cap, has a high velocity because it must move completely around the Earth, along path *E-E',* during a single revolution of the cap. Any point on the cap between points *P* and *E* has an intermediate velocity that is slower if the point is closer to *P,* faster if it is closer to *E.*

No plate is large enough at present to cover half the Earth, nor does any plate rotate around a spreading pole in the center of the plate. But the principle is the same for a small plate as it is for a hemispherical cap. Consider Plate *A* in Figure 17.9, which is only a small portion of the hemispherical cap. The motion of Plate *A* is from east to west around the spreading axis. Point *A'',* close to the spreading pole, must move more slowly than point *A',* more distant from the pole.

The motion of each of the Earth's plates can be described in terms of rotation around a spreading axis, and the velocity of each point on the plate depends on its distance from the spreading pole. One consequence of differing velocities of motion is this: The width of new oceanic crust that borders a divergent margin between two plates increases with distance from the spreading pole (Fig. 17.10). A further consequence is that each segment of an oceanic ridge lies on a line of longitude that passes through the spreading pole, and each transform fault that offsets the oceanic ridge lies on a line of latitude around the spreading pole. The relation between transform faults and spreading poles can be used to determine the position of the spreading pole of each plate. Using the same property, the positions of old transform faults can be used to determine the positions of former spreading poles and, therefore, to determine if a plate has, at some time, changed its direction of motion.

Topography of the Seafloor

The topography of the seafloor is controlled by the growth and movement of plates. Two prominent features in particular are related to divergent plate margins. The first feature is the mid-ocean ridge. The shape of the ridge is strongly influenced by the rate of spreading.

Fast spreading rates, 9–20 cm/yr, mean that new

oceanic crust is created very rapidly. This in turn means that magma must rise rapidly and continually from below and that large magma chambers must lie at shallow depths below the center of the ridge. As a result, a fast spreading ridge like the East Pacific Rise is thermally inflated and stands high above the seafloor; yet even so, a fast-

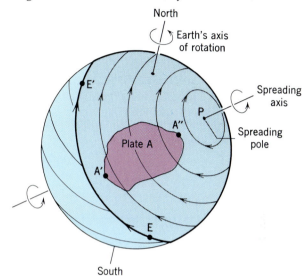

FIGURE 17.9 Rotation of a hemispherical cap that fits over a sphere. Movement of Plate A, a part of the cap, is controlled by the rate of rotation around the spreading axis. The movement of each plate of lithosphere on the Earth's surface can be described as a rotation about a spreading axis. (*Source:* Adapted from Wyllie, 1976.)

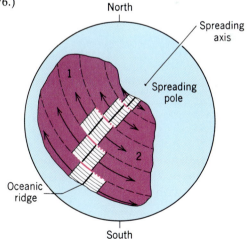

FIGURE 17.10 Relation between spreading axis, oceanic ridge, and transform faults in two adjacent plates. Plates 1 and 2 have a common spreading edge offset by transform faults (red). Each segment of the oceanic ridge lies on a line of longitude that passes through the spreading pole. Each transform fault lies on a line of latitude with respect to the spreading pole. The width of new oceanic crust (striped shading) increases away from the spreading pole.

spreading ridge has a distinct, but shallow, central graben (Fig. 17.11). By contrast, a slow spreading ridge like the Mid-Atlantic Ridge is cooler, less inflated, and has a more pronounced topography. The overall ridge still stands high above the deep ocean floor, but the central graben is wider and more pronounced.

The second prominent feature is the ocean floor itself. A large fraction of the heat that escapes from the Earth's interior does so along divergent plate margins. As a result, not only the mid-ocean ridges, but also the adjacent seafloor, are high points because the lithosphere beneath them is thermally expanded. As lithosphere moves away from a mid-ocean ridge, it cools and contracts. As contraction occurs, the depth of the seafloor in-

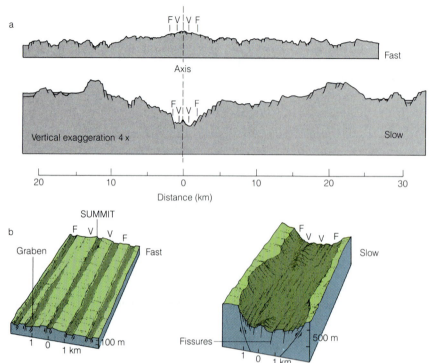

FIGURE 17.11 Topography of a mid-ocean ridge is controlled by the rate of spreading. (*a*) Profiles across a fast and a slow spreading ridge, stretching approximately 20 km on either side of the central axis, which is located in the center of the graben. (*b*) Three-dimensional diagrams of topography at the center of the spreading edges. The letters F and V show positions on the profiles in *a*. (*Source:* After Macdonald, 1982.)

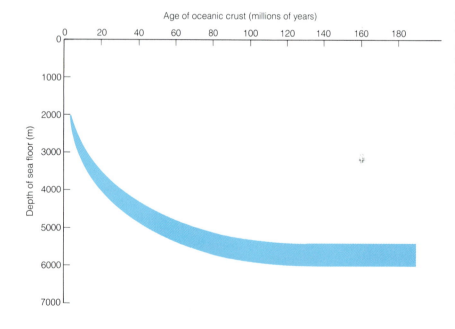

FIGURE 17.12 Average depth of the seafloor in the world oceans as a function of the age of the oceanic crust. Near the spreading ridge, young lithosphere is thermally expanded. As it moves away from the crust, the lithosphere cools and contracts. The ocean depth increases as a result.

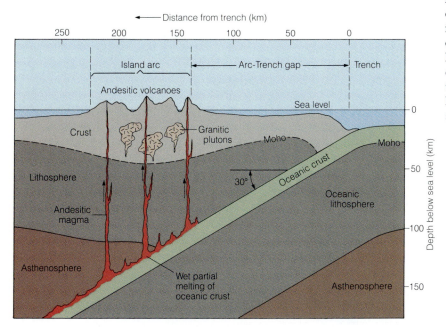

FIGURE 17.13 Formation of andesitic magma by wet partial melting of subducted oceanic crust occurs between about 100 and 150 km below the surface. If the downgoing slab has a dip of 30°, the arc-trench gap is about 150 km. If the dip is steep, the gap is shorter.

creases. A constant depth is reached after about 110 million years, by which time oceanic lithosphere has cooled and reached thermal equilibrium (Fig. 17.12). To a first approximation, therefore, the depth of the ocean floor below sea level provides an estimate of the age of the oceanic crust.

Convergent Margins

Edges of consumption or *subduction zones* were defined in Chapter 2 as *the edges along which plates of lithosphere turn down into the mantle.* They are convergent plate margins where two plates move toward each other. Convergent margins are margins of compression and the kinds of faults present are those associated with compressional forces—reverse and thrust faults (Fig. 15.14).

Island Arcs

The Earth's surface area does not seem to be changing. This means that the production of new lithosphere at divergent margins must be balanced by the destruction of old lithosphere. The process occurs at convergent margins when one plate sinks downward beneath the other at an angle of 20–60° to the horizontal (Fig. 17.5). As the plate descends into the mantle, it is heated up and eventually reaches a temperature at which wet partial melting commences. This process seems to be the one that forms andesitic magma (Chapter 4). Rising to the surface, the magma forms a chain of volcanic islands. Termed an *island arc,* the *chain of stratovolcanoes is parallel to the seafloor trench and separated from it by a distance of 150–300 km,* the distance depending on the angle of dip of the descending plate (Fig. 17.13).

The Japanese islands are part of a modern-day island arc (Fig. 4.20). Another very pronounced island arc is the Aleutian chain of islands that continues westward from the southwestern tip of mainland Alaska. There are many other island arcs around the edge of the Pacific Ocean. Examine Figure 4.19 and it becomes apparent that the Andesite Line coincides with island arcs. Note, too, that although each arc is part of a circle, some arcs are parts of a circle with a large radius and some are highly curved and are parts of a circle with a smaller radius. The radius of curvature is an indication of the angle at which lithosphere is plunging back into the mantle. If the angle of plunge were perpendicular to the Earth's surface, an island arc would be straight. If the angle of plunge is almost flat, the island arc has a pronounced curvature (Fig. 17.14).

Careful determination of the age of oceanic crust being subducted shows that old, cold, and therefore dense crust forms island arcs that have a large radius of curvature. Young oceanic crust that has still not reached thermal equilibrium, and that is less dense than older, colder crust, forms arcs with short radii of curvature. This observation is a very informative one, because it indicates that oceanic

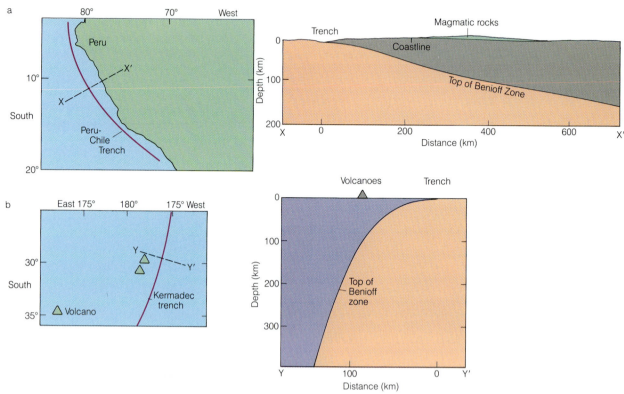

FIGURE 17.14 The steepness of the angle at which lithosphere sinks at a subduction zone controls the curvature of the trench and of the magmatic arc. (*a*) Beneath Peru the top of the Benioff zone dips at a shallow angle, and the adjacent Peru–Chile Trench has a pronounced curvature. (*b*) The Benioff zone beneath associated with the Kermadec Trench in the southwest Pacific has a steep dip and the trench has only a slight curvature.

lithosphere must actually be sinking under its own weight through the hot, weak asthenosphere. This means, further, that old, cold lithosphere, when capped by oceanic crust, must be more dense than the hot, plastic asthenosphere. The older and colder the lithosphere is, the faster the rate of sinking and the steeper the angle of the Benioff zone. As lithosphere sinks, it must start to heat up. Earthquakes can occur in the down-going slab so long as it is cool enough to be brittle. Even with a sinking rate of 8 cm/yr, calculations show that lithosphere loses its brittle properties by the time it reaches a depth of 670–700 km (Fig. 17.15). This is probably the reason that earthquake foci are never deeper than 700 km.

As old, dense lithosphere sinks, lithosphere on the adjacent plate overrides the sinking slab. If the forward motion of the overriding plate is equal to or greater than the sinking rate, a zone of pronounced compressive deformation results. A subduction zone subjected to major compressive forces is called a *coupled convergent margin* because the two plates deform each other.

Mélange

Many features occur as a result of deformation along coupled plate edges, but the most distinctive and characteristic is the development of a *mélange, a chaotic mixture of broken and jumbled rock* (Fig. 17.16). Once a subduction zone forms and a seafloor trench is created, sediment accumulates in the trench. A sinking plate drags sedimentary rock downward beneath the overriding, coupled plate. Sedimentary rock has a low density. As a result it is buoyant and cannot be dragged down very far. Caught between the overriding and sinking plates, the sediment becomes shattered, crushed, sheared, and thrust faulted to form a mélange (Fig. 17.17). As the mélange thickens, it becomes metamorphosed. The cold sedimentary rocks are dragged down so rapidly that they remain cooler than adjacent rock at the same depth. The kind of metamorphism that is common in many mélange zones, therefore, is that shown in Case *c* of Figure 6.18—a high-pressure, low-temperature metamorphism distinguished by blue schists and ec-

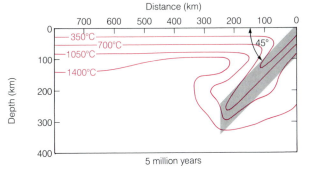

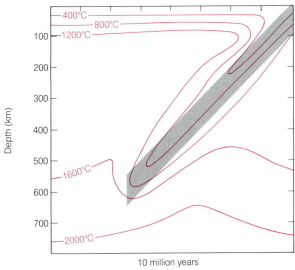

FIGURE 17.15 Computer-aided calculations of the fate of a descending slab of cool lithosphere. A plate 100 km thick, descending at an angle of 45° and a rate of 8 cm/yr, will cool the surrounding mantle but will slowly become heated as it sinks. Contours depict the temperature. Between 600 and 700 km the temperature of the tip of the descending slab reaches the temperature of the adjacent mantle and earthquakes cease. (*Source:* After Hsui and Toksöz, 1979.)

logites. The blue color comes from a bluish amphibole called glaucophane.

Back-Arc Basins

When the sinking rate of a subducting plate is faster than the forward motion of the overriding plate, part of the overriding plate can be subjected to tensional stress. Such a convergent margin is said to be an *uncoupled margin.* The leading edge of the overriding plate must remain in contact with the subduction edge or else a huge void would open. What happens is that the overriding plate grows slowly larger at a rate equal to the difference in velocities between the two plates. Most commonly, an uncoupled margin is manifested by a thinning of the crust and an opening of an arc-shaped basin behind the island arc (Fig. 17.18). Basaltic magma may rise into a so-called *back-arc basin* and a small region of new oceanic crust may even form. Because of the proximity of stratovolcanoes in the island arc, an abundant source of sediment is generally available; hence, back-arc basins tend to be filled with a mixture of volcanic rocks and clastic sediments.

Convergent margins can sometimes be coupled, sometimes uncoupled. As a result, many have both a mélange and a back-arc basin. The distinctive features of a convergent margin are shown in Figure 17.18.

Transform Fault Margins

The faults at the margins of plates are *transform faults* (Fig. 15.19). They are huge, vertical, strike-slip faults cutting down into the lithosphere. They can form when either a new divergent or a convergent margin fractures the lithosphere (Fig. 17.19). Neither compressional nor tensional forces are associated with the faults; they are simply margins along which two plates slide past each other. The sliding margins smash and abrade each other like two giant strips of sandpaper, so the faults are marked by zones of intensely shattered rocks. Where the faults cut oceanic crust, they make elongate zones of narrow ridges and valleys on the seafloor (Fig. 14.6). When transform faults cut continental crust, they do influence the topography; however, the features are less pronounced than on the seafloor. Transform faults on the land tend to be marked by parallel or nearly parallel faults in a zone that can be as much as 100 km wide.

The sliding movement of transform faults causes a great many shallow-focus earthquakes, some of them of high magnitude. Most transform faults do not have any volcanic activity associated with them. Occasionally, however, a small amount of plate separation does occur and a "leaky" transform results in a small amount of volcanism. Probably the best-known transform fault in North America is the San Andreas Fault in California. The many earthquakes that disturb California are caused by movement along it. Some of those earthquakes, including the one that devastated San Francisco in 1906, have been particularly destructive, and it is probable that future quakes will be just as devastating. As long as the plates continue

FIGURE 17.16 The Lichi Formation, near Taitung, Taiwan is a chaotic mélange of mudstone, sandstone, and shale produced by thrusting associated with subduction of the Philippine Plate beneath Taiwan.

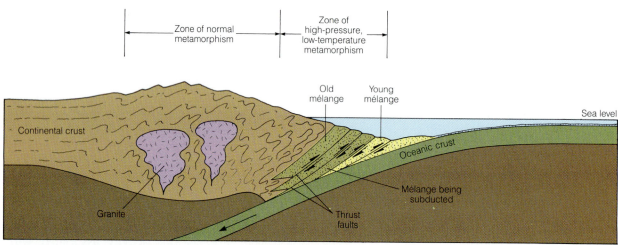

FIGURE 17.17 Mélange is formed when young sediment in a trench is smashed by moving lithosphere and dragged downward in slices bounded by thrust faults. As successive slices are dragged down, older mélange, closer to the continent, is pushed back up. The process is like lifting a deck of cards by adding new cards at the base of the deck. Mélange is characterized by blueschist metamorphism.

to move, activity must occur along the San Andreas Fault, and residents of California can expect more earthquakes.

The San Andreas Fault is the largest of several transform faults that offset the segment of oceanic ridge called the East Pacific Rise. Figure 17.20 shows how the transform faults and segments of the East Pacific Rise separate the North American Plate from the Pacific Plate. Figure 17.20 also shows that the San Andreas is only one of several faults that break the continental crust. The others are subsidiary to the San Andreas, however, and are part of a fault zone that is roughly parallel to the main fault. Movement along the San Andreas Fault arises from movement between the North American and Pacific Plates. The peninsula of Baja California and the portion of the state of California that lies west of the San Andreas Fault are on the Pacific Plate. That plate is moving northwest, relative to the North American Plate, at a rate of several centimeters per year. In about 10 million years Los Angeles will have moved far enough north so as to be opposite San Francisco. In about 60 million

years, at the present rate of movement, the segment of continental crust on which Los Angeles lies will have become separated completely from the main mass of continental crust that comprises North America.

The Plate Mosaic and Plate Motions

To a first approximation, plates of lithosphere behave as rigid bodies. This means that plates do not stretch and shrink, like rubber sheets. The distance between New York and Chicago, both on the North American Plate, remains fixed, even though

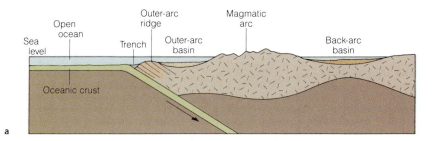

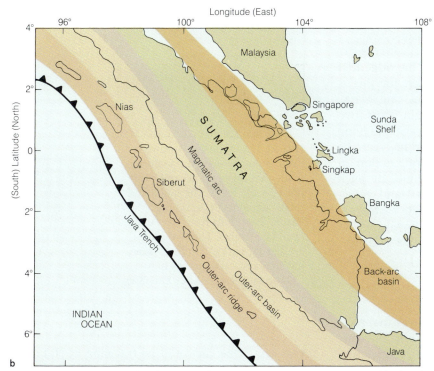

FIGURE 17.18 Topographic features of a convergent plate margin. (*a*) Idealized cross section showing the most distinctive features. The outer-arc and inner-arc basins both tend to be filled with sediment derived from the highland of the adjacent magmatic arc. The outer-arc ridge tends to be underlain by mélange; it is topographically elevated due to thrust faulting. (*b*) Map of portion of Sumatra showing the positions of the major topographic features in a present-day convergent plate boundary.

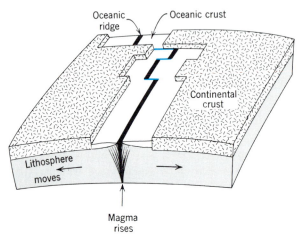

FIGURE 17.19 Transform faults (blue) form when an oceanic ridge first forms. Shapes of continental margins reflect faults now found along oceanic ridge.

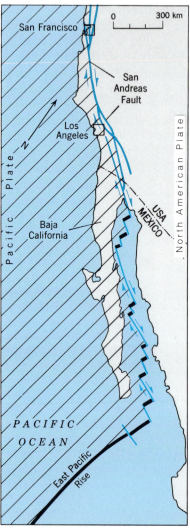

FIGURE 17.20 The San Andreas Fault is one of several transform faults that offset an oceanic ridge (East Pacific Rise). The faults and ridge segments separate the Pacific Plate (left) from the American Plate (right). The two plates are sliding past each other, causing frequent earthquakes in California. (*Source:* After Elders et al., 1972.)

the plate may flex and warp up and down. Of course, the distances between places on adjacent plates do change due to plate motions. Using velocities calculated from magnetic time lines, Figure 17.21 shows the motions of plates today. As mentioned at the beginning of this chapter, the calculated results agree with plate velocities measured directly using satellites and lasers.

Magnetic time lines are symmetrical with respect to a spreading ridge and parallel to the ridge that created them (Fig. 17.3). The reason for this is straightforward. Each magnetic time line marks the edge of an earlier divergent margin. Thus, two magnetic time lines having the same age, but lying on opposite sides of an oceanic ridge, can be brought together to show the configuration of plates as they were at an earlier time. By such means the opening of ocean basins—and movements of continents as a result of plate motions—can be reconstructed. Figure 17.22 shows a reconstruction of the opening of the southern Atlantic Ocean by such means.

Hot Spots and Absolute Motions

It was pointed out earlier in this chapter that plate motions determined from magnetic time lines are only relative motions. In order to determine absolute motions, an external frame of reference is necessary. A familiar example of absolute versus relative motion occurs when one automobile overtakes another. If observers in the two automobiles could only see each other, and could not see the ground or any fixed objects outside their cars, they could only judge the *difference* in velocity between the two cars. One car could be traveling at 50 km/h, the overtaking car at 55 km/h, but all that the observers could determine is that the *relative velocity* difference is 5 km/h. On the other hand, if the observers could measure velocity with respect to a stationary or fixed reference such as the ground surface, they could determine that the *absolute velocities* were 50 and 55 km/h, respectively.

We would be constrained to determine relative plate velocities if a fixed reference framework did not exist. Fortunately, there is a reasonable hypothesis for such a framework. During the last century, the American geologist James Dwight Dana observed that the age of volcanoes in the Hawaiian island chain increased from southeast to northwest (Fig. 17.23). As discussed earlier in this chapter, the Canadian geophysicist, J. Tuzo Wilson, suggested that a record of the movement of the seafloor was recorded by the age of the volcanic islands. Wilson postulated that a deep, long-lived magma source lies somewhere far down in the mantle. As the magma oozes up, it creates volcanoes on the surface of the lithosphere. Because the lithosphere moves, a volcano can only remain in contact with the magma source for about a million years. Wilson made his suggestion as a way of testing seafloor spreading; however, it was not long before he realized that if hot spots do exist

deep in the mantle, and do have long-continued lives, they might provide a series of fixed points on the surface of the Earth against which plate motions can be determined. More than a hundred hot spots have now been identified (Fig. 17.24). Using them for references, it is apparent that the African Plate must be very nearly stationary. Because the African Plate is almost completely surrounded by spreading edges, and because the relative velocities along the encircling ridges are closely matched (Fig. 17.21), we must conclude that the Mid-Atlantic Ridge is moving westward and that the oceanic ridge that runs up the center of the Indian Ocean is moving to the east. If the absolute motion of the African Plate is zero or nearly so, the Mid-Atlantic Ridge in the southern Atlantic Ocean must be moving westward at the rate of about 2 cm/yr, and the absolute velocity of the South American Plate must be 4 cm/yr.

The Australian-Indian Plate is moving almost di-

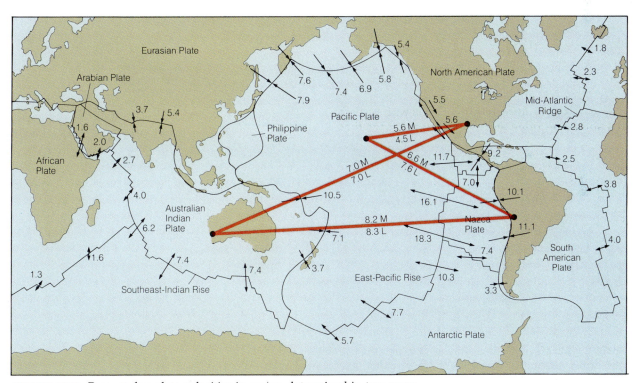

FIGURE 17.21 Present-day plate velocities in cm/yr, determined in two ways. Numbers along the mid-ocean ridges are mean velocities determined by use of magnetic reversals. A velocity of 16.1, as shown for the East Pacific Rise, means that the distance between any point on the Nazca Plate and any point on the Pacific Plate increases, on the average, by 16.1 cm each year in the direction of the arrow. The long lines connect stations that are used to determine plate motions using satellite laser ranging (L) techniques. The measured velocities between stations are very close to the average velocities estimated from magnetic reversals (M). (*Source:* Adapted from a NASA report. Geodynamic Branch, May 1986.)

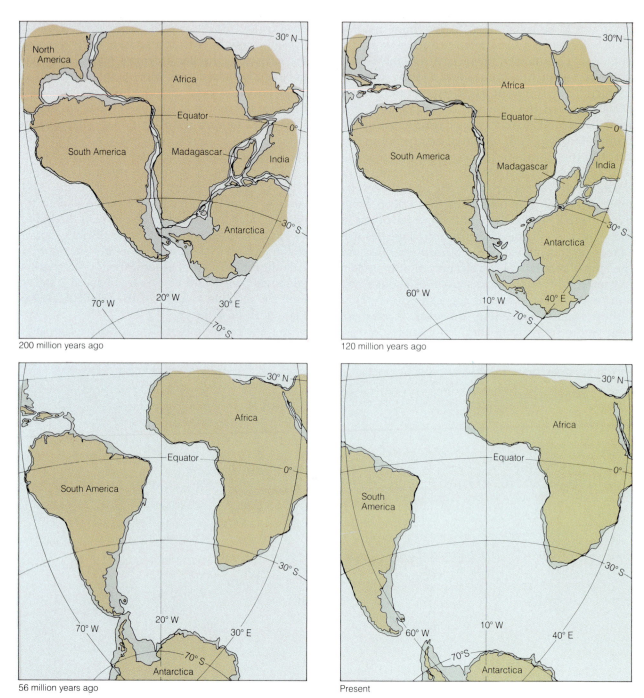

200 million years ago

120 million years ago

56 million years ago

Present

FIGURE 17.22 Magnetic data were used to plot the opening of the southern part of the Atlantic Ocean as South America and Africa drifted apart. About 200 million years ago, the present continents were joined together to form Pangaea. When the pieces are fitted back together along a line 1000 m below sea level, very few overlaps (shaded), or gaps (dark) remain. Notice how the continents move relative to the equator and the way Antarctica slowly drifts south. (*Source:* After Owen, 1983.)

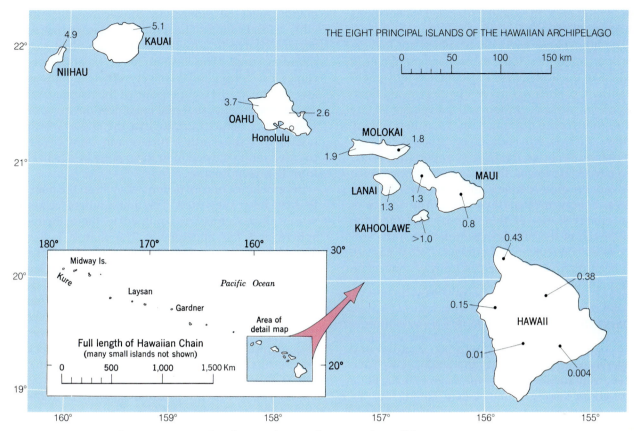

FIGURE 17.23 In the Hawaiian Archipelago, volcanism has ceased on all islands except the two most southeasterly ones, Maui and Hawaii. Activity on Maui is very infrequent. On Hawaii, three volcanoes are still alive. The tops of the most northwesterly islands, such as Midway and Laysan, have been eroded to positions below sea level and are now crowned by coral atolls. Beyond Midway are many seamounts that have now sunk below the sea. Numbers beside larger islands are K/Ar dates (in millions of years) of basalts that form the cones. Volcanism has moved steadily from northwest to southeast, each island apparently taking little more than one million years to grow from the sea floor to its ultimate height as the Pacific Plate moved over a hot spot in the mantle below. The next Hawaiian island is already forming beneath the sea. Southeast of Hawaii, a submarine volcano called Loihi, is steadily growing upward from the sea floor. (*Source:* Data from U.S. Geol. Survey.)

rectly northward. All other plates, with the exception of the stationary African Plate, are moving in approximately eastward or westward directions. Several plates do not have subduction margins and must therefore be increasing in size. Most of the modern subduction zones are to be found around the Pacific Ocean along the edge of the Pacific Plate, so much of the oceanic lithosphere that is now being destroyed is in the Pacific. It follows then, that the Indian Ocean, the Atlantic Ocean, and most other oceans must be growing larger, while the Pacific Ocean must be steadily getting smaller.

CAUSES OF PLATE TECTONICS

Just as Alfred Wegener could not explain what made continents drift, we are still unable to say *exactly why* plates of lithosphere move. Until we can explain the driving force, plate tectonics must remain a kinematic description of what occurs without knowing why it happens. The situation is analogous to knowing the details of shape, color, size, and speed of an automobile but not knowing what makes it run. But meanwhile we can hypothesize about the causes of the motion and test the hypotheses by making detailed calculations based on the laws of nature.

The lithosphere and asthenosphere are inevitably bound together. If the asthenosphere moves, it will make the lithosphere move, just as movement of sticky molasses will move a piece of wood floating on its surface. So too will movement of the lithosphere cause movement in the asthenosphere below. Such is our state of uncertainty that we cannot yet separate the relative importance of the two effects. However, there is one point on which we can be quite certain. In order for plates to move, energy must be expended. The source of that energy is the Earth's internal heat, and the way much of the heat energy must reach the surface is by convection in the mantle. What has not yet been discovered is the precise way convection and plate motions are linked.

Convection in the Mantle

The mantle is solid rock; however, it is hot and apparently weak enough so that under slow strain rates even small stresses will make it flow like a very sticky viscous liquid. Like a liquid, too, the mantle must be subject to convection currents when a local source of heat causes a mass of rock to become heated to a higher temperature than surrounding rock. The heated mass expands, becomes less dense, and rises very slowly at rates as low as 1 cm/yr. To compensate for the rising mass, cooler, more dense material must flow downward. The laws of nature indicate that the rate at which heat leaves the Earth's surface can only be accounted for if convection in the mantle brings heat from the deep interior.

Several kinds of convection cells within the mantle have been suggested. The first kind is a cell in which all movement is confined to the asthenosphere and the lithosphere. The mantle below about 700 km would have, in this mode, very little motion but would serve as a giant stove to heat the asthenosphere. Each plate of lithosphere would be the top of a giant convection cell. This means that there would have to be several convection cells; their sizes and shapes are indicated by the sizes and shapes of the plates. The masses of hot rock are postulated to rise vertically beneath oceanic ridges, then turn sideways, flowing horizontally as plates. The farther the rock moves away from the

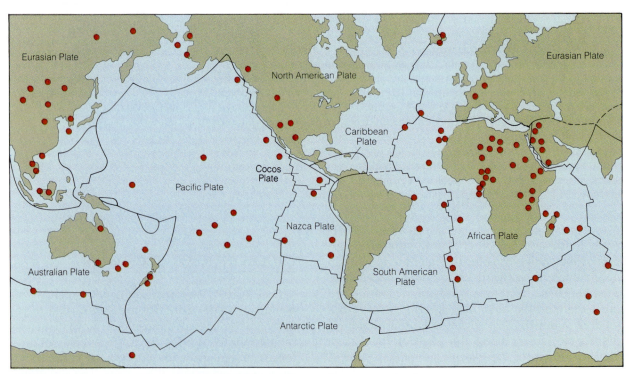

FIGURE 17.24 Long-lived hot spots at the Earth's surface, each a center of volcanism, are believed to lie above deep-seated sources of magma in the mantle. Because the magma sources lie far below the lithosphere, and do not move laterally, hot spots can be used to determine the absolute motions of plates.

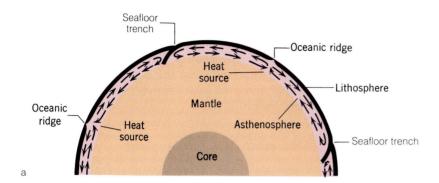

a

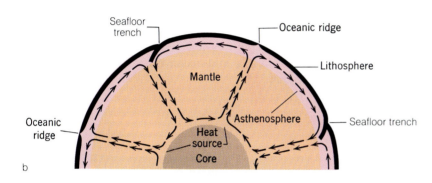

b

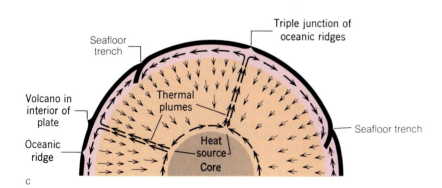

c

FIGURE 17.25 Three suggested mechanisms by which convection and flow in the asthenosphere might move plates of lithosphere. (*a*) Convection is confined to the asthenosphere. (*b*) Convection involves the entire mantle. (*c*) Thermal plumes rise from the mantle-core boundary and cause local hot spots at the Earth's surface.

ridge, the cooler it gets. Eventually, the lithosphere becomes so cool and so dense that it sinks back into the asthenosphere. The place where sinking occurs is beneath a seafloor trench (Fig. 17.25*a*). There are many problems with the suggestion that convection cells are confined to a few hundred kilometers in the mantle. One problem concerns the plates themselves. As previously discussed, ocean ridges move. Some plates are expanding, others are shrinking. It is very difficult to understand how convection cells could move and change size.

A second suggestion, one that avoids the problem of heat sources in the asthenosphere, is that convection cells involve the entire mantle, and that the heat brought up comes from the outer core (Fig. 17.25*b*). Just how localized transfer of heat takes place between core and mantle is not known. The flowing motions and sideways movement of lithosphere would be the same in a whole mantle convection pattern as in an upper-mantle-limited convection. The main differences are the sizes of the convection cells in the two cases and, in the latter case, the notion that convection can involve not only the weak asthenosphere, but also the stronger mesosphere below. These larger convection cells are even more difficult to reconcile with moving oceanic ridges and plates of varying size.

A third, and to many people a more likely possibility is that some sort of stacked convection system exists. If the two convection systems shown in Figures 17.25a and b were combined, large, deep convection cells would supply heat to drive smaller and shallower cells. The boundary between the two cell systems need not be the base of the asthenosphere. Some scientists have argued that it is more likely to be at a depth of 670 km where there is a pronounced seismic discontinuity.

Even stacked convection cells leave many unanswered problems. One concerns the existence of long-lived, deep-seated sources of magma indicated by the hot spots. Some scientists suggest that if local hot regions do occur on the core/mantle boundary, they will be small, roughly circular spots. Instead of producing a large convection cell, they maintain, a small hot spot will cause a long cylinder of hot rock, a few hundred kilometers in diameter, to rise. They refer to the vertically rising cylinder as a *thermal plume* and suggest that plumes are the sites of the hot spots that cause long-continued volcanism (such as the Hawaiian Islands). All upward motion could be accounted for by no more than 20 thermal plumes, although more probably exist. When a plume reaches the base of the lithosphere, it creates a local hot spot from which a little magma rises. However, most of the convecting plume would spread laterally and flow horizontally in all directions beneath the lithosphere. By this suggestion, the asthenosphere becomes, in essence, the top of a whole series of plume-driven convection cells. Return flow to balance the concentrated upward flow in the plumes would not necessarily involve well-defined, down-flowing plumes, but could be accomplished by slow downward movement of the entire mantle. Movement of plates of lithosphere by thermal plumes is more difficult to visualize because flow of the asthenosphere should be equal in all directions away from the hot spot. This means that plate motion must somehow involve the lithosphere itself.

All of the preceding discussion about convection is speculation. There is evidence from seismic tomography and heat flow that convection of some sort does occur beneath the lithosphere. However, it is difficult to see how plate motion can be due entirely to convection. For this reason, most scientists agree with the hypothesis that the motion of

the lithosphere is due to a combination of processes, and that convection is only one of the processes. One important thing that convection must do is keep the asthenosphere hot and weak by bringing up heat from the deep mantle and core. In this sense at least, convection is essential for plate tectonics.

Movement of the Lithosphere

There are three forces that might play a role in making the lithosphere move. The first is a push away from a spreading edge. Rising magma at a spreading edge creates new crust, and in the process it pushes the lithosphere sideways (Fig. 17.26a). Once the process is started, it would tend to keep itself going. The problem is that pushing involves compression, whereas the structure of the crust along a mid-ocean ridge indicates a state of tension.

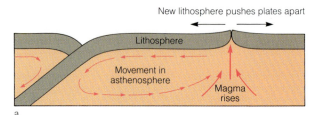

a

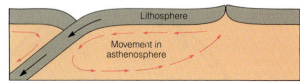

b

c

FIGURE 17.26 Three suggested mechanisms by which lithosphere might move over the asthenosphere. (a) Magma, rising at a spreading edge, exerts enough pressure to push the plates of lithosphere apart. (b) A tongue of cold, dense lithosphere sinks into the mantle and pulls the rest of the plate behind it. (c) A plate of lithosphere slides down a gently inclined surface of asthenosphere.

A second way by which lithosphere could be made to move is by dragging rather than pushing. Proponents of the dragging idea point out that a descending tongue of old, cold lithosphere must be more dense than the hot asthenosphere surrounding it. Because rock is a poor conductor of heat, they urge, the temperature at the center of a descending slab can be as much as 1000° C cooler than the mantle at depths of 400–500 km. The dense slab of lithosphere must then sink under its own weight and exert a pull on the entire plate. This is somewhat like a heavy weight that hangs over the side of a bed and is tied to the edge of a sheet. The weight falls and pulls the sheet across the bed. To compensate for the descending lithosphere, there must be a slow flow in the asthenosphere back to the spreading edge (Fig. 17.26*b*).

However, both the pushing and the dragging mechanisms have problems. Plates of lithosphere are brittle and they are much too weak to transmit large-scale pushing and pulling forces without major deformation occurring. We do not see the deformation.

The third possible mechanism for movement of a plate of lithosphere is for it to slide downhill away from the spreading edge. The lithosphere grows cooler and thicker away from a spreading edge. As a consequence, the boundary between the lithosphere and the asthenosphere must slope away from the spreading edge. If the slope is as little as 1 part in 3000, the weight of the lithosphere could cause it to slide down the slope at a rate of several centimeters per year (Fig. 17.26*c*).

At present there is no way to choose between the three lithosphere mechanisms. Calculations suggest that each operates to some extent, so that the entire process is possibly more complicated than we now imagine. The prevailing idea at present is that the sinking of old, cold, lithosphere starts the process and then the others combine to keep it going. Only future research will resolve the question. Nevertheless, without any waiting we can answer many questions about the role played by plate tectonics in the history of continents. So we turn at once to the evolution of continental crust.

SUMMARY

1. Abundant evidence proves that continents have not remained fixed on the Earth's surface, but have moved repeatedly from place to place.

2. The lithosphere is broken into six large and many smaller plates, each about 100 km thick, and each slowly moving over the top of the solid but fluid-like asthenosphere beneath it.

3. Each plate is bounded by three different kinds of margins. Divergent margins are those where new lithosphere forms. Plates move away from them. Convergent margins are lines along which plates compress each other and where lithosphere capped by oceanic crust is subducted back into the mantle. Fault margins are lines where two plates slide past each other.

4. Movement of a plate can be described in terms of rotation across the surface of a sphere. Each plate rotates around a spreading axis. The spreading axis does not coincide with the Earth's axis of rotation.

5. Because plate movement is a rotation, the velocity of movement varies from place to place on the plate.

6. Each segment of oceanic ridge that marks a divergent margin of a plate lies on a line of longitude passing through the spreading pole. Each transform fault margin of a plate lies on the line of latitude of the spreading pole.

7. The mechanism that drives a moving plate is not known, but apparently it results from a combination of convection in the mantle plus forces that act on a plate of lithosphere.

SELECTED REFERENCES

Anderson, D. L., 1971, The San Andreas Fault: Sci. American, v. 225, no. 5, p. 52–68.

Macdonald, K. C., 1982, Mid-Ocean Ridges: Fine scale tectonic, volcanic and hydrothermal processes within the plate boundary zone: Ann. Rev. Earth and Planetary Sci., v. 10, p. 155–190.

Raymond, L. A., ed., 1984, Mélanges: Their nature, origin and significance: Geol. Soc. America, Special Paper 198.

Wilson, J. T., ed., 1976, Continents adrift and continents aground: San Francisco, W. H. Freeman and Co.

Wyllie, P. J., 1976, The way the Earth works: New York, John Wiley & Sons.

CHAPTER 18

History of
the Continental Crust

The region around Bastak, southern Iran, is a landscape formed by continental collision. Sedimentary strata were thrown into symmetrical folds when Africa pushed northward against Asia. The dark, circular patches are masses of salt which rose as diapirs from a buried evaporite horizon. The rainfall is so low the salt has not washed away.

PLATE MOTIONS AND CONTINENTAL SCARS

Seafloor spreading and plate tectonics were proven correct using evidence from the oceanic crust. This result should hardly be surprising. Spreading edges are found beneath the sea. The velocity, direction of motion, and age of a plate are most convincingly established by the thermoremanent magnetism of the oceanic crust. This system is balanced by the sinking of oceanic crust along subduction zones. Plate tectonics would probably operate even if there were no continental crust at all. In a sense, continental crust is simply a passenger rafted on large plates of lithosphere. But it is a passenger that has been buffeted, stretched, fractured, and altered by the ride. Someone once characterized continental crust as the product of bump-and-grind tectonics. Each bump between two fragments forms an orogenic belt, each grind a strike-slip fault, each stretch a thinning of continental crust and a rift valley. Scars left on the continental crust by bump-and-grind tectonics are evidence of former plate motions. This is fortunate because the most ancient crust known to exist in the ocean dates only from the mid-Jurassic Period, about 165 million years ago. Indeed, direct evidence concerning geological events more ancient than the mid-Jurassic comes from the continental crust. For this reason, this chapter is focused on the history and development of the continental crust. It is a history that is increasingly obscure and uncertain as we peer further and further back in time. We start, therefore, with the most recent and best preserved evidence, then, following the tenet that the present is the key to the past, we examine successively more ancient evidence.

CONTINENTAL MARGINS

Fragmentation, drift, and the welding together of continental crust are inevitable consequences of plate tectonics. Evidence of fragmentation and welding is most strikingly preserved along the stretched or compressed margins of the fragments. It is helpful to review briefly the features associated with the five principal kinds of margins that bound continental crust. They are:

1. *Passive continental margins,* of which the Atlantic Ocean margins of the Americas, Africa, and Europe are examples.

2. *Continental subduction margins,* of which the Andean coast of South America is an example.

3. *Continental collision margins,* for which the Alpine-Himalayan mountain chain provides an example.

4. *Transform fault margins,* which are exemplified by the San Andreas Fault in California and the Alpine Fault that slices through the South Island of New Zealand.

5. *Accreted terrane margins,* consisting of small allocthonous terranes (Chapter 15) added to an existing continental margin. An example is the northwest margin of North America from northern California to Alaska.

Passive Continental Margins

A new ocean basin forms by the rifting of continental crust following the sequence illustrated in Figure 18.1. Rifting arises from tensional faults that break the crust along a complex jumble of normal faults. This process can be seen in the Red Sea, which is a very young ocean with an active spreading ridge running down its axis (Figs. 18.2 and 18.3). Indeed, the Red Sea is such a young ocean that the ridge is still partially covered by sediment.

The sequence of sediments deposited in the Red Sea rift starts with the deposition of clastic nonmarine sediments. These nonmarine sediments are followed by evaporites and then marine shales. This sequence is distinctive and apparently arises in the following manner. Basaltic magma, associated with formation of the new spreading edge, heats and expands the lithosphere so that a plateau forms with an elevation of as much as 2.5 km above sea level. When tensional forces split the crust and form a rift, there is a pronounced topographic relief between the plateau and the floor of the rift. The earliest rifting of the Red Sea must have been very much like the African Rift Valley today. Before the rift floor sank low enough for seawater to enter, clastic, nonmarine sediments such as conglomerates and sandstones were shed from the steep valley walls and accumulated in the rift. Associated with these sediments are basaltic lavas, dikes, and sills, formed by magma rising up the normal faults. As the rift widened, a point was reached where seawater entered. The early flow was apparently restricted and the water was shallow, resembling a shallow lake more than an ocean. The rate of evaporation would have been high and as a result strata of evaporite salts were laid down on top of the clastic, nonmarine sedi-

MODERN EXAMPLE

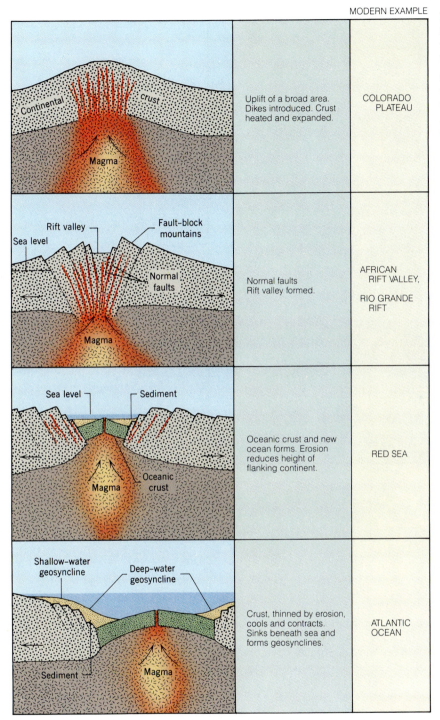

	Uplift of a broad area. Dikes introduced. Crust heated and expanded.	COLORADO PLATEAU
	Normal faults. Rift valley formed.	AFRICAN RIFT VALLEY, RIO GRANDE RIFT
	Oceanic crust and new ocean forms. Erosion reduces height of flanking continent.	RED SEA
	Crust, thinned by erosion, cools and contracts. Sinks beneath sea and forms geosynclines.	ATLANTIC OCEAN

FIGURE 18.1 Sequences of events in the rifting of continental crust to form a new ocean basin bounded by passive continental margins. The process of rifting can cease at any stage. It is not correct to conclude, therefore, that the African Rift Valley will open to form a new ocean.

ments. Finally, as rifting continued and the depth of the seawater increased, normal clastic marine sediments were deposited. This is the stage the Red Sea is in today. Eventually, if further rifting exposes new oceanic crust, the Red Sea will evolve into a younger version of the Atlantic Ocean. Sedimentation will form a pair of shallow- and deep-water geosynclines (Chapter 15) along both sides of the ocean.

Notice in Figure 18.2 that the Gulf of Aden, the Red Sea, and the northern end of the African Rift Valley meet at angles of 120°. The meeting point is a plate triple junction formed by three spreading edges. Two of the edges, the Gulf of Aden and the

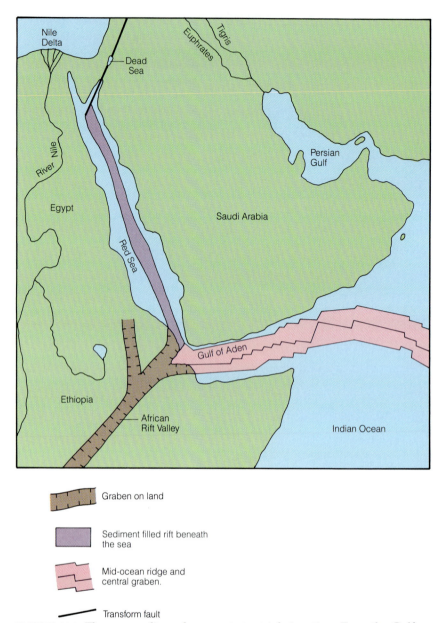

Graben on land

Sediment filled rift beneath the sea

Mid-ocean ridge and central graben.

Transform fault

FIGURE 18.2 Three spreading edges meet at a triple junction. Two, the Gulf of Aden and the Red Sea, are actively spreading, whereas the third, the African Rift appears to be a failed rift that will not develop into an open ocean. The spreading edge down the center of the Red Sea is covered by sediment.

FIGURE 18.3 Normal faults and down-dropped fault blocks along the western margin of the Red Sea near Danakil, Ethiopia. The faulting resulted from tensional forces when a new spreading edge split the crust. The white material in the foreground is an outcrop of evaporite deposits.

Red Sea, are active and still spreading. The third, the African Rift Valley is apparently no longer spreading and probably will not evolve into an ocean. What will remain on the African continent is a long, narrow sequence of grabens filled primarily with nonmarine sediment. The formation of three-armed rifts with one of the arms not developing into an ocean is apparently a characteristic feature of continental rifting caused by new spreading edges. This can be seen from Figure 18.4, which shows the reassembled positions of the continents flanking the Atlantic Ocean prior to breakup. Note that some of the world's largest rivers flow down valleys formed by rifts associated with the opening of the Atlantic Ocean. The pattern of rifts is a distinctive feature arising from plate tectonics, as is the sequence of sediment layers and the kind of igneous rock (Fig. 18.5a,b) found in failed rifts and along the flanks of rifts that open to form new oceans bounded by passive continental margins.

Continental Subduction Margins

Subduction of oceanic crust beneath continental crust produces deformation of a continental margin (together with a distinctive style of metamorphism and deformation of sediments deposited in the trench, and characteristic magmatic activity). The only tectonic setting in which sediments are subjected to high-pressure, low-temperature metamorphism is a subduction zone with a rapid rate of subduction. Sediment subjected to such metamorphism forms a mélange (Chapter 17). Adjacent to and parallel with the mélange, the thickened edge of continental crust is also metamorphosed, but in that case it is normal, regional metamorphism (Chapter 6). One distinctive feature of a continental subduction margin, therefore, is a pair of parallel metamorphic belts.

A second characteristic feature of a continental subduction margin is the arc of stratovolcanoes built on top of the continental crust. Modern examples can be seen in the chains of volcanoes in the Andes and the Cascade Range (Fig. 18.6). Where a magmatic arc has been eroded, granitic batholiths can be observed. They are remnants of the magma chambers that once fed stratovolcanoes far above. The strings of huge, elongate batholiths that run from southern California to northern British Columbia (Fig. 4.27) provide striking examples of deeply eroded magmatic arcs formed along a fossil subduction margin (Fig. 18.7).

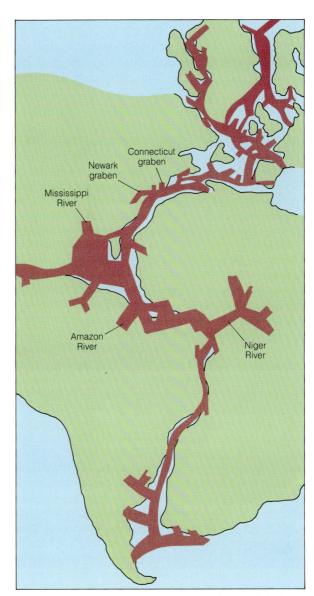

FIGURE 18.4 Map of a closed Atlantic Ocean showing the rifts formed when Pangaea was split by a spreading edge. The rifts on today's continents are now filled with sediment. Some of them serve as the channelways for large rivers. (*Source:* After Burke, 1980.)

Continental Collision Margins

When continental crust is part of a plate of lithosphere that is being subducted beneath the margin of a second piece of continental crust, the two continental fragments will eventually collide. The collision sweeps up the sediment accumulated along the leading edges of both continents and forms a fold-and-thrust mountain system. The suture zone between the two masses of deformed sediment is commonly marked by the presence of serpentinites formed by alteration of sheared and deformed fragments of ophiolites caught up in the collision (Fig. 15.42). Fold-and-thrust mountain

a

b

FIGURE 18.5 (*a*) Coarse, poorly sorted floodplain sediments deposited in the Connecticut graben. The sediments were derived from the highland that lay to the east of the graben. Sediments are graded, each graded unit representing a single flood event. Coarser sediments are light colored, finer sediments are dark. The red color of the sediments indicates a terrestrial deposition. (*b*) A half-graben formed at the time of the breakup of Pangaea. Looking north from New Haven, Connecticut a highland of metamorphosed Paleozoic-aged rocks is visible to the west (left). The unconformity separating Triassic and Jurassic sediments deposited in the graben (*a*) is marked by a ridge running from left center to upper right. The unconformity dips at about 30° to the east (right). The prominent curved ridge in the center is West Rock, a gabbro sill that intrudes the sediments. The sill and the metamorphic rocks resist weathering and form topographic highs. The sedimentary rocks weather readily and form topographic lows.

systems may also have stratovolcanoes, batholiths, and paired metamorphic belts associated with them, because a collision margin must be a continental subduction margin prior to collision.

One distinctive feature of a fold-and-thrust mountain system formed by collision is that the new mountain system lies in the interior of a major landmass. A modern example is the great Himalayan mountain chain formed by the collision of India with Asia. Another can be seen in the Alps, which were formed by the collision of Africa and Europe starting in early Mesozoic time. Older examples are provided by the Ural and Verkhoyansk mountains in the U.S.S.R. (Fig. 15.26), and by the Appalachians (Fig. 18.8), each of which was formed by Paleozoic collisions.

There are several differences between orogenic belts formed along continental subduction margins and those formed by collision. The paired metamorphic belts are asymmetric with respect to the subduction zone. A collision zone, on the other hand, is roughly symmetrical because there is deformed continental crust on both sides of the collision zone (Fig. 15.42).

It seems reasonable to conclude that all of the now deeply eroded orogenic belts around the world were once fold-and-thrust mountain chains. A further conclusion is that such orogenic belts were formed either along continental subduction

FIGURE 18.6 Looking south down the Cordillera Blanca, a chain of giant stratovolcanoes in northern Peru. The volcanoes formed as a result of subduction of the Nazca Plate beneath South America. Erosion is starting to dissect the old volcanoes. On the west side of Huascaran, the large mountain in the center of the photograph, a great mudslide can be seen. The slide, which occurred in 1970, buried the town of Yungay and claimed many thousands of victims (see Fig. 9.27).

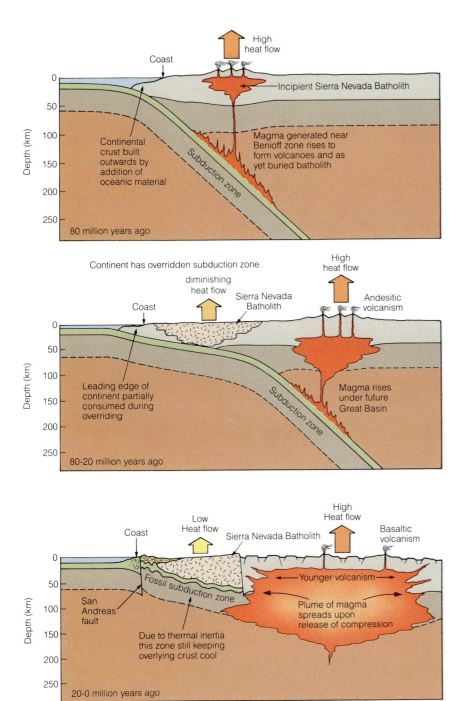

FIGURE 18.7 One interpretation of the way the Sierra Nevada Batholith formed. Prior to 80 million years ago an active chain of volcanoes lay along the western edge of the continent. Beneath the volcanoes the Sierra Nevada Batholith was formed. Between 80 and 30 million years ago, North America overrode the subduction zone and the portion of the subducted crust undergoing partial melting moved easterly beneath the Basin and Range Province. Between 30 million years and the present, North America overrode part of the mid-ocean ridge. Subduction ceased and the plate boundary became a transform fault (the San Andreas Fault). The Basin and Range Province is disrupted by volcanism and block faulting. (*Source:* After Henyey and Lee, 1976.)

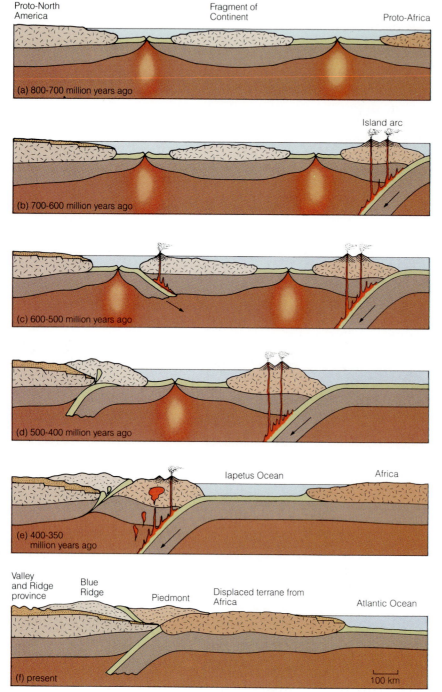

Proto-North America

Fragment of Continent

Proto-Africa

(a) 800-700 million years ago

Island arc

(b) 700-600 million years ago

(c) 600-500 million years ago

(d) 500-400 million years ago

Iapetus Ocean

Africa

(e) 400-350 million years ago

Valley and Ridge province

Blue Ridge

Piedmont

Displaced terrane from Africa

Atlantic Ocean

(f) present

100 km

FIGURE 18.8 A suggested sequence of events explaining the evolution of the southern Appalachians in terms of a plate tectonic model. A sequence of subduction zones, collisions, and thrusting produces the present-day structure. (*Source:* After Cook et al. 1979.)

or continental collision edges and that they mark former plate margins.

Transform Fault Margins

Giant strike-slip faults provide the most direct and convincing evidence that large lateral motions have occurred in the past. It is rarely possible to prove that ancient strike-slip faults were actually transform faults that connected spreading edges or subduction edges, but the inference is strong that they were. This inference arises from the manner in which the presently active large, strike-slip faults originated. As seen in Figure 18.9, the San Andreas Fault apparently formed when the westward-moving North American continent overrode portions of the East Pacific Rise. The San Andreas Fault is the transform fault that connects the two remaining segments of spreading ridge.

Accreted Terrane Margins

Plate motions can raft small fragments of crust tremendous distances. Eventually, any fragment that is not consumed by subduction will be added (accreted) to a larger continental mass. Some of the small fragments are volcanic island arcs formed by subduction of oceanic crust beneath oceanic crust. Other fragments form when they are sliced off the margin of a large continent, much as the San Andreas Fault is slicing a fragment off North America today. Other combinations of volcanism, rifting, faulting, and subduction can also form small fragments of crust that are too buoyant to be subducted. In the western Pacific Ocean there are many such small fragments of continental crust. Examples are the island of Taiwan, the Philippine islands, and the many islands of Indonesia. The ultimate fate of all of these small fragments, each of which is called a *terrane* because it has its own distinctive geological character, is to be accreted to a larger continental mass. Accreted terranes, then, modify a preexisting subduction, collision, or transform fault margin by the addition of rafted-in, exotic blocks of crust.

The western Pacific is a place where the accretion process is apparently in progress today (Fig. 18.10). The western margin of North America, where the importance of accreted terranes was first recognized, is an example of a young, accreted margin. Using a combination of lithologic and paleontologic studies, structural analysis, and paleomagnetism, many allocthonous terranes have now been identified. They have all been accreted since the early Mesozoic, about 200 million years ago (Fig. 18.11). One of the terranes is an ancient seamount, another is an older limestone platform formed on some other continental margin. Still others are volcanic island arcs and even fragments of old metamorphic rocks. Paleomagnetic studies suggest that some terranes have moved 5000 km or more and that once accreted, the process of movement did not necessarily cease. Later motion along transform faults caused still further reorganizations.

The recognition of accreted terrane margins is a relatively new discovery. In a sense, it is a second stage of complexity in the plate tectonic revolution. A great deal still remains to be discovered concerning terranes. How to recognize a terrane, how to

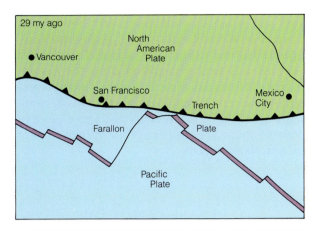

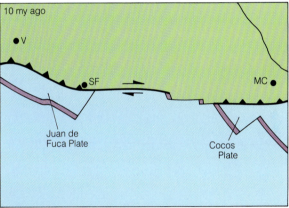

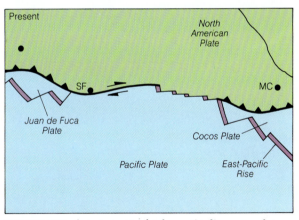

FIGURE 18.9 A sequence of schematic diagrams showing the proposed origin of the San Andreas Fault. Twenty-nine million years ago the edge of North America overrode a portion of the Farallon Plate, creating two smaller plates, the Coco Plate and the Juan de Fuca Plate, in the process. The San Andreas Fault is the transform fault that connects the remaining pieces of the severed spreading ridge (Figure 17.20). (*Source:* From Stewart, 1978.)

work out where it came from, and how it was moved are all challenges facing geologists. Only recently has the existence of accreted terranes of Paleozoic age been recognized along the eastern margin of North America. Although it is a new and exciting concept, accreted terrane margins of continents seems to be yet another distinctive piece of evidence that can be used to prove the existence of ancient plate motions.

THE PHANEROZOIC EON

Plate motion during the past 200 million years of the Phanerozoic Eon can be reconstructed with considerable accuracy. Based on evidence from still existing oceanic crust, such reconstructions demonstrate the former existence of the giant supercontinent, Pangaea. The present phase of

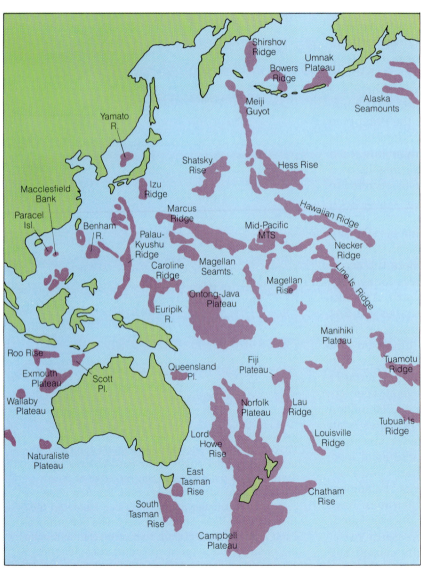

FIGURE 18.10 Regions of high sea-floor topography and islands in the western Pacific Ocean. As the Pacific Plate moves westerly, and the Australian-Indian Plate continues to move north, it is probable that most of these oceanic islands and plateaus will be accreted to the moving continents, increasing the sizes of the continents in the process and causing orogenic deformation. (*Source:* After Ben-Avraham et al. 1981.)

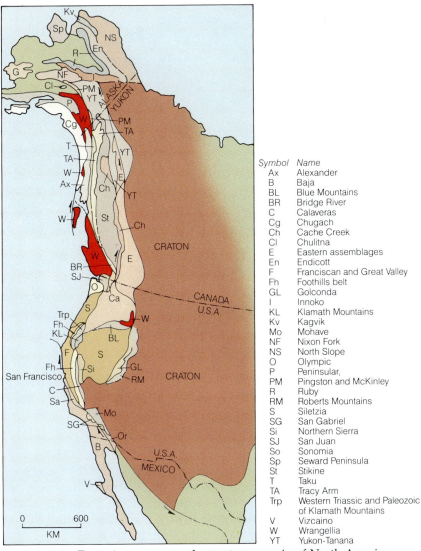

Symbol	Name
Ax	Alexander
B	Baja
BL	Blue Mountains
BR	Bridge River
C	Calaveras
Cg	Chugach
Ch	Cache Creek
Cl	Chulitna
E	Eastern assemblages
En	Endicott
F	Franciscan and Great Valley
Fh	Foothills belt
GL	Golconda
I	Innoko
KL	Klamath Mountains
Kv	Kagvik
Mo	Mohave
NF	Nixon Fork
NS	North Slope
O	Olympic
P	Peninsular,
PM	Pingston and McKinley
R	Ruby
RM	Roberts Mountains
S	Siletzia
SG	San Gabriel
Si	Northern Sierra
SJ	San Juan
So	Sonomia
Sp	Seward Peninsula
St	Stikine
T	Taku
TA	Tracy Arm
Trp	Western Triassic and Paleozoic of Klamath Mountains
V	Vizcaino
W	Wrangellia
YT	Yukon-Tanana

FIGURE 18.11 Tectonic terranes on the western margin of North America. During the Mesozoic and Cenozoic numerous tectonic terranes were accreted to the western margin of the craton. Each terrane is fault bounded and is a geological entity characterized by a distinctive stratigraphic sequence and/or a structure history that differs markedly from adjoining terranes. Some terranes, such as Wrangellia (W) were fragmented during the accretion process and now occur in several different fragments. (*Source:* Adapted from Beck et al. 1980.)

plate motions involves the fragmented remains of Pangaea.

By using the kinds of evidence just discussed for former plate margins it has been possible to identify the major fragments of continental crust from which Pangaea was assembled by continental collisions during the Paleozoic Era. By further using the paleomagnetism of lavas of appropriate age, it has been possible to show where some of the blocks of continental crust were at different times during the Paleozoic Era. One attempt at such a construction is shown in Figure 18.12.

The evidence is convincing that plate tectonics as we know it today has been operating throughout the 575 million years of the Phanerozoic Eon. Consider the Urals, for example. The Urals are a spectacular north–south mountain chain in the U.S.S.R. Blueschists formed by high-pressure, low-temperature metamorphism prove ancient subduction, and ophiolites mark the Paleozoic col-

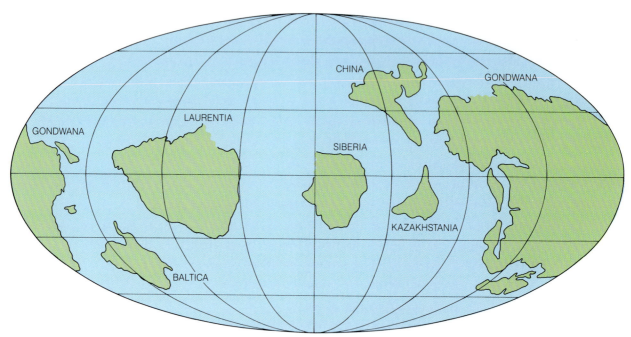

FIGURE 18.12 Reconstruction of the map of the world 540 million years ago, in Mid Cambrian time. Gondwana, a supercontinent made up of present-day Australia, Africa, India, and South America dominates the map. Note that the portion of Laurentia where New York City is situated today was then in the Southern Hemisphere, about where Rio de Janeiro is today. (*Source:* After Bambach et al. 1980.)

lision margin. Paleomagnetic measurements on both sides of the collision zone confirm that the two blocks of crust once moved as separate pieces. The Urals, then, are the place where a large ocean was eliminated by subduction during the late Paleozoic. So convincing is this evidence that names have been given to some of the ancient oceans. The ocean that disappeared when North America and Europe collided to form the Appalachians on one side and the Caledonide and Variscide mountains on the other (Fig. 2.13) is termed the Iapetus Ocean.

THE PROTEROZOIC EON

The Phanerozoic Eon commenced with the Cambrian Period 570 million years ago. The boundary between the Phanerozoic Eon and the preceding Proterozoic Eon is defined by fossils. It is the place in the geological record where fossils with hard shells first appear. The reason for their appearance is unsolved but it does not seem to be due to a dramatic tectonic event, such as the breakup of Pangaea, which occurred at approximately the

Paleozoic–Mesozoic boundary. But the paleomagnetic evidence is clear that plate tectonics were operating at the beginning of the Phanerozoic and that continental crust was moving around during the later part of the Proterozoic Eon.

The Proterozoic Eon commenced with the close of the Archean, 2500 million years ago, and ended 570 million years ago. Evidence from geological mapping and from a consideration of the rock cycle is ambiguous, but it appears that by the beginning of the Proterozoic, the volume of continental crust was close to its present value. Growth of new continental crust from that time on seems to have occurred by the accretion of andesitic island arcs. Accretions that are younger than 2500 million years make up only a small percentage of the present-day continents. So the growth rate has been slow.

The most convincing evidence that the continental crust was in motion throughout the Proterozoic Eon is provided by the remanent magnetism of Proterozoic-aged rocks. Figure 18.13 shows the apparent polar wandering curves for the North American continental masses through a long part of the Proterozoic. The curve suggests that individual fragments of continental crust were moving, relative to the poles, at rates of a few centimeters a year, similar to present rates. The paleomagnetic data are not sufficiently precise to show how the continents moved relative to each other. The evidence to prove that collisions occurred and that oceans opened and closed must come from ancient plate margins. Many rifted margins (some associated with swarms of basaltic dikes) and a number of collision margins can be identified, but the evidence is still insufficient for the preparation of paleogeographic maps of the kind shown in Figure 18.12. The fragmentary evidence that has been deduced suggests that there might have been earlier periods of continental aggregation, so supercontinents like Laurasia might have formed during the Proterozoic. The evidence also suggests that fragmentations and collisions occurred repeatedly and in approximately the same place because parallel collision zones tend to overlap, much as the Ap-

palachian Orogen overlaps the Grenville Orogen (Fig. 15.30). This seems to indicate that oceans, whether large or small, were formed and destroyed when continents moved apart and then returned close to their previous positions. The pattern suggested is similar to that of the Iapetus Ocean which disappeared during the Paleozoic, but was succeeded by the North Atlantic in the Mesozoic. Why many collision zones are parallel remains an unanswered question.

The Wopmay Orogen

The oldest well-preserved remains of a mountain system that matches the details of young Phanerozoic mountain systems is found in northwestern Canada. The Wopmay Orogen was formed between 2000 and 1800 million years ago on the northwestern flank of the Slave Craton (Fig. 15.30). The Slave Craton is a small continental frag-

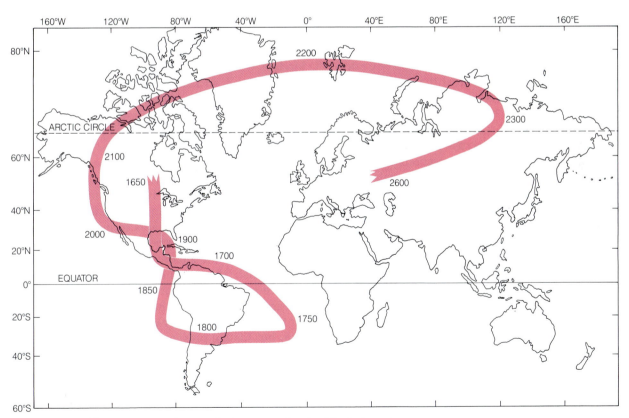

FIGURE 18.13 The apparent polar wandering curve for the North American Craton during the lower Proterozoic. Numbers beside the curve are millions of years before the present. Similar curves have been constructed for all cratons. They provide evidence that the continental crust was subjected to lateral motions as far back in time as 2500 million years ago. (*Source:* Adapted from Van der Voo, 1981.)

ment that apparently collided with, and was accreted to, the now larger Churchill Craton sometime before 2000 million years ago. The Wopmay Orogen and the Slave Craton are well exposed today because repeated glaciations during the past two million years have scoured and polished the surface. The evidence is fascinating.

The geology of the Slave Craton and the Wopmay Orogen are depicted in Figure 18.14. The boundary between the Slave and Churchill Cratons is marked in part by two rifts. The rifts, now filled by sedimentary rocks, are evidence that tensional forces once tugged at the region but failed to reopen the collision line between the two cratons. The northwestern edge of the Slave Craton was rifted, however, and more than 2000 million years ago an ocean basin opened. The age can be deduced because the oldest rocks associated with the breakup are approximately 2000 million years old. It is possible, but not proven, that the sediment filled rifts and the rifted edge of the Slave Craton formed at the same time. The bottom layers of sediment deposited along the rifted margin are clastic, nonmarine sediments; they pass upward into marine sediments. This is the typical sequence of a passive margin. The Wopmay Orogen came into being through a collision between the Slave Craton and a mass of continental crust that once lay further to the northwest. We can deduce this because thrusting caused by the collision is toward the east.

Compare the map 2A across the Wopmay Orogen in Figure 18.14 with the sections shown in Figures 15.37 and 15.40 across the Appalachians and Canadian Rockies, respectively. The Wopmay is remarkably like its younger counterparts, even though it is 1500 million years older. The thin sequence of rocks deposited on the old continental shelf is now the Wopmay fold-and-thrust belt. Even the rock types in the fold-and-thrust belt are similar to those in the Appalachians—clean quartz-sandstones and limestones which are characteristic of shallow, marine deposition on a continental shelf. Flanking the fold-and-thrust belt to the west is a belt of metamorphic rocks. They were once mudstones and coarse clastic sediments, many of which still display graded bedding. The graded bedding proves the sediments were deposited by turbidity currents that swept down the face of the continental slope. The metamorphic rocks were obviously derived from sediments deposited in the deep-water geosyncline that formed over the continental slope and continental rise. The igneous rocks that intrude the metamorphic rocks apparently mark the place where the sedimentary section was thickest and where wet partial melting occurred at the base of the thickened pile.

The kinds of sedimentary rocks found in the Wopmay Orogen provide convincing evidence that the continental crust during the early part of the Proterozoic Eon behaved very like the crust today. Deformation in the Wopmay Orogen, together with the failed rifts between the Slave and Churchill Cratons, prove that some kind of plate motions were causing collisions between crustal units, at least as far back in time as 2000 million years. Beyond that point the picture becomes murkier and much more difficult to understand.

THE ARCHEAN EON

The Archean Eon was a time of tremendous change. By the time it ended, 2500 million years ago, most of the continental crust had been formed. But just when the crust started is still unknown. So far, the most ancient rocks discovered are boulders of lava in a conglomerate near Isua, Greenland that are 3800 million years old. The kind of weathering that produces boulders takes place on land, so a land surface must have existed 3800 million years ago. With a single, fragmentary exception, no evidence has yet been discovered that dates from the 800-million-year period between 3800 and 4600 million years ago when the Earth was formed (Chapter 8). The single exception is some detrital grains of zircon (Zr_2SiO_4) in a clastic sedimentary rock from Western Australia. Using uranium-lead dating methods, it has been shown that the zircon grains formed 4200 million years ago. What cannot be discerned is the parent rock in which the zircons formed. But zircons are characteristically found in granitic rocks so the inference is strong that at least a small amount of continental crust might have existed at such a distant time. Research may eventually fill in the time gap, but for the present most of the available evidence concerns rocks formed between 3800 and 2500 million years ago.

High-Grade Terranes

Archean geology is dominated by huge areas of coarse-grained granitic rocks called *high-grade terranes*. Not all of the rocks in high-grade terranes

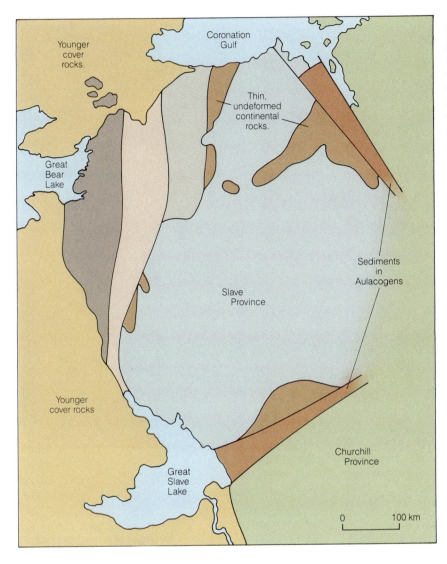

FIGURE 18.14 The Wopmay Orogen in Canada. See Figure 15.30 for the location. Of the three principal belts of the orogen, one, a magmatic belt along the western margin, was formed by magmatism arising from subduction. The central belt, now deformed and metamorphosed, consists of sediments that accumulated along the foot of the continental slope. The eastern belt of folded and thrust rocks is comprised of sediments that accumulated on the continental shelf. Separating the Slave and Churchill Provinces are two aulacogens, now filled with a thick sequence of terrestrial and marine sediments of the Proterozoic Age. An aulacogen is a large graben that opens outward toward the margin of a craton. (*Source:* Adapted from Hoffman, 1973.)

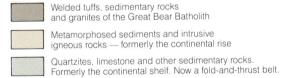

Welded tuffs, sedimentary rocks and granites of the Great Bear Batholith

Metamorphosed sediments and intrusive igneous rocks — formerly the continental rise

Quartzites, limestone and other sedimentary rocks. Formerly the continental shelf. Now a fold-and-thrust belt.

are primary igneous rocks. Some are ancient sedimentary rocks that have been subjected to such high-grade metamorphism that they have textures resembling those of granites.

The granitic intrusions in high-grade terranes differ from younger granites. Granite batholiths formed during the Phanerozoic tend to be elongate in shape and to occur in linear belts. The Archean granites are approximately round in shape (Fig. 15.29) and occur in regions that are as broad as they are long.

The metamorphic rocks in high-grade terranes are difficult to explain. Some seem to have been

volcanic or pyroclastic rocks, whereas others were sedimentary rocks. Where it is possible to assign a provenance to the sedimentary rocks they often resemble shallow-water sediments formed on sedimentary shelves. But the rocks have been complexly deformed (Fig. 18.15) and often subjected to more than one period of metamorphism. The last period of metamorphism, which occurred between 2800 and 2600 million years ago, subjected some of the high-grade terranes to pressures equivalent to burial depths of 50 km and temperatures as high as 900° C. How continental crust could have been buried to such depths and then brought back to

FIGURE 18.15 Complexly deformed gneisses that were subjected to extremely high-grade metamorphism about 2800 million years ago. This exposure, in the Sand River, South Africa, is a belt of deformed rocks known as the Limpopo mobile belt. It lies between the Zimbabwe and Kaapvaal Cratons.

FIGURE 18.16 Texture in a magnesium-rich, peridotitic lava. Erupted during the Archean Eon, the lava was highly fluid and extremely hot. As it cooled, elongate, needlelike crystals of olivine and pyroxene grew in the lava producing the patterns known as *spinifex texture.* The outcrop is in Munro County, Ontario, and is approximately 25 cm across. Individual crystals as long as 8 cm are present.

the surface is unknown, because it suggests exceptionally thick sections of crust. This in turn suggests that the Archean crust must have been somewhat ductile and readily deformed.

Low-Grade Terranes

Between the high-grade terranes there are long, thin, highly deformed belts of lavas and sediments that have only been subjected to low-grade metamorphism. These areas are known as *low-grade terranes*. The lavas are generally basaltic in character and commonly show pillow structure. Metamorphism has converted the lavas to low-grade, chlorite schists. They are the greenstones of granite-greenstone terranes. High-grade terranes correspond roughly to the granites, low-grade terranes to the greenstones in Figure 15.29.

Some of the lavas in low-grade terranes are remarkable because they have the compositions of peridotites. Peridotites are olivine-rich rocks and in younger rocks they are only formed by magmatic segregation. They don't form peridotitic magmas because the melting temperatures are too high. Temperatures of modern lavas don't exceed 1250° C. Peridotite lavas must have been hotter than 1500° C and they must have been completely liquid, not a mixture of crystals and liquid, because they usually display unusual quench textures that can only form in a liquid (Fig. 18.16).

Peridotite lavas are known only from Archean rocks. Their presence confirms a deduction concerning the Earth's internal temperature. The Earth's radioactivity must have been much higher during the Archean. It is estimated that three times as much heat had to be dissipated during the Archean than is dissipated today. A higher rate of heat loss probably meant higher temperatures in parts of the Archean mantle. Higher temperatures meant that higher-temperature magmas could form. Higher rates of heat loss possibly also caused the lithosphere to be thinner and more flexible, and if plates moved, the rates of motions may have been faster. Paleomagnetic research has not yet provided answers for such questions.

The structural relationship between blocks of high-grade metamorphic rocks and granites in direct contact with lavas and sediments that have only been subjected to low-grade metamorphism is rarely clear. The evidence has usually been destroyed. In some cases, it is apparent that the sediments and lavas are confined to ancient rifts in high-grade terranes and, thus, sit unconformably on them. In other cases it appears as if the sediments and lavas have been invaded by the granites.

Formation of Archean Crust

How might the ancient cratons have formed? There are probably as many answers to this question as there are specialists who study Archean rocks. The subject is an intensely interesting one for future research. If it is accepted that the Principle of Uniformity is a guide to processes operating during the Archean, a reasonable plate tectonic model for crustal development can be made, as shown in Figure 18.17.

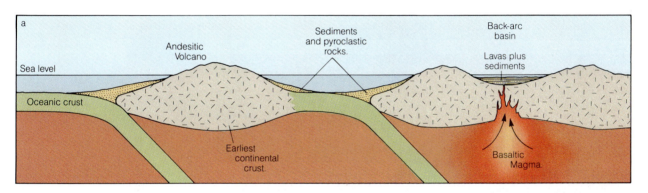

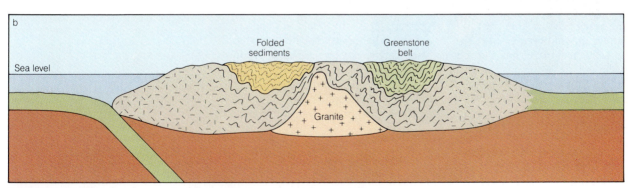

FIGURE 18.17 The development of the earliest continental crust following a uniformitarian model based on plate tectonics. (*a*) Subduction of oceanic crust produces andesitic magma and small island arcs of andesitic volcanoes. Sediment aprons form around the volcanoes and mixtures of sediment and basaltic lavas fill back-arc basins. (*b*) Collisions between arcs build larger masses of continental crust. Ancient sediments and lavas are folded into greenstone belts. Granites form by partial melting at the base of the crust.

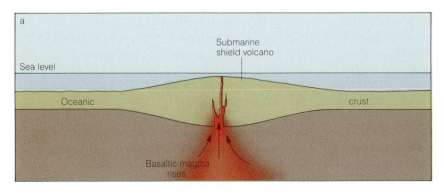

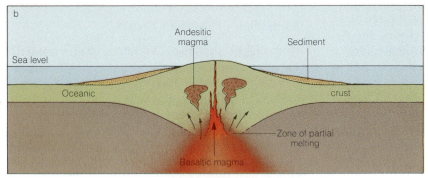

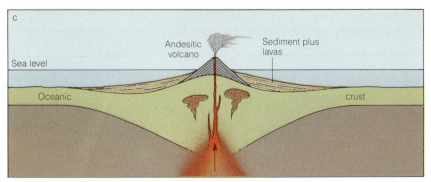

FIGURE 18.18   The development of the earliest continental crust by sag-duction. This is a non-uniformitarian model, which presumes that plate tectonics did not operate at the time. (*a*) A submarine shield volcano grows and causes the base of the oceanic crust to be depressed. (*b*) As the shield volcano grows larger, the thin oceanic crust sags to a depth where partial melting produces andesitic magma. When the volcano stands above sea level, erosion produces sediment and sediment aprons form around the base. (*c*) An andesitic stratovolcano eventually grows above the submerged and eroded shield volcano. Aprons of sediment and lava surround the volcano. When plate tectonics finally commence, the earliest continental crust grows larger through collisions.

Growth of continental crust from a plate-tectonic viewpoint starts with a world covered with oceanic crust beneath a globe encircling ocean. Volcanic-island arcs of andesitic composition, formed by subduction of oceanic crust, would be the first solid rocks to rise above the sea and these would then be subjected to erosion. The oldest sediments have affinities to younger continental shelf sediments and probably formed on narrow shelves that surrounded the island arcs. Accretion of island arcs by subduction-driven collisions would have produced high-grade terranes, and formed granites. The granite batholiths are not in long linear belts because the ancient plates were probably much smaller than those of the Phanerozoic. By the uniformitarian model the low-grade terranes were back-arc basins in which sediments and basaltic lavas were deposited. Collisions would have deformed the rocks in the back-arc basins but would not subject them to high-grade metamorphism because they were too far from the actual collision edges.

The model just outlined for growth of the Archean crust is just that, a model. It explains many facts but has not been proven. Some scientists believe that nonuniformitarian models are more reasonable. Higher-heat flow would produce a thinner and more ductile lithosphere they argue. Instead of breaking into plates, the lithosphere would stretch and thin down in some spots, thicken in others. Beneath a submarine volcano, for example, the crust and lithosphere would

stretch and sag (Fig. 18.18). When the pile of volcanic rocks was sufficiently thick, and had sagged deeply enough, partial melting would form andesitic magma. The sagging process, which has been termed *sag-duction*, would lead to the same kinds of rock that are produced by subduction, but it leaves unanswered the question of accretion into larger continental masses. Perhaps sag-duction and subduction both happened. The Archean continues to be an enigma and its rocks continue to be intriguing items for research. Perhaps some of the answers about the formation of ancient crust will be found on other planets where crust more ancient than 3800 million years has been preserved. Before addressing that question, however, we return, in the next chapter, to the question of the influence of plate tectonics and external processes on the development of the Earth's landscapes.

SUMMARY

1. There are five kinds of continental margins: passive, subduction, collision, transform fault, and accretion margins.

2. Passive margins develop by rifting of the continental crust. The Red Sea is an example of a young rift.

3. A characteristic sequence of sediments forms along a passive margin, starting with clastic nonmarine sediments, followed by marine evaporites, and then marine clastic sediments.

4. Continental subduction margins are the locale of paired metamorphic belts, chains of stratovolcanoes, and linear belts of granitic batholiths.

5. Collision margins are the locations of fold-and-thrust mountain systems.

6. Accreted terrane margins arise from the addition of allocthonous blocks of crust brought in by subduction and transform fault motions.

7. Plate tectonics must have operated as far back in time as 2000 million years ago. The proof is well displayed in the ancient collision margin represented by the Wopmay Orogen.

8. The continental crust grew during the Archean Eon. The volume of continental crust was close to the present volume of the continental crust by 2500 million years ago.

9. The way the Archean crust grew is open to question. It may have formed by the subduction of oceanic crust to produce andesitic island arcs and accretion of the arcs by collision.

SELECTED REFERENCES

Ayres, L. D., Thurston, D. C., Card, D. K., and Weber, W., eds., 1985, Evolution of Archean supracrustal sequences: Geol. Assoc. of Canada, Special Paper 28.

Hoffman, P., 1973, Evolution of an early Proterozoic continental margin: The Coronation geosyncline and associated aulocogens of the northwestern Canadian Shield: Phil. Trans. Royal Soc. London, Ser. A, v. 273, p. 547–581.

McElhinny, M. W., ed., 1979, The Earth: Its origin,

structure and evolution: London and New York, Academic Press.

Moorbath, S., 1977, The oldest rocks and the growth of continents: Sci. American, v. 236, no. 3, p. 92–105.

Tarling, D. H., ed., 1978, Evolution of the Earth's crust: London and New York, Academic Press.

Windley, B. F., ed., 1976, The early history of the Earth: New York, John Wiley & Sons.

CHAPTER 19

The Evolution of Landscapes

Dramatic landscape of Iceland, a large volcanic island lying astride the Mid-Atlantic Ridge, reflects the combined activity of volcanoes, glaciers, streams, and mass-wasting processes operating over millions of years.

DYNAMICS OF LANDSCAPE EVOLUTION

One of the most fascinating and unique attributes of our planet is its amazing variety of natural landscapes. Who could fail to be impressed by a view of the majestic snow-covered Himalaya rising abruptly from the plains of India, the lofty Andes of South America with their array of active volcanic cones, or the massive glacier system of Antarctica rising nearly as high above sea level as the highest mountains of Europe. Equally impressive are the vast subtropical deserts of northern Africa, central Australia, and the Arabian Peninsula, the broad fertile expanses of the North American Great Plains, the dense jungle terrain of South America's Amazon basin, or the undulating glaciated landscape of eastern Canada and north-central United States with its myriad lakes and streams.

Even a casual look at landscapes is likely to raise questions in our mind. Why, for example, can we see clear evidence of the sculptural effects of running water in areas now so dry that they have no flowing streams? Why does the surface of the Dead Sea in the Middle East lie hundreds of meters below sea level? Why do some islands in the Pacific Ocean rise to heights of 4 km above sea level while others barely reach the height of an average man?

These and other questions lead to more fundamental ones: How can such a diversity of landscapes be explained? Are landscapes eternal, as many of our forebears once thought, or are they transient, undergoing evolutionary change with the passage of time? What can landscapes tell us about the processes which have shaped the land, and about the nature and history of the Earth's mobile lithosphere?

Factors Controlling Landform Development

We have already seen that distinctive landforms result from the activity of various surface processes. A sand dune has a form that is different from that of a moraine. We can distinguish, as well, between an alluvial fan and a delta on the basis of their form. In each case the active process and depositional environment lead to a unique end product, or landform. *Process*, then, is one factor that helps dictate the character of landforms.

Climate, in turn, helps determine which processes are active in any area. As we have noted, in humid climates, streams may be the primary agency for moving and depositing sediment, whereas in an arid region wind may locally assume

a dominant role. Similarly, glaciers and periglacial phenomena are largely restricted to high latitudes and high altitudes where frigid climates prevail. Because climate also controls vegetation cover, it further controls the effectiveness of some important erosive processes (Fig. 11.41). In the same way that we can identify distinctive climatic regions of the Earth, it is also possible to identify distinctive landscape regions that are dominated by landforms resulting from one or more surface processes. However, because climates have changed through time, the active surface processes in some regions also have changed. As a result, certain landscapes largely reflect former conditions rather than those of the present.

Within any climatic zone, a given surface process may interact with surface materials differently, depending on their *lithology*. Rock types that are less erodible than others commonly have greater relief and produce more-prominent landforms than those more susceptible to erosion. A given rock type, however, can behave differently under different climatic conditions. For example, limestone may underlie valleys in moist climatic zones where dissolution is effective, but in dry desert areas such rocks may form bold cliffs.

Structure also plays an important role. With a little experience, a person trained in geology can rather easily identify structural features at the Earth's surface visible from an airplane. Due to differential erosion, certain folded or faulted beds stand in relief or control the drainage pattern in such a way that they impart a grain or pattern to the landscape, thereby disclosing the underlying structure (Fig. 15.25). A well-jointed or fractured rock is likely to be more susceptible to weathering, mass-wasting, and erosion than massive rock of the same composition, and typically will form more subdued terrain.

Initial relief of the land (its average altitude above sea level) is another primary control on landscape development which, in turn, is determined by the tectonic environment. Tectonically active regions are susceptible to high rates of uplift leading to high summit altitudes, steep slopes, large potential energy for erosion, and corresponding high sediment yields. Such landscapes tend to be extremely dynamic. Measured rates of *denudation, the sum of the weathering, mass-wasting, and erosional processes that result in the progressive lowering of the Earth's surface*, are generally high. In areas far removed from active tectonism, where relief is low, erosional processes tend to operate at slower rates, and changes take place more gradually. However, even in nontectonic areas, relatively rapid changes of sea level or regional isostatic movements, resulting from changing ice and water loads, may initiate and control significant changes in the landscape over broad areas.

Finally, the concept of landscape evolution necessarily involves the element of *time.* Although some landscape features can develop rapidly, even catastrophically, others quite obviously develop only over long geologic intervals. We know this, or at least we infer this, from measurements of surface processes now operating and by dating deposits that place limits on the ages of specific landforms or land surfaces.

Landscape Equilibrium

Change is implicit in the concept of landscape evolution. Presumably, landforms or landscapes will experience change if there is a change in any of the controlling variables. Change may be started by a tectonic event that causes a landmass to be uplifted, or by a drop of sea level that causes streams to assume new gradients. It can be initiated by a shift in climate that may modify the relative effectiveness of different surface processes. A change may also result as a stream, eroding downward through weak rock, suddenly encounters massive hard rock beneath.

Over short intervals of time, rates of change may vary due to natural fluctuations of the magnitude and intensity of surface processes. Over longer intervals, the rate of change may increase due to more rapid tectonic uplift, or experience a gradual decrease as a land surface is progressively worn down and approaches the level of the sea.

Does a landscape, then, ever achieve a state of equilibrium in which no change takes place? The answer apparently is no, for we have abundant evidence that the Earth's surface is now, and very likely always has been, a dynamic surface, constantly experiencing changes in response to the natural motions of the lithosphere, hydrosphere, and atmosphere. Nevertheless, it is apparent that conditions of near-equilibrium, or *quasi-equilibrium,* can be achieved. Therefore, it is convenient to think of equilibrium as a condition toward which landscapes evolve over geologically long intervals of time.

Several types of equilibrium can be envisaged. One involves a steady decline in the rate of change of a landscape through time until the rate is extremely slow (Fig. 19.1*a*). The end result is a landscape in *quasi-equilibrium.* A *steady-state equilibrium*

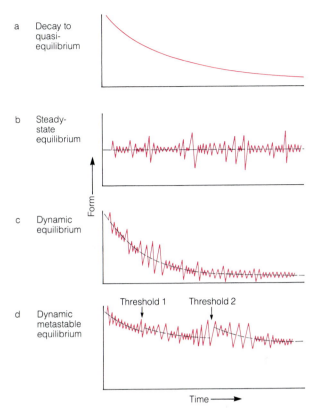

FIGURE 19.1 Graphs showing types of equilibrium conditions in landscape evolution. (*Source:* After Chorley et al., 1984.)

exists when landscape form fluctuates about some average, or mean, condition (Fig. 19.1*b*). An example is a stream that experiences changes in the form of its channel and banks as discharge varies around some average value. If landscape form oscillates about a mean value which is itself declining with time, a condition of *dynamic equilibrium* is said to exist (Fig. 19.1*c*). Such a condition may occur in a river that is cutting downward in its valley, but simultaneously is experiencing variations in discharge or sediment load that lead to minor alternating episodes of cutting and filling.

Complex Responses

One reason why a state of perfect equilibrium in landscape systems is difficult to achieve is that the responses to change are often complex. Within a landscape unit such as a large drainage basin, a change in one of the controlling factors may at first impact only a small part of the basin. For example, a sudden vertical movement along a fault that crosses the basin near its mouth will affect the long profile of the stream at that point and begin a series of compensating adjustments both in the

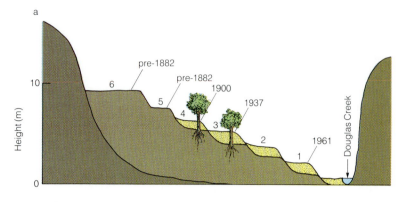

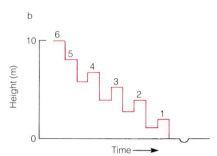

FIGURE 19.2 Evolution of Douglas Creek valley, western Colorado. (*Source: After Womack and Schumm, 1977.*) (*a*) Sketch showing section through Douglas Creek Valley, western Colorado, and terrace surfaces standing above the stream. Ages of the surfaces are based on tree-ring dating and historical information. Surfaces 5 and 6 were present before erosion began in 1882. (*b*) Behavior of Douglas Creek through time. Vertical segments indicate incision (downward) or deposition (upward); horizontal segments indicate periods of relative stability.

upstream and downstream directions. Areas in the headward part of the basin may initially be unaffected. However, as the stream system adjusts, changes will progressively affect other parts of the basin, including tributary streams, valley sides, and ultimately valley heads, which lead the system once more toward equilibrium. The change may move through the system in a complex manner, with many lags and minor adjustments taking place.

The example of Douglas Creek in western Colorado illustrates such a complex response (Fig. 19.2). Adjacent to the creek are six terraces, the highest two of which predate a period of downcutting that began in 1882. The lower discontinuous, nonpaired terraces resulted from pauses in downcutting during which deposition occurred. When incision of the main channel began, upstream tributaries were rejuvenated and began cutting downward as well. The resulting supply of sediment from the tributaries swamped the system downstream causing the main stream to stop its downward cutting and deposition to occur. This interval of deposition continued until the sediment supply from above decreased and the main stream could continue to incise its valley floor.

If several changes affect a basin simultaneously (for example, a sudden intense rainstorm concentrated in one tributary and a massive landslide in another part of the basin), the lagged responses to these events may interact in an even more complex manner. It is not surprising, therefore, that geolo-

gists working with the depositional and erosional products of such changes often have difficulty in sorting out the events that caused them.

Threshold Effects

In many natural systems, sudden changes can take place without any outside stimulus when a critical *threshold* condition is reached. Sand grains in a

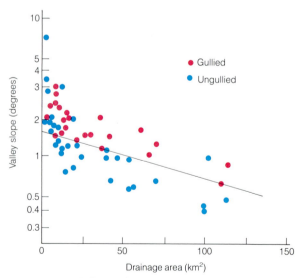

FIGURE 19.3 Relationship between valley slope and drainage area in the Piceance Creek basin of Colorado. The line marks the threshold slope above which gullying of the valley floor is observed. (*Source: After Patton and Schumm, 1975.*)

stream channel may remain at rest until a critical current velocity is reached, at which point they begin to move (Fig. 11.21). Certain glaciers apparently start to surge when the buildup of water pressure at their base reaches a critical threshold value that forces the ice to float off its bed (Chapter 13).

The concept of thresholds is also applicable to landscape evolution. It implies that the development of landscapes, rather than being progressive and steady, can be punctuated by occasional abrupt changes. A landscape in dynamic equilibrium may be subjected to a sudden change in form if a process operating on it reaches some threshold level (Fig. 19.1*d*). The condition of dynamic equilibrium is thereby affected, a change in the landscape takes place, and a new equilibrium condition is established. This has been referred to as *dynamic metastable* (changing) *equilibrium*. Field studies in the western United States have shown, for example, that the development of gullies on the floors of some alluvial valleys is related to valley slope (Fig. 19.3). Where valley-floor slopes reach a certain critical value, which depends on the area of the drainage basin, they become unstable and gullying begins. Where their slopes are gentler, valley

floors remain ungullied. Other studies have shown that sediments stored within some alluvial systems become unstable at critical threshold slopes, leading to episodes of accelerated erosion.

Scales of Space and Time

Landscapes can be examined at different spacial scales ranging from very large to very small (Table 19.1). The largest obvious features of the Earth are continents and ocean basins, but we can also recognize major structural elements such as lithospheric plates, the boundaries of which do not always coincide with continental landmasses. Within the continents we can further identify major topographic features, most of which are related to broad-scale structure. These include mountain ranges, sedimentary basins, plateaus, and major rift systems. At smaller scales we can recognize further subdivisions that reflect specific structural, eruptive, lithologic, erosional, or depositional units. At the smallest scales of subdivision, single processes may be operative and changes may take place very rapidly. At the largest scales, however, a complex interplay of many different processes and factors occurs and landscapes may appear to

TABLE 19.1 *Spacial Classes of Landscape Elements*

Order	Class	Examples	Typical Width
1	Major Earth features	Continents, ocean basins, lithospheric plates	> 1000 km
2	Major landform provinces	Mountain ranges, plateaus, sedimentary basins	100–1000 km
3	Major structural or constructional geologic features	Volcanoes, structural domes, fault-block mountains	10–100 km
4	Large-scale erosional and depositional units	Large valleys, ice-sheet moraines, small deltas	1–10 km
5	Medium-scale erosional and depositional units	Floodplains, cirques, drumlins, cinder cones, alluvial fans,	100–1000 m
6	Small-scale erosional and depositional units	Sand dunes, eskers, taluses, beaches, small deflation basins	10–100 m
7	Minor terrain features	Solifluction lobes, small gullies, channels of braided streams, rockfall boulders	1–10 m
8	Micro-terrain features	Small slumps, small-scale patterned ground	10–100 cm
9	Minor roughness features	Glacially grooved bedrock, sand ripples	1–10 cm
10	Micro-roughness features	Glacial striations, differentially weathered minerals in a rock	< 1 cm

change very slowly. Thus, a landslide deposit or a lava flow are formed almost instantaneously in the geologic sense, whereas a mountain range will evolve only over many millions of years.

In the case of small-scale features, geologists can observe and measure the active process, and see the resulting landform evolve. On much larger spacial scales, direct observation becomes impossible because of the extremely long spans of time over which major landscape features develop. In these cases we are left with only two main options. On the one hand we can observe processes operating under natural conditions and extrapolate their measured rates back through time. However, this is an uncertain approach at best, for we know that surface environments, and the magnitude and effectiveness of various surface processes, have differed greatly in the past (Chapter 20). Alternatively, we can develop theories and construct

models to explain the origin of the large-scale landforms and regional landscapes that are observed today.

Cycle of Erosion

The most influential theory of landscape evolution was proposed by an American geographer, W. M. Davis, in the late nineteenth century. Davis called his model the *cycle of erosion*, implying that it had a beginning and an end. A cycle was initiated by rapid uplift of a landmass, with little accompanying erosion, so that the initial relief was large. Erosion then progressively sculpted the land and reduced its altitude until it was worn down close to sea level (Fig. 19.4). Davis deduced that a landscape passed through a series of stages. During the earliest stage, streams cut down vigorously into the uplifted landmass and produced sharp, V-

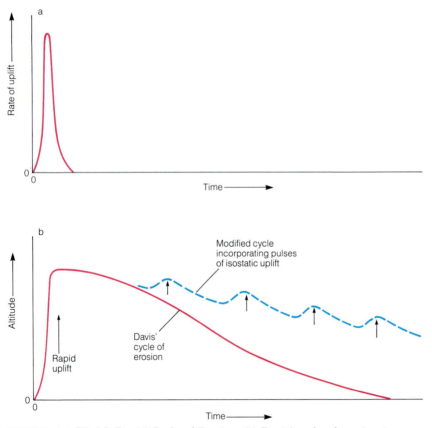

FIGURE 19.4 W. M. Davis' Cycle of Erosion. (*a*) Davis' cycle of erosion involves a pulse of rapid uplift during which uplift rate increases sharply, then declines. (*b*) As uplift rate increases, the land increases rapidly in altitude. As uplift rate drops, the uplifted land is gradually reduced by erosion. If isostatic adjustments to the removal of sediment are incorporated, a longer time is required to erode the land surface to low altitude.

shaped valleys, thereby increasing the local relief. Gradually, the original gentle upland surface is consumed as the drainage system expands and valleys become deeper and wider. During the next stage the land achieves its maximum local relief. Streams having reached a graded condition begin to meander in their valleys, and valley slopes are gradually worn down by mass-wasting and erosion. In the final stage, the landscape consists of broad valleys containing wide floodplains, stream divides are low and rounded, and the landscape is worn down ever closer to sea level.

Davis' theory attracted wide attention and formed the basis for most interpretations of landscape evolution during the following decades. Elaborations of the theory subsequently were made to account for landscapes strongly influenced by lithologic, structural, and climatic controls. Davis also visualized interruptions of erosion cycles due to climatic fluctuations or to renewed uplift. Others later pointed out that isostatic response of the crust to the progressive transfer of sediment from the land to the ocean should lead to periodic uplifts, requiring that a cycle be longer than Davis envisaged (Fig. 19.4*b*).

Landform Evolution and Plate Tectonics

The major landscape features of the Earth have developed over long intervals of time as the lithosphere has evolved and continents have been continually rearranged. The lateral motions and resulting collisions of lithospheric plates, leading to the generation of orogenic belts, have provided much of the driving force for landscape change over hundreds of millions of years. With acceptance of the theory of plate tectonics, it has become increasingly difficult to reconcile Davis' concept of an erosion cycle with what is known about global tectonics and Earth history.

Davis visualized a normal erosion cycle as beginning with a brief, sharp pulse of uplift, followed by a long interval of crustal quiet as the land was gradually worn down toward sea level (Fig. 19.4*a*). However, when a new belt of crustal convergence develops, uplift is likely to begin slowly and increase gradually to some maximum average value that is dependent on absolute rates of plate motion and on compensating isostatic adjustments (Fig. 19.5*a*). Such rates could continue as long as the relative plate motion is maintained, perhaps for many millions or tens of millions of years. Under such conditions, a landmass would be likely to increase gradually in altitude as the rate of uplift

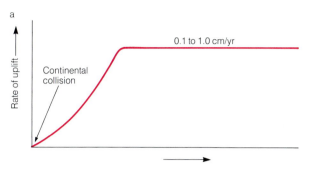

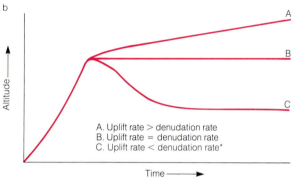

*Denudation rate decreases with declining altitude, so a balance is ultimately achieved between uplift and denudation.

FIGURE 19.5 Model of landscape evolution adjacent to a convergent plate boundary. (*a*) Rate of uplift increases following continental collision and is maintained at a level that is determined by the rate of convergence. (*b*) Altitude of the land surface increases following collision and then, depending on the relationship of uplift rate to denudation rate: (1) continues to rise slowly; (2) maintains a relatively constant altitude; or (3) declines in altitude to some steady-state condition.

increases. Once a maximum uplift rate is reached, the land would slowly continue to increase in altitude if the uplift rate exceeded the denudation rate. Alternatively, it might maintain a certain average altitude if uplift and denudation rates were balanced, or it might decline in altitude if the denudation rate exceeded the uplift rate. Over time, the rising landmass might become deeply dissected by streams, resulting in extreme relief, as in the case of the Himalaya in Nepal (see below).

The model presented above is not completely realistic, for spreading rates and average uplift rates vary through time. Studies have shown, for example, that since late in the Cretaceous Period, the rate of seafloor spreading in the Indian Ocean has undergone significant change (Fig. 19.6). Prior to about 40 million years ago, India moved pro-

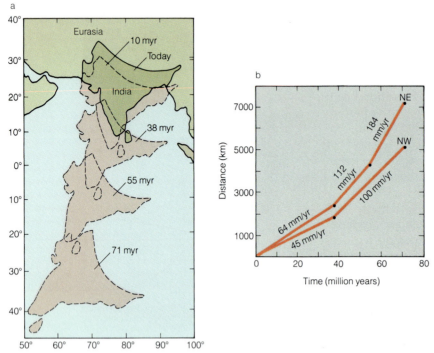

FIGURE 19.6 Variations in spreading rates in the Indian Ocean. (*Source:* After Molnar, Chen, and Tapponnier, 1981.) (*a*) Position of India with respect to Eurasia at different times since the Late Cretaceous Period, based on magnetic anomalies on the ocean floor. Collision of the two landmasses is thought to have occurred between about 50 and 40 million years ago. (*b*) Distance of the northeast and northwest tips of India from their present positions as a function of time. About 40–50 million years ago, before collision, the rate of movement was more than 100 mm/yr, whereas afterward it decreased to about half as much.

gressively toward the Eurasian continent at rates averaging 100–200 mm/yr. As the two landmasses collided, the average rate of convergence slowed to only about 45–64 mm/yr (Fig. 19.6*b*).

Measured uplift rates in orogenic belts are also variable. The highest rates range between 1 and about 10 mm/yr, averaged over intervals of several thousand to several million years. However, it appears likely that average values have changed through time as rates of seafloor spreading and plate convergence varied. Each such change is likely to lead to a compensating adjustment in landscapes as they begin to evolve toward a new condition of equilibrium.

The example of landscape development shown in Figure 19.5 is but one of many that could be postulated. By varying the rates of uplift, denudation, and isostatic adjustment to account for various interactions of lithospheric plates on different time scales, many different models can be developed that represent landscape evolution under a range of tectonic conditions.

Landscapes of Low Relief

The ultimate reduction of a landmass to low altitude, as envisioned by Davis, is likely to occur only if changes in plate motion lead to diminishing orogeny and tectonic uplift ceases. Denudational processes can then gradually lower the relief. Examples of such landscapes can be found in the world's shield areas where the roots of ancient mountain systems have been exposed as the crust has thinned through the action of long-continued erosion and compensating isostatic adjustment (Fig. 15.27).

Widespread erosional landscapes having low relief and low altitude are not commonplace. This must either mean that the Earth's crust has been very active in the recent geologic past or that such landscapes take an extremely long time to develop. Estimates have been made of the time it would take to erode a landmass to or near sea level by extrapolating current denudation rates into the past. Such estimates must take into account two important factors. Studies have shown that rates of denudation are strongly related to altitude (Fig. 19.7), implying that as the land is lowered the rate of denudation will decline. Furthermore, the eroding land will rise isostatically as the crust adjusts to transfer of sediment from the land to the sea. Both factors tend to increase the time it takes for the final reduction of a landmass to low altitude. Assuming that no tectonic uplift is taking place, estimates of the time it would take to reduce a landmass about 1500 m high to near sea level range from approximately 15 to 110 million yr.

While one might argue that the Earth's crust has been unusually active during the last 15 million years or more, could there have been earlier intervals of relative crustal quiet when low-relief surfaces did develop? Many ancient land surfaces are preserved in the geologic record as unconformities. Some can be traced over thousands of square kilometers and can be shown to possess only slight relief (Fig. 7.1). Associated with such surfaces are weathering profiles that imply long intervals of continuous weathering at relatively low altitude. Such buried paleolandscapes offer evidence of times when broad areas were eroded to low relief and may provide important clues about the history of lithospheric plates in the remote past.

LANDSCAPES OF PLANET EARTH

Every landscape has a story to tell. Each is the product of a past history of geologic events and climatic change which together have shaped the present surface. The following examples provide but a brief glimpse of the varieties of landscapes on our planet and the geologic forces that have generated them.

The Roof of the World

The lofty glacier-clad Himalaya, rising steeply from the plains of India, includes within its array of peaks the highest summits in the world (Figs. 19.8 and 19.9). To the north of the range lies the vast upland of the Qinghai-Xizang (or Tibetan) Plateau (Fig. 19.9a) which with average altitudes of 4000–5000 m is the highest extensive land area on Earth (Fig. 19.9b). Both the Himalaya and the adjacent plateau are geologically young features, still in the process of formation. Together they provide an important example of what happens when converging lithospheric plates cause continents to col-

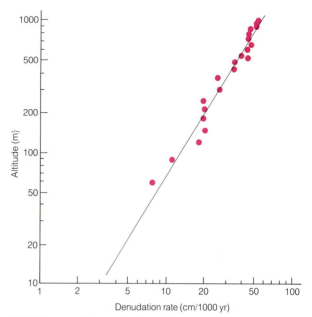

FIGURE 19.7 Change in denudation rate with increasing altitude in eastern Papua New Guinea. (*Source:* After Ruxton and McDougall, 1967.)

lide and thereby produce a major mountain system. In this part of central Asia we can see young orogenic landscapes that are literally changing before our eyes as denudational processes attack the rapidly rising crustal rocks.

Most of the interior of the high, cold plateau is dry, for monsoon rains moving inland across southern Asia are blocked by the steep front of the Himalaya. Therefore, rivers draining the uplands mainly collect water from the high mountains and from the eastern, lower quarter of the plateau which is more easily penetrated by moist air masses from the south.

Some streams that drain the high mountains and adjacent plateau follow belts of erodible rock or prominent structures to which they have become adjusted (subsequent streams; Fig. 11.38). Others flow down the regional slope of the land (consequent streams). However, segments of four of the largest streams cut sharply across the trend of major structures as they flow from the southern plateau through deep valleys in the high mountains and out onto the Indian plains (Fig. 19.9a).

The northwest-flowing Indus River drains the northern slope of the western Himalaya, as well as the lofty Karakoram and other nearby ranges beyond the western extremity of the plateau; however, it then changes course to pass directly across the mountains in a deep canyon beside the high massif of Nanga Parbat (8131 m). The Sutlej River

FIGURE 19.8 Oblique view of the Himalaya and the Qinghai-Xizang Plateau rising above the plains of India, as seen from an orbiting spacecraft.

originates north of the Himalaya, and after paralleling the crest for nearly 300 km turns abruptly southwest to cross the range axis between peaks that reach altitudes of more than 6000 m. The Arun River also begins north of the range crest, but crosses it through a deep canyon that lies 6500 m below the nearby summits of Chomolungma (Mt. Everest) (8850 m) and Kanchenjunga (8590 m). The north slope of the central and eastern Himalaya, as well as the southern plateau, are drained by the Tsangpo which flows eastward more than 1000 km before it turns sharply southward across the range and then travels west along its southern base as the Brahmaputra River.

Two hypotheses have been proposed to explain the cross-structural courses of these rivers. The first assumes the drainage to have originated as a series of consequent streams flowing down the north and south flanks of the mountain range as it began to rise. This initial pattern was modified because the most powerful south-flowing streams, due to greater discharge and steeper slopes on that flank of the range, were able to cut vigorously headward past the range crest and capture drainage on the northern side. The alternative hy-

pothesis proposes an antecedent origin for these streams, the courses of which are believed to predate the Himalaya. When major uplift began, they were able to deepen their channels at a rate that matched or exceeded the rate of uplift and so could maintain their courses.

Several lines of evidence favor the hypothesis of antecedent origin. For example, as the upper Arun River turns south toward the Himalaya it flows across a body of hard gneiss in a remarkable deep gorge. Yet adjacent to it is a belt of weak, erodible schist. Had the river acquired its course by progressive headward erosion, it most reasonably would have cut back into the more erodible rock instead of through the hard gneiss, which could have been avoided. On the other hand, if the mountains rose across its ancestral course, the river would have become intrenched in the hard rocks as its channel was deepened by vigorous downward erosion.

Chinese geologists have suggested that the Himalaya did not exist as a major topographic feature before the middle of the Miocene Epoch (about 20 million years ago), and that the plateau remained at altitudes of less than about 2000 m

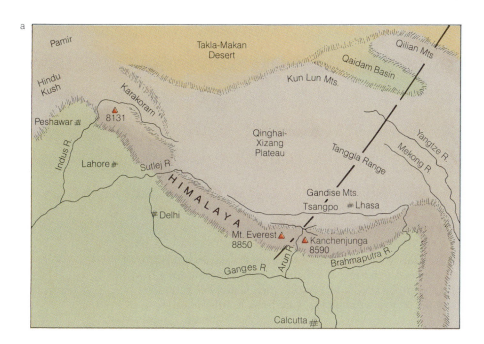

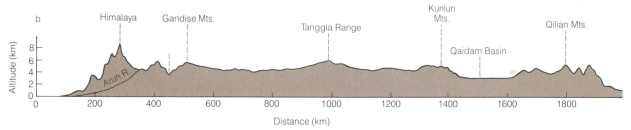

FIGURE 19.9 Himalayan drainage systems. (*a*) Major streams draining the Himalaya and southern Qinghai-Xizang Plateau. The Indus, Sutlej, Arun, and Tsangpo-Brahmaputra Rivers originate north of the Himalaya and flow across the structural axis of the Himalaya in deep valleys. (*b*) Topographic profile showing the high altitude of the plateau and the course of the Arun River as it crosses the axis of the Himalaya.

throughout most of the Pliocene Epoch (circa 5.3 to 1.6 million years ago). Pliocene plant fossils have been found on the plateau at altitudes of 4000–6000 m in what are now alpine and cold desert environments. They include many subtropical forms, modern examples of which live at altitudes of less than 2000 m. Remains of *Hipparion,* a primitive horse adapted to subtropical forests, further point to a mild, humid climate in the plateau region at that time. *Hipparion* fossils are widespread throughout Asia suggesting that these animals could migrate freely over the whole continent at a time when no major mountain barriers existed.

Ancient karst topography has also been discovered high on the plateau in the frigid periglacial zone. It is similar to karst now developing at low altitudes in the warm monsoonal zone of southern China and is regarded, therefore, as further evidence of substantial uplift since its formation during the Pliocene.

The rate of uplift apparently increased at the beginning of the Quaternary Period when forests on the plateau were replaced by a treeless cool-climate vegetation and when thick alluvial deposits began to accumulate as the rising mountains were eroded.

If the varied evidence has been correctly interpreted, the plateau and the bordering Himalaya must have experienced uplift of at least 3000–4000 m during the last 2 million years or less, with especially rapid uplift occurring during the last few hundred thousand years (Fig. 19.10*a*). Relatively

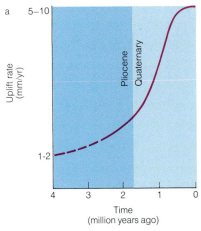
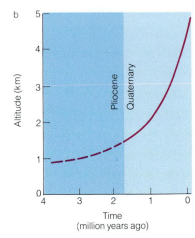

FIGURE 19.10 Changes in uplift rate and altitude of central Asia since the Pliocene Epoch. (*a*) Estimated uplift rate of Qinghai-Xizang Plateau during the last 4 million years, showing increasing rate during the Quaternary Period. (*b*) Increase in altitude of the plateau during the Quaternary, as inferred from geologic evidence.

low altitudes during the Pliocene should correlate with low rates of uplift (Fig. 19.10*b*), whereas high uplift rates during the Quaternary elevated both the Himalaya and the interior plateau to great altitude. Prior to large-scale Quaternary uplift, major rivers may have flowed south, draining the low-lying plateau. As uplift accelerated, only the largest streams were able to maintain their courses across the rapidly rising mountain system.

Are the Himalaya and the plateau still growing today? Estimates of current uplift rates in the high mountains range from 2 to more than 5 mm/yr. Maximum denudation rates for drainage basins that lie entirely or partly within the Himalaya are between 0.5–2.5 mm/yr. Even if these figures are not very accurate, they nevertheless suggest that in some sectors the mountain system is increasing in altitude more rapidly than erosional forces can tear it down. For mountaineers, this means ever-greater challenges as the highest summits continue to rise. It also means that the Himalaya and the great plateau to the north will remain areas of dynamic change as the landscapes continue to evolve.

The Andean Cordillera

The Andes are a majestic system of parallel, closely adjacent mountain ranges that bound the western margin of South America. The system extends across 65° of latitude, from 10° N in Venezuela to 55° S in southern Chile and Argentina, a distance

of 7500 km. In southern Peru and Bolivia, the Andes achieve a maximum width of 900 km (Fig. 19.11).

Many peaks in the Andes reach altitudes of more than 5000 m, with the highest soaring to 7000 m. The combination of high peaks and the north–south trend of the ranges is a severe obstacle to travel between the eastern and western portions of South America. There are no large, transverse valleys through which traffic can easily move. As a result, travelers must cross the formidable crest through high and generally rugged passes. Not surprisingly, rather different cultures developed on each side of the mountains.

The Andes are impressive indeed, but they would be even more impressive if it were possible to view the range from the lowest parts of the adjacent ocean floor. Running parallel to the coast, and only a few hundred kilometers from the Andes, lies the Peru-Chile Trench (Fig. 19.11). Off the coast of Peru and northern Chile this great linear trench reaches depths of more than 7000 m. The vertical distance between the bottom of the trench and the crest of the adjacent mountains is, therefore, as much as 14 km. This is the most pronounced change in height, over a short horizontal distance, of any place on the Earth.

The drainage pattern of South America is the most asymmetrical of all the continents. Streams flowing down the steep western slope of the Andes generally travel less than 200 km before reaching the Pacific. By contrast, some major streams draining the eastern slopes flow 4000 km or more before reaching the Atlantic. Through most of the southern Andes, the drainage divide forms the political boundary between Chile and Argentina. The marked asymmetry of the drainage pattern, therefore, explains why Chile is such a long and narrow country compared to its neighbor to the east.

The western margin of South America coincides with the edge of the American Plate (Fig. 19.11). Along a major part of this margin, the Nazca Plate (which is capped by oceanic crust) is being sub-

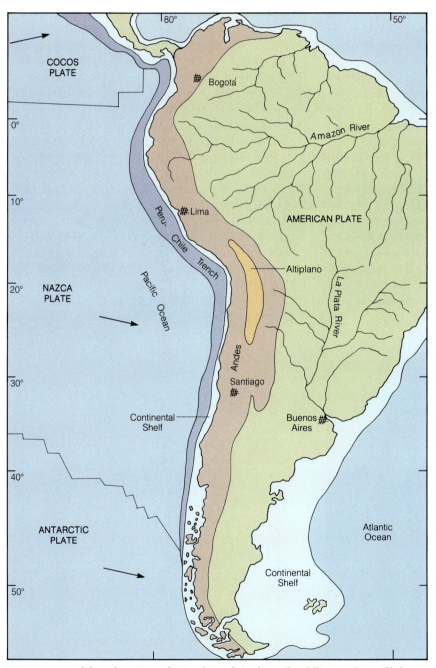

FIGURE 19.11 Map showing relationship of Andean Cordillera to Peru-Chile Trench and the adjacent subducting Nazca and Antarctic plates. The Altiplano is a high intermontane basin between western and eastern ranges of the cordillera.

ducted beneath continental crust of the American Plate. In southernmost Chile, a small portion of the Antarctic Plate is being subducted. The dip of the subducting plates is toward the east and the angle of dip is between 10° and 35°, as determined from earthquake foci. Throughout most of the Andes, geophysical data suggest that the continental crust, from the edge of the Peru-Chile Trench eastward for a distance of at least 700 km, is in a state of compression. Not surprisingly, the prominent geological features of the Andes are mainly of two types. Either they are compressional in origin, such as folded and thrust-faulted mountain ranges, or magmatic in origin, as represented by prominent volcanoes and elongate batholiths. The magma that produced the igneous features was presumably generated by melting processes associated with the subducting lithosphere.

The great topographic variations in the Andes and the tremendous latitudinal extent of the ranges means that the Andean cordillera (Spanish for "chain") is exposed to a wide range of climatic influences. These include the humid tropics in the north, the world's most arid desert (the Atacama), and the glacial and periglacial environments of southern Chile and Argentina. The vegetation cover and the type and intensity of surface processes are similarly varied. Superimposed on the basic tectonic and magmatic landscape of the Andes, therefore, are the effects of climate-related surface processes.

The northern Andes, because they lie largely within the belt of trade winds, receive their precipitation mainly from moist westward-flowing tropical air. As a result, the eastern, windward slopes at the headwaters of the Amazon Basin are clothed in dense vegetation, while the western, leeward slopes tend to be much drier and less-densely vegetated. Along the crest, both here as well as farther south, rise numerous stratovolcanoes, the highest of which are capped by snow and ice (Fig. 19.12). Many are recently active. Both their lower slopes and the adjacent terrain are mantled with pyroclastic debris, and lahars often extend down the major valleys.

In southern Peru and northern Chile and Argentina, upwelling cold ocean waters off the coast and descending dry air give rise to arid landscapes marked by salty playa lakes, desert alluvial fans, and fields of yardangs. High in the central Andes lies the Altiplano, an arid plain at an altitude of about 4000 m which is the surface of an intermontane tectonic basin that has filled with sediment shed from adjacent ranges. On the Altiplano lies Lake Titicaca, highest of the world's large lakes (Fig. 19.13).

Farther south, the Andes pass into the belt of westerly winds where the Pacific slope receives in-

FIGURE 19.12 Osorno Volcano, an historically active Andean stratovolcano in southern Chile. Lava flows and lahars from this and an adjacent volcano have helped to raise the level of Lago Llanquihue (foreground), one of several large moraine-dammed lakes along the front of the Andes.

FIGURE 19.13 Lake Titicaca, the world's highest large lake, occupies the floor of the Altiplano along the boundary between Peru and Bolivia. Two large stratovolcanoes lie along its southern shore and ice-clad peaks of the high Andes rise to the northeast (upper right) in this satellite view.

creasingly greater amounts of precipitation. There the lush rain-forest vegetation of southern Chile contrasts with the dry pampas of Patagonia that lie in the rain shadow east of the mountains.

The relatively smooth western coastline of South America ends abruptly in the southern Lake District of Chile at 40° S. At this latitude, ice-age Andean glaciers reached the Pacific coast, reflecting the increasingly wetter and colder conditions of southern latitudes. Their erosive action sculpted the deformed rocks into a complex system of long, deep fjords and numerous islands that extends south to Cape Horn. Two of the longest of these submerged glacial valleys form the famous Strait of Magellan, separating the continent from the large island of Tierra del Fuego, and Beagle Channel, through which H.M.S. *Beagle* carried Charles Darwin on his famous scientific voyage in 1833–34 (Fig. 19.14).

Also at about 40° N the offshore trench ceases to be a recognizable topographic feature (Fig. 19.11). This change partly reflects the increasing influence of glaciation toward the southern end of the Andes. Much of the great volume of sediment eroded by glaciers during successive glacial ages was dumped into the adjacent ocean where it gradually filled the trench and eliminated it as a prominent topographic form.

Intense glaciation has resulted in an eastward shift of the continental drainage divide at several places in the southernmost Andes. Because the ice apparently was thickest over the now partly submerged fjord region, it flowed eastward across saddles in the main Andean crest, progressively deepening and ultimately eliminating these topographic barriers. In several sectors, the Pacific/Atlantic divide now lies along the crest of end moraines marking the limits of former glaciers that

FIGURE 19.14 Beagle Channel, along southern margin of Tierra del Fuego, has been eroded along a major fault system. Ice-age glaciers carved this large fjord along which Charles Darwin made geological observations during the voyage of H. M. S. *Beagle.*

terminated near or beyond the eastern front of the mountains (Fig. 19.15).

The Dead Sea Depression

The Dead Sea, lying between Israel and Jordan in the Middle East, earned its name because fish and other animals cannot survive in its salty waters. The waters of the sea contain 25 percent dissolved salts by weight. As a result, the water is so dense it is impossible for swimmers to sink. The average water level is 360 m below the surface of the nearby Mediterranean, making the Dead Sea the lowest body of water in the world, but it varies several meters during the year due to seasonal variations in inflow from the Jordan River and from small streams that rise in the adjacent hills.

The depression that holds the Dead Sea has the same origin as the depression of the Gulf of Aqaba to the south, and the Sea of Galilee to the north (Fig. 19.16). Each is a long, narrow basin bounded by faults. The basins formed as a result of motions along a transform fault boundary separating two small tectonic plates. The Dead Sea Fault is actually a fault zone, rather than a single continuous fracture. The fault splits and branches into several subparallel fragments, and slippage occurs along each fragment. On its southern end, the fault zone joins the divergent plate boundary that runs up the center of the Red Sea. On its northern margin

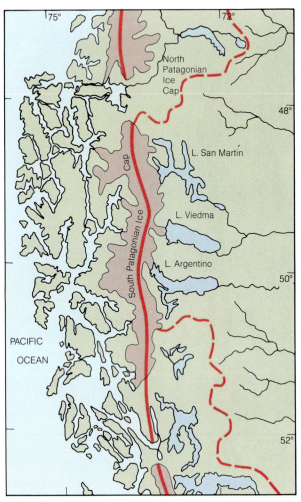

FIGURE 19.15 Map of southernmost Andes showing coastal fjords of Chile, Patagonian Ice Caps, and topographic crest of the range (bold solid line). In two sectors, the continental drainage divide (dashed line) lies well east of the crest and partly coincides with a belt of Pleistocene end moraines.

the fault zone joins the convergent plate boundary that extends eastward through southern Turkey, along the edge of the Taurus Mountains.

Relative motions on the segments of the Dead Sea fault system are all the same—strike-slip movement with a left-lateral motion. That is, the relative motion of the plate to the west of the fault is toward the south, while that to the east is toward the north (Fig. 19.16). Where segments of the fault system overlap, they do so in such a way that the northern segment is stepped to the west.

The landscape of the Dead Sea region is strongly influenced by geologic structure (Fig. 19.17). Slippage on the overlapping fault segments subjects the small blocks of crust trapped between the ends to tensional stretching. The small, stretched blocks of crust subside, creating steep-walled, pull-apart basins bounded by steeply dipping strike-slip fault escarpments. Pull-apart basins formed in this manner share similar features with the rift valleys that form when continental crust is split by a new spreading edge. Both kinds of basin tend to have steep walls, and both tend to become filled rapidly with sediment (Fig. 19.18).

The Dead Sea has a length of 75 km, a width of 16 km, and a maximum depth of 365 m. The principal inflow of water is from the Jordan River. The river first flows into the Sea of Galilee, then follows the trace of the fault into the Dead Sea. There is no outflow from the Dead Sea. Local rainfall is very low, temperatures are high, and the only way water can escape is through evaporation. Thus, any dissolved salts brought in by the river become concentrated in the waters of the sea. Both the Jordan River and small streams that flow directly into the Dead Sea carry suspended particles as well as dissolved salts. The suspended matter is the principal source of the sediment that is accumulating in the lake basin.

If sediment is continually being deposited in this enclosed basin, then why does it not fill up? The reason that the sea is so deep is that the rate of subsidence exceeds the present rate of sedimentation. This has not always been so. During periods of higher rainfall and greater runoff in the geologic past, the lake stood well above its present level, and the rate of sedimentation was higher. Averaged over a long period, however, the rate of subsidence has nearly equaled the rate of infilling, so the sea and its underlying basin of sediment probably can be regarded as being in, or close to, a steady-state condition. The present rate of sediment accumulation apparently represents a fluctuation below the long-term average.

Volcanoes in the Sea

Although we cannot directly observe the development of major second- and third-order landscape features (Table 19.1) because of the immense time involved, it would be instructive to examine a simple series of landforms that represent successive stages of landscape development. On the continents such landform series are not easy to find or interpret because of lithologic and structural complexities. However, nearly ideal examples do exist in mid-ocean volcanic island chains. The Hawaiian Islands and the associated Emperor Seamounts, which together extend 6000 km across the Pacific Ocean Basin, are among the most instructive (Fig. 19.19).

The American geologist James Dwight Dana first pointed out, in the mid-nineteenth century, that the Hawaiian Islands appear to increase in age up the chain from Hawaii to Kauai (Fig. 17.23). Dana based his judgment on the degree to which the islands have been dissected by erosion. The island of Hawaii, which includes several active volcanoes

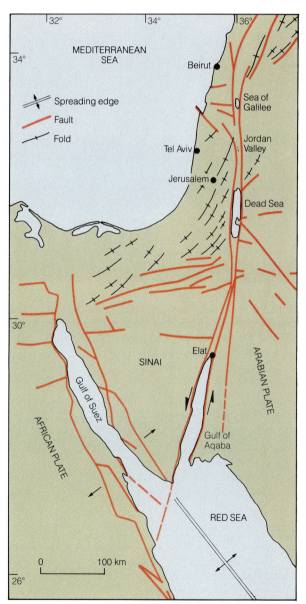

FIGURE 19.16 Map showing transform plate boundary along which lie the Sea of Galilee, the Dead Sea, and the Gulf of Aqaba. Arrows indicate sense of relative motion of the different plates. (*Source:* After Garfunkel, 1981.)

FIGURE 19.17 Geologic structure controls the major landscape features near the Sinai Peninsula (center) which is bounded to the west (left) by the Gulf of Suez and to the east by the Gulf of Aqaba. The Dead Sea, to the north of Gulf of Aqaba, occupies the floor of the Dead Sea Rift Zone.

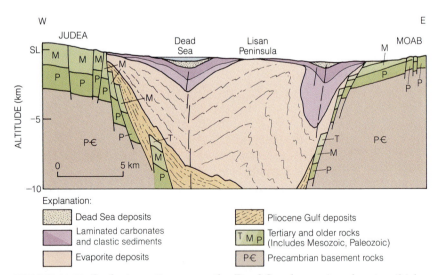

FIGURE 19.18 Geologic section across the Dead Sea depression showing thick fill of sediment that has accumulated since the end of the Tertiary Period. (*Source:* After Zak and Freund, 1981.)

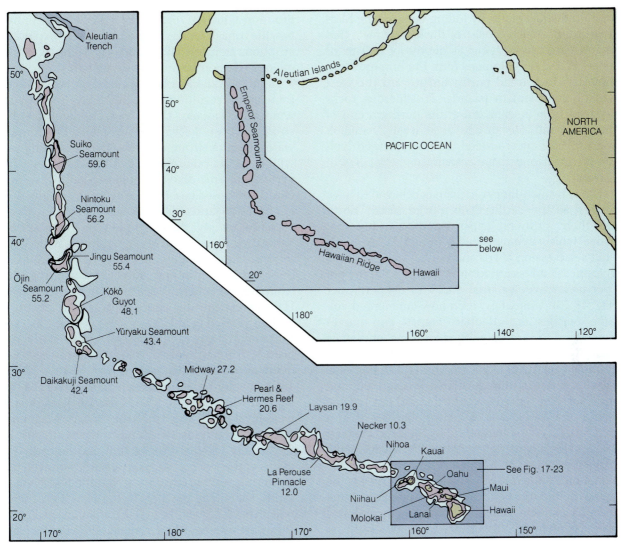

FIGURE 19.19 Hawaiian-Emperor chain, showing oldest reliable ages (in millions of years) for basaltic rocks of the volcanic shields. (*Source:* After Clague and Dalrymple, 1987.)

among the five that comprise it, is relatively unaffected by erosion except along part of its windward seacoast where some of the oldest lavas are found. The volcanoes of Maui and Molokai have been strongly eroded by streams and by wave action, while significant parts of the summits of Oahu, Kauai, and Niihau have been eroded away.

Dana's hypothesis was later proved correct when lavas from the islands were dated by the K/Ar method and their magnetic polarity was measured. The oldest rocks exposed on Hawaii are less than 730,000 years old, the age of the last major magnetic reversal. The best available ages for the volcanoes of Maui, Lanai, and Kahoolawe are 0.8–1.9 million yr, of Molokai 1.8–1.9 million yr, of Oahu 2.6–3.7 million yr, and of Niihau and Kauai 4.9–5.1 million yr (Fig. 17.23). Geologic mapping has shown that the volcanoes have had rather similar histories, with broad, relatively simple shields of basalt being constructed as the islands emerged from the sea. By examining the eroded landscapes on successively older volcanoes, one can obtain a reasonable understanding of how the most ancient Hawaiian landscapes probably evolved.

The emergent islands of the Hawaiian group represent only the youngest and highest part of the chain. To the northwest lie a series of small islets and rocks that are remnants of former larger volcanic islands (Fig. 19.19). Farther up the chain are found small atolls, including those of the Midway group, which consist of thick coral reefs that cap volcanic rocks lying at depths of 150 m or more. Still older volcanoes form the Emperor Seamounts which trend northward toward the west end of the Aleutian Islands. Fossiliferous rocks dredged from a seamount at the northern end of the chain indicate an age of at least 70 million years. These now-eroded and submerged volcanoes must once have resembled the islands of Hawaii.

The Hawaiian-Emperor chain is believed to have formed as the Pacific Plate moved slowly north and then northwest across a mid-ocean hot spot above which frequent and voluminous eruptions built a succession of volcanoes. Once formed, each volcanic island is carried slowly away from the hot spot toward cooler crust and into deeper water (Fig. 19.20). Ultimately, each now-submerged volcano reaches the Aleutian Trench and moves downward toward the subduction zone, ending its history.

Initial construction of a volcanic edifice proceeds rapidly. Based on the calculated average eruption rate (0.05 km^3/yr) and volume (42,500 km^3) of the island of Hawaii, it is estimated that less than a million years was required for its construction (Fig. 19.21). This is consistent with the known age of the oldest rocks on the island. During this constructional phase the rate of upbuilding far exceeds the rate of denudation, and the landscape assumes the form of coalescing shield volcanos. As the volcanic pile accumulates, the localized added weight on the seafloor causes isostatic subsidence of the ocean crust. This helps to limit the maximum altitude to which a volcanic island can rise. As the basaltic phase ends, and subsequent explosive eruptions of more silica-rich lavas decrease in frequency, erosion begins to take its toll.

High Hawaiian volcanoes intercept the moist tradewinds, resulting in abundant precipitation on windward slopes. Deep canyons are cut by streams whose discharge is enhanced by groundwater issuing from aquifers that are confined within the porous sequence of thin lava flows by vertical volcanic dikes. Simultaneously, waves attack the exposed coasts, producing steep cliffs that are prone to landsliding. In this manner an island becomes deeply dissected and its margins are worn back by wave attack. As volcanic activity diminishes and finally ceases, denudational processes and continuing submergence bring the landmass ever closer to sea level. In the case of the Hawaiian chain, this entire process takes about 10 million years (Fig. 19.21).

During the emergent phase of a tropical island's history, fringing reefs develop like those seen on many of the Hawaiian islands (Fig. 14.10*a*). These are gradually transformed into barrier reefs as the land slowly subsides. The Hawaiian Islands lack well-developed barrier reefs and enclosed lagoons, but good examples are found to the southwest in the Society Islands (Fig. 14.10*b*). As the rocky remains of the volcano subside beneath sea level, the bordering reef becomes an atoll dotted with small sandy islets (Fig. 14.10*c*). In the Hawaiian chain, this phase takes another 15–20 million years before slow northwestward movement of the Pacific Plate carries an atoll out of the tropical zone of coral growth and into ever-deeper water.

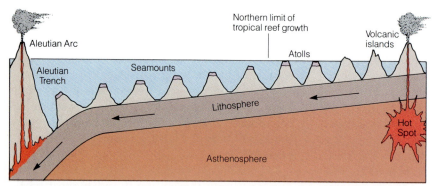

FIGURE 19.20 Diagrammatic profile (not to scale) along Hawaiian-Emperor chain illustrating how emergent volcanic islands are transformed into atolls and then seamounts as the Pacific Plate moves slowly away from the mid-Pacific hot spot toward the Aleutian Trench, where it is subducted.

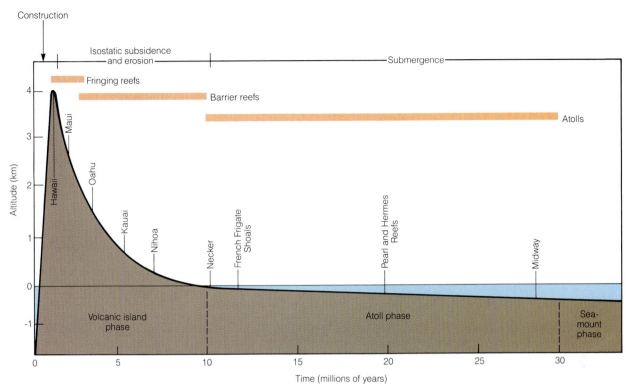

FIGURE 19.21 Change in altitude of a volcanic island as it is built up from the seafloor and then, through isostatic subsidence and erosion, is gradually reduced to sea level. Continued submergence occurs as the island moves away from the hot spot on the slowly cooling and subsiding lithosphere and becomes first an atoll and then a seamount. Mid-Pacific islands, from Hawaii to Midway, provide examples of different stages in the evolutionary sequence.

At this point the subaerial island landscape ceases to exist, except as a relict beneath the ocean where it may continue to experience slow modification with passing time. It has taken some 25–30 million years since initial outpouring of lava on the seafloor for the island to be transformed into a seamount, and it will take more than twice this long for the volcano and its coralline cap to complete its northward ride and be carried into the deep-ocean trench.

This simplified model for the development of Hawaiian landscapes omits several complications that are likely to influence evolution of the island chain. Some studies suggest, for example, that volcanic activity has been episodic and that eruption rates have varied through time. The measured eruption rate for Kilauea Volcano during the present century, about 0.1 km^3 of lava per year, greatly exceeds the long-term average rate for the Hawaiian chain as a whole (0.018 km^3/yr) and for the Emperor Seamounts (0.012 km^3/yr). This means that the volume of individual volcanoes,

their height above sea level, and their denudational history can be quite variable.

Erosion of the volcanoes is also strongly affected by climate and sea level. During glacial ages, world sea level is low, stream runoff apparently is high, and valleys are actively deepened. During interglacial times, high sea level causes coastal cliffs to erode back by wave attack and valley floors to fill with sediment. This cyclic variation in the relative effectiveness of stream erosion and marine erosion is superimposed on the long-term denudational trend (Fig. 19.21). It implies that erosion is episodic and that Hawaiian landscapes exist in a state of dynamic equilibrium (Fig. 19.1c).

The Western Australia Shield

The principal concentration of rocks of Archean age in Australia is found in the Western Australia Shield. Rocks older than 2.5 billion years are known to crop out in an area of about a million square kilometers (Fig. 19.22); however, two cra-

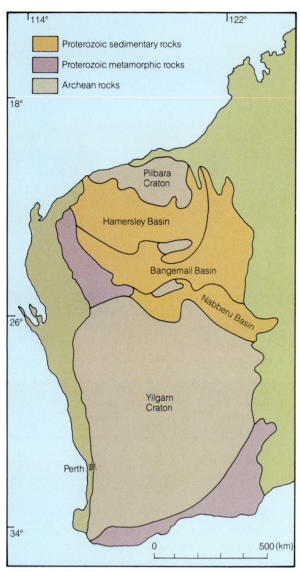

FIGURE 19.22 Map of the Western Australia Shield, showing both Archean cratons and the overlying Proterozoic basins.

tons, the Pilbara and the Yilgarn, account for most of the outcrop. It has not been proven that the Pilbara and Yilgarn Cratons are continuous at depth, but it is suspected they may be because an area of Archean rock crops out between them in a small dome.

Most of the region between the Pilbara and Yilgarn Cratons is occupied by sedimentary basins of Proterozoic age—the Hamersley, Bangemall, and Nabberu Basins. Sedimentary rocks in these basins are characteristic of the kinds of sediments known to be deposited under shallow marine conditions. The rocks have been neither extensively metamor-

phosed nor severely deformed. These facts suggest that the basins were formed by shallow downwarping or rifting of the shield and that the basins have not been significantly changed by subsequent tectonic activity. Both clastic and chemical sedimentary rocks are present in the basins, but the most distinctive rock type is called *banded iron formation*. These finely laminated sequences of chert and magnetite are chemical sediments.

Rocks within the cratons are either high-grade metamorphic derivatives of earlier sedimentary rocks, or they are igneous. The most abundant igneous rock is granite which is present as numerous small batholiths that intrude basalts which have been converted by low-grade metamorphism into greenstones. The assemblage of igneous rocks found in the Western Australia Shield is typical of the granite-greenstone terrains found in most shield areas of the world (Fig. 19.23). Associated with the greenstones are the famous gold-mining districts of Kalgoorlie, Coolgardie, and Norseman.

Most major shield areas of the Earth are low in altitude and in relief. The Western Australia Shield is no exception, for its surface generally lies between 100- and 500-m altitude. The form of the landscape is strongly controlled by rock type, structure, and relative erodibility. Granites tend to form low, slightly rounded outcrops. The intervening greenstones display little relief except where thin bands of iron-rich chert are present. The cherts are more resistant to weathering and they form low, sharp ridges. The most distinctive features of the landscape are formed by quartzites and banded iron formations of the Proterozoic basins. Because these rocks are very resistant to erosion, they form bold outcrops (Fig. 19.24).

Except for its extreme southwestern corner, the Western Australia Shield receives very little precipitation, generally only about 200–300 mm/yr. As a result, the region is characterized by sparse vegetation and poorly developed stream systems. Most rainfall soon evaporates, and the little that does run off accumulates in shallow, elongated, salt-encrusted playas. No large rivers flow off the shield.

Weathering is very deep and in places is known to reach hundreds of meters. This suggests that the shield must have been exposed to erosion for a very long time, that it has experienced no significant uplift or downwarping, and that at some time in the past it must have had a wetter climate. The land surface has now reached a state of quasi-equilibrium and the shield is a region with a denudation rate close to zero.

At many places on the shield it is possible to find remnants of an old lateritic surface (Fig. 19.25). As discussed in Chapter 9, laterites are characteristically formed on landscapes having low relief, and in warm regions with high rainfall. Laterites are forming today, for example, in such places as southeastern Asia and the Amazon Basin of South America. The Western Australia laterites cannot be dated exactly, but it is possible to determine that they are not forming in today's arid climate. They are known to date from at least the early Tertiary Period, and may be as old as the Jurassic. They indicate that the shield has been a region of low relief for more than 30 million years, and possibly as long as 200 million years. Furthermore, the landscape must have experienced a significant change of climate during that time.

Today the shield lies between 20° and 35° S latitude. These are the latitudes where, as explained in Chapter 12, dry subtropical air causes major deserts to form. However, Australia is slowly drifting directly north. The spreading rate between the Australian and Antarctic Plates is about 7 cm/yr. The absolute motion of the Antarctic Plate (relative to a fixed point on the Earth) is believed to be small, probably no more than 1 to 2 cm/yr. This suggests that Australia is moving north with an absolute motion of about 5 cm/yr. The climate of the world in the early Tertiary was much warmer than it is today, so at the measured average rate of movement, 40 million years ago the Western Australia Shield would have been some 2000 km south of its present position in a temperate zone of high rainfall like that of southern Chile today. A warm early Tertiary climate and high rainfall probably explains the origin of the old laterite. Like a ship slowly sailing north, the Western Australia Shield is passing through different climatic zones. Today it lies in the dry subtropical belt. However, if the present direction and rate of motion continue for another 20 million years, it will have entered a zone of tropical climate.

FIGURE 19.23 View from an orbiting spacecraft showing granite-greenstone terrain of Pilbara Craton. Light areas are underlain by granitic rocks and greenish areas by greenstone.

FIGURE 19.24 Bold outcrops of quartzite and banded iron formation in the Hamersley Basin.

Glacial Landscapes of Northeastern North America

An airline passenger traveling the polar route from London to Vancouver or Seattle, and who can arrange for a window seat, will be treated to one of the most scenic and geologically spectacular flights in the world. After passing Iceland astride the Mid-Atlantic Ridge, seas littered with icebergs, and the rugged highlands and enveloping ice sheet of Greenland, the plane takes a course across a landscape that has only recently emerged from beneath a vast continental glacier. First comes Baffin Island with its deep fjords (Fig. 19.26a) and ice caps that contain the last remnants of a thick ice sheet that covered northeastern North America during the last glacial age. On the west coast of the island, and on many other smaller islands north of Hudson Bay, can be seen multitudes of raised beaches that record isostatic uplift of the land as the ice sheet thinned and retreated (Fig. 13.39). West of Hudson Bay the plane passes over large areas of bare, scoured bedrock and elongate ridges and giant grooves that mark the passage of the glacier many millennia ago (Fig. 19.26b). Ahead in the distance, the sun reflects off immense glacial lakes, their basins eroded along the contact between hard metamorphic rocks of the Canadian Shield and more erodible sedimentary strata that overlie them. Before reaching the Rocky Mountains, the airplane passes over rolling morainal topography, dotted with numerous kettle lakes, that

FIGURE 19.25 Subdued landscape of Western Australia Shield with Tertiary laterite at surface.

The Evolution of Landscapes

a

b

FIGURE 19.26 Landscapes resulting from ice-sheet glaciation in North America. (*a*) Fjords along the northeastern coast of Baffin Island in the Canadian Arctic. (*b*) Large-scale linear topography west of Hudson Bay produced by erosion of bedrock at the base of the Laurentide Ice Sheet. (*c*) Pitted morainal topography dotted with kettle lakes formed near the margin of the retreating ice sheet west of Winnepeg, Manitoba.

c

formed near the margin of the retreating ice sheet (Fig. 19.26*c*).

The different landscapes along the plane's route are identifiable along other transects across the continent. Traced laterally, they are seen to form concentric zones (Fig. 19.27). In the north, the outermost zone is marked by no obvious glacial erosion or by local concentrated erosion that produced deep glacial troughs. Next comes an extensive interior zone in which most landscape features are attributable to ice scour. The belt of deepest scour lies near the outside of this zone. Beyond lies a broad depositional zone in which

end moraines and dead-ice deposits largely mask the underlying bedrock.

A model that attempts to explain this concentric landscape zonation relates each zone to conditions at the base of the former ice sheet (Fig. 19.28). In the far north, where little or no glacial erosion is evident, the ice sheet is presumed to have been very cold and frozen to its bed. Selective linear erosion is associated mainly with uplands where ice on high plateaus was cold and frozen at its base, but thicker ice in adjacent troughs was melting at its base and was able to erode. In the interior of the glacier, the ice was so thick that its great weight led to pressures high enough to produce basal melting. This warm basal ice was able to flow across and scour its bed. The ice became colder in the outer part of this zone where the glacier was thinner. As it began to freeze, debris was plucked off the bed, leading to increased erosion. In the more temperate environments along the western and southern margin of the ice sheet, the basal ice was again largely warm and melting. In this zone transported debris was deposited to form morainal topography.

The glacial landforms of the various zones we see are related largely to effects of the last glacia-

tion and are comparatively young features. The gross form of the land, however, is due to repeated glaciations during the past 2–3 million yr. Within the outer depositional zone are superimposed layers of glacial drift separated by weathering horizons. Each layer gives evidence of ice-sheet expansion and subsequent contraction. However, there are large gaps in this depositional record, so the exact number of ice-sheet invasions is not known. Some former ice sheets were as large as the last, but others apparently were smaller. The boundaries between erosional and depositional zones, therefore, are likely to have shifted from one glaciation to the next.

How much has this landscape been changed as a result of glaciation? Attempts have been made to estimate the total erosion since glaciation began, but the results vary considerably depending on

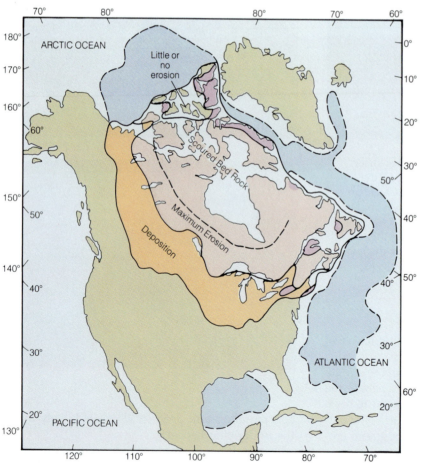

FIGURE 19.27 Map showing limit of last great Pleistocene ice sheet in eastern North America and concentric zones of deposition and erosion inside it. Zones offshore outlined by dashed lines represent principal regions where glacially eroded fine-grained sediments accumulated on the seafloor. (*Source:* After Bell and Laine, 1985.)

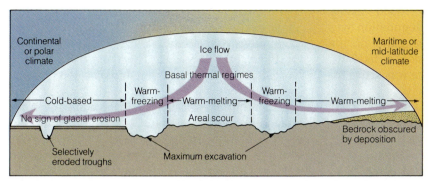

FIGURE 19.28 Diagrammatic profile through a continental ice sheet showing path of ice flow (bold arrows), conditions at the base of the glacier, and generalized erosional effects. Note the contrast in conditions along a margin having polar or continental climate with those at a margin marked by more temperate or maritime climate. (*Source:* After Chorley et al., 1984.)

what is measured. Estimates based only on the volume of sediment found within the depositional zone of the ice sheet clearly underestimate the total, for sediment was also carried beyond the glacial limit by meltwater streams and some was reworked by the wind. The bulk of the fine load reached the adjacent ocean basins where it accumulated on the seafloor (Fig. 19.27). If all the known sediment eroded from the area of the former ice sheet is summed, it totals close to 1.6 million km^3. This is the equivalent of eroding away a layer of solid rock 120 m thick from the glaciated

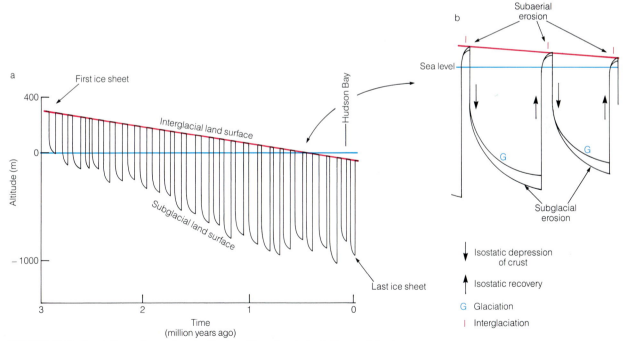

FIGURE 19.29 Isostatic and erosional effects of ice-sheet glaciation in North America. (*a*) Schematic model showing changes in altitude of land surface in Hudson Bay region due to isostatic depression of the crust, subglacial erosion, and isostatic recovery as ice sheets repeatedly form, reach a maximum size, and disappear. Net erosion over several million years has lowered land surface by several hundred meters. (*b*) Expanded view of two glacial–interglacial cycles. Subaerial erosion during interglaciations is minimal compared to erosion by ice during glaciations.

area. Of course, the rock was not eroded equally from all the landscape. Instead, denudation was probably greatest in the interior parts of the glaciated zone where glacial erosion is likely to have been most effective and the ice was active for the longest time. How much more than the average was removed from there is not known, but it is possible that the landscape has been eroded down at least several hundred meters since the first ice sheet formed (Fig. 19.29).

The present landscape of this large region of North America has developed, therefore, very late in geologic history. Although the average rate of denudation over the last 3 million years has been quite low, the actual rate has been quite variable, because long intervals marked by active glacial erosion and deposition alternated with shorter intervals when less-effective nonglacial agencies were operating (Fig. 19.29). If the last glacial cycle is representative of earlier ones, then during each a large portion of central Canada may have been continuously covered by glacier ice for many tens of thousands of years. The outer parts of the glaciated region were covered for shorter intervals when the ice sheets expanded to their greatest size. During interglacial ages, which may have averaged no more than about 10,000 years long, the land was largely free of ice. At such times, nonglacial processes did little to modify the landscape which remained, as it is today, largely a relict of the previous glaciation.

Landscape development in this region of fluctuating ice cover thus has been marked by pulses of erosion and deposition alternating with brief intervals of relative landscape stability. A landscape in long-term dynamic equilibrium is implied (compare Figs. 19.1 and 19.29a), but one in which locally significant changes in surface altitude are

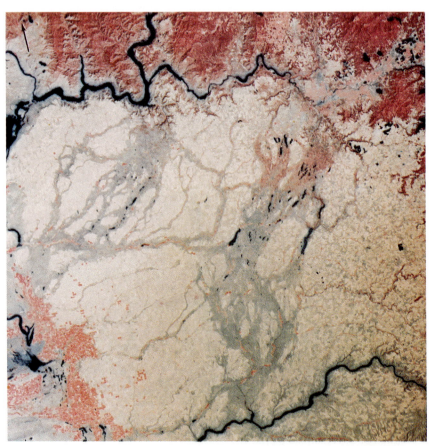

FIGURE 19.30 Satellite view of Columbia Plateau showing anastamosing channel system of Channeled Scabland where floodwaters have stripped away cover of light-colored loess from underlying dark-gray basaltic lava flows. Floodwaters traveled from the upper right to lower left. The scabland terrain is bordered on north by the Columbia River and on south by the Snake River.

FIGURE 19.31 Features attributable to catastrophic flooding in the Columbia Plateau region. (*a*) Dry Falls cataract system mid-way along Grand Coulee, one of the largest flood channels. (*b*) Enormous flood bar composed of coarse poorly sorted gravel. (*c*) Giant current ripples formed along a bend of the Columbia River. Composed of coarse gravel, the ripples are up to several meters high and their crests are as much as 100 m apart. (*d*) Large boulder transported by floodwaters and deposited beyond the mouth of Grand Coulee.

related to the isostatic effects of waxing and waning continental ice sheets.

Catastrophic Floods and the Channeled Scabland

Much of the Columbia River Basin in northwestern United States is underlain by numerous basalt flows that collectively form the extensive Columbia Plateau. During and since the final extrusive episode which ended about 6 million years ago, the plateau lavas have been regionally tilted, locally deformed into linear anticlinal and monoclinal ridges, and partly eroded away as the Cascade Range was uplifted along their western margin.

A rich topsoil developed in loess that overlies the basalt is artificially irrigated, making the plateau a prime wheat-producing region—but not all of it. A glance at a satellite image of the plateau in eastern Washington State shows that the pattern of checkered wheatfields is disrupted by an array of dark channel-like features where lava lies at the surface, stripped of its cover of loess (Fig. 19.30). Although they resemble a large braided stream system, the channels are largely dry and covered by desert scrub vegetation. In the 1920s, geologist J Harlen Bretz began a study of this curious landscape, which he named the Channeled Scabland. His investigation led to some startling conclusions and embroiled him in a heated controversy that lasted for many decades.

Bretz carefully documented the character and distribution of an assemblage of landscape features that provided evidence about the origin of the Scabland. They include dry coulees (canyons) with abrupt cliffs marking former huge cataract systems, plunge pools, potholes, and deep rock basins, all carved in the basalt (Fig. 19.31). Associated with the dry channelways are huge gravel bars containing enormous boulders, deposits of gravel in the form of gigantic current ripples, and upper limits of water-eroded land that lie hundreds of meters above valley floors (Figs. 19.31 and 19.32). Tributaries to many of the major channels contain gravel beds that slope upstream and fine-grained silts providing evidence of ponding.

Although Bretz considered alternative hypotheses to explain this array of erosional and depositional features, he was led inescapably to conclude that they could only be accounted for by a catastrophic event—a truly gigantic flood. The proposed flood, however, must have been far larger than any documented historic flood. The widespread erosional and depositional features attributed to it could be traced across the broad plateau and down the valley of the Columbia River to its mouth (Fig. 19.33). A simple calculation pointed to an enormous volume of water, the source of which was not initially apparent. However, the problem was resolved with the discovery that the continental ice sheet which covered western Canada during the last glaciation had advanced across the Clark Fork River and dammed a huge lake in the vicinity

of Missoula, Montana. Glacial Lake Missoula, as it came to be known, contained between 2000–2500 km^3 of water when at its maximum level and remained in existence only so long as the ice dam was stable. When the glacier retreated or began to float in the rising lake, the dam failed and the water emptied rapidly from the basin, as if a plug were pulled from a gigantic bathtub. The only possible exit route lay across the Columbia Plateau and down the Columbia River canyon to the sea.

There was much resistance to the flood hypothesis in the years following its introduction, but alternative hypotheses were discarded one by one as evidence steadily accumulated and Bretz' idea gained support. Much additional study throughout the Columbia River basin has generated new ideas and controversy. It is now generally agreed that not one, but many floods crossed the plateau; however, the exact number, timing, and geologic effects of these events have yet to be worked out in detail.

The probable sequence of events in the erosion of Scabland terrain has been reconstructed based on the landform characteristics and on experimental studies of stream erosion in simulated bedrock

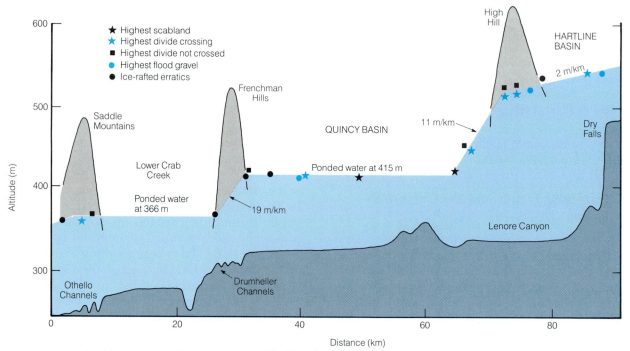

FIGURE 19.32 Profile across southwestern part of Scabland region showing reconstructed water depths during passage of floodwaters based on various kinds of field evidence. Level surface of water in Lower Crab Creek and in Quincy Basin indicates temporary ponding upstream from topographic constrictions at Saddle Mountains and Frenchman Hills. (*Source:* After Baker, 1973.)

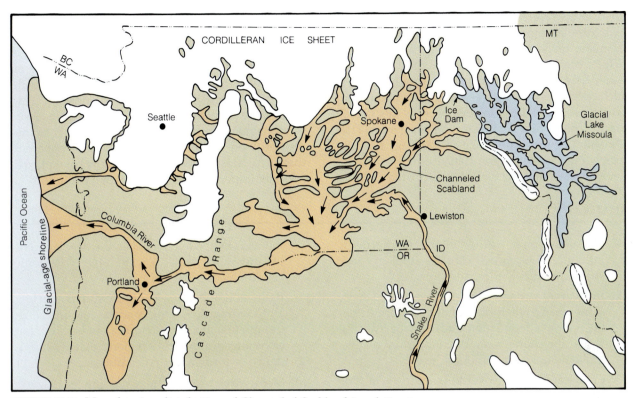

FIGURE 19.33 Map showing distribution of Channeled Scabland in relation to Glacial Lake Missoula, the ice dam on the Clark Fork River, and areas inundated by floodwaters along and adjacent to the lower Columbia River. At the time of flooding, the edge of the continental ice sheet lay south of the international boundary and the Pacific shore lay farther west due to worldwide lowering of sea level. (*Source:* After Baker and Bunker, 1985.)

that consists of a mixture of sand and clay. As high-velocity floodwater overtopped a drainage divide, it quickly removed the erodible loess that caps the basalt. The upper part of a lava flow, consisting of small, well-jointed columns, was probably eroded rapidly into elongate grooves oriented in the direction of water flow. As the rock was plucked out, underlying large columns of basalt were exposed and removed, forming huge potholes. With continued quarrying, prominent channels were eroded. In some of these, large cataract systems formed and migrated headward, causing further deepening of the channels. Most of the erosional landscape features of the plateau fall into one of these erosional stages.

The depositional landforms provide additional clues about the nature of the floodwaters. The giant ripples, for example, have been carefully measured and their height and width has been related to water depth and water-surface slope, as reconstructed from field measurements of upper

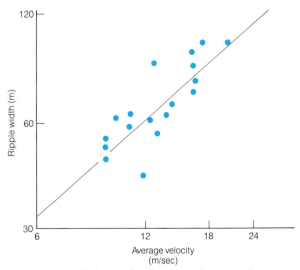

FIGURE 19.34 Relationship between the size of giant current ripples produced by Scabland floods and the average velocity of floodwaters. By measuring ripple width, an estimate of flood velocity can be obtained. (*Source:* After Baker, 1973.)

flood limits (Fig. 19.32). Standard equations used by hydraulic engineers then permitted estimates to be made of flow velocities during flooding. The results indicate that the giant ripples formed in water depths of 12–150 m, in places where water surface slope was between 1.3–5.6 m/km, and under average flow velocities of 9–18 m/s (32–65 km/h). The further demonstration of a general relationship between ripple size and average flow velocity (Fig. 19.34) means that estimates of flood velocity at a site can be obtained in the field by determining the average dimensions of ripples there.

Such investigations hold promise for gaining a more complete understanding of the dynamics of catastrophic flooding and the manner in which giant floods can radically alter a landscape. Although the prehistoric Scabland floods were of exceptional magnitude, the discovery of other examples of flood-sculptured terrain make it apparent that such catastrophic geologic events have played a locally important role in landscape evolution.

SUMMARY

1. The development of distinctive landforms is influenced by process, climate, lithology, rock structure, initial relief, and time.

2. In tectonic settings, high uplift rates and steep slopes lead to high rates of denudation. In stable regions, denudation rates may be low, but landscape changes can be initiated by relative movements of land and sea level.

3. Landscapes evolve over long intervals of time toward conditions of equilibrium. However, because landscapes experience continuous change, they tend to exist under conditions of steady-state or dynamic equilibrium.

4. The adjustment of a landscape to changing geologic or climatic conditions may be complex and involve many lags and minor readjustments.

5. A landscape in dynamic equilibrium may experience sudden change if a process operating on it achieves some critical threshold level, thereby leading to a new equilibrium condition.

6. Small landforms that develop over short time periods can be observed in the process of formation. The evolution of major landscape elements over long time intervals can be inferred by extrapolating modern rates of processes or by developing theories and models.

7. W. M. Davis' "cycle of erosion" attempted to explain landscape evolution as a series of stages through which an uplifted landmass was progressively reduced to a surface of low relief. Elaborations of this theory invoked the specific influences of lithology, structure, climate, and isostatic crustal movements on landscape development.

8. Landscape evolution is now being reassessed in terms of plate tectonics theory. Different models are needed to explain the development of landscapes under a wide range of tectonic conditions.

9. Reduction of landmasses by denudation to low altitude requires long periods of time. Although few modern examples can be cited, ancient unconformities of regional extent are evidence of former low-relief landscapes.

10. Major streams that drain the Qinghai-Xizang Plateau and cross the structural axis of the Himalaya apparently were able to maintain their courses by vigorous erosion and keep pace with the rapidly rising mountain system.

11. Major contrasts in landscape elements along the trend of the Andean Cordillera can be related to differences in climatic influences through 65° of latitude. Glaciation in the southern Andes resulted in an extensive fjord system, local eastward shift of the continental divide, and filling of the adjacent Peru-Chile Trench with sediments.

12. The Dead Sea occupies a linear depression resulting from motion along a transform fault that separates two small tectonic plates. Although present sedimentation rates are low, long-term subsidence has approximately equaled the rate of infilling of the basin.

13. The Hawaiian-Emperor chain of volcanoes has been built as the Pacific Plate traveled north, then northwest across a mid-plate hot spot. Island landscapes reflect the complex effects of volcanism, isostatic subsidence, and fluctuating climate and sea level. Volcanic islands ultimately subside beneath sea level to become coral atolls and then submerged seamounts as they move toward the Aleutian Trench.

14. Landforms of the Western Australia Shield are strongly controlled by lithology and structure.

Deep weathering and a widespread laterite of Tertiary age developed when the shield lay farther south in a zone of moister climate.

15. Concentric landscape zones in northeastern North America reflect conditions at the base of a vast ice sheet during the last glacial age. Landscape development over the last several million years has been marked by episodes of glacial erosion and deposition that alternated with intervals of relative landscape stability during interglacial times.

16. Large interconnected channels, huge abandoned cataract systems, giant current ripples, and other evidence of deep, flowing water show that the surface of the Columbia Plateau was sculptured by floodwaters released catastrophically during the late Pleistocene from a large ice-dammed lake in western Montana.

SELECTED REFERENCES

Landscape Evolution

Chorley, R. J., Schumm, S. A., and Sugden, D. E., 1984, Geomorphology: London, Methuen.

Embleton, C., 1984, Geomorphology of Europe: New York, John Wiley & Sons.

Hunt, C. B., 1974, Natural regions of the United States and Canada: San Francisco, W. H. Freeman and Co.

The Roof of the World

Liu Dong-sheng, ed., 1981, Geological and ecological studies of the Qinghai-Xizang Plateau: Beijing, Science Press (New York, Gordon and Breach), 2 vols.

Wager, L. R., 1937, The Arun River and the rise of the Himalaya: Geographical Journal, v. 89, p. 239–250.

Andean Cordillera

James, D. E., 1973, The evolution of the Andes: Sci. American, v. 229, p. 60–69.

Zeil, Werner, 1979, The Andes, a geological review: Berlin, Gebrüder Borntraeger.

Dead Sea Depression

Freund R. and Z. Garfunkel, eds., 1980, The Dead Sea Rift: Tectonophysics, v. 80, p. 1–303.
Gerson, R., Grossman, S., and Bowman, D., 1985, Stages in the creation of a large rift valley—geomorphic evolution along the southern Dead Sea Rift, *in* Morisawa, M. and J. T. Hack, eds., Tectonic geomorphology: Boston, Allen and Unwin, p. 53–73.

Volcanoes in the Sea

Clague, D. A., and Dalrymple, G. B., 1987, The Hawaiian-Emperor volcanic chain. Part I. Geologic evolution, *in* Decker, R., T. L. Wright, and P. H. Stauffer, eds., Volcanism in Hawaii: U.S. Geol. Survey Prof. Paper 1350, p. 5–54.

Macdonald, G. A., Abbott, A. T., and Peterson, F. L., 1983, Volcanoes in the sea: the geology of Hawaii, 2nd ed: Honolulu, Univ. Hawaii Press.

Western Australia Shield

Goode, A. D. T., 1981, Proterozoic geology of Western Australia, *in* Hunter, D. R., ed., Precambrian of the Southern Hemisphere: New York, Elsevier, p. 105–204.

Rutland, R. W. R., 1981, Structural framework of the Australian Precambrian, *in* Hunter, D. R., ed., Precambrian of the Southern Hemisphere: New York, Elsevier, p. 1–32.

Glaciated Landscapes of Northeastern North America

Bell, M., and Laine, E. P., 1985, Erosion of the Laurentide region of North America by glacial and glaciofluvial processes: Quaternary Research, v. 23, p. 154–174.

Sugden, D. E., 1978, Glacial erosion by the Laurentide Ice Sheet: Journal of Glaciology, v. 20, p. 367–391.

Glacial Floods and the Channeled Scabland

Baker, V. R., and Bunker, R. C., 1985, Cataclysmic late Pleistocene flooding from Glacial Lake Missoula: A review: Quaternary Science Reviews, v. 4, p. 1–41.

Baker, V. R., and Nummendal, D., 1978, The Channeled Scabland: Washington, D.C., National Aeronautics and Space Administration.

C H A P T E R 2 0

Climate and Climatic Change

Evidence of changing climate in Kyoto, Japan, since 812 A.D. is documented in records of the annual cherry blossom festival, for the time of blooming is closely related to spring temperature.

PRESENT AND FORMER CLIMATES

Scientists have long puzzled over the occurrence of geologic features that seem out of place in their present climatic environment. Abundant fossil bones and teeth of hippopotamus—of the same kind that lives in East Africa today—have been found in interglacial deposits of southeastern England. Herds of hippos lived there about 100,000 years ago under conditions that may have been something like those of modern subtropical Africa. At many sites beyond the margin of the Great Lakes in north-central United States, plant remains have been found that show this region formerly resembled arctic landscapes like those now found in far northern Canada. In ancient limestones thrust up into rugged mountain peaks in Arctic Alaska are found fossil corals whose modern descendants live in warm tropical waters. Geologists exploring the small scattered ice-free areas of Antarctica have discovered beds of coal containing fossils of plants that must have required a warm, moist climate for their survival. In each of these cases, a significant change in local climate apparently has taken place so that the present biota living in these areas is quite different from the fossil forms we see preserved in the geologic record.

One of the chief differences between present and former conditions on the Earth is **climate,** *the average weather conditions of a place or area over a period of years.* Climate is determined by a variety of factors, but those of greatest interest are temperature, precipitation, cloudiness, and windiness. The biosphere adjusts closely to climate, so that in each climatic region is found a distinctive assemblage of animals and plants (Fig. 20.1). Although in some areas the animals and plants apparently are in equilibrium with the climate, in others a recent change of climate has caused some species to leave, to die out, or to invade. In such areas the present population of plants and animals may not represent an equilibrium assemblage, but rather one that is still adjusting to environmental change.

Besides fossils, other sorts of anomalous geologic features give evidence of climatic change. A list would include (1) glacial features in warm

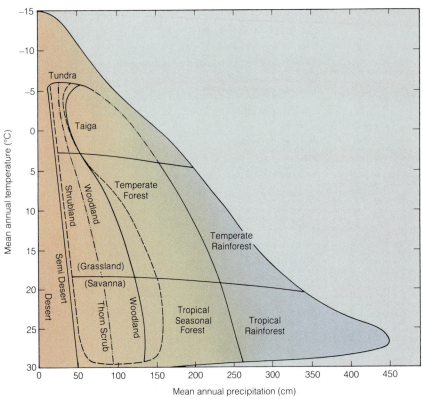

FIGURE 20.1 Principal world vegetation communities determined by temperature and precipitation. Boundaries between communities are approximate. Dashed line encloses a wide range of environments in which grassland or woody plants may form the dominant vegetation. (*Source:* After Whittaker, 1975.)

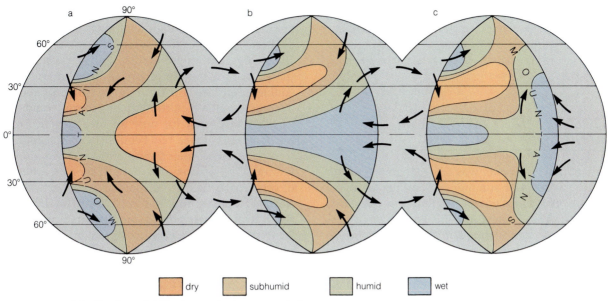

FIGURE 20.2 Idealized model Earth in which principal climatic zones are controlled by latitude and geography. Continents are shown without mountains (*b*), and with mountains on west (*a*) and east (*c*) coasts. Arrows show directions of prevailing winds. (*Source:* After Bambeck et al., 1980.)

lands, (2) desert sand dunes now covered with stablizing vegetation, (3) beaches and other features of extensive former lakes in dry desert basins, (4) systems of stream channels that are now dried up, (5) remains of dead trees at positions above the present upper treeline in a mountainous area, and (6) surface soils that are incompatible with the present climate.

Climate and climatic change have become important subjects of scientific study, especially because of the potential impact that changes of climate can have on human affairs. Study of climatic change involves scientists from many different fields—atmospheric sciences, geology, geophysics, biology, and oceanography, to name only a few. Because most of the evidence is geologic, the study of former climates is largely a geologic problem. If we hope to predict future changes of climate, then we must try to gain an understanding of the pattern and causes of climatic changes in the past.

As we have seen, solar heat warms the continents and oceans, and drives winds and ocean currents that distribute the heat (Fig. 12.2). Surface seawater in the subtropics is evaporated and carried by winds toward higher latitudes. There the vapor condenses, forms clouds, and is precipitated as rain and snow. The arrangement of climatic zones in belts that approximately parallel lines of latitude is interfered with and distorted by the pat-

tern of oceans, landmasses, high mountains, and plateaus (Fig. 20.2). Therefore, average temperature and precipitation vary greatly from one place or region to another. When a change, however slight, occurs in this vast and complex system, local climates in many areas are likely to be affected.

The understanding of climatic change is made especially difficult because of the complexity of its interacting components (Fig. 20.3). The Sun's heat drives the system and constitutes a primary external factor that can perturb it, but at the same time such internal components as the atmosphere, biosphere, hydrosphere, cryosphere, and lithosphere can all experience changes that will likewise upset the climatic balance. Some factors operate on short time scales whereas others may have a noticeable effect only over long periods of time. A single factor, operating independently, may have little impact, but in concert with others it may trigger a change or enforce a trend that can lead to a significant climatic effect.

In this chapter the focus will be on evidence from the recent geologic past. Such evidence is more abundant and easier to decipher than that from more ancient times. The discussion emphasizes the glacial ages, which provide dramatic proof that the Earth has experienced very different environmental conditions during the period when the human species was developing and spreading across the surface of the land. This record of changing climate provides the basis for testing physical mechanisms responsible for climatic changes on the Earth, changes that influence and control the importance of many surface geologic processes.

EVIDENCE THAT CLIMATES ARE CHANGING

The fact that last winter was colder than the winter before, or that last summer was rainier than the previous one does not prove that the climate is changing. The recognition of climatic variations must be based on changes in average conditions over a number of years, of which no two are ever identical. Several years of abnormal weather may not mean that a change is occurring, but trends that persist for decades may signal a shift to a new climatic state.

Figure 20.4 shows a continuous record of temperature and annual snowfall measured at Great St. Bernard Pass in the Swiss Alps during the past 100 years. Both temperature and snowfall fluctuated approximately in phase, with intervals of cool temperature corresponding to times of above-average snowfall. Although short-term trends in this record persist for only about a decade, over the long term there has been a slight but general trend toward warmer temperature and less snowy winters. The temperature pattern is representative of other parts of the Northern Hemisphere where average temperature also rose after the 1880s to reach a peak between about 1940 and 1950 (Fig. 20.5); thereafter, temperature began to fall. The amplitude of the change, amounting to some 0.6° C, is about a tenth of that which occurred between the peak of the last glacial age and the present interglaciation, yet its effects were seen widely. During those six decades, mountain glaciers in

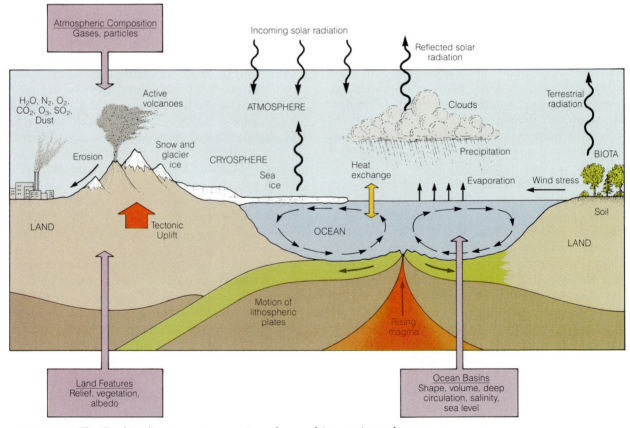

FIGURE 20.3 The Earth's climate system consists of several interacting subsystems: the atmosphere, cryosphere, oceans, land, and biota. Solar energy drives the system, although some of the incoming radiation is reflected back into space from clouds, snow, ice, and atmospheric pollutants. Tectonic movements affect surface relief and the geometry of continents and ocean basins, while volcanic and industrial gases affect atmospheric composition. (*Source:* Modified from Gates, 1979.)

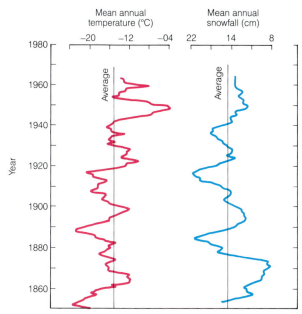

FIGURE 20.4 Variations of mean annual temperature and snowfall recorded at Great St. Bernard Pass in the Alps since the middle of the nineteenth century. Vertical lines through the plots show long-term average values. (Data from Janin, 1970.)

most parts of the world shrank in size, some conspicuously (Fig. 20.6). Sea ice was observed less frequently off the coast of Iceland.

The biosphere, as one might expect, was also affected. From 1880 to 1940, the summer growing season increased in length and crop yields generally improved. Territories normally occupied by various plant and animal species expanded slightly toward the poles. The common codfish, for example, was almost unknown in Greenland before the twentieth century, but it then began to migrate northward. Year by year it was observed at places farther north until within a period of only 27 years its limit had extended northward through 9° of latitude.

That the world's climates can change detectably within a human lifetime is a relatively new realization. With it has come increasing concern about the impact of such changes on nature and on society, as well as the possible impact of human activities on the Earth's climate.

CLIMATIC CHANGES DURING THE LAST THOUSAND YEARS

Historic Records

In trying to reconstruct climatic fluctuations of the last millennium, instrumental records can be used

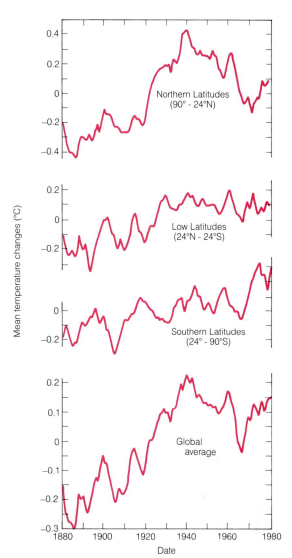

FIGURE 20.5 Surface air temperature trends for three latitude bands and for the entire Earth during the last 100 years. Plots have been smoothed mathematically to remove "noisy" year-to-year fluctuations. Because the curves were compiled mainly from land-based records, they probably provide only a general approximation of actual temperature trends for this period, especially in the dominantly oceanic Southern Hemisphere. (*Source:* After Hansen et al., 1981.)

only for the very latest part of that time. To extend the reconstruction farther back we must rely on what are termed *proxy records, information that can act as a substitute for data that may be unobtainable.* Such records, although lacking the precision of instrumental data, often confirm one another and collectively provide a good general picture of climatic trends (Fig. 20.7). However, these records are not uniformly distributed geographically but come mainly from the densely populated middle latitudes of the Northern Hemisphere. Some of the longest and most informative series include, for example, (1) dates for the blooming of cherry trees in Kyoto, Japan, since 812 A.D., (2) the number of severe winters in China since the sixth century A.D., (3) variations in the occurrence of sea ice at the coast of Iceland since 860 A.D., (4) the frequency of dust falls in China since 300 A.D., (5) the height of the Nile River at Cairo since 622 A.D., (6) wheat prices (a reflection of climatic adversity) in England, France, the Netherlands, and northern Italy since 1200 A.D., and (7) the quality of wine harvests in Germany since the ninth century A.D. Each of these phenomena reflects prevailing climate to a greater or lesser degree, and therefore, each is regarded as a useful proxy indicator of climate.

Isotopes and Tree Rings

A further source of paleoclimatic information comes from ice cores collected from polar glaciers. Careful laboratory measurements of the ratio of

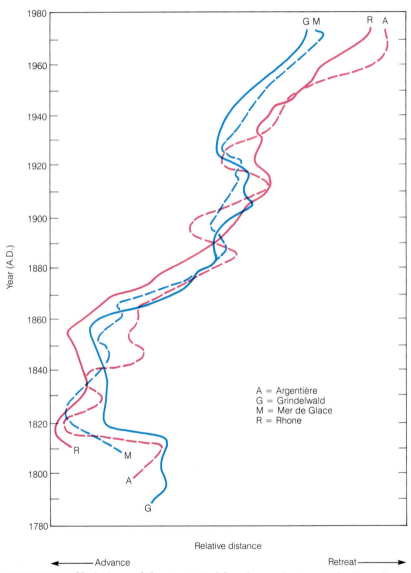

FIGURE 20.6 Variations of the termini of four large glaciers in the French and Swiss Alps since the beginning of the nineteenth century. Main episodes of advance and retreat are common to all the glaciers. Minor differences in the curves reflect local topographic and climatic factors that influence the mass balance of each glacier. Pronounced glacier recession since the middle 19th century coincides with a significant warming trend during this same interval (compare with Fig. 20.5).

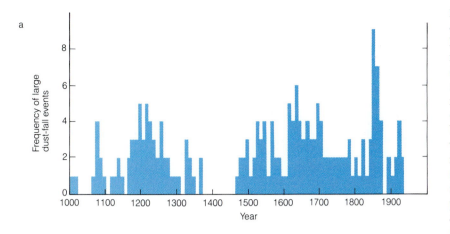

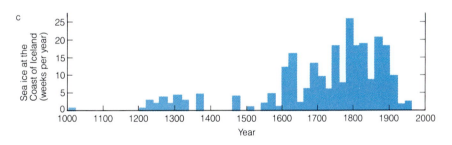

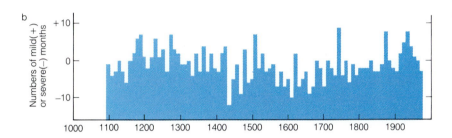

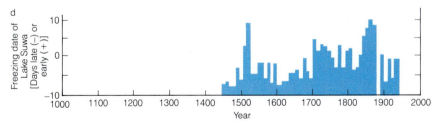

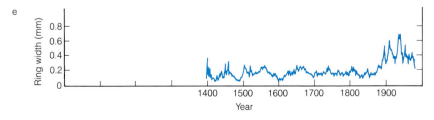

FIGURE 20.7 Climate proxy records spanning all or part of the last 1000 years: (*a*) frequency of major dust-fall events in China (*Source:* After Zhang, 1982); (*b*) severity of winters in England, recorded as frequency of mild or severe months (*Source:* After Lamb, 1977.); (*c*) number of weeks per year during which sea ice reached the coast of Iceland (*Source:* After Lamb, 1977.); (*d*) freezing date of Lake Suwa, Japan, relative to long-term average (*Source:* After Lamb, 1966.); (*e*) ring-width variations for trees in northern Labrador (*Source:* After Payette et al., 1985.).

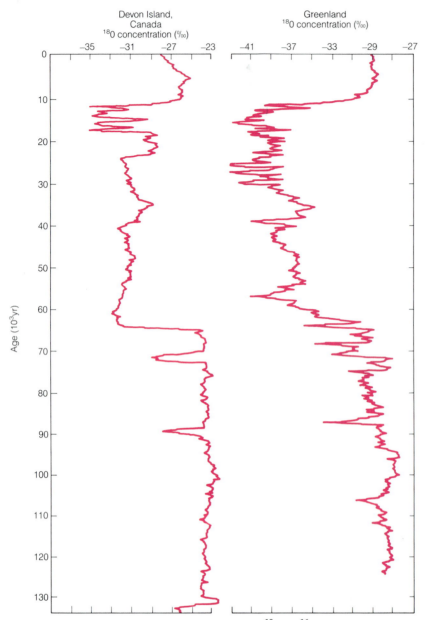

FIGURE 20.8 Variations in ratio oxygen isotopes ^{18}O to ^{16}O through the Devon Island Ice Cap and the Greenland Ice Sheet. Intervals of strong negative values between about 65,000 and 10,000 years ago in each glacier encompass the last glacial age. The sharp rise in the curves about 10,000 years ago marks an abrupt change from glacial to interglacial climatic conditions. (*Source:* After Paterson et al., 1977)

two isotopes of oxygen (^{18}O and ^{16}O) that make up part of the molecules of glacier ice can tell us the air temperature at the time when the ice reached the glacier surface as snow, for the isotope ratio depends on air temperature. Cores obtained from the Greenland and Antarctic ice sheets, as well as from several smaller ice caps, provide continuous records of fluctuating temperatures near the surface

of these glaciers, in some cases extending back many tens of thousands of years (Fig. 20.8).

Trees offer additional important information about past climates. Trees living in middle and high latitudes typically add a growth ring each year, the width and density of which reflect the local climate (Fig. 20.9). Many species live for hundreds of years. A few, like the Giant Sequoia and

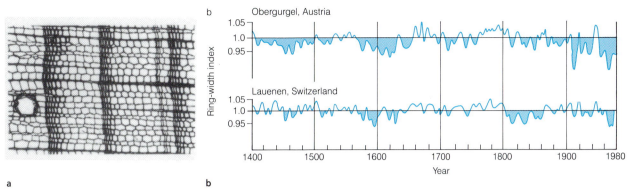

b Obergurgel, Austria

Ring-width index

1.05
1.0
0.95

Lauenen, Switzerland

1.05
1.0
0.95

1400 1500 1600 1700 1800 1900 1980

Year

a   b

FIGURE 20.9 (*a*) Enlarged cross section of a 1500-year-old fossil larch tree found in the moraine of a Swiss glacier and showing annual growth rings. Early wood of each year consists of large, well-formed cells. Late wood contains smaller, closely spaced cells. (*b*) Tree-ring chronologies based on wood density measurements of spruce trees at two sites in the Alps that are 200 km apart. The index is a measure of growth relative to long-term average growth (1.0). A general similarity can be seen among periods of low growth (shaded) that correspond with times of cold climate and glacier expansion.

Bristlecone pine of the California mountains, can live for thousands of years. Mathematical techniques permit specialists in tree-ring analysis to reconstruct temperature and precipitation patterns from tree rings over broad geographic areas for specific years in the past. These provide a picture of changing weather patterns for regions of continental or subcontinental size.

The Little Ice Age

Old lithographs and paintings of Alpine valleys from the eighteenth and early nineteenth centuries show glaciers terminating far beyond their present positions (Fig. 20.10), sometimes close to farmland and buildings that date to earlier centuries. Documents from the sixteenth and seventeenth centuries describe glaciers that advanced over small villages and became larger than at any time in human memory. Similar ice advances took place in other parts of the world and in many cases led to the greatest expansion of glaciers since the end of the last glacial age, some 10,000 years earlier. The climatic change that heralded these conditions apparently occurred in the thirteenth century, bringing an end to an era of relatively mild climate during the Middle Ages. *The interval of generally cool climate between the middle thirteenth and middle nineteenth centuries, during which mountain glaciers expanded worldwide*, is commonly referred to as the **Little Ice Age.** Throughout much of western Europe the climate of this interval was punctuated by frequent episodes marked by snowy winters

a

b

FIGURE 20.10 (*a*) Lithograph showing Lower Grindelwald Glacier in 1808 advancing during the last phase of the Little Ice Age. (*b*) Lithograph of same glacier in 1826. Ice now reaches floor of main valley and is beginning to retreat from its maximum stand. (*Source:* After Röthlisberger, 1980.)

and cool, wet summers that led to many crop failures. In Iceland, widespread abandonment of farms occurred throughout the eighteenth century, large numbers of people died of starvation and disease related to malnutrition, and the crime rate rose sharply during severe years.

During the coldest parts of the Little Ice Age, temperatures in middle latitudes averaged about 1°–2° C colder than now. As mountain glaciers advanced, the sea-ice cover expanded in the North Atlantic, and there were series of winters when rivers and canals in Holland and England froze over, enabling large numbers of people to skate on them. In one winter the western Baltic Sea froze so solidly that travel between Germany and Sweden was possible by wheeled vehicles.

Little Ice Age conditions persisted until the middle of the last century when a general warming trend began that caused mountain glaciers to re-

FIGURE 20.11 Skeletons of Greenland colonists who probably died early in the fourteenth century. Excavated from a churchyard near Godthåb at latitude 64° North.

treat and many coastal glaciers to calve back rapidly, exposing long, deep fjords (Fig. 13.16). At the same time, the edge of the sea ice retreated northward in the North Atlantic Ocean. Although minor fluctuations of climate continued to take place, the overall trend of increasing warmth brought conditions that were increasingly favorable for crop production in middle latitudes at a time when the human population was expanding rapidly and entering the Industrial Age.

The Greenland Colonies

The onset of the Little Ice Age is illustrated dramatically by the history of the Norse colonies in Greenland. In the summer of 985, Erik the Red, father of Leif Ericson, sailed up a fjord in southwest Greenland with several hundred emigrants from Iceland who brought with them their livestock and household goods. They established two colonies that later numbered 4000 or more people. They raised sheep and cattle, exported butter and cheese, and even tried unsuccessfully to grow grain. Trade routes were well established between Greenland and Iceland, and voyages were made to Norse settlements in the New World. In the earlier years, floating ice was not mentioned, but later, ice began to appear. Sea travel grew increasingly difficult and sailing directions were altered. The growing season became shorter and cooler, with serious affects on farming. By 1350 the more northerly colony could not hold out and was abandoned. The final reference to the colonies was made in 1410 by an Icelander who took what may have been the last ship from Greenland.

The stone houses and churches of the colonies remain. Scientific excavation of more than 100 burials (Fig. 20.11), in ground now perennially frozen, has helped to reconstruct the story. Early burials were deep, and in coffins made of imported woods; as the colony became poorer, later burials were shallow, and in shrouds only. The medieval cloaks that served as shrouds were pierced by roots of plants that in the subsequent cold years could not have grown there because the burials and the ground enclosing them were solidly frozen. The later buried skeletons are of young people, suggesting a short life expectancy; all are less than 160 cm (5 ft 3 in) in height and are often misshapen, suggesting a deficient diet; teeth are worn down extraordinarily, indicating that the people probably ate coarse vegetable food. The skeletons are not at all like those of their healthy Norse ancestors who died during the twelfth century. Overtaken by the Little Ice Age, the later people must have endured appalling living conditions that were marginal at best. Apparently the late survivors died of starvation or were so weakened that they could not withstand illness. With the passing of the Little Ice Age, conditions in southern Greenland have improved and parts of the land are once again suitable for farming.

Physical Changes During the Last Millennium

Assembling the varied evidence of Northern Hemisphere climatic fluctuations over the past thousand years, we can say that mild temperatures early in the present millennium gave way about 700 years ago to temperatures that averaged about 1° to 2° C lower. Various paleoclimatic information shows that the change to colder climates was accompanied by other recognizable changes that affected the atmosphere, oceans, and cryosphere. For example, the general atmospheric circulation was intensified (winds became stronger). In the belt of westerly winds, established storm tracks shifted toward the equator by 5° to 10° of latitude, bringing a wetter climate to western Europe. The area of floating sea ice in the Arctic Ocean expanded southward into the North Atlantic, and the path of the Gulf Stream shifted southward. The snowline fell about 100 to 150 m causing mountain glaciers throughout the world to expand. These changes fit a consistent pattern and are what might be expected to occur during the onset of a glacial age.

POSTGLACIAL CLIMATIC CHANGES

Thus far we have looked at climatic change on time scales of decades and centuries. We come now to changes more difficult to reconstruct, those measured in thousands of years. Such changes have taken place since the end of the last glacial age, an interval referred to as postglacial time. A date of about 10,000 years is a convenient one to use for the end of the last glaciation. Of course the date is arbitrary, for the transition from glacial to postglacial conditions was gradual and took place at different times in different latitudes. By 10,000 years ago, however, the world was clearly at the threshold of an interglacial age, and many environments were undergoing rapid and significant changes.

Evidence from Plant Fossils

In the late nineteenth century, at a time when the glacial ages were still very poorly understood, it was vaguely supposed that the cold climate of glacial times must have given way to milder climate that warmed continuously until temperatures reached those of today. That this supposition was wrong was first revealed by studies of fossil plant remains in Scandinavia. Plant leaves and stems, preserved in postglacial sediments of ponds and bogs, were discovered and identified. Later, tiny fossil pollen grains were systematically analyzed. Although large plant fragments permit identification to the level of individual species, they are not common in geologic deposits. By contrast, fossil pollen grains are numerous and possess a hard waxy coating that resists destruction by chemical weathering. Most pollen is transported by the wind and settles in lakes, ponds, and bogs where it is trapped and protected from destructive oxidation in the wet environment. Mixed with other plant remains and mineral sediment, the pollen forms part of the strata that slowly accumulates year by year (Fig. 20.12).

Pollen has several advantages for paleoclimatic studies. It permits identification of the plants from which it comes, although commonly only to the level of genus. Furthermore, generally being enclosed in sediment rich in organic matter, the age of pollen-bearing layers often can be determined by radiocarbon dating. A sample taken by a tubular coring device driven into the sediments yields a vast number of pollen grains that can be identified by types, counted, and the results treated statistically. In this way, the pollen grains reveal the particular assemblage of plants that flourished nearby when the enclosing sediment layers were deposited. If a modern vegetation assemblage can be found which has a composition like that suggested by a pollen count, then the precipitation and temperature at the site of the modern assemblage can be used to estimate climatic conditions represented by the fossil assemblage.

Botanists who have studied sediments from lakes and bogs in Scandinavia and in the glaciated region of the northern United States and southern Canada have found that at most sites the organic-rich layers rest on till or stratified drift left after retreat of the last ice sheet. The postglacial pollen-bearing sediments in many cases extend to the surface of the deposit where they are still accumulating. Samples taken at close intervals from bottom to top provide a continuous history of environ-

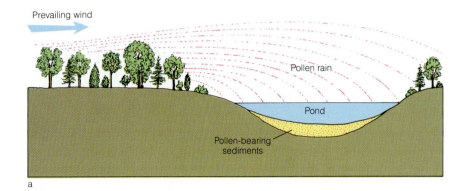

Prevailing wind

Pollen rain

Pond

Pollen-bearing
sediments

a

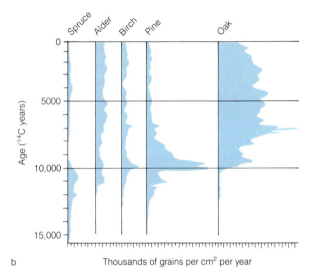

b

Thousands of grains per cm² per year

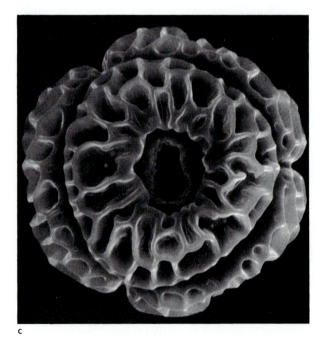

c

FIGURE 20.12 Fossil pollen used to reconstruct past vegetation and climate. (*a*) Fallout of windborne pollen grains derived mostly from nearby trees into a pond where they are incorporated as part of the accumulating strata. (*b*) Simplified pollen diagram prepared from data collected from Rogers Lake, Connecticut. (*Source: After Davis, 1983.*) Variations in pollen influx as a function of time show changes in forest composition. A major change occurred about 10,000 years ago when early spruce/pine forest was replaced by a forest dominated by pine and deciduous trees. (*c*) Scanning electron microscope photograph of a grain of *Drymis winterii* pollen, having a diameter of 42 microns.

mental change since the glacial age. The results, when first deciphered, were unexpected. Instead of a slow, progressive postglacial warming, several important fluctuations of climate were detected, including a time when average temperatures were even higher than today. This ***Hypsithermal interval*** (Greek for "high heat") was *a period of warm postglacial climate that reached its peak in many areas between about 8000 and 6000 years ago* (Fig. 20.13).

An interesting yet perplexing feature of many pollen records is that reconstructed vegetation assemblages at some levels have no modern counterparts. Apparently many forests in postglacial Europe and North America looked quite different from the natural forests of today. The reason seems to lie in the different rates at which various tree species expanded into new territory from their ice-age refuges. This resulted in continual changes in forest composition (Fig. 20.14). As a result, it is

not always possible to reconstruct former climates confidently from fossil plant assemblages, for we may lack modern examples for comparison.

The most sensitive records of vegetation changes generally come from boundary zones where two types of vegetation meet, or from near the latitudinal or altitudinal limits of trees. In northern Canada, west of Hudson Bay, radiocarbon-dated samples of pollen-bearing lake sediments show that the boundary between forest and treeless tundra shifted well north of its present position during the Hypsithermal interval, but then moved south as the climate again cooled. A similar and contemporaneous shift of vegetation occurred in Minnesota where prairie species expanded eastward replacing forest. After reaching its greatest extent about 6000–7000 years ago, the prairie then retreated westward as the climate became cooler and moister (Fig. 20.15).

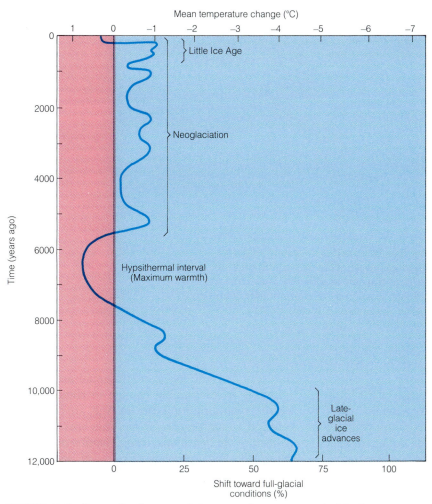

FIGURE 20.13 Generalized curve of changes of temperature in Northern Hemisphere inferred from fluctuations of snowline and of vegetation communities in the mountains.

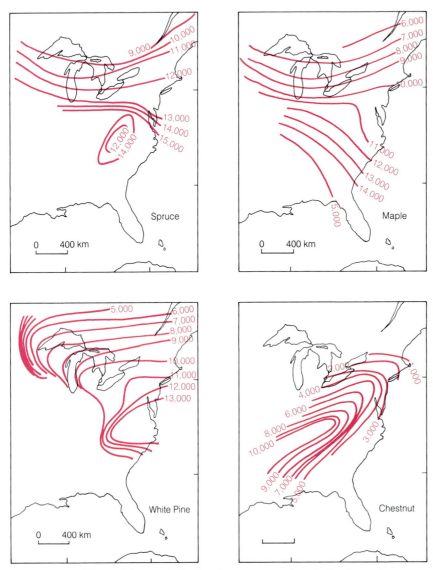

FIGURE 20.14 Time of first arrival of various tree species at sites in eastern North America (years ago) during and after retreat of the last continental ice sheet. Forest species migrating generally northward at different rates crossed parallels of latitude at quite different times, giving rise to continuously changing vegetation assemblages. (*Source:* After Bernabo and Webb, 1977.)

Neoglaciation

Moraines often seen just beyond those of the Little Ice Age represent earlier glacier expansions. Generally, they can be distinguished from the younger deposits by their denser vegetation cover, deeper soils, and more subdued form. Radiocarbon dating of buried wood and soils shows that such episodes occurred about 1200–1000 years ago, 3300–2400 years ago, and 5800–4900 years ago (Fig. 20.13). At those times glaciers were about the same size as during the Little Ice Age, indicating that the snow-line was lowered by a nearly equal amount and implying a comparable change in climate. These events, including the Little Ice Age, are collectively referred to as *Neoglaciation.* This was *an interval postdating the time of maximum Hypsithermal warmth when, following a period of ice recession, many glaciers were regenerated and others expanded in size.* Because Neoglacial moraines have been found in nearly all the world's mountain ranges, the climatic changes that led to their formation must have been worldwide events.

Lake-Level Fluctuations

Lakes in closed basins rise and fall in response to changes of climate. Chronologies of lake-level fluctuations based on radiocarbon dating indicate dissimilar histories for different climatic zones. Most lakes in tropical Africa were low during the last glaciation, implying arid conditions (Fig. 20.16). They then rose to their highest levels during early postglacial time, suggesting that precipitation increased more than enough to offset higher evaporation rates in a time of warmer climate. By contrast, lakes in the temperate zone of the western United States were high during glacial times, probably a result of reduced evaporation rates under cooler climatic conditions. Their levels subsequently fell during the Hypsithermal interval as the climate of that region became warmer and drier.

THE LATEST GLACIAL AGE

Glaciers and Permafrost

Before 10,000 years ago, the Earth for many millennia was in the last glacial age. It was a world that was very different from the one with which we are familiar. As the climate cooled about 25,000 years ago, an extensive ice sheet that had formed over north-central Canada began to expand south toward the United States and west toward the Rocky Mountains (Figs. 13.37 and 20.17). Moving across the Great Lakes region, the ice margin overwhelmed spruce trees that grew in scattered groves in a frigid tundra landscape and buried their remains in till. Dozens of ancient logs found exposed in the sides of stream valleys are bent and twisted, indicating that they were alive when the glacier destroyed them. Some retain their bark, and some lie pointing in the direction of ice flow, like large aligned arrows. A radiocarbon date of the outermost wood in a log tells us the approximate time when the ice arrived and the tree was killed. These dates are consistent with the places of occurrence of the dated logs; that is, they become younger from north to south. Dates for wood found near the extreme southern limit of the ice sheet fall between 21,000 and 18,000 years. Dividing the distance between the localities of two successive samples by the difference between their dates yields an average rate of advance of the ice margin across that distance. Results from many such pairs of dates suggest that the ice was moving forward at between 25 and 100 m/yr. These rates are comparable to those of some existing glaciers, which gives us confidence that the calculated results are reasonable.

Similar events were occurring simultaneously in northern Europe where another great ice sheet spread southward and overwhelmed the landscape (Fig. 20.18). Other large ice sheets expanded over arctic regions of North America and Eurasia, including some areas now submerged by shallow

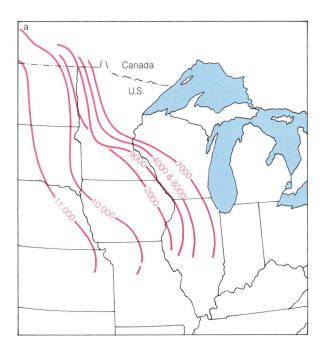

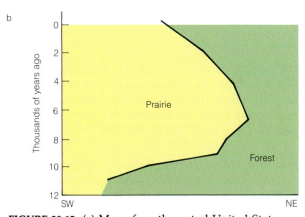

FIGURE 20.15 (*a*) Map of north-central United States showing changes in position of prairie/forest border during the last 11,000 years (ages are in years before present). (*b*) Curve showing northeastward expansion and subsequent retreat of prairie southwest of the Great Lakes. (*Source:* After Bernabo and Webb, 1977.)

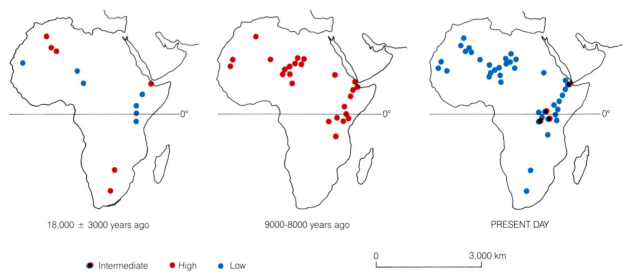

18,000 ± 3000 years ago   9000-8000 years ago   PRESENT DAY

0   3,000 km

●◐ Intermediate   ● High   ● Low

FIGURE 20.16 Lake levels in tropical Africa were low during dry full-glacial times about 18,000 years ago, but rose to high levels during the early Holocene as climate became wetter. A subsequent shift to drier conditions caused lakes to fall to present low levels. (*Source:* After Street and Grove, 1976.)

polar seas, and over the mountainous cordillera of western Canada (Fig. 13.37). The ice sheets in Greenland and Antarctica expanded as falling sea level exposed areas of the surrounding continental shelves. Glacier systems also grew in the world's major mountain ranges, including the Alps, Andes, Himalaya, and Rockies, as well as in numerous smaller ranges and on isolated peaks scattered widely through all latitudes.

We can assume that ice shelves must also have existed under full-glacial conditions, but their size and distribution are not easy to determine. Some geologists postulate that an ice shelf may have covered all of the Arctic Ocean and extended south into the northern reaches of the Atlantic Ocean, thereby linking the major northern ice sheets into a continuous glacier system that mantled nearly all of the arctic and much of the subarctic regions of our planet. If this view is correct, a well-equipped and hardy traveller in those days might have stepped onto the glacier in southern Illinois and, heading north across the pole, journeyed continuously over thick glacier ice for more than 9000 km, all the way to its limit in northern Germany or Poland. Other geologists concede that ice shelves very likely were present locally in favorable places, just as they are today in Antarctica, but suggest that the polar sea was largely covered by thin sea ice that may have extended far south of its present limit into the North Atlantic.

With the southward spread of ice sheets on the

northern continents, periglacial zones were displaced to lower latitudes and lower altitudes. Ice-wedge casts and other indicators of frozen ground that are found widely in Britain south of the glacial limit point to widespread permafrost and a cold periglacial climate. Similar features noted throughout central Europe between the limit of the northern ice sheet and the Alpine glaciers indicate harsh full-glacial conditions. In the Soviet Union, permafrost extended far south of its present limit in a zone reaching 1000 km or more beyond the ice sheet. However, in North America evidence of full-glacial permafrost is restricted largely to Alaska, to a generally narrow belt adjacent to the southernmost limit of glacial drift in the Great Plains and Great Lakes regions, and to the high mountains of the American West, especially the Rockies (Figs. 20.17 and 13.32). This contrast may largely reflect the fact that the massive Eurasian glacier lay north of 50° latitude, whereas the ice sheet over central North America extended south of 40° into more temperate latitudes. The periglacial zone was therefore much narrower in the United States because the north-to-south gradient of climate there was far steeper.

Vegetation Zones

Much of our knowledge of environmental conditions outside the great ice sheets during full-glacial times is based on interpretation of fossil vegetation, but the picture is far from complete. The number of sites from which plant fossils of this age have been sampled is reasonably large for Europe, but reconstructions for North America are based on a far smaller sample. In many parts of the world, sites are so few that useful conclusions are difficult to draw.

The vegetation pattern in eastern North America, before European immigrants changed it radically, consisted of several subparallel belts that mainly reflect the gradual increase of temperature from the pole toward the equator. Superimposed

on this latitudinal pattern is a change from moist forest in the east to dry grassland in the west. In the Far West, a complex of vegetation assemblages exists, with patterns determined by latitude, altitude, topography, and distance from the Pacific Ocean. Dense conifer forests of the Pacific Northwest give way gradually to chaparral and shrublands in the coastal Southwest, while sagebrush vegetation dominates in the arid interior basins.

The vegetation distribution of glacial times was apparently quite different (Fig. 20.17). Just south of the ice sheets that covered most of Canada and

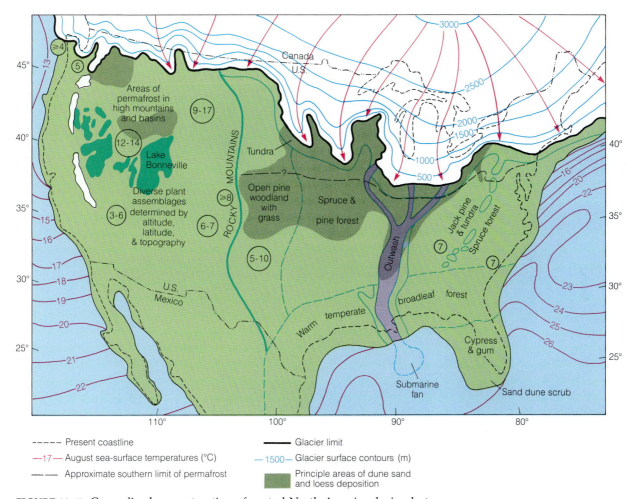

FIGURE 20.17 Generalized reconstruction of central North America during last glacial maximum about 18,000 years ago. Contours depict surface of ice sheets in Canada, with principal ice-flow directions indicated by arrows. Lake Bonneville and other large pluvial lakes are shown in western United States. Coastlines lie beyond present coast (dashed line) due to fall of sea level of about 100 m. Sea-surface temperatures (in °C) are based on analysis of microfossils from deep-sea cores. Encircled numbers are inferred amount of temperature lowering based on various kinds of proxy data. Dominant vegetation zones are shown east of Rocky Mountains. To west, mountainous terrain gave rise to complicated pattern of vegetation assemblages.

the northern United States lay a narrow strip of tundra with local clumps of spruce. An extension of that zone projected southward along the high, frigid crest of the Appalachians. Beyond the tundra was a belt of forest dominated by spruce and Jack pine. This passed southward into a broad belt of warm-temperate forest dominated by oak and hickory, along with pine. To the west, where there are grassy plains today, there grew an open woodland consisting of scattered pines with grass between them. Thus, in the glacial age, tundra and spruce forest grew much farther south than they do now, consistent with a colder climate. Today's dry grassland country was then mostly woodland, suggesting that in this zone more moisture was

available to support trees. It was once supposed that as the ice sheets spread, vegetation zones seen on today's map crept gradually southward, each maintaining its own character. But the fossil pollen show that the change was more complicated, and that various plants were displaced in different directions and at different rates, forming new mixtures of species.

In the west, mountain ranges and intervening basins break up the simple latitudinal zonation of vegetation, because each highland has its own belts of vegetation arranged according to altitude, temperature, and moisture. Although sample sites are too few to enable us to produce a regional map of ice-age vegetation like that for the east, local

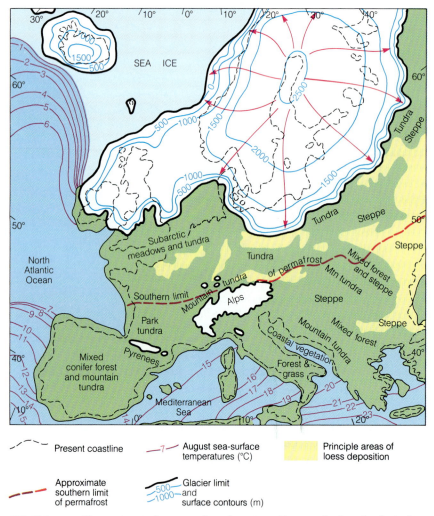

FIGURE 20.18 Reconstructed geography of western Europe during the last glacial maximum. A vast continental ice sheet extended across the North Sea between Britain and Scandinavia and was separated from the glacier-covered Alps by a frigid periglacial zone with tundra vegetation. Cold polar waters extended far south of their present limit in the North Atlantic. Due to lower sea level, the English Channel and North Sea were emergent land areas.

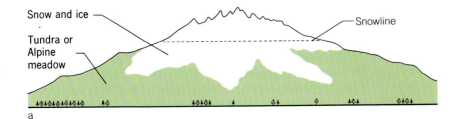

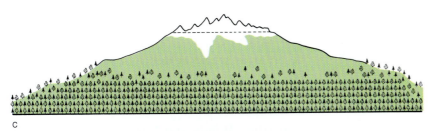

FIGURE 20.19 Idealized sketch showing altitudinal shift of vegetation zones on a high middle-latitude mountain during the last 20,000 years (not to scale). (*a*) About 20,000 years ago. (*b*) About 6000 years ago. (*c*) Present day.

studies indicate that vegetation zones shifted downward during glacial times in response to cooler temperatures (Fig. 20.19). Some of the best and most interesting information comes from deposits left by ancient wood rats of the American Southwest who subsisted on vegetation collected close to their dens. Analysis of the fossil plant remains and fecal pellets shows that vegetation boundaries moved downward some 1000 m during the coldest interval.

In Europe a comparable vegetational response took place (Fig. 20.18), but with one major difference. In North America migrating plants driven south by the ice could inhabit the relatively warm lowlands that extended to the Gulf of Mexico. However, in Europe, the Alps, with their own ice cover some 800 km long and 150 km wide, constituted a high, cold barrier north of the Mediterranean Sea. Many species were trapped between the large ice sheet in the north and the Alpine glaciers to the south and were driven to extinction. Thus western Europe, which before the glacial ages had

an abundance of tree types, now has only 30 naturally occurring species. In the north only six species of broad-leaf trees survive. North America, with no Alpine barrier standing between the Great Lakes and the Gulf of Mexico, has 130 species.

Pluvial Lakes

In the western Americas and in many other regions now dry, glacial-age climates resulted in the creation or enlargement of lakes (Fig. 20.16). For example, the basin of Great Salt Lake, Utah, was then occupied by the gigantic Lake Bonneville (Fig. 20.17), more than 300 m deep and having a volume comparable to that of Lake Michigan. Beaches, deltas of tributary streams, and lake-bottom sediments are all there to tell the story. Such a lake is commonly called a ***pluvial lake;*** by definition, it is *a lake that existed when, under a former climate, rainfall in the surrounding region was greater than today's.* Although the name implies a rainy (pluvial) climate, we now realize that reduced evaporation caused

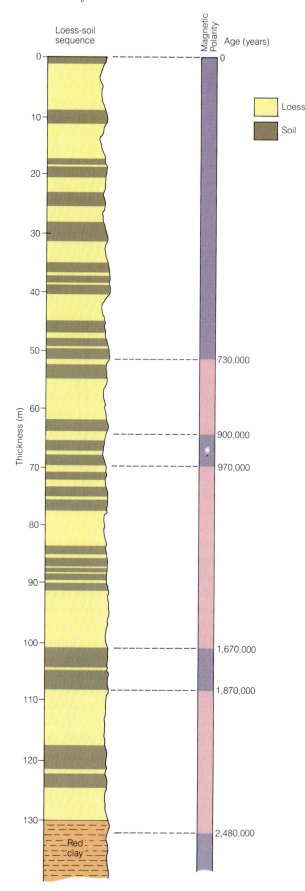

by lower temperatures in many cases was a major factor in raising water levels in these basins, for other types of paleoclimatic information from the same areas often point to reduced precipitation during glacial times. These lakes, then, were likely the result of (1) lower temperature that led to reduced water loss by evaporation, and/or (2) increased rainfall in the regions where they existed.

Direct geologic evidence and radiocarbon dates show that high stands of some pluvial lakes were contemporaneous with expansion of glaciers during the most recent glacial age. This does not mean, however, that such lakes were fed by glacial meltwater. Some, like Lake Bonneville, were, but only in a very limited way. Most of the lakes occupied basins to which glacial waters had no access.

Pluvial lakes were abundant also in many other dry regions of the world. High lake levels during glacial times have been recorded in the Saharan region of North Africa (Fig. 20.16), in the Middle East, and in southern Australia. There are other desert areas where evidence of such former lakes has been found, but in most cases we do not yet know when the highest levels were attained.

Prevailing Winds and Wind-Blown Dust

At the height of the glacial age, conditions were both windy and dusty throughout much of the middle latitudes. We infer this from several lines of evidence. Loess deposits of glacial age found south of the ice limit in midwestern United States become both thinner and finer as one travels east from major outwash deposits, implying that the dust was distributed by prevailing westerly winds which deflated the active outwash surfaces. Thick loess deposits in central China lie east of desert basins in central Asia that were swept by cold, dry winds during glacial times (Fig. 12.33). Loess deposits in eastern Europe lie downwind from extensive outwash surfaces lying between the Alps and the limit of the northern ice sheet, and they contain fossil plants and animals that suggest cold,

FIGURE 20.20 Sequence of loess deposits and soils exposed near Lochuan in the Loess Plateau of central China. The loess rests unconformably above a deposit of red clay and spans approximately the last two and a half million years. Paleomagnetic reversals identified in the section permit dating of specific levels. Loess units represent dry, windy glacial ages whereas soil formation occurred during moister, less-windy interglacial times. (*Source:* Livetal, 1980.)

FIGURE 20.21 *". . . and the record low for this date is 147° below zero, which occurred 25,000 years ago during the Great Ice Age."*

dry conditions. In each of these regions successive sheets of loess are separated by soils of interglacial character, indicating that loess deposition was a feature of glacial times (Fig. 20.20).

That full-glacial periods were both windy and dusty is also shown by studies of fine dust retrieved from Greenland ice-sheet cores. The percentage of wind-blown dust in the cores increases sharply at levels corresponding to the last glacial age. Because Greenland and much of northern North America were ice-covered at that time, the dust very likely originated in dry, windy periglacial zones elsewhere in the Northern Hemisphere.

Temperature Estimates

In the popular imagination, glacial ages were times when temperatures were very cold, perhaps rivaling those in the middle of Antarctica today (Fig. 20.21). While not denying that such extreme cold may have existed locally, it is clear that *average*

temperatures were not very much lower during glaciations. The evidence is varied, and it comes both from the ocean basins and the land.

The upper parts of cores taken from seafloors throughout the world's oceans consist of soft sediments that commonly contain multitudes of tiny fossils. Most of the fossils are of microorganisms that live in the surface waters and whose shells rain down on the seafloor in vast numbers (Fig. 5.27). The rate of sedimentation is extremely slow, however, so that it may take more than a thousand years for a single centimeter of sediment to accumulate (Table 5.2). Because the assemblage of organisms that live in the surface waters depends largely on the temperature there, fossil remains in sediment provide a record of changing conditions at the ocean surface.

The fossil content of the sediments changes downward through a core, typically shifting back and forth from predominantly warm-water (interglacial) to cold-water (glacial) types. Assemblages of living forms are associated with a certain range of surface-water temperature. By identifying the species present at any level in a sediment core and comparing the assemblage with modern ones it is possible to infer what the surface ocean temperature must have been when the shells were settling to the seafloor. In practice, geologists can select a level in a core that represents the last glacial maximum and determine the former surface-water temperature at the core site from the contained fossils. Information from hundreds of cores scattered widely over the oceans has been used to derive a global map of sea-surface temperature for the last glacial maximum (Fig. 20.22b) which can be compared with temperatures in today's oceans (Fig. 20.22a).

Surprisingly, the *average* global difference between present and ice-age sea-surface temperatures amounts to only about 2.3° C, but this figure is misleading. In some large regions, like the subtropics, little or no change is detected. In others, such as the North Atlantic, sea-surface temperatures were locally 14° C or more colder than now. In this region, cold polar water that is now largely restricted to areas north of latitude 60° descended far south at the glacial maximum to reach the shore of northeastern United States in North America and the Iberian Peninsula in western Europe (Fig. 20.22b). The greatest changes occurred in northern middle latitudes where the large northern ice sheets were located, and near the equator, where cold water welled up off the coasts of Africa and South America and spread westward across the

a   TODAY

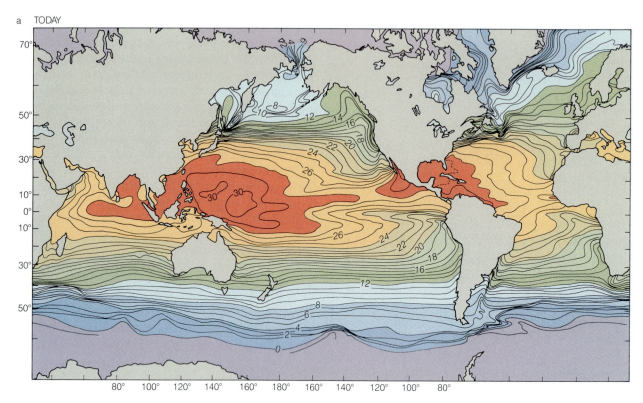

b   LAST GLACIAL MAXIMUM

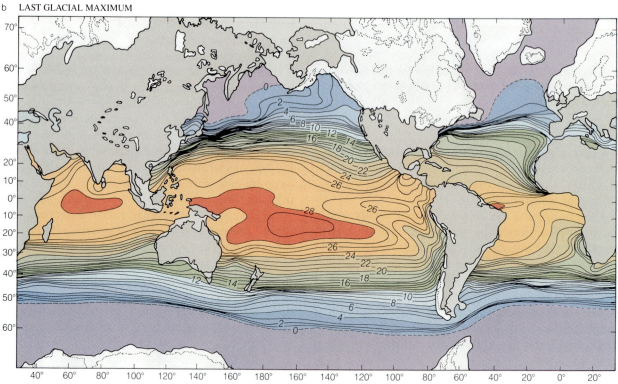

FIGURE 20.22 (*a*) Map showing modern sea-surface temperatures (°C) during August. (*b*) Map showing reconstructed August sea-surface temperatures during the last glacial maximum, about 18,000 years ago. Cold polar water extended far south of its present limit in the North Atlantic, and plumes of cold water flowed westward in the equatorial Pacific and Atlantic. (*Source:* After CLIMAP Project Members, 1976.)

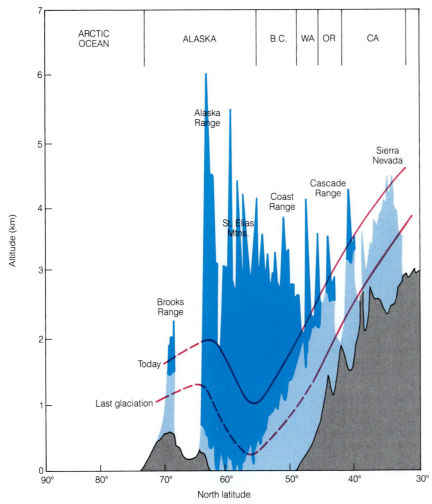

FIGURE 20.23 Transect along the coastal mountains of western North America showing the relationship of the present snowline to existing glaciers (blue) and of the ice-age snowline to expanded glaciers of that time (light blue). Difference between present and ice-age snowline was about 1000 m along the southern part of the transect and about 600 m in northern Alaska.

equatorial zone. Thus, the changes accompanying a shift to glacial conditions did not affect the whole world equally. Some zones saw little change while others experienced climatic environments like those of present-day Antarctica.

Estimates of temperature lowering on the land also display a range of values. In mid-latitude coastal regions temperatures were generally reduced by about 5–8° C whereas in continental interiors reductions of 10–15° are indicated (Fig. 20.17). Such estimates are arrived at in several ways, of which the following are the most important:

1. By reconstructing the equilibrium-line altitudes of former glaciers and comparing the results

with present-day values, a figure for snowline depression can be obtained (Fig. 20.23). An estimate of temperature lowering can then be arrived at by assuming that the **lapse rate,** *the rate of decrease in temperature with altitude in the atmosphere,* is 6° C/km, an average contemporary value. Applied to the calculated snowline depression, the resulting temperature lowering can be found. Such estimates are not likely to be precise, for glacial-age lapse rates may have been different from those of today. Furthermore the estimate is made by assuming that there was no change in precipitation, a factor that also controls the altitude of the snowline.

2. By using information about ice-age vegetation

assemblages derived from pollen studies and applying it to the range of temperatures within which the same plants now live, estimates of temperature change can be derived. However, this approach works only where modern examples can be found of the particular assemblages that occur in the fossil record. A similar approach can be taken for animal fossils, such as beetles, with comparable results.

3. By obtaining measurements of the oxygen-isotope ratio in ice-sheet cores that penetrate to the level of the last glacial age, ice-age air temperature at the glacier surface can be found (Fig. 20.8). Available records show a marked change in isotope values at a level coinciding with the transition from mild interglacial climate recorded in the upper parts of the cores to cold ice-age temperatures below.

4. By plotting the distribution of periglacial features that indicate the former presence of permafrost in areas now lacking it, an estimate can be made of the minimum temperature change that has taken place. At present, permafrost exists mainly in areas where the mean annual air temperature is below about −5° C. If evidence of former permafrost is found at a place where the annual temperature is now 4° C, then the former periglacial climate is inferred to have been at least 9° colder.

EARLIER GLACIAL AGES

Until rather recently, it was thought that the Earth had experienced four glacial ages during the Pleistocene Epoch. This assumption was based on studies of ice-sheet and mountain-glacier deposits and had its roots in early studies of the Alps where geologists identified stream terraces they thought were related to four ice advances. This traditional view was discarded when studies of deep-sea sediments disclosed a long succession of glaciations, the most recent of which was shown by radiocarbon dating to equate with the youngest glacial drift on land. Paleomagnetic dating of the cores (Chapter 8) showed that glacial-interglacial cycles average about 100,000 years long and that during the last million years alone there had been about 10 such episodes. For the Pleistocene Epoch as a whole, between 15 and 20 glacial ages are recorded, rather than the traditional four. The implications are clear: whereas seafloor sediments provide a continuous record of climatic cycles,

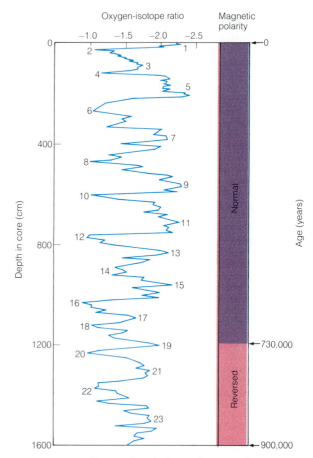

FIGURE 20.24 Curve of variations of oxygen-isotope ratio in an equatorial Pacific deep-sea core representing global changes in ice volume during the last 900,000 years. Each glacial/interglacial cycle is about 100,000 years long. (*Source:* After Imbrie and Imbrie, 1979.)

evidence of glaciation on land generally is discontinuous, interrupted by many unconformities, and includes only a partial record of past glacial events.

The seafloor evidence is of three kinds. First, the biologic component of the sediments shows repeated shifts of surface-water animal and plant populations, from warm interglacial forms to cold glacial forms, with increasing depth in a core. Second, the percentage of calcium carbonate in cores from some oceanic regions fluctuates in much the same manner. Third, the ratio of the isotopes ^{18}O to ^{16}O fluctuates with a pattern similar to that shown in the biologic and mineral fractions of the sediments. Whereas the isotopic variations in ice cores are believed to represent fluctuations in air temperature at the glacier surface, in Pleistocene marine sediments they are thought primarily to reflect changes in global ice volume. During glacial ages, when water is evaporated from the oceans and precipitated on land to form glaciers, water containing the light isotope ^{16}O is more easily evaporated than water containing the heavier ^{18}O. As a result, Pleistocene glaciers contained more of the light isotope, while the oceans became enriched in the heavy isotope. Isotope curves derived from the sediments therefore give us a continuous reading of changing ice volume on the planet (Fig. 20.24). Because glaciers wax and wane in response to climatic changes, the isotopes also give a generalized view of global climatic change.

Over most of the last million years, peaks in the isotope curve that represent times of high global ice volume are nearly equal in amplitude. This suggests that during most glaciations the buildup of ice on land was about the same. The interglacial peaks are also nearly equal, indicating that earlier interglacial ages were probably broadly similar in character to the present one. Nevertheless, evidence that sea level was some 5–6 m higher during the last interglaciation (about 120,000 years ago) than it is today suggests that there must have been less ice on the planet at that time.

Measurements of oxygen isotopes in long deep-sea cores that extend through the entire Cenozoic section show that the oceans have grown colder over the last 66 million years (Fig. 20.25). During one pronounced cooling event at the end of the Eocene Epoch, ocean temperatures declined by nearly 5° C within only about 100,000 years. In concert with the long-term cooling trend, glaciers spread from highlands in Antarctica and reached

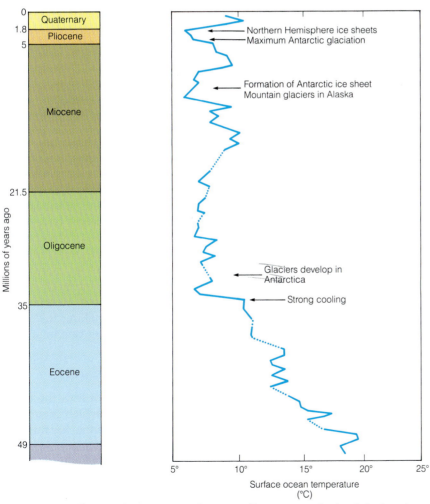

FIGURE 20.25 Oxygen-isotope curve from a sediment core obtained during the Deep Sea Drilling Project in the western Pacific Ocean. Warm surface waters in the Eocene Epoch gave way to colder conditions after about 35 million years ago, with major ice-sheet glaciation beginning about 10 million years ago. (*Source:* After Shackleton and Kennett, 1975.)

the sea. Then, in the Miocene Epoch, some 12 to 10 million years ago, ice volume increased and an ice sheet formed over the continent as temperatures continued to fall. The presence of such a large polar ice mass reduced average temperatures on the Earth still further and caused a substantial drop in sea level. From that time onward large glaciers began to form in the high mountains of Alaska and the southern Andes. Although the evidence is still sketchy, it appears that large ice sheets did not form in northern middle latitudes until about 2.5–3 million years ago. If this sequence of events is correct, glaciation has gradually affected more and more of the Earth's land surface during the Cenozoic: first the Antarctic, then high-latitude mountain systems, and more recently the northern middle latitudes.

Ancient glaciations, identified mainly by tillites and striated rock surfaces, are known from still older parts of the geologic column (Fig. 5.12). The earliest recorded glaciation dates to about 2.3 billion years ago in the early Proterozoic. Evidence of others has been found in rocks of late Proterozoic, early Paleozoic, and late Paleozoic age. During the latest of these intervals 50 or more glaciations are believed to have occurred. The geologic record is fragmentary and not always easy to interpret, but evidence from such low-latitude regions as South America, Africa, and India, as well as from Antarctica, suggest that the Earth's land areas must have had a very different relationship to one another during the late Paleozoic glaciation than they do today. In the Mesozoic Era, glaciation of similar magnitude apparently did not occur, for geologic evidence points to a long interval of mild temperatures both on land and in the oceans.

THE WARM MIDDLE CRETACEOUS

Several lines of evidence indicate that the climate of the middle Cretaceous Period, about 100 million years ago, was unusually warm. Coral reefs, which are now confined to tropical latitudes, are found in Cretaceous rocks as much as 1500 km closer to the poles. Nonseasonal land plants had similar distributions, and tropical and subtropical vertebrates, like the alligator and crocodile, ranged as far as present-day Labrador, within about 30° of the North Pole. This was a time of high organic productivity in the ocean basins and more than half of all known petroleum reserves exist in rocks of Cretaceous age. Of these, as many as 60 percent were formed during the warm middle Cretaceous.

Sea level stood much higher at that time, as shown by the distribution of Cretaceous marine rocks on any geologic map of the world. They extend far inland on most of the continents, overlapping older rocks and reaching altitudes well above the average level of younger marine rocks. These relationships show that the ratio of land to ocean was then much smaller than today (Fig. 7.22). The higher sea level, and the apparent absence of a permanent ice cover, is consistent with a warmer climate. Isotopic studies of ocean-floor sediments of that age suggest that deep ocean water was as much as 15° C warmer than now.

Based on such evidence, it has been estimated that average world temperature was at least 6° C milder than today and possibly as much as 14° C, with the greatest difference being in the polar regions. Whereas today the difference in temperature between the poles and the equator amounts to some 41° C, it is estimated that during the middle Cretaceous it was no more than 26° C and possibly as little as 17° C.

WHAT CAUSES CLIMATES TO CHANGE?

What factors cause the climate to warm and cool at irregular intervals, thereby bringing about great changes in the Earth's surface processes and environments? The search for an answer is a huge undertaking, in pursuit of which tens of thousands of pages of reasoning and discussion have been published. Yet we still have achieved only a part of the answer. Because climates change on different time scales, ranging from decades to many millions of years, several quite different mechanisms are likely to be responsible. These mechanisms may interact in complex ways, making the problem of climatic variability a difficult one to solve.

Glacial Eras and Shifting Continents

Let us first examine the question of how the Earth can pass from a climatic state in which few or no glaciers exist and surface temperatures are mild, to one in which it enters a long succession of glacial-interglacial ages marked by the waxing and waning of large continental ice sheets and numerous smaller glaciers. Then we can ask what controls the glacial-interglacial cycles, the little ice ages, and the even shorter changes that we see evidence of in the geologic record.

Successions of glacial ages, often encompassing many millions of years, can be identified in the geologic record of the last two and a half billion years (Fig. 20.26). They were separated by long periods of mild climate, that of the Mesozoic Era being the most recent example. What seems to be the only reasonable explanation for their pattern is suggested by slow but important geographic changes that affect the crust of the planet. These changes include (1) the movement of continents as they are carried along with shifting plates of lithosphere, (2) the large-scale uplift of continental crust where one plate overrides another, (3) the creation of high mountain chains where continents collide, and (4) the opening or closing of ocean basins and seaways between moving landmasses.

The effect of such earth movements on climate is illustrated by the fact that low temperatures are found, and glaciers tend to form and persist, in two kinds of situations: high latitudes and high altitudes—especially in places where winds can supply abundant moisture evaporated from a nearby ocean. The Earth's largest existing glacier is centered on the South Pole where temperatures are constantly below freezing and the land is surrounded by ocean. The only glaciers found at or close to the equator lie at extremely high altitudes.

Abundant evidence now leads us to conclude that the positions, shapes, and altitudes of landmasses have changed with time (Chapter 17), in the process altering the paths of ocean currents and of atmospheric circulation. Where evidence of ancient ice-sheet glaciation is now found in low latitudes, we infer that such lands were formerly located in higher latitudes where large glaciers could be sustained. Although this conclusion appears to explain the pattern of glaciation during and since the late Paleozoic, information about earlier glacial eras is very fragmentary and more difficult to evaluate.

Can such reasoning also help us to explain the much warmer conditions of the middle Cretaceous? It seems apparent that the significantly different positions of landmasses at that time must have played an important role in determining the global climate. In the Cretaceous Period continents were concentrated in low latitudes, an arrangement that permitted warm ocean currents to transfer heat poleward efficiently. Neither polar region had a large landmass, like the Antarctic has today. Consequently, no large continental ice sheets could form there. Furthermore, large ice-covered seas surrounded by land, like the Arctic Ocean, did not exist. Instead, the polar oceans, being less-reflective than either bare land or ice-covered land, would have been able to absorb more heat, further warming the climate.

Although geographical factors apparently were important in explaining Cretaceous climate, mathematical climate models indicate that geography alone is insufficient to explain fully the elevated temperatures of that time. A variety of evidence suggests that a high concentration of carbon dioxide in the atmosphere (Fig. 2.22; see below) may also have been important in raising temperatures well above those of today.

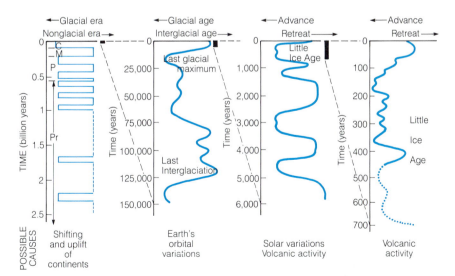

FIGURE 20.26 Time scales of climatic variations and possible causes. In first column: Pr = Proterozoic, P = Paleozoic, M = Mesozoic, and C = Cenozoic.

Ice-Age Periodicity and the Astronomical Theory

As initially discovered through studies of glacial deposits, and later verified by studies of deep-sea cores, glacial and interglacial ages have alternated for more than a million years (Fig. 20.24). The cause of this basic climatic pattern, which produced worldwide geographic changes and major relocations of plants and animals, has long been a fundamental challenge to the development of a comprehensive theory of climate. A preliminary answer was provided by Scottish geologist, John Croll in the middle nineteenth century, and was later elaborated by Milutin Milankovitch, a Serbian astronomer of the early twentieth century.

Croll and Milankovitch recognized that minor variations in the path of the Earth in its orbit around the Sun and in the inclination, or tilt, of the Earth's axis cause slight but important variations

in the amount of radiant energy reaching the top of the atmosphere at any given latitude, even though the total heat received by the planet remains little changed. Three movements are involved (Fig. 20.27*a*). First, the axis of rotation, which now points to the North Star, wobbles like a spinning top, moving slowly in a circular path. Simultaneously, the elliptical orbit of the Earth is also rotating, but much more slowly. These two motions together result in the *precession of the equinoxes* which is a progressive change in the Earth–Sun distance for a given date (Fig. 20.27*b*). At the spring and autumn equinoxes the day and night are of equal length. At the summer and winter solstices, days are longest and shortest, respectively. These points move, or precess, slowly around the orbital path, completing one full cycle in about 23,000 years. Second, the *tilt* of the axis, which now averages 23.5°, shifts about 1.5° to either side during a span of about 41,000 years

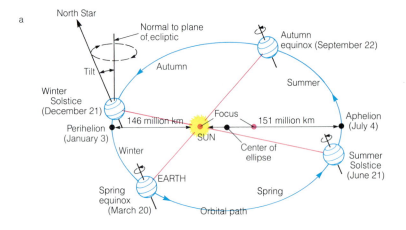

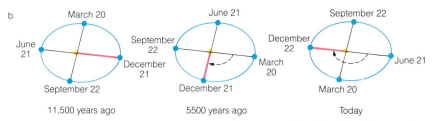

FIGURE 20.27 (*a*) Geometry of the Sun–Earth system. The Earth's orbit, an ellipse with the Sun at one focus, defines the plane of the ecliptic. The Earth moves around its orbit in the direction of the arrows, while spinning about its axis, which is tilted to the plane of the ecliptic at 23.5° and points toward the North Star. (*b*) Precession of the equinoxes causes the position of the equinoxes and the solstices to shift slowly around the Earth's elliptical orbit. About 11,500 years ago, the winter solstice occurred near the aphelion (the most distant point on the orbit from the Sun), whereas today it occurs at the opposite end of the orbit near perihelion (the closest approach to the Sun).

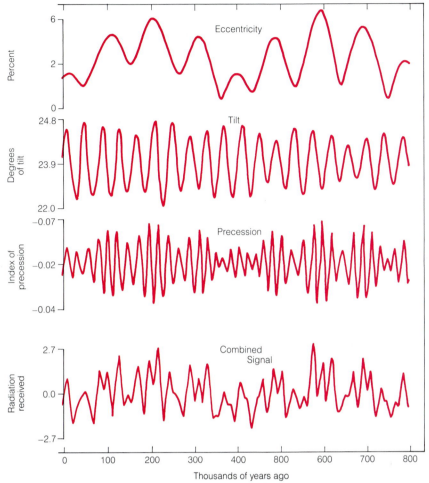

FIGURE 20.28 Pattern of orbital eccentricity, axial tilt, and precession during the last 800,000 years. When combined, these factors produce an irregular curve that portrays variations in the amount of radiation reaching the Earth as a function of time. (*Source:* After Imbrie et al., 1984.)

(Fig. 20.27*a*). Finally, the shape, or *eccentricity*, of the orbit changes over an average period of nearly 100,000 years. About 100,000 years ago the orbit was highly eccentric (less circular) compared to what it has been for the last 10,000 years.

Each of these astronomical factors varies on a different time scale, so when they interact the resulting pattern is complicated (Fig. 20.28). The variations at any latitude can be expressed as an irregular curve showing that the solar heat received fluctuates continuously by a slight amount. Careful analysis of paleoclimatic data from deep-sea cores shows clearly that the factors controlling the observable variations fluctuate on the same time scales as those of axial tilt and precession. This persuasive evidence has provided considerable support for the theory that astronomical factors control the timing of the glacial-interglacial cycles.

Solar Variations, Volcanic Activity, and Little Ice Ages

The major glacial eras in Earth history can be thought of as primary, or first-order, climatic events, and the glacial-interglacial cycles as second-order events. Superimposed on them is a series of third-order events having a time scale similar to the Little Ice Age (Fig. 20.26). These latter climatic fluctuations are measured in centuries or decades, periods too short to be affected either by movements of continents or by the three primary astronomical variations, and require us to seek other explanations for their cause. Among a

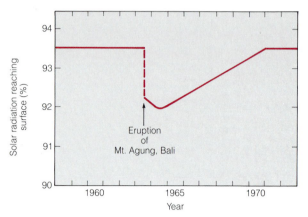

FIGURE 20.29 Effect of volcanic particles and gases from the explosive 1963 eruption of Mount Agung in the East Indies on radiation reaching the Earth's surface at the Mauna Loa Observatory in Hawaii. The sudden drop in the curve signals the arrival of the volcanic cloud. After seven years the atmosphere has returned to its pre-eruption state. (*Source:* After Ellis and Pueschell, 1972.)

number of suggested explanations, two have received special attention.

One obvious hypothesis is based on the concept that the energy output of the Sun fluctuates through time. The idea is appealing because it might explain climatic variations on many different time scales. However, it is a difficult idea to test. Although measurements of solar energy reaching the Earth have been made for about 75 years, they show only a very slight change over this interval. Fluctuations in the production of radiocarbon in the upper atmosphere, which are detected through isotopic measurements of tree rings, are known to be related to solar activity. However, comparison of such fluctuations with climatic records covering all or part of the last 1000 years has revealed no significant relationship. Therefore, although possibly the Sun has an influence on climate on short time scales, a relationship has not yet been clearly demonstrated.

Fine ash thrown into the atmosphere during a large explosive volcanic eruption can create a veil of dust that will circle the globe. The fine particles tend to scatter incoming solar radiation, resulting in a slight cooling at the Earth's surface. Although the dust settles out rather quickly, generally within several months or a year, tiny droplets of sulfuric acid, produced by the interaction of volcanically emitted SO_2 gas and water vapor, also scatter the Sun's rays and may remain in the upper atmosphere for several years (Fig. 20.29). They too reflect incoming radiation back into space. During the major eruptions of El Chichón volcano in

southern Mexico during late 1982 and early 1983, direct solar radiation reaching the surface was reduced by about 25 percent. It has been calculated that this would result in a maximum cooling of surface air by about 0.4° C. While isolated volcanic events are unlikely to affect climate over the long term, if several major eruptions occur within a decade or two, their combined effect will be greater and last for a longer time. Might such episodes of increased volcanic activity explain climatic variations on the scale of decades and centuries? The idea is attractive, but we still know few details about the history of volcanic activity prior to the present century. By measuring the variations in acidity with increasing depth in polar ice cores, which is related to fallout of precipitation rich in volcanic acids, the intensity of volcanism over many hundreds of years can be detected. Such a record from the Greenland Ice Sheet points to higher levels of volcanic activity during the Little Ice Age than during periods before or after, which lends support to this hypothesis.

FUTURE CLIMATES

As our expanding civilization becomes increasingly vulnerable to changes of climate, even of small magnitude, the need for reliable prediction of future climatic changes becomes obvious. Any who have planned a weekend outing on the basis of a favorable weather report, only to be met with an unexpected rainstorm, will probably guess that scientific predictions of future climate are likely to be even less reliable than our attempts to forecast the weather. At present, prediction of future climatic trends is based largely on extrapolation—that is, estimation of the probable future continuation of a trend shown by past events of a similar kind. If, in 1940, we had extrapolated the Northern Hemisphere temperature curve seen in Figure 20.5, we might have extended the 1880–1940 trend upward. Later events would have proved us wrong, but therein lies the danger of extrapolation. However, if we have a longer record, and the curve goes up and down in a regular manner, if its peaks and troughs are spaced in a repeating pattern, then extrapolation of the pattern may be justified. Yet, if we do not understand the basic causes of the fluctuations in a climatic record, then our predictions may turn out to be as unreliable as those of the forecaster who predicted the favorable weekend weather.

What then can be said about probable future cli-

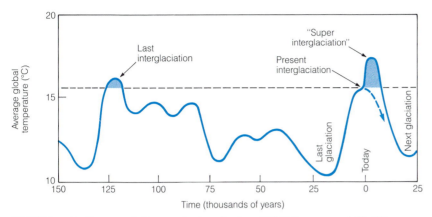

FIGURE 20.30 Course of climate during the last 150,000 years and 25,000 years into the future. Natural course of future climate (dashed line) would involve declining temperatures until the next glacial maximum, about 23,000 years from now. With CO_2-induced greenhouse effect, continued warming may lead to a super interglaciation within the next several hundred years. During such an interval, temperature may rise above that of the last interglaciation. The decline toward the next glaciation would thereby be delayed by several millennia. (*Source:* After Imbrie and Imbrie, 1979.)

mates? If the Croll-Milankovitch astronomical variations do control the timing of the glacial ages, the climatic curves can be extended into the future by calculating the interacting effects of the Earth's predictable movements. The results suggest that although we may be nearing the end of the present interglaciation, the next full-scale glaciation will not culminate for about another 23,000 years (Fig. 20.30). Obviously, it will be a long time before this prediction is proved right or wrong!

The near future is of greater interest to us. Can we say for certain that the climate of the Little Ice Age is behind us and that we can look forward to more equable climates in coming centuries? Some scientists have suggested that the cooling trend which began in the middle 1940s may be the start of a return to Little Ice Age conditions (some alarmists even claim that it is the beginning of a rapid plunge into the next glacial age). Others think that this may be but a minor reversal in an otherwise longer-term trend toward still warmer climates, to be brought about by the ever-increasing release of CO_2 and other industrial gases into the atmosphere.

Measurements show that the concentration of CO_2 in the atmosphere has increased from 316 ppm (parts per million) in 1959 to 338 ppm in 1980, a 7 percent increase over 21 years (Fig. 2.19). The progressive increase, totaling about 25 percent since 1850, has been taking place since large-scale combustion of fossil fuels began at the start of the

industrial revolution. Estimates of future use of fossil fuels indicate a high probability that the CO_2 concentration will reach twice the preindustrial value before the year 2100. The CO_2 has the same effect on incoming solar radiation as the glass covering a greenhouse: it allows short-wave radiation from the Sun to reach the ground, but it prevents long-wave radiation emitted by the Earth from getting back into space. The surface, therefore, heats up to a new equilibrium condition that depends on the concentration of CO_2.

The increasing buildup of atmospheric CO_2 is expected to cause significant worldwide warming within the next century or two. Computer simulations using various models suggest that a doubling of CO_2 will lead to a surface warming of about 4° C. Although in theory the Earth should have warmed by a degree or more due to the CO_2 buildup since 1850, the observed increase amounts to perhaps only half that amount. The reason may lie in the fact that the oceans absorb heat from the atmosphere, and therefore delay the surface warming. Models show that even if fossil-fuel consumption and related CO_2 buildup were suddenly to cease (a highly unlikely scenario), the surface air temperature would continue to rise by as much as 0.7° C as the atmosphere moves toward a new equilibrium condition.

The "greenhouse effect" is not restricted to buildup of CO_2 alone. Other industrial gases, especially several chlorofluorocarbons (CF_2Cl_2 and

$CFCl_3$) used as spray-can propellants and refrigerants, have a very strong greenhouse effect. Nitrous oxide (N_2O), a product of fossil fuel combustion and the use of nitrogen fertilizers, and methane (CH_4) are also concentrating in the atmosphere. The combined effect of these "trace gases" in promoting surface warming is estimated to be approximately equal to that of CO_2.

Observations and modeling results have led some scientists to predict that the Earth may be about to enter a "Super Interglaciation" in which temperatures will rise to levels not experienced since the last interglaciation about 120,000 years ago, or even since the warm Cretaceous Period (Fig. 20.30). The ultimate results of this huge man-made atmospheric "experiment" we have embarked upon are uncertain, but they could be very dramatic in some regions. It seems quite possible that a warmer Earth will lead to extensive reduction in polar ice bodies, with a consequent rise of world sea level, perhaps by many meters. The resulting slow inundation of coastal lands, where a large percentage of the Earth's inhabitants dwell, would lead to large-scale dislocations of populations in coming centuries. Society would be affected in other ways as well. Agriculture, forestry, fisheries, water supplies, transportation, and energy production would all be impacted, with social and political repercussions that are difficult to foresee.

Although we now know enough about mechanisms of climatic change to make some intelligent projections about the timing of future glacial and interglacial ages, we quite naturally are more concerned about climate in the near future and the effect it will have on us and our children. However, the exact course of future climate over the short term remains uncertain. This uncertainty poses a major challenge to Earth scientists who have at their disposal an array of geologic information about past climates and surface conditions that can provide important clues about the mechanisms of climate change and environmental changes that the future may hold in store for us.

SUMMARY

1. Climate is the average weather condition over a period of years. Former climates are determined mainly from fossil plants and animals, relict physical features, and from isotopic studies.

2. Changes in climates within the last 100–200 years are established by instrumental records. After rising to a high level in the 1940s, the average temperature of the world has fallen.

3. Climatic changes during the last 1000 years are established (less accurately) by proxy records. In northern middle latitudes an interval of mild climate during the Middle Ages was followed by cold Little Ice Age conditions.

4. Changes during the last 10,000 years are established mainly from studies of fossil pollen, of isotopes in ice cores, and of fluctuations of glaciers, lake levels, and tree limits. In many areas, postglacial temperatures rose to a peak about 6000–8000 years ago during the Hypsithermal interval, then fell during Neoglaciation.

5. During the last glacial age land-surface temperatures were from 5–15° C lower than today's. Sea-surface temperatures fell as much as 14°, with the greatest changes occurring in the North Atlantic, the North Pacific, and in the equatorial zone. Temperature changes are determined from fossils on land and in deep-sea cores, from isotopic measurements of ice cores, from evidence of lowered snowlines, and from distribution of periglacial features.

6. Glacial ages have alternated with interglacial ages in which temperatures approximated those of today. Studies of marine cores indicate that glacial-interglacial cycles were about 100,000 years long, and that during the Pleistocene there have been 15 or more such cycles. In most glaciated regions we see evidence only of the 3 or 4 most extensive glaciations.

7. Glacial ages are discerned in many parts of the geologic column; they extend back at least 2.3 billion years.

8. The occurrence of glacial ages probably is related to favorable positioning of continents and ocean basins, brought about by movements of lithospheric plates. The timing of the glacial-interglacial cycles appears to be closely controlled by changes in the orbital path and axial tilt of the Earth, which affect the distribution of solar radiation received at the surface.

9. Climatic variations on the scale of centuries and decades may result from fluctuations in energy output from the Sun, or from injections of volcanic dust and gases into the atmosphere.

10. Predictions of future climate rely largely on extrapolation of measured trends, but should improve as more is learned about the specific causes of short-term climatic variations.

SELECTED REFERENCES

Chorlton, W., 1983, Ice ages: Alexandria, Va., Time-Life Books.

Flint, R. F., 1974, Three theories in time: Quaternary Research, v. 4, p. 1–8.

Frakes, L. A., 1979, Climates through geologic time: Amsterdam, Elsevier.

Global Atmospheric Research Project, 1975, Understanding climatic change: A program for action: Washington, D.C., National Research Council/ National Academy of Sciences.

Imbrie, J., and Imbrie, K. P., 1979, Ice ages: Solving the mystery: Short Hills, N.J., Enslow.

Lamb, H. H., 1972, Climate—present, past and future, v. 1, Fundamentals and climate now: London, Methuen.

Lamb, H. H., 1977, Climate—present, past and future, v. 2, Climatic history and the future: London, Methuen.

Le Roy Ladurie, E., 1971, Times of feast, times of famine: A history of climate since the year 1000: Garden City, N.J., Doubleday.

Skinner, B. J., ed., 1981, Climates past and present: Los Altos, Calif., William Kaufmann.

Wigley, T. M. L., Ingram, M. J., and Farmer, G., eds., 1981, Climate and history: Studies in past climates and their impact on man: Cambridge, Cambridge University Press.

Miners swarm all over the diggings at Serra Pallada, a fabulously rich gold deposit in Brazil.

The Earth's Resources

CHAPTER 21

Sources
of Energy

Geothermal bores, as much as 1,000 m deep, supply volcanically heated steam to power plants at Wairakei, New Zealand. The project produces about 10 percent of all electricity used in New Zealand.

MATERIALS AND ENERGY

When we speak of the present western civilization as being in "the industrial age," we mean the age of enormously increased use of fuels and metals. Another term would be "the age of intensive use of energy in industry." Stone Age people had an industry too: They chipped and flaked mineral substances (mostly pieces of flint) to make tools and weapons. However, theirs was an industry with a low input of energy because the energy was supplied by human muscle. Although they were limited by this fact and by a very narrow choice of materials to work with, we must not underestimate them. They were as intelligent as we are, lacking only the skill that comes from our accumulated experience.

This lack of skill was gradually overcome by Stone Age people and their descendants. Here are a few of their accomplishments, discovered and dated by archeologists:

4000 B.C.   Chaldeans, who lived at the head of the Persian Gulf in a province of ancient Babylonia, had become skilled workers in metals such as gold, silver, copper, lead, tin, and iron.

3000 B.C.   Eastern Mediterranean people were making glass, glazed pottery, and porcelain.

2500 B.C.   Babylonians were using petroleum instead of wood for fuel.

1100 B.C.   Chinese were mining coal and were drilling wells hundreds of feet deep for natural gas.

These accomplishments were arts learned by experience, and they implied the substitution of metals, glass, and other substances for stone. However, the making of a copper implement used more energy than the manufacture of a stone tool. Yet more was needed to make objects of glass. So with each advance, more energy and, therefore, more fuel were needed. Most of the energy for working iron, copper, lead, and other materials came from wood fuel and from the muscles of men and animals. But as time passed, other fuels came to be used, such as coal, oil, and natural gas. The new fuels were more efficient than the old. By using them, each man could produce many more implements than he could have produced with wood fuel or with muscle. Therefore, the new fuels sparked rapid growth of the amounts of metals used.

Use of Energy

Metals to build machines and the energy needed to power these machines are won from the Earth. Supplies of both metals and energy are found in certain kinds of rocks—unfortunately, not very common rocks—and the supplies have been formed through geological processes. Through our use of metals and energy, geological processes play a major role in the daily life of each and every one of us.

Metals and energy are complementary. Machines made of metal require fuels to supply the energy to run them. As more machines are built (in order to develop transportation systems, to build houses, or to till fields and grow crops) so does the use of energy increase. It is humbling to stop and consider the inadequacies of personal muscle power by comparison with our present-day energy needs.

A healthy, hard-working person can produce just enough muscle energy to keep a single 100 W light bulb burning during an 8 h working day. When we purchase the same amount of energy from our local electrical source, we pay about 10 cents. Viewed strictly as machines, we aren't worth much. By comparison, the amount of supplementary energy used each 8 h working day in North America could keep 300 of those bulbs burning for every person living there. Think of all this supplementary energy, if you will, as 300 silent "energy slaves" working eight hours a day to keep us well fed and enjoying a high standard of living. It is the "energy slaves" that produce our high crop yields, keep industry productive, power our automobiles, and heat our homes. Every person on the Earth now relies on at least some supplementary energy, which means everyone has some "energy slaves" working for them. In India the use of supplementary energy is now equivalent to 15 "energy slaves" for every man, woman, and child. In South America the figure is 30, in Japan, 100, in the U.S.S.R., 120, and in Europe, almost 200.

To see where all the energy is used we have to look at society as a whole and sum up all the energy that is employed to grow and transport food, make clothes, cut lumber for new homes, light streets, heat and cool office buildings, and to do myriad other things. The uses can be grouped into three categories: transportation, home and industry (meaning all manufacturing and raw material processing, plus the growing of foodstuffs), plus commercial uses. The present-day uses of energy

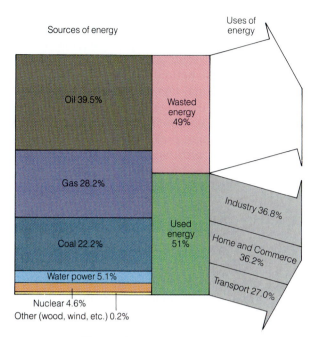

FIGURE 21.1 Uses and sources of supplementary energy in the United States in 1985. Wasted energy arises both from inefficiences of use and from the fact that there is a limit to the efficiency of an engine, and therefore of the fraction of energy that can be usefully employed, that is imposed by the fundamental laws of thermodynamics.

in the United States are summarized in Figure 21.1.

How much energy do all the peoples of the world use? The total is enormous. In 1985, the supplementary energy drawn from the major fuels such as coal, oil, natural gas, and nuclear power plants was 6.2×10^{19} cal. Nobody keeps accurate accounts of all the wood and animal dung burned every day in the cooking fires of Africa and Asia, but the amount has been estimated to be so large that when wood and dung are added, the world's total energy consumption rises to about 7.2×10^{19} cal/yr. This is equivalent to the burning of 2 tons of coal or 10 barrels of oil for every living man, woman, and child each year!

Supplies of Energy

The chief sources of energy consumed in highly industrialized nations are few: the fossil fuels (coal, oil, natural gas), hydroelectric power, nuclear energy, wood, wind, and a very small amount of muscle energy. As recently as a century ago, wood was an important fuel in industrial societies, but now it is used mainly for space heating in some dwellings. Wood and wind together supply only 0.2 percent of the energy consumed in the United States today. Excluding muscle (a resource now getting very little exercise in industrial coun tries), the sources of energy consumed in the United States are shown in Figure 21.1.

For Europe, the breakdown of energy supply and use is different: Coal accounts for about half the energy consumed, and industry for more than 40 percent of the use. For the world as a whole, with its hundreds of millions of agricultural workers, the breakdown is different again; wood is a more substantial source of energy, and fossil fuels, especially oil and gas, account for less.

Because our ability to produce and use the Earth's mineral supplies depends on our having energy available to do the necessary work, we first discuss the principal sources of energy, and then, in Chapter 22, the sources of mineral supplies.

FOSSIL FUELS

The term fossil fuel refers to the remains of plants and animals trapped in sedimentary rock. Fossil fuels occur in many ways, depending on the kind of sediment, the kind of organic matter trapped, and the changes that have occurred during the long geological ages since the organic matter was trapped.

Essentially all living organisms derive their energy from the Sun. The only known exceptions to this statement are a few animals that live around submarine hot springs on mid-ocean ridges; they derive their energy from the Earth's internal heat. The principal energy-trapping mechanism of living organisms that derive their energy from the Sun is photosynthesis. Plants combine water and carbon dioxide to make carbohydrates and oxygen as shown in Figure 21.2.

This combination process uses energy and plants get the energy from sunlight. The oxygen formed during photosynthesis is passed into the atmosphere and in this way plants and sunlight control the composition of the atmosphere.

The organic compounds in plants is the fuel that keeps animals alive and active; animals are, therefore, secondary consumers of trapped solar energy. When one animal eats another a little bit of trapped solar energy is once again passed along. When plants or animals die and decay, oxygen from the atmosphere combines with carbon and hydrogen in the organic compounds to form H_2O and CO_2 once again. In the process a small amount

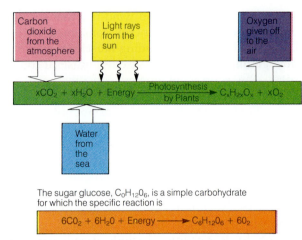

The sugar glucose, $C_6H_{12}O_6$, is a simple carbohydrate for which the specific reaction is

$$6CO_2 + 6H_2O + Energy \longrightarrow C_6H_{12}O_6 + 6O_2$$

FIGURE 21.2 Photosynthesis is the process by which plants combine carbon dioxide and water, using the Sun's energy to produce carbohydrates which have the general chemical formula $C_xH_{2x}O_x$, where x always is a whole number.

of energy is released, so the photosynthesis reaction is reversed.

The rates at which organic matter is formed through photosynthesis, and broken down by decay, are essentially the same; if they were not essentially equal, the world would soon be covered by increasingly deep piles of organic matter. However, the growth and decay rates are not *exactly* the same. In many sediments, a little organic matter is trapped and buried before it is completely removed by decay. In this way some of the solar energy becomes stored in rocks—hence, the term fossil fuel. The amount of trapped organic matter is far less than 1 percent of the organic matter formed by growing plants and animals. However, from the late Proterozoic (about 600 million years ago) to the present, through which time the size of the biomass seems to have been as large as it is today, the total amount trapped has grown to be very large.

The kind of organic matter that is trapped in sediments plays an important role in the kind of fossil fuel that forms. In the ocean, tiny photosynthetic phytoplankton and bacteria are the principal sources of trapped organic matter. Shales are the sedimentary rocks that do most of the trapping. Bacteria and phytoplankton contribute mainly organic compounds called proteins, lipids, and carbohydrates and it is these compounds that are transformed (mainly by heat) to oil and natural gas. On land, it is higher plants such as trees, bushes, and grasses that contribute most of the trapped organic matter; they are rich in resins,

waxes, and lignins, as well as carbohydrates in the form of cellulose. The trapped organic matter tends to remain solid and form coals although a certain amount of natural gas can be formed too. In many shales, burial temperatures never reach the levels at which the original organic molecules are completely broken down. Instead, what happens is that an alteration process occurs in which wax-like substances with large molecules are formed. This material, called *kerogen*, is the substance in oil shales; it can be converted to oil and gas by applying sufficient heat.

Coal

The black, combustible sedimentary rock we call coal is the most abundant of the fossil fuels. Most of the coal that is mined is eventually burned under boilers to make steam for electrical generators, or it is converted into coke, an essential ingredient in the smelting of iron ore and the making of steel. In some parts of the world coal is still used as a fuel for steam locomotives, but this use is declining. When coal is heated under suitable conditions, a fraction of it can be converted to a liquid fuel. Because internal combustion engines are more efficient than steam engines, the future will probably see more and more coal used in this way. In addition to its use as a fuel, coal is a raw material from which nylon and many other plastics, plus a multitude of organic chemicals, can be made.

Origin of Coal

Coal occurs in strata (miners call them seams) along with other sedimentary rocks, mostly shale and sandstone. A look through a magnifying glass at a piece of coal reveals the shapes of bits of fossil wood, bark, leaves, roots, and other parts of land plants, chemically altered but still identifiable. This observation leads at once to the conclusion that coal is fossil plant matter (Fig. 21.3). This conclusion we can incorporate into our definition of *coal* as *a black sedimentary rock consisting chiefly of decomposed plant matter and containing less than 40 percent inorganic matter.*

Many coal seams include fossil tree stumps rooted in place in the underlying shale, evidently a former clay-rich soil. Unlike the material in most sedimentary rock, the sediment we now find in coal seams was not eroded, transported, and deposited; instead it accumulated right where the plants grew. It was recognized long ago that places where coal accumulated were ancient swamps, because (1) a complete physical and chemical gradation exists from coal to peat, which today accumulates mainly in swamps, and (2) only under swamp conditions is the conversion of plant matter to coal chemically probable. On dry land and in running water, oxygen is abundant and dead plant matter gradually rots away. However, under stagnant or nearly stagnant swamp water, oxygen is used up and not replenished. Instead, the plant matter is

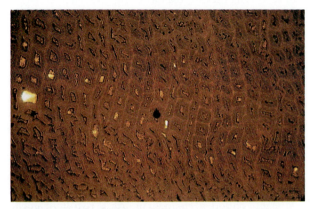

FIGURE 21.3 Coal of Pennsylvanian age from Nashville, Illinois, showing cellular structure in a fossilized fragment of wood from a plant called *Lycopod periderm.* The photograph, taken in reflected light, is of an etched surface. The field of view is approximately 0.3 cm across.

attacked by anaerobic bacteria which partly decompose it by splitting off some of the oxygen and hydrogen. These two elements escape, combined in various gases, and the carbon gradually becomes concentrated in the residue. Although they work to destroy the vegetal matter, the bacteria themselves are destroyed before they can finish the job, because the poisonous acid compounds they liberate from the dead plants kill them. This could not happen in a stream because the flowing water would bring in new oxygen to decompose the plants and would also dilute the poisons and permit the bacteria to complete their destructive process.

With the destruction of bacteria, the plant matter has been converted to peat. But as the peat is buried beneath more plant matter and beneath accumulating sand, silt, or clay, both the temperature and pressure increase. These bring about a series of continuing changes called *coalification* (Fig. 21.4). The peat is compressed, water squeezed out, and the volatile organic compounds such as methane (CH_4) escape leaving an increased proportion of carbon. The peat is converted successively into lignite, subbituminous coal, and bituminous coal. These coals are sedimentary rocks. However, a still-later phase, *anthracite*, is a metamorphic rock. Since anthracite occurs in folded strata that have been subjected to low-grade

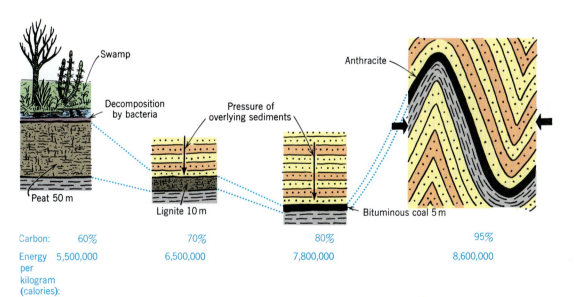

Carbon:	60%	70%	80%	95%
Energy per kilogram (calories):	5,500,000	6,500,000	7,800,000	8,600,000

FIGURE 21.4 Accumulating plant matter is converted into coal by decomposition, pressure, and heat. By the time it has become bituminous coal, a layer of peat has decreased to one tenth of its original thickness. During the same time the proportion of carbon has increased and the calorific value (the amount of heat energy obtained when a kilogram of coal is burned) has risen continually.

metamorphism, we infer that it has undergone a further loss of volatiles and carbon concentration, as a result of the pressure and heat that accompany folding and metamorphism. Because of its low content of volatiles, anthracite is hard to ignite, but once alight, it burns with almost no smoke. In contrast, lignite, rich in volatiles, ignites so easily that it is dangerously subject to spontaneous ignition (in chemical terms, rapid oxidation), and burns smokily. In certain regions where metamorphism has been intense, coal has been changed so thoroughly that it has been converted to graphite, in which all volatiles have been lost, leaving nothing but carbon. Graphite, therefore, will not burn in an ordinary fire.

Most coals are known as *humic* coals; they consist of organic debris that has passed through a peat stage. The major components of humic coals are twigs, branches, and other woody material, together with leaves and fronds. They produce a coal full of shiny, brown to black materials known as *macerals*. Much less common types of coal, known as *sapropelic* coals, consist primarily of the remains of algae, spores, and plant waxes that collect in local, oxygen deficient lakes and lagoons. These coals, commonly called *cannel* and *boghead* coals, are dull rather than lustrous like humic coals, and lack obvious signs of bedding.

Occurrence

We have said that coal occurs in layers or seams, which are merely strata. Each seam is a flat, lens-shaped body corresponding to the area of the swamp in which it accumulated originally. Most coal seams are 0.5 to 3 m thick, although some reach more than 30 m. They tend to occur in groups. In western Pennsylvania, for example, there are about 60 beds of bituminous coal. This indicates that coal formed in slowly subsiding sedimentary basins. The occurrence of coal beds in the rock layers of every period later than Devonian indicates that during the last 300 million years or so, swamps rich in vegetal growth have been recurrent features of the land. Peat is accumulating today, at an average rate of about 1 m every 100 years in swamps on the Atlantic and Gulf Coastal Plains of the United States. The swamps now represented by coal beds must have been much the same as these. Presumably, too, the old coal swamps flourished in a climate somewhat like that of the Atlantic and Gulf Coast swamps of today.

Coal swamps seem to have formed in many sedimentary environments, of which two predominated. One consists of slowly subsiding basins in continental interiors and the swampy margins of shallow inland seas formed at times of high-sea stands. This is the home environment of the bituminous and subbituminous coal seams in Utah, Montana, Wyoming, and the Dakotas. The other consists of continental margins with wide continental shelves that were also flooded at times of high sea level. This is the environment of the bituminous coals of the Appalachian region. That same environment exists along the east coast of North America today, and it is there that one of the largest modern coal swamps is to be found. The Dismal Swamp, in Virginia and North Carolina, with an area of 5700 km^2, contains an average thickness of 2 m of peat.

In many coal basins, swamps were formed repeatedly as the sea level rose and fell, or as the local area was uplifted and submerged. What is observed today is a repetitive sequence of sedimentary rocks—sandstone, shale, coal, sandstone, shale, coal, etc.,—reflecting cyclic deposition. Each unit is *a repeated sequence of peat and clastic sediments* called a **cyclothem**. As a result of cyclic deposition, most of the world's coal basins contain many coal seams that overlie one another.

Although peat can form under even subarctic conditions, it is clear that the luxuriant plant growth needed to form thick and extensive coal seams happen most readily under a tropical or semitropical climate. Even the dense growth in Dismal Swamp is probably insufficient to produce, ultimately, a coal seam as thick as some of the seams in Pennsylvania. This means that either the global climate was very much warmer, or the swamps in which most of the world's coal seams formed were within 30° of the equator, when the plant matter accumulated. Probably both effects were involved. In Figure 2.22, it is apparent that one of the two great coal-forming periods, about 90 million years ago during the Cretaceous Period, coincided with a time of exceptionally high CO_2 in the atmosphere. As a result the global climate was probably much hotter and coal swamps may have flourished in high latitudes. The second coal-forming period, the Carboniferous (about 300 million years ago), was also a time of higher CO_2 content than today, but nothing like the warm Cretaceous period. Continental reconstruction suggests that the Carboniferous coal swamps did form within 30° north and south of the equator and that they have reached their present latitudes by continental drifting.

Peat formation has been widespread and more

or less continuous from the time land plants first appeared about 450 million years ago during the Silurian Period. The size of the swamps in which the peat accumulated, however, has varied greatly and so, as a consequence, has the amount of coal formed. By far, the greatest period of coal swamp formation occurred during the Carboniferous and Permian, when Pangaea existed. The great coal beds of Europe and the eastern United States formed at this time, when the plants of coal swamps were giant ferns and the so-called scale trees (gymnosperms). The second great period of coal deposition peaked during the Cretaceous, but commenced in the early Jurassic and continued until the mid-Tertiary. The plants of the coal swamps during this period were flowering plants (angiosperms), much like flowering plants today.

Distribution

Coal is not only abundant, it is also rather widely distributed. Experts believe that most of the world's major coal seams have been discovered. In that case a good estimate of the amount of mineable coal can be made. Coal seams that are thinner than 0.3 m are too thin to be mined, and seams deeper than 2000 m make mining dangerous as well as expensive. Taking 0.3 m as the lower limit of thickness and 2000 m as the depth limit for mining, experts estimate that 13,800 billion tons of coal can be mined. Most of this reserve is in Asia (Fig. 21.5) (principally in the U.S.S.R.), but North America is well endowed too (Fig. 21.6).

Most of the world's coal resources are in the Northern Hemisphere. Land plants first evolved during the Silurian Period, so no coal is older than Silurian. Evidently the southern continents have not spent much time close to the equator since the Silurian.

Mining

The average thickness of all seams now mined in the United States is about 1.6 m. In the past, most coal has been obtained by mining underground, and historically, miners have recovered no more than about 50 percent of the coal present. Underground mining has always been a dangerous and unpleasant occupation. Increasingly, underground mining is being modernized by automatic coal-cutting and coal-loading machines, which have increased miners' safety, increased production of coal per man-hour by factors of 10 to 20

times, and increased the percentage of coal recovered from some seams to 60 percent or more.

Much of the coal mined today is recovered by surface mining. When coal lies less than 60 m beneath the surface, it can be mined by stripping away the overlying surface rock, and then digging out the exposed coal with huge power shovels that take 10 to 100 m^3 at a single bite (Fig. 21.7). Referred to as *strip mining*, this method greatly reduces the danger to miners and permits recoveries of 90 percent or more of the coal in the ground. One drawback, of course, is that the surface is disturbed. Because much of the shallow, strippable coal occurs in rich farming areas, the farming is disturbed. The land can be restored after mining is completed, but reclamation is slow and sometimes expensive. The big complaint is that it is often done in a shoddy manner!

A "rule of thumb" concerning strip mining is that extraction is economic if the ratio of the thickness of overlying rock to the thickness of the coal seam is less than 20:1. It is estimated that less than 10 percent of all coal in the United States is in thick enough seams and is close enough to the surface to be reached by strip mining.

Another method of mining coal from the surface, without even going underground, employs huge augers. This method is used mainly in hilly areas of the Appalachian states. First a bench is cut with bulldozers to expose the coal seam, then two or more augers, about 2 m in diameter, are installed. They bore horizontally into the seam and move the coal outward between their spiral blades to loading belts (Fig. 21.8). Although the method is

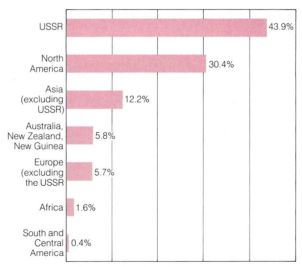

FIGURE 21.5 Geographic distribution of the world's coal resources. (*Source:* Data from U.S. Geol. Survey, 1974.)

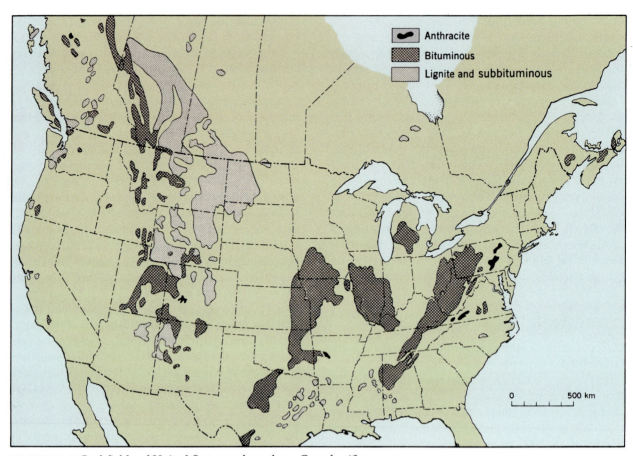

FIGURE 21.6 Coal fields of United States and southern Canada. (*Source:* Adapted from maps by U.S. Geol. Survey and Geol. Survey of Canada.)

FIGURE 21.7 This coal strip-mine at Wyodak, Wyoming, produces, each day, 300,000 tons of subbituminous coal from a seam 20 m to 30 m thick.

FIGURE 21.8 A huge multiple auger in eastern Kentucky bores into a coal seam from an artificially excavated bench. Coal is cut, moved out, and loaded in a single operation.

cheap and efficient, only a very small percentage of all coal can be recovered by augers. Unfortunately, however, the cutting of benches in hilly country creates serious environmental problems such as acid stream waters and local landslides.

Experiments are now being made with burning coal while it is still in the ground to produce gas for industrial use. This technique would eliminate mining altogether, and if successfully developed, would greatly reduce environmental damage.

Petroleum: Oil and Natural Gas

Rock oil is one of the earliest products our ancestors learned to use. In the valleys of the Tigris and Euphrates Rivers in Iraq, natural oil seeps occur. The people living in those regions have used the oil since before recorded history. Some of the ways Babylonians, Assyrians, and other ancient people used oil were: (1) to make bitumen glue to hold arrowheads on spears; (2) to set tiles; (3) to make a mortar for holding bricks together in a building; (4) to waterproof boats; and (5) to embalm bodies. In later times, people in the Arab world and China used oil for many additional purposes. However, the major use of oil really started about 1847 when a merchant in Pittsburgh, Pennsylvania, started bottling and selling rock oil from natural seeps to be used as a lubricant. Five years later, in 1852, a Canadian chemist discovered that distillation of both rock oil and coal yielded kerosene, a liquid that could be used in lamps. This discovery spelled doom for candles and whale-oil lamps. Wells were soon being dug by hand near Oil Springs, Ontario in order to increase the yield of oil. In Romania, using the same hand-digging process, oil production in 1857 was 2000 barrels[1] a year. In 1859, the first oil well was drilled at Titusville, Pennsylvania. On August 27, 1859, at a depth of 21.2 m, oil-bearing strata were encountered and up to 35 barrels of oil a day were pumped out. Oil was soon discovered in West Virginia (1860), Colorado (1862), Texas (1866), California (1875), and many other places.

The earliest known use of natural gas was in China, where more than 3000 years ago, gas seeping out of the ground was collected and transmitted through bamboo pipes to be used to evaporate salt water in order to recover the salt. Modern uses of gas started in the early seventeenth century in Europe, where gas made from wood and coal was used for illumination. Commercial gas companies were founded as early as 1812 in London, and 1816 in Baltimore. The stage was set for the exploitation of an accidental discovery at Fredonia, New York, in 1821. A water well drilled in that year produced not only water, but bubbles of a mysterious gas. The gas was accidentally ignited and produced such a spectacular flame that a new well was drilled on the same site and wooden pipes were installed to carry the gas to a nearby hotel where 66 gas lights were installed. By 1872, natural gas was being piped as much as 40 km from its source.

Oil and gas are the two chief kinds of petroleum. We define **petroleum** as *gaseous, liquid, and semisolid substances, occurring naturally and consisting chiefly of chemical compounds of carbon and hydrogen.* Oil and gas occur together and are searched for in the same way. Therefore, we can follow general practice and talk about oil pools, oil exploration, and the origin of oil with the understanding that we mean not only oil but gas as well. In discussing oil, we shall deal with its occurrence, its origin, and distribution, in that order.

Occurrence

The accumulated experience of more than a century of exploration, drilling, and producing has taught us much about where and how oil and gas occur. Oil possesses two important properties that affect its occurrence. It is fluid, and it is generally lighter than water. Oil is produced from pools (an **oil pool** is *an underground accumulation of oil and gas in a reservoir limited by geologic barriers*). The word pool sometimes gives a wrong impression because an "oil pool" is not a lake of oil. It is a body of rock

---

[1] A barrel is equal to 42 U.S. gal and is the volume generally used when oil is discussed.

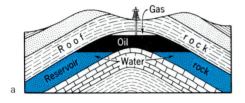

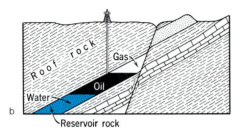

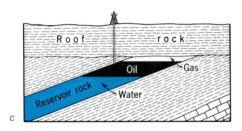

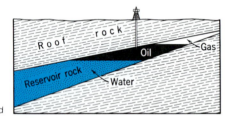

FIGURE 21.9 Four of the many kinds of oil traps. Here *A* and *B* are structural traps; *C* and *D* are stratigraphic traps—in *C,* an unconformity marks the top of the reservoir; in *D,* a porous stratum thins out and is overlain by an impermeable roof rock. Gas (white) overlies oil (black), which floats on groundwater (blue), saturates reservoir rock, and is held down by roof of claystone. Oil fills only the pore spaces in the rock.

in which oil occupies the pore spaces. *A group of pools, usually of similar type, or a single pool in an isolated position* constitutes an **oil field.** The pools in a field can be side by side or one on top of the other.

For oil or gas to accumulate in a pool, five essen-

tial requirements must be met. (1) There must be a *reservoir rock* to hold the oil, and this rock must be permeable so that the oil can percolate through it. (2) The reservoir rock must be overlain by a layer of impermeable *roof rock,* such as shale, to prevent upward escape of the oil, which is floating on groundwater. (3) The reservoir rock and roof rock must form a *trap* that holds the oil and prevents it from moving any farther under the pressure of the water beneath it (Fig. 21.9). These requirements are much like those of an artesian-water system, but with the essential difference that the artesian aquifer connects with the surface, whereas the oil pool does not. Although these three features—reservoir, roof, and trap—are essential, they do not guarantee a pool. In many places where they occur together, drilling has shown that no pool exists, generally because of lack of a source from which oil could enter the trap. So, to the foregoing requirements for a pool we must add two others: (4) there must be *source rock* to provide oil; and (5) the *deformation* that forms the trap must occur *before all the oil has escaped* from the reservoir rock.

Origin

Oil consists of scores of different hydrocarbon compounds, and no two oils have exactly the same mix of compounds. Even so, the bulk chemical compositions of most oils are very similar (Table 21.1). Natural gas, by contrast with oil, is dominantly methane (CH_4), though minor amounts of other gases such as ethane (C_2H_6), propane (C_3H_8), and butane (C_4H_{10}) may also be present.

Petroleum is believed to be a product of the decomposition of organic matter of both plant and animal origin. This belief arises principally because of two observations: (1) oil possesses optical properties known only in hydrocarbon compounds derived from organic matter; and (2) oil contains nitrogen and certain compounds (porphyrins) that scientists believe can only originate in living matter.

Oil is nearly always found in marine sedimentary strata. Indeed, in places on the seafloor, particularly on the continental shelves and at the bases of the continental slopes, sampling has shown that fine-grained sediment now accumulating contains up to 8 percent organic matter that is chemically good potential oil substance. Thus, on the basis of the Principle of Uniformity, we can draw the conclusion that the oil we now find in rock originated as organic matter deposited with marine sediment.

TABLE 21.1 *Composition of Typical Petroleum*

Element	Oil (percent)	Natural Gas (percent)
Carbon	82.2–87.0	65–80
Hydrogen	11.7–14.7	1–25
Sulfur	0.1– 5.5	Trace to 0.2
Nitrogen	0.1– 1.5	1–15
Oxygen	0.1– 4.5	—

Source: Levorson, 1967.

This suggestion is only a general one. Analyzing the organic matter in modern sediments, chemists have found significant differences between its composition and that of petroleum. Therefore, in order to effect the change from organic matter to petroleum, a complex series of changes must occur. The following simplified theory about how the steps occur is widely held and is supported by enough facts to be at least somewhere near the truth.

1. The raw material consists mainly of microscopic marine organisms, mostly phytoplankton, living in multitudes at and near the sea surface. Measurements show that the sea grows at least 31.5 g of organic matter per square centimeter per year, and the most productive inshore waters sometimes grow as much as six times this amount. The latter value represents more organic matter than could be harvested in a year from the richest farmland.

2. When the microscopic organisms die, they fall to the bottom. Some of the dead plant matter is devoured by scavengers, some is oxidized by oxygen in the water, and the rest becomes buried in the accumulating pile of sediment. There it is attacked and decomposed by bacteria, which split off and remove oxygen, nitro-gen, and other elements, leaving residual carbon and hydrogen. Sediment that is rich in organic matter teems with bacteria.

3. Deep burial beneath fine sediment destroys the bacteria and provides pressure and heat. Millions of years of further chemical changes convert the substance into droplets of liquid oil and minute bubbles of gas.

4. Gradual compaction of the enclosing sediments, under the pressure of their own increasing weight, and any deformation that might occur by folding, reduces the space between the rock particles and squeezes out oil and gas substance into nearby layers of sand or sandstone, where open spaces are larger.

5. Aided by their buoyancy and by the circulation of artesian water, oil and gas migrate (generally upward) through the sand until they reach the surface and are lost, or until they are caught in a trap and form a pool.

We repeat that this statement is oversimplified. A long and complex chain of chemical reactions is involved in the conversion of the original organic constituents to crude petroleum. Also, chemical changes may occur in oil and gas even after they have migrated into their reservoirs. This may help explain why chemical differences exist between the oil in one pool and that in another.

The migration of oil needs more explanation. The sediment in which oil substance is accumulating today is rich in clay minerals, whereas most of the strata that constitute oil pools are sandstones (consisting of quartz grains), limestones and dolostones (consisting of carbonate minerals), and much-fractured rock of other kinds (Fig. 21.10). It seems obvious, therefore, that oil forms in one kind of material and at some later time migrates to another. The migration process is essentially anal-

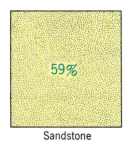
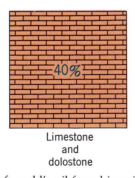
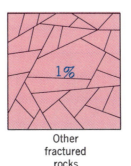

Sandstone   59%

Limestone and dolostone   40%

Other fractured rocks   1%

FIGURE 21.10 Percentage of world's oil found in principal kinds of reservoir rocks.

ogous to the movement of groundwater. When, as we mentioned above, oil is squeezed out of the clay-rich sediment in which it originated and enters a body of sandstone or limestone somewhere above, it can migrate more easily than before. One reason is that sandstone, as explained in Chapter 5, is far more permeable than any clay-rich rock. Another reason is that the force of molecular attraction between oil and quartz or carbonate minerals is less strong than that between water and

quartz or water and carbonate minerals. Hence, because oil and water do not mix, water remains fastened to the quartz or carbonate grains while oil occupies the central parts of the larger openings. Because it is lighter than water, the oil tends to glide upward past the carbonate- and quartz-held water. In this way it becomes segregated from the water; when it encounters a trap, it can form a pool.

Most of the oil that forms in sediments does not find a suitable trap and eventually makes its way, along with the artesian water, to the surface. It is estimated that no more than 0.1 percent of all the organic matter originally buried in a sediment is eventually trapped in an oil pool. Most of it escapes to the surface. It is not surprising, therefore, that the highest ratio of oil pools to volume of sediment is found in rock no older than 2.5 million years, and that nearly 60 percent of all the oil so far discovered has been found in strata of Cenozoic age (Fig. 21.11). This does not mean that older rocks produced less oil. It simply means that oil in older rocks has had a longer time in which to escape. Another interesting discovery by oil drillers is that the amount of oil decreases as drilling proceeds deeper and deeper (Fig. 21.12). This too might be expected because deeper rocks are more highly compacted and the pressure that drives the oil and water upward is greater there.

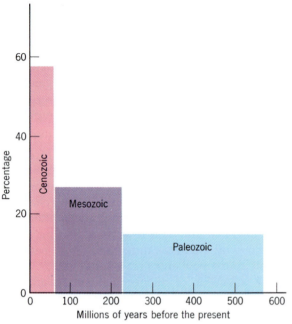

FIGURE 21.11 Percentage of world's total oil production from strata of different ages.

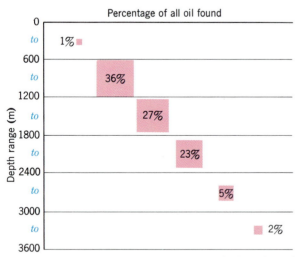

FIGURE 21.12 Relation between depths of oil pools and percentage of world's oil production.

Distribution

Petroleum, like coal, is widespread but distributed unevenly. The reasons for the uneven distribution are not so obvious as they are with coal. Suitable source sediments for oil are very widespread and seem as likely to form in subarctic waters as in tropical regions. The critical controls seem to be a supply of heat to effect the conversion of solid organic matter to liquid and gaseous forms, and the formation of a suitable trap before the petroleum has leaked away.

Conversion of solid organic matter to oil and gas happens within a specific range of depth and temperature defined by the geothermal gradients shown in Figure 21.13. If a thermal gradient is too low—that is, less than 1.8° C/100 m—conversion does not occur. If the gradient is above 5.5° C/100 m, conversion to gas starts at such shallow depths that very little trapping occurs. Once oil and gas have been formed, they will only accumulate in pools if suitable traps are present. Most oil and gas pools are found beneath anticlines; the timing of the folding event is therefore a critical part of the trapping process. If folding occurs after petroleum

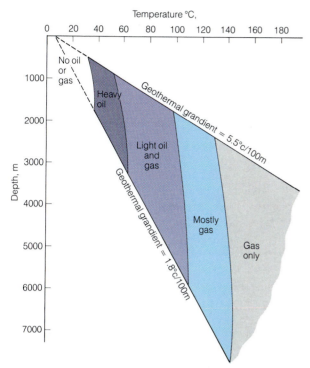

FIGURE 21.13 Regions of depth and temperature within which the generation and trapping of oil and gas occur.

generation and migration has occurred, pools cannot form. The great oil pools in the Middle East arose through the fortunate coincidence of a high thermal gradient and development of anticlinal traps due to the collision of Europe and Asia with Africa.

How much oil is there in the world? This is an extremely controversial question. Approximately 600 billion barrels of oil have been discovered, but a great deal remains to be found by drilling. Unlike coal, for which the volume of strata in a basin of sediment can be accurately estimated, the volume of undiscovered oil can only be guessed at. The way guesses are made is to use the accumulated experience of a century of drilling. Knowing how much oil has been found in an intensively drilled area, such as eastern Texas, experts make estimates of probable discoveries in other regions where rock types and structures are similar. Using this approach, and considering all the sedimentary basins of the world (Fig. 21.14), experts estimate that somewhere between 1500 and 3000 billion barrels of oil have already been, or eventually will be, discovered and produced. So far, over 500 billion barrels of oil have been produced. The estimates include gas, which is added in on the basis of its calorific value, 1470 m³ of gas being taken as equal

in heating capacity to one barrel of oil. Considering that the rate at which the world uses oil has now risen to between 20 and 30 billion barrels each year, and that consumption continues to grow despite embargoes, high prices, and political difficulties, the total amount of oil left seems highly inadequate for probable long-term future needs.

Not only does the supply seem inadequate, but also it is distributed unevenly on a geographic basis. The estimate takes into account all the sediments on the continental shelves around the world and the great piles of sediment at the bases of continental slopes. Most of the promising sediments occur along passive continental margins. Along convergent continental margins, the sediment piles and potential for oil seem to be less. However, the sediments that accumulate in both outer and back arc basins have high oil potential. Because they are just being explored, the offshore regions offer the greatest potential for future discoveries. These are difficult and expensive places to explore and drill, and many problems of ownership remain to be solved. Wells are already being drilled as far as 250 km offshore, and oil and gas are being produced at water depths as great as 300 m. Eventually, much more distant and deeper-water sites will be drilled. The geographic regions, both onshore and offshore, where petroleum is known or estimated to occur are shown in Figure 21.15. The same figure also shows where the world's oil presently comes from. North America and Africa produce more petroleum by comparison with their total resources than other continents do. This worries some people because it means these two continents are likely to be the first to run out of local oil.

When petroleum flows or is pumped from an oil pool, as much as 60 percent of the oil originally present remains trapped as coatings on mineral grains, and in innumerable tiny holes and fractures in the reservoir rock. Some of this trapped oil can be recovered by secondary processes such as blasting and underground heating, but as much as half of the original total still remains trapped and is not recoverable by any presently known methods. Trapped oil thus constitutes a potential resource, as large as flowing oil, but a great technological advance is needed before it can be recovered.

Heavy Oils and Tars

Oil that is exceedingly viscous and thick will not flow and cannot be pumped. Colloquially called *tar*

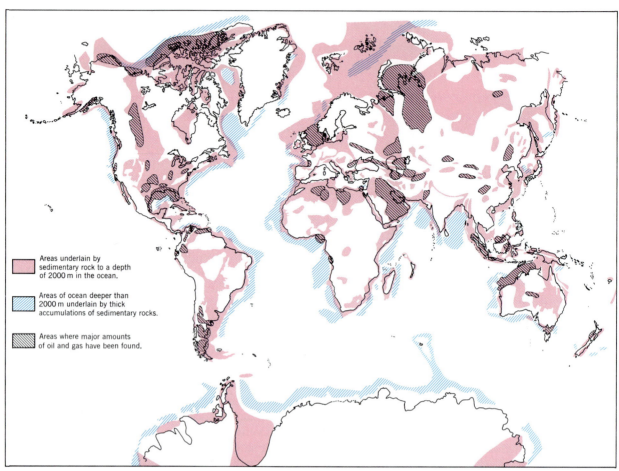

FIGURE 21.14 World map showing areas underlain by sedimentary rock and regions where large accumulations of oil and gas have been located. Where the ocean is deeper than 2000 m, sedimentary rock has yet to be tested for its oil and gas potential.

or *asphalt*, heavy, viscous oil acts as a cementing agent between mineral grains. The tar can be recovered from the tar sands only if the sand is heated to make the tar flow. The resulting tar must then be processed to recover the valuable gasoline fraction. The heating can be done underground, somewhat as in the secondary recovery of trapped oil. Some tars are recovered by steam heating in California. Most present practice, however, relies on mining and surface processing. This means that deposits must be close enough to the surface to be worked by inexpensive surface-mining methods. By far the largest occurrence of tar sands that meets these requirements in North America is in Alberta, Canada, where the *Athabasca Tar Sand* covers an area of 5000 km² and reaches a thickness of 60 m (Fig. 21.16). Mining of the Athabasca Tar Sand already exceeds a rate of 100,000 tons a day. Assuming 50 percent recovery, as much as 500 bil-

lion barrels of viscous tar might eventually be recovered. Deposits of tar sand almost as large as the Athabasca deposit are known in Venezuela and in the U.S.S.R. A great many smaller deposits of tar sand are also known around the world. Some experts suggest that the total reserve of tar might be as large as the sum of all the trapped and flowing oil. If just half this amount were recovered, tar sands would ultimately yield as much oil as would flowing oil and gas from oil pools.

Oil Shale

Another source of petroleum consists of solid organic matter (kerogen) enclosed in fine-grained sedimentary rock. If the organic matter is heated, the solid breaks down and liquid and gaseous hydrocarbons, similar to those in oil, can be distilled out. All sedimentary rock contains some organic matter, but to be considered an energy resource the organic matter must yield more energy than that required for the processes of mining and distillation. The only kind of sedimentary rock that contains sufficient solid organic matter to be given any attention is shale, and only those shales that yield 40 or more liters of distillate per ton can be considered, because the energy needed to mine and process a ton of shale is equivalent to that created by burning 40 l of oil.

Unfortunately, most shales will not yield 40 l of oil per ton, but a few particularly rich oil-shale deposits are known. In two places in the world (Estonia, U.S.S.R., and China), shales that yield as much as 320 l of oil per ton are being worked. The world's largest deposit of rich oil shale, however, is in the United States. During the Eocene Epoch many large, shallow lakes existed in basins in Colorado, Wyoming, and Utah; in three of them was deposited a series of rich organic sediments that are now the *Green River Oil Shales* (Figs. 5.8 and

21.16). The richest shales were deposited in the lake in Colorado (Fig. 21.17). These shales are capable of producing as much as 240 l of oil per ton. Scientists of the U.S. Geological Survey estimate that oil-shale resources capable of producing 40 l or more oil a ton, in the Green River Oil Shales alone, total about 2000 billion barrels of oil. It is most unlikely that all this oil could ever be recovered. Many low-grade areas hardly warrant treatment, and because mining and processing the shale is expensive and creates tremendous environmental disturbance, probably even some of the richest areas will not be touched. Nevertheless, experts suggest that 50 percent of the oil from the Green River Oil Shales may someday be recovered. Many of the experts believe that large-scale mining and processing of the shale will be under way before the year 2000.

Rich resources of oil shale in other parts of the world have not been adequately explored, but another huge deposit occurs in Brazil, in the Irati

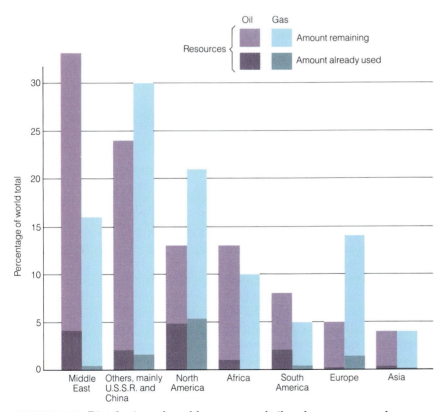

FIGURE 21.15 Distribution of world resources of oil and gas, expressed as a percentage of the total, and the amount used up by 1982. North America produces more than its share, compared to its resources. The U.S.S.R. and China produce less than their relative share. Note that North America and the U.S.S.R. tend to be gas-rich, while the Middle East and Africa tend to be oil-rich.

a

b

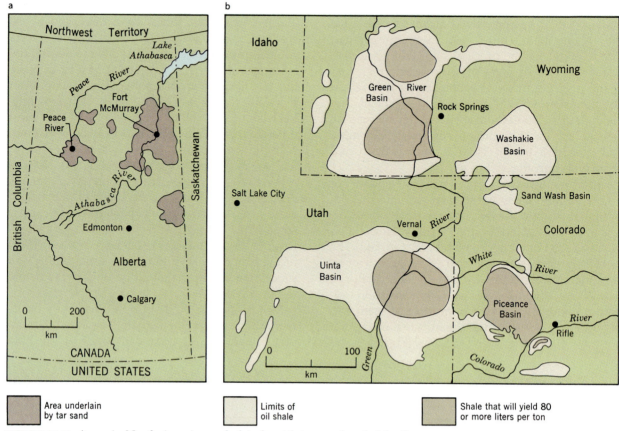

| | Area underlain by tar sand | | Limits of oil shale | | Shale that will yield 80 or more liters per ton |

FIGURE 21.16 Areas in North America underlain by rich tar sand and rich oil shale. (*a*) Athabasca Tar Sand in Alberta; (*b*) Green River Oil Shale, originally deposited as sediment rich in organic matter. The sediment accumulated in freshwater lakes that formed in shallow basins in Colorado, Wyoming, and Utah. Compaction and cementation of the sediment formed the shale we see today. (*Source:* U.S. Geol. Survey, 1968.)

Shale. Another very large deposit is known in Australia, and others have been reported in such widely dispersed places as Brazil, South Africa, and China. At this stage it is not possible to make an accurate worldwide estimate of the total amount of distillable oil in oil shale. What is clear, however, is that oil shale is by far the largest of the fossil fuel resources and is possibly a hundred or more times larger than coal. The trouble is that most of the oil shale is too lean ever to be worked, and as a consequence the amount of economically recoverable shale oil around the world may not even be as large as recoverable resources of flowing oil.

How Much Fossil Fuel?

Because of their "one crop" availability, are the supplies of fossil fuels adequate to meet future demands? If we use a barrel of oil as our unit of measurement, we can compare quantities of all fossil fuels directly. Thus, approximately 0.22 ton of coal produces the same amount of heat energy as one barrel of oil; so the $13{,}800 \times 10^9$ tons of coal stated earlier to be the world's recoverable coal reserves are equivalent to about 62,730 billion barrels of oil.

Considering the 1984 world-use rate of 30 billion barrels of oil a year, and comparing the estimated recoverable amounts of fossil fuels (Table 21.2), it is apparent that only coal seems to have the capacity to meet long-continued demands. The two fuels on which we now rely most heavily, oil and gas, are very efficient and the least expensive of all the fossil fuels in terms of manpower and disruption of environments. As we shift, inevitably, to increased use of other fuels, a great many problems can be anticipated. One is a rapid increase in the CO_2 content of the atmosphere. Environmental problems may even make it necessary to decide that we should not mine and use fossil fuels on an

a

b

FIGURE 21.17 Green River Oil Shale. (*a*) The richest outcroppings are in the Parachute Creek area of western Colorado. The chocolate-colored, thin-bedded strata crop out prominently. (*b*) Specimen of oil shale from Parachute Creek, Colorado. The specimen is 15 cm high. The alternating organic-rich (dark), and carbonate-rich (light) bands are believed to be varves. Disturbance of strata is considered to have occurred during the compaction of sediments.

ever-larger scale. Therefore, we shall consider other possible sources of energy.

OTHER SOURCES OF ENERGY

Three sources of energy other than fossil fuels have already been developed to some extent: the Earth's biomass, hydroelectric energy, and nuclear energy. Several others—such as the Sun's heat, winds, tides, and the Earth's internal heat—have been tested and developed on a limited basis. None has yet been developed on a large scale in the United States, but the day may not be far in the future when each will become locally important.

Biomass Energy

Scientists working for the United Nations estimate that wood and animal dung used for cooking and heating fires now amounts to energy production of 10^{19} cal annually. This is approximately 14 percent of the world's total use of supplemental energy. The greatest use of wood as a fuel occurs in developing countries where the cost of fossil fuel is very high in relation to income.

Measurements made on living plant matter indicate that new plant growth on land equals 1.5×10^{11} metric tons of dry plant matter each year. If all of this were burned, or used in some other way as

TABLE 21.2 *Amounts of Fossil Fuels Possibly Recoverable Worldwide. (Unit of comparison is a barrel of oil)*

Fossil Fuel	Total Amount in Ground (billions of barrels)	Amount Possibly Recoverable (billions of barrels)
Coal	About 100,000	62,730[a]
Oil and gas (flowing)	1500 to 3000	1500 to 3000
Trapped oil in pumped-out pools	1500 to 3000	0 to ?
Viscous oil (tar sands)	3000 to 6000	500 to ?
Oil shale	Total unknown. Much greater than coal.	1000 to ?

[a] 0.22 ton of coal = 1 barrel of oil.

an energy source, it would produce 62×10^{19} cal—almost nine times more supplemental energy than the world uses today. Obviously, it is not possible to do this because forests would have to be destroyed, plants could not be eaten, and agricultural soils would be devastated. The danger in using biomass energy on a large scale is that food supplies could suffer. Nevertheless, controlled cropping of fuel plants could increase the fraction of the biomass now used for fuel and in several parts of the world, such as Brazil and China, experiments are already underway to develop this obvious energy source.

Hydroelectric Power

Hydroelectric power is recovered from the potential energy of water in streams and rivers as they flow downward to the sea. Water ("hydro") power is an expression of solar power because it is the Sun's heat energy that drives the water cycle. That cycle is continuous; so energy obtained from flowing water is also continuous (unlike coal and oil, it cannot be used up). However, to convert the power of flowing water into electricity efficiently, it is necessary to dam streams. Because a stream carries a suspended load of sediment, the reservoir behind the dam eventually will become filled with deposited sediment. Depending on the sediment load, a reservoir could be completely filled up within 50 to 200 years. Lake Nasser, the reservoir held behind the great Aswan Dam on the river Nile, will be almost half filled with silt by the year 2025. Thus, although water power is continuous, the reservoirs needed for the conversion of water power to electricity have limited lifetimes.

Water power has been used in small ways for thousands of years, but only in the twentieth cen-

tury has widespread use been made for generating electricity. All the water flowing in the streams of the world has a total amount of recoverable energy that has been estimated as 2.2×10^{19} cal/yr. This is an amount of energy equivalent to burning 15 billion barrels of oil.

The distribution of water power around the world, and the extent to which it is already developed, is shown in Figure 21.18. There we see that the greatest undeveloped potential lies in two continents of the southern hemisphere, Africa and South America, both of which lack large coal reserves.

Nuclear Energy

Nuclear energy has a very large potential but it also has some difficulties. The use of nuclear energy by industry already plays an important part in the world's economy. Nuclear energy in a nuclear reactor is the heat energy produced during controlled fission of suitable radioactive isotopes. Three of the same radioactive atoms that keep the Earth hot by spontaneous radioactive decay—^{238}U, ^{235}U, and ^{232}Th—can be mined and used to produce energy by inducing fission under controlled conditions. Fission is accomplished by bombarding the radioactive atoms with neutrons, thus accelerating the rate of disintegration and the release of heat energy. The device in which this operation is carried out is called a *pile*.

When ^{235}U fissions, it not only releases heat and forms new elements but also ejects some neutrons from its nucleus. These neutrons can then be used to induce more ^{235}U atoms to fission, and a continuous chain reaction occurs. The function of a pile is to control the flux of neutrons so that the rate of fission can be controlled. When a chain reaction proceeds without control, an atomic explosion occurs. Controlled fission, therefore, is the method used by nuclear power plants, and a tremendous amount of energy can be obtained in the process. During fission, one gram of ^{235}U produces as much heat as the burning of 13.7 barrels of oil. Unfortunately, however, ^{235}U is the only natural radioactive isotope that will maintain a chain reaction, and is the least abundant of the three radioactive isotopes. Only one atom of each 138.8 atoms of uranium in nature is ^{235}U. The remaining atoms are ^{238}U which will not sustain a chain reaction. All the present nuclear power plants use ^{235}U. However, if ^{238}U is placed in a pile with ^{235}U that is undergoing a chain reaction, some of the neutrons will bombard the ^{238}U and convert it to a new isotope, plutonium-239 (^{239}Pu). This new isotope can, under suitable conditions, sustain a chain reaction of its own. The pile in which the conversion of ^{238}U takes place is called a *breeder reactor*. The same kind of device can be used to convert ^{232}Th into a new isotope, ^{233}U, that also will sustain a chain reaction. However, a number of technolog-

ical problems remain before breeder reactors move from the experimental stage into widespread practical use.

Already there are nearly two hundred nuclear plants operating around the world. They utilize the heat energy from fission to produce steam, which in turn drives turbines and generates electricity. Approximately 7.6 percent of the world's electrical power is derived from nuclear power plants. In France, more than half of all the electrical power comes from nuclear plants; the fraction is rising sharply in some other European countries and Japan too. The reason for the increase is obvious. Japan and most European countries do not have adequate supplies of fossil fuels in order to be self-sufficient. There are many problems, however. If a nuclear power plant breaks down, it can release dangerous radioactive substances into the environment, as vividly exemplified by the disaster at Chernobyl in the U.S.S.R. in 1986. Furthermore, the waste left after the fission process is complete can remain toxic for many thousands of years. How to bury and isolate the wastes so that

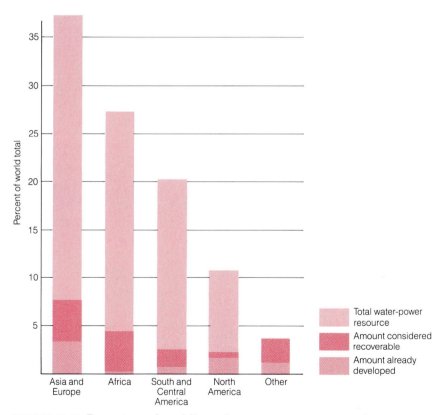

FIGURE 21.18 Percentage of world's total water-power resources, the amount estimated to be recoverable, and the extent to which the recoverable potential has already been developed. (*Sources:* From F. C. Adams, after Hubbert, 1969; and from The World Environment, 1972–1982, a report by the United Nations Environment Programme, 1982.)

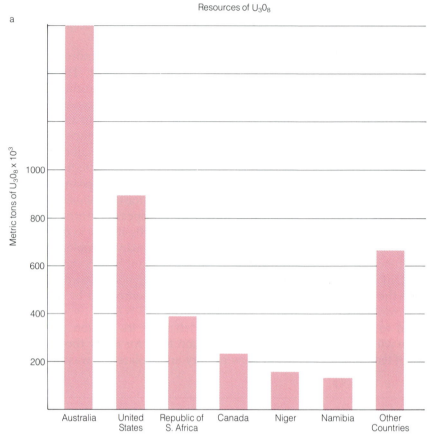

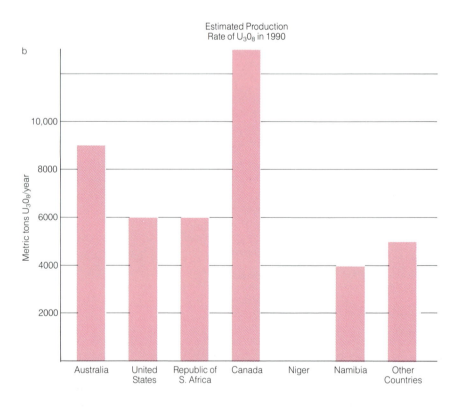

FIGURE 21.19 Uranium ore which is rich enough for use in ^{235}U power plants, together with estimated rates of production by the year 1990. Data are reported as the oxide, U_3O_8. No data are available for the communist countries. (*Sources:* Production rate estimates from Mining Annual Review, 1983; resource data from Am. Assoc. of Petroleum Geologists Bulletin, 1983, and Mining Annual Review, 1980.)

they do not contaminate the air or groundwater is a continuing and massive problem. Concerns with plant safety and waste disposal have led to a slowdown in the rate of new plant installations in the United States. Around the world, however, new plants are continuing to be built. This leads to the question of the availability of uranium and thorium to serve as fuel.

Uranium is present in the continental crust in very small amounts, constituting only 0.00016 percent of average crustal rock. Thorium is about four times more abundant, but neither uranium nor thorium is a very common element. Fortunately, both uranium and thorium do occur in local enrichments—ore deposits—that can be mined. Uranium is much more susceptible to the processes that produce ore deposits; consequently, most interest has centered on it. If we were to use only ^{235}U atoms, supplies of nuclear fuel would be very limited because only the richest of uranium ores could be worked. This is so because the cost of separating ^{235}U atoms from ^{238}U (or at least of raising the $^{235}U/^{238}U$ ratio so the fuel will sustain a chain reaction) is so high that it is not feasible also to pay high costs for mining and recovering the raw uranium ore. At present the ^{238}U is of no use, and so is simply a by-product. If, however, breeder reactors can be perfected and if ^{238}U can be used also, 138.8 times more uranium would be available. By avoiding the expensive atomic enrichment process, it would be possible to spend more for uranium ore to be used in breeder reactors than we can as long as we are limited to ^{235}U alone. Higher prices mean that less-concentrated ores can be worked, and this in turn means a vast increase in the amount of uranium fuel that is available.

Uranium ores rich enough to be worked and used solely for ^{235}U plants are very limited. The cost of uranium before separation should be no more than a few cents per gram. Estimates of the total ore in Western countries amounts to only 3,463,000 tons (Fig. 21.19). If all this uranium were used in ^{235}U plants, and the plants were 50 percent efficient, the heat energy produced would only be 23×10^{19} cal. But if all the ^{238}U could be used in a breeder reactor, the equivalent figure would be 3200×10^{19} cal.

Apart from rich uranium deposits, it is difficult to make an assessment of uranium resources because the needed work has not been done. Within the United States alone, a general evaluation of low-grade source materials indicates that, at a cost of up to \$1.00/g, as much as two billion tons of

uranium might be recovered. It is little wonder, therefore, that many people view nuclear power as a possible answer to the decline of resources of oil and gas.

Tidal Energy

Tidal energy resembles hydroelectric energy in that it is recovered from flowing water and cannot be used up. People have harnessed tides to drive water wheels for many centuries; much of the flour used in Boston in the seventeenth and eighteenth centuries was ground in a tidal mill. But today's interest in tides lies in their use as a source of water to drive electrical generators. Water in a restricted bay, retained behind a dam at high tide and allowed to flow out again at low tide, can drive a generator in the same way that river water can. One important difference between hydroelectric power and tidal power is that rivers flow continually, whereas tides can only be impounded twice a day, and electricity can only be produced periodically.

Along most of the world's coasts the differences in height between high tide and low tide are too small—about 2 m—to drive generators. In some restricted bays, however, tidal ranges exceed 15 m, and here it is possible to recover tidal energy. In France, at the mouth of the River Rance, a tidal-power plant has been operating successfully for several years. Another plant has been built in the U.S.S.R. near Murmansk on the White Sea. But no tidal-power plants have been built in the United States. When they are, the most likely sites are in Passamaquoddy Bay in Maine, and in some bays and inlets of Alaska.

Experts estimate that if all sites with suitable tidal ranges were developed to produce power, the total potential would only be equivalent to the burning of 170 million barrels of oil a year. Furthermore, about half the potential is along the sparsely inhabited Kimberley coast in northwestern Australia. Thus, tidal power could have significant local uses but could never be important on a worldwide scale.

Geothermal Power

Geothermal power, as the Earth's internal heat flux is called, has been used for more than 50 years in Italy and Iceland and more recently in other parts of the world, including the United States. The steam produced in hot-spring areas can be used to power generators in the same way as

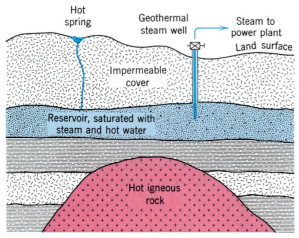

FIGURE 21.20 A typical geothermal steam reservoir. Water in a permeable aquifer, such as sandstone, is heated by magma or hot igneous rock. As steam and hot water are withdrawn through the well, cold water flows into the reservoir through the aquifer.

steam produced in coal- and oil-fired boilers. To capture steam in hot-spring areas, it is necessary to drill into the hot underground reservoirs that feed the springs, bring the steam or hot water to the surface in pipes, and feed it into a power plant (Fig. 21.20). The places where sources of heat are close enough to the surface to produce steam, and therefore where geothermal energy can be developed, are mostly in areas of current or recent volcanic activity, where magma or hot intrusive igneous rock is close to the surface and can serve as a source of heat. In the United States these areas include California, Nevada, Montana, Wyoming, New Mexico, Utah, Alaska, and Hawaii. In other countries, in addition to Italy and Iceland, large resources of geothermal energy exist in Japan, U.S.S.R., New Zealand, several countries in Central America, Ethiopia, and Kenya. Not surprisingly, most of the world's geothermal steam reservoirs are around the margins of plates, because plate margins are where most of the recent volcanic activity has occurred.

A depth of 3 km seems to be a rough lower limit for big geothermal steam and hot-water pools. Experts from the U.S. Geological Survey have reported that down to 3 km the world's reservoirs of geothermal energy contain about 1.9×10^{19} cal that can be recovered—equivalent to burning 13 billion barrels of oil. This estimate incorporates the observation that in New Zealand and Italy only about one percent of the energy in a geothermal reservoir is recoverable. If the recovery efficiency

were to rise, the estimate of recoverable geothermal resources would also rise.

Interesting geothermal experiments are now being conducted in New Mexico. In the Jemez Mountains on the edge of an extinct (but still hot) volcano, scientists have drilled deep into the hot rock. They then shattered the hot rock with explosives to create an artificial reservoir, and pumped water through to produce steam. The test was only partly successful. A major difficulty was that water did not flow uniformly through the hot rock but instead followed narrow flow paths, and the rocks that lined them soon cooled down. Further tests are now planned, not only in the Jemez Mountains, but in France, England, and other countries also.

THE FUTURE

We really do not face an energy shortage. Vast amounts of energy are available—more than we can ever use. But first we must learn how to use energy in different forms. The only crises we face are of our own making because we have, for too long, relied on limited supplies of oil and gas. But as we have just shown, many alternatives to oil

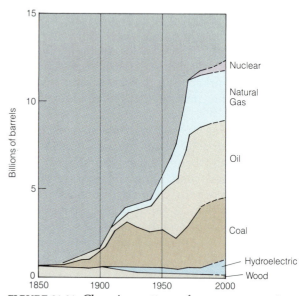

FIGURE 21.21 Changing pattern of energy sources in the United States. In 1850 firewood supplied most needs, but as energy demands increased, coal, oil, and natural gas became increasingly important. The rate at which energy is used, declined in the early 1970s following the oil embargo. Dashed lines show projections into the future. Energy is expressed in barrels of oil.

and gas are already in use, at least in a small way. Still others are available and are hardly used at all. The Sun's heat rays, the energy of the wind to turn windmills, the waves that pound all seashores, the giant ocean currents, and even the temperature difference between warm water at the ocean surface and cold water at depth are all potential energy resources. Still other potential sources are to be found in the possibility of duplicating the process of nuclear fusion that creates the Sun's heat energy. We can release the energy in a hydrogen bomb; can we learn to do so in a slower controlled way?

As they have done in the past, energy sources will surely change in the future (Fig. 21.21). Changing energy sources will inevitably bring changes in life-styles. As fossil fuels are depleted, it seems inevitable that present systems of transportation will change too. As energy sources alter, so will the machines that use the energy. Demand patterns for metals to build the machines must also change. How the demand patterns change will be conditioned, in part, by the metals available and how much it costs to mine and process them. We turn next, therefore, to a discussion of mineral resources.

SUMMARY

1. Ancient industry, based largely on wood and muscle, has given way to industry that uses energy intensively and is based mainly on fossil fuels.

2. Coal originated as plant matter in ancient swamps, and is both abundant and widely distributed.

3. Oil and gas probably originated as organic matter sedimented on seafloors and decomposed chemically. Later these fluids moved through reservoir rocks and were caught in geologic traps to form pools.

4. Oil and gas are limited in abundance; the world's supplies cannot long sustain our present rate of use.

5. As much as half the oil in a reservoir remains trapped, as coatings on mineral grains and in pockets between grains, after flowing oil has been pumped out.

6. Heavy, nonflowing oil (tar) can be recovered by mining techniques. The amount of tar in the world probably equals the amount of flowing oil.

7. When heated, part of the solid organic matter found in shale will convert to oil and gas. Oil from shales is the world's largest resource of fossil fuel. Unfortunately, most shales contain so little solid organic matter that more oil must be burned to heat the shale than is produced by the conversion process.

8. Nuclear energy is derived from atomic nuclei of unstable elements, chiefly uranium. The amount of nuclear energy available from naturally occurring radioactive elements is the single largest energy resource available.

9. Other sources of energy currently used to some extent are geothermal heat, the tides, and energy from flowing streams.

SELECTED REFERENCES

Averitt, P., 1975, Coal resources of the United States, January 1, 1974: U.S. Geol. Survey Bull. 1412.

Darmstadter, J., Landsberg, H. H., and Morton, H. C., 1983, Energy today and tomorrow: Englewood Cliffs, N.J., Prentice-Hall.

Levorsen, A. I., 1967, Geology of petroleum, 2nd ed.: San Francisco, W. H. Freeman.

Perry, H., 1983, Coal in the United States: A status report: Science, v. 222, p. 377–384.

Skinner, B. J., 1985, Earth resources, 3rd ed.: Englewood Cliffs, N.J., Prentice-Hall.

Sources of Materials

A striking specimen of malachite from Zaire. Malachite, a hydrous copper carbonate mineral, is sometimes mined for its copper content but more commonly is used for jewelry.

USES OF MATERIALS

Having discussed the principal sources of *energy*, we now turn to the mineral substances that provide *materials* from which engines and a myriad of other necessary things can be made. The number and diversity of such substances is so great that to make a simple classification covering all of them is almost impossible. Nearly every rock and mineral can be used for something. A society such as ours, that possesses an energy-intensive industry, not only requires a diverse group of metals for machines, but also demands a host of nonmetallic mineral products [such as shale and limestone for making cement, gypsum for making plaster, salt for making chemical compounds, and calcium phosphate (apatite) for making fertilizer]. To supply all needs, the people of the world now mine several hundred kinds of mineral products. The amounts used are truly enormous, and are still increasing. In the United States in 1985, the quantity of iron and steel used was equivalent to 500 kg for every man, woman, and child in the country. In the same year the use of sand and gravel amounted to 3000 kg per person; of crushed stone, 3900 kg; of aluminum, 20 kg; and of copper, 8 kg. When the mineral substances mined for energy (petroleum and coal) are added to all the rest, we get a total of 15,000 kg, or 15 metric tons of mineral substances that are mined and used *each year* for every person living in the United States. Because living standards vary around the world, the quantity of material mined annually for every person in the world is considerably less, 3750 kg. However, the world consumption is slowly rising because living standards around the world are slowly getting better. For the year 1985, it is estimated that the 4.8 billion people of the world consumed an enormous 18 billion metric tons of newly mined mineral substances. It is difficult to imagine where all these materials go, but they are used in every corner of our lives. They build roads and skyscrapers, generate electricity, make automobiles and ashtrays, grow and transport food, and make clothes and indeed all the familiar objects around us. Even the paper on which these words are printed contains mineral substances in addition to woody vegetable matter.

Supplies of Minerals

Many industrialized nations possess a strong mineral base; that is, they are rich in many kinds of *mineral deposits* (any volume of rock containing an enrichment of one or more minerals) which they are exploiting vigorously. Yet no nation is entirely self-sufficient in mineral supplies, and so each must rely to some extent on other nations to fulfill its needs (Fig. 22.1).

Mineral resources have distinctive aspects that differ from plant and animal resources such as those produced through agriculture, grazing, forestry, and fisheries. First, occurrences of usable minerals are limited in abundance and distinctly localized at places within the Earth's crust. This is the main reason why no nation is self-sufficient where mineral supplies are concerned. Because usable minerals are localized, they must be searched out, and the search ranges over the entire globe. The special branch of geology concerned with discovering new supplies of usable minerals is called, appropriately, *exploration geology.*

Second, the quantity of a given mineral available in any one country is rarely known with accuracy, and the likelihood that new deposits will be discovered is difficult to assess. As a result, production over a period of years can be difficult to predict. Thus, a country that today can supply its need for a given mineral substance may face a future in which it will become an importing nation. A good example is Britain. A little more than a century ago, Britain was a great mining nation, producing and exporting such materials as tin, copper, tungsten, lead, and iron. Today, most of the known deposits have been worked out and Britain, once self-sufficient in most minerals, is almost entirely an importing country.

Third, unlike plants and animals that are cropped yearly or seasonally, deposits of minerals are depleted by mining, which eventually exhausts them. Therefore, minerals have only a "one-crop" availability per occurrence; this disadvantage can be offset only by finding new occurrences or by making use of scrap—that is, by reusing the same material repeatedly.

These peculiarities of the mineral industry place a premium on the skills of the geologists, prospectors, and engineers who play the essential roles in the finding and mining of mineral substances used by society. The task of finding and mining is accomplished through the application of the basic principles that have been set forth in this book, and by the use of additional specialized knowledge concerning the origin and distribution of mineral deposits. Much ingenuity has been expended in bringing the production of minerals to its present state. Because known deposits are being rapidly exploited, while demands for minerals continue to grow, we can be sure that even more ingenuity will be needed in the future.

A few mineral substances can be used just as they are found in the ground. Examples are gemstones such as diamond and emerald, common salt, and sand and gravel. The form in which most mineral substances can be used, however, is not the form in which they occur. Most minerals must be processed before use. Because processing costs are high and vary from mineral to mineral, some minerals are more desirable than others. Iron, which is used mainly as a metal, occurs in hundreds of different minerals. But only the oxide minerals hematite, magnetite, and limonite, and the carbonate mineral siderite ($FeCO_3$), are sought and mined. From each of these minerals metallic iron can be separated by straightforward smelting processes in which an oxide mineral is mixed with carbon and the mixture is heated in a blast furnace. Carbon is usually added as *coke*, which is prepared by heating bituminous coal to drive off gases such as methane. The iron oxide is chemically reduced to metallic iron when the oxygen combines with the carbon to form carbon dioxide gas. An example

of such a reaction is given in the chemical equation below:

$$2Fe_2O_3 + 3C \xrightarrow{\text{heat}} 4Fe + 3CO_2$$

Hematite   Carbon   Iron   Carbon
   from   dioxide
   coke

By comparison, the cost of recovering iron from common silicate minerals, such as garnet, biotite, or olivine, is much more difficult because silicon and other chemical elements must also be removed.

Therefore, minerals for industry are sought in deposits from which the desired substances can be recovered least expensively. The more concentrated the preferred minerals, the more valuable the deposit. In some deposits the desired minerals are so highly concentrated that even very rare substances such as gold and platinum can be seen with the naked eye. For every desired mineral substance there is a level of concentration below which the deposit cannot be worked economically. To distinguish between profitable and unprofitable mineral deposits, we use the word **ore,** meaning *an aggregate of minerals from which one or more minerals*

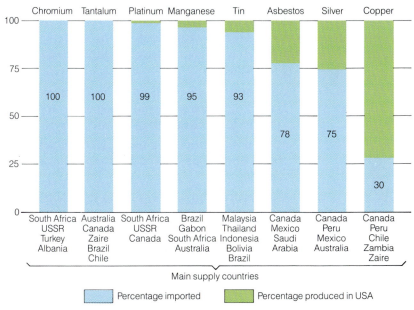

FIGURE 22.1 Selected mineral substances for which United States consumption exceeds production. The difference must be supplied by imports. Data are plotted for 1980, but the percentage changes little from year-to-year. (*Source:* Data from U.S. Bureau of Mines, 1980.)

can be extracted profitably. It is not always possible to say exactly how much of a given mineral must be present in order to constitute an ore. For example, two deposits may contain the same iron mineral and be the same size, but one is ore and the other is not. The reasons one deposit is not ore are many. For example, it could be too deeply buried, or so remote that the costs of mining and transport are so high that the final product, metallic iron, is not competitive with iron from other deposits. Furthermore, as both costs and market prices fluctuate, a particular aggregate of minerals may be an ore at one time but not at another.

Along with ore minerals from which the desired substances are extracted, there are other minerals, collectively termed *gangue* (pronounced *gang*). These are *the nonvaluable minerals of an ore.* Familiar minerals that commonly occur as gangue are quartz, feldspar, mica, calcite, and dolomite.

The ore problem has always been twofold: (1) to find the ores (which altogether underlie an infinitesimally small proportion of the Earth's land area); and (2) to mine the ore and get rid of the gangue as cheaply as possible. Getting rid of gangue and the mining itself are both technical problems; engineers have been so successful in solving them that some deposits now considered ore are only one-sixth as rich as were the lowest-

grade ores 100 years ago (Fig. 22.2). Notice, however, that the curve in Figure 22.2 has risen up from its lowest value, reached during the 1970s. The reason for this is that overproduction of copper around the world, combined with an economic recession, produced a glut of newly mined copper. This, in turn, drove the price of copper down and led to the closing of many mines and, in particular, those mines with ore containing the lowest percentage of copper.

ORIGIN OF MINERAL DEPOSITS

Ores are mineral deposits because each of them contains a local enrichment of one or more minerals or mineraloids. The reverse is not true, however. Not all mineral deposits are ores. Ore is an economic term, while mineral deposit is a geological term. How, where, and why a mineral deposit forms is the result of one or more geological processes. Whether or not a given mineral deposit is an ore is determined by how much we humans are prepared to pay for its contents. Fascinating though the economics of ore deposits and mining are, we cannot explore the topic in this volume. Instead, we will limit discussion to the origin of mineral deposits without necessary regard to questions of economics.

In order for a deposit to form, some process or combination of processes must bring about a localized enrichment of one or more minerals. A convenient way to classify mineral deposits is through the principal concentrating process. Minerals become concentrated in five ways:

1. Concentration by hot, aqueous solutions flowing through fractures and pore spaces in crustal rock to form *hydrothermal mineral deposits.*

2. Concentration by magmatic processes within a body of igneous rock to form *magmatic mineral deposits.*

3. Concentration by precipitation from lake water or seawater to form *sedimentary mineral deposits.*

4. Concentration by flowing surface water in streams or along the shore to form *placer* or *detrital mineral deposits.*

5. Concentration by weathering processes to form *residual mineral deposits.*

Hydrothermal Mineral Deposits

Many of the most famous mines in the world contain ores that were formed when their essential

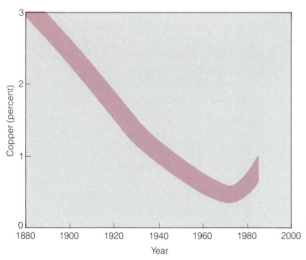

FIGURE 22.2 Declining percentage of copper needed for a mineral deposit to be ore. The large reduction occurred because large-volume mining with efficient machines led to a steady reduction of mining costs. The sharp upturn, which started in the mid-1970s and continues into the 1980s, is due to increased fuel costs of the 1970s and the worldwide economic recession of the 1980's.

minerals were deposited from hot-water solutions, commonly called *hydrothermal solutions*. It is probable that more mineral deposits have been formed by deposition from hydrothermal solutions than by any other mechanism. However, despite the importance of such deposits, the origins of the solutions are often difficult to decipher. Deposition occurs deep underground where we cannot see it happen; by the time a deposit is finally uncovered by erosion, the solution that formed it is no longer present. Nevertheless, enough clues have been found so that many details of the process of deposition are now understood.

Composition of the Solutions

The principal ingredient of hydrothermal solutions is water, although small amounts of gases such as carbon dioxide and methane are often present also. This information is obtained by analyzing the tiny bubbles of ore fluid that are trapped inside crystals of quartz and other minerals present in the ore (Fig. 22.3). The water is never pure and always contains (dissolved within it) salts such as sodium chloride, potassium chloride, calcium sulfate, and calcium chloride. The amounts of such solutes vary, but most solutions range from about the saltiness of seawater (3.5 percent dissolved solids by weight) to about 10 times the saltiness of seawater (Table 22.1). A hydrothermal solution is therefore a brine, and brines, unlike pure water, are capable of dissolving minute amounts of seemingly insoluble minerals such as gold, chalcopyrite, galena, and sphalerite. This happens because certain of the constituents in a brine, particularly chlorine, bond strongly to metals such as zinc,

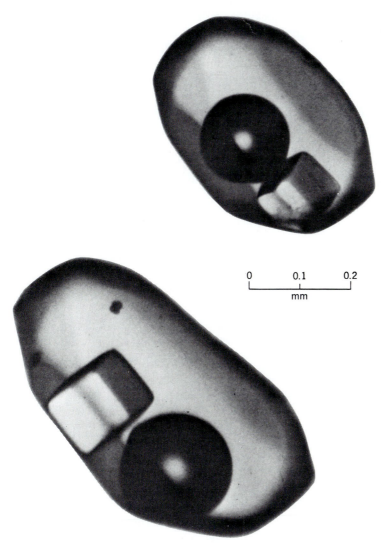

0   0.1   0.2
mm

FIGURE 22.3 Tiny inclusions of the solution that formed a small mineral deposit in the Swiss Alps were trapped inside a quartz crystal. Similar inclusions, visible only through a microscope, can be found in minerals from many deposits. Dark, circular objects are gas bubbles. When first formed the inclusions were full of solution and no gas bubbles were present. Later, as temperatures dropped, the solution contracted and the gas bubbles formed. The small cubes are crystals of sodium chloride. They also formed as temperatures dropped and the originally high temperature solution became saturated in salt. If the quartz crystal is heated back to its original temperature of formation, the solution again becomes homogeneous and both gas bubbles and salt crystals disappear.

forming complex ions of which $(ZnCl_4)^{-2}$ is an example. Complex ions tend to keep zinc in solution and depress its tendency to be precipitated as the ore mineral sphalerite (ZnS), even though sulfide ions (S^{-2}) are present in the solution.

Origins of the Solutions

Brines have many sources. One way they form is by the cooling and crystallization of magma formed by wet partial melting (Chapter 4). Most of the water that causes the wet partial melting is released when such a magma solidifies. Instead of being pure water, however, it carries in solution the most soluble constituents in the magma such as NaCl, and elements that do not readily enter quartz, feldspar, and other common minerals by atomic substitution. The solution is a brine and examples of the elements carried in solution are gold, silver, copper, lead, zinc, mercury, and molybdenum.

High temperatures increase the effectiveness of brines to form hydrothermal mineral deposits. Hot solutions dissolve more material than cold solutions do. Also, reactions between circulating brines and enclosing rock are facilitated greatly by high temperature. It is not surprising, therefore, that many mineral deposits are associated with sources of magmatic heat and that their source materials came from cooling magmas or from hot volcanic rocks that were invaded by circulating fluids. Nor is it surprising that a great many mineral deposits are found in the upper portions of volcanic piles where they were deposited when upward-moving hydrothermal solutions became cool and precipitated the ore minerals.

Submarine volcanic eruptions are common along mid-ocean ridges and above subduction zones. One very important class of mineral deposit, called *volcanogenic massive sulfide deposits*, forms as a result. The hydrothermal solutions in this case are believed to start out as seawater rather than as water in magma, because basalt is a "dry" magma. All the cracks and openings in volcanic rocks in the oceanic crust are saturated by seawater. As magma rises through the oceanic crust toward a site of submarine eruption, the crust is heated and its contained seawater evolves into a hydrothermal solution.

The ore-mineral constituents in massive sulfide deposits apparently come from the igneous rocks in the oceanic crust. Heated seawater reacts with the rocks it is in contact with, causing small changes in mineral composition. For example, feldspars are changed to clays and epidote, and pyroxenes are changed to chlorites. As the minerals are transformed, trace metals such as copper and zinc, present by atomic substitution, are released and become concentrated in the hot seawater.

Hydrothermal solutions formed beneath the sea can become so hot that they rise rapidly through fractures and form jet-like eruptions of hot, hydrothermal solutions into cold seawater (Fig. 22.4).

TABLE 22.1 *Compositions of Some Hydrothermal Solutions (Concentrations Are in Parts per Million)*

Ion in Solution	Seawater	Modern Solutions		Ancient Solutions	
		1	2	3	4
Cl^{-1}	19,400	155,000	157,000	87,000	46,500
Na^{+1}	10,800	50,400	76,140	40,400	19,700
Ca^{+2}	411	28,000	19,708	8600	7500
K^{+1}	392	17,500	409	3500	3700
Sr^{+2}	8.1	400	636	—	—
Ba^{+2}	0.021	235	—	—	—
Li^{+1}	0.17	215	7.9	—	—
Mg^{+2}	1290	54	3080	5600	570
SO_4^{-2}	904	5	309	1200	1600
Fe^{+2}	0.0034	2290	14	—	—
Mn^{+2}	0.0004	1400	46.5	450	690
Zn^{+2}	0.005	540	3.0	10,900	1330
Pb^{+2}	0.00002	102	9.2	—	—
Cu^{+2}	0.0009	8	1.4	9100	140

Sources: Seawater, after Handbook of Geochemistry, K. H. Wedepohl, ed.; Solution 1, Salton Sea Geothermal brine, California, after White and Muffler, 1969; Solution 2, Cheleken Geothermal brine, U.S.S.R., after Lebedev and Nikitina, 1968; Solution 3, Fluid inclusion in fluorite, Cave-in-Rock, Illinois, after Roedder et al., 1963; Solution 4, Fluid inclusion in sphalerite, Creede, Colorado, after Skinner and Barton, 1973.

FIGURE 22.4 A so-called "black smoker" and the chimney-shaped deposit formed around it. This "smoker" was photographed from a submersible at a depth of 2500 m below sea level on the East Pacific Rise, at 21° N latitude. Seawater descends along fractures in rocks of the oceanic crust, becomes heated by magma chambers beneath the mid-ocean ridge, and then rises convectively again to the sea floor. The "smoker" seen here has a temperature of 320°C. The rising hot water is actually clear; the black color is due to fine particles of iron sulfide and other minerals precipitated from solution as the plume is cooled through contact with cold seawater.

Many such eruptions have recently been observed along the East Pacific Rise. When a jetting hydrothermal solution cools, it deposits minerals such as pyrite, chalcopyrite, sphalerite, and galena in a massive blanket around the erupting vent (Fig. 22.5). Volcanogenic massive sulfide deposits have been found in rocks as old as 3 billion years and can be observed to be forming today; they are possibly the most common of all kinds of hydrothermal mineral deposits.

Some hydrothermal solutions start out as rainwater. All rocks contain at least trace amounts of soluble salts. Rainwater that percolates through a sufficiently large volume of rock can eventually be-

come salty. The effect is more pronounced when water percolates through sedimentary rocks of marine origin, or when it comes into contact with evaporites. By interaction with surrounding rocks, therefore, underground water can become a brine. Just as reactions between heated seawater and volcanic rocks of the oceanic crust will generate a hydrothermal solution, so too will reactions between salty underground water and the sedimentary rocks it passes through form a hydrothermal solution. Modern hydrothermal solutions formed in this way have been encountered in deep drill holes in several places such as the geothermal fields near the Salton Sea, in California, and the Cheleken Peninsula, on the eastern side of the Caspian Sea, in the U.S.S.R. (Table 22.1).

Hydrothermal solutions having similar compositions can apparently form in many different ways. Which ore constituents are carried in solution depends on the kinds of rocks involved in the formation of the solution. For example, copper and zinc are present in pyroxenes by atomic substitution, so that the pyroxene-rich rocks of the oceanic crust yield solutions charged with copper and zinc. Most massive sulfide deposits are copper- and zinc-rich as a result.

The important questions concerning hydrothermal solutions are not *where* the water came from or *how* it became a brine, but rather *where did the ore constituents come from* and *what made the solution precipitate its soluble mineral load?*

FIGURE 22.5 A sample of the massive blanket of sulfide minerals that form around the sea-floor vents at 21° N. This specimen is part of a system of chimneys. Lining the chimney is pyrite and scattered through the pyrite are grains of chalcopyrite and sphalerite. Surrounding the massive pyrite is a mixture of gypsum, pyrite, and chalcopyrite. The specimen is 28 cm across.

FIGURE 22.6 Quartz vein cutting across fine-grained metamorphic rock from West Cummington, Massachusetts. Such veins are formed when dissolved minerals are deposited from hot, watery solutions. The ruler is scaled in inches.

FIGURE 22.7 Dark-colored wall-rock alteration produced in a calcareous siltstone by hydrothermal solutions. When the hot solutions flowed through joint-controlled openings, they introduced iron and other chemical elements, allowing an assemblage of amphiboles, pyroxenes, and garnets to grow.

Causes of Precipitation

When a deposit-forming solution moves slowly upward, as with groundwater percolating through a confined aquifer, the solution cools very slowly. If dissolved minerals were precipitated from such a slow-moving solution, they would be spread out over great distances and would not be sufficiently concentrated to form a useful mineral deposit. But when a solution flows rapidly, as in an open fracture or a mass of shattered rock or any other place where flow is less restricted, cooling can be sudden and happen over short distances. Rapid precipitation and a concentrated mineral deposit are the result. Other effects such as boiling, a rapid decrease in pressure, composition changes of the solution caused by reaction with adjacent rock, and dilution by mixing with groundwater can also cause rapid precipitation and form concentrated deposits. When valuable minerals are present, an ore is the result.

Examples

As noted earlier, deposits formed by precipitation from hydrothermal solutions are very common. In most areas of metamorphic rock, quartz veins can be observed (Fig. 22.6). Such veins are deposited by hot salty solutions that have slowly escaped upward as they responded to the pressure and heat involved in metamorphism. Most of the veins that are products of metamorphism are not ore deposits, because precipitation is spread out over long distances and, as a result, the deposits are too small and too lean.

Many veins containing valuable minerals are found in regions of volcanic activity, particularly where the volcanism is rhyolitic or andesitic. This is so because volcanism heats solutions very near the surface, making them effective ore-formers. The famous gold deposit at Cripple Creek, Colorado, was formed in a small caldera and the huge tin deposits in Bolivia are all localized in and around stratovolcanoes. In each case, the magma chamber that fed the volcano also served as the source of the hydrothermal solutions that rose up and formed the mineralized veins in the overlying rocks. Formation of hydrothermal solutions by volcanism also occurs, as we have seen, when volcanism takes place beneath the sea. The gold deposits at Kirkland Lake in Ontario and Kalgoorlie in

Western Australia were formed apparently during submarine volcanic activity. So too were the famous copper deposits of Cyprus, and those in New Brunswick, Canada, and many other places.

When a body of granitic magma cools, it is a source of heat just as the magma chamber beneath a volcano is—and it is also a source of hydrothermal solutions that are released by crystallization. Such solutions move outward from a cooling intrusive body. They will flow through any fracture or channel, altering the surrounding rock in the process and commonly depositing valuable minerals (Fig. 22.7). Many famous ore bodies are associated with shallow intrusive rocks. The tin deposits of Cornwall, England, and the copper deposits at Butte, Montana; Bingham, Utah; and Bisbee, Arizona are examples.

Many of the ore bodies formed by hydrothermal solution are alike in that they are related in one way or another to igneous activity. We have seen already that igneous activity tends to be concentrated along both spreading edge and subduction edge boundaries of plates. It is not surprising, therefore, that certain kinds of ore deposits are also concentrated near present or past plate boundaries (Fig. 22.8). This realization, which is one of the many new insights arising from plate tectonics, is still being studied in detail, because it may help in the discovery of new deposits, especially if former plate boundaries can be identified. The relation between plate boundaries and ore deposits provides a possible explanation for the existence of **metallogenic provinces.** These are *limited regions of the crust within which mineral deposits occur in unusually large numbers.* A striking example is the metallogenic province that runs along the western margin of the Americas. Within the province is the world's greatest concentration of large copper deposits (Fig. 22.9). These deposits are associated with porphyritic igneous rock (Chapter 4) and are called *porphyry coppers.* The igneous bodies (and, therefore, the deposits themselves) are believed to have formed as a consequence of subduction because they are in, or adjacent to, old stratovolcanoes. Although some details of the association remain obscure, the close parallelism between convergent plate margins and metallogenic provinces strongly supports the notion that porphyry copper deposits must somehow be connected with subduction.

Magmatic Mineral Deposits

The processes of partial melting and fractional crystallization in magmas (Chapter 4) are, by their very definitions, ways of separating some substances from others. These special circumstances lead sometimes to the creation of large and potentially valuable mineral deposits.

Pegmatites

When a magma undergoes differentiation by fractional crystallization, the residual melt becomes

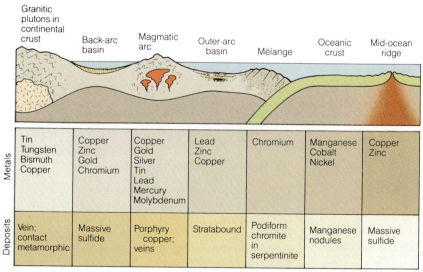

	Granitic plutons in continental crust	Back-arc basin	Magmatic arc	Outer-arc basin	Mélange	Oceanic crust	Mid-ocean ridge
Metals	Tin Tungsten Bismuth Copper	Copper Zinc Gold Chromium	Copper Gold Silver Tin Lead Mercury Molybdenum	Lead Zinc Copper	Chromium	Manganese Cobalt Nickel	Copper Zinc
Deposits	Vein; contact metamorphic	Massive sulfide	Porphyry copper; veins	Stratabound	Podiform chromite in serpentinite	Manganese nodules	Massive sulfide

FIGURE 22.8 Diagram showing the kinds of mineral deposits and the most important metals concentrated in relation to tectonic plates.

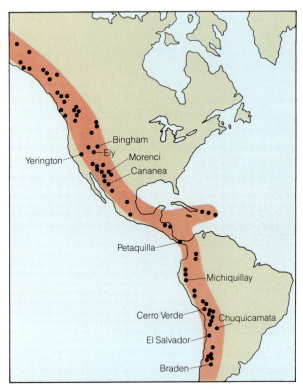

FIGURE 22.9 Metallogenic province along the western edge of the Americas. The province parallels the edge of the North and South American Plates, and is due to the igneous activity caused by subduction. Many of the world's largest copper deposits occur in the province. Some important deposits are named.

progressively enriched in chemical elements that are not present in the early-crystallizing minerals. Separation and crystallization of the remaining melt produces an igneous rock that contains the concentrated elements that do not go off in the hydrothermal solution that forms at the same time. The most important example of this kind of concentration process is found in pegmatites, an especially coarse-grained kind of igneous rock discussed more fully in Appendix C. Apparently, these unusual rocks form by extreme differentiation of deep-seated granitic magma. Commonly, they contain significant enrichments of rare elements such as beryllium, tantalum, niobium, uranium, and lithium.

Crystal Settling

As noted in Chapter 4, one form of magmatic differentiation occurs when dense minerals accumulate on the floor of a magma chamber. The process is called *crystal settling*, and in some cases the segregated minerals make desirable ores. Most of the

world's chromium ores were formed in this manner by accumulation of the mineral chromite ($FeCr_2O_4$) (Fig. 22.10). The largest known chromite deposits are in South Africa, Zimbabwe, and the U.S.S.R. Similarly, vast deposits of ilmenite ($FeTiO_3$), a source of titanium, were formed by magmatic differentiation. Large deposits occur in the Adirondack Mountains.

Immiscible Magmas

Concentration also occurs by a kind of magmatic differentiation that has not been mentioned previously, but that is closely related to crystal settling. In the first case, crystals settle; in the second, droplets of sulfide magma settle.

FIGURE 22.10 Perfection of layering in an igneous rock. Dark layers consisting of grains of chromite ($FeCr_2O_4$) are separated by layers of nearly pure anorthite (plagioclase). The layers were formed by crystal settling during cooling and the crystallization of the Bushveld Igneous Complex, South Africa, the largest known layered intrusion. The specimen is 8 cm across and 14 cm high.

When ore that contains sulfide minerals (such as chalcopyrite and pyrite, together with some silicate gangue minerals) is melted in the smelter, the molten sulfides, being heavy, sink to the bottom; the molten silicates, being light, rise to the top as a sort of scum called *slag*. The sulfide and silicate melts are liquids that will not mix. As with oil and water, instead of mixing, the two liquids separate into distinct layers. They are said to be *immiscible*. The property of immiscibility forms the basis of the processing methods used in the smelting of many ores. A similar form of concentration occurs in nature when, for reasons not clearly understood, certain magmas separate into two immiscible magmas. One, a sulfide liquid that is rich in copper and nickel, sinks to the floor of the magma chamber because it is denser. The world's greatest known concentration of nickel ore, at Sudbury, Ontario, is believed to have formed in this fashion (Fig. 22.11). Other great nickel deposits in Canada, Australia, and Zimbabwe formed in the same manner.

Sedimentary Mineral Deposits

The term **sedimentary mineral deposit** is applied to any *local concentration of minerals formed through processes of sedimentation*. Any process of sedimentation can form localized concentrations of minerals, but it has become common practice to restrict use of the term *sedimentary* to those mineral deposits formed through precipitation of substances carried in solution. Such sediments are also known as chemical sediments (Chapter 7), because some chemical process must occur in order to produce the precipitate.

Evaporite Deposits

The most important way in which sedimentary mineral deposits form is by evaporation of lake water or seawater. The *layers of salts that precipitate as a consequence of evaporation*, are called **evaporite deposits.**

Examples of salts that precipitate from lake waters of suitable composition are sodium carbonate (Na_2CO_3), sodium sulfate (Na_2SO_4), and borax ($Na_2B_4O_7 \cdot 10H_2O$). Such salts are used in many manufacturing processes such as production of paper, soap, certain detergents, antiseptics, and chemicals for tanning and dyeing.

Much more common and important than non marine evaporites are the marine evaporites formed by evaporation of seawater. The most important salts that precipitate from seawater are gyp-

FIGURE 22.11 Photograph of a polished surface on ore formed by the cooling of an immiscible sulfide liquid. The dark grains are olivine crystals, the light material is pyrrhotite (FeS), plus chalcopyrite. The specimen is approximately 0.5 cm across and comes from the Alexo Mine, Ontario.

sum ($CaSO_4 \cdot 2H_2O$), halite (NaCl), and carnallite ($KCl \cdot MgCl_2 \cdot 6H_2O$). Low-grade metamorphism of evaporite deposits causes another important mineral, sylvite (KCl), to form. Marine evaporite deposits are widespread; in North America, for example, strata of marine evaporites underlie as much as 30 percent of the entire land area (Fig. 22.12). From marine evaporites we recover much of the salt we use (Fig. 22.13), the gypsum used for plaster, and potassium used in plant fertilizers.

Iron and Phosphorus Deposits

Salt, gypsum, and other nonmetallic substances have been concentrated by evaporation. Iron minerals and the important phosphatic fertilizer-mineral apatite have also been concentrated from solution in the sea, but the abundance of iron and phosphorus in average seawater is so small that thick beds of iron and phosphorus minerals could not have been concentrated solely by evaporation; some other process must have been responsible.

An example of such beds is the Clinton iron formation which occurs as discrete strata in several sedimentary units—one of them locally more than 10 m thick, that extends from New York State southwestward for 1100 km into Alabama. The

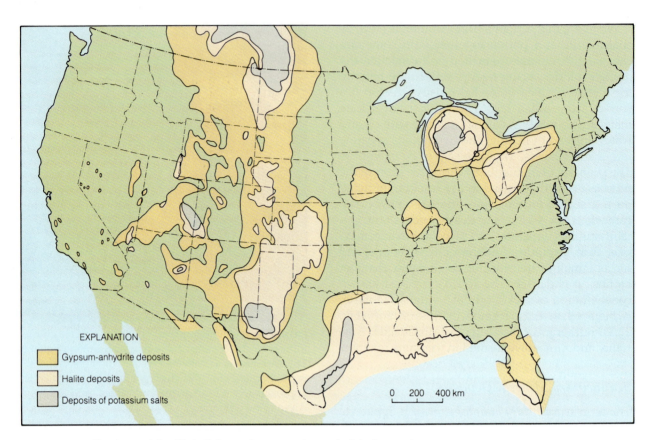

FIGURE 22.12 Portions of the United States known to be underlain by marine evaporite deposits. The areas underlain by gypsum and anhydrite do not contain halite. Areas underlain by halite are also underlain by gypsum and anhydrite. Areas underlain by potassium salts are also underlain by halite and by gypsum and anhydrite. (*Source:* After U. S. Geological Survey, 1973.)

FIGURE 22.13 Mining halite near Windsor, Ontario. Horizontal bedding is visible; each pair of dark and light bands represents a single year's cycle of evaporitic precipitation, the dark band representing the winter months, the wider light band, the summer.

steel industry of Birmingham, Alabama, was based on the convenient proximity of iron-rich sediments to both the coal and limestone needed for smelting. Fossils show that the ore strata were deposited in a shallow sea during the Silurian Period. Ripple marks, mud cracks, and oolites further indicate shallow water. The iron mineral is hematite, and the content of iron is 35–40 percent.

Just how the Clinton deposits formed is still open to question. There are no modern deposits to serve as examples. Iron occurs in two oxidation states and as a result forms two kinds of ions. They are ferrous iron, Fe^{+2}, and ferric iron, Fe^{+3}. Many compounds containing Fe^{+2} are soluble in sea, lake, and river water. Compounds containing Fe^{+3} are essentially insoluble in the same waters. The problem is that when water is in contact with oxygen in the atmosphere, only Fe^{+3} compounds can form. Therefore, how can surface waters concentrate iron? One suggestion is that the iron was dissolved from iron-bearing minerals in mafic igneous rock, and was carried perhaps as a ferrous bicarbonate $[Fe(HCO_3)_2]$ by groundwater to the shallow sea. There it precipitated as a ferric oxide or ferric hydroxide when the groundwater came into contact with seawater containing dissolved oxygen. No satisfactory explanation has been given that covers all aspects of iron sediments. The precipitate is a chemical sediment and the iron concentrated in Clinton-type deposits amounts to billions of tons, far more than seems likely for concentration by groundwater. The enormous iron-rich strata in the Alsace region of eastern France, and similar strata in England, are also Clinton-type deposits.

Large and impressive as the Clinton deposits and others like them are, they are all tiny by comparison with the class of deposits characterized by the Lake Superior iron deposits. The deposits were long the mainstay of the United States steel industry but are declining in importance today as imported ores displace them. The deposits are of early Proterozoic age, about 2 billion years old, and are found in sedimentary basins on every craton, particularly in Labrador, Venezuela, Brazil, the U.S.S.R., India, South Africa, and Australia. Lake Superior-type deposits are one of the most unusual kinds of chemical sediment known. Like the Clinton-type deposits, the important ore minerals are iron oxides. But they differ from the Clinton deposits in two important respects. First, whereas the Clinton deposits contain fossils and other detritus, the Lake Superior deposits are sediments of wholly chemical origin and are free of

FIGURE 22.14 Lake Superior-type iron formations formed from iron-rich chemical sediment deposited in early Proterozoic seas. The Brockman Iron Formation of the Hamersley Range, Western Australia, consists of numerous cherty iron-rich sedimentary beds, here making vertical cliffs, interbedded with softer and more easily eroded shales. The shales underlie the sloping surfaces between each iron-rich unit.

detritus. Lake Superior-type iron formations are commonly interbedded with clastic sedimentary rocks, although the iron deposits themselves are devoid of clastic grains. Second, whereas the Clinton deposits contain calcite or detrital quartz grains as a gangue, the Lake Superior deposits contain very fine bands of recrystallized chert (Fig. 22.21). Every aspect of the Lake Superior deposits indicates they are chemical precipitates. Because the deposits are so large, we infer that transportation of the iron must have taken place in surface water. This inference forces us to a further conclusion. Because all of the Lake Superior-type deposits formed about two billion years ago, the composition and properties of the atmosphere or the seawater, or both, must have been different from those of today. The most important difference probably involved the atmosphere: It must have contained less oxygen, so the surface waters were less oxygenated than those of today and they could transport large amounts of dissolved ferrous iron. Sufficient oxygen must have been present to allow precipitation of ferric oxide and ferric hydroxide, but not so much that solution in surface waters was entirely prevented. The differences in the atmosphere and seawater would have produced different styles of weathering, transport, and precipitation from those we see today and this could possibly account for the unusual composition of the iron-rich sediments. A more likely explanation, however, involves the release of iron into seawater by submarine volcanoes.

However the unusual Lake Superior-type deposits formed, they are so large they will be sources of iron for centuries ahead, although the deposits are not rich enough to be smelted directly. Secondary enrichment of iron formations formed many of the ores worked today. But a similar kind of enrichment can be produced by metallurgical treatment. *Taconite*, a Lake Superior-type iron formation that has been metamorphosed, is sufficiently coarse-grained so that the iron minerals and chert can be separated by metallurgy to yield an almost pure concentrate of hematite or magnetite. Taconite concentrate is now the preferred blast-furnace feed at many iron foundries.

Precipitation from seawater is also the origin of most of the world's manganese deposits. The two largest deposits occur in the U.S.S.R. They are of early Cenozoic age and seem to have formed by groundwater flow, as suggested for the origin of Clinton-type iron ores. Manganese, like iron, has two oxidation states, Mn^{+2} and Mn^{+4}. The Mn^{+2} state forms soluble compounds, while Mn^{+4} forms insoluble compounds. All the ore minerals are oxides containing Mn^{+4}. Manganese minerals are precipitating, even today, on the deep-sea floor. Nodular growths of manganese oxides occur on

the floors of all the oceans (Fig. 22.15). Ranging in diameter from 1 cm to as much as 30 cm, the nodules form by very slow precipitation from seawater. Individual nodules may take millions of years to grow. But the seafloor is so extensive, and in places the frequency of nodules is so high, that literally trillions of nodules are present. Dredging tests to recover the nodules from the floor of the Pacific Ocean have already been made. In addition to the manganese minerals, the nodules contain 1 percent or more of copper, nickel, and cobalt present in atomic substitution which may prove to be even more valuable than the manganese.

Sedimentary mineral deposits of phosphorus form through the precipitation of apatite $[Ca_5(PO_4)_3(OH,F)]$ from seawater. The surface waters of the ocean are depleted in phosphorus because fish and other animals that live in the water extract phosphorus to make bone, scales, and other parts of their bodies. When the animals die, and sink to the seafloor, their bodies slowly decay and release the contained phosphorus to the deep ocean water. If such phosphorus-rich waters are brought to the surface by rapid upwelling, precipitation of apatite can occur. Phosphorus-rich sediments are forming today off the west coasts of Africa and South America, but the process has been much more common at times in the past—particularly when shallow epicontinental seas were prev-

FIGURE 22.15 Internal structure of a manganese nodule from the floor of the Pacific Ocean. The nodule is 11 cm across and 5 cm high. The larger nodule apparently formed when two smaller nodules grew together. The variation in color of the layers is due to the variation in composition and to changes in porosity from layer to layer. The nodule contains approximately 1 percent copper and 0.5 percent nickel, which are incorporated into the structure of the manganese oxide minerals.

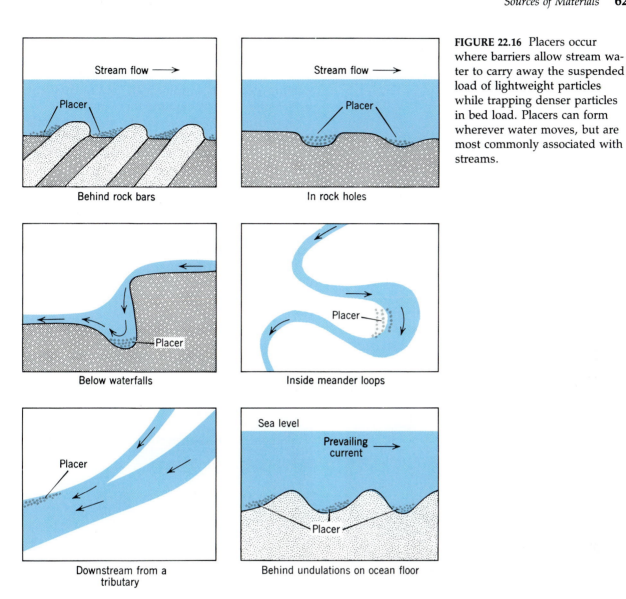

Stream flow ⟶

Placer

Behind rock bars

Stream flow ⟶

Placer

In rock holes

Placer

Below waterfalls

Placer

Inside meander loops

Placer

Downstream from a tributary

Sea level

Prevailing current ⟶

Placer

Behind undulations on ocean floor

FIGURE 22.16 Placers occur where barriers allow stream water to carry away the suspended load of lightweight particles while trapping denser particles in bed load. Placers can form wherever water moves, but are most commonly associated with streams.

alent, as was the case, for example, along the margins of the old Tethys Sea during the Mesozoic and Cenozoic Eras.

Placer Deposits

The famous California goldrush of 1849 followed the discovery that the sand and gravel in the bed of a small stream contained bits of gold. Similar gold-bearing gravels are found in many parts of the world. The gravels themselves are sometimes rich enough to be ores, but even when they are too lean the gold is a clue that a source for the gold must lie upstream. Indeed, many mining districts have been discovered by following trails of gold and other minerals upstream to their sources in veins in bedrock. Because pure gold is heavy (density, 19

g/cm^3), it is deposited from the bed load of a stream very quickly, while quartz, with a density of only 2.65 g/cm^3, is washed away. As most silicate minerals are light by comparison with gold, grains of gold become mechanically concentrated in places where the velocity of stream flow is least (Fig. 22.16). *A deposit of heavy minerals concentrated mechanically is a **placer**.* Besides gold, other heavy, durable metallic minerals form placers. These include minerals that occur as pure metals, such as platinum and copper, as well as tinstone (cassiterite, SnO_2) and nonmetallic minerals such as diamond and ruby and sapphire (both gem forms of corundum). Even if a vein contains a low percentage of gold or cassiterite, the placer it yields may be quite rich. In order for minerals to become concentrated in placers they must not only be dense,

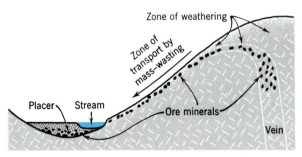

FIGURE 22.17 Chemically resistant minerals from weathered fissure veins creep down slopes by mass-wasting and are redeposited by streams as placers.

they must be resistant to chemical weathering and not readily susceptible to cleaving as the mineral grains are tumbled in the stream.

Every phase of the conversion of gold in a fissure vein into placer gold has been traced.

Chemical weathering of the exposed vein releases the gold, which then moves slowly downslope by mass-wasting (Fig. 22.17). In some places mass-wasting alone has concentrated gold or cassiterite sufficiently to justify mining these metals.

More commonly, however, the mineral particles get into a stream, which concentrates them more effectively than mass-wasting. Most placer gold occurs in grains the size of silt particles, the "gold dust" of placer miners. Some of it is coarser; pebble-sized fragments are *nuggets* (Fig. 22.18), of which the largest ever recorded weighed 80.9 kg. In following placers upstream, prospectors have learned that rounding and flattening (by pounding) increase downstream, just as does rounding of ordinary pebbles. When they find angular nuggets, prospectors know the primary source is close.

Mining was done first by hand, simply by swirl-

a

b

c

FIGURE 22.18 Formation of a nugget. (*a*) A vein of metallic gold cutting through a pebble of vein quartz. Stream abrasion causes the brittle quartz to chip and be reduced in size, while the malleable gold deforms but is not reduced. The specimen is 4 cm long. (*b*) The ratio of gold to quartz increases as the quartz is abraded away. Eventually, a nugget of almost solid gold results. The specimen is 4 cm long. (*c*) A nugget of metallic gold from California. No quartz remains. The specimen is 4.5 cm long.

ing stream sediment around in a small pan of water. Later it was done by jetting water under high pressure against gravel banks and washing the sediment through troughs that caught the heavy grains of gold behind cleats. Nowadays, the more efficient method of dredging is used. The platinum placers of the Ural Mountains in the U.S.S.R., the rich diamond placers in Zaire, and the hundreds of tin placers in Malaysia and Indonesia are examples of other mechanical concentrations by streams.

Not all placers are to be found in modern stream channels. Indeed, the most famous and largest gold deposits ever found are the Witwatersrand deposits in South Africa. These ancient placers are more than 2 billion years old. The gold grains, which are microscopic in size, are found in conglomerates in which quartz pebbles and pyrite are the main constituents (Fig. 22.19). The Witwatersrand placers are found where ancient river systems formed deltas in near-shore portions of a Proterozoic basin. The Witwatersrand ores are remarkable for many reasons, but especially for the enormous amount of gold they have yielded. It is estimated that as much as 30 percent of all the gold that has ever been mined has come from the Witwatersrand ores.

Gold, diamonds, and several other minerals have been concentrated in beach sands by surf and long-shore currents. Diamonds are being obtained in large quantities from gravelly beach placers, both above and below present sea level, along a 350-km strip of the coast of Namibia in southwestern Africa. Weathered from deposits in the interior, the diamonds were transported by the

a

b

FIGURE 22.20 Residual mineral deposits rich in iron and aluminum are typically formed under tropical or semitropical conditions. (*a*) Red laterite enriched in iron, near Djenné, Mali. The upper two meters of the laterite consist of rounded concretions of limonite cemented by more limonite to create a hardened mass. The texture is developed through repeated solution and redeposition of the limonite. The exposure is a quarry; the hard laterite is used for buildings as well as for roads and runways. (*b*) Bauxite from Weipa in Queensland, Australia. Long-continued leaching of clastic sedimentary rocks under tropical conditions has removed all original constituents, such as silica, calcium, and magnesium, leaving a rich bauxite consisting largely of the mineral gibbsite ($Al(OH)_3$). Nodules of gibbsite form by repeated solution and redeposition.

FIGURE 22.19 Detrital gold is recovered from ancient placer deposits of the Witwatersrand, South Africa. The gold is found at the base of conglomerate layers interbedded with finer-ground sandstone, here seen in weathered outcrop at the site where gold was first discovered in 1886.

Orange River to the coast and were spread southward by long-shore drift. Later, some beaches were lifted up tectonically, and others submerged by rise of sea level.

Residual Mineral Deposits

Weathering is the combination of physical and chemical changes in rocks exposed at the Earth's surface. The changes are brought about by the atmosphere, the hydrosphere, and the biosphere. As discussed in Chapter 9, weathering occurs because newly exposed rocks are not chemically stable in contact with rainwater and the atmosphere. Chemical weathering, in particular, leads to many important mineral concentrations by the removal

FIGURE 22.21 Chemical weathering of an iron-rich sedimentary rock removes silica and leaves a secondarily enriched capping of iron ore. (*a*) Siliceous iron-rich sediments of the Brockman Iron Formation, Hamersley Range, Western Australia. Typical of the Lake Superior-type iron formations, the white layers are largely chert, the darker reddish and bluish colored layers consist largely of iron-rich silicate, oxide, and carbonate minerals. (*b*) Secondarily enriched iron ore above the Brockman Iron Formation, Hamersley Range, Western Australia. The sedimentary structure is still visible, but the cherts have been leached out (open spaces) and the iron minerals have all been converted to limonite. The original sedimentary rock had an iron-content of above 25 percent, by weight. The secondarily enriched capping contains approximately 60 percent iron.

of soluble materials in solution and the concentration of a less-soluble residue. The process is called ***residual concentration,*** defined as *the natural concentration of a mineral substance by removal of a different substance with which it was associated.*

A common example of a deposit formed through residual concentration is laterite (Chapter 9). Limonite (ferric-iron hydroxide) is among the least soluble of the many minerals formed during chemical weathering. Under conditions of high rainfall in a warm, tropical climate, other minerals are slowly leached out of a soil, leaving an iron-rich limonitic crust of laterite at the surface (Fig. 22.20*a*). In a few places, laterites have developed such high contents of iron they can even be mined.

While iron-rich laterite is by far the most common kind of residual mineral deposit, the most important deposits so far as human exploitation are concerned are the aluminous laterites called *bauxites.* Bauxite is a mixture of several aluminum hydroxide minerals and is the preferred source for aluminum.

Bauxite is a product of special chemical weathering. Aluminum is present as a constituent of original, primary silicate minerals such as feldspars, micas, and clays. When the minerals are subjected to chemical weathering, silica is carried away in solution, gradually increasing the concentration of aluminum in the residue. Bauxite can be observed grading downward into the underlying rock. From the locations of bauxite ores we know that the weathering takes place beneath erosion surfaces of low relief where the water table is close to the surface and the circulation of groundwater is slow. Bauxites only develop under tropical and subtropical climates. However, the chemistry of the process is complex, and there is still much to be learned about just what happens. Apparently groundwater sometimes has just the acidity necessary to decompose the original silicate minerals and to carry silica away in solution, but not to remove the aluminum and iron oxides. Even though a parent rock may originally have contained no more than a few percent aluminum, the resulting bauxite may reach a concentration of 40 percent aluminum (Fig. 22.20*b*).

Laterites developed on serpentine-rich rocks are almost invariably iron-rich, but because the parent rocks from which serpentines develop (peridotites), tend to be nickel-rich, many of the laterites are nickel-rich also. Indeed, nickel-rich laterites are mined in such places as New Caledonia, Cuba, and Oregon. Peridotites are most commonly found as components of ophiolites. Because ophiolites

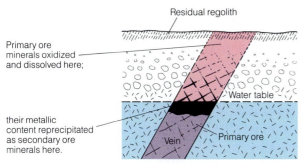

Residual regolith

Primary ore
minerals oxidized
and dissolved here;

Water table

their metallic
content reprecipitated
as secondary ore
minerals here.

Vein — Primary ore

FIGURE 22.22 Descending groundwater oxidizes and impoverishes copper ore above the water table, forming an acid solution that removes soluble copper compounds and leaving a residue of limonite. The descending acid solution deposits copper at and below the water table, producing secondary enrichment below.

tend to occur along convergent plate margins, nickeliferous laterites also tend to form along modern or ancient subduction zones.

Secondary Enrichment

When chemical weathering and residual enrichment change a preexisting mineral deposit, the result can sometimes be a spectacular upgrading in the metal content of the deposit. The process is called *secondary enrichment,* and it is defined as *the natural enrichment of an existing mineral deposit through processes of chemical weathering and groundwater transport.* Iron-rich sedimentary rocks sometimes contain as much as 25 percent iron by weight. Such rocks are too lean to be ores, but if further enrichment occurs through removal of gangue minerals in solution, a secondarily enriched ore containing as much as 60 percent iron can result (Fig. 22.21). Many of the world's great iron-ore deposits have arisen through secondary enrichment. The process is very important in the formation of rich manganese deposits too.

Secondary enrichment has also been important in many of the great copper-ore deposits of the world and, in particular, those found in the arid regions of the southwestern United States. The primary mineral deposits, which are hydrothermal in origin, contain pyrite (FeS_2) and chalcopyrite ($CuFeS_2$). These minerals are oxidized by rainfall and the atmosphere to form sulfuric acid (H_2SO_4). The acid then dissolves some of the remaining copper minerals, forming copper sulfate and leaving an insoluble residue of limonite. When the copper-rich acid solution percolates down and reaches the water table, it reacts by depositing some of the

dissolved copper as chalcocite (Cu_2S), forming a secondarily enriched capping above the leaner primary ore below (Fig. 22.22).

USEFUL MINERAL SUBSTANCES

The way minerals are concentrated into ore deposits has little if anything to do with the way we use the mined products. Some minerals are very abundant and so the products made from them are inexpensive; brick, which is produced from common clay minerals, is an example. Other minerals are rare, so the mined products are expensive; gold, silver, and diamonds are examples. Because mineral substances are so important in our daily lives we must question the adequacy of supply, especially for those expensive materials that only occur in small amounts and that are hard to find. In order to consider the question of adequacy it is convenient for discussion to group mineral products on the basis of the way we use them rather than the way they occur. Excluding substances used for energy, there are two broad groups: (1) minerals from which metals such as iron, copper, and gold can be recovered by special processes of smelting; and (2) minerals such as salt, gypsum, and clay used not for the metals they contain but for their properties as chemical compounds. The nonmetallic substances can be further subdivided on the basis of more specialized uses (Table 22.2).

TABLE 22.2 *Principal Mineral Substances, Grouped According to Use*

1. Metals
   a. Geochemically Abundant Metals:
      Iron, aluminum, magnesium, manganese, titanium
   b. Geochemically Scarce Metals:
      Copper, lead, zinc, nickel, chromium, gold, silver, tin, tungsten, mercury, molybdenum, uranium, platinum, palladium, and many others
2. Nonmetallic Substances
   a. Used for Chemicals:
      Sodium chloride (halite), sodium carbonate, sulfur, borax, fluorite
   b. Used for Fertilizers:
      Calcium phosphate (apatite), potassium chloride (sylvite), sulfur, calcium carbonate (limestone), sodium nitrate
   c. Used for Building:
      Gypsum, limestone, clay (for brick and tile), asbestos, sand, gravel, crushed rock of various kinds, shale (for cement)
   d. Used for Ceramics and Abrasives:
      Ceramics: Clay, feldspar, quartz
      Abrasives: Diamond, garnet, corundum, pumice

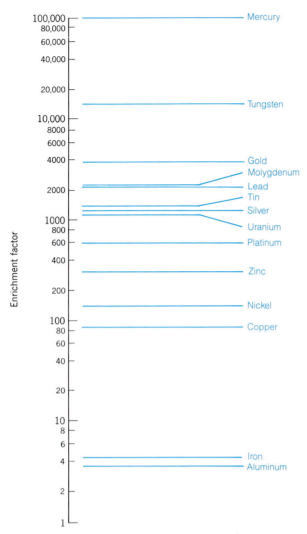

Enrichment factor

Mercury — 100,000, 80,000, 60,000, 40,000

Tungsten — 20,000, 10,000, 8000, 6000

Gold, Molygdenum — 4000

Lead, Tin — 2000

Silver

Uranium — 1000, 800

Platinum — 600

Zinc — 400, 200

Nickel — 100, 80

Copper — 60, 40, 20, 10, 8, 6

Iron — 4

Aluminum — 2, 1

FIGURE 22.23 Before a mineral deposit can be called an ore, the percentage of valuable metal in the deposit must be greatly enriched above its average percentage in the Earth's crust. The enrichment is greatest for metals that are least abundant in the crust, such as gold and mercury. As mining and mineral processing become more efficient and less expensive, it is possible to work leaner ore and enrichment factors decline. Note that the scale is a magnitude (logarithmic) scale, in which the major divisions increase by multiples of ten.

Metals

Without exception, the useful metals are present in the crust in such low amounts that we can only mine and recover them when rich deposits can be located. Just how much richer than ordinary rock a mineral deposit must be before it is an ore deposit depends on the metals present and the market value of the material. For example, at present iron must be enriched five times above its average value in the crust, while copper requires an enrichment of 75 times, platinum 600, and mercury an enormous 100,000 times (Fig. 22.23) before a deposit is rich enough to be an ore.

Metals can be usefully subdivided on the basis of their average percentage in the crust. Those present in such abundance that they make up 0.1 percent by weight, or more, of the crust are considered to be *geochemically abundant*. These are the metals iron, aluminum, manganese, magnesium, and titanium. Geochemically abundant metals require comparatively small enrichment factors to form extremely large deposits. The minerals that are concentrated in deposits of the geochemically abundant metals tend to be oxides and hydroxides. The most important kinds of deposits are residual, sedimentary, and magmatic deposits (Table 22.3). The size and number of ore deposits of the geochemically abundant metals is so great that we can consider the supplies to be virtually inexhaustible. Even if preferred ores, such as the bauxites now used as sources of aluminum, should someday be used up, there are other kinds of deposits to which we could turn without a great increase in the cost of the final product. For example, there are huge deposits of minerals such as kaolinite and the calcic-plagioclase anorthite that are almost as much enriched in aluminum as bauxite is—and these deposits are far more abundant.

Metals that make up less than 0.1 percent by weight of the crust are said to be *geochemically scarce*. They behave very differently from the geochemically abundant metals. Every common rock contains one or more minerals of iron and aluminum and a great many contain the minerals of the other abundant metals—magnesium, titanium, and manganese. By contrast, with the exception of copper, zinc, and chromium, minerals of the scarce metals are never present in common rocks. However, chemical analysis of any rock reveals that even though the minerals are absent, geochemically scarce metals are certainly present. Further research reveals that the scarce metals are present exclusively by atomic substitution (Chapter 3) in the common rock-forming minerals. Atoms of the scarce metals (such as nickel, cobalt, and copper) can readily substitute for more common atoms (such as magnesium and calcium). In order for a mineral deposit to form, therefore, some gathering and concentrating agent such as a hot brine must react with the rock-forming minerals and leach the scarce metals from them. The brine must then transport the metals in solution and deposit them as separate minerals in a lo-

TABLE 22.3 *Geochemically Abundant Metals: Principal Ore Minerals and Kinds of Deposits*

Metal	Important Minerals	Metal Content of an Average Ore (Wt.%)	Classes of Deposits	Examples of Deposits
Iron	Magnetite, Fe_3O_4 Hematite, Fe_2O_3 Limonite, $FeO(OH)$ Siderite, $FeCO_3$	60%	Sedimentary Magmatic segregation	Mesabi Range (Minn.) Hamersley Range (W. Australia) Alsace (France) Kiruna (Sweden)
Aluminum	Gibbsite, $Al(OH)_3$ Diaspore, $AlO(OH)$	35%	Residual (bauxites)	Jamaica Weipa (Queensland), Guyana
Magnesium	Dolomite, $CaMg(CO_3)_2$ Magnesite, $MgCO_3$	15%	Sedimentary	Common worldwide
Titanium	Ilmenite, $FeTiO_3$ Rutile, TiO_2	25%	Magmatic segregation Placer	Allard Lake (Quebec) Queensland (Australia), North Carolina
Manganese	Pyrolusite, MnO_2 Psilomelane, $BaMn_9O_8 \cdot 2H_2O$	50%	Sedimentary with secondary enrichment Deep-sea nodules	Chiatura and Nikopol (U.S.S.R.) Brazil, Mexico All oceans

calized place. With such a complicated chain of events, it is not surprising that deposits of geochemically scarce metals are rarer and very much smaller than deposits of the abundant metals.

Most minerals that are concentrated to form deposits of the scarce metals are sulfides; a few, such as the most important minerals of tin and tungsten, are oxides. In the case of gold, platinum, palladium, and a few less common elements, the metal itself is the most important mineral. Most scarce metal deposits form as hydrothermal and magmatic mineral deposits. In the case of the precious metals gold and platinum, placer concentration is also important (Table 22.4).

The number and size of mineral deposits that contain concentrations of scarce metals seems to be proportional to the percentage of the metals in the crust. In Figure 22.24 the sum of all the ore discovered in the United States is plotted against the crustal abundance of selected metals. Clearly, the economically recoverable amount of each scarce metal also seems to be related to its average content in the crust. Even though iron is concentrated by many different processes, it is apparent that iron also seems to fit the relationship depicted in Figure 22.24. This suggests that if all the mineral deposits, whether ore or not, could be found, the amount of each metal concentrated in them would be proportional to the abundance of each metal in the crust. From this we might conclude that a prudent society should mine and use metals at rates proportional to the supplies available and, hence, proportional to their geochemical abundances. Worldwide rates of use are in fact far from this ideal balance. By comparison with iron, which we have seen is very abundant, we use copper at a rate 13 times faster than we should, lead 41 times faster, and mercury 91 times faster. The scarcer the metal, the faster we seem to be mining out the ore deposits. This leads to the hypothesis that within the lifetimes of some people living today, demands for some of the geochemically scarcer metals such as mercury, silver, and gold might cease to be met by the mining of new metal. If that hypothesis is correct, our society will either have to develop ways of near-perfect recycling of scarce metals or learn to do without them and to substitute more abundant materials instead.

Nonmetallic Substances

The great diversity of nonmetallic substances makes it impossible to devise an unambiguous scheme of classification. Some substances have several different uses. Although far from perfect, the list in Table 22.2 includes the major uses of nonmetallic substances.

TABLE 22.4 *Selected Geochemically Scarce Metals: Principal Ore Minerals and Kinds of Deposits*

Metal	Important Minerals	Metal Content of an Average Ore (Wt.%)	Classes of Deposits	Examples of Deposits
Copper	Chalcocite (Cu_2S) Chalcopyrite ($CuFeS_2$) Bornite (Cu_5FeS_4) Enargite (Cu_3AsS_4)	0.8%	Hydrothermal (veins and porphyry coppers) In shales	Bingham Canyon (Utah) Butte (Montana) Kupferschiefer (Germany, Poland) Zambia, White Pine (Mich.)
Lead	Galena, PbS	4%	Hydrothermal veins and volcanogenic massive sulfide deposits	Mississippi Valley types, such as Viburnum (Missouri), and Pine Point (Canada). Broken Hill (New South Wales), Mt. Isa (Queensland)
Zinc	Sphalerite, ZnS	3%	Same as lead	
Gold	Metallic gold, Au	0.003%	Hydrothermal veins Placers	Homestake (So. Dakota) Carlin, Getchell, and Cortez (Nevada) Siberian rivers, Alaska, Witwatersrand (So. Africa)
Molybdenum	Molybdenite, MoS_2	2%	Hydrothermal deposits	Climax and Henderson, Colorado
Platinum	Metallic platinum, Pt	0.001%	Magmatic segregation	Bushveld Igneous Complex (So. Africa) Urals
Tin	Cassiterite, SnO_2	0.4%	Hydrothermal deposits Placers	Llallagua (Bolivia) Marine placers in Malaya, Thailand, Indonesia
Uranium	Uraninite, UO_2 Carnotite, $K_2(UO_2)_2 \cdot 3H_2O$	0.18%	Hydrothermal deposits	Colorado Plateau, Jabiluka and Roxby Downs (Australia)

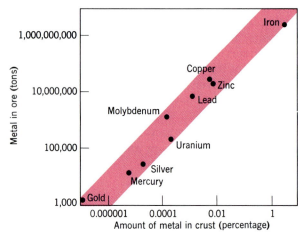

FIGURE 22.24 Relation between the average content of a metal in the crust (geochemical abundance) and the metal present in all the ore deposits discovered in the United States. (*Source:* After McKelvey, 1973.)

Chemical Materials

Most of the materials used by the chemical industry are organic substances (coal, oil, and gas). They are the raw materials used to make petrochemical products such as plastics, drugs, pesticides, synthetic fibers, and countless other products. Many inorganic substances are used also. The most important are sodium chloride, sodium carbonate, sodium sulfate, and minerals such as borax ($Na_2B_4O_7 \cdot 10H_2O$) that contain boron. Each of these is recovered from evaporites. Fortunately, the supplies of all of the chemical substances are very large.

Fertilizer Materials

Growing plants draw the chemical elements they need from the soil. Nature has its own way of recycling the necessary elements. When a plant dies and decays, rainwater washes the chemicals back into the soil. The same chemicals can then be used during the next growing season. Some of the useful chemicals may be removed by groundwater, but the loss is slowly made up by the addition of new materials through chemical weathering of minerals in the bedrock. When we interfere with this process by growing crops and removing the plant matter for food, the soil becomes progressively depleted in the needed elements. The elements needed for growth must then be replaced in the form of soluble fertilizers added to the soil.

The three essential fertilizer elements are: (1) nitrogen, which is recovered by chemical means from the atmosphere; (2) potassium, which is recovered as the soluble mineral sylvite (KCl) from marine evaporite deposits; and (3) phosphorus, recovered as apatite $[Ca_5(PO_4)_3(OH,F)]$ from a special class of marine chemical sediments known as *phosphorites*. For each essential fertilizer element the world's supplies are very large, and probably sufficient for all foreseeable needs. But there are problems associated with the geographic distribution of fertilizer raw materials. Fertilizer must be used in large amounts and the costs of treatment are high. Deposits of phosphorus and potassium minerals are geographically restricted, and costs of transport to all the places the materials are needed become significant. It is difficult for poorer countries such as India to purchase all the fertilizers they need.

Calcium carbonate and sulfur must also be added to soils to keep a balance between acidity and alkalinity favorable for maximum plant growth. Calcium carbonate is produced from the abundant limestone strata around the world, but sulfur is more restricted, forming mainly in fumaroles (Chapter 4), in certain gypsum deposits in marine evaporites, and in the H_2S recovered from many natural gas wells. Supplies of sulfur are very large, but unfortunately for poor countries that need the sulfur, the supplies are geographically restricted.

Building Materials

Besides cut stone, crushed stone, and sand and gravel, of which the world has enormous resources, the main building materials are: cement (manufactured from shale and limestone), gypsum (used for plaster), clay (used for tile and brick), and asbestos (a variety of serpentine used for wallboards, siding, and insulations). With the exception of asbestos, which is formed principally as an alteration product of peridotites by hot groundwater, supplies of building materials are enormous. Problems associated with their use tend to be problems of transportation, siting of quarries in environmentally acceptable areas, and costs of treatment and preparation. It has been suggested that there will be a shortage of standing room on the Earth before there is a shortage of building materials.

Ceramic and Abrasive Materials

We sometimes forget how extensively ceramic materials (including glass) are used in our everyday lives. Some ceramics have properties that make them desirable substitutes for certain metals; so it is reasonable to anticipate that the uses of ceramics will grow. Supplies of the essential raw materials for ceramics—certain kinds of clays, feldspar, and quartz—are all abundant and widespread.

Abrasives also play a widespread and vital part in our lives, although their importance is not widely appreciated. Modern industry employs machines that work accurately and efficiently at high speeds. Such machines require precision grinding and exact shaping and polishing of the machine parts; for these purposes a wide variety of abrasives is needed. The abrasives must be hard enough to cut metals and tough enough so they will not rapidly fracture during the grinding process. The principal abrasive minerals are: quartz and garnet, both of which are common rock-forming minerals; corundum, which is rare as a mineral but easily synthesized, and diamond. It is not commonly appreciated that about 80 percent of all the diamonds produced are used as abrasives, and only 20 percent are cut as gems. Nor is it widely realized that although the primary source of diamonds is kimberlites, a rare kind of igneous rock erupted from the mantle, most diamonds are recovered from placers. This is so because when diamonds do reach the surface, they are so hard and tough that even pounding in streams does not quickly wear them down. Because they are more dense than the silicate minerals with which they occur, diamonds are readily concentrated in placers. Most of the world's production has, for many years, come from placers in African countries such as Zaire, Central African Republic, and Namibia, and from Siberia. However, recent discoveries in northwestern Australia mean that that country will also be a major producer in the future.

Not all the abrasive materials used today are natural. Two of the toughest and most useful abrasive substances, diamond and corundum, can be readily produced in synthetic forms.

FIGURE 22.25 Two miners digging out a vein of silver ore, Potosi, Bolivia. The width of the vein corresponds to the width of the opening the miners are standing in.

FUTURE SUPPLIES OF MINERALS

Mineral deposits are exploited in the least expensive way possible. For some deposits this means underground mines in which miners drill and blast away at narrow veins (Fig. 22.25); the quantities of material dug from the ground in this fashion are truly enormous. For other deposits, exploitation means huge open pits from which the ore is removed by truck or train (Fig. 22.26). Yet for still others, such as some salt bodies, exploitation may mean solution in hot water pumped down a drill hole and recovery of salt by evaporation from the resulting brine. All kinds of mining are becoming increasingly efficient. Once a deposit is discovered, it can be worked quickly and efficiently. But how can we discover enough deposits? Minerals for industry, like fossil fuels, are being used at ever-increasing rates; and mineral deposits, as we have stressed, are a one-crop-only resource.

At present, new mineral deposits are being found about as fast as old ones are being ex-

FIGURE 22.26 Mining copper from an open pit. An aerial view of the Sierrita Pit, near Tucson, Arizona. Mining proceeds by moving each bench back, thus making the overall pit wider, and allowing a new bench to be started at the bottom of the pit. Each bench is about 40 m high. The size can be judged from the large trucks visible at the bottom of the pit.

hausted. But the new finds are mostly in places such as Chile, Australia, and Siberia, where intensive prospecting has begun rather recently. Antarctica is the only continent where intensive prospecting has yet to begin. In Europe, which has been intensively prospected for many years, new deposits are found rarely; indeed, in the areas conquered and occupied by the Romans, no new deposits of the metals used by them have been discovered since they departed more than 1500 years ago. Extensions of ore bodies in known mineralized ores have been discovered in many places in Europe, but not in totally new areas of mineralization. This is not to say that there are not more ore deposits to be found in what was once Roman-occupied Europe. Old-time prospectors could only sample outcrops and find mineral deposits that were not covered by younger sediments. Modern prospectors who are aided by sophisticated geophysical and geochemical devices can sense mineral deposits buried by as much as 150 m of soil and sediments. They can only do so under ideal conditions, however, and so far they have been unsuccessful where cover rocks are more than 150 m deep. Considering that about half of the Earth's surface is covered by sediment that is more than 150 m deep, we must conclude that there are still a great many mineral deposits to be found. First we must discover how to find them. Then we must ask whether the cost of finding and mining such buried deposits will be worthwhile. The answer to that question is one for the future. Even so, the number of mineral deposits in the crust is fixed. Every time one is mined out, there is one less left to serve as a future supply. The rate at which we use mineral resources is growing and the number of deposits left to be found is shrinking. Perhaps, therefore, when all parts of the Earth's surface have been prospected intensively, people will have to look to kinds of deposits not presently exploited. For geochemically scarce metals (such as mercury, silver, gold, and molybdenum) that time may come soon. For these scarcer materials, careful conservation of supplies will some day have to be practiced, substitutes will have to be found for some of their applications, and efficient recycling will become essential. For materials that are abundant in the crust (such as iron, aluminum, common salt, and sulfur), sup-

plies will probably always be sufficient to meet demand.

Probably the future of mineral supplies will be similar to the future of energy supplies. A sufficiency will be found, but present patterns of use will change. Some materials will become more important, others less important, and yet other substances, not used today, will assume major importance. But one prediction can safely be made: Mining operations will get bigger, more materials will be used, and pollution will probably be intensified. Environments will come under even heavier pressure than that which endangers them today. In short, human beings will continue to be an increasingly important factor in geological processes.

SUMMARY

1. The mineral industry is based mainly on local concentrations of useful minerals in mineral deposits.

2. When a mineral deposit can be worked profitably it is called an ore deposit.

3. Mineral deposits form when minerals become concentrated in one of five different ways: (1) by concentration through weathering processes to form residual deposits; (2) by concentration in flowing water to form placers; (3) by concentration from lake water or seawater to form sedimentary mineral deposits; (4) by concentration through magmatic cooling and crystallization processes to form magmatic mineral deposits; and (5) by concentration through hydrothermal solutions to form hydrothermal mineral deposits.

4. Deposits of geochemically abundant metals, which make up 0.1 percent or more of the crust,

form in several ways. The amounts available for exploitation are enormous.

5. Deposits of geochemically scarce metals, present in the crust in amounts less than 0.1 percent, form mainly as hydrothermal and magmatic deposits. Amounts of scarce metals available for exploitation are limited and geographically restricted.

6. Secondary enrichment by weathering and groundwater has been important in further concentrating many copper and iron deposits.

7. Gold, platinum, tinstone, diamonds, and other minerals are commonly found mechanically concentrated in placers.

8. Nonmetallic substances are used mainly in the chemical industry for fertilizers, for building materials, and for ceramics and abrasives.

9. The most critical mineral substances, as far as future supplies are concerned, seem to be the geochemically scarce metals.

SELECTED REFERENCES

Bates, R. L., 1960, Geology of the industrial minerals and rocks: New York, Harper and Row.

Brobst, D. A., and Pratt, W. P., eds., 1973, United States mineral resources: U.S. Geol. Survey Prof. Paper 820.

Hutchison, C. S., 1983, Economic deposits and their tectonic setting: New York, Wiley-Interscience.

Jensen, M. L., and Bateman, A. M., 1981, Economic mineral deposits, revised printing: New York, John Wiley & Sons.

Sawkins, F. J., 1984, Metal deposits in relation to plate tectonics: Berlin, Springer-Verlag.

Skinner, B. J., ed., 1981, Economic Geology Seventy-Fifth anniversary volume, 1905–1980: El Paso, The Economic Geology Publishing Co.

Skinner, B. J., 1986, Earth resources, 3rd ed.: Englewood Cliffs, N.J., Prentice-Hall.

Skinner, B. J., and Turekian, K. K., 1973, Man and the ocean: Englewood Cliffs, N.J., Prentice-Hall.

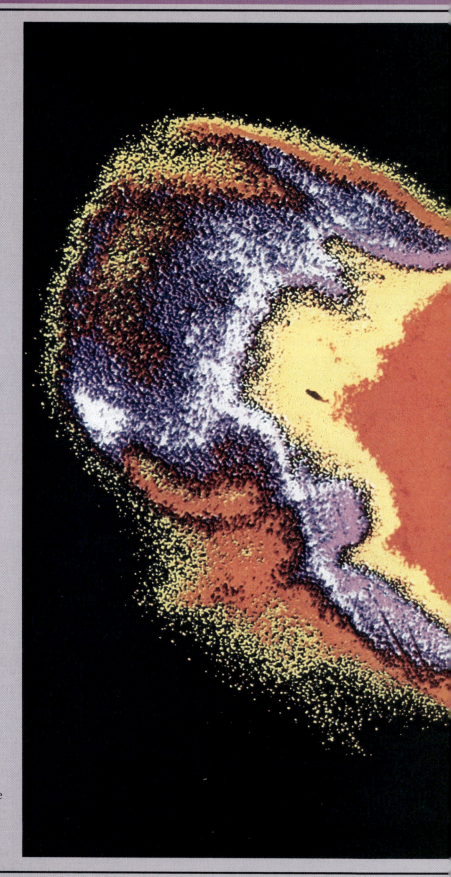

When the visible portion of the Sun is hidden from view during a solar eclipse by the Moon, a striking corona of intensely hot, atomic particles can be detected. The shape of the corona is controlled by the Sun's magnetic field.

Beyond
Planet Earth

The Planets: A Review

The four largest moons of the giant planet, Jupiter (upper right). Io (upper left) is closest to Jupiter, followed by Europa (center), Ganymede and Callisto (lower right). Nine smaller moons also circle Jupiter.

THE SOLAR SYSTEM

The first person to view the sky through a telescope was Galileo. The year was 1609 and his homemade device was crude by comparison with a modern child's toy telescope. Galileo was astonished to see mountains on the Moon and large flat areas that looked to him like seas. He observed several moons circling Jupiter and he saw disc-shaped rings around Saturn. He also saw that Venus, like the Moon, had phases—full, half, quarter—and that huge, dark spots appeared every now and then on the surface of the Sun. The discoveries electrified Galileo's contemporaries, just as visits to the Moon by astronauts and images of distant planets sent back by unmanned spaceships have electrified the present generation.

Interest in visible stars dates back to prehistory. Our ancestors noticed that a few stars seem to wander in completely different paths from the annual progression of most of their fellows across the sky. The Greeks mapped the paths of these strange objects and called them *planetai*, or wanderers. The Romans named the objects after their gods: Saturn, Jupiter, Mars, Venus, and Mercury. We now know that the curious wandering paths of the planets result from orbital motions around the Sun and that the other objects in the heavens are other suns, so far distant that they seem to occupy fixed places in the sky. Three other planets—Uranus, Neptune, and Pluto—were unknown to our ancestors because they are visible only through telescopes. Finally, we now know that other planets besides the Earth and Jupiter have moons of their own. A few vital statistics of the planets and their larger moons are given in Table 23.1.

In 1957, when the first artificial satellite was placed in orbit around the Earth, a brand-new scientific specialty emerged. *Planetology,* devoted to *a comparative study of the Earth with the Moon and other planets*, has taught us much about the Earth's earliest days—that time, before 3.8 billion years ago, for which most of the record seems to have been erased by the operation of the rock cycle. In the present chapter we can give only a brief account of the successes of planetology. However, they have been enormous and have motivated sci-

TABLE 23.1 *Properties of the Planets and Some of Their Moons*

	Diameter (km)	Mass		Density (g/cm^3)
		(g)	(Earth = 1)	
Terrestrial Planets				
Mercury (no moons)	4,880	3.30×10^{26}	0.055	5.4
Venus (no moons)	12,104	4.87×10^{27}	0.815	5.3
Earth (1 moon)	12,756	5.97×10^{27}	1.000	5.5
Moon	3,476	7.35×10^{25}	0.012	3.3
Mars (2 moons)	6,787	6.42×10^{26}	0.107	3.9
Phobos	~22	9.55×10^{19}	1.6×10^{-9}	1.9
Deimos	~13	1.97×10^{18}	3.3×10^{-10}	1.5
Giant Planets				
Jupiter (15 moons)	142,800	1.90×10^{30}	317.9	1.3
Io	3,632	8.93×10^{25}	0.015	3.5
Europa	3,126	4.86×10^{25}	0.008	3.2
Ganymede	5,276	1.49×10^{26}	0.030	2.0
Callisto	4,820	1.06×10^{26}	0.018	1.8
Saturn (15 moons)	120,000	5.69×10^{29}	95.2	0.7
Mimas	390	3.76×10^{22}	7.6×10^{-6}	1.2
Enceladus	500	7.40×10^{22}	1.4×10^{-5}	1.2
Tethys	1,050	6.26×10^{23}	1.3×10^{-4}	1.2
Dione	1,120	1.05×10^{24}	1.8×10^{-4}	1.4
Rhea	1,530	2.28×10^{24}	4.2×10^{-4}	1.3
Titan	5,120	1.34×10^{26}	2.3×10^{-2}	1.9
Iapetus	1,440	1.88×10^{24}	3.2×10^{-4}	1.2
Uranus (15 moons)	51,800	8.66×10^{28}	14.6	1.2
Titania	1,000?	1.10×10^{24}	1.8×10^{-4}	2.1?
Oberon	802?	2.96×10^{23}	4.9×10^{-5}	1.1?
Neptune (2 moons)	49,500	1.03×10^{29}	17.2	1.7
Pluto (no moons)	6,000?	1.5×10^{25}	0.003	1.1

entists to seek a unifying theory of origin for all suns and their planets. If the scientists are successful, it may one day be possible to say which of the billions of other suns, visible in the heavens, have planets like the Earth, and perhaps even to say which, if any, of those planets may support biospheres of their own.

The Solar System consists of the Sun, nine planets, 50 known moons, a vast number of asteroids, millions of comets, and innumerable small fragments of rock called meteorites. All of the objects in the Solar System move through space in smooth, regular orbits, held in place by gravitational attraction. The planets, asteroids, and comets circle the Sun while the moons circle the planets. *Meteorites,* which are *small stony or metallic objects from interplanetary space that impact a planetary surface,* can only be observed in motion when they flash into view as they plunge through the Earth's atmosphere. However, they, too, are believed to follow orbits around the Sun.

The orbits of the planets around the Sun are elliptical. The planets all revolve around the Sun in the same direction and, except for Pluto, the orbits lie in nearly the same plane as the orbit of the Earth. The orbit of Pluto is tilted at 17° to the ecliptic, which is the name given to the plane of the Earth's orbit (Fig 20.27a). Most of the moons revolve around the planets in the same direction as the planets revolve around the Sun.

The Solar System occupies a region in space that is at least 12 billion km in diameter. Measuring outward from the Sun, the **Titius-Bode rule** states that *the distance to each planet is approximately twice as far as the next inner one.* Thus, the distance from the Sun to Mercury is 58 million km, while the distance to Venus, the next planet, is 107 million km. The rule seems to break down between Mars and Jupiter. Mars is 226 million km from the Sun, Jupiter is 775 million km. This rule suggests that there should be a planet about 400 million km from the Sun. There is no planet at this place, but there is, instead, a very large number—at least 100,000—of **asteroids** which are *irregularly shaped rocky bodies that have orbits lying between the orbits of Mars and Jupiter.* These asteroids are either fragments of a planet that once existed and was somehow broken up or, more likely, they are rocky fragments that failed to gather into a planetary mass.

The distances between the planets are immense. Vast distances, like vast stretches of time, are difficult to comprehend. To put the Solar System into perspective, think of the Sun as a basketball. The nearest planet, Mercury, would be a speck of dust about 12 m away. Earth would be a grain of sand about 1 mm in diameter at a distance of 30 m, Saturn a pebble the size of a grape nearly 300 m away, and Pluto, the most distant planet, would be another grain of sand 1200 m away. The comets, which are believed to be mixtures of rock, ice, and dust, lie farther still, beyond the orbit of Pluto, and they would form a cloud of miniscule dust specks about 1600 m away!

The planets can be separated into two groups based on their densities and compositions. The innermost planets, Mercury, Venus, Earth, and Mars, are small, rocky, and dense. Each has a density of 3 g/cm^3 or more. They are similar in composition, and we call them the *terrestrial planets* because they are similar to *terra* (Latin for the Earth). The asteroids and meteorites are also dense, rocky objects. They are believed to have formed in the inner regions of the Solar System, in the same way and time as the terrestrial planets.

The planets more distant from the Sun are much larger, yet much less dense than the terrestrial planets. The masses of Jupiter and Saturn are 317 and 95 times the mass of the Earth, but their densities are only 1.3 and 0.7 g/cm^3, respectively. The *jovian* planets, which take their name from *Jove,* an alternate designation for the Roman god Jupiter, probably have rocky cores that resemble terrestrial planets, but most of their planetary masses are contained in thick atmospheres of hydrogen and helium. It is this thick atmosphere that keeps the densities of the giant planets low.

We know that rocky matter exists in the outer regions of the Solar System because some of the moons that circle the giant, gassy planets have high densities like that of the terrestrial planets. Two of Jupiter's moons, Io and Europa, have densities between 3.2 and 3.5 g/cm^3, and are close to the density of the Earth's Moon. Space exploration has revealed that two of Jupiter's low-density moons, Callisto and Ganymede, have thick blankets of ice surrounding small, rocky cores. As we shall discuss later in this chapter, the progression outward from the Sun, from dense, rocky planets to less-dense, giant gassy planets with ice-shrouded moons, provides an important clue to the way the Solar System formed.

Cratered Surfaces

The processes that shape the surface of the Earth can be divided into three groups: tectonic processes, magmatic processes, and the surficial processes of weathering, mass-wasting, and erosion.

FIGURE 23.1 Meteor Crater, near Flagstaff, Arizona. The crater was created by the impact of a bolide about 20,000 years ago. It is 1.2 km in diameter and 200 m deep. Note the raised rim of the crater wall and the blanket of ejecta thrown out of the crater.

These processes have been the principal topics discussed in previous chapters. To varying degrees each group of processes plays, or has played, a role in shaping the surfaces of all the rocky planets and moons in the Solar System. However, in a planetary context, a fourth, more important process must be added—the process of impact cratering.

Impact cratering is *the process by which a planetary surface is deformed as a result of a transfer of energy from a bolide to the planetary surface. A **bolide*** is defined as *an impacting body; it can be a meteorite, an asteroid, or a comet.* The velocities of meteorites, which are small bolides, have been measured between 4–40 km/s. If a bolide were large, and had such a velocity, the amount of energy released on impact would be enormous. It has been calculated, for example, that a bolide 30 m in diameter and traveling at a speed of 15 km/s would, on impact, release as much energy as the explosion of 4 million tons of TNT. The resulting impact crater would be the size of Meteor Crater in Arizona—1.2 km across and 200 m deep (Fig. 23.1). Cratering is a very rapid geological process; the Meteor Crater event is estimated to have lasted about 1 min.

Approximately 200 impact craters have been identified on the Earth. However, impact events must have been much more common than this small number implies. The reason so few craters have been found is that weathering and the rock cycle continually erase the evidence. On most

planetary surfaces in the Solar System a very different situation prevails. On most of the terrestrial planets, and on many of the rocky moons, atmospheres are absent, and tectonic activity has either ceased or is so slow that a rock cycle probably does not operate. As a result, the most striking features to be seen on most of the solid surfaces of the Solar System are impact craters, some of which are more than 4 billion years old (Fig. 23.2). The largest craters are 1000 km or more in diameter. They range down in size to tiny craters, the size of a pin's head, produced by impact of dust-sized bolides.

No large, natural impact crater has been observed as it formed. What is known about the process of cratering comes largely from laboratory experiments. The sequence of events is illustrated in Figure 23.3. As a high-speed bolide impacts and penetrates the surface it causes a high-velocity jet of debris to be ejected away from the point of impact. At the same time, the impact compresses the underlying rocks and sends intense shock waves outward. The pressures produced by the shock waves from a large bolide are so great that the strength of the rock is exceeded and a large volume of crushed and brecciated material results. In very large impact events, local melting and vaporization may even occur. Once the compressive shock waves have passed, a rapid expansion or decompression occurs. Expansion causes more material to be ejected from the impact crater and produces a blanket of ejecta that surrounds the crater and thins away from the rim. Ejecta blankets can be seen in Figures 23.1, 23.2, and 23.4. Beyond the ejecta blanket, immediately adjacent to the crater rim, there are rays of discontinuous ejecta and even small secondary craters. Note, in Figure 23.3, that the stratigraphy of layers in the ejecta blanket adjacent to the crater rim is overturned. In the very largest impact structures, the central crater is circled by one or more raised rings of deformed rock. The outer rings are presumed to form as a result of the initial compression (Fig. 23.5). Following the immediate impact event, a number of post-impact events tend to modify the crater. Crater walls may slump (Fig. 23.6), isostatic rebound may produce changes in the floor and rim of the crater, erosion may fill the crater with debris, and in some instances, magma may rise along fractures produced by the impact and lava may fill the crater.

THE TERRESTRIAL PLANETS

Besides impact cratering, volcanism has been widespread on each of the terrestrial planets and the Moon. On Mercury, the Moon, Mars, and

Venus, it is clear that volcanism is predominantly basaltic. Besides the Earth, only on Venus is it possible that other kinds of volcanism may have occurred. However, Venus is cloud-covered and

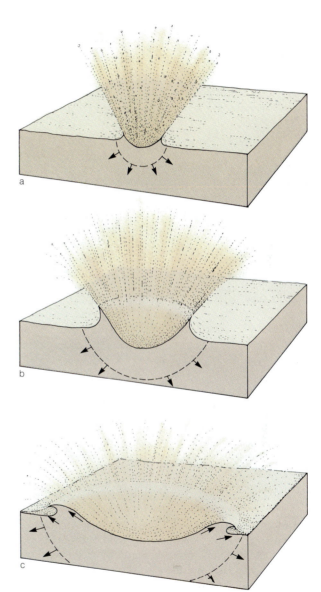

FIGURE 23.3 The features and shape of an impact crater are the same regardless of the angle of entry of the impacting bolide. Stages in the development are: (*a*) The initial bolide contact ejects a high velocity jet of near-surface material. (*b*) The passage of shock waves through the bedrock produces high pressures and the compression of strata. In places the rock strength is exceeded and fracturing and brecciation result. As the compressive wave passes, decompression throws broken rock out of the crater. (*c*) Strata along the rim of the crater are folded back and overturned by decompression. The ejected debris forms a circular blanket around the crater. (*Source:* After Gault et al, 1968.)

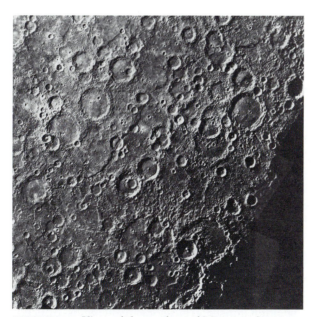

FIGURE 23.2 View of the surface of Mercury showing a densely cratered surface produced by bolide impacts. The high density of craters indicates that the surface is very ancient—as much as 4 billion years—and that the rock cycle is not active on Mercury. The image was taken from a distance of 76,000 km by the spacecraft *Mariner 10*.

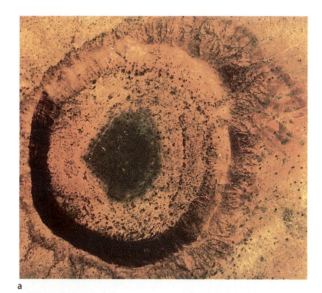

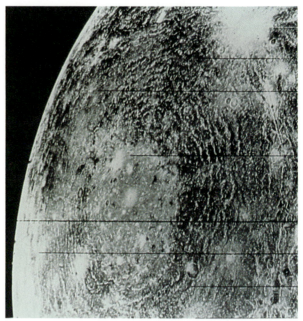

FIGURE 23.4 (*a*) Wolf Creek Crater, Western Australia. The exact age of the crater is not known but it is inferred to be relatively young because the crater has not been deeply eroded and the ejecta blanket is still present. Scale can be judged from the two single-lane roads that approach it from the upper right. (*b*) Impact craters on the Moon's surface. Note the raised crater rims and the prominent ejecta blankets.

FIGURE 23.5 (*a*) Multiple concentric rings surrounding Valhalla, a crater on Callisto, a moon of Jupiter. The crater (central bright spot) is 600 km in diameter. The rings were probably formed as a result of circular thrust faults caused by the compression due to the giant impact. The surface of Callisto is ice. Rings are also observed around giant craters on rocky planetary bodies, but they are neither so pronounced nor so numerous as the rings around Valhalla.

difficult to study, so the question remains open. What space exploration has made clear, and what has been confirmed by analysis of lunar and martian rocks, is that basaltic volcanism is a very common process in the Solar System. This means, presumably, that each planet has gone through (or is still going through) a stage in its developmental history when internal heating caused partial melting. The product of this partial melting is basaltic magma. The products of partial melting are apparently similar on all of the rocky bodies in the Solar System; this is suggestive evidence that the parent bodies all have similar compositions. This means, in turn, that major differences between the rocky planets must reflect other factors in addition to composition.

Mercury

Mercury, the innermost planet, is so close to the Sun that it can only be seen just before sunrise or just after sunset. Telescopic viewing is difficult under such circumstances and as a result, almost everything that is known about Mercury comes from observations made during fly-by missions of unmanned spacecraft. Mercury has a diameter of 4880 km and is just a little larger than the Moon. It rotates slowly about its axis, and has a density of

FIGURE 23.6 A large impact crater on the Moon. The crater is more than 200 km in diameter; the oblique photo was taken from a manned spacecraft. The highlands in the distance are part of the ejecta blanket. The stepped terraces in the middle distance were formed as a result of post-cratering collapse of the crater rim. The hills in the foreground lie at the center of the crater and were formed by rebound of the crater floor during decompression.

5.4 g/cm^3. The high density is puzzling. On the Earth, the density of the mantle increases with depth due to compression from the weight of rock above. The same effect must occur in all planets. Because Mercury is so much smaller, the increased density of its interior layers must be much less than the equivalent effect on the Earth. To account for the high overall density, therefore, it is necessary to conclude that Mercury has a metallic core that is about 3600 km in diameter, and that this core accounts for 80 percent of the planet's mass. Mercury's core alone is the size of the Moon. Images of the surface, sent back by spacecraft, show that Mercury is heavily pockmarked by ancient impact craters, that it lacks an atmosphere, and that there is no evidence of moving plates of lithosphere. The largest impact basins are filled with what are apparently basaltic lava flows; this means that Mercury had a period of magmatic activity. However, the lava plains are not crumpled and deformed, so the magmatic activity was not followed by tectonic activity.

The largest impact structure on Mercury is the Caloris Basin, 1300 km in diameter (Fig. 23.7). On the far side of Mercury, exactly opposite Caloris, the surface is jumbled into a weird, hilly terrain (Fig. 23.8). Scientists who have studied Mercury believe that the weird terrain was produced by compressive shock waves from the bolide that

formed Caloris. The waves passed completely through the planet and disrupted the far side (Fig. 23.9).

The most extraordinary feature about Mercury is the presence of a magnetic field that is about 1/100 as strong as the Earth's. The field is dipolar and the

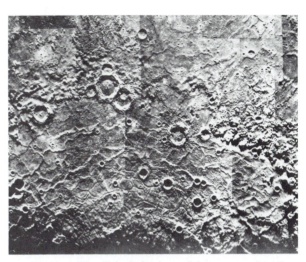

FIGURE 23.7 The Caloris basin on Mercury, a giant impact crater 1300 km in diameter. Note the irregularity of the concentric ridges and the radial fractures. The cause of the radial fractures is not known but presumably they arise from impact.

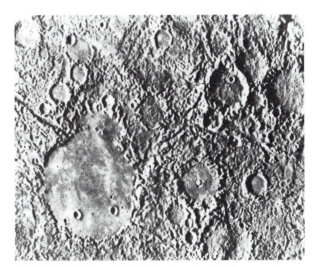

FIGURE 23.8 The weird hilly terrain that lies exactly opposite Caloris, the giant impact structure on Mercury. The smooth floors in some of the impact craters probably result from basaltic lava that was erupted after development of the "weird" terrain. A possible origin for the weird terrain is shown in Figure 23.9. The area shown is 540 km across.

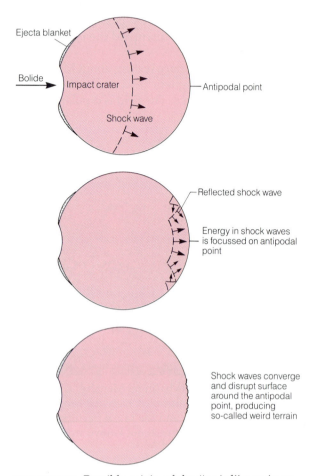

FIGURE 23.9 Possible origin of the "weird" terrain on Mercury (see Fig. 23.8). The terrain is exactly opposite the giant impact crater, Caloris. Compressive waves caused by the impact travelled through the body of the planet, and surface waves travelled around the edge. Both sets of waves were focused at the antipodal point, and they produced severe disruption of the surface. (*Source*: After Schultz and Gault, 1975.)

magnetic axis coincides with the axis of rotation. The origin of the magnetic field is a puzzle. If Mercury's core is molten, the magnetic field could be caused by the fluid motions of the planet's rotation. There are two objections to this idea. The first is that Mercury rotates so slowly—once every 59 days—it is difficult to see how fluid motions strong enough to produce such a magnetic field could exist in the core. The second concerns tectonism. If Mercury has a huge core of molten iron, the interior must be very hot. Why then, is Mercury not tectonically active? The puzzle of the magnetic field remains unanswered. Even though it seems far-fetched, some scientists are considering the possibility that the magnetic field is caused by

remanent magnetism of basalts produced by a former magnetic field from the core. The most important fact we have learned about Mercury is that there is another terrestrial planet besides the Earth with a magnetic field. We can look forward to much research in the attempt to solve the puzzle of the source of the field.

Venus

Venus (diameter, 12,104 km; density, 5.3 g/cm^3) is the planet most like the Earth in size and mass. However, the similarities are fewer than the differences. Venus is enveloped in a cloudy atmosphere of carbon dioxide that is a hundred times more dense than the Earth's atmosphere. The clouds prevent direct observation of the Venusian surface. Information comes from several spacecraft that have landed on the surface and radioed back information, and from radar measurements of the surface topography.

Landing spacecraft have reported that the surface temperature of Venus is astonishingly hot, about 500° C. At this temperature metals such as lead, zinc, and tin are in a molten state. The explanation of the high temperature is evident. First, Venus is closer to the Sun than the Earth. However, more important is the fact that the carbon dioxide in Venus's atmosphere acts like the glass

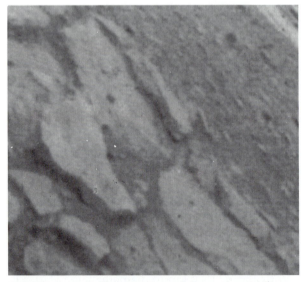

FIGURE 23.10 A close-up view of the surface of Venus from the Soviet spacecraft *Venera 13*. The slabby rock surfaces are suggestive of pahoehoe-like basaltic lava flows. The smaller particles visible offer proof that chemical weathering is causing the rock to disintegrate.

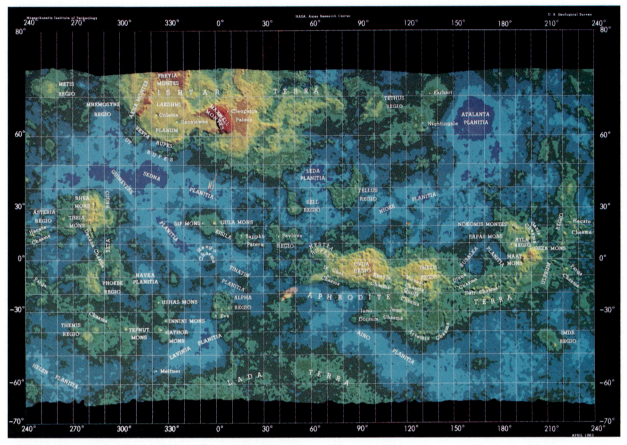

FIGURE 23.11 Topography of Venus as determined by radar. Blue areas are lowlands, green, yellow, and red, in that order, represent increasingly high topography. If continental crust exists on Venus, it is represented by the yellow and red highlands.

of a greenhouse: It lets the Sun's rays through to heat the surface, but serves as a barrier that prevents heat from leaving (Chapter 20). Russian spacecraft have operated in this intense heat long enough to send back a few small-scale images of the surface of Venus. The images show masses of broken rock fragments, each about 20 cm across, covering the surface (Fig. 23.10). Unfortunately, they don't tell us much about the composition of the rocks. However, several of the Russian spacecraft have also carried out rough chemical analyses before they were overcome by the high temperatures. In each case, the analyses suggest that the rocks analyzed were basaltic in composition, but features such as high-potassium contents support the idea that Earthlike magmatic differentiation may have been operating.

Sensitive radar measurements of Venus's shape show that large circular structures (presumably, impact craters) are present, and that much of the planet is nearly smooth (Fig. 23.11), with a total relief from the bottom of the deepest chasm to the top of the highest peak of 13 km. By comparison, the total relief on the Earth is 20 km. There is no evidence on Venus of two topographic levels, as there is on the Earth (Fig. 2.7). The Earth's topography is a result of the differentiation processes that produce oceanic and continental crust, respectively. Apparently, this has not happened on Venus, leading to the conclusion that if plate tectonics has operated there, it must have been on a much smaller scale than on the Earth.

Visiting spacecraft discovered that Venus lacks a magnetic field. Because the density of Venus is very close to the Earth's density, Venus must have an iron core and (presumably, like the Earth's) at least part of it is molten. The reason for the lack of a magnetic field is probably to be found in the slow rate of rotation of Venus about its axis—once every 243 days. The rotation is apparently too slow to cause fluid motion in the core.

We can speculate that the history of Venus may

be like the early part of the Earth's history. Apparently magmas were created, convection probably occurred in the mantle, and possibly there was even a period of moving lithospheric plates. Some have suggested that Venus is actually more advanced in its history than the Earth, that erosion has proceeded further, and that future study of Venus will provide clues to the Earth's future. The observation that large impact craters are present does not support this idea. All such suggestions are at best speculation, and we must await future spacecraft missions and more reliable observations before it is possible to draw any definite conclusions about the Earth's twin.

The Moon

The Earth's Moon is unique in the Solar System. Because its diameter is 3476 km, only a little less than that of Mercury, the Moon is often described as a small terrestrial planet, and the Earth–Moon pair is described as a double planet. The moons of Jupiter and Saturn, although as big as or even bigger than our Moon, are tiny in comparison with the sizes of their giant neighbors. The moons of Mars are only tiny bodies, a few kilometers in diameter.

The exact place in the Solar System where the Moon originated is not known with certainty. It may have formed where it is now (close to the Earth and attracted to it by gravity). However, it may have condensed as a separate planet, moving in its own orbit around the Sun. Then, when the separate paths of the Earth and the Moon brought them momentarily close together, the Earth's gravitational pull may have plucked the Moon from its path, causing it to move into a new path and to orbit the Earth endlessly. Although we still cannot choose between these two possibilities, this fact does not influence an important conclusion we *can* draw: Because the Moon is a small, dense, rocky planet, it must have formed in the inner regions of the Solar System, just as the other terrestrial planets did. If ideas about the way planets form are even partly correct, the Moon's structure and composition should be similar to those of the other terrestrial planets. Information about the Moon, therefore, should help us in our understanding of the Earth and the other terrestrial planets.

Structure

Each time astronauts have visited the Moon, the measurements made have yielded clues about its structure. When they departed, they left behind instruments that continued the measurements and that transmitted the results to the Earth. The most informative measurements are of four kinds; measurements of seismic waves, magnetism, the heat that flows out of the Moon, and gravity.

Seismic Evidence. Compared to earthquakes, moonquakes are weak and also infrequent. Moonquakes large enough to be detected by instruments placed on the Moon by astronauts number fewer than 400/yr.; on the Earth the same instruments would record nearly a million quakes per year. Despite the infrequency of moonquakes and the weakness of the seismic waves they generate, the quakes reveal a good deal about the lunar structure. Some of the most important points are: (1) on the side of the Moon that faces the Earth (at least on those parts that have been visited by astronauts) there is a crust about 65 km thick; (2) the Moon is layered (Fig. 23.12); (3) covering the surface is a layer of regolith that ranges from a few meters to a few tens of meters thick; (4) below the regolith is a layer (about 2 km thick) of shattered and broken rock produced by the continual rain of large and small bolides; (5) below the broken-rock zone is about 23 km of basalt, then 40 km of a feldspar-rich rock; and (6) at a depth of 65 km the velocities of seismic waves increase rapidly, indicating that the lunar crust overlies a mantle.

The exact composition of the lunar mantle is not known. Wave velocities suggest it is similar to that of the Earth's mantle, but that the lunar mantle is colder and more rigid than the Earth's mantle. Whether or not the lunar mantle is layered is not known. Nor is it yet resolved if the Moon has a core, because moonquakes are too weak to pass through the Moon and resolve the question.

The comparative scarcity of moonquakes (and their weakness) immediately suggests that processes such as volcanism and plate tectonics, the causes of most earthquakes, are not happening on the Moon. Some moonquakes are caused by meteorites hitting the Moon. However, most quakes occur in groups, and at the times when its elliptical orbit brings the Moon closest to the Earth—the very moment when gravitational forces between the Earth and the Moon are strongest. This suggests that most moonquakes result from the gravitational pull that the Earth exerts on the Moon. The pull causes slight movement along cracks, each one causing a tiny moonquake. Foci of moonquakes have been recorded as deep as 1000 km; this observation, too, is informative. It means

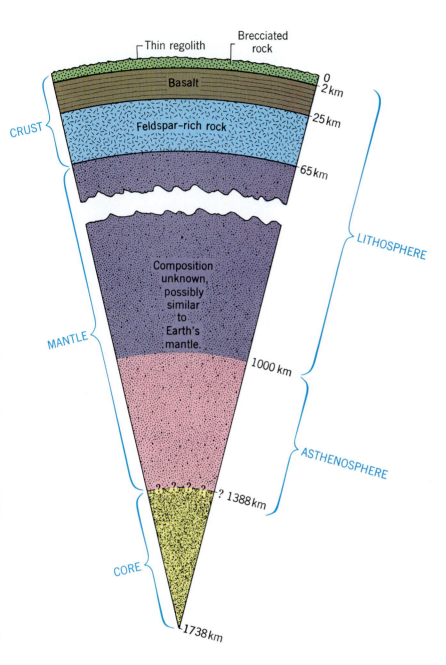

Figure 23.12 A section through the Moon showing the probable layered structure. The structure of the crust is known with certainty only in the vicinity of the astronauts' landing sites, and the existence of the small core is still uncertain. The depth of the boundary between the lithosphere and the asthenosphere is known only approximately.

that unlike the Earth, the Moon is rigid enough at 1000 km for elastic deformation to store energy and for brittle failure to occur. This in turn means that the lunar lithosphere must be at least 1000 km thick and that the asthenosphere lies very deep within the Moon's body. The thick, rigid lithosphere makes it most unlikely that there is any present-day tectonic activity on the Moon.

Magnetism. Although the Moon lacks a dipolar magnetic field such as that of the Earth, it is weakly magnetic. The magnetism comes from a surprising source: bodies of igneous rock, all of which retain remanent magnetism. Therefore, although there is no lunar magnetic field today, a field must have existed at the time when the igneous rocks solidified from magma. Unless a magnetic field can be generated in some way that we know nothing about, we can infer that the Moon probably has a core, that the core once created a magnetic field, but that it is no longer capable of doing so. We can also infer that the core must be small. The Moon's density is 3.3 g/cm^3, while that of the Earth is 5.5 g/cm^3. If we assume that the Moon's core, like the Earth's, is mainly iron and is, therefore, very dense, we can state that its radius can be no larger than about 700 km. If the core

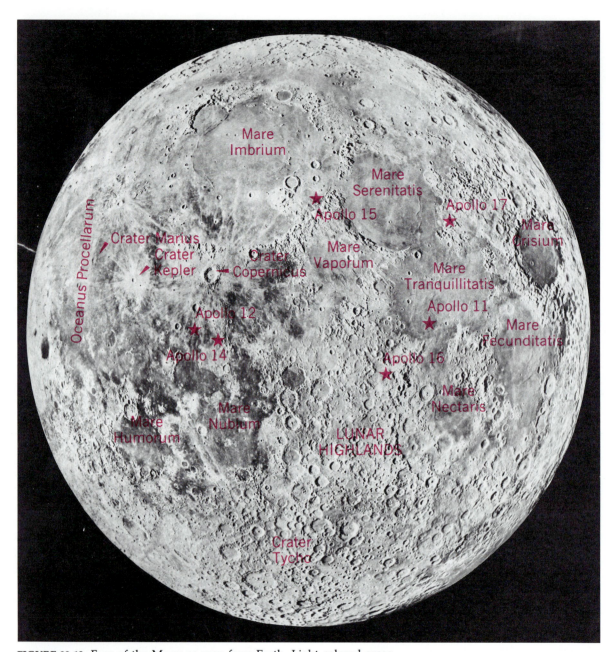

FIGURE 23.13 Face of the Moon as seen from Earth. Light-colored areas, pitted with craters caused by impacts of meteorites, are lunar highlands. Dark-colored, lowland areas are maria (plural of mare), formed when basaltic lava flowed out to fill the craters made by exceptionally large impacts. Copernicus, Kepler, and Tycho are three prominent, young meteorite craters that formed after the mare basins were filled. The impacts that made them splashed bright-colored rays of rocky debris over ancient basalt. The sites of the six Apollo lunar landing missions are indicated. (*Source:* Mosaic LEM-1, 3rd edition, 1966, made from telescopic photographs by U. S. Air Force.)

were larger, the Moon's overall density would have to be greater than it is.

Heat Flow. The amount of heat flowing outward from the Moon was measured during the *Apollo 15* and *Apollo 17* missions and was found to be surprisingly high—about half the rate of the Earth. A high rate of heat flow and a thick lunar lithosphere seem at first to be contradictory. However, they need not be so. They lead us to conclude that the lunar crust contains a high concentration of heat-producing radioactive elements. From this we can infer that some kind of magmatic differentiation must have been involved in the formation of the lunar crust, as a result of which the radioactive elements became concentrated in a thin, near-surface layer.

Indirect measurement also yields information about interior temperatures. The Sun continually emits electrically charged particles which stream away into space. Those streams are similar to electrical currents, and as they reach the Moon they cause secondary electrical currents to flow within it. The currents are weak—so weak that they cannot be felt—but they are detected by sensitive measuring devices. The strength of an electrical current depends on the composition and temperature of the substance in which it flows; for rock material, the principal control is temperature. The strength of the Moon's electrical currents indicates that at a depth of 1000 km the temperature of the lunar mantle can be no greater than 1000–1100° C. Therefore, this means that the temperature at the base of the lunar lithosphere can only be 1100° C. This in turn suggests that temperatures in the underlying lunar asthenosphere are probably too low for partial melting to occur.

Gravity. We can divide the Moon's surface into two general categories: (1) *highlands,* mountainous areas that appear to us as light-colored patches; and (2) *maria* (the "seas" seen by Galileo), smooth lowland areas that appear to us as dark-colored regions (Fig. 23.13). The highlands are regions of intense cratering. The maria are covered by basaltic lava flows (Fig. 23.14). In places, the lavas can be seen to fill and cover the impact craters. We conclude, therefore, that the highlands are older than the maria. Note, in Figure 23.13, that the maria are circular or nearly so. Detailed study shows that they are actually the sites of giant impacts that must have predated the basaltic flows that now fill them.

In the highlands, mountains soar tens of

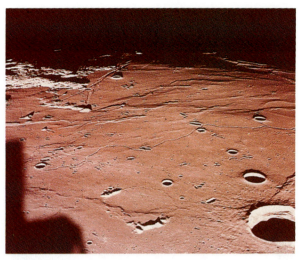

FIGURE 23.14 A photograph of the surface of a mare. Taken by an astronaut during the *Apollo 11* mission the photograph shows portions of Mare Tranquillitatis. The irregular ridges, looking like long sand dunes, are ancient basaltic lava flows. The impact crater on the lower righthand side is called Maskelyne.

thousands of meters above the maria, and in places stand even higher than the Earth's mountains. Because lunar mountains are big, one way to obtain clues about the Moon's interior regions is to learn whether isostasy operates as it does on the Earth, and whether the great lunar mountains have roots. The test is best made by measuring variations in the Moon's gravitational pull. Such measurements were made continually while spacecraft orbited the Moon; they indicate that the mountains must have roots because the highlands are isostatically balanced. We can infer, then, that when the lunar mountains formed, at least one of the Moon's outer layers must have been sufficiently fluid-like and plastic to enable the highlands to float.

However, the Moon's gravity also has an unexpected surprise and an additional clue. Some of the maria are not isostatically balanced; they have too much mass near the surface, and they produce *a region with an anomalously high gravity pull* called a **mascon** (abbreviated from *mass concentration*). Apparently mascons resulted from flows of dense basalt piling-up on the surface. The excess mass on the surface leads us to draw an interesting inference. When the maria formed, the Moon was not capable of attaining isostatic balance. The highlands formed before the maria did. Thus, the Moon had once, but later lost, the capacity of isostatic adjustment.

Astronauts not only measured the Moon's structures; they returned to the Earth with a treasure trove of samples of rock and regolith. From these, too, we have obtained many clues about the Moon's composition and history.

Rock and Regolith

Astronauts brought back three kinds of material: (1) a variety of igneous rocks; (2) breccias; and (3) samples of the regolith popularly called Moon dust (Fig. 23.15).

Igneous Rock. The most interesting samples are the igneous rocks; in terms of age and composition, there are three different kinds. The first and oldest consists of feldspar-rich rocks such as anorthosite, a variety of igneous rock formed by extreme magmatic differentiation, and consisting largely of calcium-rich plagioclase. Radiometric dates of these oldest rocks, which come from the highlands, indicate they could have been formed as long ago as 4.5 billion years, only 100 million years after the formation of the Moon. The second kind of igneous rock is basalt that contains high concentrations of potassium and phosphorus; this too is 4 billion or more years old. The potassium-rich basalts likewise come from highland areas and they are lava flows that seem to represent the last igneous activity in the highlands. They may possibly be the rocks that cause the high heat flow, because they are richer in radioactive elements than any of the other rocks found on the Moon. The third kind of igneous rock is also basalt, but it is rich not in potassium but in iron and titanium. It has been found only in the maria, and is dated radiometrically at 3.2–3.8 billion yr. We infer it is the material that underlies each mare to a depth of about 25 km. Mare basalt, then, formed several hundred million years after the highlands had formed. Even though magmas do not seem to occur on the Moon today, the mare basalts prove that magmas similar to those on the Earth have existed on the Moon for a long time.

Regolith and Breccias. What clues can be extracted from samples of the other kinds of Moon rock—regolith and breccia? The lunar regolith is a mixture of gray pulverized rock fragments and small particles of dust, many of which are glassy. Its composition is essentially that of the lunar igneous rocks. Regolith covers all parts of the lunar surface like a gray shroud (Fig. 23.16), as if giant hammers had crushed the surface rock. Indeed the surface has been hammered, but by the bolides

a

b

FIGURE 23.15 Two kinds of lunar samples brought back by astronauts. (*a*) Basalt, containing numerous vesicles formed when gases escaped during cooling and crystallization of lava. Collected during the *Apollo 12* mission, this sample is typical of the kinds of basalt found in the maria. (*b*) Regolith, a mixture of many rock and mineral types, together with glassy fragments produced by the bombardment of the lunar surface by meteorites. Only coarser grains from the regolith, from 1 to 2 mm across, are shown in the picture.

that continually strike its surface. The impacts are unhindered by the moderating influences of an atmosphere. Presumably, the layer of shattered rock, 2 km deep, beneath the regolith was created in the same way. The breccias brought back by astronauts are samples of the layer. They consist of compacted aggregates of rock fragments, mineral chips, and regolith. They, too, seem to have been formed by the impact of bolides. Apparently, some bolides compress the regolith and broken rock so greatly that the particles are welded together again to form new breccia. The process is one by which new rock is created from regolith.

The samples of regolith and breccia, then, provide clues about erosion on the lunar surface—erosion caused by the impacts of bolides.

History of the Moon

From the clues mentioned above it is possible to construct a history of the Moon. The story begins about 4.6 billion years ago, by which time the Moon had formed as a solid body. The evidence does not prove exactly how the Moon was formed; however, it does indicate that the final stages of formation involved the impacting of innumerable bolides, large and small, because ancient impact craters are still present. Probably, the entire growth of the Moon occurred by *accretion, the process by which solid bodies gather together to form a planet.*

By 4.6 billion years ago the Moon had accreted to about its present size. However, as it accreted and grew larger, the strength of its gravitational attraction increased, so that the speeds at which accreting bolides reached the lunar surface grew ever greater. Eventually, speeds became so great that each impact generated a large amount of heat. Near the end of the accretion process, so much heat was generated that an outer layer, 150–200

FIGURE 23.16 Photograph of the lunar surface near the landing site of *Apollo 15*, adjacent to the Hadley Rille. The fragment-littered surface is being physically weathered by impact cratering. Small craters can be seen on the slopes toward the rear of the photo.

km thick, of the Moon's body was apparently melted. Thus, the Moon seems to have had a solid interior with a molten outer shell. The period of rapid accretion soon ended, and the outer layer of magma began to cool and crystallize. It has been suggested that the first minerals to crystallize were plagioclase feldspar; and that because they were lighter than the parent magma, they floated in it. Thus, magmatic differentiation was operating, and soon a thick crust, rich in feldspar crystals, floated in the remaining liquid. The dense residue became the upper part of the lunar mantle. Bolides fall on the lunar surface today; so we can infer that bolides must also have been falling while the crust was forming. The largest impacts must have broken the crust, letting liquid from below ooze out to form the ancient potassium-rich lava flows. Those ancient products of magmatic differentiation are the materials the astronauts found in the lunar highlands.

The layer of magma, 200 km thick, would have cooled and crystallized within about 400 million years. By about 4 billion years ago, therefore, the highland crust had formed, and the major activity on the Moon had become the incessant rain of bolides that pitted and pocked the surface. Some bolides, larger than others, made exceptionally large impact scars. These scars, huge circular cavities, eventually became the mare basins, and the tilted crust around their margins became the Moon's highest mountains. If we study closely the photograph of the "front" side of the Moon (Fig. 23.13) we see that all the mare basins are roughly circular and have raised rims.

While the Moon's surface was being bombarded by bolides, its interior regions were slowly heating up. If we judge from the composition of meteorites that fall on the Earth and presume that the Moon has a similar composition, the amount of radioactivity was small, but nevertheless it was sufficient to cause partial melting in the upper mantle, beginning about 3.8 billion years ago. So formed, the magma worked its way up to the surface along fractures caused by impacts of the largest meteorites. When it reached the surface, the magma filled the impact basins and formed the basalt flows now seen in the maria (Fig. 23.14). About 3 billion years ago, the extrusion of mare basalt flows ceased and the Moon probably looked as it is depicted in Figure 23.17b. We infer that from 3 billion years ago to the present no further magmatic activity has occurred. Except for the continuing rain of bolides, the Moon has remained a tectonically and magmatically dead planet.

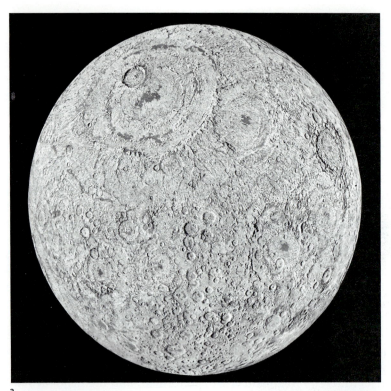

a

b

FIGURE 23.17 How the "front" side of the Moon probably looked at two times in the past. The figures, which should be compared with the Moon's aspect today (Fig. 23.13), were prepared by starting with a photograph of the Moon in its present appearance, and painting in a surface reconstructed from geologic maps. (*a*) As the Moon probably looked about 3.9 billion years ago, before basalt flows filled the mare. The crater that later became Mare Imbrium looks like a giant bulls-eye at the upper left. Note the multiple rings around the impact. Virtually the entire Moon was then covered with anorthosite. The few dark areas represent old potassium-rich lavas. (*b*) As the Moon appeared about 3.0 billion years ago. The basaltic flows that filled the mare basins between 3.8 and 3.2 billion years had ceased. Except for the young impact craters, such as Copernius and Kepler, the Moon looked much as it does today.

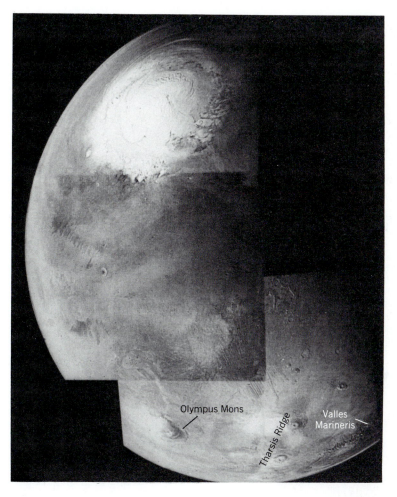

Olympus Mons

Tharsis Ridge

Valles Marineris

FIGURE 23.18 The northern hemisphere of Mars in the late spring seen from a distance of 13,700 km. The figure is a mosaic of images from *Mariner 9*. The northern polar cap is greatly shrunken from its winter maximum. The origin of the spiral-shaped structure is uncertain but is probably related to spiral wind patterns. Near the base of the image are several shield volcanoes, including the giant Olympus Mons. The great canyonlike feature at lower right is Valles Marineris, named for the *Mariner* spacecraft.

Lessons for the Earth

Did the Earth ever look like the Moon? It probably did. The Moon supports the idea that the terrestrial planets formed by accretion of solid bodies. The Earth is larger than the Moon, so its gravitational pull is stronger. Presumably, therefore, an even thicker layer of magma covered the Earth at the end of the accretion process. Perhaps the earliest crust began to form through cooling of that magma. However, all traces of the Earth's primitive crust have been lost. The reason is not hard to find. Because radioactive heating of the Earth has continued to create magma, probably the earliest crust has been remelted and reabsorbed.

Mars

Mars, with a diameter of 6787 km, possesses a mass only one tenth of the Earth's mass. Yet despite its small size, Mars is earthlike in many ways. It rotates once every 24.6 h, so the length of the Martian day is nearly the same as the length of an

Earth day. Also, Mars has an atmosphere, although it is only one one-hundredth as dense as the Earth's and consists largely of carbon dioxide. Mars has polar "ice" caps (Fig. 23.18), consisting mostly of frozen carbon dioxide ("dry ice"), although a small amount of water ice is also present. Like the Earth, Mars has seasons, and the diameters of the ice caps alternately grow and shrink with the coming of winter and summer.

Like the Moon, Mars lacks a dipolar magnetic field. Like the Moon, too, it has a density somewhat less than that of the Earth. However, when allowance is made for the fact that Mars is smaller than the Earth, and that internal pressures are less, the difference in density is small. This means that the composition of the Earth and Mars must be similar and that Mars has a core. The conclusion that Mars has a core and is therefore compositionally layered is supported by another observation. Mars wobbles as it rotates, and the way it wobbles is characteristic of a layered body, not a homogeneous one. Unfortunately, no further in-

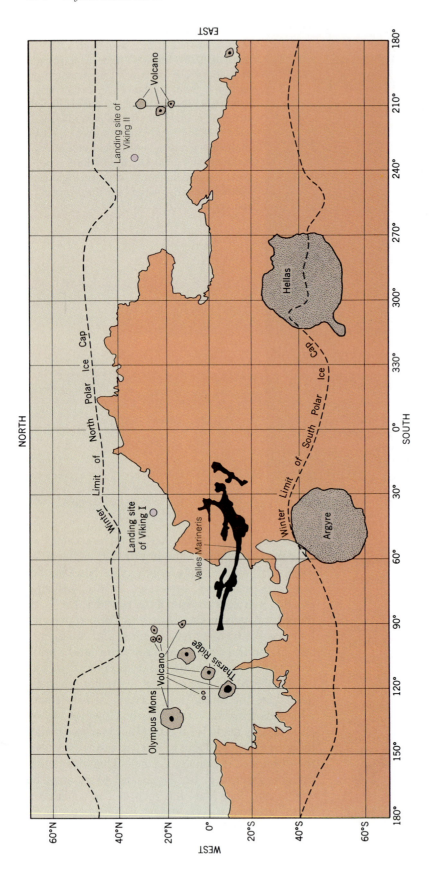

FIGURE 23.19 The principal features of Mars shown on a Mercator projection (Appendix D). Some features on this map are visible also in Figure 23.18. The southern hemisphere is occupied by densely cratered terrain (orange) believed to be ancient crust. The northern hemisphere is smoother and less cratered; the crust there is believed to be younger. It was on the edge of the smooth, northern hemisphere in a region called Chryse Planitia, that *Viking I* landed on July 20, 1976. *Viking 2*, which also landed on the northern hemisphere, came down a few months later on Utopia Planitia. Several large shield volcanoes, each crowned by a caldera, occur in the younger crust, while two very large impact structures, Argyre and Hellas, occur in the ancient crust. The feature labeled Valles Marineris is the largest canyon in a region of vast canyons and gorges (Fig. 23.18). The winter limit of the northern polar cap is much farther south than the edge of the cap seen in later spring in Figure 23.18. (*Source:* Adapted from a map, based on *Mariner 9* photographs, prepared by U. S. Geological Survey.)

formation is available about the layering. One of the spacecraft that landed on Mars in 1976 carried a seismometer with it, but no marsquakes occurred during the months the seismometer was active, so seismic evidence of layering is unavailable.

Surface Features

The topography of Mars is extraordinary, and many features have yet to be adequately explained. The most obvious surface features are shown in Figure 23.19. Approximately half the Martian surface, the southern half, is densely cratered and resembles the surfaces of the Moon and Mercury. The largest impact crater discovered in the Solar System, Hellas, with a diameter of nearly 2000 km, is on Mars. Scientists who have studied the images of the Martian surface returned by unmanned spacecraft conclude that the southern hemisphere is covered by ancient crust similar to the crust of the lunar highlands, and presumably of the same age. The dense population of craters records a rain of bolides prior to 4 billion years ago.

The northern hemisphere presents a very different picture. Craters are sparse and large areas are relatively smooth. The obvious conclusion to draw is that some process (or processes) has produced a much younger surface. The probable answer to the puzzle is the widespread evidence of volcanism in the northern hemisphere. At least 20 huge shield volcanoes and many smaller cones have been discovered. The giant among them is Olympus Mons (Fig. 23.20), whose basal diameter is 600 km, approximately the distance from Boston to Washington, D.C. Olympus Mons stands 27 km above the surrounding plains and is capped by a complex caldera that is 80 km across (Fig. 23.21). Mauna Loa, the largest volcano on the Earth, is also a shield volcano, but it is only 225 km across and 9 km high above the adjacent seafloor. It is estimated that the amount of volcanic rock in Olympus Mons exceeds all of the volcanic rock in the entire Hawaiian chain of volcanoes. The presence of a huge volcanic edifice such as Olympus Mons implies several things. First, in order for such a huge volcano to form, there must be long-lived

FIGURE 23.20 Olympus Mons, giant shield volcano on Mars, is the largest volcano known on any planet. Its diameter is about 600 km, it is capped by a caldera 80 km across, and its base is marked by steep, near vertical cliffs that were possibly cut by intense wind erosion as a result of fierce Martian sandstorms. Within the main caldera, smaller calderas and a circular crater can be seen. The two small circular structures on the flanks of the volcano, the caldera, are believed to be volcanic craters.

FIGURE 23.21 Detail of the summit caldera of Olympus Mons. Several phases of caldera collapse have occurred. The irregular ridges in the center foreground are probably basaltic flows similar to those seen on the lunar mare (Fig. 23.14).

sources of magma in the Martian interior. Second, the magma source must remain connected to the volcanic vent for a very long time. This in turn means that the Martian lithosphere must be stationary. In short, plate tectonics is not operating on Mars. A third implication is that the Martian lithosphere must be thick and strong. If it were not, it would be bowed down by the weight of Olympus Mons. Isostasy is apparently not operating, or if it is operating, the isostatic response is very slow. Are volcanoes active on Mars today? This intriguing question cannot be answered exactly, but it is likely that some volcanism persists. The evidence comes from the volcanic cones. If they were old features, they would be pitted by impact craters. Craters are rare on the volcanic slopes, leading some experts to conclude that the youngest flows on Olympus Mons are less than 100 million years old.

Adjacent to Olympus Mons, cutting both young and ancient terrain and running generally parallel to the Martian equator, is a region of extraordinary canyons. Here *Mariner 9* photographed Valles Marineris (Fig. 23.22), the longest canyon discovered on any planet. At least 4000 km long, 100 250 km wide, and 7 km deep, Valles Marineris is a system of canyons that dwarf the Grand Canyon in Arizona. If the same feature were present on the Earth it would stretch from San Francisco to New York. What formed the great canyons is not

known, but they are a series of giant grabens. Possibly they formed when great crustal up-warping occurred, or when the crust subsided into openings left empty when magma was extruded to build the huge volcanoes. Whatever the reason for Valles Marineris, the giant canyons have been modified in various ways after their formation. Walls have slumped, large landslides and mudflows seem to have occurred, features that look surprisingly like beaches from old lakes have been reported, evidence of running water can be seen, and there is abundant evidence of wind erosion.

Erosion

As remarkable as the giant volcanoes and canyons are, even more remarkable features can be seen in the images returned by orbiting spacecraft, and by the two Viking landers. *Viking 1* landed in a region of the northern hemisphere called Chryse Planitia, and *Viking 2* landed in a somewhat similar region called Utopia Planitia (Fig. 23.19). In both cases the images sent back showed a red-brown surface covered by loose stones and windblown sand (Fig. 23.23). Both landers made chemical analyses of the regolith on which they landed, and in both cases the results indicated clays and a sulfate mineral (probably gypsum). Weathering has obviously influenced the Martian surface and wind continually blows the weathering products around. Indeed, when the spacecraft *Mariner 9* was nearing Mars in 1971, a sandstorm of spectacular proportions was raging. Dust and sand particles were observed as high as 55 km above the surface, winds reached velocities of 300 km/h and the storm raged unabated for several months. Under conditions such as these, sand particles can abrade rock effectively; based on the widespread occurrence of yardangs (Chap. 12), it seems likely that on Mars, wind-driven sand is now the main agent of erosion. Peering into one of the impact craters near Hellas, *Mariner 9* photographed a series of spectacular sand dunes (Fig. 23.24) that suggest what probably happens when the great storms abate. Probably all craters and all low places on Mars contain piles of sand. Every time a storm blows up, the sand is swirled into the lower atmosphere. When the storms die down, gentle winds slowly sweep the sand back into all the sheltered hollows on the surface.

Yet still more spectacular than the Martian sand is evidence that water or some other liquid has influenced the surface of Mars (Fig. 23.25). There are valleys that look much like those cut by inter-

FIGURE 23.22 A portion of the southern margin of Valles Marineris, a system of giant canyons near the Martian equator. The area in view is 150 km wide. The canyon wall, with its giant landslide, is 2 km high.

FIGURE 23.23 An image of the Martian surface from *Viking 2.* The red color indicates that the surface rocks have undergone oxidation as a result of weathering.

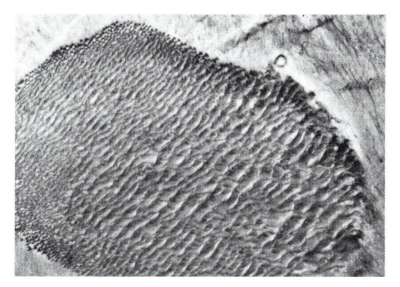

FIGURE 23.24 Dark-colored area of sand dunes on Mars, near Hellas. The distance from crest to crest of adjacent dunes is about 1.5 km. Numerous dune areas have been observed on Mars, many of them inside large craters, indicating that dunes are common features of the Martian landscape.

mittent desert streams on the Earth. They meander, they branch, and they have braided patterns and other features characteristic of valleys made by running water. Some of the features look as if they were caused by gigantic floods. Yet Mars now lacks rainfall, streams, lakes, and seas. The sparse water that does occur on Mars is in the atmosphere, or exists near the poles condensed as ice. The Martian surface is too cold for water to exist as

a liquid. Some have suggested that ice might be present beneath the surface dust as permafrost, and that in former times the Martian climate warmed up, causing the ice to melt and creating torrential floods. Others suggest that Mars simply cannot have sufficient ice in polar caps or as permafrost to melt and produce sufficient water to have cut great stream channels. The origin of the Martian channels, then, remains unsolved.

 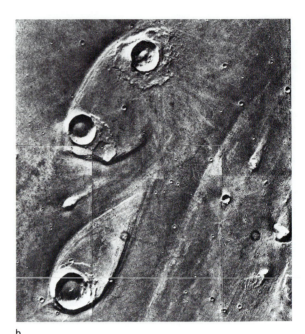

a   b

FIGURE 23.25 Part of the surface of Mars, showing features that have been interpreted as ancient stream channels. (*a*) Channels that branch in dendritic patterns. The channels are most prominently developed in the old, cratered terrain of the southern hemisphere. The width of the image is about 320 km. (*b*) Part of a channel system in Chryse Planitia. The teardrop-shaped plateaus formed behind resistant craters when floodwaters moved from lower left to upper right. Largest crater is about 10 km in diameter.

FIGURE 23.26 Chaotic terrain on Mars. A mosaic of Viking orbiter images showing a pronounced channel (left) emerging from a region of slumping and collapse. The channel is about 20 km wide. It eventually connects with other channels and then flows into Chryse Planitia. The chaotic terrain probably formed when local melting led to the collapse of permafrost. The mud and water released by the collapse cut the channel.

Geological History

The early history of Mars must have been much like that of the Earth and the Moon. The terrain of the southern hemisphere, formed in the early days, is probably equivalent to the lunar highlands. Then, as in both the Earth and the Moon, radioactive heating inside Mars started to create magma by partial melting. Because Olympus Mons and its mates are shield volcanoes, we can infer that the magma they extruded was of low viscosity and, therefore, was basalt. This means that magmatic differentiation has occurred, but we cannot yet say when the volcanism started. To be able to do that we must have radiometric-age measurements. But the fresh-looking surfaces of the big cones suggest that the volcanoes might be active still, and that Mars (like the Earth) is still continuing to make magma.

Planetary differentiation and volcanism would have released volatiles such as H_2O and CO_2. Because Mars is compositionally like the Earth, it can be calculated that the amount of H_2O released would have been sufficient to cover a smooth, uniform Mars with water to a depth of 50–100 m. It is possible, therefore, that very early in Martian history—a time corresponding to the Archean on the Earth—rains may have fallen, lakes and streams may have existed, and water erosion may have occurred. For some reason the Martian atmosphere became thinner. Probably this happened because weathering and erosion locked up H_2O and CO_2 in clay and carbonate minerals. It was a one-way process. H_2O and CO_2 were not released to the atmosphere again because the rock cycle did not operate and therefore did not cause recycling. The temperature would have dropped as the atmosphere thinned. Liquid water eventually became unstable on the surface of Mars. However, H_2O could still exist as ice, and this is presumably the form it is in today, possibly existing as a cement between grains in the regolith. In short, it is suggested that Mars became a planet covered by permafrost. The torrential floods that probably caused some of the most striking erosion features (Fig. 23.26) could have occurred when regions of permafrost were subjected to sudden melting, perhaps by near intrusion of magma or sudden changes in climate.

Summary of the Terrestrial Planets

The Earth, the Moon, and the other terrestrial planets seem to have had similar early histories.

Each seems to have formed at the same time. Where ancient surfaces still exist, as on the Moon and Mercury, the end stage of formation involved a violent rain of infalling bolides (Fig. 23.27). Each planet then seems to have experienced a period of radioactive heating during which a core was formed (Fig. 23.28) as well as the basaltic magmas similar to those on the Earth. Mercury and the Moon are too small to have held, gravitationally, the gases given off during differentiation, so they no longer have atmospheres. Venus, the Earth, and Mars are larger and they retain atmospheres.

Plate tectonics seems to be an Earth process. The other terrestrial planets have single, continuous plates of lithosphere and there is no evidence to suggest that the plates move. Tectonics on the Earth involves lateral forces due to plate motions. On the other planets tectonics involves vertical forces, presumably caused by local expansions and contractions. To what extent those vertical forces are caused by internal convection remains an unanswered question.

Despite the differences between the terrestrial planets, they share many features in common. The differences apparently arise through the interaction of several major factors. First, the planet size

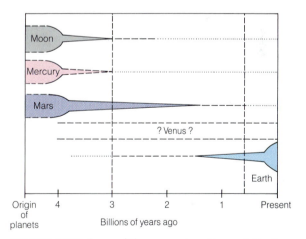

FIGURE 23.27 Ages of the present surfaces of the terrestrial planets. The width of the bar at a given time indicates the percentage of the surface having the indicated age. For the Moon and Mercury, 80% of the present surfaces were already in place 4 billion years ago, and close to 100% were emplaced by 3 billion years ago. The Martian surface has a longer history, while the evidence for Venus is uncertain. On the Earth, more than 65% of the surface is covered by oceanic crust that is less than 200 million years old. (*Source:* After Head and Solomon, 1981.)

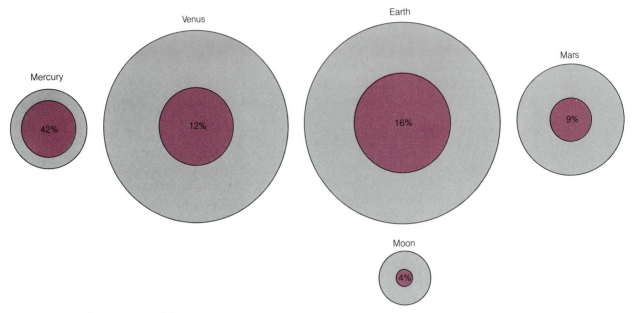

FIGURE 23.28 Comparison of the sizes and radii of the cores of the terrestrial planets. The percentage of the total planetary volume represented by the core is shown. (*Source:* After Carr, 1984.)

controls not only the atmosphere, but also the thermal properties. Small planets cool rapidly and magmatic activity soon ceases. Large planets cool more slowly and remain active. Second, the distance of a planet from the Sun determines whether or not H_2O can exist as a liquid. The third factor is the presence or absence of life. The Earth's atmosphere is the way it is because living plants and animals play essential roles in the geochemical cycles that control the composition. If life had developed on Venus, that planet would probably have an atmosphere like the Earth's. However, life apparently did not develop, so all of the CO_2 is still in the atmosphere. On the Earth, plants and animals have been the means whereby carbon dioxide has been removed from the atmosphere and the carbon is locked up in rocks as fossil organic matter and as calcium carbonate.

THE JOVIAN PLANETS AND THEIR MOONS

Jupiter

The giant, gassy planets tell us little about the evolution of the Earth, but they provide the best preserved samples of the gases from which all planets are believed to have formed. Thus, they reveal much about how the Solar System may have

formed. Jupiter is the largest and best studied of the giant planets. It has a dense atmosphere composed of hydrogen, helium, ammonia, and methane, plus other trace constituents. It is presumed that there is a rocky core inside the dense atmosphere.

Jupiter has about twice the mass of all of the other planets combined. Had it been slightly larger, it probably would have reached an internal temperature high enough for nuclear burning to start, and as a result it would have been a sun. Jupiter is unusual in many ways. One unusual feature is that it gives off twice as much energy as it receives from the Sun. The reason for this seems to be that Jupiter is still undergoing gravitational contraction which gives off heat energy.

One of the most interesting things about Jupiter is its moons, four of which are as large, or larger, than the Earth's Moon. The moons closest to Jupiter, Io and Europa, have densities of 3.5 and 3.2 g/cm^3, respectively, indicating that they are rocky bodies.

Io, the large moon that is closest to Jupiter, is extraordinary. It is a highly colored body with shades of yellow and orange predominating, suggesting it is covered by sulfur and sulfurous compounds (Fig. 23.29). Impact craters are absent and the reason is not hard to find—Io is volcanically active. Not only is Io volcanically active, it is by far the most volcanically active body so far discovered

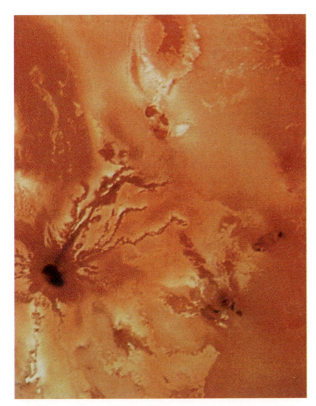

FIGURE 23.29 The bright colors on the surface of Io, the innermost moon of Jupiter, are believed to be caused by sulfur and sulfurous compounds given off during volcanism. The feature in the lower left of the image is a volcanic cone; numerous lava flows, probably basaltic in composition, radiate out from the volcano. The width of the field of view is 1000 km. The image was taken by *Voyager I* in 1979, at a distance of 128,500 km.

in the Solar System. Impacts by bolides certainly must occur, but the craters are quickly covered up by volcanic debris.

Io's volcanism seems to be of two kinds. The first is the familiar basaltic volcanism found so widely through the Solar System. Lava plains and shield volcanoes are the result. One of the shield volcanoes, Ra Patera, is almost as large as Olympus Mons on Mars. Fresh lava flows can be seen on its slopes. The second kind of volcanism seems to involve sulfur and sulfur dioxide (SO_2). Huge orange-yellow flows of what is presumed to be molten sulfur have been seen—some are as much as 700 km long. Most striking, however, are active volcanic plumes that throw sprays of sulfurous gases and entrained solid particles as high as 300 km above the surface of Io. Nine active plumes were observed by the two spacecraft, *Voyager I* and *Voyager II*, as they flew by Io (Fig. 23.30). The volcanic plumes seem to be geyser-like in origin, but

the fluid that boils and erupts is SO_2, not H_2O. It has been estimated that the plumes eject 10^{16} g of fine, solid particles each year. This quantity is sufficient to bury the surface of Io with a layer of pyroclastic debris 100 m thick in a million years. No wonder there are no impact craters to be seen. The process of surface renewal is much faster than it is on the Earth.

The amount of heat energy needed to drive Io's volcanoes is much greater than the heat that could be produced through radioactive decay in a stony planet. Io's volcanic heat comes from a different source—the gravitational pull exerted by the huge mass of Jupiter. As Io moves around Jupiter in an elliptical orbit it is periodically stretched more or less by the gravitational pull—more during close approach, less when far away. The bending and stretching due to the fluctuating gravitational pull generate heat, just as a copper wire becomes hot if it is bent back and forth. No other object in the Solar System demonstrates the effect of tidal stresses so dramatically as Io.

Standing further away from Jupiter than Io are the three large moons Europa, Ganymede, and

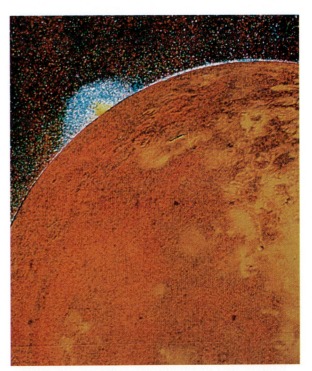

FIGURE 23.30 A volcanic eruption on Io. The volcanic plume is mostly gas, but small solid particles are also distributed by the gas. The plume rises to a height of 100 km above the surface of Io and is believed to be largely sulfur dioxide (SO_2). Several sites of active volcanism have been discovered on Io.

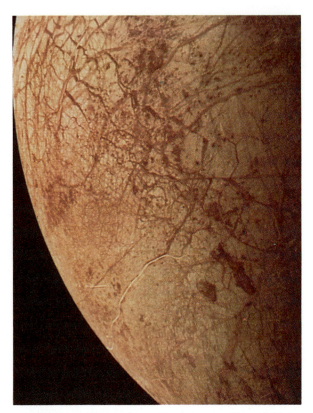

FIGURE 23.31 The surface of Europa, smallest of the four large moons of Jupiter. Europa has a density below that of Io indicating that it contains a substantial quantity of ice. The surface is mantled by ice to a depth of 100 km. The fractures indicate that some internal process must be renewing the surface on Europa. The dark material in the fractures apparently rises up from below. The cause of the fracturing is not known. The image was taken by *Voyager 2* in July, 1979.

Callisto. Their densities decrease the further they are away from Jupiter. Europa has a density of 3.2 g/cm^3; Ganymede, 2.0 g/cm^3; and Callisto, 1.8 g/cm^3. The reason for the lowered densities was discovered during the Voyager missions; each of the moons is sheathed with a layer of ice. The outer moons, and especially Europa, may have small metallic cores, but their densities suggest that their main masses lie in thick mantles consisting of ice and silicate minerals. Above the mantles are crusts of nearly pure ice a hundred kilometers or more thick.

Craters are rare on Europa; however, the surface is split and criss-crossed by an intimate network of fractures (Fig. 23.31). Presumably some tectonic process renews Europa's ice surface through fracture and upwelling. The upwelling may involve melting of the ice. The source of energy that results

in melting is probably tidal, caused, as with Io, by the gravitational pull of Jupiter.

Ganymede has a much thicker sheath of ice, it is too far from Jupiter to be influenced by tides, and it is pitted by craters. However, on Ganymede some slow-acting process is apparently renewing and reworking the ice surface, because some regions have few if any impact craters. The surface is divided into dark and light areas (Fig. 23.32). The dark areas are more heavily cratered than the light areas and, presumably, are older. Within the lighter, younger areas there are striking grooves, fractures, and grabenlike structures. It is possible that a unique kind of plate tectonics may be operating on Ganymede. The dark regions are ancient ice-continents, the lighter areas are the places

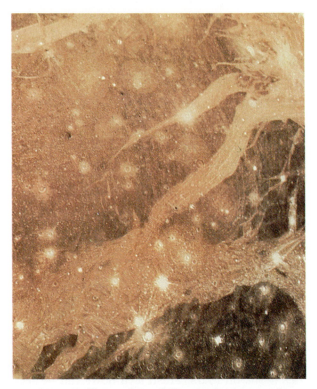

FIGURE 23.32 The surface of Ganymede, largest of Jupiter's moons, viewed from a distance of 312,000 km by *Voyager 2* in 1979. Ganymede is covered by a thick crust of ice. The dark surface is ancient ice, presumably covered by dust and impact debris. It is split into continent sized fragments that are separated by light-colored, grooved terrains of younger ice. Ganymede is apparently tectonically active and the grooved terrains seem to be the places where new ice rises from below, but how this happens, and what causes the grooves, is not known. The field of view is approximately 1300 km across.

where ice rises convectively from the depths to create new ice crust.

The most distant moon, Callisto, must contain the greatest proportion of ice because its density is the least. The surface is again icy but it must be very ancient ice as the density of impact craters is great (Fig. 23.33). No evidence has yet been found to suggest that the surface of Callisto is being renewed and reworked.

Saturn

Saturn has a composition like that of Jupiter but it is not quite as large. Like Jupiter, Saturn radiates more energy than it receives from the Sun. The most striking feature of Saturn is its immense ring system (Fig. 23.34). The rings are a disc about 65,000 km in diameter and less than 10 km thick. They are composed of ice particles, icy snowballs, rocks, and dust ranging in diameter from a fraction of a millimeter to about a meter. The rings are thought to be either the remains of a moon that got too close to Saturn and was broken up by tidal

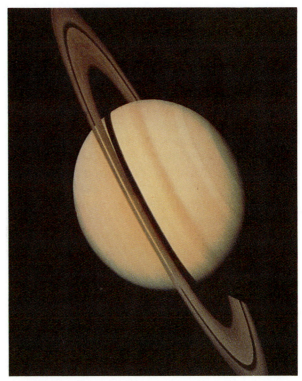

FIGURE 23.34 The remarkable ring structure that surrounds Saturn. The width of the rings at the center of the disk is 10,000 km. Note the shadow cast by the rings on the surface of Saturn.

FIGURE 23.33 A photomosaic of Callisto, outermost of Jupiter's moons. Viewed from a distance of 390,000 km, the surface can be seen to be riddled with impact craters. The craters appear bright because Callisto has a thick outer layer of ice; hence, impact craters expose new, clean ice below the old, dust covered surface.

forces, or to be material that never did aggregate to form a moon.

Before the Voyager mission to Saturn in 1980, little was known of Saturn's moons. Less is known about them still than is known about the moons of Jupiter. Most of the moons are small and have low densities. In composition and structure they seem to resemble Ganymede and Callisto. They are ice-covered and cratering is extensive. Two of the moons display evidence, like that on Ganymede, suggesting that ice flow may be renewing the surfaces. The most distinctive among the Saturnian moons is Titan, a body larger than Mercury. It is the only moon in the Solar System large enough to retain a substantial atmosphere. Unfortunately, the atmosphere is an opaque, orange-colored smog that shrouds the surface from view. The composition of Titan's atmosphere is mostly nitrogen; however, ethane, acetylene, ethylene, hydrogen cyanide, and other unpleasant substances are also present.

The density of Titan is 1.9 g/cm^3, which suggests that it contains about 45 percent ice and 55 percent rocky matter. When first formed, Titan probably incorporated some ammonia (NH_3) and methane (CH_4) in its mass. Through radioactive heating and differentiation, an H_2O and CO_2 atmosphere that was both ammonia-rich and methane-rich probably formed around Titan. Sunlight working on the atmosphere then caused other compounds such as ethylene and acetylene to form through photochemical reactions, and it is those compounds that produce the smog that covers Titan. Very little sunlight can penetrate the smog so the surface of Titan must be a cold and unfriendly place. Scientists who have studied the data sent back by *Voyager II* suggest that the surface temperature is $-180°$ C, that Titan is covered by an ocean of liquid ethane and methane, and that great islands of H_2O and CO_2 ices probably rise up from the ocean floor. Titan is stranger even than science fiction. Who could have imagined a planetary body with oceans of liquid hydrocarbons and continents of ice?

ORIGIN OF THE SOLAR SYSTEM

How did the Solar System form? An exact answer may never be forthcoming, but the outlines of the process can be discerned from evidence obtained by astronomers, from our knowledge of the Solar System today, and from the laws of physics and chemistry.

The birth throes of our Sun and its planets were the same as those of any other sun. Birth began with space that seemed to be empty, yet was not entirely empty. Atoms of various elements were present everywhere, even though thinly spread; they formed a tenuous, turbulent, swirling gas. When the gas thickened by a slow gathering of all the thinly spread atoms, the Sun was formed. The kinetic energy of those turbulent gas swirls eventually gave rise to the rotations of the Sun and planets. The gathering force of the gas was gravity, and as the atoms slowly moved closer together, the gas became hotter and denser. As one part of the gathering process, the Earth and the other planets formed.

More than 99 percent of all atoms in space are atoms of hydrogen and helium, the two lightest kinds of atoms. Samples of the gas are still preserved in the atmospheres of Jupiter and Saturn. Near the center of the gathering cloud of gas, the atoms became so tightly pressed and so hot that atoms of hydrogen and helium began to fuse together to form heavier elements. The gas cloud rotated because it inherited the turbulent energy of the original gas cloud. Fusion of light elements to form heavier ones causes heat energy to be released; the hydrogen and helium undergo *nuclear burning*. When, in the gas cloud that formed the Solar System, nuclear burning commenced, the Sun was born. The time was about 6 billion years ago. However, nuclear burning was confined to the center of the gas cloud. A vast, rotating envelope of less-compressed gas still surrounded the Sun.

We noted in Chapter 2 that rotation gives rise to a centrifugal force. While gravity tended to pull the gaseous envelope inward toward the Sun, the centrifugal force tended to pull it outward. As a result of the two opposing forces, the gas cloud slowly became *a flattened, rotating disk of gas* surrounding the hot Sun. Such a disk is a **planetary nebula** (Fig. 23.35).

At some stage the cool outer portions of the planetary nebula became compacted enough to allow solid objects to condense, in the same way that ice condenses from water vapor to form snow. The solid condensates eventually became the planets. Planets nearest the Sun, where the temperature was highest, contain only compounds that are capable of condensing at high temperatures. Those compounds consist of elements such as iron, silicon, magnesium, and aluminum; we call them *refractory elements*. Planets distant from the Sun, where temperatures were lower, contain not only refractory elements but also *volatile elements,* such as hydrogen and sulfur, that do not condense at high temperatures but will do so at low temperatures. The gradual decrease in the size of the metallic cores of terrestrial planets apparently reflects the condensation of the planetary nebula. Mercury, the closest planet to the Sun has the largest core and, therefore, the largest percentage of iron. Mars, the most distant terrestrial planet has the smallest core and, therefore, the lowest amount of iron. As the fraction of refractive elements in a planet decreases, the fraction of volatile elements increases. The further away from the Sun the condensation process occurred, the lower was the temperature and the greater the fraction of volatile elements. The most striking demonstration of this fact is the increasing amount of ice in moons, and the consequent lowering of density, the further an object is from the Sun (Table 23.1).

The size of a planet is also related to its distance from the Sun. A large-diameter ring of gas contains more atoms than a small-diameter ring. Planets close to the Sun formed from comparatively small amounts of gas and are, therefore, smaller than more distant planets that condensed from large rings of gas. Near the outermost margin of the nebula, where the gas cloud was very tenuous and thin, the size of the planets becomes smaller again.

Condensation of a Planetary Nebula

Condensation is analogous to the cooling of magma. First one solid compound condenses,

then another. A common example of condensation of a solid is the formation of snowflakes from water vapor. In the case of snow, only one kind of solid (ice) condenses from the gas. In a planetary nebula there are many different gases present, so many different solids condense. The earliest compounds to condense were probably metallic iron and water-free silicate minerals. Just how the planetary snowflakes were aggregated to form planets is a continuing subject for research. The main process seems to have involved aggregation of small flakes to form larger ones by gravitational attraction. Particles aggregate together to form

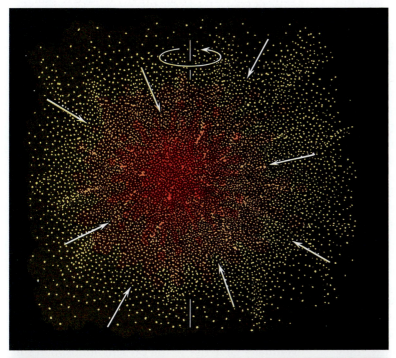

FIGURE 23.35 The gathering of atoms in space created a rotating cloud of dense gas that eventually became the Sun, and a surrounding disk-shaped planetary nebula. The planets formed by condensation in the planetary nebula.

larger and larger particles. Eventually bodies the size of Mars were colliding together to form larger planets. Indeed, one widely held hypothesis is that a late collision between the proto-Earth and a Mars-sized bolide splashed off enough material to form the Moon. The collision must have come after differentiation of the proto-Earth had started and after its core had separated. The material splashed off by the inferred collision came from the mantle of the proto-Earth. According to the hypothesis, that is why the Moon is so similar in composition to the Earth's mantle, and why the Moon has a tiny core, or no core at all.

The condensation of the Earth and the other planets was completed by 4.6 billion years ago. But there still remained around the Earth an atmosphere of hydrogen and helium. These gases are no longer with us. How the terrestrial planets lost their primordial atmospheres is still in doubt. One widely held theory involves the Sun. Astronomers have observed that early in a sun's life cycle the sun undergoes a short period of violent turbulence during which intense winds of charged atomic particles are blown out into space. According to the theory, the wind from our Sun, about 4.6 billion years ago, swept the remaining gases of the planetary nebula away into outer space, leaving the terrestrial planets free of atmosphere. Because of their great masses and huge gravitational attractions, the giant planets retained their atmospheres of hydrogen and helium.

Probably all of the terrestrial planets once had a molten outer layer due to the early, violent accretion process. On the Earth, all evidence of that ancient magma has been destroyed. The interiors of the earliest planets were certainly hot as a result of collision during accretion, but apparently, they were not hot enough to melt and form magma. Yet radioactive elements trapped within the planets must immediately have started to cause internal heating and to raise temperatures. As time passed, rising temperatures reached a point at which magma formed (Fig. 23.36). Opinions differ as to whether entire planets melted. Probably they did not, but partial melting certainly occurred. Any metallic iron in the planet would melt and sink toward the center, forming a core in the process. The sinking iron would release more gravitational energy, speeding up the heating process still further. The fraction of mantle with the lowest melt-

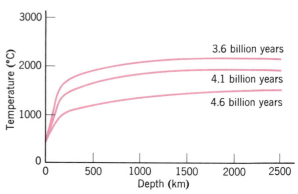

FIGURE 23.36 Earth's changing geothermal gradient. The initial geothermal gradient, 4.6 billion years ago when the Earth was aggregated from small solid bodies to form a larger solid body, was slowly raised by heat released during radioactive decay. Eventually iron melted and sank toward the core and magmas formed in the mantle. (*Source:* After Hanks and Anderson.)

ing temperature and lowest density would form magma and rise upward. In this fashion, the planets underwent an early profound differentiation that produced the compositional layering.

Metallic iron, being heavy, concentrated in the core, whereas magma, being light, moved upward, carrying with it most of the volatile elements in the mantle. This made it possible for a crust to form. Convection in the mantle must have begun and possibly plates of lithosphere started to move—at least they did so on the Earth. As internal heating continued and as more and more magma was extruded, volcanoes gave off gases. It was these later gases, mainly water vapor, carbon dioxide, methane, and possibly ammonia, that eventually gave rise to the present atmospheres of the terrestrial planets. From the same source came the water we now find in the Earth's hydrosphere. The next great event in the Earth's history was the appearance of life. How, when, and where it happened is still unknown. It seems to have happened at least 3.5 billion years ago, at a time when basaltic volcanism was still active on the Moon, but before the main mass of continental crust had formed on the Earth. Unknown, too, is the reason that life arose on the Earth but apparently not on the other terrestrial planets. Perhaps future research will provide answers to even these most complex of all questions.

SUMMARY

1. The Sun formed when atoms in space became sufficiently compacted for nuclear burning to begin.

2. The planets formed by condensation from a disk-shaped envelope of gas (the planetary nebula) that rotated around the Sun. All planets condensed at the same time, about 4.6 billion years ago.

3. Planets close to the Sun, such as Mercury, Mars, the Earth, and Venus, are small, dense, rocky bodies. Planets farther away, such as Saturn and Jupiter, are large, low-density bodies.

4. The Earth, the Moon, and the terrestrial planets have layered structures. The layered structures probably formed by differentiation soon after the planets formed from the planetary nebula.

5. Each of the terrestrial planets went through a period of internal radioactive heating that led to generation of basaltic magma.

6. The Earth and possibly Mars and Venus are still producing magma from radioactive heating. Mars and Venus may be tectonically active but not in the same way as the Earth. Mars and Venus appear to be one-plate planets, not multiplate planets like the Earth.

7. The Moon probably has a small core surrounded by a thick mantle, and is capped by a crust 65 km thick.

8. On the Moon, magma was formed early, but is no longer generated. The Moon is a magmatically dead planet.

9. The highlands of the Moon are remnants of ancient crust built by magmatic differentiation more than 4 billion years ago.

10. The maria (lunar lowlands) are vast basins created by the impacts of giant meteorites and later filled in by lava flows.

11. Mars seems to be geologically active. Olympus Mons, a shield volcano on Mars, is the largest volcano yet found in the Solar System.

12. The principal eroding agent on Mars is wind-driven sand and dust. Water or some other flowing liquid cut stream channels at some time in the past.

13. Venus has about the same size and density as the Earth. It has a dense atmosphere of carbon dioxide and a surface temperature of about 500° C.

14. None of the other terrestrial planets have climates hospitable to human life.

SELECTED REFERENCES

Basaltic Volcanism Study Project, 1981, Basaltic volcanism on the terrestrial planets: New York, Pergamon Press.

Carr, M. H., 1981, The surface of Mars: New Haven, Yale University Press.

Carr, M. H., ed., 1984, The geology of the terrestrial planets: NASA SP-469, Washington, D.C., National Aeronautics and Space Administration.

Glass, B. P., 1982, Introduction to planetary geology: Cambridge, Cambridge University Press.

Greeley, R., 1985, Planetary landscapes: London, Allen and Unwin.

Greeley, R., and Iverson, J. D., 1985, Wind as a geological process: Cambridge, Cambridge University Press.

Head, J. W., and Solomon, G. C., 1981, Tectonic evolution of the terrestrial planets: Science, vol. 213, p. 62–76.

Hunten, D. M., Colin, L., Donahue, T. M., and Moroz, V. I., eds., 1983, Venus: Tucson, University of Arizona Press.

Morrison, D., ed., 1982, Satellites of Jupiter: Tucson, University of Arizona Press.

Murray, B. M., Malin, C., and Greeley, R., 1981, Earthlike planets: New York, W. H. Freeman and Co.

Scientific American, 1983, The Planets: New York, W. H. Freeman and Co.

Veverka, J., 1985, Planetary geology in the 1980's: NASA SP-467, Washington, D.C., National Aeronautics and Space Administration.

Organization of Matter

WHAT IS MATTER?

Every object in the universe is composed of matter. However, can we really say precisely what *matter* is? Ancient Greek philosophers first attempted the question and came close to a correct answer. They saw that matter occurs in three different states—solid, liquid, and gaseous—and that different forms of matter combine and react with each other, giving rise to new forms. Seeking some unifying principle, they examined properties such as color, hardness, taste, and odor of common materials, including rocks and minerals of many kinds. In a sense, some of those Greek philosophers were the first geologists, and the names they gave to some forms of matter are with us to this day. The modern word *copper* comes directly from the Greek work *cyprus*, and words such as *sapphire* and *magnetite* are little changed from the Greek originals.

Although the ancient philosophers learned much about matter, they were unable to resolve the question of what matter is. That question remained unanswered until the present era because no one was able to decide between two equally plausible possibilities, which can be stated in terms of questions that they themselves posed: Can matter be endlessly subdivided into tiny pieces, all of which retain the properties of the whole? Or is matter built up from submicroscopic particles of only a few kinds, so that the different properties of matter reflect different arrangements of the particles? There is some truth in both lines of thinking, but the second is essentially the correct one. Matter can be subdivided into atoms, and all atoms of a given chemical element have identical chemical properties. However, atoms themselves are composed of still smaller subatomic particles (protons, neutrons, and electrons), and these are the fundamental particles the Greek philosophers were seeking. Modern physicists continue the search, discovering still smaller particles like quarks which combine to make protons and neutrons. It is the protons, neutrons, and electrons that determine chemical properties.

Structure of Atoms

Atoms are held together by electrical forces; therefore, we can say that matter is basically electrical in character. Physicists have found that all atoms have a structure that can be described in a simplistic way as possessing a central portion called the *nucleus* which has a *positive electrical charge.*

Spinning around the nucleus in defined paths, or orbits, are *negatively charged particles* called **electrons**. The positive charge of the nucleus exactly balances the negative charge of the orbiting electrons. Of all atoms, the simplest is hydrogen, its nucleus having a single positive charge. Only one orbiting electron, therefore, is required to provide electrical balance. The nucleus of the hydrogen atom gets its positive charge from a single particle called a proton. A **proton** is *a positively charged particle with a mass 1832 times greater than the mass of an electron.*

The second simplest atom is helium. Its nucleus has two positive charges, indicating that it contains two protons and, therefore, requires two orbiting electrons to maintain electrical neutrality (Fig. A.1). Sensitive measurements show, however, that the mass of a helium nucleus is very slightly greater than that of four protons. The increase in mass results from the presence of yet another type of particle in the nucleus. The newcomer is *an electrically neutral particle with a mass 1833 times greater than that of the electron,* called a **neutron.**

Protons, neutrons, and electrons may be regarded for our purposes as the three principal subatomic particles that combine to form all atoms,

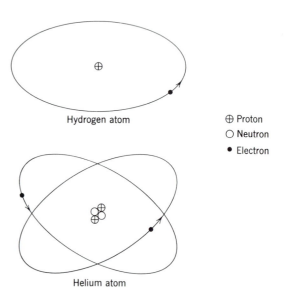

Hydrogen atom

⊕ Proton
○ Neutron
● Electron

Helium atom

FIGURE A.1 Hydrogen and helium atoms, shown schematically. Hydrogen is the simplest atom with one proton (positive) electrically balanced with one electron (negative). Helium has two electrons in orbit around a nucleus with two protons and two neutrons. The *atomic number* of helium is 2 (the number of protons); its *mass number* is 4 (the sum of protons and neutrons).

TABLE A.1 *The Fundamental Particles of Matter*

Name	Electric Charge	Relative Mass
Proton	+1	1832
Neutron	0	1833
Electron	−1	1

and they are always the same, no matter in what atoms they occur. The way they fit together in a given atom determines the chemical properties of the atom. Therefore, protons, neutrons, and electrons are the fundamental particles of matter about which the Greeks speculated and their different arrangements determine the different chemical properties of matter.

Because protons and neutrons are so much heavier than electrons (Table A.1), 99.99 percent of the mass of an atom resides in the nucleus. Atoms are tiny particles: The radius of an atom is about 10^{-8} cm. The radius of the atomic nucleus, however, is very much smaller—about 10^{-12} cm. Therefore, as Figure A.1 reveals, an atom is mostly open space. It has a tiny but heavy nucleus, surrounded by tiny electrons that move in distant orbits.

Elements are organized on the basis of the numbers of protons and electrons present: hydrogen, which has one proton and one electron, then helium, with two protons and two electrons, and so on. *The number of protons in the nucleus of an atom is the **atomic number.*** The known elements, with their atomic numbers and their abundance in the Earth's crust, are listed in Table A.2.

ISOTOPES AND RADIOACTIVITY

Although the number of neutrons that fit into the nucleus of an atom does not follow a simple pattern, it nevertheless is an important property of the atom because neutrons and protons together determine its mass. *The sum of the protons and neutrons in the nucleus of an atom is called the **mass number.*** Both atomic number and mass number are such important properties of atoms that they are recorded as subscripts and superscripts, respectively, usually on the left-hand side of the chemical symbol for the element. For example, $^{40}_{19}K$ means an atom of potassium with mass number 40 and atomic number 19. The atom, therefore, contains 21 neutrons and 19 protons. *Atoms having the same atomic number but differing numbers of neutrons in the nucleus are called **isotopes.*** Thus, isotopes of an

element have identical atomic numbers but different mass numbers. All isotopes that are known to occur in the Earth are listed in Table A.3.

Hydrogen has three isotopes. Each has one proton and one electron. As shown in Figure A.2, the three isotopes of hydrogen are $^{1}_{1}H$, $^{2}_{1}H$, and $^{3}_{1}H$, known in scientific language as protium, deuterium, and tritium, respectively. Most elements have two or more isotopes, and for one element (tin) as many as 10 isotopes have been discovered. The number of isotopes that are possible for any given element is limited, however, because the mass numbers of elements can vary only within certain limits. Many combinations of protons and neutrons are unstable and break up spontaneously. *The decay process by which an unstable atomic nucleus spontaneously transforms to another nucleus, emitting energy in the process, is called **radioactivity.***

The rates at which radioactive isotopes disintegrate are inherent properties of each isotope and are unaffected by changes in temperature or pressure. The constancy of decay rates is very important because decay rates of certain isotopes, such as $^{14}_{6}C$ and $^{40}_{19}K$ are so slow that we can use them as accurate clocks for timing events in the Earth's history.

When a radioactive isotope decays, it may form either a new isotope of the same element or an isotope of a different element. The decay products are called *daughters* which may, in turn, be either stable or radioactive. If they are radioactive, then further disintegrations will eventually occur, and still newer daughters will form.

When an atom disintegrates radioactively, some of the energy that holds the nucleus together is released. The disintegrating atom may also lose some of the subatomic particles. These may be either *alpha particles* (α-particles) or *beta particles* (β-particles). Alpha particles are $^{4}_{2}He$ nuclei, stripped of their electrons so that their loss reduces the mass number of an atom by 4 and the atomic number by 2. Beta particles are electrons expelled from the nucleus; their loss converts a neutron to a proton and thus increases the atomic number of an atom by 1 but leaves the mass number unchanged.

The results of radioactive emissions of α- and β-particles are illustrated in Figure A.3, which depicts a sequence of decays beginning with the most common isotope of uranium, $^{238}_{92}U$. The uranium atom emits an α-particle to form a daughter atom that is an isotope of thorium, $^{234}_{90}Th$. The thorium isotope then emits a β-particle to become $^{234}_{91}Pa$, and the new daughter product in turn emits a sec-

TABLE A.2 *Alphabetical List of the Elements*

Element	Symbol	Atomic Number	Crustal Abundance, Weight Percent	Element	Symbol	Atomic Number	Crustal Abundance, Weight Percent
Actinium	Ac	89	Man-made	Mercury	Hg	80	0.000002
Aluminum	Al	13	8.00	Molybdenum	Mo	42	0.00012
Americium	Am	95	Man-made	Neodymium	Nd	60	0.0044
Antimony	Sb	51	0.00002	Neon	Ne	10	Not known
Argon	Ar	18	Not known	Neptunium	Np	93	Man-made
Arsenic	As	33	0.00020	Nickel	Ni	28	0.0072
Astatine	At	85	Man-made	Niobium	Nb	41	0.0020
Barium	Ba	56	0.0380	Nitrogen	N	7	0.0020
Berkelium	Bk	97	Man-made	Nobelium	No	102	Man-made
Beryllium	Be	4	0.00020	Osmium	Os	76	0.00000002
Bismuth	Bi	83	0.0000004	Oxygen[b]	O	8	45.2
Boron	B	5	0.0007	Palladium	Pd	46	0.0000003
Bromine	Br	35	0.00040	Phosphorus	P	15	0.1010
Cadmium	Cd	48	0.000018	Platinum	Pt	78	0.0000005
Calcium	Ca	20	5.06	Plutonium	Pu	94	Man-made
Californium	Cf	98	Man-made	Polonium	Po	84	Footnote[d]
Carbon[a]	C	6	0.02	Potassium	K	19	1.68
Cerium	Ce	58	0.0083	Praseodymium	Pr	59	0.0013
Cesium	Cs	55	0.00016	Promethium	Pm	61	Man-made
Chlorine	Cl	17	0.0190	Protactinium	Pa	91	Footnote[d]
Chromium	Cr	24	0.0096	Radium	Ra	88	Footnote[d]
Cobalt	Co	27	0.0028	Radon	Rn	86	Footnote[d]
Copper	Cu	29	0.0058	Rhenium	Re	75	0.00000004
Curium	Cm	96	Man-made	Rhodium[c]	Rh	45	0.00000001
Dysprosium	Dy	66	0.00085	Rubidium	Rb	37	0.0070
Einsteinium	Es	99	Man-made	Ruthenium[c]	Ru	44	0.00000001
Erbium	Er	68	0.00036	Samarium	Sm	62	0.00077
Europium	Eu	63	0.00022	Scandium	Sc	21	0.0022
Fermium	Fm	100	Man-made	Selenium	Se	34	0.000005
Fluorine	F	9	0.0460	Silicon	Si	14	27.20
Francium	Fr	87	Man-made	Silver	Ag	47	0.000008
Gadolinium	Gd	64	0.00063	Sodium	Na	11	2.32
Gallium	Ga	31	0.0017	Strontium	Sr	38	0.0450
Germanium	Ge	32	0.00013	Sulfur	S	16	0.030
Gold	Au	79	0.0000002	Tantalum	Ta	73	0.00024
Hafnium	Hf	72	0.0004	Technetium	Tc	43	Man-made
Helium	He	2	Not known	Tellurium[c]	Te	52	0.000001
Holmium	Ho	67	0.00016	Terbium	Tb	65	0.00010
Hydrogen[b]	H	1	0.14	Thallium	Tl	81	0.000047
Indium	In	49	0.00002	Thorium	Th	90	0.00058
Iodine	I	53	0.00005	Thulium	Tm	69	0.000052
Iridium	Ir	77	0.00000002	Tin	Sn	50	0.00015
Iron	Fe	26	5.80	Titanium	Ti	22	0.86
Krypton	Kr	36	Not known	Tungsten	W	74	0.00010
Lanthanum	La	57	0.0050	Uranium	U	92	0.00016
Lawrencium	Lw	103	Man-made	Vanadium	V	23	0.0170
Lead	Pb	82	0.0010	Xenon	Xe	54	Not known
Lithium	Li	3	0.0020	Ytterbium	Yb	70	0.00034
Lutetium	Lu	71	0.000080	Yttrium	Y	39	0.0035
Magnesium	Mg	12	2.77	Zinc	Zn	30	0.0082
Manganese	Mn	25	0.100	Zirconium	Zr	40	0.0140
Mendelevium	Md	101	Man-made				

Source: After K. K. Turekian, 1969.

[a]Estimate from S. R. Taylor (1964).

[b]Analyses of crustal rocks do not usually include separate determinations for hydrogen and oxygen. Both combine in essentially constant proportions with other elements, so abundances can be calculated.

[c]Estimates are uncertain and have a very low reliability.

[d]Elements formed by decay of uranium and thorium. The daughter products are radioactive with such short half-lives that crustal accumulations are too low to be measured accurately.

TABLE A.3 *Naturally Occurring Elements Listed in Order of Atomic Numbers, Together with the Naturally Occurring Isotopes of Each Element, Listed in Order of Mass Numbers*

Atomic Number	Name	Symbol	Mass Numbers[b] of Natural Isotopes
1	Hydrogen	H	1, 2, 3[c]
2	Helium	He	3, 4
3	Lithium	Li	6, 7
4	Beryllium	Be	9, 10
5	Boron	B	10, 11
6	Carbon	C	12, 13, 14
7	Nitrogen	N	14, 15
8	Oxygen	O	16, 17, 18
9	Fluorine	F	19
10	Neon	Ne	20, 21, 22
11	Sodium	Na	23
12	Magnesium	Mg	24, 25, 26
13	Aluminum	Al	27
14	Silicon	Si	28, 29, 30
15	Phosphorus	P	31
16	Sulfur	S	32, 33, 34, 36
17	Chlorine	Cl	35, 37
18	Argon	A	36, 38, 40
19	Potassium	K	39, 40, 41
20	Calcium	Ca	40, 42, 43, 44, 46, 48
21	Scandium	Sc	45
22	Titanium	Ti	46, 47, 48, 49, 50
23	Vanadium	V	50, 51
24	Chromium	Cr	50, 52, 53, 54
25	Manganese	Mn	55
26	Iron	Fe	54, 56, 57, 58
27	Cobalt	Co	59
28	Nickel	Ni	58, 60, 61, 62, 64
29	Copper	Cu	63, 65
30	Zinc	Zn	64, 66, 67, 68, 70
31	Gallium	Ga	69, 71
32	Germanium	Ge	70, 72, 73, 74, 76
33	Arsenic	As	75
34	Selenium	Se	74, 76, 77, 80, 82
35	Bromine	Br	79, 81
36	Krypton	Kr	78, 80, 82, 83, 84, 86
37	Rubidium	Rb	85, 87
38	Strontium	Sr	84, 86, 87, 88
39	Yttrium	Y	89
40	Zirconium	Zr	90, 91, 92, 94, 96
41	Niobium	Nb	93
42	Molybdenum	Mo	92, 94, 95, 96, 97, 98, 100
44	Ruthenium	Ru	96, 98, 99, 100, 101, 102, 104
45	Rhodium	Rh	103
46	Palladium	Pd	102, 104, 105, 106, 108, 110
47	Silver	Ag	107, 109
48	Cadmium	Cd	106, 108, 110, 111, 112, 113, 114, 116
49	Indium	In	113, 115
50	Tin	Sn	112, 114, 115, 116, 117, 118, 119, 120, 122, 124
51	Antimony	Sb	121, 123
52	Tellurium	Te	120, 122, 123, 124, 125, 126, 128, 130
53	Iodine	I	127
54	Xenon	Xe	124, 126, 128, 129, 130, 131, 132, 134, 136
55	Cesium	Cs	133
56	Barium	Ba	130, 132, 134, 135, 136, 137, 138
57	Lanthanum	La	138, 139
58	Cerium	Ce	136, 138, 140, 142
59	Praseodymium	Pr	141
60	Neodymium	Nd	142, 143, 144, 145, 146, 148, 150

Table A.3 *(Continued)*

Atomic Number[a]	Name	Symbol	Mass Numbers[b] of Natural Isotopes
62	Samarium	Sm	144, 147 , 148 , 149 , 150, 152, 154
63	Europium	Eu	151, 153
64	Gadolinium	Gd	152 , 154, 155, 156, 157, 158, 160
65	Terbium	Tb	159
66	Dysprosium	Dy	156, 158, 160, 161, 162, 163, 164
67	Holmium	Ho	165
68	Erbium	Er	162, 166, 167, 168, 170
69	Thulium	Tm	169
70	Ytterbium	Yb	168, 170, 171, 172, 173, 174, 176
71	Lutetium	Lu	175, 176
72	Hafnium	Hf	174, 176, 177, 178, 179, 180
73	Tantalum	Ta	180, 181
74	Tungsten	W	180, 182, 183, 184, 186
75	Rhenium	Re	185, 187
76	Osmium	Os	184, 186, 187, 188, 189, 190, 192
77	Iridium	Ir	191, 193
78	Platinum	Pt	190, 192 , 195, 196, 198
79	Gold	Au	197
80	Mercury	Hg	196, 198, 199, 200, 201, 202, 204
81	Thallium	Tl	203, 205
82	Lead	Pb	204, 206, 207, 208
83	Bismuth	Bi	209
84	Polonium	Po	210
86	Radon	Rn	222
88	Radium	Ra	226
90	Thorium	Th	232
91	Protactinium	Pa	231
92	Uranium	U	234 , 235 , 238

[a]Atomic number = number of protons.
[b]Mass number = protons + neutrons.
[c]□ indicates isotope is radioactive.

ond β-particle to become $^{234}_{92}U$. The decay process continues through a total of 14 radioactive daughter products until a stable atom, $^{206}_{82}Pb$, is formed. We infer from this decay scheme that the uranium content of the Earth must be slowly decreasing and, conversely, that the total lead content must be slowly increasing. The rate of decay of uranium is quite slow, so there is plenty left for us to mine if we wish to do so. The rate of decay is also slow enough so that it has been possible to measure the

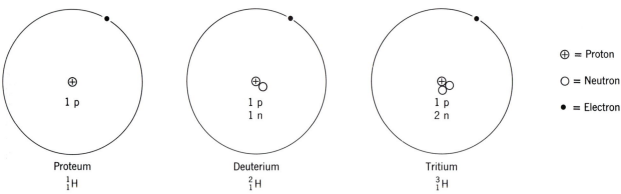

FIGURE A.2 Three isotopes of hydrogen, each of which has one proton in the nucleus and one electron in orbit. The isotopes differ in the number of neutrons each nucleus contains. All isotopes of an element have the same atomic number, but differ in their mass numbers.

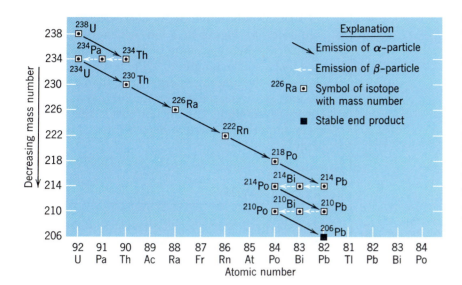

FIGURE A.3 Example of a radioactive-decay series. It commences with the most common isotope of uranium, $^{238}_{92}$U. The stable end product is an isotope of lead, $^{206}_{82}$Pb, but in order to reach it a number of different isotopes of different elements are produced as intermediate steps. Release of an α-particle from a nucleus reduces the atomic number by 2 and the mass number by 4. Release of a β-particle raises the atomic number by 1 but leaves the mass number unchanged.

way in which the Earth's lead content has changed through its history. This measurement allows us to infer a great deal about the age of the Earth.

When an atom decays radioactively, the principal way energy is released is as electromagnetic rays of very short wavelength, called *gamma rays* (γ-rays), and as heat which is generated whenever a nuclear disintegration occurs. The generation of heat by spontaneous nuclear decay has played a vitally important part in Earth history as well as in the evolution of other bodies in the Solar System, because radioactive decay of natural elements is believed to be one of the most important fuels that drive the great internal processes of planets.

MATTER AND ENERGY

We stated earlier that the nature of matter is essentially electrical. We also know that electricity is a form of energy, so we can modify our earlier statement to say that matter is essentially energy. The first person to understand this important concept was Albert Einstein who, in 1905, showed that matter and energy are connected by the equation

$$E = mc^2$$

where E is energy, m is mass, and c is the velocity of electromagnetic waves in a vacuum.

When we consider the structure of atoms, the equivalence of matter and energy becomes vitally important. For example, the nucleus of an atom of $^4_2$He contains two protons and two neutrons. Expressed in atomic mass units (AMU), a proton weighs 1.00758 AMU and a neutron weighs

1.00893 AMU. Therefore, the mass of $^4_2$He should be $(2 \times 1.00758) + (2 \times 1.00893) = 4.03303$ AMU. When the helium nucleus is weighed, however, it is found to be only 4.00260 AMU; obviously some of the mass has been lost. Explaining the mystery of the lost mass was one of the greatest triumphs of atomic physics. When the four fundamental particles join together to form a helium nucleus, some of their mass is converted to energy, which appears partly as electromagnetic radiation and partly as heat energy. The fusion of light atomic particles to form heavier ones proceeds continuously in the Sun. Since nuclei of hydrogen (which are simply protons) are abundant in the Sun, astronomers believe that hydrogen fusion to helium is the main source of solar energy. As a result of fusion, the Sun radiates a continuous stream of electromagnetic waves into space, and is slowly converting more of its matter into energy. *Fusion of light atomic particles to form heavy ones does not happen naturally on the Earth, because the high temperatures needed to trigger it do not occur, but it is the process that goes on in the hydrogen bomb.*

A nucleus can be broken down into its constituent particles only by the addition of sufficient energy to replace that lost during the joining process. The missing mass of the particles in the nucleus acts, in a sense, as a sort of "glue" that binds the nucleus together. The energy given off when protons and neutrons combine to form an atomic nucleus, therefore, is called the *binding energy* of the nucleus. As we can see in Figure A.4, the relation between binding energy and mass numbers is not a simple one. The curve dips down for ele-

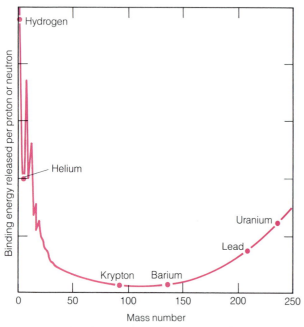

FIGURE A.4 Binding energy, the amount of energy produced by conversion of some of the mass of each proton and neutron combined in a nucleus, varies with the mass number of the nucleus (see text for further explanation). The lower an element is on the curve, the more energy will be given off when protons and neutrons combine to form its nucleus. The fusion of four hydrogen nucleii to form helium releases a vast amount of energy. This is the process that produces the Sun's energy and that indirectly drives most of the Earth's external activities. The splitting of a heavy uranium atom into krypton and barium atoms also releases energy. This is the process that occurs in the atom bomb. Lead sits lower on the curve than uranium. Natural radioactive decay of uranium to lead releases energy, and is one of the principal sources of energy for Earth's internal activities.

ments with intermediate mass numbers, but rises again as elements with high mass numbers are reached. This means that if the nucleus of a heavy atom such as uranium is split into one or more elements with intermediate mass numbers in the range 75–150, energy is released, just as it is when light nuclei fuse to form heavier ones. When the nucleus of a $^{235}_{92}$U atom is struck by a neutron moving within a certain range of velocities, the uranium nucleus splits into lighter atoms of barium and krypton, plus several neutrons, and a great deal of energy (Fig. A.5). The process of splitting heavy atoms is called *fission*, and it is the process that occurs in the atom bomb and that is used in a controlled manner in nuclear power plants.

Energy released by atomic fission appears both as electromagnetic waves and as heat. The process

of natural radioactive decay also releases heat. If we refer to Figure A.3, in which the natural radioactive decay scheme of $^{238}_{92}$U is depicted, we see that a stable daughter, $^{206}_{82}$Pb, results. Lead sits lower on the binding-energy curve than uranium; so energy is released during the process of natural decay of uranium. The fraction that is heat energy is an important source of Earth's internal heat energy.

COMPOUNDS

The chemical properties of an element (that is, the way elements combine together to form compounds) are determined by the orbiting electrons. Electrons are confined to specific orbits which are arranged at predetermined distances from the nucleus. Because the electrons in each orbit have a specific amount of energy characteristic for that orbit, the orbit distances are commonly called *energy-level shells*. The maximum number of electrons that can occupy a given energy-level shell is fixed. Shell 1, closest to the nucleus, is small and can accommodate only 2 electrons; shell 2, however, can accommodate 8 electrons; shell 3, 18; and shell 4, 32.

When an energy-level shell is filled with electrons, it is very stable, like an evenly loaded boat. To fill their energy-level shells and so reach a stable configuration, atoms share or transfer electrons among themselves. The movement of electrons naturally upsets the balance of electrical forces, for an atom that loses an electron has lost a negative electrical charge and, therefore, has a net positive charge, while one that gains an electron has a net negative charge. The sharing and transfer of electrons gives rise to electronic forces, or *bonds*, that bind atoms together into *compounds*. For example, as can be seen in Figure A.6, lithium has shell 1 filled, but has only one electron in shell 2. The lone outer electron is loosely held and easily transferred to an element such as fluorine, which has 7 electrons in shell 2, needing only one more to be completely filled. In this fashion both the lithium and fluorine finish with filled shells, and the resulting positive charge on the lithium (written Li^{+1}) and the negative charge on the fluorine (F^{-1}) bind the two atoms together. Lithium and fluorine form the compound lithium fluoride, which is written LiF to indicate that for every Li atom there is a counterbalancing F atom. Properties of compounds are quite different from the properties of their constituent elements. The elements of sodium (Na) and chlorine (Cl) are highly toxic, for example, but

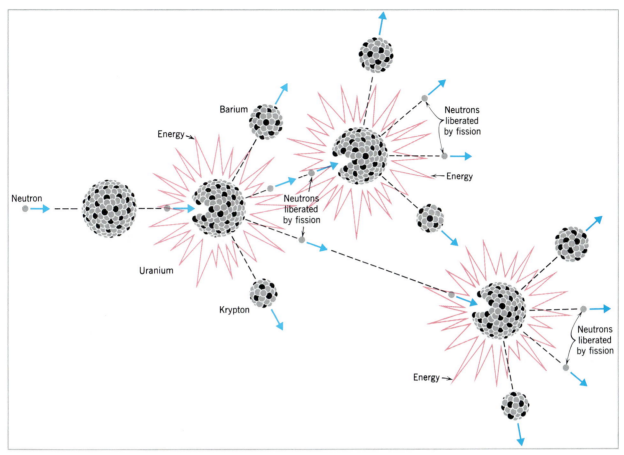

FIGURE A.5 Release of energy by a chain reaction in uranium. A neutron strikes the nucleus of a $^{235}_{92}U$ atom, causing it to fission into lighter elements such as barium and krypton, releasing energy and more neutrons. The new neutrons cause more $^{235}_{92}U$ atoms to disintegrate, producing a continuous chain reaction.

the compound sodium chloride (NaCl) is a compound that is essential for human health. A single LiF pair is called a molecule of lithium fluoride. A *molecule* is *the smallest unit that retains all the properties of a compound.*

Bonds

Some elements occur naturally with their energy-level shells completely filled. They occur as individual, electrically neutral atoms and show little or no tendency to react with other elements and form compounds. Because they are so unreactive, elements with normally filled outer shells are called *noble gases*. The noble-gas elements are helium, neon, argon, krypton, xenon, and radon. All elements other than the noble gases readily bond with other atoms and form compounds. Atoms form bonds with like or unlike atoms. However, regardless of the pairing, when atoms transfer or

share electrons, they finish with a net positive or negative charge, depending on whether they give up or receive the transferred electrons. The manner in which electrons are transferred or shared leads to several distinctive kinds of bonds. When electrons are transferred completely from one atom to another, the transferring atom becomes a *cation*, and the receiving atom becomes an *anion*. *The electrostatic attraction between negatively and positively charged ions is called an* **ionic bond.** LiF, NaCl, and CaF_2 are examples of ionically bonded compounds.

The force that arises when two atoms share one or more electrons is called **covalent bonding.** One common substance in which covalent bonding occurs is water, H_2O. The outer shell of an oxygen atom has six electrons but requires eight for maximum stability. A hydrogen atom has one electron but requires two for maximum stability. How this is accomplished is shown in Figure A.7. The silicate

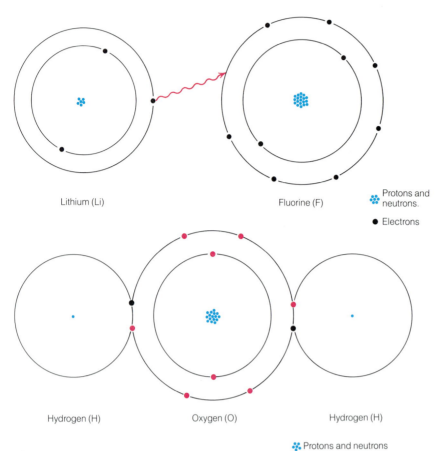

FIGURE A.6 To form lithium fluoride, an atom of lithium combines with an atom of fluorine. The lithium atom transfers its lone outer-shell electron to fill the fluorine atom's outer shell. The electrostatic force that holds the lithium and fluorine together is an ionic bond.

Lithium (Li)

Fluorine (F)

⚛ Protons and neutrons.

● Electrons

FIGURE A.7 Two atoms of hydrogen form covalent bonds with an oxygen atom through sharing of electrons. The oxygen atom thereby has its most stable configuration with eight electrons in the outer shell and each of the hydrogen atoms fill their outer shells with two electrons.

Hydrogen (H)   Oxygen (O)   Hydrogen (H)

⚛ Protons and neutrons
● Electrons belonging to oxygen
● Electrons belonging to hydrogen

tetrahedron is also covalently bonded. A large number of common minerals have either covalent or ionic bonding or a combination of the two. A smaller number of minerals display a third kind of bonding called the *metallic bond,* which is really *a variation of covalent bonding in which there are more electrons than needed to satisfy bond requirements.* The atoms are held together by covalent bonds and the additional electrons are free to diffuse through the structure, sometimes drifting, sometimes replacing an electron forming a covalent bond. The drifting electrons are very loosely held and are the reason that metals have special properties such as high electrical and thermal conductivity, opacity, and ductility.

There is, lastly, a special kind of bond, the *van der Waals bond,* that does not involve the transfer or sharing of electrons but is instead *a weak electrostatic attraction that arises because certain ions and atoms are distorted from a spherical shape.* The van der Waals bond is a weak bond—much weaker than ionic or covalent bonds—but it plays an important role in the structure of certain minerals, of which graphite is a good example. Graphite has a sheet-like structure; carbon atoms are bonded covalently within the sheets but the sheets are held together by van der Waals bonds. These van der Waals bonds are so weak they are easily broken. As a result, graphite and all other minerals held together by such bonds have very pronounced cleavage.

Identification of Common Minerals

IDENTIFICATION OF COMMON MINERALS

Minerals that are abundant in rocks, or common as ores, number only a few dozen, and most can be identified without complicated equipment if sizable pieces are available. The techniques to be described consist of direct observations and simple tests. With these it is possible to recognize groups of rock-forming minerals and the common ore minerals. The most helpful accessory equipment to aid in making the tests is: (1) a pocketknife; (2) a 10-power magnifying lens; (3) a piece of white, broken porcelain with a rough, unglazed surface; and (4) a small hand magnet or compass (the same tests can be performed by magnetizing the blade of your pocketknife).

PHYSICAL PROPERTIES

The properties of minerals are determined by their compositions and their crystal structures. Once we know which properties are characteristic of which minerals, we can use those properties to identify the minerals. Simple property tests are reliable and unambiguous because the number of common minerals is small. It is not necessary, therefore, to analyze a mineral chemically or to determine its crystal structure in order to identify most common ones. The characteristics most often used in identifying minerals are the obvious physical properties, such as color, shape of crystal, and hardness—plus some less obvious properties, such as cleavage and specific gravity. Each property is discussed below.

Crystal Form and Habit

When a mineral grows freely (without obstruction from adjacent minerals), it forms a characteristic *geometric solid that is bounded by symmetrically arranged plane surfaces.* The characteristic solid is called the **crystal form** of a mineral. The plane faces of a crystal are an external expression of the strict, internal geometric arrangement of the constituent atoms. Each plane surface corresponds to a plane of atoms in the crystal structure. The individual atoms are too small to be observed directly, even with the best microscopes, but their ordered array can be sensed and measured in a number of ways. The most obvious expression of the internal structure is the crystal form. Unfortunately, crystals are rare in nature because minerals do not usu-

FIGURE B.1 Two quartz crystals with the same crystal forms. Although the size of the individual faces differ markedly between the two crystals, it is clear that each face on one crystal is parallel to an equivalent face on the other crystal. It is a fundamental property of crystals that arises as a result of the internal crystal structure, that the angles between adjacent faces are identical for all crystals of the same mineral.

FIGURE B.2 Crystals of stibnite (Sb_2S_3) almost invariably occur as long, thin, needlelike crystals. The white, tabular-shaped crystals intergrown with the needles are calcite ($CaCO_3$).

ally grow into open, unobstructed space. When nice crystals are found, however, an examination of the crystal form tells much about the crystal structure and immediately aids us in identification.

The sizes of individual crystal faces differ. Under some circumstances a mineral may grow a long, thin crystal; under others, it may grow a short, fat one. Superficially, the two crystals may look very different; however, the unique characteristic of crystals is not the relative sizes of the individual crystal faces, but the angles between the faces. The angle between any designated pair of crystal faces in a mineral is constant and is the same for all specimens of the mineral, regardless of the overall shape. Two crystals of quartz (SiO_2) are shown in Figure B.1. One is flattened, the other elongate, but it is clear that the same sets of crystal faces occur on both minerals. The sets of faces are parallel; therefore, the angle between any two equivalent faces must be the same on each crystal.

Every mineral has a characteristic crystal form. Some have such distinctive forms we can use the property as an identification tool without having to measure angles between faces. For example, the mineral pyrite (FeS_2) is commonly found as intergrown cubes (Fig. 3.10) with markedly striated faces, while the mineral stibnite (Sb_2S_3) almost invariably forms long, needle-like crystals (Fig. B.2).

As noted earlier, most minerals do not grow freely into open spaces and, therefore, do not develop well-shaped crystals. Instead, the growing minerals usually encounter other minerals and other obstructions that prevent the development of crystal faces. Usually, then, we cannot use crystal form to identify minerals, but we can sometimes make use of distinctive growth habits to aid identification. For example, Figure B.3 shows asbestos, a variety of the mineral serpentine that characteristically grows as fine, elongate threads. Another example of a distinctive growth habit is shown in Figure B.4. This is psilomelane, a common manganese oxide that possesses *botryoidal* structure, a collection of smooth, rounded surfaces together resembling grapes closely bunched. Other common habits are the *earthy* form of many iron oxides, which are crumbly like soil, and *micaceous*, which is a habit of many silicate minerals which cleave into thin platelike fragments.

Cleavage and Fracture

If we break a mineral with a hammer, or drop the specimen on the floor so that it shatters, many of the broken fragments are seen to be bounded by rough fractures. However, careful examination will reveal that some surfaces are smooth, plane faces. In exceptional cases, such as calcite, most of the breakage surfaces will be smooth, plane faces so that the fragments resemble small crystals. A closer look shows that all fragments break along similar planes. *The tendency of a mineral to break in preferred directions along bright, reflective plane surfaces is called* **cleavage.** The plane surfaces along which cleavage occurs is governed by the crystal structure (Fig. 3.11). They are planes along which the bonds between atoms are relatively weak. Be-

FIGURE B.3 Some minerals have distinctive growth habits even though they do not develop well-formed crystal faces. Asbestos, a variety of the mineral serpentine, grows as fine, cottonlike threads that can be separated and woven into fireproof fabric.

cause the cleavage planes are direct expressions of the crystal structure, they are valuable guides for the identification of minerals.

Many minerals have distinctive cleavage planes. One of the most distinctive is found in the mineral *muscovite* (Fig. B.5). Clay minerals also have distinctive cleavage, and it is this easy cleavage direction that makes them feel smooth and slippery when rubbed between the fingers. Another mineral with a highly distinctive cleavage is calcite, which breaks into perfect rhombs (Fig. B.6). Besides micas, clays, and calcite, a number of other common minerals such as feldspar, amphibole, pyroxene, and galena have distinctive cleavages. Indeed, it is cleavage that allows us readily to distinguish amphibole from pyroxene—two groups of minerals that, in most properties, are nearly

FIGURE B.4 Psilomelane, a manganese oxide, commonly displays botryoidal surfaces. This habit is also common in various iron oxide minerals.

FIGURE B.5 Perfect cleavage of the mica mineral, muscovite, shown by very thin, plane flakes into which this platy crystal has been split. The cleavage flakes suggest leaves of a book, a resemblance embodied in the name "books of mica" for crystals elongated in a direction perpendicular to the cleavage flakes.

identical. The reason why cleavage is distinctive is because cleavages accurately reflect the ways in which silicate tetrahedra polymerize in the structures (Fig. 3.16).

Because of a distinctive habit of growth within the crystals, one of the two cleavage surfaces of plagioclase nearly always appears to be *striated* (Fig. B.7). Striations are a reliable means of distinguishing plagioclase from potassium feldspar. They are seen to best advantage with a hand lens when the cleavage surface reflects a bright light.

Not all minerals have distinctive cleavages. A few lack cleavage planes altogether, and thus are distinctive in the opposite sense. Minerals lacking perceptible cleavage include: garnet, which breaks along irregular fractures; and quartz (Fig. B.8) and olivine, which fracture irregularly or display **conchoidal fracture,** *breakage resulting in smooth, curved surfaces.* Some minerals break along splintery surfaces that resemble those of wood.

Luster

The quality and intensity of light reflected from a mineral produce an effect known as **luster.** Two miner-als with almost the same color can have totally different lusters. The more important are described as *metallic*, like that on a polished metal surface; *vitreous*, like that on glass; *resinous*, like that of resin; *pearly*, like that of pearl; *greasy*, as if the surface were covered with a film of oil; and *adamantine*, having the brilliance of a diamond.

Color and Streak

The color of a mineral is one of its striking properties, but unfortunately is not a very reliable means of identification. Color commonly results from impurities, which are present in only small amounts. Some minerals display various colors. Quartz, for example, can be clear and colorless, milky white, rose-colored, violet, and dark gray to black. Calcite, likewise, can be clear, milky white, pink, green, and gray. Among the feldspars, flesh-colored, cream-colored, pink, and light green are characteristic of potassium feldspar and its relatives, whereas dead white, gray, and light blue typify plagioclase.

Color in minerals is determined by several factors, but the main cause is variation in composition. As we have seen, atomic substitution causes the compositions of minerals to vary within small ranges. Some elements can create strong color effects even when they are present in very small amounts. For example, the mineral corundum (Al_2O_3) is commonly white or grayish in color, but when small amounts of Cr have replaced Al by solid solution, this mineral is blood red, forming *ruby*, a prized gem variety of corundum. Similarly, when small amounts of Fe and Ti are present, the corundum is deep blue and another prized gemstone, *sapphire*, is the result (Fig. B.9).

Color in opaque minerals with metallic and semimetallic luster can be very confusing because the color is partly a property of grain size. One way to reduce errors of judgment where color is concerned, is to prepare a **streak,** which is *a thin layer of powdered mineral made by rubbing a specimen on a nonglazed porcelain plate.* The powder diffuses light and gives a reliable color effect that is independent of the form and luster of the mineral specimen. Red streak characterizes hematite whether the specimen itself is red and earthy, like the streak, or black and metallic, like magnetite. Limonite streaks brown, and magnetite streaks black. Many minerals, particularly those with vitreous luster, streak an undiagnostic white.

a

b

FIGURE B.6 (*a*) Perfect rhombs of *calcite* (calcium carbonate) formed by cleavage planes in three directions. (*b*) Photograph of a broken surface of calcite as it appears through a 10-power magnifying glass. Two directions of cleavage are obvious; the third direction is the flat surface perpendicular to the direction of viewing.

FIGURE B.7 Characteristic striations on the cleavage surface of plagioclase. The striations arise through a property of crystal growth called twinning, and can only be observed on one of the two prominent cleavage directions in plagioclase.

FIGURE B.8 Irregular fracture of quartz. Curved fracture surfaces at end of quartz crystal from Arkansas.

Hardness

Relative resistance of a mineral to scratching is **hardness,** another distinctive property of minerals. Hardness, like crystal form and cleavage, is governed by crystal structure and by the strength of the bonds between atoms. The stronger the bonds, the harder the mineral. Degree of hardness can be decided in a relative fashion by determining the ease or difficulty with which one mineral will scratch another. Talc, the basic ingredient of most body ("talcum") powders, is the softest mineral known, and diamond is the hardest. A relative hardness scale between talc (number 1) and diamond (number 10) is divided into 10 steps, each marked by 1 of 10 common minerals (Table B.1). These steps do not represent equal intervals of

hardness, but the important feature of the hardness scale is that any mineral on the scale will scratch all minerals below it. Minerals on the same step of the scale are just capable of scratching each other. For convenience, we often test relative hardness by using a common object such as a penny, or a penknife, as the scratching instrument.

In tests of hardness, several precautions are necessary. A mineral softer than another may leave a mark that looks like a scratch, just as a soft pencil leaves its mark. A real scratch does not rub off. The physical structure of some minerals may make the hardness test difficult. If a specimen is powdery or in fine grains, or if it breaks easily into splinters, an apparent scratch may be deceptive.

Density and Specific Gravity

The final obvious physical property of a mineral is its density, which in practical terms means how heavy it feels. We know that equal-sized baskets of feathers and of rocks have different weights: feathers are light, rocks are heavy. The property that causes this difference is **density,** or *the average mass per unit volume.* The units of density are numbers of grams per cubic centimeter (g/cm^3). Minerals with a high density, such as gold, have their atoms closely packed. Minerals with low density, such as ice, have loosely packed atoms.

Minerals are divided into a heaviness or density scale. Gold has the highest density of all minerals, 19.3 g/cm^3, but many others such as galena (7.5),

FIGURE B.9 The many colors of corundum (Al_2O_3), here cut as gemstones, are caused by tiny amounts of trace elements entering the crystal structure by atomic substitution. The red color of ruby is produced by Cr^{+3} substituting for Al^{+3}, while the blue color of sapphire is produced by Fe^{+3} and Ti^{+4}.

TABLE B.1 *Scale of Hardness*

	Relative Number in the Scale	Mineral	Hardness of Common Objects
Decreasing	10.	Diamond	
	9.	Corundum	
	8.	Topaz	
	7.	Quartz	
	6.	Potassium feldspar	Pocketknife; glass
	5.	Apatite	
	4.	Fluorite	
	3.	Calcite	Copper penny
	2.	Gypsum	Fingernail
	1.	Talc	

magnetite (5.2), and hematite (5.3) feel heavy by comparison with most silicate minerals which have densities between 2.5–3.0 g/cm³.

Density is not commonly measured directly. Instead, unit weights of different minerals are compared against the unit weight of a standard substance. In comparing unit weights, we commonly use water as a standard. The *specific gravity* of any substance is expressed as *a number stating the ratio of the weight of a substance to the weight of an equal volume of pure water*. Specific gravity can be approximated by comparing different minerals held in the hand. Metallic minerals such as galena feel "heavy," whereas nearly all others feel "light." Specific gravity is a ratio of two weights, so it does not have any units. Because the density of pure water is 1 g/cm³, the specific gravity of a mineral is numerically equal to its density.

Magnetism

Of the common minerals only magnetite and pyrrhotite are strongly magnetic. They can be singled out at once by their strong attraction to a small magnet. A convenient way to test for magnetism is to magnetize one of the blades of your pocketknife.

CHEMICAL PROPERTIES

Only two chemical tests are commonly used in the beginning study of minerals: (1) taste test for halite; and (2) acid test for calcite and dolomite. The salty taste of halite is distinctive. Carbonate minerals effervesce (make bubbles) in dilute hydrochloric acid. Calcite effervesces freely no matter what the size of the particles. Dolomite may not effer-

vesce at all unless the specimen is powdered or the acid is heated. Dolomite powder effervesces slowly in cool dilute acid.

Caution. Hydrochloric acid is hard on teeth and has an unpleasant taste. Where many students share mineral specimens caution in the use of acid is necessary. Acid should be applied in small drops and when the test is finished the specimen should be blotted dry. The next user of the specimen may decide to try the taste test.

SOME TIPS ABOUT MINERAL IDENTIFICATION

Success in identifying minerals requires practice and the systematic application of tests. Learn to use a 10-power magnifying lens and always employ it in your examination of minerals. Use it when looking for cleavage, testing hardness, seeking crystal form, or any other test.

When faced with an unknown mineral, you will apply the first test automatically by recognizing whether the specimen has a metallic or nonmetallic luster. Minerals with metallic luster are usually ore minerals and should immediately be tested for streak. Other tests such as those for hardness, cleavage, specific gravity, and magnetism follow. Minerals with nonmetallic, and especially those with vitreous luster, have very pale streaks which are not diagnostic. For such minerals you must rely on properties other than color or streak. Cleavage and hardness are especially useful.

For convenience, the common minerals with metallic lusters are arranged alphabetically in Table B.2, and the common rock-forming minerals with nonmetallic lusters are listed in Table B.3. The chemical formulas in the second column of both tables are there for reference only, not to aid in identifications. The final column in Tables B.2 and B.3 list the physical properties most characteristic for each mineral, together with suggestions for distinguishing between similar-looking minerals. Most ore minerals will be encountered only in mineral deposits. A few, notably chalcopyrite, hematite, ilmenite, limonite, magnetite, pyrite, pyrrhotite, and rutile, may also be encountered as accessory minerals in common igneous, sedimentary, or metamorphic rocks.

Gemstones are minerals with desirable properties of color and wearing ability. A list of the commonly observed gem minerals, together with the names of the gem varieties, is given in Table B.4.

TABLE B.2 *Properties of the Common Minerals with Metallic Luster*

Mineral	Chemical Composition	Form and Habit	Cleavage	Hardness / Specific Gravity	Other Properties	Most Distinctive Properties
Bornite	Cu_5FeS_4	Massive. Crystals very rare.	None. Uneven fracture.	3 / 5	Brownish bronze on fresh surface. Tarnishes purple, blue, and black. Gray-black streak.	Color, streak.
Chalcocite	Cu_2S	Massive. Crystals very rare.	None. Conchoidal fracture.	2.5 / 5.7	Steel-gray to black. Dark gray streak.	Streak.
Chalcopyrite	$CuFeS_2$	Massive or granular.	None. Uneven fracture.	3.5–4 / 4.2	Golden yellow to brassy yellow. Dark green to black streak.	Streak. Hardness distinguishes from pyrite.
Chromite	$FeCr_2O_4$	Massive or granular.	None. Uneven fracture.	5.5 / 4.6	Iron black to brownish black. Dark brown streak.	Streak and lack of magnetism distinguishes from ilmenite and magnetite.
Copper	Cu	Massive, twisted leaves and wires.	None. Can be cut with a knife.	2.5–3 / 9	Copper color but commonly stained green.	Color, specific gravity, malleable.
Galena	PbS	Cubic crystals, coarse or fine-grained granular masses.	Perfect in three directions at right angles.	2.5 / 7.6	Lead-gray color. Gray to gray-black streak.	Cleavage and streak.
Gold	Au	Small irregular grains.	None. Malleable.	2.5 / 19.3	Gold color. Can be flattened without breakage.	Color, specific gravity, malleability.
Hematite	Fe_2O_3	Massive, granular, micaceous	Uneven fracture.	5–6 / 5	Red-brown, gray to black. Red-brown streak.	Streak, hardness.
Ilmenite	$FeTiO_3$	Massive or irregular grains.	Uneven fracture.	5.5–6 / 4.7	Iron-black. Brown-red streak differing from hematite.	Streak distinguishes hematite. Lack of magnetism distinguishes magnetite.
Limonite (*Goethite* is most common.)	A complex mixture of minerals, mainly hydrous iron oxides.	Massive, coatings, botryoidal crusts, earthy masses.	None.	1–5.5 / 3.5–4	Yellow, brown, black, yellow-brown streak.	Streak.

TABLE B.2 *(Continued)*

Mineral	Chemical Composition	Form and Habit	Cleavage	Hardness	Specific Gravity	Other Properties	Most Distinctive Properties
Magnetite	Fe$_3$O$_4$	Massive, granular. Crystals have octahedral shape.	None. Uneven fracture.	5.5–6.5	5	Black. Black streak. Strongly attracted to a magnet.	Streak, magnetism
Pyrite ("Fool's gold")	FeS$_2$	Cubic crystals with striated faces. Massive.	None. Uneven fracture.	6–6.5	5.2	Pale brass-yellow, darker if tarnished. Greenish-black streak.	Streak. Hardness distinguishes from chalcopyrite. Not malleable, which distinguishes from gold.
Pyrolusite	MnO$_2$	Crystals rare. Massive, coatings on fracture surfaces.	Crystals have a perfect cleavage. Massive breaks unevenly.	2–6.5	5	Dark gray, black or bluish black. Black streak.	Color, streak.
Pyrrhotite	FeS	Crystals rare. Massive or granular.	None. Conchoidal fracture.	4	4.6	Brownish-bronze. Black streak. Magnetic.	Color and hardness distinguish from pyrite, magnetism from chalcopyrite.
Rutile	TiO$_2$	Slender, prismatic crystals or granular masses.	Good in one direction. Conchoidal fracture in others.	6–6.5	4.2	Red-brown (common), black (rare). Brownish streak. Adamantine luster.	Luster, habit, hardness.
Sphalerite (zinc blende)	ZnS	Fine to coarse granular masses. Tetrahedron shaped crystals.	Perfect in six directions.	3.5–4	4	Yellow-brown to black. White to yellow-brown streak. Resinous luster.	Cleavage, hardness, luster.
Uraninite	UO$_2$ to U$_3$O$_8$	Massive, with botryoidal forms. Rare crystals with cubic shapes.	None. Uneven fracture.	5–6	6.5–10	Black to dark brown. Streak black to dark brown. Dull luster.	Luster and specific gravity distinguish from magnetite. Streak distinguishes from ilmenite and hematite.

TABLE B.3 *Properties of Rock-Forming Minerals with Nonmetallic Luster*

Mineral	Chemical Composition	Form and Habit	Cleavage	Hardness / Specific Gravity	Other Properties	Most Distinctive Properties
Amphiboles. (A complex family of minerals, *Hornblende* is most common.)	$X_2Y_5Si_8O_{22}(OH)_2$ where X = Ca, Na; Y = Mg, Fe, Al.	Long, six-sided crystals; also fibers and irregular grains	Two; intersecting at 56° and 124°	5–6 / 2.9–3.8	Common in metamorphic and igneous rocks. *Hornblende* is dark green to black; *actinolite*, green; *tremolite*, white.	Cleavage, habit.
Andalusite	Al_2SiO_5	Long crystals, often square in cross-section.	Weak, parallel to length of crystal.	7.5 / 3.2	Found in metamorphic rocks. Often flesh-colored.	Hardness, form.
Anhydrite	$CaSO_4$	Crystals are rare. Irregular grains or fibers.	Three, at right angles.	3 / 2.9	Alters to gypsum. Pearly luster, white or colorless.	Cleavage, hardness.
Apatite	$Ca_5(PO_4)_3(F, OH, Cl)$	Granular masses. Perfect six-sided crystals.	Poor. One direction.	5 / 3.2	Green, brown, blue, or white. Common in many kinds of rocks in small amounts.	Hardness, form.
Aragonite	$CaCO_3$	Massive, or slender, needle-like crystals.	Poor. Two directions.	3.5 / 2.9	Colorless or white. Effervesces with dilute HCl.	Effervescence with acid. Poor cleavage distinguishes from calcite.
Asbestos			See Serpentine			
Augite			See Pyroxene			
Biotite			See Mica			
Calcite	$CaCO_3$	Tapering crystals and granular masses.	Three perfect; at oblique angles to give a rhomb-shaped fragment.	3 / 2.7	Colorless or white. Effervesces with dilute HCL.	Cleavage, effervescence with acid.
Chlorite	$(Mg, Fe)_5(Al, Fe)_2 Si_3O_{10}(OH)_8$	Flaky masses of minute scales.	One perfect; parallel to flakes.	2–2.5 / 2.6–2.9	Common in metamorphic rocks. Light to dark green. Greasy luster.	Cleavage—flakes not elastic, distinguishes from mica. Color.
Dolomite	$CaMg(CO_3)_2$	Crystals with rhomb-shaped faces. Granular masses.	Perfect in three directions as in calcite.	3.5 / 2.8	White or gray. Does not effervesce in cold, dilute HCl unless powdered. Pearly luster.	Cleavage. Lack of effervescence with acid.

TABLE B.3 (*Continued*)

Mineral	Chemical Composition	Form and Habit	Cleavage	Hardness / Specific Gravity	Other Properties	Most Distinctive Properties
Epidote	Complex silicate of Ca, Fe and Al	Small elongate crystals. Fibrous.	One perfect, one poor.	6–7 / 3.4	Yellow-green to dark green. Common in metamorphic rocks.	Habit, color. Hardness distinguishes from chlorite.
Feldspars: Potassium feldspar (*orthoclase* is a common variety)	$KAlSi_3O_8$	Prism-shaped crystals, granular masses.	Two perfect, at right angles.	6 / 2.6	Common mineral. Flesh-colored, pink, white, or gray.	Color, cleavage.
Plagioclase	$NaAlSi_3O_8$ (albite) and $CaAl_2Si_2O_8$ (anorthite) and all compositions between.	Irregular grains, cleavable masses. Rarely as tabular crystals.	Two perfect, not quite at right angles.	6–6.5 / 2.6–2.7	White to dark gray. Cleavage planes may show fine parallel striations.	Cleavage. Striations on cleavage planes will distinguish from orthoclase.
Fluorite	CaF_2	Cubic crystals, granular masses.	Perfect in four directions.	4 / 3.2	Colorless, blue-green. Always an accessory mineral.	Hardness, cleavage, does not effervesce with acid.
Garnets	$X_3Y_2(SiO_4)_3$; X = Ca, Mg, Fe, Mn; Y = Al, Fe, Ti, Cr.	Perfect crystals with 12 or 24 sides. Granular masses.	None. Uneven fracture.	6.5–7.5 / 3.5–4.3	Common in metamorphic rocks. Red, brown, yellow-green, black.	Crystals, hardness, no cleavage.
Graphite	C	Scaly masses.	One, perfect. Forms slippery flakes.	1–2 / 2.2	Metamorphic rocks. Black with metallic to dull luster.	Cleavage, color. Marks paper.
Gypsum	$CaSO_4 \cdot 2H_2O$	Elongate or tabular crystals. Fibrous and earthy masses.	One, perfect. Flakes bend but are not elastic.	2 / 2.3	Vitreous to pearly luster. Colorless.	Hardness, cleavage.
Halite	NaCl	Cubic crystals.	Perfect to give cubes.	2.5 / 2.2	Tastes salty. Colorless, blue.	Taste, cleavage.
Hornblende			See Amphibole			
Kaolinite	$Al_2Si_2O_5(OH)_4$	Soft, earthy masses. Submicroscopic crystals.	One, perfect.	2–2.5 / 2.6	White, yellowish. Plastic when wet; emits clayey odor. Dull luster.	Feel, plasticity, odor.

TABLE B.3 *(Continued)*

Mineral	Chemical Composition	Form and Habit	Cleavage	Hardness	Specific Gravity	Other Properties	Most Distinctive Properties
Kyanite	Al_2SiO_5	Bladed crystals.	One perfect. One imperfect.	4.5 parallel to blade, 7 across blade	3.6	Blue, white, gray. Common in metamorphic rocks.	Variable hardness, distinguishes from sillimanite. Color.
Mica: Biotite	$K(Mg, Fe)_3$-$AlSi_3O_{10}$-$(OH)_2$	Irregular masses of flakes.	One, perfect.	2.5–3	2.8–3.2	Common in igneous and metamorphic rocks. Black, brown, dark green.	Cleavage, color. Flakes are elastic.
Muscovite	$KAl_3Si_3O_{10}(OH)_2$	Thin flakes.	One, perfect.	2–2.5	2.7	Common in igneous and metamorphic rocks. Colorless, pale green or brown.	Cleavage, color. Flakes are elastic.
Olivine	$(Mg, Fe)_2SiO_4$	Small grains, granular masses.	None. Conchoidal fracture.	6.5–7	3.2–4.3	Igneous rocks. Olive green to yellow-green.	Color, fracture, habit.
Orthoclase			See Feldspar				
Plagioclase			See Feldspar				
Pyroxene (A complex family of minerals. *Augite* is most common.)	$XY(SiO_3)_2$ $X = Y = Ca,$ Mg. Fe	8-sided stubby crystals. Granular masses.	Two, perfect, nearly at right angles.	5–6	3.2–3.9	Igneous and metamorphic rocks. *Augite,* dark green to black; other varieties white to green.	Cleavage
Quartz	SiO_2	6-sided crystals, granular masses.	None. Conchoidal fracture.	7	2.6	Colorless, white, gray, but may have any color, depending on impurities. Vitreous to greasy luster.	Form, fracture, striations across crystal faces at right angles to long dimension.
Serpentine (Fibrous variety is *asbestos*)	$Mg_3Si_2O_5(OH)_4$	Platy or fibrous.	One, perfect.	2.5–5	2.2–2.6	Light to dark green. Smooth, greasy feel.	Habit, hardness.
Sillimanite	Al_2SiO_5	Long needle-like crystals, fibers.	Breaks irregularly, except in fibrous variety.	6–7	3.2	White, gray. Metamorphic rocks.	Hardness distinguishes from kyanite. Habit.
Talc	$Mg_3Si_4O_{10}(OH)_2$	Small scales, compact masses.	One, perfect.	1	2.6–2.8	Feels slippery. Pearly luster. White to greenish.	Hardness, luster, feel, cleavage.

TABLE B.3 *(Continued)*

Mineral	Chemical Composition	Form and Habit	Cleavage	Hardness / Specific Gravity	Other Properties	Most Distinctive Properties
Tourmaline	Complex silicate of B, Al, Na, Ca, Fe, Li and Mg.	Elongate crystals, commonly with triangular cross section.	None.	7–7.5 / 3–3.3	Black, brown, red, pink, green, blue, and yellow. An accessory mineral in many rocks.	Habit.
Wollastonite	$CaSiO_3$	Fibrous or bladed aggregates of crystals.	Two, perfect.	4.5–5 / 2.8–2.9	Colorless, white, yellowish. Metamorphic rocks. Soluble in HCl.	Habit. Solubility in HCl and hardness distinguish amphiboles, kyanite, sillimanite.

TABLE B.4 *Properties of Some Common Gemstones*

Mineral and Variety	Composition	Form and Habit	Cleavage	Hardness / Specific Gravity	Other Properties	Most Distinctive Properties
Beryl: *Aquamarine* (blue) *Emerald* (green) *Golden beryl* (golden-yellow)	$Be_3Al_2Si_6O_{18}$	Six-sided, elongate crystals common.	Weak.	7.5–8 / 2.75	Bluish green, green, yellow, white, colorless. Common in pegmatites.	Form. Distinguished from apatite by its hardness.
Corundum: *Ruby* (red) *Sapphire* (blue)	Al_2O_3	Six-sided, barrel-shaped crystals.	None, but breaks easily across its crystal.	9 / 4	Brown, pink, red, blue, colorless. Common in metamorphic rocks. Star sapphire is opalescent with a six-sided light spot showing.	Hardness.
Diamond	C	Octahedron-shaped crystals.	Perfect, parallel to faces of octahedron.	10 / 3.5	Colorless, yellow; rarely red, orange, green, blue or black.	Hardness, cleavage.
Garnet: *Almandite* (red) *Grossularite* (green, cinnamon-brown) *Demantoid* (green)	A rock-forming mineral—See Table B.3					

TABLE B.4 *(Continued)*

Mineral and Variety	Composition	Form and Habit	Cleavage	Hardness / Specific Gravity	Other Properties	Most Distinctive Properties
Opal (A mineraloid)	$SiO_2 \cdot nH_2O$	Massive, thin coating. Amorphous.	None. Conchoidal fracture.	5–6 / 2–2.2	Colorless, white, yellow, red, brown, green, gray, opalescent.	Hardness, color, form.
Quartz: (1) Coarse crystals *Amethyst* (violet) *Cairngorm* (brown) *Citrine* (yellow) *Rock crystal* (colorless) *Rose quartz* (pink) (2) Fine-grained *Agate* (banded, many colors) *Chalcedony* brown, gray) *Heliotrope* (green) *Jasper* (red)		A rock-forming mineral—See Table B.3				
Topaz	$Al_2SiO_4(OH, F)_2$	Prism-shaped crystals, granular masses.	One, perfect.	8 / 3.5	Colorless, yellow, blue, brown.	Hardness, form, color.
Tourmaline		A rock-forming mineral—See Table B.3				
Zircon	$ZrSiO_4$	Four-sided elongate crystals, square in cross-section.	None.	7.5 / 4.7	Brown, red, green, blue, black.	Habit, hardness.

SELECTED REFERENCES

Berry, L. G., Mason, B., and Dietrich, R. V., 1983, Mineralogy, 2nd ed.: San Francisco, W. H. Freeman.

Deer, W. A., Howie, R. A., and Zussman, J., 1966, An introduction to the rock-forming minerals: New York, John Wiley.

Dietrich, R. V., 1969, Mineral tables—Hand specimen properties of 1500 minerals: New York, McGraw Hill.

Dietrich, R. V., and Skinner, B. J., 1979, Rocks and rock minerals: New York, John Wiley.

Klein, C., and Hurlbut, C. S., Jr., 1985, Manual of mineralogy: New York, John Wiley.

Pough, F. H., 1976, A field guide to rocks and minerals, 4th ed.: Boston, Houghton Mifflin.

Identification
of Common Rocks

DIAGNOSTIC FEATURES

The three major classes of rocks (igneous, sedimentary, and metamorphic are defined and discussed in Chapters 4, 5, and 6. In each class there are many distinct kinds, and each has a specific name. Fortunately, fewer than 30 make up the great bulk of the visible part of the Earth's crust. These common kinds must be learned well if we are to read correctly the history recorded in the crust. By studying the representative specimens of all the important kinds, we can learn the properties some have in common, and the distinctive features by which each kind is identified.

Ideally, we should see each specimen in the field as part of an exposure in which the larger features and relations are clearly shown. The sedimentary layering of sandstone or limestone and the intrusive relation of a dike tell us at once that the rock is sedimentary or igneous. Some hand specimens in a laboratory may not have clear indications of their general classification. However, any systematic description of common rocks lists them according to class, and we welcome any clue that may tell us at the start whether an unknown specimen is igneous, sedimentary, or metamorphic. A number of such clues are found with practice, and even without them a specimen can ordinarily be traced quickly to its class by the process of elimination.

Texture

Examine a rock specimen closely for the pattern of visible constituents, just as you inspect the weave (texture) in cloth. Phaneritic rocks have visible mineral grains. If these can be made out with the unaided eye over the entire surface, the rock has *granular texture*. This texture may be categorized as follows: *coarse* if the average grains are 5 mm or more across; *medium* if the average is 1–5 mm across; and *fine* if less than 1 mm across. Aphanitic rocks have such small grains they cannot be distinguished by the naked eye, or even the eye aided by a 10-power hand lens. Some igneous rocks have *glassy* texture; the cooling from magma was too rapid for any grains to form.

Many igneous rocks are porphyritic, with distinct phenocrysts (Fig. 4.10*d*); this proves igneous origin. As a first step toward their identification certain textures, then, tell us the general classification of some rock specimens.

In examining texture, we are interested not only in the size of the grains, but also in their shapes and the way they fit together. If the grains are angular, and dovetail one into another to fill all the space, they must have been formed by crystallization, and the rock is probably either igneous or metamorphic. If the grains are separated by irregular spaces filled with fine cementing material, they are probably fragments and the rock may be of either sedimentary or volcanic origin.

Mineral Assemblage

Many rocks are identified by their component minerals. The critical minerals are determined by their physical properties as explained in Appendix B. Grains large enough for clear visibility may be studied without a microscope. A hand lens that magnifies about 10 times is a useful aid in studying the mineral grains, even in coarse-textured rocks.

Other Properties

Some limestones may resemble fine-grained igneous rocks but are much softer. Every rock specimen under study should be tested for hardness.

Some rock types show characteristic forms on fracture surfaces. Other tests are mentioned in the descriptions of specific rocks.

The principal group of rocks discussed in this appendix is the igneous group. They offer more trouble to a beginning student than sedimentary or metamorphic rocks. Therefore, the following discussion and Table C.1 present the necessary data for identifying common igneous rocks. Identification of sedimentary and metamorphic rocks can be effected by studying Chapters 5 and 6 and using Tables C.2 and C.3, respectively.

KINDS OF IGNEOUS ROCK

In an introductory study, igneous rocks can be classified more systematically than others, and details required for satisfactory analysis of laboratory specimens are here separated from the general treatment in the body of the book. Table C.1 supplements Figure 4.11. Igneous rocks are grouped in this table according to (1) texture and (2) composition. Any classification for use with hand specimens must be general, and the number of names in this table is reduced to a minimum.

Nature does not always draw sharp boundaries; there are all conceivable gradations in texture and

TABLE C.1 *Aids in Identification of Common Igneous Rock*[a]

Texture	Minerals			Rock Name	Helpful Distinguishing Features
	Quartz	Feldspar	Other		
Coarse-grained Grains uniform in size	Abundant	Abundant. Potassium-feldspar exceeds plagioclase	Muscovite and/or biotite common. Hornblende sometimes present	**Granite**	Quartz and feldspar predominant. Light-colored rock, commonly pink, white shades of gray. Make sure potassium-feldspar exceeds plagioclase. Easily confused with granodiorite
Coarse-grained Grains uniform in size	Abundant	Abundant. Plagioclase exceeds potassium-feldspar	Muscovite and/or biotite common. Hornblende sometimes present	**Grandior-ite**	Quartz and feldspar predominant. Shades of gray
Coarse-grained Grains uniform in size	Sparse or absent	Abundant plagioclase. Potassium-feldspar rare or absent	Biotite and/or hornblende common. Pyroxene sometimes present	**Diorite**	About equal amounts of light- and dark-colored minerals. A darker rock than grandiorite. Absence or sparsity of quartz is diagnostic
Coarse-grained Grains uniform in size	Sparse or absent	Abundant. Potassium-feldspar exceeds plagioclase	Biotite, hornblende, nepheline may be present	**Syenite**	Commonly pink or red. Distinguish from granite by quartz content
Coarse-grained Grains uniform in size	Absent	Common. Plagioclase only	Pyroxene abundant. Olivine may be present	**Gabbro**	Dark minerals exceed light. A dark-colored rock. Distinguish from peridotite and pyroxenite by common plagioclase
Coarse-grained Grains uniform in size	Absent	Plagioclase is abundant	Pyroxene and olivine as minor constituents	**Anortho-site**	A light-colored rock consisting very largely of plagioclase
Coarse-grained Grains uniform in size	Absent	Rare or absent	Pyroxene abundant. Olivine may be present	**Pyroxe-nite**	A dark-colored rock consisting very largely of pyroxenes
Coarse-grained Grains uniform in size	Absent	Rare or absent	Olivine abundant. Pyroxene common to abundant	**Peridotite**	Dark-colored rock. Olivine is commonly a clear green and grains rounded
Medium-grained Grains uniform in size	Rare or absent	Abundant. Plagioclase only	Pyroxene common. Olivine may be present	**Diabase**	A common medium-grained, dark-gray-colored rock. Look for pyroxene and plagioclase. Distinguish from basalt by grain size and lack of extrusive volcanic features. Often called trap rock
Fine-grained Grains uniform in size	Abundant. Hard to see because of grain size	Abundant. Potassium-feldspar exceeds plagioclase	Hornblende, biotite may be present	**Rhyolite**	A light-colored volcanic rock. White, gray, red, purple. May contain some glass. Often shows signs of flowage
Fine-grained Grains uniform in size	Sparse or absent	Abundant. Plagioclase exceeds potassium-feldspar	Pyroxene, hornblende, biotite may be present	**Andesite**	A dark-colored volcanic rock. Shades of gray, brown, green, purple. Glass is not common

[a]To use the table, first determine the texture, and find which entries in Column 1 best describe it. Then determine if quartz is present, and if it is sparse or abundant. These options are listed in Column 2. Finally, determine which feldspars are present (Column 3), and what the remaining minerals are (Column 4). These four sets of observations uniquely determine the rock type, listed in Column 5. Column 6 is a review column in which the key identification features are mentioned together with suggestions on general rock features that may be helpful.

TABLE C.1 (*Continued*)

Texture	Minerals			Rock Name	Helpful Distinguishing Features
	Quartz	Feldspar	Other		
Fine-grained Grains uniform in size	Absent	Abundant. Plagioclase only	Pyroxene common. Olivine often present	**Basalt**	A common dark-colored volcanic rock. No quartz present. Often rings like a bell when struck with a hammer
Glassy	—	—	—	**Obsidian**	A dense glass. May contain some vesicles
Glassy	—	A few feldspar crystals may be present	—	**Pumice**	A glassy froth
Phenocrysts in a coarse- or medium-grained groundmass	Determine overall composition of rock. If a granite composition, rock is a **granite porphyry,** if a gabbro composition, rock is a **gabbro porphyry.** Common varieties are granite, granodiorite, and diorite porphyries. Phenocrysts commonly quartz, feldspar, hornblende				
Phenocrysts in a fine-grained groundmass	Determine overall composition of rock. Texture indicates a porphyritic volcanic rock. Most common varieties are **rhyolite porphyry** and **andesite porphyry.** Phenocrysts commonly quartz, feldspar, biotite				
Fragmental texture	Determine overall composition and size of fragments. Texture indicates a pyroclastic rock. Most common varieties are **rhyolitic ash tuff** and **andesitic ash tuff**				
Fragmental texture with glassy fragments flattened to give a texture that resembles flow structure	Determine overall composition. Flattened glassy fragments indicate a pyroclastic rock that has been welded. Most common variety is a **rhyolitic welded tuff**				

TABLE C.2 *Aids in Identification of Sedimentary Rock*

Rock Name	Composition	Critical Tests
1. Clastic Sedimentary Rock		
Conglomerate	Cemented particles, somewhat rounded, considerable percentage of pebble size	Larger particles more than 2 mm in diameter; smaller particles and binding cement in interstices
Breccia	Fragments conspicuously angular, with binding cement	Large particles of pebble size or larger
Sandstone	Rounded fragments of sand size, 0.02 to 2 mm; binding cement	Grains commonly quartz, but other minerals qualify in general classification
Arkose	Important percentage of feldspar grains, sand size or larger	Essential that feldspar grains make 25 percent or more of rock; some may be larger than sand size
Graywacke	Fragments of quartz, feldspar, rock fragments of any kind, with considerable clay	Poor assortment of several kinds of ingredients, with considerable clay in matrix
Siltstone	Chiefly silt particles, some clay	Surface is slightly gritty to feel
Shale	Chiefly clay minerals	Surface has smooth feel, no grit apparent
2. Chemical Sedimentary Rock		
Limestone	Calcite; may be even-grained and crystalline	Easily scratched with knife; effervesces in cold dilute hydrochloric acid
Dolostone	Dolomite; may be even-grained and crystalline	Harder than limestone, softer than steel; requires scratching or powdering for effervescence in cold dilute hydrochloric acid

TABLE C.3 *Aids in Identification of Metamorphic Rock*

Rock Name	Distinguishing Characteristics	Rock Name	Distinguishing Characteristics
1. Foliated Metamorphic Rock		*2. Nonfoliated Metamorphic Rock*	
Slate	Cleaves into thin, plane plates that have considerable luster; commonly the sedimentary layers of parent rock make lines on plates; thin slabs ring when they are tapped sharply	**Quartzite**	Consists wholly of quartz sand cemented with quartz; outlines of sand grains show on broken surfaces; the breaks passing through the grains; wide range in shades of color
Phyllite	Surfaces of plates highly lustrous; plates commonly wrinkled or sharply bent; grains of garnets and other minerals on some plates	**Marble**	Wholly crystallized limestone or dolostone; grain varies from coarse to fine; responds to hydrochloric acid test, as do calcite and dolomite; accessory minerals have developed from impurities in original rock
Schist	Well foliated, with visible flaky or elongate minerals (mica, chlorite, hornblende); quartz a prominent ingredient; grains of garnet and other accessory minerals common; foliae may be wrinkled	**Granofels**	Coarse-grained rock, commonly with the composition of a granite, but without mineral layering
Gneiss	Generally coarse-grained, with imperfect but conspicuous foliation; lenses and layers differ in mineral composition; feldspar, quartz, and mica are common ingredients	**Hornfels**	Hard, massive, fine-grained rock, commonly with scattered grains or crystals of garnet, andalusite, staurolite, or other minerals that are common in zones of contact metamorphism

composition in igneous rocks. The separating lines in the table, therefore, are somewhat fictitious. We can find a series of specimens that will bridge the gaps in composition between granite and granodiorite, or granodiorite and diorite.

Minerals of igneous rocks are divided generally into a light-colored group (including the feldspars, quartz, and muscovite) and a dark-colored group (including biotite, pyroxene, hornblende, and olivine).

Quartz grains in an igneous rock indicate a surplus of silica in the parent magma. Therefore, the presence or absence of quartz grains is a logical basis for drawing a line between granitic and dioritic rocks. In rocks that have no quartz, the feldspars are often as abundant as the dark-colored minerals; the boundary between diorite and gabbro is drawn where the dark minerals exceed 50 percent of the total. This boundary is carried through between andesite and basalt on the basis of color.

Granite

Feldspar and quartz are the chief minerals in granite. Some mica, either biotite or muscovite, is usually present. Many granites contain scattered grains of hornblende. Commonly, the dark minerals are in nearly perfect crystals; this suggests that they formed first, while most of the mass was molten. Feldspar formed next and the grains crowded

against and hampered each other in growth. Quartz, the surplus silica, crystallized last and so is molded around the angular grains of the earlier minerals. This *interlocking arrangement of visible mineral grains characteristic of granite* is called **granular texture.**

Technically, the term granite is applied only to quartz-bearing rocks in which potassium feldspar is predominant. The name granodiorite applies to similar rocks in which plagioclase is the chief feldspar. Without special equipment, the differences in feldspars are not always easily recognized, and in a general study the term granite sometimes is extended to this whole group of rocks. We sometimes recognize the variation in mineral composition by speaking of the *granitic* rocks. They are widespread in all the continents.

Pegmatite, or "giant granite," is a special kind of *granite which has abnormally large mineral grains.* The term is generally used for granitic rocks in which average grain diameters are 2 cm or larger. Individual mineral grains up to several meters long have been discovered in some pegmatites. Pegmatite quarries produce in commercial quantities large sheets of mica and minerals that yield the valuable elements lithium and beryllium.

Diorite

The chief mineral in diorite is feldspar, mainly plagioclase. This may not be evident to the unaided eye. Quartz is usually absent and when present it is only in minor amounts. Generally, the dark minerals are more abundant than in granite. Diorite forms many large masses, but it is not nearly so abundant as granitic rocks.

Gabbro

Dark diorite grades into *gabbro* as the dark minerals exceed 50 percent of the rock and plagioclase becomes subordinate. The chief dark mineral in gabbro is pyroxene, commonly with some olivine. These minerals are more dense than feldspar, and gabbro is distinctly more dense than granite and average diorite.

Diabase has fine-grained, intermediate texture between gabbro and basalt.

Anorthosite, Pyroxenite, and Peridotite

Under some circumstances plagioclase may comprise more than 90 percent of the volume of an igneous rock; associated minerals are usually pyroxene or, less commonly, olivine. Instead of calling such a rock a plagioclase-rich gabbro, it is given a special name. A granular rock composed largely or entirely of plagioclase is *anorthosite*.

If it is one of the dark minerals that comprises the bulk of a gabbroic igneous rock, special names are given to them too. A granular rock composed 90 percent or more of pyroxene in *pyroxenite*. If 90 percent of the rock is olivine the rock is *peridotite*. Both these rocks are very dark-colored and dense, and are commonly associated with ores containing the metals nickel, platinum, chromium, and iron. In many masses, the pyroxene and olivine have been partly or completely altered to serpentine.

Porphyritic Rock

Commonly, both coarse- and fine-grained igneous rocks contain prominent phenocrysts. If the larger grains make up less than about 25 percent of the rock mass, we say that the rock is *porphyritic* and give it the name suited to its groundmass (for example, porphyritic granite, porphyritic diorite, and porphyritic andesite). If the proportion of phenocrysts is more than 25 percent, we call the rock a *porphyry* and combine this term with the name that is proper for the groundmass (for example, granite porphyry, diorite porphyry, and rhyolite porphyry).

Rhyolite

A fine-grained rock with phenocrysts of quartz is *rhyolite*. The quartz indicates an excess of silica and, therefore, a close chemical kinship to granite. Rhyolites usually have phenocrysts of potassium-feldspar and biotite as well as quartz. Colors of the groundmass range from nearly white to shades of gray, yellow, red, or purple. Rhyolite commonly has irregular bands made by flowage of stiff magma shortly before it becomes solid.

Andesite

A fine-grained rock generally similar to rhyolite but lacking the quartz phenocrysts is *andesite*. Usually it has phenocrysts of plagioclase and dark minerals. Common colors are shades of gray, purple, and green, but some andesites are very dark, even black. Freshly broken, thin edges of dark andesites transmit some light and appear almost white when held before a bright source of light. In this way they are distinguished from basalt, which is opaque even on thin edges. The lighter-colored andesites commonly have irregular banding similar to that of rhyolite.

Andesite is extremely abundant as a volcanic rock, especially around the margins of the Pacific Ocean. The name comes from the Andes of South America.

Basalt

Basalt is a fine-grained rock, often porphyritic, that appears dark even on freshly broken thin edges. Common phenocrysts are plagioclase, pyroxene, and olivine. Common colors are black, dark brown or green, and very dark gray. In the upper and lower parts of lava flows basalt is commonly *vesicular*, meaning it is filled with small openings or *vesicles* made by escaping gases (Fig. C.1). In many flows these openings have been filled with calcite, quartz, or some other mineral deposited from solution. Vesicles filled by secondary minerals are called *amygdules;* a basalt containing them is called an *amygdaloidal* basalt (Fig. C.2).

Glassy Rock

Quick chilling of magma forms natural glass. *Obsidian* is a highly lustrous glassy rock. Obsidian displays a conchoidal pattern when broken.

Clear natural glass is not common; most obsid-

FIGURE C.1 Vesicular basalt. Note the phenocrysts of yellowish-green olivine. The specimen is 4.5 cm across.

FIGURE C.2 Amygdaloidal basalt formed when secondary minerals such as calcite and zeolites fill vesicles. The specimen is 4.5 cm across.

ians appear dark, even black. Because many of them correspond in chemical composition to rhyolite and granite, they seem to contradict the rule that rocks with a high silica content are light-colored. But obsidian chipped to a thin edge appears white, even transparent. The dark coloring results from a small content of dark mineral matter distributed evenly in the glass. Basalt glass, by contrast with rhyolite glass, is opaque or very dark-colored even on a thin edge.

Pumice is glass froth, full of cavities made by gases escaping through stiff, rapidly cooling magma. Because many of the cavities are sealed, pumice will often float on water for a long time. As the thin walls of the cavities transmit light, pumice is almost white, though it may form the cap of a black sheet of obsidian.

Pyroclastic Rock

The size of pyroclastic fragments, cemented to form a rock, determines the name applied to pyroclastic rock. When particles are tiny (less than 2 mm across) they are called *ash* and the rock is an *ash tuff*. Larger particles, between 2 and 64 mm in diameter, are called lapilli and the resulting rock is a *lapilli tuff*. The largest pyroclastic particles, with diameters in excess of 64 mm, are called bombs and a rock containing them is an *agglomerate*.

Pyroclastic rocks are largely comprised of volcanic fragments and, hence, they have the same compositions as igneous rocks. The compositions are determined in the same way as they are for igneous rocks, and the names are used as adjectives; for example, we refer to a lapilli tuff with an andesitic composition as an *andesitic lapilli tuff*.

The particles of ash in some falls are very hot, and when they land, the temperature is high enough so the particles weld together. A glassy or fine-grained rock formed by fusion of volcanic ash during deposition is *welded tuff* (Fig. 4.14). Some natural glasses are welded tuffs rather than chilled lava flows. Volcanic rocks in southeastern Arizona, Utah, and Nevada that strongly resemble flows of rhyolite are really welded tuffs. Many tuffs and agglomerates are stratified and look much like sedimentary rocks. Successive layers of ash are spread by air currents and become solidified as distinct beds. Furthermore, many explosive eruptions are accompanied by heavy rains and the ash is swept out in thin uniform beds. The loose volcanic debris on steep slopes becomes saturated and masses of it move down as mudflows and debris flows. Pyroclastic rocks, therefore, are hybrid in classification—partly igneous and partly sedimentary.

KINDS OF SEDIMENTARY ROCK

A study of the common sedimentary rocks, using good specimens in a laboratory, is a convenient way to become familiar with the different kinds of sedimentary rock. To aid this study, a brief description of common sedimentary rock types is given below, and is summarized in Table C.2.

Clastic Sedimentary Rock

All clastic sedimentary rock is composed of particles of broken rock transported, deposited, compacted, and then cemented to form a coherent mass. *A clastic sedimentary rock that contains numerous rounded pebbles or larger particles is* **conglomerate.** The pebbles, cobbles, and boulders have been more or less rounded during transport by streams or glacier ice or in buffeting by waves along a shore. They may consist of any kind of rock, but most commonly of those rich in the durable mineral quartz. Usually, the spaces between pebbles contain sand cemented with silica, clay, limonite, or calcite.

Sedimentary breccia is a *clastic sedimentary rock that resembles conglomerate, but most of whose fragments are angular instead of rounded.* We find all gradations between conglomerate and breccia.

A clastic sedimentary rock consisting of cemented sand grains is **sandstone.** With progressive change in size of grain, fine sandstone grades into *siltstone.* In many rocks, grain sizes are mixed; so we speak of conglomeratic sandstone or sandy siltstone.

In most sandstones the grains consist almost entirely of quartz. The cementing material varies, as in conglomerate; calcium carbonate is common, but silica makes a more durable rock. Color in sandstone, produced partly by the color of the grains and partly by that of the cementing material, varies within a wide range.

Arkose is *a variety of sandstone with a large proportion of feldspar grains.* A composition consisting of feldspar and quartz suggests granite, so that arkose might be mistaken for it. In arkose the grains do not interlock; they are rounded and separated by fine-grained cementing material.

Graywacke is a *poorly sorted variety of sandstone containing grains of quartz and feldspar, plus fragments of broken rock, and with considerable amounts of clay in the matrix.*

Shale (also called **claystone**) is a *fine-grained clastic sedimentary rock consisting largely of clay and silt.* It is so fine grained that to the unaided eye it seems homogeneous. Shale is soft and generally feels smooth and greasy, but some fine sand or coarse silt may make it feel gritty. Shales generally split into thin layers or flakes parallel to the sedimentary layering. Rocks of similar composition but with thick, blocky layers are termed *mudstone.*

The color of claystones and mudstones typically ranges through shades of gray, green, red, and brown. Some layers that contain considerable carbon are black.

Chemical Sedimentary Rock

The difference between chemical and clastic sedimentary rock is principally one of transportation. In chemical sedimentary rock the constituents were transported in solution and precipitate by inorganic or organic chemical reactions.

Limestone is *chemical sedimentary rock which consists chiefly of calcite.* Limestones may have many impurities and may vary greatly in appearance from occurrence to occurrence. Limestone belongs as a chemical sedimentary rock because even if there are clastic particles (such as shell fragments and corals) present, these were formed first by chemical and biological activity. Some limestones that are uniformly fine-grained were formed purely as chemical precipitates, aided more or less by tiny organisms. Some of the sediment on today's seafloors probably represents an early stage in the formation of fine-grained limestone. By contrast, many limestones are coarse-grained, either from recrystallization of the calcium carbonate or because they are made largely of detrital shell fragments.

Dolostone, like limestone, is either *a detrital or a chemical sedimentary rock which consists chiefly of the carbonate mineral dolomite, $CaMg(CO_3)_2$.* Dolomite looks like calcite, which is why dolostone looks like limestone. But dolomite is slightly harder than calcite and only effervesces with acid on a scratched surface or in powdered form.

KINDS OF METAMORPHIC ROCK

Metamorphic rocks are classified on the basis of texture (Chapter 6). Where a particular mineral is very obvious, its name may be used as a prefix. Examples of the seven common metamorphic rocks are given below and identifying features are summarized in Table C.3.

Slate is *fine-grained metamorphic rock with a pronounced cleavage.* Cleavage planes separate slates into thin, flat plates, commonly cutting across original sedimentary layering. Although surfaces of the cleavage slabs have considerable luster, mineral grains can be seen only with very high magnification. (Fig. 6.6). A common color is dark bluish gray, generally known as "slate color," but many slates are red, green, or black.

Phyllite is *an exceptionally lustrous rock formed by a higher stage of metamorphism than slate.* The mica flakes responsible for the luster can sometimes be seen without magnification. Some phyllites have visible grains of garnet and other minerals (Fig. 6.9). The cleavage plates commonly are wrinkled or even sharply bent.

Schist, a *well foliated metamorphic rock in which the component platy minerals are clearly visible,* represents a higher stage of metamorphism than phyllite. Mica schist is rich in mica (either biotite, muscovite, or both together). Chlorite schist and hornblende schist are also common. Quartz is abundant in all kinds of schists and many are studded with garnets and other metamorphic minerals.

Gneiss is a *coarse-grained, foliated metamorphic rock, always with marked layering but with imperfect cleavage* (Fig. 6.11). Gneisses have a streaky, roughly banded appearance, caused by alternating layers that differ in mineral composition. Feldspar, quartz, mica, amphibole, and garnet are common minerals in gneisses. *Granitic gneiss* is a *distinctly banded rock with the mineral composition of granite.* *Granofels* is a *coarse-grained metamorphic rock with the composition of a granite but without mineral layering.*

Marble is created by metamorphism of limestone, during which the calcite grains grow until the entire rock mass has become coarse grained (Fig. 6.13). Dolomite marble is formed in the same way from dolostone. *Marble,* then, is merely *coarsely crystalline limestone or dolostone.* Impurities in the original rock may cause the growth of small amounts of pyroxene, amphibole, and other minerals, which in many marbles make striking patterns.

Some commercial "marbles" are actually limestones and dolostones that "take a good polish," and have not been metamorphosed at all.

Quartzite is a *metamorphic rock developed from sandstone by introduction of silica into all spaces between the original grains of quartz* (Fig. 6.13). When quartzite is broken, the fracture passes through the original quartz grains, not around them as in ordinary sandstone. Usually, quartzites have no obvious patterns of foliation. Quartzite and marble are examples of typically nonfoliated metamorphic rocks, in contrast to gneiss, schist, phyllite, and slate, which are all foliated.

Hornfels is a rock formed near intrusive igneous bodies where the invaded rock is greatly altered by high temperature. Shale and some other fine-grained rocks are thus changed to *hornfels,* a *very hard, nonfoliated metamorphic rock, commonly studded with small crystals of mica and garnet.*

SELECTED REFERENCES

Dietrich, R. V., and Skinner, B. J., 1979, Rocks and rock minerals: New York, John Wiley.

Ehlers, E. G., and Blatt, H., 1982, Petrology. Igneous, sedimentary and metamorphic: San Francisco, W. H. Freeman.

Tennissen, A. C., 1974, Nature of Earth materials: Englewood Cliffs, New Jersey, Prentice-Hall.

APPENDIX D

Maps,
Cross Sections,
Field Measurements,
Interpretations

USES OF MAPS

An important part of the accumulated information about the geology and morphology of the Earth's crust exists in the form of maps. Nearly everyone has used automobile road maps in planning a trip or in following an unmarked road. A road map of a state, province, or county does what all maps have done since their invention at some unknown time more than 5000 years ago: it reduces the pattern of part of the Earth's surface to a size small enough to be seen as a whole. Maps are especially important for an understanding of geologic relations because a continent, a mountain chain, or a major river valley are of such large size that they cannot be viewed as a whole unless represented on a map.

A map can be made to express much information within a small space by the use of various kinds of symbols. Just as some aspect of physics and chemistry use the symbolic language of mathematics to express significant relationships, so many aspects of geology use the simple symbolic language of maps to depict relationships too large to be observed within a single view. Maps made or used by geologists generally depict either of two sorts of things:

1. The shape of the Earth's surface, on which are shown hills, valleys, and other features. Maps of this kind are *topographic maps*.

2. The distribution and attitudes of bodies of rock or regolith. Maps showing such things are *geologic maps*. They are often plotted on a topographic base map.

BASE MAPS

Every map is made for some special purpose. Road maps, charts for sea or air navigation, and geologic maps are examples of three special purposes. However, whatever the purpose, all maps have two classes of data: base data and special-purpose data. As base data, most geologic maps show a latitude-longitude grid, streams, and inhabited places; many also show roads and railroads and details of topography. Geologists may take an existing base map containing such data and plot geologic information on it, or they may start with blank paper and plot on it both base and geologic data—a much slower process if the map is made accurately.

Two-Dimensional Base Maps

Many base maps used for plotting geologic data are two-dimensional; that is, they represent length and breadth but not height. A point can be located only in terms of its horizontal distance, in a particular direction, from some other point. Hence, a base map always embodies the basic concepts of direction and distance. Two natural reference points on the Earth are the North and South Poles. Using these two points, a grid is constructed by means of which any other point can be located. The grid we use consists of lines of *longitude* (half circles joining the poles) and *latitude* (parallel circles concentric to the poles and perpendicular to the axis connecting the poles) (Fig. D.1). The longitude lines (*meridians*) run exactly north–south, crossing the east–west *parallels* of latitude at right angles. Since the circumference of the Earth at its equator (and the somewhat smaller circumference through its two poles) is known with fair accuracy, it is possible to define any point on the Earth in terms of direction and distance from either pole or from the point of intersection of any parallel with any meridian.

For convenience in reading, most maps are drawn so that the north direction is at the top or upper edge of the map. This is an arbitrary convention adopted mainly to save time. The north direction could just as well be placed elsewhere, provided its position is clearly indicated.

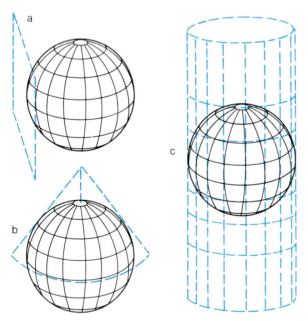

FIGURE D.1 The Earth's latitude–longitude grid can be projected onto a plane A, cylinder C, or cone B that theoretically can be cut and flattened out.

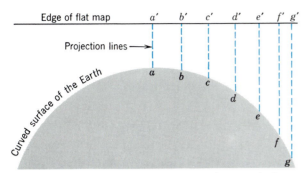

FIGURE D.2 Equally spaced points (*a, b, c, . . .*) along a line in any direction on the Earth's surface become unequally spaced when projected onto a plane. This is why all flat maps are distorted.

Map Projections

The Earth's surface is nearly spherical, whereas maps (other than globes) are two-dimensional planes, usually sheets of paper. It is geometrically impossible to represent any part of a spherical surface on a plane surface without distortion (Fig. D.2). The latitude-longitude grid has to be projected from the curved surface to the flat one. This can be done in various ways, each of which has advantages, but all of which represent a sacrifice of accuracy in that the resulting scale on the flat map will vary from one part of the map to another. The most famous of these is the Mercator projection, prepared by projecting all points radially onto a cylinder (Fig. D.1*c*), then unfolding the cylinder; although it distorts the polar regions very greatly, compass directions drawn on a Mercator projection are straight lines. Because this is of enormous value in navigation, the Mercator projection is widely used in navigators' charts.

Other kinds of projections are shown in Figure D.1. A commonly used projection for small regions of the Earth's surface is the conic projection (Fig. D.1*b*). Figure 4.34 is an example of conic projection. Some commonly used varieties are polyconic, in which not one cone, as in Figure D.1*b*, but several cones are employed, each one tangent to the globe at a different latitude. This device reduces distortion.

Map Scales

The accuracy with which distance is represented determines the accuracy of the map. *The proportion between a unit of distance on a map and the unit it represents on the Earth's surface* is the **scale** of the map. It is expressed as a simple proportion, such as 1:1,000,000. This ratio means that 1 meter, 1 foot, or other unit on the map represents exactly 1,000,000 meters, feet, or other units on the Earth's surface. It is approximately the scale of many of the road maps widely used by motorists in North America. Scale is also expressed graphically by means of a numbered bar, as is done on most of the maps in this book. A map with a latitude-longitude grid needs no other indication of scale (except for convenience), because the lengths of a degree of longitude (varying from 110.7 km at the equator to 0 at the poles) and of latitude (varying from 109.9 km at the equator to 110.9 km at the poles) are known.

The most commonly used scale for both topographic and geologic maps prepared in the United States by the Geological Survey is 1:24,000. Many older maps employ a scale of 1:62,500, approximately equal to 1 in. = 1 mile. When maps of larger regions are prepared scales of 1:100,000, 1:250,000, and 1:1,000,000 are employed. Use of scales at 1:24,000 and 1:62,500 arises from the practice of preparing maps that cover a quadrangular segment of the surface that is either 15 minutes (15′) of longitude by 15′ of latitude (scale, 1:62,500), or 7½′ × 7½′ (scale, 1:24,000).

Contours and Topographic Maps

A more complete kind of base map is three-dimensional; it represents not only length and breadth but also height. Therefore, it shows **relief** (*the difference in altitude between the high and low parts of a land surface*) and also **topography,** defined as *the relief and form of the land. A map that shows topography is a* **topographic map.** Topographic maps can give the form of the land in various ways. The maps most commonly used by geologists show it by contour lines.

A **contour line** (often called simply a **contour**) is *a line passing through points having the same altitude above sea level.* If we start at a certain altitude on an irregular surface and walk in such a way as to go neither uphill nor downhill, we will trace out a path that corresponds to a contour line. Such a path will curve around hills, bend upstream in valleys, and swing outward around ridges. Viewed broadly, every contour must be a closed line, just as the shoreline of an island or of a continent returns upon itself, however long it may be. Even on maps of small areas, many contours are closed lines, such as those at or near the tops of hills. Many, however, do not close within a given map area; they extend to the edges of the map and join the contours on adjacent maps.

Imagine an island in the sea crowned by two prominent isolated hills, with much steeper slopes on one side than on the other and with an irregular shoreline. The shoreline is a contour line (the zero contour) because the surface of the water is horizontal. If the island is pictured as submerged until only the two isolated peaks project above the sea, and then raised above the sea 5 m at a time, the successive new shorelines will form a series of contour lines separated by 5 m contour intervals. (A **contour interval** is *the vertical distance between two successive contour lines,* and is commonly the same throughout any one map.) At first, two small islands will appear, each with its own shoreline, and the contours marking their shorelines will have the form of two closed lines. When the main mass of the island rises above the water, the remaining shorelines or contours will pass completely around the landmass. The final shoreline is represented by the zero contour, which now forms the lowest of a series of contours separated by vertical distances of 5 m.

The following rules apply to contours:

1. All points on a contour have the same *elevation,* (also called *altitude*), which is *the vertical distance above mean sea level.*

2. A contour separates all points of higher elevation from all points of lower elevation.

3. In order to facilitate reading the contours on a map, certain contours (usually every fifth line) are drawn as a bolder line. Contours are numbered at convenient intervals for ready identification. The numbers are always multiples of the contour interval. For example, contour intervals of 5 m mean that successive contours are drawn at 10, 15, 20 m, etc.

4. Contours do not split or cross over, but at vertical cliffs they merge.

5. Because the contours that represent a depression without an outlet resemble those of an isolated hill, it is necessary to give them a distinctive appearance. Therefore, depression contours are *hatched;* that is, they are marked on the downslope side with short transverse lines called *hachures.* An example is shown on one contour in Figure D.3*b.* The contour interval employed is the same as in other contours on the same map.

6. Closely spaced contours indicate steep slopes, and widely spaced contours indicate gentle slopes.

7. Contours crossing a valley form a V-shape pointing *up* the valley, while contours on a ridge form a V-shape pointing *down* the ridge.

Idealized Example of a Topographic Map

Figures D.3*a* and *b* show the relation between the surface of the land and the contour map representing it. Figure D.3*a,* a perspective sketch, shows a stream valley between two hills, viewed from the south. In the foreground is the sea, with a bay sheltered by a curving spit. Terraces in which small streams have excavated gullies border the valley. The hill on the east has a rounded summit and sloping spurs. Some of the spurs are truncated at their lower ends by a wave-cut cliff, at the base of which is a beach. The hill on the west stands abruptly above the valley with a steep scarp and slopes gently westward, trenched by a few shallow gullies.

Each of the features on the map (Fig. D.3*b*) is represented by contours directly beneath its position in the sketch.

Topographic Profile

The *outline of the land surface along a given line* is called a **topographic profile.** A profile can be drawn along any line on a topographic map. Both the horizontal and the vertical scales must be designated for a profile. Most commonly, the map scale is chosen for the horizontal scale. The vertical scale is commonly made somewhat larger than the horizontal scale in order to exaggerate, or emphasize, the topography. The *ratio of the horizontal scale to the vertical scale in a topographic profile* is called the **vertical exaggeration.** If the horizontal scale is 1 cm = 1,000 m, and the vertical scale is 1 cm = 200 m, the vertical exaggeration is 1000/200 = 5X. A topographic profile of Figure D.3*b* is shown at a vertical exaggeration of 5X in Figure D.4.

To prepare a topographic profile perform the following steps:

1. Select the vertical and horizontal scales.

2. On a sheet of graph paper, select one of the horizontal lines as a baseline. Choice of a base varies from profile to profile; it can be sea level or any convenient height, such as the contour below the lowest point on the profile. Then mark in elevations on the graph paper, choosing the spacing appropriate to the vertical scale.

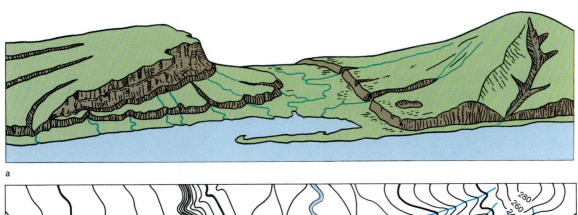

a

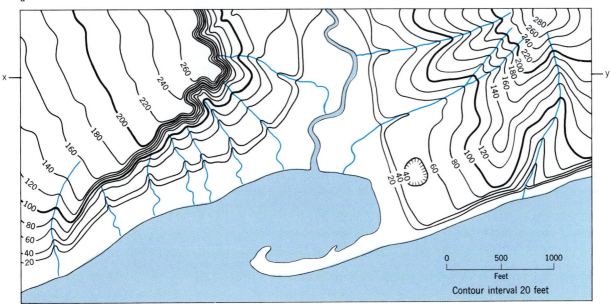

b

FIGURE D.3 (*a*) Perspective sketch of a landscape. (*Source:* Modified from U.S. Geological Survey.) (*b*) Topographic map of the area shown in Figure D.3*a*. Note that this map is scaled in feet and the contour interval is 20 ft. (*Source:* Modified from U.S. Geological Survey.)

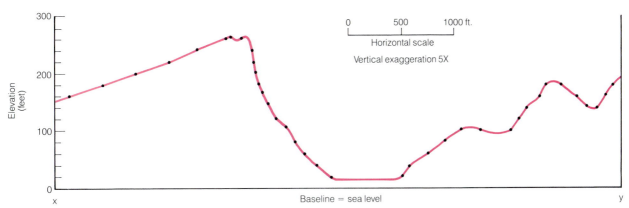

FIGURE D.4 Topographic profile along the line X–Y in Figure D.3*b*.

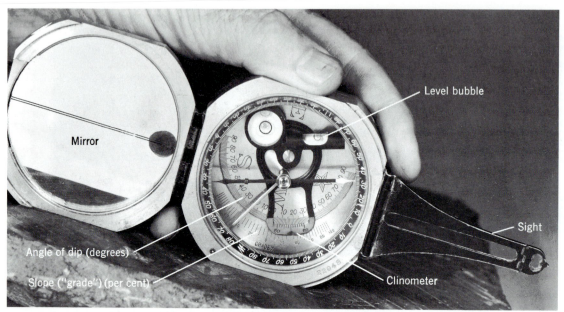

FIGURE D.5 Use a brunton (pocket transit) to measure the angle of dip of a stratum. The sides of the brunton are plane, parallel surfaces. With one side placed on a sloping plane surface in the direction of dip, the level bubble of the clinometer is centered by means of a lever on the back of the case (not visible). The angle is read on one of the two arcs below. The inner semicircular arc, calibrated in degrees from 0 to 90°, shows a reading of 17.5°, the outer, shorter scale, calibrated in percent, reads 31 percent. Compass directions are read on the dark outer scale when the instrument is held face up and is leveled by use of the circular level bubble. The hinged mirror and sight aid in taking bearings on selected points, and in using the brunton as a hand level (Fig. D.8), with the clinometer set to read zero degrees.

3. Place the edge of the graph paper on the line of the profile marked on the map.

4. Wherever a contour crosses the line of the profile, make a line on the graph paper at the appropriate elevation.

5. Join the points on the graph paper with a smooth line.

GEOLOGIC MAPS AND CROSS SECTIONS

The most common kind of *geologic map* is *a map that shows the distribution, at the surface, of rocks of various kinds or of various ages.* Examples are shown in Figure 15.30 (a map of basement rocks of successive ages), Figure 15.38 (a map of folded and faulted strata in the Alps), and Figure 4.34 (a sketch map showing the distribution of basalt compared with that of other rocks).

Field Equipment

A first essential for making a geologic map in the field is a base map, preferably a topographic map with a contour interval no greater than 3–5 m. Other probable necessities are a hammer, a steel tape for measuring the thickness of strata, and a pocket transit (geologist's compass; the most widely used variety in the United States and Canada is a Brunton compass, or simply a brunton). A brunton is a compact instrument (Fig. D.5) that is not only a compass, but also a clinometer and a sighting device used in reading compass directions and in hand leveling.

Strike and Dip

One of the commonest kinds of field measurement is determination of the attitude of a contact between strata, or any other planar surface. To represent the orientation of an inclined plane, we need to remember two principles of geometry: (1) the intersection of two planes defines a line; and (2) in an inclined plane only one horizontal direction exists. The horizontal line formed by the intersection of the inclined plane with the horizontal plane can be visualized as the waterline on a boat-launching

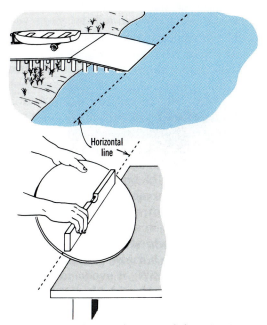

FIGURE D.6 The one horizontal direction in an inclined plane is illustrated by the water line against a boat-launching ramp, and by a carpenter's level held against an inclined board. The two horizontal lines shown are the directions of strike of the two inclined planes (see text).

ramp or by placing the edge of a carpenter's level in a horizontal position on a sloping plane (Fig. D.6). Instead of the level we can place one edge of a brunton, in the level position, against the inclined surface of a stratum.

The compass direction of the horizontal line in an inclined plane is the **strike** of the plane. There are two common ways of expressing strike and other compass directions. In the United States and Canada, directions are often expressed as angles between 0° and 90° east or west of true north. A strike trending 20° east of north would be written N20°E (Fig. D.7b); one trending 72° west of north would be written N72°W. In Europe and elsewhere, directions are expressed as *azimuths*, meaning they are measured as angles clockwise from true north (0°), through a full circle of 360°. N20°E, expressed as an azimuth is 20°; N20°W is 340°.

Once we know the strike we need only one more measurement to fix the orientation of the plane. That is the **dip,** *the angle in degrees between a horizontal plane and the inclined plane, measured down from horizontal in a plane perpendicular to the strike.* Dip is measured with a *clinometer,* usually a brunton in the position seen in Figure D.5. Not only the angle but the direction of dip (always at right angles to the strike) must be noted.

When plotted on a map as symbols (Fig. D.7a), orientation measurements made at each locality graphically convey the significant structural features of an area (Fig. D.7c). In a similar manner dip directions of cross strata can be plotted on a map to indicate directions of flow of ancient currents (Fig. 5.31).

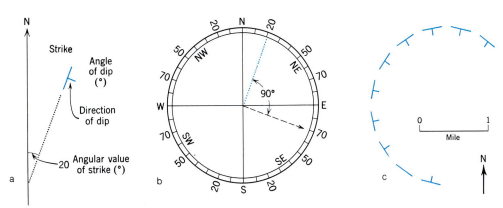

FIGURE D.7 Plotting strike and dip. (a) Strike and dip are plotted on maps with the T-shaped symbol shown in blue. (b) Strike and direction of dip shown in a are indicated by blue lines on the face of a compass. Strike is N20°E, direction of dip S70°E. Angle of dip, being a vertical angle, can not be shown. (c) Strike and dip symbols, from measurements at several localities on the same stratum, plotted on a geologic map. Evidently the structure is a syncline plunging southeast, the northeast limb dipping more steeply than the southwest. Compare Figure 15.25.

Thickness of Strata

The perpendicular distance between the upper and lower surfaces of a stratum, the *thickness* of the stratum, can be measured directly with a steel tape or other scale or with a hand level (Fig. D.8*a*) or altimeter. Thickness of an inclined stratum is usually calculated from simple trigonometric data (Fig. D.8*b*).

Patterns Made by Strata

Horizontal Strata. Figure D.9*a* is a geologic sketch map and a combined topographic profile and geologic cross section showing three units: two shale units with a unit of limestone between them. Each is a *formation* as defined in Chapter 7. The sketch map has been constructed on a topographic base representing two rounded hills with a stream between them. The two black lines represent the traces, along the land surface, of the interface of contact between the base of the limestone and underlying shale and of that between the top of the limestone and overlying shale. Symbols (Fig. D.10*a*) at five places indicate that the strata exposed at those places were found to be horizontal; that is, their dip is zero.

The geologist who drew the black lines, or "contacts," on the map shown in Figure D.9*a* determined the altitude of the lower one by hand level from a known point farther down the slope and then walked around each hill, following by eye, and then plotting on his map the change in type of rock from shale below to limestone above. His circuit of each hill brought him back to his starting point, and the contact he had plotted maintained a constant altitude around both hills. Therefore, the line representing it extends between the same two contours around both hills and is parallel with the contours.

The geologist then repeated the process with the higher contact, thereby completing the map. By hand leveling (Fig. D.8*a*) he could measure the vertical distance between base and top of the limestone and thereby find that the stratum is 8 m thick. For the thicknesses of each of the two layers

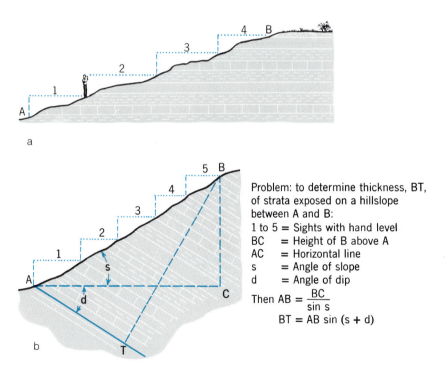

Problem: to determine thickness, BT, of strata exposed on a hillslope between A and B:

1 to 5 = Sights with hand level
BC = Height of B above A
AC = Horizontal line
s = Angle of slope
d = Angle of dip

Then $AB = \dfrac{BC}{\sin s}$

$BT = AB \sin (s + d)$

FIGURE D.8 Use of hand level in measuring heights, and thicknesses of strata exposed in slopes. (*a*) Thicknesses of horizontal strata are measured directly by hand level. The geologist determines accurately the height of his eyes above the ground, and multiplies this figure by the number of "sights" (1, 2, 3, . . .) he makes. If, instead of a hand level, he uses an altimeter, he computes differences between readings at critical points. (*b*) Where strata are inclined, vertical distance between base and top is measured, up the slope, by hand level. Angle of dip is measured with a clinometer. Thickness is determined by the trigonometric computation given above.

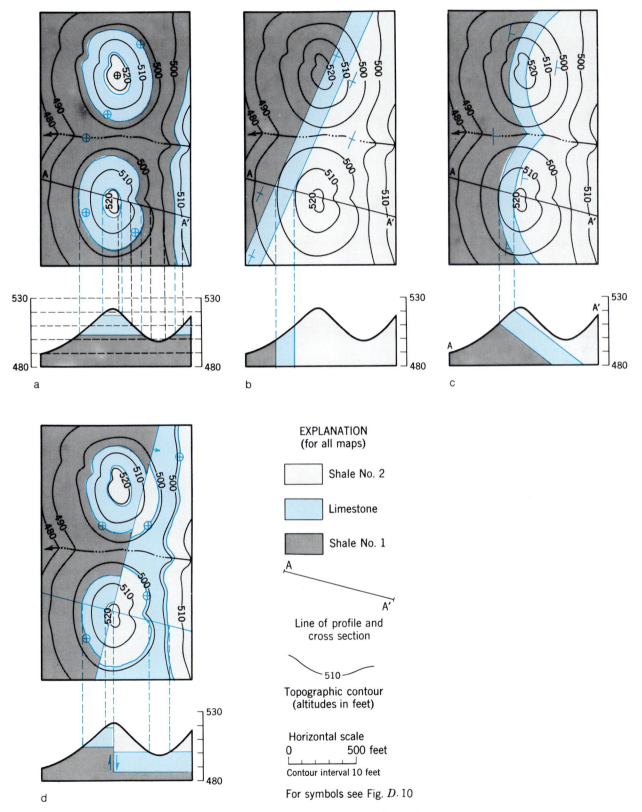

EXPLANATION
(for all maps)

Shale No. 2

Limestone

Shale No. 1

A ———— A'

Line of profile and
cross section

——— 510 ———

Topographic contour
(altitudes in feet)

Horizontal scale

0   500 feet

Contour interval 10 feet

For symbols see Fig. *D*.10

FIGURE D.9 Simple geologic sketch maps, each with topographic profile and geologic cross section below it, showing relation between pattern on map and geometry of section in four situations. Scale is in feet. (*a*) Strata horizontal. (*b*) Strata vertical. (*c*) Strata dipping east at 30° angle. (*d*) Horizontal strata cut by a fault.

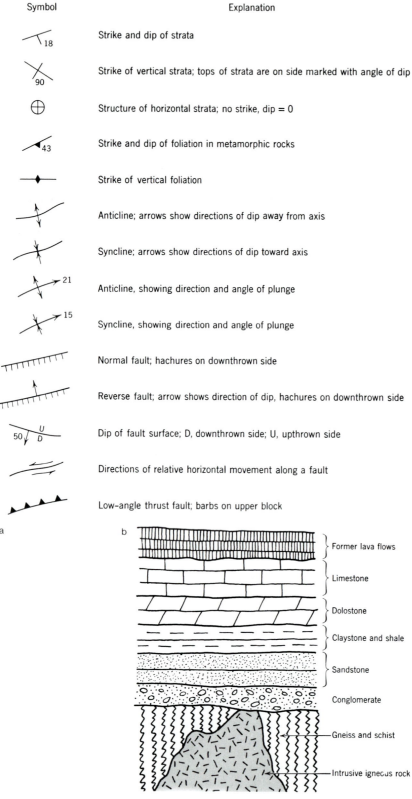

Symbol   Explanation

Strike and dip of strata

Strike of vertical strata; tops of strata are on side marked with angle of dip

Structure of horizontal strata; no strike, dip = 0

Strike and dip of foliation in metamorphic rocks

Strike of vertical foliation

Anticline; arrows show directions of dip away from axis

Syncline; arrows show directions of dip toward axis

Anticline, showing direction and angle of plunge

Syncline, showing direction and angle of plunge

Normal fault; hachures on downthrown side

Reverse fault; arrow shows direction of dip, hachures on downthrown side

Dip of fault surface; D, downthrown side; U, upthrown side

Directions of relative horizontal movement along a fault

Low-angle thrust fault; barbs on upper block

a

b

Former lava flows

Limestone

Dolostone

Claystone and shale

Sandstone

Conglomerate

Gneiss and schist

Intrusive igneous rock

FIGURE D.10 (*a*) Symbols commonly used to show structure on geologic maps. (*b*) Representative patterns commonly, but not universally, used to show kinds of rock in geologic cross sections.

of shale, however, he could measure only minimum values, because neither the base of one nor the top of the other is exposed within the area of the map.

The area, on a geologic map, shown as occupied by a particular rock unit is the **outcrop area** of that unit. An outcrop area, therefore, is the area of the Earth's surface in which some particular rock unit constitutes the highest part of the underlying rock, whether exposed at the surface or covered by regolith derived from it.

Vertical Strata. The lines representing contacts between strata whose dip is vertical appear as straight lines on a geologic map (Fig. D.9b). They change direction only where the strike of an interface of contact changes, or where the contact is offset by a fault.

Inclined Strata. On a geologic map, the lines of contact between strata that are inclined cross the contours in directions that vary with angle of dip of the strata and with slope of the surface as represented by trend and spacing of contours. In Figure D.9c we note that the lines representing the top and base of the limestone layer swing west through broad arcs in each of the two hills and bend east in a chevron-like pattern as they cross the valley. Measurements of strike and dip at five localities show that all three strata are dipping east. If the dip were west instead of east, the pattern would be reversed, with arcs swinging east and the chevron pointing west.

Faults

On a geologic map, faults (which are, of course, one kind of surface of contact) are represented by lines thicker than those used for all other kinds of contact. Where faults displace rock units they also displace other contacts between the units; these appear on a map as offsets (Fig. D.9d). The details of faults and other structures are brought out on geologic maps by means of special symbols (Fig. D.10a).

Geologic Cross Sections

Once a geologic map has been constructed it can be used as the basis for constructing cross sections along planes that extend downward below the surface, at right angles to the plane of the map. A **geologic cross section** is *a diagram showing the arrangement of rocks in a vertical plane.* It represents what would be revealed, as outcrops, in the vertical wall of a deep trench. In a natural trench (such as the Grand Canyon) we could construct a geologic cross section by measuring the exposed rocks directly, with an accuracy as great as would be possible in making a geologic map of a comparable area. Most sections, however, are made by geometric projection of the contacts already drawn on a map. The accuracy of a section constructed in this way varies with the kinds of rock bodies present. Where strata of originally wide extent have been deformed, surfaces of contact are generally parallel and can be projected downward with fair accuracy. However, where rock bodies having irregular surfaces of contact are present and where exposures of the rocks are few, sections are less accurate and some are little more than speculative. Sections of this kind can be checked and corrected only by comparing them with subsurface data obtained from drilling records or geophysical exploration.

The method of constructing geologic cross sections is illustrated in Figure D.9. A distinctive unit of limestone, shown in the four geologic sketch maps, is horizontal in *a*, vertical in *b*, dipping at 30° in *c*, and cut by a fault in *d*. What would be the geometry of this unit as seen in a vertical plane?

Below the map on Figure D.9a we construct a topographic profile along the line A–A'. Next we draw lines from the borders of the outcropping unit to intersect the profile on the profile grid. By connecting points on the profile that represent, respectively, the top and bottom of the unit, the cross section of the layer is completed.

Because the topography on all four base maps is the same, construction lines for making the topographic profile are shown for *a* only. On the identical profiles for *b*, *c*, and *d*, the lines, dropped from lower and upper boundaries of the outcropping limestone unit, give the points required for completing the sections. Patterns commonly used in geologic cross sections to represent rocks of various kinds are illustrated in Figure D.10b.

MAPS OF IGNEOUS AND METAMORPHIC ROCKS

Because sedimentary rocks are commonly stratified with rather distinct upper and lower surfaces of contact and because many of them are extensive, they present fewer mapping problems than do most igneous and metamorphic rocks. An exception is volcanic rocks that occur in widespread sheets like sedimentary strata and may be inter-bedded with layers of volcanic ash (Fig. 8.3). In some districts it is possible to map these units separately just as sedimentary formations are mapped. But in many volcanic fields, bodies of extrusive igneous rock and bodies of volcanic ash (erupted from a number of centers at irregular intervals) are mixed in great confusion. Such complexes are often mapped only as general assemblages, except at small scales, without regard to the various kinds of volcanic rock that compose them.

Intrusive igneous bodies exposed at the surface are highly varied in form and in their relation to associated rock. Commonly, they cut across older sedimentary units and their outcrop areas range widely in shape and size. Those large enough to be shown clearly on the scale of the map are outlined and marked with distinctive colors or patterns.

Most metamorphic rocks have complex structure and are not easily divisible into distinctive units. Therefore, many such rocks are represented on maps by a single color or pattern. On the other hand, in some areas, metamorphic rocks are mapped in great detail. Various kinds of schists and gneisses are mapped individually, as formations; attitudes of foliation are recorded at many points by symbols (Fig. D.10*a*), and faults are mapped, many of them through long distances. In such areas the various intensities of metamorphism that have affected the rocks are identified through critical minerals. With such identification at a large number of points, it is possible to draw a series of isograds (Figs. 6.16 and 6.17). The trends and patterns of isograds give clues to the directions in which the former pressures were applied, and also suggest former temperature gradients.

ISOPACH MAPS

Somewhat analogous to an isograd is an *isopach,* a *line on a map that connects points of equal thickness of a rock unit.* Isopachs are usually shown on special maps of limited areas where thickness is particularly significant. For example, an isopach map might show the thickness of a surface cover of loess, volcanic ash, alluvium, or glacial drift. Again, it might represent the thinning, in some direction, of a buried unit similar to the one appearing in cross section in Figure 21.9*d*. An isopach map of that unit probably would be useful in a local search for petroleum.

SELECTED REFERENCES

General

Thompson, M. M., 1979, Maps for America: Washington, D.C., U.S. Government Printing Office.

Base Maps

Raisz, E., 1948, General cartography, 2nd ed.: New York, McGraw-Hill.

Robinson, A. H., 1960, Elements of cartography, 2nd ed.: New York, John Wiley.

Geologic Maps and Cross Sections

Blyth, F. G. H., 1965, Geological maps and their interpretation: London, Edward Arnold.

Field Techniques

Compton, R. R., 1961, Manual of field geology: New York, John Wiley.

Lahee, F. H., 1961, Field geology, 6th ed.: New York, McGraw-Hill.

Units and Their Conversions

PREFIXES FOR MULTIPLES AND SUBMULTIPLES

$$
\begin{aligned}
1{,}000{,}000{,}000 &= 10^9 &&\text{giga} \\
1{,}000{,}000 &= 10^6 &&\text{mega} \\
1{,}000 &= 10^3 &&\text{kilo} \\
100 &= 10^2 &&\text{hecto} \\
10 &= 10 &&\text{deka} \\
0.1 &= 10^{-1} &&\text{deci} \\
0.01 &= 10^{-2} &&\text{centi} \\
0.001 &= 10^{-3} &&\text{milli} \\
0.000001 &= 10^{-6} &&\text{micro} \\
0.000000001 &= 10^{-9} &&\text{nano} \\
0.000000000001 &= 10^{-12} &&\text{pico}
\end{aligned}
$$

COMMONLY USED UNITS OF MEASURE

Length

Metric Measure

1 kilometer (km)	= 1000 meters (m)
1 meter (m)	= 100 centimeters (cm)
1 centimeter (cm)	= 10 millimeters (mm)
1 millimeter (mm)	= 1000 micrometers (μm) (formerly called microns)
1 micrometer (μm)	= 0.001 millimeter (mm)
1 angstrom (Å)	= 10^{-8} centimeters (cm)

Nonmetric Measure

1 mile (mi)	= 5,280 feet (ft) = 1,760 yards (yd)
1 yard (yd)	= 3 feet (ft)
1 fathom (fath)	= 6 feet (ft)

Conversions

1 kilometer (km)	= 0.6214 mile (mi)
1 meter (m)	= 1.094 yards (yd) = 3.281 feet (ft)
1 centimeter (cm)	= 0.3937 inch (in)
1 millimeter (mm)	= 0.0394 inch (in)
1 mile (mi)	= 1.609 kilometers (km)
1 yard (yd)	= 0.9144 meter (m)
1 foot (ft)	= 0.3048 meter (m)
1 inch (in)	= 2.54 centimeters (cm)
1 inch (in)	= 25.4 millimeters (mm)
1 fathom (fath)	= 1.8288 meters (m)

Area

Metric Measure

1 square kilometer (km^2)	= 1,000,000 square meters (m^2)
	= 100 hectares (ha)
1 square meter (m^2)	= 10,000 square centimeters (cm^2)
1 hectare (ha)	= 10,000 square meters (m^2)

Nonmetric Measure

1 square mile (mi^2)	= 640 acres (ac)
1 acre (ac)	= 4840 square yards (yd^2)
1 square foot (ft^2)	= 144 square inches (in^2)

Conversions

1 square kilometer (km^2)	= 0.386 square mile (mi^2)
1 hectare (ha)	= 2.471 acres (ac)
1 square meter (m^2)	= 1.196 square yards (yd^2) = 10.764 square feet (ft^2)
1 square centimeter (cm^2)	= 0.155 square inch (in^2)
1 square mile (mi^2)	= 2.59 square kilometers (km^2)
1 acre (ac)	= 0.4047 hectare (ha)
1 square yard (yd^2)	= 0.836 square meter (m^2)
1 square foot (ft^2)	= 0.0929 square meter (m^2)
1 square inch (in^2)	= 6.4516 square centimeter (cm^2)

Volume

Metric Measure

1 cubic meter (m^3)	= 1,000,000 cubic centimeters (cm^3)
1 liter (l)	= 1000 milliliters (ml)
	= 0.001 cubic meter (m^3)
1 centiliter (cl)	= 10 milliliters (ml)
1 milliliter (ml)	= 1 cubic centimeter (cm^3)

Nonmetric Measure

1 cubic yard (yd^3)	= 27 cubic feet (ft^3)
1 cubic foot (ft^3)	= 1728 cubic inches (in^3)
1 barrel (oil) (bbl)	= 42 gallons (U.S.) (gal)

Conversions

1 cubic kilometer (km^3)	= 0.24 cubic miles (mi^3)
1 cubic meter (m^3)	= 264.2 gallons (U.S.) (gal)
	= 35.314 cubic feet (ft^3)
1 liter (l)	= 1.057 quarts (U.S.) (qt)
	= 33.815 ounces (U.S. fluid) (fl. oz)
1 cubic centimeter (cl^3)	= 0.0610 cubic inch (in^3)
1 cubic mile (mi^3)	= 4.168 cubic kilometers (km^3)
1 acre-foot (ac-ft)	= 1,233.46 cubic meters (m^3)
1 cubic yard (yd^3)	= 0.7646 cubic meter (m^3)
1 cubic foot (ft^3)	= 0.0283 cubic meter (m^3)
1 cubic inch (in^3)	= 16.39 cubic centimeters (cm^3)
1 gallon (gal)	= 3.784 liters (l)

Mass

Metric Measure

1000 kilogram (kg)	= 1 metric ton (also called a tonne) (m.t)
1 kilogram (kg)	= 1000 grams (g)

Nonmetric Measure

1 short ton (sh.t)	= 2000 pounds (lb)
1 long ton (l.t)	= 2240 pounds (lb)
1 pound (avoirdupois) (lb)	= 16 ounces (avoirdupois) (oz) = 7000 grains (gr)
1 ounce (avoirdupois) (oz)	= 437.5 grains (gr)
1 pound (Troy) (Tr. lb)	= 12 ounces (Troy) (Tr. oz)
1 ounce (Troy) (Tr. oz)	= 20 pennyweight (dwt)

Conversions

1 metric ton (m.t)	= 2205 pounds (avoirdupois) (lb)
1 kilogram (kg)	= 2,205 pounds (avoirdupois) (lb)
1 gram (g)	= 0.03527 ounce (avoirdupois) (oz) = 0.03215 ounce (Troy) (Tr. oz) = 15,432 grains (gr)
1 pound (lb)	= 0.4536 kilogram (kg)
1 ounce (avoirdupois) (oz)	= 28.35 grams (g)
1 ounce (avoirdupois) (oz)	= 1.097 ounces (Troy) (Tr. oz)

Pressure

1 pascal (Pa)	= 1 newton[a]/square meter (n/m^2)
1 kilogram/square centimeter (kg/cm^2)	= 0.96784 atmosphere (atm)
	= 14.2233 pounds/square inch (lb/in.^{2}) = 0.98067 bar
1 bar	= 0.98692 atmosphere (atm)
	= 10^{10} pascals (Pa) = 1.02 kilograms/square centimeter (kg/cm^2)

Energy and Power

Energy

1 erg = the work done by a force of 1 dyne when its point of application moves through a distance of 1 cm in the direction of the force.

1 calorie (cal) = the amount of heat that will raise the temperature of 1 g of water 1° C with the water at 4° C.

1 erg	= 9.48 × 10^{-11} British thermal unit (Btu)
1 erg	= 7.367 × 10^{-8} foot-pounds (ft-lb)
1 erg	= 2.778 × 10^{-14} kilowatt-hours (kWh)
1 kilowatt-hour (kWh)	= 3413 Btu = 3.6 × 10^{13} ergs
1 Btu	= 2.928 × 10^{-4} kilowatt-hours (kWh)
	= 1.0548 × 10^{10} ergs
1 joule (J)	= 10^7 ergs
1 calorie (cal)	= 3.9685 × 10^{-3} Btu
	= 4.186 × 10^7 ergs

Power; (energy per unit time).

1 watt (w)	= 3.4129 Btu/h
1 watt (w)	= 1.341 × 10^{-3} horsepower (hp)
1 watt (w)	= 1 joule per second (J/s)
1 watt (w)	= 14.34 calories per minute (cal/min)

Temperature

To change from Fahrenheit (F) to Celsius (C)

$$°C = \frac{(°F - 32°)}{1.8}$$

To change from Celsius (C) to Fahrenheit (F)

$$°F = (°C \times 1.8) + 32°$$

[a] a newton is the unit of force that gives a mass of 1 kg an acceleration of 1 m/s/s.

Glossary

Some definitions are not included in the glossary; *mineral names* can be found in Appendix B, Tables B.1, B.2, and B.3; *rock names* are defined in Appendix C, Tables C.1, C.2, and C.3, and Figure 4.11; *units of measurement* are defined in Appendix E.

A good professional reference for definitions of geological terms is the *Glossary of Geology*, second edition, 1980, edited by R. L. Bates and J. A. Jackson, and published by the American Geological Institute, Falls Church, Virginia.

Ablation. The loss of mass to a glacier.

Ablation area. A region of net loss on a glacier characterized by a surface of bare ice and old snow from which the last winter's snowcover has melted away.

Abyssal plain. A large flat area of deep-sea floor having slopes less than about 1 m/km.

Accreted terrane margins. Continental margins formed by the addition of allochthonous terranes.

Accretion. The process by which solid bodies gather together to form a planet or a continent.

Accumulation. The addition of mass to a glacier.

Accumulation area. An upper zone on a glacier, covered by remnants of the previous winter's snowfall and representing an area of net gain in mass.

Active layer. A thin surface layer of permafrost that thaws in summer and refreezes in winter.

Aggradation. Depositional upbuilding, as by a stream.

Allochthonous terranes. Blocks of crust moved laterally by strike-slip faulting or by a combination of strike-slip faulting and subduction, then accreted to a larger mass of continental crust.

Alluvial fan. A fan-shaped body of alluvium typically built where a stream leaves a steep mountain valley.

Alluvium. Sediment deposited by streams in nonmarine environments.

Alpha particle (α-particle). An atomic particle expelled from an atomic nucleus during certain radioactive transformations, equivalent to an $^4_2$He nucleus stripped of its electrons.

Altitude (also called *elevation*). The vertical distance above mean sea level.

Amplitude (of a wave). The distance from the bottom of a wave trough to the top of a wave crest.

Amygdule. A vesicle filled by secondary minerals deposited by groundwater.

Angle of repose. The steepest angle, measured from the horizontal, at which rock debris remains stable.

Angular unconformity. An unconformity marked by angular discordance between older and younger rocks.

Anion. A negative ion.

Annular pattern. Stream pattern in which streams follow nearly circular or concentric paths along belts of weak rock.

Antecedent stream. A stream that has maintained its course across an area of the crust that was raised across its path by folding or faulting.

Anthracite. A metamorphic rock derived from coal by heat and pressure.

Anticline. An upfold in the form of an arch.

Aphanites. Igneous rocks in which the component grains cannot be readily distinguished with the naked eye or even with the aid of a simple hand lens.

Apparent polar wandering. The apparent motions of the magnetic poles derived from measurements of pole directions using paleomagnetism.

Aquifer. A body of permeable rock or regolith saturated with water and through which groundwater moves.

Arête. A jagged, knife-edged ridge created where glaciers have eroded back into a ridge.

Artesian spring. Natural springs that draw their supply of water from a confined aquifer.

Artesian well. A well in which water rises above the aquifer.

Asphalt. See *tar.*

Asteroids. Irregularly shaped rocky bodies that have orbits lying between the orbits of Mars and Jupiter.

Asthenosphere. The region of the mantle where rocks become plastic, like toffee or tar, and are easily deformed. It lies at a depth of 100 to 350 km below the surface.

Atmosphere. The air sphere, consisting of the mixture of gases that together we call air.

Atoll. A coral reef, often roughly circular in plan, that encloses a shallow lagoon.

Atom. The smallest individual particle that retains all the properties of a given chemical element.

Atomic number. The number of protons in the nucleus of an atom.

Atomic substitution. See *ionic substitution.*

Authigenic sediment. Sediment formed in place.

Axial plane. An imaginary plane that divides a fold as symmetrically as possible, and that passes through the axis.

Axis (of a fold). The median line between the limbs, along the crest of an anticline or the trough of a syncline.

Back-arc basin (also called *inner-arc basin*). An arc-shaped basin formed behind a magmatic arc, by crustal thinning, in an overriding plate of lithosphere at a subduction zone.

Backshore. A zone extending inland from a berm to the farthest point reached by waves.

Bajada. A broad alluvial apron composed of coalescing adjacent fans.

Bar. An accumulation of alluvium formed where a decrease in stream velocity causes deposition.

Barchan dune. A crescent-shaped sand dune with horns pointing downwind.

Barrier island. A long island built of sand, lying offshore and parallel to the coast.

Barrier reef. A reef separated from the land by a lagoon.

Base level. The limiting level below which a stream cannot erode the land.

Batholith. A very large, discordant, intrusive igneous body of irregular shape.

Bay barrier. A ridge of sand or gravel that completely blocks the mouth of a bay.

Beach. Wave-washed sediment along a coast, extending throughout the surf zone.

Beach dune. A hummocky or elongate mound of sand bordering a beach.

Bed load. Coarse particles that move along a stream bed.

Bedrock. The continuous mass of solid rock that makes up the crust.

Benioff zone. A narrow, well-defined zone of deep-earthquake foci beneath a seafloor trench.

Berm. A nearly horizontal or landward-sloping bench formed of sediment deposited by waves.

Beta particle (β-particle). An electron expelled from an atomic nucleus during certain radioactive transformations.

Biosphere. The totality of the Earth's organisms and, in addition, organic matter that has not yet been completely decomposed.

Blowout. A small shallow deflation basin excavated in loose sand.

Body waves. Seismic waves that travel outward from an earthquake focus and pass through the Earth.

Bolide. An impacting body; it can be a meteorite, an asteroid, or a comet.

Bond (chemical). The electrostatic force that holds atoms together to form compounds by the sharing and transfer of electrons.

Bornhardts. A special type of inselberg having rounded or domelike form.

Bottomset layer. A gently sloping, fine, thin part of each layer in a delta.

Boudinage. A style of deformation in layered rocks in which the brittle layers break into elongate blocks while the enclosing ductile layers stretch and flow.

Bouguer correction. A correction applied to measurements of the Earth's gravitational pull to account for the mass of material between the gravimeter and the reference surface used for free-air corrections.

Boulder train. A group of erratics spread out fanwise.

Bowen's reaction series. A schematic description of the order in which different minerals crystallize during the cooling and progressive crystallization of a magma. See *continuous* and *discontinuous reaction series.*

Branching decay. A radioactive parent decays to produce either of two different daughter products.

Brittle fracture. Rupture of a solid body that is stressed beyond its elastic limit.

Burial metamorphism. Metamorphism caused solely by the burial of sedimentary or pyroclastic rocks.

Caldera. A roughly circular, steep-walled volcanic basin several kilometers or more in diameter.

Caliche. A solid, almost impervious layer of whitish calcium carbonate in a soil profile.

Calorie. The amount of heat energy needed to raise the temperature of one gram of water by one degree Celsius.

Calving. The progressive breaking off of icebergs from a glacier that terminates in deep water.

Capacity. The potential load a stream can carry.

Carbonate compensation depth. The level in the sea below which the rate of dissolution of calcium carbonate exceeds the rate of its deposition.

Cataclastic metamorphism. Metamorphism caused solely by mechanical effects such as crushing, flattening, and elongation.

Cation. A positive ion.

Cave. A natural underground opening, generally connected to the surface and large enough for a person to enter.

Cavern. A large cave or system of interconnected cave chambers.

Cenote. Name for a sinkhole on the Yucatan Peninsula, Mexico.

Centripetal pattern. Stream pattern in which streams converge toward a central depression.

Chain decay. A radioactive parent decays to form a radioactive daughter, which in turn forms another radioactive daughter, and so forth until a stable daughter product is reached.

Chemical elements. The most fundamental substances into which matter can be separated by chemical means.

Chemical remanent magnetism. Remanent magnetism acquired through chemical precipitation and growth of magnetic minerals in a sediment.

Chemical sediment. Sediment formed by precipitation of minerals from solution in water.

Chemical weathering. The decomposition of rocks.

Chrons. See *magnetic epochs.*

Cirque. A bowl-shaped hollow on a mountainside open downstream and bounded upstream by a steep slope (headwall) and excavated mainly by frost wedging and by glacial abrasion and plucking.

Cirque glacier. A glacier that occupies a bowl-shaped hollow on the side of a mountain.

Clastic sediment. See *detritus.*

Cleavage. The tendency of a mineral to break in preferred directions along bright, reflective plane surfaces.

Climate. The average weather conditions of a place or area over a period of years.

Clinometer. A device for measuring dip.

Col. A gap or pass in a mountain crest where the headwalls of two cirques intersect.

Colluvium. Loose, incoherent deposits on or at the base of slopes and moving mainly by creep.

Column. A stalactite joined with a stalagmite, forming a connection between the floor and roof of a cave.

Columnar joints. Joints that split igneous rocks into long prisms or columns.

Competence. The size of particles a stream can transport under a given set of hydrologic conditions.

Composite volcanoes. See *stratovolcanoes.*

Compound. A combination of atoms of different elements bonded together.

Compressional waves. See *P waves.*

Conchoidal fracture. Breakage resulting in smooth, curved surfaces.

Concretion. A hard, localized body having distinct boundaries enclosed in sedimentary rock, and consisting of a substance precipitated from solution, commonly around a nucleus.

Conduction. The means by which heat is transmitted through solids.

Cone of depression. A conical depression in the water table immediately surrounding a well.

Consequent stream. A stream whose pattern is determined solely by the direction of slope of the land.

Contact metamorphism. (Also called *thermal metamorphism*). Metamorphism adjacent to an intrusive igneous rock.

Continental crust. The crust that comprises the continents.

Continental drift. The slow, lateral movements of continents across the surface of the Earth.

Continental rise. A region of gently changing slope where the floor of the ocean basin meets the margin of the continent.

Continental shelf. A submerged platform of variable width that forms a fringe around a continent.

Continental shield. The portion of a craton where rocks are exposed at the surface.

Continental slope. A pronounced slope beyond the seaward margin of the continental shelf.

Continuous reaction series. The continuous change of mineral composition, through solid solution, as a magma crystallizes. See *discontinuous reaction series.*

Contour. See *contour line.*

Contour currents. Bottom currents in the ocean that flow horizontally, following bathymetric contours.

Contour interval. The vertical distance between two successive contour lines.

Contour line (= contour). A line passing through points having the same altitude.

Contourites. Sediments deposited by contour currents.

Convection. The process in liquids and gases by which hot, less dense materials rise upward, being replaced by cold, downward flowing fluids to create a convection current.

Convergent margin. See *edge of consumption.*

Coral reef. A ridge of limestone built by colonial marine organisms.

Core. The spherical mass, largely of metallic iron, with admixtures of nickel, sulfur, silicon, and other elements, at the center of the Earth.

Coriolis effect. An effect that causes any body that moves freely with respect to the rotating solid Earth to veer toward the right in the Northern Hemisphere and toward the left in the Southern Hemisphere, regardless of the direction in which the body may be moving.

Correlation. Determination of equivalence, in geologic age and position, of the succession of strata found in two or more different areas.

Covalent bonding. The force that arises when two atoms share one or more electrons.

Crater. A funnel-shaped depression at the top of a volcano from which gases, fragments of rock, and lava are ejected.

Craton. A portion of the Earth's crust that has attained tectonic stability and has been little deformed for a prolonged period.

Creep. The imperceptibly slow downslope movement of regolith.

Crevass. A deep, gaping fissure in the upper surface of a glacier.

Cross section. See *geologic cross section.*

Cross-strata. Strata that are inclined with respect to a thicker stratum within which they occur.

Crust. The outermost and thinnest of the Earth's layers, which consists of rocky matter that is less dense than the rocks of the mantle below.

Cryosphere. The portion of the hydrosphere that is ice, snow, and frozen ground.

Crystal. A solid compound composed of ordered, three-dimensional arrays of atoms or ions chemically bonded together and displaying crystal form.

Crystal form. A geometric solid that is bounded by symmetrically arranged plane surfaces.

Crystal structure. The geometric pattern that atoms assume in a solid.

Curie point. A temperature above which all permanent magnetism is destroyed.

Cycle. A sequence of recurring events.

Cycle of erosion. A theory of landscape evolution in which a landmass is uplifted and is gradually worn down by erosion to a surface of low relief.

Cyclothem. A repeated sequence of peat and clastic sediments.

Daughter product (= daughter). The product arising from radioactive decay. Compare *parent.*

Debris flow. The downslope movement of a mass of unconsolidated regolith more than one-half of which is coarser than sand.

Décollement. A body of rock above the detachment surface of a thrust fault.

Deep-sea fans. Huge fan-shaped bodies of sediment at the base of the continental slope that spread downward and outward to the deep-sea floor.

Deflation. The picking up and removal of loose rock particles by wind.

Deflation armor. See *desert pavement.*

Degradation. Downcutting, as by a stream.

Delta. A body of sediment deposited by a stream where it flows into standing water.

Dendritic pattern. Stream pattern consisting of irregular branching of channels ("treelike") in many directions.

Density. The average mass per unit volume.

Density current. A localized current, within a body of water, caused by dense water sinking through less-dense water.

Denudation. The sum of the weathering, mass-wasting, and erosional processes that result in the progressive lowering of the Earth's surface.

Depositional remanent magnetism. Remanent magnetism acquired through processes of sedimentation.

Depth of compensation. That depth in the Earth above which segments of crust and upper mantle act as blocks that rise or sink depending on the mass and density of the individual blocks.

Deranged pattern. Stream pattern in which streams show a complete lack of adjustment to underlying structural or lithologic control.

Desert. Arid land, whether "deserted" or not, in which annual rainfall is less than 250 mm (10 in.) or in which the evaporation rate exceeds the precipitation rate.

Desertification. The invasion of desert into nondesert areas.

Desert pavement (deflation armor). A surface layer of coarse particles concentrated chiefly by deflation.

Desert varnish. A thin, dark, shiny coating, consisting mainly of manganese and iron oxides, formed on the surfaces of stones and rock outcrops in desert regions after long exposure.

Detachment surface. The surface along which a large-scale thrust fault moves.

Detritus. (Also called *clastic sediment*). The accumulated particles of broken rock and skeletal remains of dead organisms.

Diachronous boundaries. Boundaries that vary in age in different areas.

Diagenesis. Changes that affect sediment after its initial deposition and during and after its slow transformation into sedimentary rock.

Diamictite. Sedimentary rock formed by lithification of diamicton.

Diamicton. Nonsorted sediment, regardless of origin.

Diastem. A short break in sedimentation resulting from normal variations about an average condition, but without a major change in the regular sedimentary pattern.

Differential weathering. Weathering that occurs at different rates or intensity as a result of variations in the composition and structure of rocks.

Dikes. Tabular sheets of intrusive igneous rock cutting across the layering of the intruded rock.

Dip. The angle in degrees between a horizontal plane and an inclined plane, measured down from horizontal in a plane perpendicular to the strike.

Dip-slip fault. A normal or reverse fault on which the only component of movement lies in a plane normal to the strike of the fault surface.

Discharge. The quantity of water that passes a given point in a stream channel per unit time.

Discharge area. Area where subsurface water is discharged to streams or to bodies of surface water.

Disconformity. Parallel layered strata separated by an irregular surface of erosion having appreciable relief.

Discontinuous reaction series. The discontinuous sequence of reactions by which early-formed minerals in a crystallizing magma react with residual liquid to form new minerals. See *continuous reaction series.*

Dissolution. The chemical weathering process whereby minerals and rock material pass directly into solution.

Dissolved load. Matter dissolved in stream water.

Divergent margin. See *spreading edge.*

Divide. The line that separates adjacent drainage basins.

Doline. See *sink.*

Dowsing. The practice of locating groundwater by means of a divining rod.

Drainage basin. The total area that contributes water to a stream.

Drainage density. The ratio of the total length of all stream segments in a drainage basin to the area of the basin.

Dripstone. A deposit chemically precipitated from water in an air-filled cavity.

Drumlin. A streamlined hill consisting of glacially deposited sediment and elongated parallel with the direction of ice flow.

Ductile deformation. The irreversible deformation induced in a solid that is stressed beyond its yield point, but before rupture occurs.

Dune. A mound or ridge of sand deposited by wind.

Dynamic equilibrium (in landscape evolution). A condition in which a landscape form oscillates about a mean value which is itself declining with time.

Earthflow. A downslope movement, transitional between debris flows and mudflows, in which the debris contains 80 percent or more of sand and finer particles, and has a higher water content than debris flows.

Earthquake focus. The point of the first release of energy that causes an earthquake.

Earth's gravity. An inward-acting force with which the Earth tends to pull all objects toward its center.

Economy of groundwater system. A measure of the input of water to and outflow of water from the entire system.

Edge of consumption. The same as a convergent margin, a subduction zone, or an edge of subduction. The edge along which a plate of lithosphere turns down into the mantle. Coincident with trenches in the sea floor.

Edge of subduction. See *edge of consumption.*

Elastic deformation. The reversible or nonpermanent deformation that occurs when an elastic solid is stretched and squeezed and the force is then removed.

Electrons. Negatively charged atomic particles.

Elevation. See *altitude.*

End moraine. A ridgelike accumulation of drift deposited along the margin of a glacier.

Energy. The capacity to produce activity.

Epicenter. That point on the Earth's surface that lies vertically above the focus of an earthquake.

Equilibrium line. A line that marks the level on a glacier where net mass loss equals net gain.

Erg. See *sand sea.*

Erosion. The complex group of related processes by which rock is broken down physically and chemically and the products moved.

Erratic. A glacially deposited rock fragment whose composition differs from that of the bedrock beneath it.

Esker. A long narrow ridge, often sinuous, composed of stratified drift.

Estuary. A semi-enclosed body of coastal water within which seawater is diluted with fresh water.

Evaporite. Sedimentary rock composed chiefly of minerals precipitated from a saline solution through evaporation.

Evaporite deposits. Layers of salts that precipitate as a consequence of evaporation.

Exfoliation. The spalling off of successive shells, like the ''skins'' of an onion, around a solid rock core.

Exploration geology. The special branch of geology concerned with discovering new supplies of usable minerals.

Exposure (also called an *outcrop*). A place where solid rock is exposed at the Earth's surface.

External processes. All the activities involved in erosion, and also in the transport and deposition of the eroded materials.

Extraterrestrial material. Material originating outside the Earth.

Extraterrestrial sediment. Microscopic meteoritic particles that are mixed randomly in other types of deep-sea sediment.

Extrusive igneous rock. Rock formed by the solidification of magma poured out onto the Earth's surface.

Facies. A distinctive group of characteristics, within a rock unit, that differs as a group from those elsewhere in the same unit.

Fan. See *alluvial fan.*

Fault. A fracture along which the opposite sides have been displaced relative to each other.

Fault breccia. Crushed and broken rock adjacent to a fault.

Fault drag. Structures created by bending adjacent to faults.

Faunal succession (law of). Fossil faunas and floras succeed one another in a definite, recognizable order.

Fiord. See *fjord.*

Firn. Snow that survives a year or more of ablation and achieves a density that is transitional between snow and glacier ice.

Fission tracks. Regions of a crystal damaged by passage of an α-particle released by radioactive decay.

Fissure eruption. Extrusion of volcanic rock or pyroclasts and associated gases along an extended fracture.

Fjord. A deep glacially carved valley submerged by the sea. Also spelled *fiord.*

Flocculation. A process whereby many minute suspended particles form clotlike accumulations that become massive enough to settle.

Flood. A discharge great enough to cause a stream to overflow its banks.

Floodplain. The part of any stream valley that is inundated during floods.

Flowstone. A deposit chemically precipitated from flowing water in the open air or in an air-filled cavity.

Foliation. A plane defined by any planar set of minerals, or bonding of minerals, found in a metamorphic rock.

Footwall block. The surface of the block of rock below an inclined fault.

Foreset layer. The coarse, thick, steeply sloping part of each layer in a delta.

Foreshore. A zone extending from the level of lowest tide to the average high-tide level.

Formation. A stratum or collection of strata distinctive enough on the basis of physical properties to constitute a basic unit for geologic mapping.

Fossil. The naturally preserved remains or traces of an animal or a plant.

Free-air correction. A correction applied to measurements of the Earth's gravitational pull in order to remove effects caused by differences in elevation.

Frequency (of a wave). The number of wave crests that pass a given point each second.

Fringing reef. A coral reef attached to or bordering the adjacent land.

Frost heaving. The lifting of regolith by freezing of contained water.

Frost wedging. The formation of ice in a confined opening within rock, thereby causing the rock to be forced apart.

Fumarole. A volcanic vent that emits only gases.

Gamma rays (γ-rays). Very short wavelength electromagnetic radiation given off by an atomic nucleus during certain radioactive transformations.

Gangue. The nonvaluable minerals of an ore.

Gauging station. Place on a stream where systematic measurements are made of the discharge.

Gelifluction. The slow downslope movement of saturated sediment, associated with frost action, in cold-climate regions.

Geochemically abundant elements. Those chemical elements that individually comprise 0.1 percent or more, by weight, of the crust.

Geochemically scarce elements. Those chemical elements that individually comprise less than 0.1 percent by weight of the crust.

Geologic column. A composite diagram combining in chronological order the succession of known strata, fitted together on the basis of their fossils or other evidence of relative or actual age.

Geologic cross section. A diagram showing the arrangement of rocks in a vertical plane.

Geologic map. A map that shows the distribution, at the surface, of rocks of various kinds or of various ages.

Geologic time scale. A sequential arrangement of geologic time units, as currently understood.

Geology. The science of the Earth.

Geosyncline. A great trough that has received thick deposits of sediment during its slow subsidence through long geologic periods.

Geothermal gradient. The rate of increase of temperature downward in the Earth.

Geyser. A hot spring equipped with a system of plumbing and heating that causes intermittent eruptions of water and steam.

Glacial drift. Sediment deposited directly by glaciers or indirectly by meltwater in streams, in lakes, and in the sea.

Glacial marine drift (also referred to as *glacial marine sediment*). Terrigenous sediment dropped onto the sea floor from floating ice shelves or from ice bergs.

Glacial striations. Long subparallel scratches inscribed

on a rock surface by rock debris embedded in the base of a glacier.

Glaciation. The modification of the land surface by the action of glacier ice.

Glacier. A body of ice, consisting largely of recrystallized snow, that shows evidence of downslope or outward movement due to the stress of its own weight.

Gneiss. A coarse-grained, foliated metamorphic rock, always with marked layering but with imperfect cleavage.

Gondwanaland. The southern half of Pangaea, consisting of present-day Australia, India, Madagascar, Africa, and South America.

Graben. A trenchlike structure bounded by parallel normal faults. See *half-graben.*

Graded layer. A layer in which the particles grade upward from coarse to finer.

Graded stream. A stream in which the slope has become so adjusted, under conditions of available discharge and prevailing channel characteristics, that the stream is just able to transport the sediment load available to it.

Gradient. A measure of the vertical drop over a given horizontal distance.

Granite-greenstone terrane. A distinctive grouping of metamorphosed pillow basalts, known as greenstones, and small granitic plutons, found in many cratons.

Granofels. A coarse-grained metamorphic rock with the composition of a granite, but without mineral layering.

Granular texture. The interlocking arrangement of visible mineral grains characteristic of granite.

Gravimeter (also called a *gravity meter*). A sensitive device for measuring the pull of gravity at any locality.

Gravity meter. See *gravimeter.*

Groin. A low wall, built on a beach, that crosses the shoreline at a right angle.

Ground moraine. Widespread drift with a relatively smooth surface topography consisting of gently undulating knolls and shallow closed depressions.

Groundwater. All the water contained in spaces within bedrock and regolith.

Guyot. A seamount with a more-or-less flat top well below sea level.

Half-graben. A trenchlike structure formed when the hanging-wall block moves downward on a curved fault surface. See *graben.*

Half-life. The time required to reduce the number of parent atoms by one-half.

Hanging wall. The surface of the block of rock above an inclined fault.

Hardness. Relative resistance of a mineral to scratching.

Hiatus. The lapse in time recorded by an unconformity.

High grade of metamorphism. Metamorphism under conditions of high temperature and high pressure.

High-grade terranes. Geologic provinces dominated by huge areas of coarse-grained, granitic rocks of both igneous and metamorphic parentage.

Hinge fault. A fault on which displacement dies out perceptibly along strike and ends at a definite point.

Horn. A sharp-pointed peak bounded by the intersecting walls of three or more cirques.

Horst. An elevated elongate block of crust bounded by parallel normal faults.

Humus. The decomposed residue of plant and animal tissues.

Hydration. The absorption of water into a crystal structure.

Hydraulic gradient. The slope of the water table.

Hydrologic cycle. The cyclic movement of water through evaporation, wind transport, stream flow, percolation, and related processes.

Hydrolysis. A chemical reaction in which the H^+ or OH^- ions of water replace ions of a mineral.

Hydrosphere. The "water sphere," embracing all the world's oceans, lakes, streams, water underground and all the snow and ice, including glaciers.

Hydrothermal mineral deposit. Any local concentration of minerals formed by deposition from a hydrothermal solution.

Hydrothermal solutions. Hot brines either given off by cooling magmas, or produced by relations between hot rock and circulating water, that concentrate ore minerals.

Hypsithermal interval. A period of warm postglacial climate that reached its peak in many areas between about 8000 and 6000 years ago.

Iapetus. The name given to the ocean that disappeared when North America and Europe collided during the Paleozoic Era.

Ice cap. A dome-shaped body of ice and snow that covers a mountain highland, or lower-lying land at high latitude, and that displays generally radial outward flow.

Ice-contact stratified drift. Stratified sediment deposited in contact with supporting glacier ice.

Ice field. A broad, nearly level area of glacier ice in a mountainous region consisting of many interconnected mountain glaciers.

Ice sheet. A continent-sized mass of ice thick enough to flow under its own weight and that overwhelms nearly all the land surface within its margin.

Ice shelf. Thick glacier ice that floats on the sea and commonly is located in coastal embayments.

Ice-wedge polygon. A large polygonal feature that forms by contraction and cracking of frozen ground.

Igneous rock. Rock formed by the cooling and consolidation of magma.

Ignimbrite. See *welded tuff.*

Impact cratering. The process by which a planetary surface is deformed as a result of a transfer of energy from a bolide to the planetary surface.

Index fossil. A fossil that can be used to identify and date the strata in which it is found, and is useful for local correlation of rock units.

Inertia. The resistance a large mass has to sudden movement.

Inner-arc basin. See *back-arc basin.*

Inselbergs. Steep-sided mountains, ridges, or isolated hills rising abruptly from adjoining monotonously flat plains.

Internal processes. All activities involved in movement or chemical and physical change of rocks in the Earth's interior.

Intrusive igneous rock. Any igneous rock formed by solidification of magma below the Earth's surface.

Ion. An atom that has excess positive or negative charges caused by electron transfers.

Ionic bond. The electrostatic attraction between negatively and positively charged ions.

Ionic radius. The distance from the center of the nucleus to the outermost orbiting electrons.

Ionic substitution. (Also called **solid solution** and **atomic substitution**). The substitution of one atom for another in a random fashion throughout a crystal structure.

Island arc. An arcuate chain of stratovolcanoes parallel to a sea-floor trench and separated from it by a distance of 150 to 300 km.

Isograd. A line on a map connecting points of first occurrence of a given mineral in metamorphic rocks.

Isopach. A line on a map that connects points of equal thickness of a rock unit.

Isopleth. A line on a map that connects points having equal values of some variable such as particle size in a sediment.

Isostasy. The ideal property of flotational balance among segments of the lithosphere.

Isotopes. Atoms having the same atomic number but differing numbers of neutrons.

Joint. A fracture on which movement has not occurred in a direction parallel to the plane of the fracture.

Joint set. A widespread group of parallel joints.

Joule. A unit of energy equal to the work done when a force of one newton is displaced a distance of one meter.

Jovian planets. Giant planets in the outer regions of the Solar System that are characterized by great masses, low densities, and thick atmospheres consisting primarily of hydrogen and helium.

Kame. A short, steep-sided knoll of stratified drift.

Kame terrace. A terrace of ice-contact stratified drift along a valley side.

Karst topography. An assemblage of topographic forms resulting from dissolution of carbonate bedrock and consisting primarily of closely spaced sinks.

Kettle. A basin within a body of drift created by melting out of a mass of underlying ice.

Kettle-and-kame topography. An extremely uneven terrain resulting from wastage of debris-mantled stagnant ice and underlain by ice-contact stratified drift.

Key bed. A thin and generally widespread bed with characteristics so distinctive that it can be easily recognized.

Kimberlite pipes. Narrow pipelike masses of igneous rocks, sometimes containing diamonds, that intrude the crust but originate deep in the mantle.

Kinetic energy. Energy that results from motion of an object.

Laccolith. A concordant, lenticular, intrusive igneous body along which the layers of the invaded country rock have been bent upward to form a dome.

Lagoon. A bay inshore from an enclosing reef or island paralleling the coast.

Lahar. A mudflow that consists chiefly of volcanic debris and originates on the flank of a volcano.

Landslide. A general term covering a variety of mass-movement processes.

Lapse rate. The rate of decrease in temperature with altitude in the atmosphere.

Lateral moraine. An end moraine built along the side of a valley glacier.

Laterite. A hardened soil horizon characterized by extreme weathering that has led to concentration of secondary oxides of iron and aluminum.

Latitude. Part of a grid used for describing positions on the Earth's surface, consisting of parallel circles concentric to the poles. The circles are called *parallels* of latitude.

Laurasia. The northern half of Pangaea, consisting of present-day Asia, Europe, and North America.

Lava. Magma that reaches the Earth's surface through a volcanic vent.

Lava dome. See *plug dome.*

Law of faunal succession. See *faunal succession.*

Law of original horizontality. See *original horizontality.*

Leaching. The continued removal, by water solutions, of soluble matter from bedrock or regolith.

Left-lateral fault. A strike-slip fault in which relative motion is such that to an observer looking directly at the fault, the motion of the block on the opposite side of the fault is to the left. A right-lateral fault has right-handed movement.

Levee. See *natural levees.*

Limbs. The sides of a fold.

Lineation. A parallel arrangement of elongate mineral grains.

Lithification. The process that converts a sediment into a sedimentary rock.

Lithology. The systematic description of rocks in terms of mineral assemblage and texture.

Lithosphere. The outer 100 km of the solid Earth, where rocks are harder and more rigid than those in the plastic asthenosphere.

Little Ice Age. The interval of generally cool climate between the middle thirteenth and middle nineteenth centuries, during which mountain glaciers expanded worldwide.

Loess. Wind-deposited silt, commonly accompanied by some clay and fine sand.

Long profile. A line drawn along the surface of a stream from its source to its mouth.

Longitude. Part of a grid used for describing positions on the Earth's surface, consisting of half circles joining the poles. The half circles are called *meridians.*

Longitudinal dune. A long, straight, ridge-shaped sand dune parallel with wind direction.

Longshore current. A current, within the surf zone, that flows parallel to the coast.

Love waves. Seismic surface waves produced by torsional oscillations of the Earth.

Low grade of metamorphism. Metamorphism under conditions of low temperature and low pressure.

Low-grade terranes. Geologic provinces dominated by thin, highly deformed belts of lavas and sedimentary rocks that have been subjected to low grades of metamorphism.

Luster. The quality and intensity of light reflected from a mineral.

Magma. Molten rock, together with any suspended crystals and dissolved gases, that forms when temperatures rise and melting occurs in the mantle or crust.

Magmatic differentiation by fractional crystallization. Compositional changes that occur in magmas by the separation of early formed minerals from residual liquids.

Magmatic differentiation by partial melting. The process of forming magmas with differing compositions by the incomplete melting of rocks.

Magnetic declination. The clockwise angle from true north assumed by a magnetic needle.

Magnetic epochs. (Also called *chrons*). Periods of predominantly normal polarity (as at present), or predominantly reversed polarity.

Magnetic events. Short-term magnetic reversals.

Magnetic field. Magnetic lines of force surrounding the Earth.

Magnetic inclination. The angle with the horizontal assumed by a freely swinging bar magnet.

Mantle. The thick shell of dense, rocky matter that surrounds the core.

Mascon. A region with an anomalously high gravity pull.

Mass balance (of a glacier). A measure of the change in total mass of a glacier during a year.

Mass number. The sum of the protons and neutrons in the nucleus of an atom.

Mass-wasting. The movement of regolith downslope by gravity without the aid of a transporting medium.

M-discontinuity. See *Mohorovičić discontinuity.*

Meander. A looplike bend of a stream channel.

Mechanical weathering. Disintegration of rocks by mechanical processes, such as frost-wedging.

Megascopic. Those features of rocks that can be perceived by the unaided eye, or by the eye assisted by a simple lens that magnifies up to 10 times.

Mélange. A chaotic mixture of broken and jumbled rock above a subduction zone.

Mercalli Scale. See *modified Mercalli Scale.*

Mercator projection. A geometric projection by which points on a sphere are projected radially onto a cylinder tangent to the sphere.

Meridians. See *longitude.*

Mesosphere. The region between the base of the asthenosphere and the core/mantle boundary.

Metallic bond. A variation of covalent bonding in which there are more electrons than are needed to satisfy bond requirements.

Metallogenic provinces. Limited regions of the crust within which mineral deposits occur in unusually large numbers.

Metamorphic aureole. A shell of metamorphic rock, produced by contact metamorphism, surrounding an igneous intrusion.

Metamorphic facies. Contrasting assemblages of minerals that reach equilibrium during metamorphism within a specific range of physical conditions belonging to the same facies.

Metamorphic rock. Rock whose original compounds, or textures, or both, have been transformed to new compounds and new textures by reactions in the solid state as a result of high temperature, high pressure, or both.

Metamorphic zones. The regions on a map between isograds.

Metamorphism. All changes in mineral assemblage and rock texture, or both, that take place in rocks in the solid state within the Earth's crust as a result of changes in temperature and pressure.

Meteorites. Small stony or metallic objects from interplanetary space that impact a planetary surface.

Microscopic. Those features of rocks that require high magnification in order to be viewed.

Mid-ocean ridges. Continuous rocky ridges on the ocean floor, many hundreds to a few thousand kilometers wide with a relief of more than 0.6 km.

Migmatite. A composite rock containing both igneous and metamorphic portions.

Mineral. Any naturally formed, solid, chemical substance having a definite chemical composition and a characteristic crystal structure.

Mineral assemblage. The variety and abundance of minerals present in a rock.

Mineral deposit. Any volume of rock containing an enrichment of one or more minerals.

Mineralogy. The special branch of geology that deals with the classification and properties of minerals.

Modified Mercalli Scale. A scale used to compare earthquakes based on the intensity of damage caused by the quake.

Moho. See *Mohorovičić discontinuity.*

Mohorovičić discontinuity. (Also called *M-discontinuity* and *Moho*). The seismic discontinuity that marks the base of the crust.

Molecule. The smallest unit that retains all the properties of a compound.

Monocline. A one-limbed flexure, on both sides of which the strata either are horizontal or dip uniformly at low angles.

Mountain chain. A large scale, elongate geologic feature consisting of numerous ranges or systems, regardless of similarity in form or equivalence in ages.

Mountain range. An elongate series of mountains forming a single geologic feature.

Mountain system. A group of ranges similar in general form, structure, and alignment, and presumably owing their origin to the same general causes.

Mud cracks. Cracks caused by shrinkage of wet mud as its surface becomes dry.

Mudflow. A flowing mass of predominantly fine-grained rock debris that generally has a high enough water content to make it highly fluid.

Natural levees. Broad, low ridges of fine alluvium built along both sides of a stream channel by water that spreads out of the channel during floods.

Neoglaciation. An interval postdating the time of maximum Hypsithermal warmth when, following a period of ice recession, many glaciers were regenerated and others expanded in size.

Neutron. An electrically neutral particle with a mass 1833 times greater than that of the electron.

Newton. A unit of force defined as that force which gives a mass of one kilogram an acceleration of one meter per second per second.

Nonconformity. Stratified rocks that unconformably overlie igneous or metamorphic rocks.

Normal fault. A fault, generally steeply inclined, along which the hanging-wall block has moved relatively downward.

Nunatak. An area of ice-free land rising above the surface of a glacier.

Original horizontality. Waterlaid sediments are deposited in strata that are horizontal, or nearly horizontal, and parallel or nearly parallel to the Earth's surface.

Outcrop area. The area, on a geologic map, shown as occupied by a particular rock unit.

Outwash. Stratified drift deposited by streams of meltwater.

Outwash plain. A body of outwash that forms a broad plain.

Overland flow. The movement of runoff in broad sheets or groups of small, interconnecting rills.

Oxbow lake. A crescent-shaped shallow lake occupying the abandoned channel of a meandering stream.

Oxic horizon. A soil horizon characterized by extreme chemical alteration of the parent material.

Paleogeography. The physical geography during past geologic times.

Paleomagnetism. Remanent magnetism in ancient rock recording the direction of the magnetic poles at some time in the past.

Paleosol. A soil that formed at the ground surface and subsequently is buried and preserved.

Pangaea. The name given to a supercontinent that formed by collision of all the continental crust during the late Paleozoic.

Parabolic dune. A sand dune of U-shape with the open end of the U facing upwind.

Paraconformity. Parallel beds separated by an unconformity marked only by a bedding-plane surface.

Parallel of latitude. See *latitude.*

Parallel pattern. Stream pattern consisting of parallel or subparallel channels that have formed on sloping surfaces underlain by homogeneous rocks.

Parallel strata. Strata whose individual layers are parallel.

Parent. An atomic nucleus undergoing radioactive decay; compare *daughter product.*

Patterned ground. More-or-less symmetrical patterned forms due to frost action.

Pediment. A sloping surface, cut across bedrock and thinly or discontinuously veneered with alluvium, that slopes away from the base of a highland in an arid or semiarid environment.

Pelagic sediment. Sediment consisting of material of marine organic origin.

Peneplain. "Almost a plain." A land surface worn down to very low relief by streams and mass-wasting.

Perched water body. Water body that occupies a basin in impermeable sediments or rocks, perched in positions higher than the main water table.

Percolation. The movement of groundwater in the saturated zone.

Periglacial zone. Land area beyond the limit of glaciers where low temperature and frost action are important factors in determining landscape characteristics.

Permafrost. Sediment, soil, or even bedrock that remains continually at a temperature below 0°C for an extended time.

Permeability. The capacity for transmitting fluids.

Petroleum. Gaseous, liquid, and semisolid substances, occurring naturally and consisting chiefly of chemical compounds of carbon and hydrogen.

Petrology. The special branch of geology that deals with the occurrence, origin, and history of rocks.

Phanerites. Igneous rocks in which the component mineral grains are distinguishable megascopically.

Phase transition. Atomic repacking caused by changes in pressure and temperature.

Phenocrysts. The isolated large crystals in a porphyry.

Piedmont glacier. A broad glacier that terminates on a piedmont slope beyond confining mountain valleys and is fed by one or more large valley glaciers.

Pillow lava. Discontinuous, pillow-shaped masses of lava, ranging in size from a few centimeters to a meter or more in greatest dimension.

Pingo. A large, generally conical ice-cored mound that

commonly reaches a height of 30 to 50 m and is formed from the freezing of water within permafrost.

Piracy. See *stream capture.*

Placer. A deposit of heavy minerals concentrated mechanically.

Plagioclase. A variety of feldspar with an unbroken range of composition from albite to anorthite.

Plane of the ecliptic. The plane of the Earth's orbit around the Sun.

Planetary nebula. A flattened, rotating disk of gas surrounding a proto-sun.

Planetology. A comparative study of the Earth with the Moon and with the other planets.

Plate tectonics. The special branch of tectonics that deals with the processes by which the lithosphere is moved laterally over the asthenosphere.

Playa. A dry lake bed in a desert basin.

Plug dome. (Also called a *lava dome*). A volcanic dome characterized by an upheaved, consolidated conduit filling of lava.

Plunge (of a fold). The angle between a fold axis and the horizontal.

Plunging fold. A fold with an inclined axis.

Plutons. All bodies of intrusive igneous rock, regardless of shape or size.

Pluvial lake. A lake that existed when, under a former climate, effective precipitation in the surrounding region was greater than today's.

Point bar. A low arcuate ridge of sand or gravel along the inside of the bend of a meander loop.

Polar (cold) glacier. A glacier in which the ice is below the pressure melting point throughout, and the ice is frozen to its bed.

Polarity reversals. Changes of the Earth's magnetic field to the opposite polarity.

Polymerization. The process of linking silicate tetrahedra into large anion groups.

Polymorph. A compound that occurs in more than one crystal structure.

Porosity. The proportion (in percent) of the total volume of a given body of bedrock or regolith that consists of pore spaces.

Porphyry. Any igneous rock consisting of coarse mineral grains scattered through a mixture of fine mineral grains.

Potential energy. Stored energy.

Precession of the equinoxes. A progressive change in the Earth-Sun distance for a given date.

Pressure melting point. The temperature at which ice can melt at a given pressure.

Principle of stratigraphic superposition. See *stratigraphic superposition.*

Principle of Uniformity. The same external and internal processes we recognize in action today have been operating unchanged, though at different rates, throughout most of the Earth's history.

Prograde metamorphic effects. The metamorphic changes that occur while temperatures and pressures are rising.

Proton. A positively charged particle with a mass 1832 times greater than the mass of an electron.

Provenance. The place of origin of a sediment or sedimentary rock.

Proxy records. Information that can act as a substitute for data that may be unobtainable.

P waves. Seismic body waves transmitted by alternating pulses of compression and expansion. P waves pass through solids, liquids, and gases.

Pyroclastic cones. Cones consisting entirely of pyroclastic debris surrounding a volcanic vent.

Pyroclastic rocks. Rocks comprised of fragments of igneous material ejected from a volcano, then sedimented and either cemented or welded to a coherent aggregate.

Pyroclasts. Fragments extruded violently from a volcano.

Radial pattern. Stream pattern consisting of channels that radiate, like the spokes of a wheel, from a topographically high area.

Radioactivity. The decay process by which an unstable atomic nucleus spontaneously transforms to another nucleus.

Radiometric age. The length of time a mineral has contained its built-in clock.

Rayleigh waves. Seismic surface waves produced by spheroidal oscillations of the Earth.

Recharge area. Area where water is added to the saturated zone.

Rectangular pattern. Stream pattern in which channel systems are marked by right-angle bends.

Recurrence interval. The probable interval, in years, between floods of a given magnitude.

Regional metamorphism. Metamorphism affecting large volumes of crust, and involving both mechanical and chemical changes.

Regolith. The blanket of loose, noncemented rock particles that commonly overlies bedrock.

Regression. A retreat of the sea from the land.

Relief. See *topographic relief.*

Replacement. The process by which a fluid dissolves matter already present and at the same time deposits from solution an equal volume of a different substance.

Residual concentration. The natural concentration of a mineral substance by removal of a different substance with which it was associated.

Resurgent cauldron. The uplifting of the collapsed floor of a caldera to form a structural dome.

Retrograde metamorphic effects. Metamorphic changes that occur as temperature and pressure are declining.

Reverse fault. A fault, generally steeply inclined, along which the hanging-wall block has moved relatively upward.

Richter magnitude scale. A scale, based on the recorded amplitudes of seismic body waves, for comparing the amounts of energy released by earthquakes.

Right-lateral fault. See *left-lateral fault.*

Rock. Any naturally formed, firm, and coherent aggregate mass of mineral matter that constitutes part of a planet.

Rock avalanche. Large mass of falling rock that breaks up, pulverizes on impact, and may then continue to travel downslope, often for great distances.

Rock cleavage. The property by which a rock breaks into platelike fragments along flat planes. Also called *slaty cleavage.*

Rock cycle. The cyclic movement of rock material, in the course of which rock is created, destroyed, and altered through the operation of internal and external Earth processes.

Rockfall. The relatively free falling of detached bodies of bedrock from a cliff or steep slope.

Rock flour. Fine rock particles produced by glacial crushing and grinding.

Rock glacier. A glacierlike tongue or lobe of angular rock debris containing interstitial ice or buried glacier ice that moves downslope in a manner similar to glaciers.

Rockslide. The sudden and rapid downslope movement of newly detached masses of bedrock across an inclined surface.

Rock-stratigraphic unit. A body of rock having a high degree of overall lithologic homogeneity. Compare *time-stratigraphic unit.*

Runoff. The fraction of precipitation that flows over the land surface.

Sag-duction. A sagging process hypothesized to operate during the early Archean, by which piles of volcanic rock sagged deep enough so that partial melting commenced.

Salinity. The measure of the sea's saltiness.

Saltation. The progressive forward movement of a sediment particle in a series of short intermittent jumps along arcing paths.

Sand seas (ergs). Vast tracts of shifting sand.

Saturated zone. The groundwater zone in which all openings are filled with water.

Scale (of a map). The proportion between a unit of distance on a map and the unit it represents on the Earth's surface.

Schist. A well-foliated metamorphic rock in which the component platy minerals are clearly visible.

Schistosity. The parallel arrangement of coarse grains of the sheet-structure minerals, like mica and chlorite, formed during metamorphism under conditions of differential pressure.

Sea-floor spreading (theory of). A theory proposed during the early 1960s in which lateral movement of the oceanic crust away from mid-ocean ridges was postulated.

Seamount. An isolated submerged volcanic mountain standing more than 1000 m above the sea floor.

Secondary enrichment. The natural enrichment of an existing mineral deposit through processes of chemical weathering and groundwater transport.

Sediment. Regolith that has been transported by any of the external processes.

Sedimentary facies. A distinctive group of characteristics within a sedimentary unit that differs, as a group, from those elsewhere in the same unit.

Sedimentary mineral deposit. Any local concentration of minerals formed through processes of sedimentation.

Sedimentary rock. Any rock formed by chemical precipitation or by sedimentation and cementation of mineral grains transported to a site of deposition by water, wind, or ice.

Sedimentary stratification. A layered arrangement of the particles that constitute sediment or sedimentary rock.

Seismic belts. Large tracts of the Earth's surface that are subject to frequent earthquake shocks.

Seismic sea waves (also called *tsunami*). Long wavelength ocean waves produced by sudden movement of the sea floor following an earthquake. Incorrectly called tidal waves.

Seismic stratigraphy. A cross-sectional view of crustal rocks and sediments obtained through high-resolution seismic exploration techniques.

Seismic tomography. A way of revealing inhomogeneities in the mantle by measuring slight differences in the frequencies and velocities of seismic waves.

Seismic waves. Elastic disturbances spreading outward from an earthquake focus.

Seismograph. The device used to study the shocks and vibrations caused by earthquakes.

Seismology. The study of earthquakes.

Setting time. The moment a mineral starts accumulating a daughter product produced by radioactive decay.

Shear waves. See *S waves.*

Sheet erosion. The erosion performed by overland flow.

Sheeted flows. Thin sheets of lava with rapidly quenched, glassy surfaces.

Shield volcano. A volcano that emits fluid lava and builds up a broad, dome-shaped edifice with a surface slope of only a few degrees.

Sills. Tabular sheets of intrusive igneous rock that are parallel to the layering of the intruded rock.

Simple decay. A radioactive parent decays to produce a stable daughter product.

Sink. (Also called a *doline.*) A large solution cavity open to the sky.

Slaty cleavage. See *rock cleavage.*

Slickensides. Striated or highly polished surfaces on hard rocks abraded by movement along a fault.

Sliderock. The sediment composing a talus.

Slip face. The straight, lee slope of a dune.

Slump. A type of slope failure in which a downward and outward rotational movement of rock or regolith occurs along a concave-up slip surface.

Snowline. The lower limit of perennial snow.

Soil. The weathered part of the regolith which can support rooted plants.

Soil profile. The succession of distinctive horizons in a

soil from the surface down to the unaltered parent material beneath it.

Solid solution. See *ionic substitution*.

Solifluction. The slow downslope movement of waterlogged soil and surficial debris.

Specific gravity. A number stating the ratio of the weight of a substance to the weight of an equal volume of pure water.

Spit. An elongate ridge of sand or gravel that projects from land and ends in open water.

Spreading axis. The axis of rotation of a plate of lithosphere.

Spreading edge. (Also called a *divergent margin*.) The new, growing edge of a plate. Coincident with a mid-ocean ridge.

Spreading pole. The point where a spreading axis reaches the Earth's surface.

Spring. A flow of groundwater emerging naturally at the ground surface.

Stable platform. That portion of a craton that is covered by a thin layer of little-deformed sediments.

Stack. An isolated rocky island or steep rock mass near a cliffy shore, detached from a headland by wave erosion.

Stalactite. Icicle-like form of dripstone and flowstone, hanging from cave ceilings.

Stalagmite. Blunt "icicle" of flowstone projecting upward from cave floors.

Star dune. An isolated hill of sand having a base that resembles a star in plan.

Steady state. A condition in which the rate of arrival of material or energy equals the rate of escape.

Steady-state equilibrium. A condition in which fluctuations occur about some average value.

Stock. A small, discordant body of intrusive igneous rock.

Strain. The measure of the changes in length, volume, and shape in a stressed material.

Strain rate. The rate at which a rock is forced to change its shape or volume.

Stratification. Layered arrangement of the particles that constitute sediment or sedimentary rock.

Stratified drift. Drift that is both sorted and stratified.

Stratigraphic superposition (principle of). In a sequence of strata, not later overturned, the order in which they were deposited is from bottom to top.

Stratigraphy. The study of strata.

Stratotype. See *type section*.

Stratovolcanoes. (Also called *composite volcanoes*.) Volcanoes that emit both fragmental material and viscous lava, and that build up steep conical mounds.

Stratum (plural = strata). A distinct layer of rock that accumulated at the Earth's surface.

Streak. A thin layer of powdered mineral made by rubbing a specimen on a nonglazed porcelain plate.

Stream. A body of water that carries rock particles and dissolved substances and flows down a slope along a definite path.

Stream capture. (Also called *piracy*.) The diversion of a stream by the headward growth of another stream.

Stream-flow. The flow of surface water in a well-defined channel.

Stress. The magnitude and direction of a deforming force.

Striations (glacial). Scratches and grooves on bedrock surfaces caused by grinding of rock against rock during movement of glacier ice.

Strike. The compass direction of a horizontal line in an inclined plane.

Strike-slip fault. A fault on which displacement has been horizontal.

Structural geology. The branch of geology devoted to the study of rock deformation.

Subduction zone. See *edge of consumption*.

Submarine canyon. A steep-sided valley on the continental shelf or slope resembling a river-cut canyon on land.

Subpolar glacier. A glacier in which surface temperature reaches 0°C in summer, but beneath several meters the temperature is below the pressure melting point.

Subsequent stream. A stream whose course has become adjusted so that it occupies belts of weak rock or other geologic structures.

Superposed stream. A stream that was let down, or superposed, from overlying strata onto buried bedrock having composition or structure unlike that of the covering strata.

Surf. Wave activity between the line of breakers and the shore.

Surface ocean currents. Broad, slow drifts of surface water.

Surface waves. Seismic waves that are guided by the Earth's surface and do not pass through the body of the Earth.

Surge. An unusually rapid rate of movement of a glacier and marked by dramatic changes in glacier flow and form.

Suspended load. Fine particles suspended in a stream.

S waves. Seismic body waves transmitted by an alternating series of sideways movements in a solid. S waves cause a change of shape and cannot be transmitted through liquids and gases.

Syncline. A downfold with a troughlike form.

Taconite. A Lake Superior-type iron formation that has been metamorphosed.

Talus. The apron of rock waste sloping outward from the cliff that supplies it.

Tar. (Also called *asphalt*.) An oil that is viscous and so thick it will not flow.

Tarn. A small, generally deep mountain lake occupying a cirque.

Tectonics. The study of movements and deformation of the crust on a large scale.

Temperate (warm) glacier. A glacier in which the ice is at the pressure melting point throughout.

Tephra. A loose assemblage of pyroclasts.

Terminal moraine. An end moraine deposited at the front of a glacier.

Terrace. Abandoned floodplain formed when a stream flowed at a level above the level of its present channel and floodplain.

Terrane. A large piece of crust with a distinctive geological character.

Terrestrial planets. The innermost planets of the Solar System (Mercury, Venus, Earth, and Mars) which have high densities and rocky compositions.

Terrigenous sediment. Sediment derived from sources on land.

Tethys. The name of a narrow sea separating Gondwanaland from Laurasia.

Texture. The overall appearance that a rock has because of the size, shape, and arrangement of its constituent particles.

Thalweg. A line connecting the deepest parts of a stream channel.

Thermal metamorphism. See *contact metamorphism.*

Thermokarst. Karstlike landscape in a permafrost region produced by melting of ground ice and settling of the ground.

Thermoremanent magnetism. Permanent magnetism that is a result of thermal cooling.

Thin section. A thin slice of rock glued to a glass slide and used for microscopic examination.

Thrust faults. (Also called *thrusts*). Low angle reverse faults with dips less than 45°.

Tidal bore. A large, turbulent wall-like wave of water caused by the meeting of two tides or by the rush of tide up a narrowing inlet, river, estuary, or bay.

Tidal wave. See *seismic sea wave.*

Till. A diamicton deposited directly from glacier ice.

Tillite. A diamictite of glacial origin.

Time line. A line of constant age in a stratigraphic section.

Time-stratigraphic unit. A unit representing all the rocks, and only those rocks, that formed during a specific interval of geologic time. Compare *rock-stratigraphic unit.*

Titius-Bode rule. The distance to each planet is approximately twice as far as the next inner one, measuring from the Sun.

Tombolo. A ridge of sand or gravel that connects an island to the mainland or to another island.

Topographic map. A map that shows topography.

Topographic profile. The outline of the land surface along a given line.

Topographic relief. The difference in altitude between the highest and lowest points of a landscape.

Topography. The relief and form of the land.

Topset layer. A layer of stream sediment that overlies the foreset layers in a delta.

Transform. The junction point where one of the major deformation features—a mid-ocean ridge, a sea-floor trench, or a strike-slip fault—meets another.

Transform faults. The special class of strike-slip faults that link major structural features.

Transgression. A spreading of the sea over the land.

Transpiration. The process by which plants release water vapor from their leaves.

Transverse dune. A sand dune forming a wavelike ridge transverse to wind direction.

Trellis pattern. Stream pattern consisting of rectangular arrangement of channels in which principal tributary streams are parallel and very long.

Trenches. Long, narrow, very deep and arcuate basins in the sea floor.

Tsunami. See *seismic sea wave.*

Turbidite. Sediment deposited by a turbidity current.

Turbidity currents. Gravity-driven currents consisting of dilute mixtures of sediment and water having a density greater than the surrounding water.

Type section. (Also called a *stratotype.*) Section that displays the primary characteristics of a stratigraphic unit in a typical manner.

Unconformity. A substantial break or gap in a stratigraphic sequence that marks the absence of part of the rock record.

Unconformity-bounded sequence. A grouping of strata that is bounded at its base and top by unconformities of regional or interregional extent.

Uniform layer. A layer of sediment or sedimentary rock that consists of particles of about the same diameter.

Uniformity. See *Principle of Uniformity.*

Unsaturated zone (zone of aeration). The groundwater zone in which open spaces in regolith or bedrock are filled mainly with air.

Upwelling. Vertical movement of seawater caused by prevailing winds blowing offshore.

Valley glacier. A glacier that flows from a cirque or cirques onto and along the floor of a valley.

Valley train. A body of outwash that partly fills a valley.

van der Waals bond. A weak electrostatic attraction that arises because certain ions and atoms are restored from a spherical shape.

Varve. A pair of sedimentary layers deposited during the seasonal cycle of a single year.

Ventifact. Any bedrock surface or stone that has been abraded and shaped by windblown sediment.

Vertical exaggeration. The ratio of the horizontal scale to the vertical scale in a topographic profile.

Vesicle. Small opening, in extrusive igneous rock, made by escaping gas originally held in solution under high pressure while the parent magma was underground.

Viscosity. The internal property of a substance that offers resistance to flow.

Volcanic neck. The approximately cylindrical conduit of igneous rock forming the feeder pipe immediately below a volcanic vent.

Volcanic sediment (in the ocean). Sediment from submarine volcanoes, together with ash from oceanic and nonoceanic volcanic eruptions.

Volcano. The vent from which igneous matter, solid rock debris, and gases are erupted.

Water cycle. See *hydrologic cycle.*

Water gap. A pass, in a ridge or mountain range, through which a stream flows.

Water table. The upper surface of the saturated zone of groundwater.

Wave base. The effective lower limit of wave motion, which is half of the wavelength.

Wave-cut bench. A bench or platform cut across bedrock by surf.

Wave-cut cliff. A coastal cliff cut by surf.

Wavelength. The distance between the crests of adjacent waves.

Wave refraction. The process by which the direction of a series of waves, moving in shallow water at an angle to the shoreline, is changed.

Weathering. The chemical alteration and mechanical breakdown of rock materials during exposure to air, moisture, and organic matter.

Welded tuff. (Also called *ignimbrite*.) Pyroclastic rocks, the glassy fragments of which were plastic and so hot when deposited that they fused to form a glassy rock.

Wilson Cycle. The cycle of opening and closing of oceans, accompanied by successive fragmentations and collisions of continents.

Xenoliths. Fragments of country rock still enclosed in a magmatic body when it solidifies.

Yardang. An elongate and streamlined wind-eroded ridge.

Yield point. The point on a stress-strain curve where the elastic limit is reached and ductile deformation commences.

Zone of aeration (unsaturated zone). The groundwater zone in which open spaces in regolith or bedrock normally are filled mainly with air.

Photo Credits

Part I Opener: Ray Manley/Shostal Associates.

Chapter 1: Opener: Georg Gerster/Photo Researchers. Fig. 1.1: Joseph Viesti. Fig. 1.3: Victor Engelbert/Photo Researchers. Fig. 1.6: Bob DiNatale/The Image Bank. Fig. 1.7b: Earth Satellite Corp.

Chapter 2: Opener: WQED, Pittsburgh/Planet Earth. Fig. 2.15: Landsat/USGS/John Shelton. Fig. 2.21: K. R. Gill.

Chapter 3: Opener: Manfred Kage/Peter Arnold. Figs. 3.9b, 3.10a and b, 3.11a, 3.17, 3.18, 3.19, 3.20: B. J. Skinner.

Chapter 4: Opener: S. Maeda/The Image Bank. Fig. 4.1: B. J. Skinner. Fig. 4.3: R. S. Fiske. Figs. 4.5 and 4.10: B. J. Skinner. Fig. 4.12: S. C. Porter. Figs. 4.13, 4.14 and 4.21: B. J. Skinner. Fig. 4.22: S. C. Porter. Fig. 4.24: B. J. Skinner. Fig. 4.26: John S. Shelton. Fig. 4.28a: Jill Scheiderman. Fig. 4.28b: Bruce K. Goodwin. Figs. 4.29b and 4.30: S. C. Porter. Fig. 4.31: Bruce Marsh. Fig. 4.32: S. C. Porter. Fig. 4.33: Lyn Topinka/USGS/David A. Johnson Cascades Volcano Observatory. Fig. 4.34b: B. J. Skinner. Fig. 4.35: WHOI/Agnus 83. Fig. 4.36a: WHOI/Fred Grassle. Fig. 4.36b: E. H. Bailey/USGS. Fig. 4.38a and b: Philippines Institute of Volcanology. Fig. 4.39: S. C. Porter. Fig. 4.40: Underwood and Underwood. Fig. 4.41: S. C. Porter.

Chapter 5: Opener: Tom Bean. Fig. 5.1: Josef Muench. Fig. 5.2a and b: John S. Shelton. Fig. 5.2c: S. C. Porter. Fig. 5.2d: B. J. Skinner. Fig. 5.3: S. C. Porter. Fig. 5.4: Richard J. Stewart. Figs. 5.5, 5.6, 5.7, 5.8, 5.9 and 5.10: S. C.

Porter. Fig. 5.11: Eric Cheney. Fig. 5.12: Betty Crowell/Faraway Places. Figs. 5.13, 5.14 and 5.15a: S. C. Porter. Fig. 5.15b: Joanne Bourgois. Fig. 5.16a: Josef Muench. Fig. 5.16b: S. C. Porter. Fig. 5.17: Tom Bean 1986. Fig. 5.18: Gary Ladd. Fig. 5.19: S. C. Porter. Fig. 5.20: David H. Krinsley. Fig. 5.22a: S. C. Porter. Fig. 5.22b: J. Robert Waaland/Terraphotographics/Biological Photo Services. Fig. 5.23: U. S. Soil Conservation Service; California Institute of Technology. Fig. 5.26: Richard J. Stewart. Fig. 5.27: Deep Sea Drilling Project/Scripps Institute of Oceanography.

Chapter 6: Opener: B. J. Skinner. Figs. 6.2, 6.5 and 6.6: B. J. Skinner. Fig. 6.8: Brenda Sirois. Figs. 6.9, 6.11 and 6.12: B. J. Skinner. Figs. 6.13, 6.14 and 6.19: W. Sacco. Figs. 6.21 and 6.22: S. C. Porter.

Part II Opener: S. C. Porter.

Chapter 7: Opener: David Hiser/Photographers Aspen. Fig. 7.1: S. C. Porter. Fig. 7.5: John S. Shelton. Fig. 7.14: Robert C. Bostrom/Geological Research Corp. Fig. 7.16: Geopic/Earth Satellite Corp.

Chapter 8: Opener: Emily Harste/Bruce Coleman. Fig. 8.7: C. W. Naeser/USGS. Fig. 8.8: Ted McConnaughy. Fig. 8.9: I. Friedman/USGS. Fig. 8.10: S. C. Porter. Fig. 8.13: John Wiley Photo Library.

Chapter 9: Opener: Robert S. Anderson. Fig. 9.1a and b: William E. Ferguson. Fig. 9.1c: Llewellyn/The Picture Cube. Fig. 9.3: S. C. Porter. Fig. 9.4: Maurice

and Sally Landre/Photo Researchers. Figs. 9.5, 9.6, 9.7, and 9.8: S. C. Porter. Fig. 9.9: Robert S. Anderson. Fig. 9.11: Barrie Rokeach 1983. Figs. 9.16, 9.17, 9.19, 9.21a and 9.22: S. C. Porter. Fig. 9.24: John S. Shelton. Fig. 9.25a: W. R. Hausen/USGS. Fig. 9.26: G. R. Roberts. Fig. 9.27: Chalmers M. Clapperton. Fig. 9.28: S. C. Porter. Fig. 9.29: Shelly Katz/Black Star. Figs. 9.33 and 9.34: S. C. Porter.

Chapter 10: Opener: Hiroji Kubota/Magnum Photos. Fig. 10.14: Oriental Institute, University of Chicago. Fig. 10.17: Gary Ladd Photography. Fig. 10.18: Josef Muench. Fig. 10.19: David Hiser/Photographers Aspen. Fig. 10.21: Mike McClure/Bruce Coleman. Fig. 10.22: USGS. Fig. 10.23: John S. Shelton. Fig. 10.24: The Metropolitan Museum of Art, The Dillon Fund Gift, 1979. Fig. 10.28c: USGS. Fig. 10.29: Troy L. Péwé. Fig. 10.30: S. C. Porter.

Chapter 11: Opener: Geopic/Earth Satellite Corp. Fig. 11.1: Geopic/Earth Satellite Corp. Fig. 11.2 and 11.4: Thomas Dunne. Fig. 11.10: NASA. Fig. 11.16: S. C. Porter. Fig. 11.27: John S. Shelton. Fig. 11.30: Geopic/Earth Satellite Corp.

Chapter 12: Opener: Jim Brandenburg/Bruce Coleman. Fig. 12.3: Gary Ladd. Fig. 12.4: Gordon Wiltsie/Bruce Coleman. Fig. 12.5: Peter Fronk/Click, Chicago. Fig. 12.6: G. R. Roberts. Figs. 12.7, 12.8 and 12.9: S. C. Porter. Fig. 12.10: E. R. Degginger. Fig. 12.14: John S. Shelton. Fig. 12.16: U. S. Soil Conservation Service. Fig. 12.18: R. H. Hufnagle/Phillip Gendreau. Figs. 12.20 and 12.21: S. C. Porter. Fig.

12.23: M. J. Grolier/USGS. Fig. 12.24: Gary Ladd. Fig. 12.26: Galen Rowell/High and Wild Photography. Fig. 12.27: Eric Cheney. Fig. 12.29: S. C. Porter. Fig. 12.31: Geopic/Earth Satellite Corp. Figs. 12.32, 12.33 and 12.34: S. C. Porter.

Chapter 13: Opener: NASA. Fig. 13.1: S. C. Porter. Table 13.1a and b: Austin S. Post/USGS. Table 13.1c: Eric Cheney. Table 13.1d: Austin S. Post/USGS. Table 13.1e and f: Friedrich Röthlisberger. Table 13.1g: Chalmers M. Clapperton. Table 13.1h: Terence J. Hughes. Fig. 13.3: NASA. Fig. 13.11: S. C. Porter. Fig. 13.15: Austin S. Post/USGS. Figs. 13.17, 13.18, 13.19 and 13.20: S. C. Porter. Fig. 13.21: USGS. Fig. 13.22: Josef Muench. Fig. 13.23: Richard B. Waitt, Jr. Fig. 13.24: National Air Photo Library, Ottawa, Canada. Fig. 13.25: S. C. Porter. Fig. 13.27: John S. Shelton. Figs. 13.28, 13.29 and 13.33: S. C. Porter. Fig. 13.34: Bernard Hallet. Fig. 13.35a and b: A. Lincoln Washburn. Fig. 13.35c: Brainerd Mears, Jr. Fig. 13.36a: J. Ross Mackay. Fig. 13.36b: Troy L. Péwé. Fig. 13.36c: S. C. Porter. Fig. 13.39: S. C. Porter.

Chapter 14: Opener: Nicholas Devore III/Bruce Coleman. Fig. 14.7: D. J. Fornari, Lamont-Doherty. Fig. 14.8: S. C. Porter. Fig. 14.9a: Chesley Bonestell, in E. L. Hamilton, Geological Society of America, Mem. 63, 1951, pl. I. Fig. 14.10a and b: Nicholas Devore III/Bruce Coleman. Fig. 14.10c: NASA. Fig. 14.16a and b: G. R. Roberts. Fig. 14.16c: S. C. Porter. Fig. 14.19: John S. Shelton. Fig. 14.21a: John S. Shelton. Fig. 14.21b: Nicholas Devore III/Photographers Aspen. Fig. 14.23: S. C. Porter. Fig. 14.24: Ray Manley/Shostal Associates. Fig. 14.25: Terraphotographics/Biological Photo Services. Fig. 14.27: Earth Satellite Corp. 1980. Fig. 14.29: Earth Satellite Corp. 1979. Figs. 14.30 and 14.31: S. C. Porter. Fig. 14.33: G. R. Roberts. Fig. 14.34:

S. C. Porter. Fig. 14.35: Earth Satellite Corp. 1982. Fig. 14.36: Arthur L. Bloom. Fig. 14.37: Photri.

Part III Opener: John S. Shelton.

Chapter 15: Opener: B. J. Skinner. Fig. 15.4: R. S. Fiske. Fig. 15.5: NASA. Fig. 15.6: R. S. Fiske. Fig. 15.7: John S. Shelton. Fig. 15.8: Alaska Pictorial Service. Fig. 15.9: From Lyell, 1875, *Principles of Geology*, 12ed. Fig. 15.11: John S. Shelton. Fig. 15.13b: B. J. Skinner. Fig. 15.15d: Earth Satellite Corp. Fig. 15.16: Peter Arnold. Fig. 15.17: Paolo Koch/Photo Researchers. Fig. 15.18: John S. Shelton. Fig. 15.20: B. M. Shaub. Fig. 15.21: B. J. Skinner. Fig. 15.21b: William E. Ferguson. Fig. 15.22: Barrie Rokeach 1983. Fig. 15.25c: B. Caulfield 1977. Fig. 15.28: B. J. Skinner. Fig. 15.33: Greg Davis.

Chapter 16: Opener: Stan Wayman/Life Magazine. Fig. 16.1: Joe Cavaretta/Gamma-Liaison. Fig. 16.13: B. J. Skinner. Fig. 16.14: Asahi Shimbun. Fig. 16.21: B. J. Skinner. Fig. 16.25: A. M. Dziewonski and D. L. Anderson, *American Scientist*, v. 72, p. 483–494, 1984.

Chapter 17: Opener: G. R. Roberts. Fig. 17.16: Warren Hamilton.

Chapter 18: Opener: Geopic/Earth Satellite Corp. Fig. 18.3: E. Bonatti. Figs. 18.5 and 18.6: B. J. Skinner. Fig. 18.15: Gerald Weinbren. Fig. 18.16: A. J. Naldrett.

Chapter 19: Opener: S. C. Porter. Fig. 19.8: NASA. Fig. 19.12: S. C. Porter. Fig. 19.13: Geopic/Earth Satellite Corp. Fig. 19.14: S. C. Porter. Fig. 19.13: Geopic/Earth Satellite Corp. Fig. 19.14: S. C. Porter. Fig. 19.17: NASA. Fig. 19.23: EOSAT. Figs. 19.24 and 19.25: B. J. Skinner. Fig. 19.26: S. C. Porter. Fig. 19.30: NASA. Fig. 19.31: S. C. Porter.

Chapter 20: Opener: John Bryson/The Image Bank. Fig. 20.9a: Friedrich Röthlis-

berger. Fig. 20.21: Jørgen Meldgaard. Fig. 20.12c: Matsuo Tsukada. Fig. 20.21: Sidney Harris.

Part IV Opener: Courtesy of Dr. Glenn Allcot, USGS, Reston, Va.

Chapter 21: Opener: G. R. Roberts. Fig. 21.3: Illinois Geological Survey. Fig. 21.7: S. C. Porter. Fig. 21.8: Courier-Journal and Louisville Times. Fig. 21.17: B. J. Skinner.

Chapter 22: Opener: Peabody Museum, Yale University. Fig. 22.3: Jacques Touret. Fig. 22.4: Dudley Foster/WHOI. Fig. 22.5: B. J. Skinner. Fig. 22.6: B. M. Shaub. Fig. 22.7: John Allcock. Fig. 22.10: B. J. Skinner. Fig. 22.11: A. J. Naldrett. Fig. 22.13: Windsor Publications/FPG. Figs. 22.14, 22.15, 22.18 and 22.19: B. J. Skinner. Fig. 22.20a: J. Bahr. Fig. 22.20b: H. Murray. Fig. 22.21: B. J. Skinner. Fig. 22.25: Jeff Rotman/Peter Arnold. Fig. 22.26: B. J. Skinner.

Part V Opener: NCAR/HAO.

Chapter 23: Opener: NASA. Fig. 23.1: John S. Shelton. Fig. 23.2: NASA. Fig. 23.4a: Pickands-Mather Co. Fig. 23.4b: NASA. Figs. 23.5, 23.6 and 23.7: NASA. Fig. 23.8: Courtesy of Space Photography Laboratory, Arizona State University. Figs. 23.10 and 23.11: U. S. Dept. of the Interior, Astrogeology Dept. Figs. 23.14 and 23.15: NASA. Fig. 23.15a: B. J. Skinner. Fig. 23.15b: John A. Wood. Figs. 23.16, 23.17, 23.18, 23.20, 23.21, 23.22, 23.23, 23.24 and 23.25b: NASA. Figs. 23.26, 23.29, 23.30, 23.31, 23.32, 23.33 and 23.34: Courtesy of Space Photography Laboratory, Arizona State University.

Appendix B: B.1, B.2, B.3, B.4, B.5, B.6, B.7, B.8 and B.9: W. Sacco.

Appendix C: C.1 and C.2: B. J. Skinner.

Index

Numbers of pages on which terms are defined are in **boldface**.
Asterisks indicate illustrations.
A indicates page is in Appendices.
G indicates page is in the Glossary.